Advances in Volcanology

An Official Book Series of the International Association of Volcanology and Chemistry of the Earth's Interior

More information about this series at http://www.springer.com/series/11157

Valerio Acocella

Volcano-Tectonic Processes

Springer

Valerio Acocella
Università Roma Tre
Rome, Italy

ISSN 2364-3277 ISSN 2364-3285 (electronic)
Advances in Volcanology
ISBN 978-3-030-65967-7 ISBN 978-3-030-65968-4 (eBook)
https://doi.org/10.1007/978-3-030-65968-4

This Springer imprint is published by the registered company Springer Nature Switzerland AG
The registered company address is: Gewerbestrasse 11, 6330 Cham, Switzerland

...and the Earth becomes my throne!
(James Hetfield, 1991)

Dedicated to
Francesca, Flavia and Elena

Foreword

When Prof. Valerio Acocella asked me to write a Foreword for his book on Volcano-Tectonics (VT), I replied "... errr, OK, but I am not a volcano-tectonicist". Professor Acocella retorted that he wanted a non-VT specialist to write the Foreword as it may more-convincingly endorse the reception of the book into the general sphere of volcanology. Being a general, academic, physical volcanologist (apart from a 10-year stint on the geologic aspects of burial of high-level nuclear waste for the US Nuclear Regulatory Commission), who has spent his career on explosive eruptions and their products (except for a currently 20+-year love affair with the lava flows in large igneous provinces), this was a fair point, so I acquiesced.

The book before you by Valerio Acocella is the first of its kind. It is not that VT is new, but it has not been defined and laid out before as Valerio does here, and its importance may have been underplayed, but it should not be after this treatise. Some the world's greatest geologists have been, at times, VT specialists. I am reminded of Arthur Holmes, a believer in Alfred Wegener's continental drift hypothesis, who, in the 1930s–1940s, almost nailed what we now accept as plate tectonics and its relationship to magmatism and volcanism. Also Rein van Bemmelen in the 1940s–1950s, who, while mapping Indonesia with its overwhelmingly volcanic nature for the Dutch Government, got very close to a plate tectonic hypothesis for the controls on volcanism. But it is not just in the relationship between volcanism and tectonic plates that some of the leading twentieth-century geologists touched upon VT. In the area of applied geological science, geologists such as Jim Healy in New Zealand, while working almost solely on developing the geothermal prospects for that county in the 1950s–1960s, became intimately involved with the relationships of the Taupo Volcanic Zone fault structures, its calderas and ignimbrites, and the prospective geothermal fields, some of which came online as power-plants in his lifetime.

Despite beginnings in the early 1900s, with the instigation of the International Association of Volcanology and Chemistry of the Earth's Interior (IAVCEI) under the umbrella of the International Union of Geodesy and Geophysics (IUGG), *volcanology* was slow to develop and was only becoming a distinct discipline by the mid-1970s. It was definitely not a major discipline in my undergraduate days in the UK (1965–1970), but it had its stars ushering the science from different countries at different times, and with various approaches and purposes, for example, Sigurdur Thorarinsson in Iceland, Jim Cole and Colin Vucetich in New Zealand, Shigeo Aramaki and

Isamu Murai in Japan, Hans-Ulrich Schmincke in Germany, Franco Barberi in Italy, Bob Smith in the USA and George Walker in the UK (and, later, New Zealand and Hawai'i). In parallel, the development of a more modern approach to VT slowly gathered momentum and prominence. Notable here is David Pollard, using an approach via rock mechanics, who began publishing on the emplacement of intrusions such as laccoliths (1973). Pollard's work with his student Paul Segall on non-volcanological applications began appearing in the 1980s, but Segall's work developed into volcanic applications in the 1990s. Another early proponent of VT was Kazuaki Nakamura, who in the 1970s promoted the use of volcanoes as indicators of tectonic stress.

Mainly in the 1990s and this century, others who helped to develop the field of VT include Agust Gudmundsson, representing Iceland, Germany and the UK. Agust, initially moving along a similar path to Pollard and Segall, merging rock mechanics principles and field studies, led to improved understanding of matters such as emplacement of intrusions and calderas. Also, Cynthia Ebinger (USA) contributed much to defining the regional relationships between magmatism and tectonics. Daniel Dzurisin (USA) made important studies on volcano geodesy, mainly on Yellowstone and the Cascadian and Aleutian volcanoes. The development of GPS was important to VT as, together with InSAR, it boosted studies of volcano deformation over the last two decades. And now, Valerio is bringing all this together as a mature text on this well-developed sub-discipline of geosciences, where the main branches of volcanology and structural geology meet.

This book and its topics contribute to the development of modern volcanology, not only in terms of understanding volcanoes (there is a mini course on volcanology at the beginning), but also in explaining the role of VT in defining volcanic hazards and forecasting eruptions. Chapters 1 and 2 are introductory and provide the fundamentals. Chapters from 3 to 9 focus on the rise of magma towards the surface, describing the main volcano-tectonic processes associated with magma rise. Chapters from 10 to 13 provide regional and stylistic perspectives. Details of the contents can be found in the Preface.

Convincing the broader volcanological community about the importance not only of this book, but also of volcano-tectonics in general, is a purpose of this Foreword. A goal is to engender within the volcanological community a broader realization of volcano-tectonics. I admire tremendously those colleagues who can finish writing books—so, Valerio, congratulations! I hope that this hard-won fruit of your labour brings your readers much useful information and brings you a well-deserved round of appreciation.

Dear, generous readers, I promote to you a book on a new field which is beautifully illustrated, thoroughly researched and wide-ranging. It is a first, and I hope it will spur on many other similar studies and texts in the future.

Berkeley, CA, USA Steve Self
January 2021

Preface

Among the various university courses I took for my geoscience degree, volcanology and structural geology motivated me by far the most. Volcanoes and deformed rocks are indeed the most impressive expressions of the activity of our planet, a real manifestation of power and energy. As a student, this often came into my mind, imagining a faulted volcano pouring out blood-red lava! What can be more dramatic than faults ripping a volcano apart? It is like destroying a source of destruction.

With this intriguing image, it became natural to dive into what at that time was whispered as "volcano-tectonics". Indeed, in the early 1990s volcano-tectonics was still a rather ambiguous term, with some professors claiming that volcano-tectonics resulted from volcano stratigraphy and others mystically invoking the term anytime they could not explain a process occurring at a volcano. At that time, volcano-tectonic studies were rare, even within the international community, although I somehow realized that there was so much to study and understand on the topic. So, once I decided that research in volcano-tectonics would be my future (but without getting stuck with volcano stratigraphy), it was straightforward to look for opportunities after my university degree. In this delicate step, I was lucky enough to have the support of a couple of open-minded mentors, Renato Funiciello and Antonio Lazzarotto, who granted me total support and freedom in studying volcano-tectonic and tectono-magmatic problems during and after my Ph.D., providing me with the opportunity to express myself in this mystical topic.

Today, after two and a half decades, research in volcanology has grown to an extent that was unconceivable in the 1990s. Boosted by an impressive technological support, volcanology currently relies on continuous inputs from several disciplines within the geoscience field, as well as from independent disciplines in different fields of science, including mathematics, physics, chemistry, engineering, statistics and informatics. The result of this integration is the firm establishment of several modern research branches making up current volcanology. Among these is, indeed, volcano-tectonics, which has grown from its infancy of the late '80s-early '90s to a mature and leading topic, relying on many different techniques and tools. Today any form of confusion has definitely disappeared, and it would be appropriate to say that volcano-tectonics considers any stress-deformation process, occurring at any scale, related to the accumulation, transfer and eruption of shallow magma. These stress-deformation processes may be magma-induced or, conversely, encourage or control the accumulation, transfer and eruption of magma. In any case, focused on the conditions promoting magma accumulation, transfer and eruption, volcano-tectonics plays a fundamental role also in understanding pre-eruptive processes, providing the theoretical ground to eruption forecasting.

Therefore, (a) the impressive growth and identity acquired by this branch of volcanology, (b) which is also essential to understanding pre-eruptive processes and forecasting eruptions, (c) thus deserving to be better appreciated within the wider volcanological community, (d) especially given the lack of an accessible reference text summarizing the main achievements, are the four main motivations which, a few years ago, encouraged me to write a monograph on volcano-tectonics.

Indeed, while today there are so many academic texts on volcanoes dealing with general or particular aspects of their activity, especially eruptions, there are extremely few texts, and all released very recently (after I began preparing this monograph), which have started to consider specific aspects of volcano-tectonic processes. Distinctive from these texts, the present monograph claims to provide the first comprehensive and practical overview on volcano-tectonic processes, occurring at the local (volcano) and regional (plate boundary) scale.

Under these premises, this monograph aims at summarizing, with an integrated approach, the current (as of mid-2020) state of the art on volcano-tectonics, including its importance for volcanic hazards, as well as to provide a complementary reference for anyone interested in volcanoes and volcanic activity. *Volcano-Tectonic Processes* is in fact conceived as an accessible monograph, which may be appreciated by academics of any degree level, from upper-level undergraduate students with a general knowledge in volcanology to Ph.Ds, post-docs, researchers, lecturers and professors interested in volcanic, magmatic, structural and tectonic processes. I expect *Volcano-Tectonic Processes* to be especially appreciated by students and young researchers working on volcano-tectonic problems and experienced volcanologists wishing to throw off any residual mysticism on the topic. Although the ideal audience is supposed to have a background in geoscience (including volcanologists, structural geologists, geologists and geophysicists), this monograph may be of interest also for anyone fascinated by volcanoes and with some elementary knowledge in geoscience.

Volcano-Tectonic Processes is organized into three main parts.

Chapters 1 and 2 provide the fundamental introductory knowledge, defining many terms commonly used in the successive chapters and giving the reader the opportunity to approach volcano-tectonics with an appropriate, accessible and shared background. In both chapters, only the essential information, that is most useful for volcano-tectonics, is given, with the recommendation to rely on more specific and comprehensive textbooks for any detail.

Chapter 1 introduces the main notions about volcanoes and volcanic activity, at the same time highlighting how all the distinctive aspects of volcanoes directly or indirectly derive from their tectonic setting, a feature which is considered in detail in the third part of the book. After the overview on volcanoes and their activity, this chapter introduces and places volcano-tectonics into context.

Chapter 2 introduces the main concepts regarding deformation of the Earth's crust, merging knowledge from rock mechanics, rheology, structural geology, tectonics and geodynamics. This is a summary of the "tectonics" side of volcano-tectonics, and many volcanologists may not be very familiar with its content.

Chapters from 3 to 9 make up the second and main part, describing volcano-tectonics processes at the local (volcano) scale. These chapters follow magma in an ideal path rising through the crust, accumulating and developing magma chambers, whose activity may affect the structure and shape of the volcanic edifices, generating vertical collapses (calderas) or lateral collapses (flank collapses). Calderas and flank collapses may in turn affect the shallower transfer and eruption of magma. Therefore, knowledge of all these volcano-tectonic processes and their related signals, as revealed by monitoring systems, may allow understanding the state of active volcanoes and ultimately forecasting eruptions.

Following this ideal path, Chap. 3 discusses the rise of magma in the crust, moving from classic theories on diapirism to more modern perspectives on dikes. The latter include the fundamental factors contributing to dike propagation or arrest, which ultimately control the ascent of magma towards the surface and the possibility of an eruption.

Chapter 4 focuses on the arrest, emplacement and accumulation of the magma rising in the shallow crust, considering the development of various types of intrusions, as sills, laccoliths, bysmaliths and lopoliths, as well as larger plutons. Emphasis is then given to the formation and development of magma chambers, as deriving from the growth of the aforementioned intrusions. The nucleation of dikes from magma chambers is also considered.

Striking surface evidence of the dynamics of large and long-lived magma chambers is provided by calderas. With this regard, Chap. 5 focuses on the long- and short-term surface effects of magma withdrawal and magma accumulation within a chamber, including caldera collapse and resurgent uplift. The processes affecting shallow magma transfer and eruption at calderas are also considered.

Another major process affecting the shape of a volcanic edifice is flank collapse, a dramatic consequence of flank instability. Both flank instability and collapse are the focus of Chap. 6. Flank instability and collapse are commonly associated with eruptions, seismicity and, in coastal areas, tsunamis, playing a pivotal role in multi-hazard assessments at many active volcanoes.

Calderas and flank collapses may severely alter the shape of a volcanic edifice, so that the shallow propagation path of the magma nucleated from an underlying chamber may be affected by these topographic variations. With this regard, Chap. 7 examines the stress variations imposed by topography and other conditions associated with the shallower transfer and eruption of magma, providing the ground for a functional approach to eruption forecasting.

As magma approaches the surface, monitoring data become essential to detect shallow magma accumulation or transfer and to finally forecast eruptions. To this aim, Chap. 8 reviews the geodetic, geophysical and geochemical monitoring approaches applicable to volcano-tectonics, including the popular mathematical source models involving geodetic data. These monitoring approaches, interpreted in the frame of the previously discussed volcano-tectonic processes, allow capturing and understanding any anomalous state of the volcano, or unrest, possibly leading to an impending eruption.

Chapter 9 exploits knowledge from all the previous chapters to define hazard at volcanoes, in order to mitigate volcanic risk. The chapter first considers unrest processes at volcanoes, focusing on how the distinctive contribution of volcano-tectonics may help interpret unrest signals towards forecasting eruptions, which is the ultimate challenge for volcanology. Then, the state of the art on the delicate problem of eruption forecasting is presented.

Finally, Chaps. 10 to 13 make up the third part of the book, considering the regional tectonic framework of volcanoes, building on the message introduced in Chap. 1, that all the distinctive aspects of volcanoes directly or indirectly derive from their tectonic setting. Therefore, this part of the book changes scale, relating volcanoes and volcanic activity to the tectonic setting, which is probably the least explored topic of volcano-tectonics. Here, after a theoretical introduction (Chap. 10), key regional examples are illustrated for an in-depth understanding, highlighting general and distinctive behaviour, and offering a holistic view of tectono-magmatic processes affecting our planet.

In particular, the introductory Chap. 10 considers the distribution and general features of volcanoes in the plate tectonics frame. This chapter reviews major volcano-tectonic processes along divergent and convergent plate boundaries and hot spots, also including polygenic and monogenic volcanism. The surprisingly active role of magma in shaping plate boundaries and intraplate settings is highlighted.

Chapter 11 describes the development of divergent plate boundaries, with a progression from continental to transitional and finally oceanic domains, as currently observed in different key rift zones. A final discussion summarizes and highlights the main tectono-magmatic behaviours of rift zones, providing a general working hypothesis for their activity.

Chapter 12 focuses on representative convergent plate boundaries, which are the most hazardous areas on Earth, in terms of volcanic activity, seismicity and tsunamis. These boundaries may experience orthogonal and oblique convergence, generating volcanic arcs with extremely variable structure and activity. A final discussion summarizes and highlights the main tectono-magmatic behaviours of the arc volcanoes, including their relationships with mega-earthquakes.

Last, Chap. 13 describes the main features of hot spot volcanoes, considering examples from representative oceanic and, subordinately, continental hot spots. A final discussion summarizes and highlights similarities and differences in the tectono-magmatic behaviour of hot spot volcanoes, also providing an original working hypothesis to explain the recurrent features.

These chapters have been written with the aim of maintaining an accessible level for non-experts while trying to offer an original and comprehensive overview on volcano-tectonic processes. To achieve this, each chapter is introduced at the beginning, placing its content into context, and summarized at the end, providing an overview of the main notions. In between the introduction and the summary, an integrated approach, merging available knowledge mainly from geological, geophysical, modelling, monitoring and petrological data, has been followed, avoiding any limitation deriving from relying only on a specific method. Indeed, far from being technical, this monograph tries to summarize the general processes suggested by all types of available data, in order to be of interest to the broadest audience.

While writing the book, I soon realized that, because of the editorial "space problem" (this term pops up in several chapters), I had to make a summary of many important notions, also omitting many points. In the end, I realize that the several hundred of pages of this monograph are simply the tip of an iceberg. Nevertheless, this tip is fundamental to finding the coordinates whereby to access and place into context the related notions making up rest of the iceberg. With this aim, I referred to numerous key-studies in the references for further deepening (apology for any study I may have omitted, but the selection was among more than 10,000 papers and books). In this, I have attempted to maintain a balance between the importance of giving comprehensive and useful references to the reader and, at the same time, keeping the reading of the text as smooth as possible, without too many interruptions due to the citations. As a result, the references have been sometimes grouped together at the end of a topic. This implies that, in these cases, a specific reference may not necessarily refer to the sentence(s) immediately preceding it, but even to a previous one. This also implies that the reference supporting a given argument should, in some cases, be identified among a few other references. A quick glance at the topic of these citations in the reference list at the end of each chapter allows prompt identification of the appropriate reference for the specific sentence.

In addition, to facilitate reading of the text, key terms or definitions have been usually highlighted in bold when introduced for the first time and equations and mathematical derivations have been kept at a minimum, with the possibility to refer to the original study in case details are needed. Likewise, each chapter ends with a list of the main symbols used in the text, although this does not include any additional symbol appearing only in a specific figure, which is explained in the related caption.

Writing this monograph has been more challenging and time-consuming than expected, although it has allowed me to appreciate the impressive number of studies on volcano-tectonics and to deepen several topics outside my limited personal experience. This effort comes at what I believe is a mature stage in my career, where many scientific curiosities have been finally satisfied, although new exciting ones come into sight at the horizon. But it also comes at an extremely delicate period for the academia of Italy, with a lack of opportunities, initiatives and resources that has been ongoing for nearly a decade, worsened by a general incapacity or refusal from colleagues and institutions to rely on meritocracy. This makes the future of research in the Italian academia highly uncertain, the only certainty being that the country is losing competitiveness.

Thus, it is with a heavy heart and the legacy of this monograph that I plunge myself into the uncertain future.

Rome, Italy
September 2020

Valerio Acocella

Acknowledgements

Before the "ordinary" acknowledgment of colleagues, friends, collaborators and family, I must start from afar, with a quite eccentric note.

This monograph is in many ways a result of 25 years of hard and committed work, in which I made field campaigns in five continents, run thousands of models, read many thousands of papers, wrote a respectable (not thousands, though) amount of papers and projects, presented and convened at countless meetings, reviewed and edited hundreds of manuscripts, thought many courses, and much more... Yes, this is the usual academic life, some may say. I am not sure if it is, but surely I enjoyed doing all this in the fast lane. Odd as it may seem for an academic, as speed requires energy, I must thank James Hetfield, Lars Ulrich, Kirk Hammett, Cliff Burton, Jason Newsted and Rob Trujillo, aka Metallica, for having provided the required fuel for so many years. Without their pace, I would have certainly accomplished much less.

Now, on a different note, the "ordinary" acknowledgments.

While this book has been typed by a single Author, I consider it as the result of the direct and indirect collaboration, participation, attention, support, discussions, suggestions and comments of many colleagues and friends over 25 years.

First of all, I must thank my mentor Renato Funiciello for having given me the opportunity, mainly during my master thesis and post-doc, to dive into volcano-tectonics, always enthusiastically supportive in those early formative years. Antonio Lazzarotto then gave me the unique possibility to start my career with a Ph.D. on tectono-magmatic problems at the University of Siena, showing a similar supportive enthusiasm.

Many colleagues and friends have shared their research with me in Italy and outside. I would have missed many things without the scientific (and not just that) support of (in alphabetical order) Bekele Abebe, Yosuke Aoki, Franco Barberi, Maurizio Battaglia, Olivier Bellier, Boris Behncke, Alessandro Bonaccorso, Marco Bonini, Chiara Cardaci, Luca Caricchi, Roberto Carniel, Raffaello Cioni, Giovanni Chiodini, Jim and Christine Cole, Fabio Corbi, Giacomo Corti, Chiara Cristiani, Mauro di Vito, Rosanna De Rosa, James Dohm, Federico Galetto, Nobuo Geshi, Adelina Geyer, Anna Gioncada, Agust Gudmundsson, Grant Heiken, Sigurjon Jonsson, Derek Keir, Christopher Kilburn, Tesfaye Korme, Andrei Kozhurin, Peter Lipman, Francesco Maccaferri, Roberto Mazzuoli, Nahid Mohajeri, Genene Mulugeta, Marco Neri, Chris Newhall, Andy Nicol, Eugenio Nicotra, Ricardo

Omarini, Carolina Pagli, Danilo Palladino, Paolo Papale, Angelo Peccerillo, Adriano Pimentel, Virginie Pinel, Vera Ponomareva, Massimiliano Porreca, Giuseppe Puglisi, Eleonora Rivalta, Mauro Rosi, Joel Ruch, Laura Sandri, Federico Sani, Eugenio Sansosti, Roberto Scandone, Steve Self, Esmaeil Shabanian, Silvia Simei, Karl Spinks, John Stix, Roberto Sulpizio, Andres Tassara, Elisa Trasatti, Thor Thordarson, Alessandro Tibaldi, Daniele Trippanera, Luigina Vezzoli, Tom Walter, Ryoichi Yamada and, last but definitely not least, Takeyoshi and Yoriko Yoshida. Many of these colleagues have also shared with me days, weeks or months in the field in Italy, Iceland, Ethiopia, New Zealand, Japan, Kamchatka, Easter Island, the Azores and the Andes, where my memory goes to Ricardo Omarini.

Some of these human beings, including Jim Cole, Agust Gudmundsson, Marco Neri, Joel Ruch, Luigina Vezzoli and Takeyoshi Yoshida have been more than pleasant, kind and supportive colleagues, and I would like to remember them also for their friendship.

Many colleagues and friends helped me in reviewing early or late versions of the chapters of this book, providing constructive criticism, ideas and suggestions. In particular, Adelina Geyer, Agust Gudmundsson, Anthony Hildenbrand, Luca Ferrari, Mike Poland, Thomas Sisson, Andres Tassara, and Benjamin vanWyk deVries provided helpful reviews on an earlier version of some chapters. In addition, Falk Amelung, Bill Chadwick, Shan de Silva, Francesca Funiciello, Derek Keir, Chris Kilburn, Luis Lara, Peter Lipman, Thierry Menand, Chris Newhall, Mike Poland, Eleonora Rivalta, Joel Ruch, Laura Sandri, Steve Self, John Stix and Alessandro Tibaldi provided prompt, enthusiastic and invaluable support in reviewing the late version of one or more chapters. Professor Steve Self also kindly provided the Foreword to this monograph. Finally, Falk Amelung, Roger Buck, Michael Heap, Thierry Menand, Sigurjon Jonsson and Marlene Villeneuve provided additional helpful discussions on specific topics. Needless to say, while many ideas and suggestions in this book are borrowed from them, the errors are all mine!

I am also thankful to colleagues who enthusiastically assisted in (a) finding and providing bibliographic material and in (b) providing appropriate figures and images.

Among the former, I must thank for their patience Marco Neri and Matteo Trolese, who used very different but equally effective approaches, and the personnel at the Roma Tre BAST Library (Enza Gasbarro, Marta Izzi, Marco Muscolino).

Among those providing figures and images, I must thank Kyle Anderson, Marco Bagnardi, Francesca Bianco, Elodie Brothelande, William Chadwick, Mauro Di Vito, Cynthia Ebinger, Agust Gudmundsson, Zhong Lu, Michael Manga, Christina Neal, Marco Neri, Carolina Pagli, Adriano Pimentel, Eleonora Rivalta, Joel Ruch, Freysteinn Sigmundsson, Alessandro Tibaldi, Giuseppe Vilardo, Tom Walter, Jennifer Woods, Dapeng Zhao, as well as NASA and USGS.

In preparing this book, I have also appreciated the help and support of the Springer Team, in particular of Annette Buettner, Marion Schneider, Birke Dalia, Dinesh Natarajan and Monica Janet Michael.

Here I would also take the opportunity to thank in advance any reader who will send his/her comments to me so that improvements can be made for a next edition.

Finally, on a more personal note, I would like to remember those with whom I shared my personal life while in the academia.

First, Serena Majetta provided so much support, dedication enthusiasm and love in those carefree early days that, without her, things may have been different today: an angel in paradise.

Then, Roberta and, in particular, my parents Nicola and Anna, who supported me long before I ever started to think about volcanoes. Needless to say, without them, and their constant love, there would have been no book on Volcano-Tectonic Processes, and much, much more.

Lastly, my deepest thoughts go to those who stayed next to me for so many years and, inevitably, also experienced most of the burden imposed by writing this book: my own family. Flavia and Elena had to bear the load of an absent minded and grave father for many months. Francesca had to cope with a frantic and, at the same time, dull husband for even longer… Yes, at moments we all felt like "All work and no joy make Valerio a dull boy". Luckily for us, this book was not written in wintertime at the Overlook Hotel, so I can now focus again on our amusing journey together.

Contents

About the Author

Valerio Acocella professor of Volcano-Tectonics and Volcanic Risk at the University of Roma Tre, Rome, Italy, has been studying volcanoes worldwide for more than two decades, publishing approximately 150 scientific papers. He has coordinated large research projects on Etna and Campi Flegrei, as well as the IAVCEI Commission on Calderas. He has been serving as editor for several scientific journals and organizing sessions at many scientific meetings.

Volcanoes and Volcanic Activity 1

1.1 Introduction

Volcanoes are an astonishing manifestation of the activity of planets and their satellites, as observed in active and/or fossil examples on Earth, Mars, Venus, Mercury, the Moon and the Jovian satellite Io. On Earth, volcanoes are one of the most impressive evidence of the same imbalance in energy that is also driving plate tectonics. Active volcanoes have terrified and at the same time fascinated populations and civilizations for thousands of years. This instinctive attitude towards volcanoes has been gradually replaced, in the last centuries, by a more rational curiosity and scientific approach, heralding the birth of volcanology, although only in the last decades volcanology has grown to the point of understanding general processes. Nevertheless, this increased knowledge has been only partially paralleled by an improved capacity to forecast eruptions and mitigate the environmental impacts of volcanic activity, which remain the ultimate and exciting challenges for modern volcanology.

Any synthesis of the general knowledge gained on volcanoes would deserve a voluminous monograph, and for an overview several appropriate texts may be examined (e.g., Schmincke 2004; Lockwood and Hazlett 2010; Sigurdsson et al. 2015). Here this introductory chapter builds up on these texts to summarize the fundamental concepts on volcanoes and their activity, also pairing Chap. 2, which summarizes the basic notions on crustal deformation. Both introductory chapters provide the essential background to the remaining eleven chapters focusing on the stress-deformation processes associated with the rise, emplacement and eruption of shallow magma (i.e., in the upper crust), at the local and regional scale, which concern volcano-tectonics.

Under these premises, the main aims of this chapter are to introduce the:

- distribution, composition and types of volcanoes, as deriving from the plate tectonics frame;
- main types of volcanic activity;
- concepts of volcanic hazard and risk;
- volcano-tectonic processes, that are the focus of this book.

1.2 The Volcano Factory

Volcanoes are found exhibiting very different types of activity on several planets and satellites: in this book the focus is on the volcanoes on Earth. Herein a **volcano** is a vent that develops when **magma**, that is a high-temperature (generally >650 °C) mixture containing melt, crystals

V. Acocella, *Volcano-Tectonic Processes*, Advances in Volcanology,
https://doi.org/10.1007/978-3-030-65968-4_1

and gases, erupts on the Earth's surface, in subaerial, subglacial or subaqueous environments. The repeated eruption of magma and accumulation of volcanic deposits from one or more vents in the same location over time develops a **volcanic edifice**. This commonly has a quasi-conical shape and is from tens to thousands of metres high and several tens of metres to several tens of kilometres wide. The edifice may be characterized by a summit depression, or **crater**, up to a few hundred of metres wide and deep, from which the magma is erupted. Occasionally, magma may also be erupted from the slopes or flanks of the edifice, or even outside, through **peripheral vents** (Fig. 1.1; e.g., Schmincke 2004, and references therein).

When reaching the surface, erupted magma feeds volcanic activity, rapidly solidifying and forming **volcanic** rocks. However, most magma remains trapped at depth in the crust and solidifies more slowly, forming **plutonic** or **intrusive** rocks: taken together, volcanic and plutonic rocks constitute **magmatic** (or igneous) rocks. Therefore, while it is tempting to relate a volcano to the topographic construction built by the erupted magma, it should be understood that the volcanic edifice is just the surface part of a larger system developed at depth. Indeed, a volcano should be considered, in the broadest sense, as the surface expression of a deeper and complex network of magma, which may eventually erupt. This network defines the **plumbing system** of the volcano, which consists of magma penetrated within the surrounding or country rocks, forming shallow **intrusions** with different shapes (Burchardt 2018, and references therein). The physical conditions of these intrusions vary significantly. For example, their thermal state may range from molten (between ~700 and ~1200 °C, depending upon magma composition) to solid (that is, completely crystallized, at the country rock temperature at that depth), passing through an intermediate mushy state, where melt and crystals coexist. The geometry of the magmatic intrusions forming the plumbing system is also extremely variable, although some basic shapes may be recognized, as described below.

The shallowest portion of the plumbing system, just below the volcanic edifice, may include a subcylindrical and vertically elongated central **magmatic conduit**, which feeds the eruptions

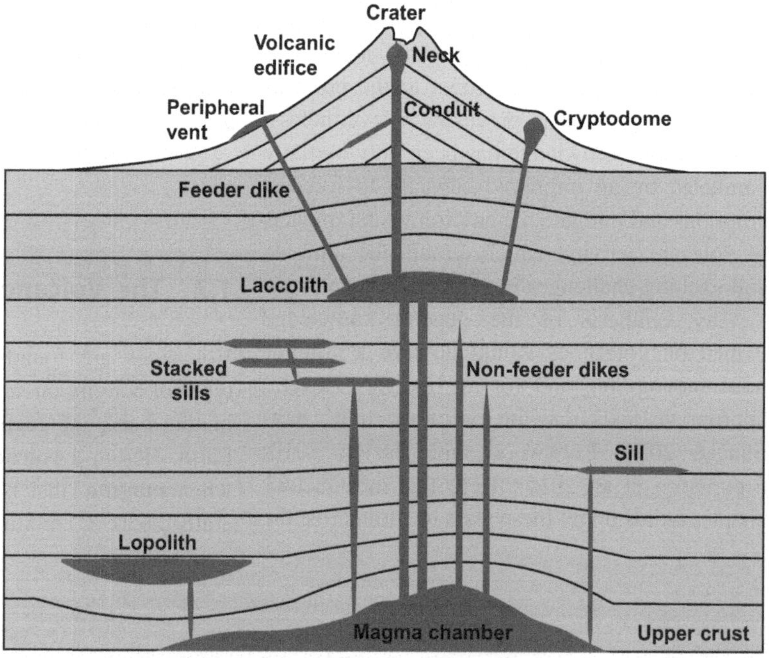

Fig. 1.1 Scheme of a volcano, consisting of the volcanic edifice, the central conduit and the shallow plumbing system, made up of different types of intrusions. Not to scale

connecting the vent(s) to a zone of deeper magma supply. The upper part of the magmatic conduit may include **necks** (vertically-elongated feeders below the vent, made up of solidified magma) or **cryptodomes** (very shallow and viscous magma stagnating and solidifying without reaching the surface). The plumbing system below the upper magmatic conduit mainly consists of intrusions with planar shape, or **magma-filled fractures**, with thickness much smaller than the lateral extent (Fig. 1.1; Burchardt 2018, and references therein). Magma-filled fractures discordant with the country rock layers are named **dikes**. These, described in Chaps. 3 and 7, are responsible for the rise of magma and, ultimately, for feeding eruptions (feeder dikes). Magma-filled fractures concordant with the country rock layers are named **sills**. Slls may thicken as a consequence of the increase in pressure of the intruded magma, eventually forming **laccoliths**, where the upper surface of the intrusion is domed, or **lopoliths,** where the lower surface of the intrusion is domed. Sills, laccoliths and lopoliths, as described in Chap. 4, are thus concordant magmatic intrusions characterized by different conditions of storage, or **emplacement**, of magma in the crust. These intrusions allow accumulating large quantities of magma (up to thousands of km^3) at shallow depth (usually a few km), promoting the development of **magma chambers** and **magma reservoirs**. These two terms are often used synonymously, although the former should refer to the melt-dominated region, and the latter to the mush-dominated (melt and crystals) region, characterized by cooling and solidification (see Chap. 4; e.g., Sparks et al. 2019, and references therein). At deeper levels magmatic intrusions include plutons and batholiths. A **pluton** is an isolated, solidified and massive intrusion of magma, with an average volume of 10^2–10^3 km^3; some plutons may represent solidified magma chambers of extinct volcanoes. A **batholith** is a larger intrusion, with an average volume of $\sim 10^4$ km^3, deriving from the assemblage of several plutons at a few tens of kilometres of depth.

Magmatic **gases** accompany the rise, emplacement and eruption of magma. Gases may be dissolved (in solution) within the magma, or exsolved (distinct) from the magma, depending on their pressure, temperature and composition, as well as on the magma composition (Fischer and Chiodini 2015, and references therein). These gases mainly consist of water vapour (H_2O), carbon dioxide (CO_2) and, subordinately, of sulphur dioxide (SO_2), hydrogen sulphide (H_2S), carbon monoxide (CO), hydrogen chloride (HCl) and hydrogen fluoride (HF). Magmatic gases released from a magma reservoir or chamber may contaminate, heat and pressurize the meteoric fluids in shallow or deep aquifers, thus forming a **hydrothermal system**. Many volcanoes show a hydrothermal system above the magma chamber, whose presence requires three components: fluids (meteoric and magmatic), permeability through rocks (fractures) so that fluids can circulate, and heat source (magma).

Magma, with its cargo of dissolved and exsolved gases, provides the energy within the volcano factory: magma transported in bursts to the surface is the most remarkable and readily observable evidence of the larger energy stored within the Earth. The energy associated with magma drives a wide range of processes: among these, the rise of magma towards the surface is largely a consequence of its heat, which makes magma lighter than the country rock, gaining the buoyancy pressure to rise.

1.3 Distribution of Volcanoes and Origin of Magmas

The distribution of volcanoes on the Earth's surface is determined by the availability and, ultimately, the generation of magma. These conditions depend on the tectonic setting, as described in the frame of the plate tectonics theory. Plate tectonics not only explains the location of volcanoes, but it also justifies the variable chemical composition of the erupted magmas. This compositional variability is in turn associated with different physical properties and finally responsible for the development of different types of volcanic edifices and eruptive styles. This section provides an introduction to the plate tectonic processes responsible for magma generation, to explain in the subsequent sections the resulting main compositional and physical variations of magmas, the different types of volcanic edifices and eruptive styles. The concepts described in this section are then discussed in more detail in Chap. 2 (concerning crustal deformation) and Chaps. 10–13 (concerning the regional context of volcanism).

The Earth consists of several shells with different physical and chemical properties (Fig. 1.2a; e.g., van der Pluijm and Marshak 2004, and references therein). The **crust** is the Earth's thinnest and outermost shell, made up of magmatic, metamorphic and sedimentary rocks. The crust can be divided into oceanic and continental: the former is found below oceans and is thinner (thinning to ~5 km) and denser (average density of ~2900 kg/m^3); the latter is found on continents, and is thicker (up to ~80 km) and lighter (average density of ~2700 kg/m^3). The boundary between the crust and the underlying layer, the mantle, is geophysically defined by the marked increase in the velocities of the seismic waves (from approximately 7 to over 8 km/s) and is called the Mohorovicic Discontinuity (or **Moho**). The **mantle** is ~2870 km thick and is predominantly solid, although in geologic time it behaves like a viscous fluid. It has average density of ~4600 kg/m^3 and its upper part mainly consist of Fe- and Mg-rich silicates, forming a rock called peridotite. Then, at ~2900 km of depth the Earth's core begins, with a radius of nearly 3500 km and average density of ~10,600 kg/m^3. The core is thought to consist mainly of nickel and iron; its outer portion is liquid, whereas the inner is solid.

Using a more dynamic subdivision of the Earth's outer portion, the crust and the uppermost mantle together form the **lithosphere**. This can reach a thickness of 150–200 km below continents and may be defined as the mechanically coupled, relatively cool (<1280 °C) and rigid outer portion of the planet (Fig. 1.2a; e.g., Stuwe 2007, and references therein). This behaviour is opposed to that of the underlying hotter (>1280 °C) and less viscous **asthenosphere**, which is the portion of upper mantle characterized by variable amount of partial melt (from ~2 to ~10%) and with variable thickness, reaching in some places ~300 km of depth. This partially molten zone is commonly explained by the depth increase in temperature approaching the melting point of the peridotite. The increase in temperature with depth derives mainly from the radiogenic heat produced by the radioactive decay of the isotopes in the Earth's interior, as well as from the primordial heat left over with its formation. Because the Earth is much hotter in its core than the outer mantle or the crust, heat migrates along thermal gradients to the Earth's surface and then radiates into space: this results in the release of heat, or **heat flow**. The mean heat flow of the Earth is estimated as of ~92 mW/m^2, although in volcanic areas magma and hot fluids may produce heat flows up to a few W/m^2 and, locally, over 100 W/m^2, as for example in parts of Yellowstone caldera (Wyoming, USA; e.g., Lucazeau 2019, and references therein). The rate of increase in temperature with regard to the depth within the Earth is expressed by the **geothermal gradient**. This gradient decreases with depth: for example, the mean geothermal gradient within a depth of 5 km in a continental crust is 25–30 °C/km, to become ~15 °C/km at a depth of 40 km; in volcanic areas the geothermal gradient may be higher than 200 °C/km, as for example in the Kenya Rift (Lagat 2003).

According to the theory of plate tectonics, the Earth's lithosphere consists of seven major plates and several minor plates which slowly move (on average with velocity of cm/yr) with regard to each other, following deeper circulation in the Earth's mantle. An assumption of plate tectonics is that the lithospheric plates are essentially rigid, implying that the deformation resulting from their relative motion focuses along the plate boundaries. As a result, the distribution of seismicity and the largest earthquakes also focus along the plate boundaries (Fig. 1.2b; e.g., Le Pichon 1968). Depending upon the predominant relative motion between neighbouring plates, three types of plate boundaries, each

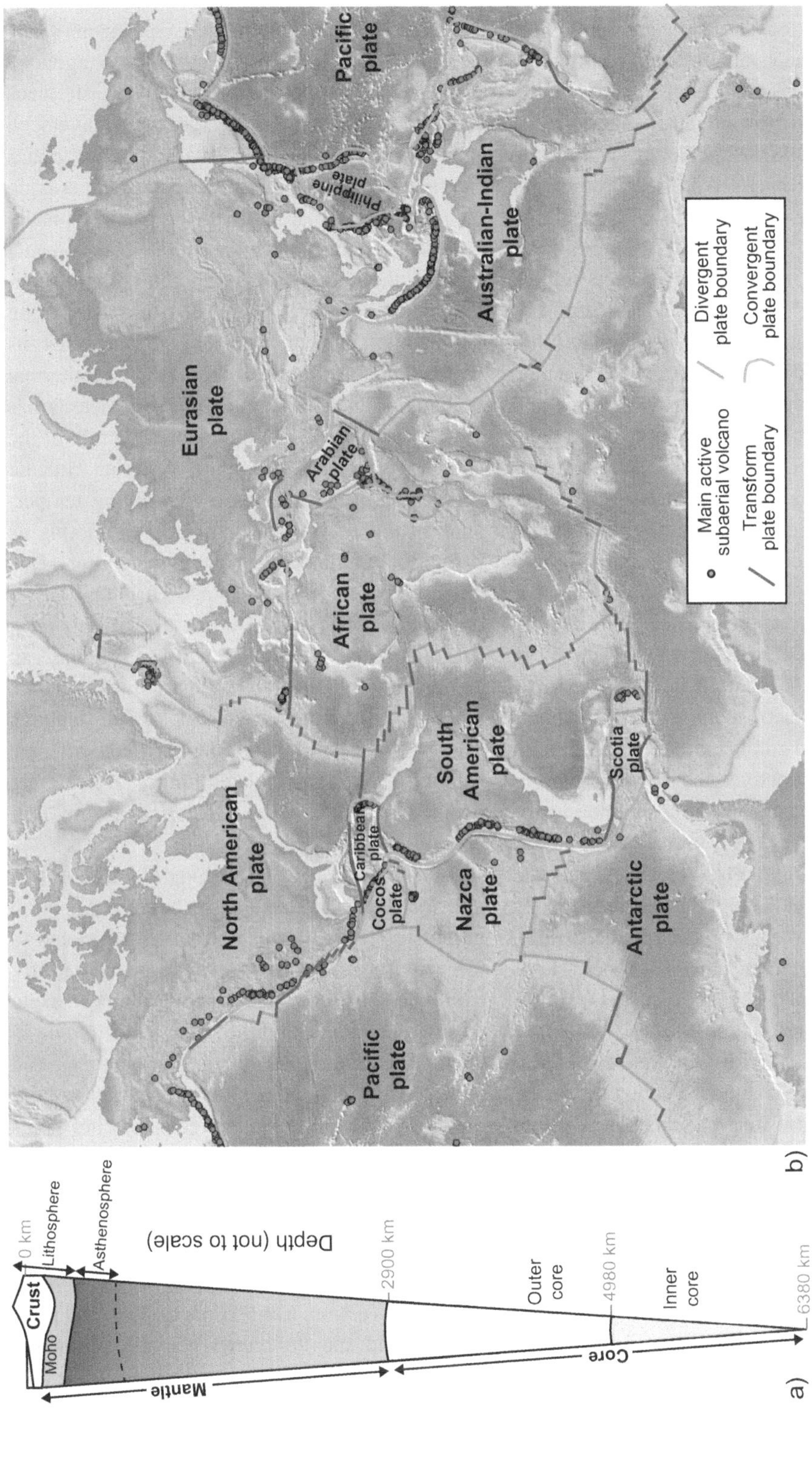

Fig. 1.2 **a** Subdivision of the internal portion of the Earth into several shells. **b** The distribution of active volcanoes on the surface of the Earth closely mimics the boundaries of the main lithospheric plates, which may be characterized by divergent, convergent or transform motions

characterized by a distinct deformation pattern, are distinguished: **divergent plate boundaries** develop along the contact between plates that move away from each other; **convergent plate boundaries** develop when two nearby lithospheric plates move toward each other; **transform plate boundaries** develop when two nearby plates move laterally past each other (for details see Chaps. 2 and 10; e.g., van der Pluijm and Marshak 2004, and references therein). Plate tectonics assumes that mantle convection (i.e., heat transfer through material flow, see Sect. 1.5) determines the motion of the overlying lithospheric plates, with upwelling regions of mantle flux underlying divergent plate boundaries and downwelling regions underlying convergent plate boundaries, In this way, the extension created along the divergent boundaries is compensated by the shortening along the convergent boundaries also at depth.

Volcanoes are not randomly distributed on the Earth's surface. The approximately 1500 terrestrial volcanoes with known or possible activity in the last 10,000 years focus in fact along narrow but continuous belts on continents and islands (Fig. 1.2b; e.g., LaFemina 2015, and references therein). This distinctive distribution closely mimics the onland portion of the boundaries of the lithospheric plates. The overall coincidence between the distribution of volcanoes and the plate boundaries indicates that the processes governing plate motion also control the generation of the magma feeding the volcanoes. However, magma-generating processes differ depending upon the type of plate boundary: while magma does not form at transform plate boundaries, divergent and convergent plate boundaries generate magma through distinct conditions resulting from their different motion. As explained in more detail in Chap. 10, divergent plate boundaries are characterized by the separation of the two nearby plates, which promotes lithospheric thinning and is accompanied by a decrease in pressure in the mantle below. This decompression generates partial melting in the asthenospheric mantle. The melt rises through buoyancy and interacts with the thinned crust, developing shallow magma chambers and feeding volcanic activity along the divergent plate boundary. This takes place predominantly on oceans, with submarine volcanism along oceanic ridges (the site of oceanic divergence), and subordinately on continents, along rift zones (the site of continental divergence). Conversely, in convergent plate boundaries the two plates moving closer commonly determine the sinking (or **subduction**) of the heavier plate (typically consisting of oceanic lithosphere) in the mantle beneath the lighter one (often consisting of continental lithosphere). The oceanic subducting plate sinks into the mantle and increases its temperature, following the local geothermal gradient. This temperature increase determines the release of fluids from the upper portion of the subducting plate into the overlying mantle wedge. The fluids decrease the melting temperature of this portion of the mantle, promoting its partial melting and thus magma generation. This magma then rises, intrudes and heats the thicker crust of the plate above. If continental, this crust has relatively low melting temperature and can be partially melted by the intruded mantle magma, generating additional crustal magma. These mantle and crustal magmas feed volcanic activity on the upper plate along a volcanic arc parallel to and at some distance from the convergent plate boundary.

As shown in Fig. 1.2b, several volcanoes are also located away from plate boundaries, in an intraplate setting. These volcanoes are related to the activity of **hot spots**, which are the surface expression of hot and narrow plumes rising through the mantle and the crust, and assumed as fixed. The fixed location of these plumes and volcanoes is thus independent of any plate configuration or motion, although some hot spots may be located along plate boundaries, most commonly divergent boundaries (Davies 1999, and references therein; Courtillot et al. 2003).

Volcanic activity is thus the expression of a much larger geodynamic process, which includes mantle convection, asthenospheric flow and plate motion, and mantle plume activity, ultimately also responsible for the generation of earthquakes. As discussed in the sections below, the specific tectonic conditions occurring at divergent and convergent plate boundaries and at hot spots are responsible for different magma compositions and properties, which ultimately affect the type of volcanic edifice and eruptive activity.

1.4 Composition of Magmas

The composition of magmatic rocks varies considerably and is usually expressed in terms of major, minor, and trace elements. Major elements have oxide abundances higher than 1 wt%, minor elements have abundances between 0.1 and 1 wt %, and trace elements are below 0.1 wt%. Major and minor elements are expressed as oxides: SiO_2, Al_2O_3, FeO, Fe_2O_3 CaO, MgO, and Na_2O are major elements, whereas K_2O, TiO_2, MnO, and P_2O_5 are minor elements. For example, volcanic rocks can be described as a function of the abundance of silica (SiO_2) against the total alkali content (Na_2O + K_2O). However, plutonic rocks, crystallizing at depth, are best described as a function of the relative percentage of the contained minerals, as quartz, plagioclase and alkali feldspar (Fig. 1.3; Le Maitre et al. 2002; Rogers 2015, and references therein). Despite their variety, magmatic rocks are dominated by two compositional types: basalts and granites. **Basalts** are volcanic rocks and, with their intrusive equivalent (gabbroes), dominate the oceanic crust. Basalts have the representative composition for **mafic** (rich in Fe and Mg oxides) rocks, also called **basic** (primitive) rocks. At the surface, the **solidus** (temperature below which the rock is completely solid) and **liquidus** (temperature above which the rock is completely liquid) of basalts are ~1000 and ~1200 °C, respectively. **Granites** are plutonic rocks and, with their less common volcanic equivalents (rhyolites), are the most abundant igneous rocks in the continental crust. Granites have the representative composition for **felsic** or **silicic** (rich in silica) rocks. The higher silica and lower iron content of granites makes them lighter than basalts, also explaining why the continents are on average 5 km higher in elevation than the ocean basins. At low crustal pressures granites have solidus temperature of ~960 °C (anhydrous solidus) that can be substantially lowered by volatile constituents, such as H_2O, reaching ~680 °C (hydrous solidus); their liquidus temperature is ~1050 °C (e.g., Brown 2013, and references therein).

Basalts are mainly generated through the separation of the plates along oceanic ridges. Here

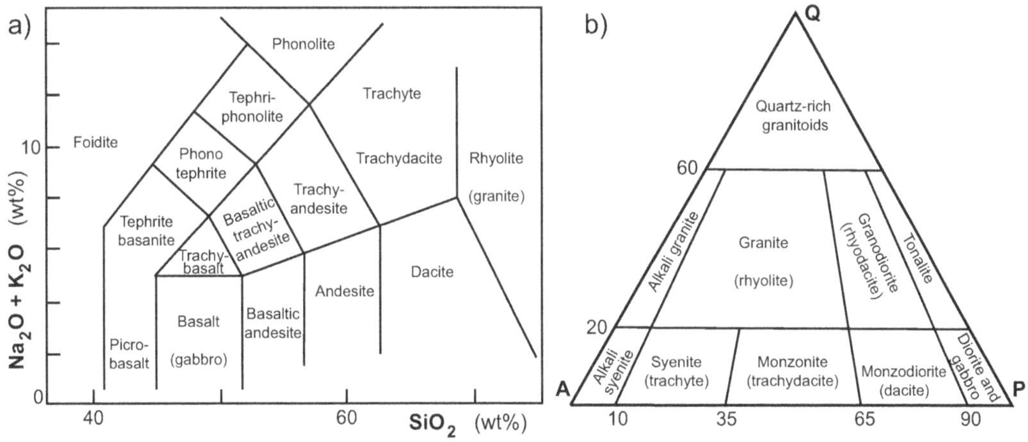

Fig. 1.3 Classification of volcanic (**a**) and intrusive (**b**) rocks. **a** Total alkali (Na_2O and K_2O) versus silica (SiO_2) diagram: superimposed are the boundaries that delineate the different types of volcanic rocks, at times including also their plutonic equivalent (in purple). **b** QAP diagram with Q representing the amount of quartz in the rock, A of alkali feldspar and P of plagioclase. Granites have more than 20% quartz, whereas rocks with less than 20% quartz are classified as intermediate; volcanic equivalents are shown in purple (modified after Le Maitre et al. 2002)

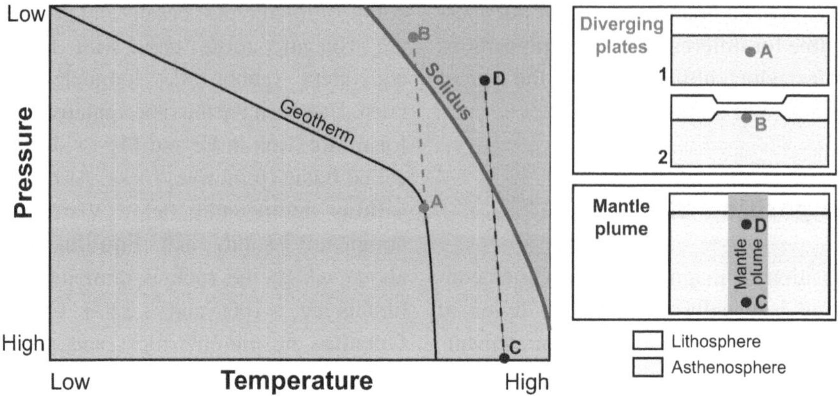

Fig. 1.4 Decompression melting of mantle rocks lying at point A on a given geotherm. If the overlying crust is extended, the rocks rise to point B, with a significant decrease in pressure and minor decrease in temperature. As point B is above the solidus, the rocks partially melt. Points C and D illustrate the decompression path of mantle rocks at higher temperatures, as within a rising plume. Here decompression occurs at higher depths and may produce a greater volume of melt. Right diagrams show schematic section views (not to scale) of decompression melting associated with plate divergence (top) and mantle plumes (bottom; modified after Rogers 2015)

plate divergence promotes thinning of the oceanic lithosphere, decompression of the asthenospheric mantle and the generation of basaltic melt from the mantle peridotite (a process named **decompression melting**; Fig. 1.4). The melt rises and intrudes along the plate boundary, originating a Mid-Oceanic Ridge Basalt (**MORB**), which is a **tholeiitic basalt** poor in sodium and potassium with ~50% of SiO_2 (Fig. 1.5; e.g., Rogers 2015, and references therein). Eventually, the basalt experiences the separation due to gravity of the minerals containing the denser elements (Mg and Fe oxides) from those containing the lighter ones (Ca, Al, alkali oxides), shifting composition from primitive (Mg- and Fe-rich minerals) to more evolved (Al-, alkali- and Si-rich minerals). The separation of the primitive and denser minerals from the evolved and lighter ones, called **fractional crystallization**, may generate a wide range of magma compositions (see below).

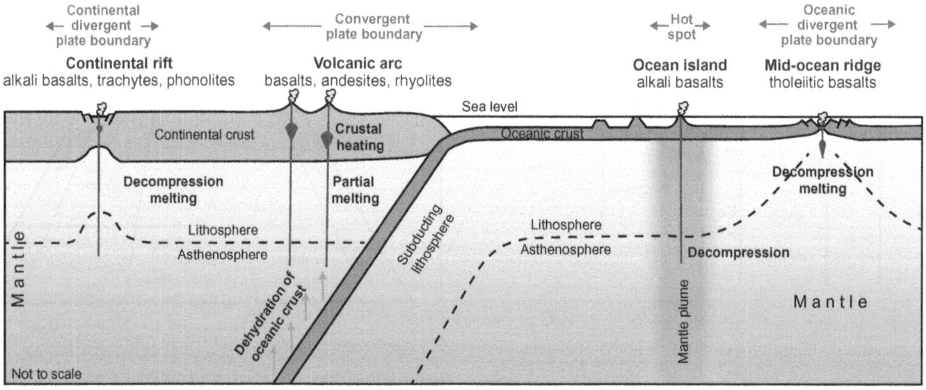

Fig. 1.5 Schematic section through continental and oceanic lithosphere, showing the major processes responsible for magma generation (in green) and the related main erupted compositions (purple) as a function of the tectonic setting (blue; modified after Rogers 2015)

Basalts may also form in oceanic intraplate settings, as in hot spot volcanic islands (as Hawaii, Tahiti, Easter Island, Reunion and Galapagos). Here melting induced by the localized upwelling and decompression of the hotter and lighter mantle plume feeding the hot spot volcanism explains the generation of basalts. As the mantle within the plume is hotter than that through which it rises, it can start melting as a consequence of decompression at a greater depth than that beneath ocean ridges (Fig. 1.4). Unlike oceanic ridge settings, here the top of the melting zone is confined by the thicker overlying lithosphere, so that the Ocean Island Basalts (**OIB**) resulting from this process derive at greater depths than MORB, with lower amount of produced melt from a given parcel of mantle. As OIB tend to be richer in alkali and iron and poorer in silica than MORB, they are termed **alkali basalts** (Fig. 1.5).

Alkali basalts may also form in correspondence of rift zones along continental divergent plate boundaries, such as the East African Rift System, or intraplate rift zones, as the Baikal Rift (central Asia), the Rhine Graben (western Europe) and the Rio Grande Rift (southern USA). As here the lithosphere experiences a lower amount of extension with regard to oceanic ridges, mantle melting is restricted to depths similar to those beneath ocean islands. Consequently, these basalts resemble those of ocean islands, dominated by alkaline compositions and their evolved derivatives, such as **trachytes** and **phonolites** (Fig. 1.5; e.g., Rogers and Hawkesworth 2000; Rogers 2015).

Basalts are also generated above subduction zones in convergent plate boundaries, where the denser oceanic lithosphere subducts into the mantle. This lithosphere carries water trapped in sediments and seafloor basalts that were hydrothermally altered at oceanic ridges. This water is released as the subducting lithosphere warms up with depth, and migrates into the overlying mantle wedge, inducing here melting by lowering the peridotite solidus of 200–300 °C: the resulting basalts are tholeiitic. If the subduction takes place beneath continents, these basalts may be trapped at the crust-mantle boundary (Moho) or within the thicker continental crust. The intrusion of basaltic magma into the continental crust heats the country rock, leading to its partial melting and remobilization. Melts of crustal rocks tend to have granitic compositions and can mix with the basaltic magma to make a hybrid rock along a "calcalkaline" trend, generating an **andesite**, whose name derives from the volcanoes of the Andes, arguably the archetypal active volcanic arc (Fig. 1.5). In addition to the mixing of basaltic and granitic melts, andesites may originate through fractional crystallization of basalts. Further fractional crystallization of andesitic magma, possibly combined with the assimilation of crustal material, leads to even more silica-rich compositions, as **dacites** and **rhyolites** (e.g., Rogers and Hawkesworth 2000; Rogers 2015).

As fractionation proceeds, the concentration of water builds up in the magma, because the crystallizing minerals are usually anhydrous. The end product is a silica-rich magma rich in water, responsible for the large and violent silicic volcanic eruptions occurring along the volcanic arcs in convergent plate boundaries. Therefore, during fractional crystallization of basalt, the progressive loss of the more primitive minerals enriches the liquid phase in the more evolved ones, reaching granitic compositions. Likewise, the composition of silicic magmas lies close to the lowest melting fraction of crustal silicate systems so that, if basalt is melted, the first melt that forms has a silica-rich composition. Thus, two possible routes to silicic magmas exist: one by partial melting of pre-existing crustal rocks, the other by extensive fractional crystallization of basaltic magma (Rogers 2015, and references therein). Silicic magmas commonly crystallize at depth forming granites, or are less frequently erupted as rhyolites. Volatiles, particularly water, play an important role in the formation of granites. Water decreases the melting temperature of crustal rocks, as it does in the mantle, and it is the dehydration of water-bearing minerals that allows the production of large volumes of granitic melt at moderate temperatures within the crust.

1.5 Physical Properties of Melts and Magmas

There is a close relationship between the evolutionary history and composition of a magma and its physical properties.

Density ρ_m is one of the most important physical properties of magma. Magma density is determined by the ratio between its mass m and volume V, as:

$$\rho_m = m/V \tag{1.1}$$

Density is strictly dependent upon pressure P, temperature T and composition: in particular, density increases as pressure rises and temperature decreases, and with anhydrous and mafic magma. As a result, densities of anhydrous melts vary from 2300 to 2900 kg/m^3 at low pressure and magmatic temperatures (~1100 °C) for silicic through intermediate to mafic and ultramafic compositions (Fig. 1.6; e.g., Lesher and Spera 2015, and references therein). Hot magma is commonly less dense than the colder country rock, gaining the **buoyancy** pressure to rise ΔP_b, defined as:

$$\Delta P_b = (\rho_r - \rho_m)gz \tag{1.2}$$

where ρ_r and ρ_m are the country (or host) rock density and magma densities, respectively, g is the acceleration due to gravity and z is the depth of the magma. Depending on the relative values of ρ_r and ρ_m, the buoyancy can be either positive (driving the lighter magma toward the surface) or negative (preventing the ascent of the denser magma); the level at which the buoyancy changes sign is the level of neutral buoyancy, that is when magma would stop ascending (e.g., Schmincke 2004, and references therein). The buoyancy of magma is also determined by its variations in density with regard to its pressure, temperature and composition. Likewise, the differentiation of magma by gravity-driven fractional crystallization critically depends on the density difference between the melt and the newly formed crystals. All these factors contribute to control the segregation, ascent, and eruption of the magma (e.g., Gregg et al. 2015). The density of magma also determines its compressibility, which is an important factor governing explosivity, especially for volatile-rich magmas. The alkali metals and water are relatively compressible and contribute significantly to the pressure dependence of density.

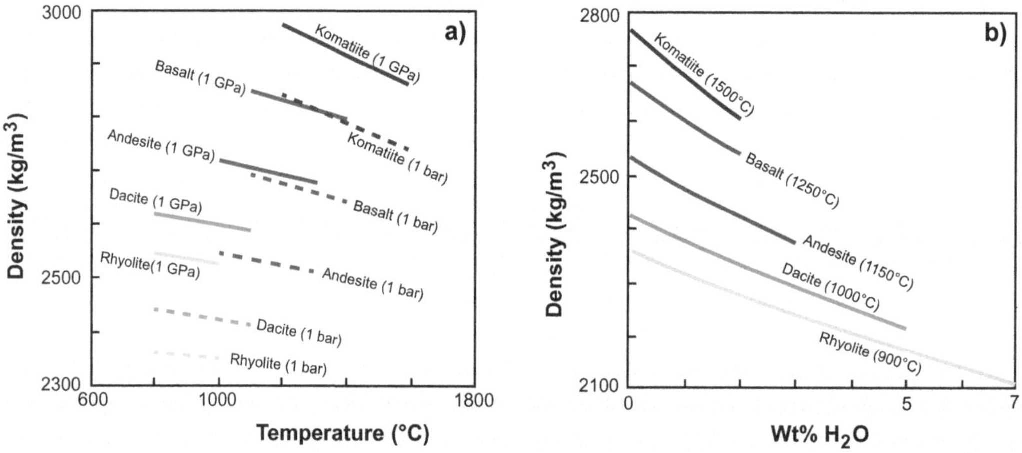

Fig. 1.6 **a** Melt density as a function of composition and temperature for natural melts spanning the volatile-free compositional range rhyolite to komatiite at 1 bar (10^{-4} GPa) and 1 GPa pressure. **b** Melt density as a function of dissolved water content for natural melts spanning the range rhyolite to komatiite. Temperatures for each melt are characteristic eruption temperatures. Densities are calculated at a pressure of 1 bar (10^{-4} GPa; modified after Lesher and Spera 2015)

Viscosity is another primary physical property of magma, as it controls different features at different scales, including the morphology of the volcanic edifice, the rates of magma flow, gas escape, convection, crystal settling or flotation, and crystal growth. The dynamic viscosity η of melt connects the shear stress σ in the fluid to its rate of strain $e^{\cdot}$, and is expressed in Pa s (see Sect. 2.4.3):

$$\eta = \sigma / e^{\cdot} \tag{1.3}$$

To a good approximation, silicate melts show a linear relationship between the strain rate and shear stress at fixed temperature and pressure, behaving as Newtonian fluids with constant viscosity (see Sect. 2.4.3). More generally, at a given pressure, viscosity depends upon the composition (including the water content) and temperature of the melt. As for the composition, melts can vary in viscosity up to several orders of magnitude, from highly viscous rhyolites (viscosity of 10^6–10^{10} Pa s) to poorly viscous basalts and komatiites (viscosity of 10^{-1}–10^2 Pa s; Fig. 1.7; Lesher and Spera 2015, and references therein). In general, the higher the silica (SiO_2) content, the higher is the viscosity of the melt. This follows from the fact that melts rich in SiO_2, as rhyolites, require higher activation energy to break the stronger Si–O bonds which link to form chains and complex three-dimensional structures within the melt (Schmincke 2004, and references therein). Volatiles (especially water) also affect the viscosity of melts (Fig. 1.7a). In molten silica, the typical framework consists of each oxygen with two nearest neighbours of silicon and each silicon surrounded by four nearest neighbours of oxygen. Increasing the concentration of non-framework components (as water, but also, to a lesser extent, alkalis) creates oxygens with only a single nearest neighbour of silicon, destroying the network structure. Small concentrations (several percent by mass) of water may have dramatic effects in lowering the viscosity of silicate liquids: while the effect on mafic compositions is moderate, granitic melts may decrease their viscosity up to 2 orders of magnitude. As for the temperature, there is an inversely proportional temperature dependence of viscosity, so that the higher the temperature, the lower the viscosity of melt, as shown in Fig. 1.7b. This

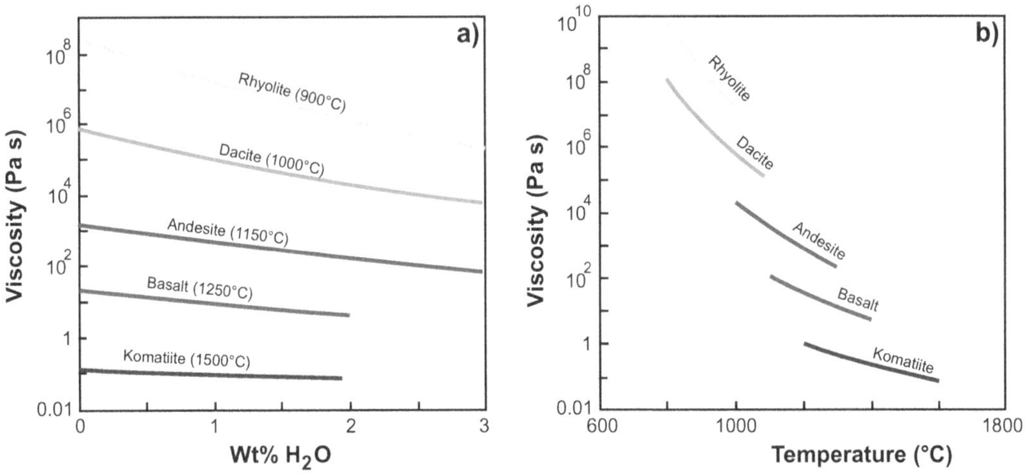

Fig. 1.7 **a** Melt viscosity (calculated at 1 bar, or 10^{-4} GPa) as a function of the composition and the dissolved water content for natural melts spanning the compositional range rhyolite to komatiite. Temperatures for each composition are characteristic eruption temperatures. **b** Viscosity as a function of temperature at 1 bar (10^{-4} GPa) for natural melts spanning the volatile-free compositional range rhyolite to komatiite. The temperature range is illustrative of typical eruption temperatures for each composition (modified after Lesher and Spera 2015)

inverse relationship is exponential, and has been approximated by different models, including the Arrhenian model, the Tammann-Vogel-Fulcher equation and more recent empirical expressions (see also Sect. 2.4.3; Giordano et al. 2008, and references therein).

Far fewer data exist on the viscosity of magmatic suspensions (where the magma is a mixture of suspended crystals and vapour bubbles in a melt matrix) compared to silicate melts. In general, magma with significant amounts of suspended crystals and bubbles exhibits a more complex rheology, departing from the Newtonian behaviour. The rheology of these multiphase (crystal-melt-vapour) fluids is the subject of experimental, analytical and numerical studies, which allow approximating the viscosity of magmatic suspensions composed of monosized particles in the creeping flow regime at very low rates of shear (Lesher and Spera 2015, and references therein; see also Sect. 2.4.6).

When two systems with different temperatures are put in contact, heat flows spontaneously from the hotter to the colder system. **Heat** is the energy, expressed in Joule, transferred from a system to the other as a result of thermal interactions. These interactions include **conduction** (that is the transfer of heat from a hotter to a colder body through contact, without flow of the body), **convection** (transfer of heat through the movement of a fluid, or bulk flow) and **radiation** (an electromagnetic phenomenon involving particle transfer). In the Earth's crust, melts and magmas transfer heat mainly through conduction.

Enthalpy is a calorimetric property of high temperature and molten silicates, bound to the thermal evolution of magma, partial melting, solidification, and heat transport. The specific enthalpy of fusion is the heat per unit mass needed at a reference pressure to transform crystals to the liquid state. There is a factor of three difference in the specific heat of fusion for a model granite ($\sim$200 kJ/kg) compared to an ultramafic composition ($\sim$600 kJ/kg). The relatively low fusion enthalpy for mineral phases comprising continental crust compositions confirms that partial melting of crust by heat exchange between mafic magma and its surroundings is thermally efficient (Lesher and Spera 2015).

Thermal conductivity, heat capacity and the resulting thermal diffusivity are the main properties controlling conductive heat flow, that is the dominant process of heat transfer in the Earth's crust. When a rock or magma gains heat, it stores some of its heat energy and transfers the remaining heat energy to the nearby rocks or magmas. Its ability to transfer heat energy is referred to as its **thermal conductivity** k, expressed in W/mK. In general, for a silicate liquid k decreases as the temperature increases, but increases as pressure increases. The heat storage capability of a rock or magma is referred to as its **heat capacity** C_p. The **thermal diffusivity** α quantifies how fast heat diffuses through a rock or magma. Expressed in m^2/s, it is defined as the ratio between the conducted and stored heat:

$$\alpha = k/\rho_m C_p \tag{1.4}$$

where ρ_m is the density of magma. Therefore, a rock or magma with higher thermal conductivity or lower heat capacity has large thermal diffusivity. The ratio expresses the rate at which the heat generated gets diffused out: the larger the thermal diffusivity, the easier the conductive propagation of heat into the medium. The thermal diffusivity is the relevant parameter in transient heat conduction problems, as well as a key parameter in determining the cooling rates of magmas (Lesher and Spera 2015).

The above considerations indicate that the most important physical properties of magma, including the viscosity, density and enthalpy of fusion, are largely affected by the tectonic setting from which the magma originates. For example, the basaltic magmas erupted at divergent plate boundaries and hot spots are denser, much more fluid and require a larger heat of fusion than the andesitic to rhyolitic magmas erupted at convergent plate boundaries. In turn, the composition and physical properties of a magma affect its rates and modes of rise and emplacement within the crust, highlighting an indirect control of plate tectonic processes also on the architecture of the plumbing system of a volcano. A more direct control on the size, shape and depth of the plumbing system results from the distribution of the regional stresses within the crust, still determined by plate tectonics. These features are discussed in detail in Chaps. 4, and 10–13.

1.6 Types of Volcanic Edifices

The previous sections have shown how a given tectonic setting determines the composition of the originated magma, which also affects its physical properties. In this section the effect of these physical properties on the size and shape of the volcanic edifice erupting the magma is considered (e.g., de Silva and Lindsay 2015, and references therein).

As regards the size, the larger and long-lived volcanic edifices growing from repeated eruptions are referred to as polygenetic, or **polygenic**, while smaller volcanic edifices resulting from a single eruptive event are called monogenetic, or **monogenic**. Polygenic or monogenic volcanoes may develop as isolated features, as the polygenic Mount Damavand volcano (Iran) or the monogenic Radicofani cone (Italy), or, more commonly, clustering in a **volcanic field**, as the polygenic volcanoes on Isabela Island (Galapagos) or the Michoacan-Guanajuato monogenic field (Mexico), with several volcanoes contemporaneously active. Polygenic and monogenic volcanoes may also display close genetic relationships, as for example indicated by the monogenic cones growing on the flanks of a polygenic volcano. The formation of polygenic or monogenic volcanoes in general depends on the tectonic setting and magma availability (see Sect. 10.6; Takada 1994).

As regards the shape, volcanic edifices are commonly conical, following the repeated accumulation of products around a central crater or

Fig. 1.8 Volcan Licancabur (Argentina-Bolivia border), here seen from northeast, is a typical Andean felsic stratovolcano and reaches 5920 m of height

Fig. 1.9 The basaltic shield volcano of Skjaldbreiður ("broad shield"), seen from Lake Thingvallavatn (Iceland)

vent. This leads to the archetype of polygenic volcano, named **stratovolcano, or composite volcano**, characterized by a conical edifice up to a few thousands of metres high, progressively steepening upward and culminating in a summit crater (Fig. 1.8; e.g., Schmincke 2004; Lockwood and Hazlett 2010, and references therein). The steepness of the flanks of stratovolcanoes (commonly with dip of 30°–35°, occasionally up to 40°–45°) results from the relatively high viscosity of the erupted lava, usually andesitic or dacitic, which does not flow for long distances and thus remains in a proximal location. The crater on the summit of a stratovolcano may include or be replaced by a **dome**, that is a steep mound of partly solidified viscous lava escaping the summit vent. Felsic stratovolcanoes are a very common type of edifice, especially in arcs along convergent plate boundaries. Conversely, volcanoes erupting low viscosity magma, usually basaltic, are characterized by a gentler edifice, called **shield** volcano as its less inclined flanks, usually dipping 10°–15°, resemble an overturned shield (Fig. 1.9). This shape results from the fact that the more fluid basaltic lavas flow for longer distances, reaching distal locations. Mafic shield volcanoes are common along oceanic divergent plate boundaries and in oceanic hot spots.

Some polygenic stratovolcano or shield volcano may lack the summit portion of the edifice, or even most of the conical topography. In this case, the volcanic edifice may show a broad depression, up to several tens of kilometres wide and very few kilometres deep, often filled with volcanic deposits and sediments. These volcanoes are named **calderas**, and are associated with the rapid or sudden withdrawal of magma, through eruption or lateral intrusion, from an underlying magma chamber (Fig. 1.10).

Calderas form independently of the viscosity of the magma, even though less viscous mafic magmas tend to form smaller calderas at divergent plate boundaries and hot spot volcanoes, and more viscous felsic magmas form larger calderas at convergent plate boundaries, highlighting distinct dynamic behaviours (see Chap. 5).

Fig. 1.10 View of the internal part of the Bromo-Tengger caldera, Java (Indonesia), approximately 8 km wide and filled with early morning mist surrounding post-caldera eruptive centres. The stratovolcano in the background is Gunung Semeru

Stratovolcanoes, shield volcanoes and calderas produce repeated eruptions and are all examples of polygenic volcanoes. However, their precise distinction may not be straightforward, as the above-mentioned typical features may coexist in a same volcanic edifice. For example, stratovolcanoes and shield volcanoes may both host a summit caldera, whereas some volcanoes may show a combination of shield and composite edifice: Mount Etna (Italy) provides an example of edifice whose base resembles a shield underlying a composite volcano hosting a summit paleo-caldera.

In map view, a volcanic edifice usually shows circular or subcircular shape, reflecting a symmetric build up, as Mayon volcano (Philippines). However, volcanic edifices are at times elongated, presenting an elliptical shape in map view. The eccentricity of the ellipse approximating the base of the cone, expressed as the ratio between its major L_M and minor L_m axes, is usually moderate (that is, $L_M/L_m < 1.4$). Nevertheless,

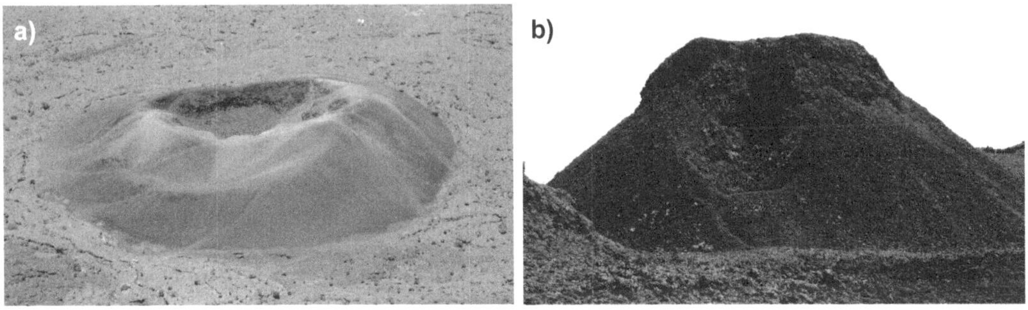

Fig. 1.11 Examples of **a** cinder and **b** spatter cones on the summit of Piton de la Fournaise volcano (La Reunion Island)

a)

b)

Fig. 1.12 Examples of eruptive fissures. **a** Fissure eruption developed along the Northeast Rift of Mount Etna (Italy) in late 2002. The fissure was fed by a dike propagating laterally from the summit conduit. Photo courtesy Marco Neri. **b** Detail of the southern portion of the ~25 km long 1783–1784 eruptive fissure of Lakagigar, Iceland, highlighted by the alignment of monogenic vents

some volcanic edifice, as Kilauea or Mauna Loa (Hawaii, USA) is strongly elongated, with eccentricity L_M/L_m larger than 1.4. The eccentricity of the base of an edifice thus quantifies its elongation, reflecting the effect of near- and/or far-field external factors on the growth of the volcano, as explained in the following chapters.

Monogenic volcanoes consist of much smaller edifices, formed during a single eruptive episode. **Cinder or scoria cones** are produced by the build up of ash or other fragments around a vent, usually during the explosive activity of low viscosity magma (Fig. 1.11a). **Spatter cones** result from a less explosive activity, when partially molten clots fall to the base of a lava fountain and enclose the vent in a ring of welded clasts (Fig. 1.11b).

Similarly to polygenic volcanoes, monogenic cones may also show preferred elongation, usually parallel to the strike of the feeder dike or of tectonic fractures in the basement. The rim of the crater of the cone may be also elongated and/or show a characteristic breaching, similarly deriving from the structural control on its development (Tibaldi 1995).

These monogenic cones may lie on the flanks of a polygenic edifice or also in a volcanic field without polygenic volcanoes. In both cases, the cones may be isolated, or, more commonly, aligned along a direction forming an **eruptive fissure**, that is the surface expression of the portion of the feeder dike reaching the Earth's surface (Fig. 1.12). The **alignment** of the cones forming the fissure, best identified remotely, may be further highlighted at a more detailed scale by the elongation of the single cones, each parallel to the alignment. Multiple and parallel eruptive fissures on the same flank of a polygenic volcano form a dike-fed **volcanic rift zone**, characterized by the repeated linear emission of magma (see Chap. 7). Multiple parallel eruptive fissures are also found outside polygenic volcanoes, identifying **rift zones** usually accompanying regional rifts or oceanic ridges along divergent plate boundaries (see Chap. 10). Therefore, volcanic rift zones and rift zones mainly differ in size, the former being restricted to the volcanic edifice (usually a few km long) and the latter extending for several tens of kilometres or more along a

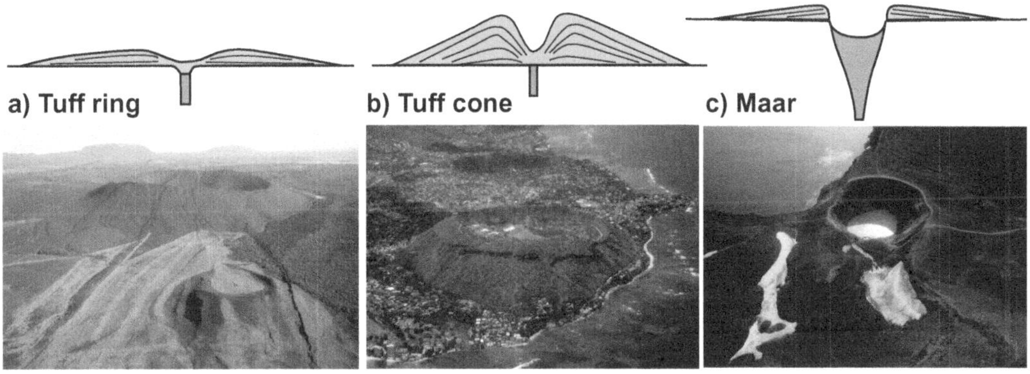

Fig. 1.13 Section (top) and areal (bottom) views of **a**: **a** tuff ring (image is of Hverfjall, Iceland), **b** tuff cone (Diamond Head, Oahu, Hawaii) and **b** maar (Viti, Askja, Iceland)

regional rift. At the scale of a regional rift zone, polygenic volcanoes also align parallel to the rift.

Monogenic volcanic edifices also derive from the sudden and explosive interaction between the magma and shallow water of meteoric origin (groundwater, lakes, glaciers, sea), forming tuff rings, tuff cones, or maars, where the magmatic products are mixed with clasts from the country rock. **Tuff rings**, commonly less than 50 m high, are defined by craters with small depth to width ratios at or above ground level and low rims with beds dipping less than 25° (Fig. 1.13a). **Tuff cones** have higher profiles than tuff rings and steeper external slopes, with bedding dips higher than 25°; their crater floors are generally above the land surface (Fig. 1.13b). **Maars** are bowl-shaped craters cutting from tens to hundreds of metres deep into the substrate surrounded by low ramparts of well-bedded ejecta (beds dipping less than 25°) that decrease rapidly in thickness away from the rim (Fig. 1.13c). Maar deposits are distinguished from those of tuff rings/tuff cones by the abundance of non-magmatic components (mainly country rock; e.g., de Silva and Lindsay 2015, and references therein).

1.7 Volcanic Activity

Volcanic activity results from the shallow emplacement, transfer and eruption of magma and manifests itself through the release of heat, hot fluids and magmatic products on the Earth's surface, in subaerial, subglacial and subaqueous environments. As mentioned, volcanic areas are characterized by a higher than average heat flow and geothermal gradient. These are commonly associated with the release of hot fluids at the surface, which may form **fumaroles** (fluids are mainly released as gases), **hot springs** (fluids mainly released as liquids), **geysers** (intermittently spouting jets of steam and hot water) and **mud pools** (acidic hot springs decomposing the country rock into clay and mud). Rather than having a direct magmatic origin, hot fluids mainly derive from groundwater, meteoric and/or surface water percolating downward and reaching depths where they are heated through convective circulation.

The most impressive manifestation of volcanic activity is however represented by **volcanic eruptions**, which occur through the effusion of lava or the explosive ejection of fragmented magma or older solidified material, in subaerial, subglacial and subaqueous settings. Of the ~1500 volcanoes which have erupted in the Holocene (last ~10,000 years), 575 have had historically documented eruptions and roughly 50–70 erupt every year (Siebert et al. 2015). Volcanic eruptions vary widely in both magnitude and duration and display a broad spectrum of eruptive styles and processes. Some eruptions may last very long, up to decades (as Kilauea). Other eruptions end much earlier: while 10% have lasted no longer than a single day, most have ended in less than 3 months, and a few last

Table 1.1 Categories of the Volcanic Explosivity Index, VEI (modified after Newhall and Self 1982; Pyle 2015)

VEI	0	1	2	3	4	5	6	7	8
Bulk tephra volume (m^3)	$<10^4$	$<10^6$	$<10^7$	$<10^8$	$<10^9$	$<10^{10}$	10^{11}	$<10^{12}$	$>10^{12}$
Plume column height (km)	<0.1	0.1–1	1–5	3–15	10–25	>25	>25	>25	>25
Qualitative description	Gentle	Effusive	Explosive	Explosive	Cataclysmic	Paroxysmal	Paroxysmal	Paroxysmal	Paroxysmal
Stratospheric injection	None	None	None	Possible	Certain	Certain	Certain	Certain	Certain
Percentage of known eruptions in the last 10,000 years	13	16	49	14	5	2	<1	<0.1	0
Typical recurrence interval	Days-weeks	Days-weeks	Days-weeks	0.3 years	3 years	20 years	80 years	500 years	7×10^5 years

longer than 3 years. The median duration of an eruption is approximately 7 weeks. The median interval between successive eruptions of a same volcano is about a dozen years; however, many volcanoes have been quiescent for significantly longer periods.

The parameters most widely used to define the scale of an eruption are the magnitude, that is the mass of erupted material, and the intensity, that is the mass eruption rate: as both parameters may vary by many orders of magnitude, logarithmic scales are used (e.g., Pyle 2015, and references therein). The concepts of magnitude and intensity are both included in the definition of the **Volcanic Explosivity Index** (VEI), which is commonly used to quantify the size of an eruption (Table 1.1; Newhall and Self 1982). The VEI uses an integer scale from 0 to 8, where each step corresponds to an increase in energy of one order of magnitude, to broadly describe the erupted volume and the eruption column height. Although this index is defined by both the eruption magnitude (volume) and intensity (column height), it does not really describe how the energy is distributed during the eruption. The VEI scale is not useful for effusive eruptions, which are predominantly non-explosive and therefore receive a default classification of 0 or 1; nor can the VEI scale be readily applied to very small eruptions. The VEI provides a scale that has been applied to more than three-fourths of reported Holocene eruptions (Fig. 1.14; Siebert et al. 2015; Papale 2018). Although VEI 2 is used as a default assignment for explosive eruptions in the absence of other criteria, it remains the dominant VEI for eruptions during the past 50 years, when more quantitative data are available. Eruptions with VEI $\geq$ 2 show a distribution characterized by a decrease in frequency with increasing size, while eruptions with VEI $\geq$ 3 highlight a power law distribution. This behaviour implies that smaller eruptions (VEI 3) have a frequency of approximately once every few weeks, while the largest eruptions

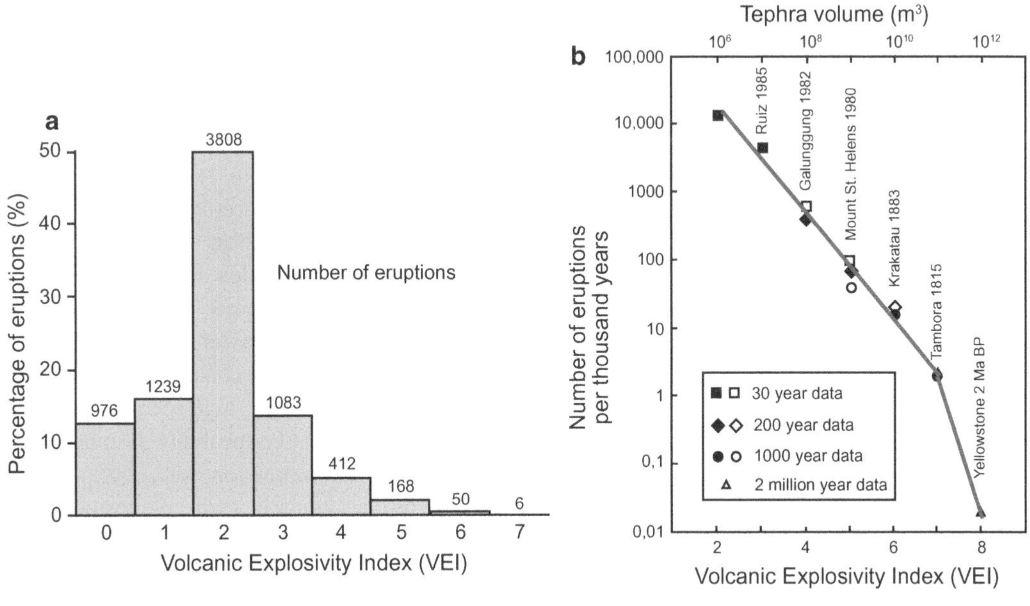

Fig. 1.14 **a** Percentage and number of Volcanic Explosivity Index (VEI) assignments for all Holocene eruptions (modified after Siebert et al. 2015). **b** Magnitude and frequency of Holocene eruptions. Cumulative number of eruptions during various intervals (before 1994) for each VEI class and normalized to "per 1000 years." Eruptions of low explosivity (VEI 0 and 1) are not shown. The best-fit line is determined by an exponential regression model for VEI 2–7 using data points that are filled. Data points for VEI 7 and 8 are shown by open triangles. Well-known eruptions are labelled to illustrate each VEI class (modified after Simkin and Siebert 2000)

(VEI 8) are believed to occur once every few hundred of thousands of years. Therefore, most eruptions are fairly modest, although proximal effects can still be considerable. Only 5% of documented Holocene eruptions are VEI 4 (the size of the devastating 1902 eruption of Pelée, Martinique), and only 2.4% of explosive eruptions are VEI 5 or larger, the size of the 1980 eruption of Mount St. Helens (Washington, USA). Eruptions the size of Mount St. Helens in 1980 occur perhaps once a decade, and those such as Krakatau (Indonesia) in 1883 (VEI 6) once a century. The known Holocene record implies that VEI 7 eruptions, such as at Tambora (Indonesia) in 1815, occur less than once a millennium, although this record is incomplete.

One assumption implicit in the VEI is that the magnitude and intensity of eruptions are in some way related, so that a single integer can fully describe the different elements of the size of an eruption. Instead, if the full spectrum of effusive and explosive volcanic activity is considered, there cannot be a single simple relationship between eruption intensity and magnitude. Magnitude and intensity are not independent variables (both are a function of the erupted mass), but using two measures of eruption size rather than one makes it possible to describe and compare both explosive and effusive eruptions and both short- and long-lived eruptions (Pyle 2015, and references therein). Therefore, to compare different types of eruptions, two separate scales are needed: one for magnitude, the other for intensity. The eruption **magnitude scale** is defined as follows:

$$\text{Magnitude} = \log_{10}(\text{erupted mass}) - 7 \quad (1.5)$$

where the erupted mass is expressed in kg. For most eruptions, magnitudes are numerically similar to their VEI, even though the magnitude scale is continuous and the VEI scale is discrete. The **intensity scale** is defined as follows:

$$\text{Intensity} = \log_{10}(\text{mass eruption rate}) + 3 \quad (1.6)$$

where the mass eruption rate is expressed in kg/s: on this scale, a very vigorous eruption has intensity of 10–12, while minor eruptions have intensity of 4–6. The intensity of eruptions scales with the rate of the released thermal energy, which represents by far the greatest usable energy contained in magma.

One useful property of the magnitude and intensity scales is that both can be used to describe historic, prehistoric, and geological events. Moreover, since the scales are logarithmic, large errors in estimates of erupted mass or mass eruption rate do not translate into large errors in magnitude or intensity. One weakness of the intensity scale is that it is still quite hard to measure (Pyle 2015). To make comparisons between different deposits, volumes are often corrected to a Dense Rock Equivalent (**DRE**) volume. This is an estimate of the volume of dense, non-vesicular magma that erupted to form the deposit.

1.8 Effusive Eruptions and Their Products

An **effusive eruption** is characterized by the flow of molten rock (lava) from a vent: this occurs when the magma from which the lava originates becomes volatiles-poor upon ascent to the surface. Deep in the crust volatiles (mainly H_2O and CO_2) are dissolved into the magma because of the high pressures, but upon ascent pressure drops rapidly and these volatiles begin to exsolve, forming bubbles. For an effusive eruption to occur, the pressure within the bubbles should remain low and/or the excess volatiles should be removed from the ascending magma. The first condition may be met by the slow ascent or decompression of magma, whereas the second condition requires efficient gas loss, with high permeability (or bubble connection). Permeability is moderately sensitive to decompression rate, but strongly affected by variations in melt composition (viscosity) and crystallinity, with high permeability being related to low viscosity magma with high melt fraction (Mueller et al. 2008; Cassidy et al. 2018; Okumura et al. 2020). Basaltic magmas normally produce effusive eruptions because they contain less volatiles and their lower viscosity favours the relaxation of the overpressure

Fig. 1.15 Hornito (small vent emitting degassed magma) feeding a minor basaltic lava flow at Mount Etna (Italy) in 2001

generated by volatile **exsolution**, that is when the dissolved volatiles in magma separate from the magma and form bubbles. In addition, basaltic magmas require a lower supersaturation pressure to overcome the energy barrier provided by surface tension. This allows bubble nucleation and coalescence with a minor amount of decompression, further increasing magma permeability and gas release (Gonnermann and Manga 2007). For all these reasons, effusive eruptions are most common in mafic volcanoes, which predominate along divergent plate boundaries and at hot spots.

Effusive eruptions are characterized by the emplacement of **lava flows** (Fig. 1.15).

Lava flows dominate the surface of most volcanoes and ocean floors. The length of a lava flow depends upon its emission rate and viscosity. As lava flows downslope, it cools, crystallizes and becomes rheologically stiffer. Therefore, while the rear and inner portions of the flow fill with fluid lava, its cooled frontal and external portions commonly stall. Lava may be transported in open channels and/or in enclosed tube systems, where cooling is significantly slower. Basaltic lava flows have typical viscosities of 10^2–10^3 Pa s, increasing towards 10^5 and 10^6 Pa s for cooler and more crystallized basalts. Lava viscosities increase with silica content, so that an andesite flow may have viscosity of 10^5–10^7 Pa s, increasing to 10^7–10^{10} Pa s for a rhyolite (Harris and Rowland 2015).

Three types of lava are traditionally distinguished: pahoehoe, ‘a’a and block lava (e.g., Harris and Rowland 2015). Pahoehoe lava has lobate forms with smooth, glassy, coherent surface, sometimes with surface folds, or “ropes”. An active pahoehoe lobe consists of a fluid core with outer crust that progressively thickens. Pahoehoe lava flows may often form positive topographic features, known as **tumuli**, where the inflating fluid mass causes the brittle crust to crack into plates which are then uptilted. The resulting tumulus is typically dome-like in section, rounded to oval in plan, with a main axial fracture running parallel to the long axis. ‘A’a lava is characterised by brecciated surface and basal crusts, separated by a coherent fluid interior. Pahoehoe and ‘a’a lavas can coexist during the same eruption and differ in viscosity (‘a’a being more viscous) and amount of stirring, or shear stress, experienced by the down slope flow (‘a’a experiencing more shear; Lockwood and Hazlett 2010). Block lava is typical of thick (thickness of tens to hundreds of meters) dacitic and rhyolitic lava flows, with relatively smooth clasts.

Some volcanoes host persistently active **lava lakes**, as Kilauea, Erta Ale (Ethiopia; Fig. 1.16), Mount Erebus (Antarctica), and Nyiragongo (Democratic Republic of the Congo). Lava lakes usually fill craters and result from conduits filled with molten magma and open (connected) to the surface (**open conduits**). This condition is opposed to that, more common, of **closed conduit** volcanoes, where solidified magma fills the conduit, confining and pressurizing the molten magma below. Lava lakes constantly churn as new lava wells up from below and cooler surface lava becomes denser and founders, triggering shallow convection (Lockwood and Hazlett 2010).

Submarine eruptions are dominated by basaltic compositions with low volatile contents, thus mainly leading to lava flows. Submarine lavas that advance slowly relative to the rate of crust formation produce pillow lavas, whereas lavas that advance quickly relative to crust formation rates produce sheet lavas. Pillow lavas are composed of meter-scale, poorly connected bulbs of lava not extending significantly from the vent. At the highest eruption/flow rates, sheet lavas are produced; their interiors are fully interconnected and the surfaces lack any indication of separate units. These features can be used as a proxy for lava flow rate and eruption rate for submarine basalts (White et al. 2015). A further product of eruption of basaltic lava into water results from large-scale fragmentation occurring when the

Fig. 1.16 Lava lake, with diameter of ~10 m, within a pit crater at Erta Ale volcano, Afar, Ethiopia, in December 2003

lava is rapidly chilled by seawater. This produces a deposit composed of glassy basaltic particles of sand size mixed with separated pillows, termed **hyaloclastite**.

Lava domes are mounds of viscous magma that usually pile up and accumulate around a vent as magma degasses and cools (Fig. 1.17; Calder et al. 2015, and references therein). Over the last decades, dome-building eruptions have been repeatedly occurring at Mount St. Helens, Soufriere Hills (Montserrat) and Unzen (Japan) volcanoes. Domes are a few tens of metres to a few kilometres wide, and up to 1 km high. Their morphology can vary from steep-sided to tabular in cross-section and circular or elliptical to irregular in plan view, mainly depending on the topography of the basement, composition of magma and mode of growth. Domes restricted to crater areas are nearly circular, while domes on the flanks of volcanic edifices tend to elongate downhill. Dome lavas cover a wide compositional range, from basaltic through rhyolitic, although the majority are andesitic and dacitic. Lava domes show highly variable crystal contents, with some constituting the most crystal-rich lavas erupted. The growth of the dome may be endogenous or exogenous: endogenous growth refers to the expansion caused by intrusion of new magma, while exogenous growth refers to dome enlargement as a result of magma flowing from the vent and forming discrete lobes of lava (e.g., Calder et al. 2015, and references therein). Active lava domes are made of an inner ductile lava core and a cooler, brittle outer rock carapace flanked by a talus composed of fragmental blocks resulting from brittle failure. Internally, lava domes reveal strain related to shear localization along the conduit and dome margins. Away from the vent, the carapace may exhibit compressional ridges, generated by folding of the dome surface due to internal shear traction from a spreading lava core. Lava domes can form relatively rapidly, over days to weeks, but eruptive episodes can extend for years to

Fig. 1.17 Aerial view of viscous lava domes forming within the Mount St. Helens (Washington, USA) crater in September 2001. Credit: U.S. Geological Survey, USGS

decades, with effusion rates varying widely, from 0.01 to over 100 m^3/s. As a result, eruptions vary from unthreatening effusion to highly unpredictable and hazardous activity. Hazards associated with lava domes include the collapse of portions of the dome and moderate to major explosive activity. The discharge rate associated with lava dome eruptions is generally low when compared to lava flow activity and explosive eruptions, with cyclicity on multiple timescales (Calder et al. 2015, and references therein).

1.9 Explosive Eruptions and Their Products

Explosive magmatic eruptions experience magma **fragmentation**, in which magma passes from a continuous liquid phase with dispersed gas bubbles and crystals to a gas phase with dispersed magma fragments (pyroclasts). The explosive force of magmatic eruptions is indeed generated during fragmentation, where the potential energy of expanding magma is converted to the kinetic energy of the gas phase and individual pyroclasts, and then enhanced by thermal expansion of the gas-particle mixture within volcanic plumes (e.g., Cashman and Scheu 2015). In particular, the decompression of a rising magma induces volatile exsolution from the melt, forming bubbles. If bubbles grow faster than they either rise individually, a condition usually associated with faster decompression, the magma may ultimately fragment in the liquid or solid state, with viscous or brittle behaviour. Conversely, if bubbles rise through a slowly decompressing magma, or if bubble networks are sufficiently interconnected to allow permeable gas escape, the magma may reach the surface without fragmenting, feeding an effusive eruption (Okumura et al. 2020). The most abundant volatile phase is water. Even small initial amounts of dissolved water will expand to very large volume fractions (>99%) at atmospheric pressure. Therefore, the amount of water retained or lost during the transit to the surface contributes substantially to the explosivity of the eruption. The fragmentation process is promoted by the

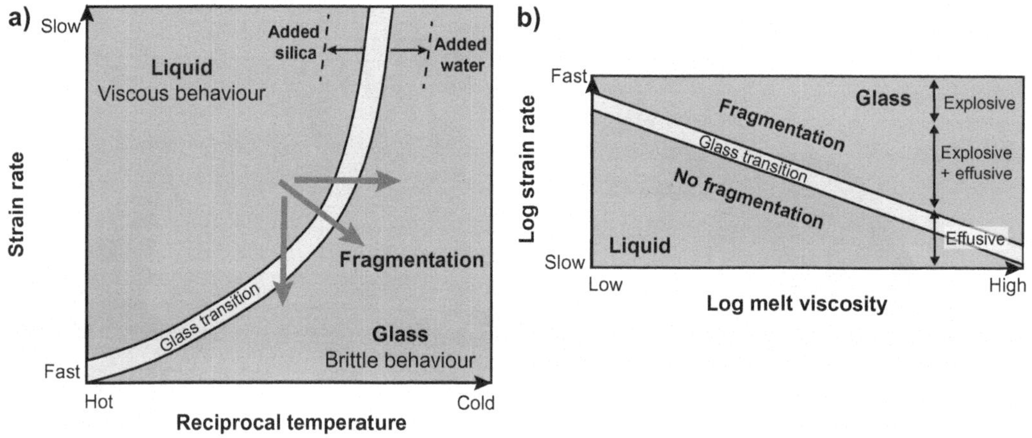

Fig. 1.18 Melt fragmentation. **a** Strain rate-temperature diagram illustrating the glass transition (in yellow), which divides the behaviour of silicate melts into viscous liquid and brittle glass. The transition is very sensitive to variations in H_2O and SiO_2 content of the melt (see shift of transition zone indicated by the black arrows in top right of diagram), highlighting the importance of degassing and crystallization. Melt fragmentation occurs by crossing the glass transition, by increasing the strain rate and/or cooling the melt (see grey arrows). **b** Glass transition in a log–log strain rate-viscosity diagram. For low strain rates the melt reacts as a liquid, and no fragmentation occurs. High strain rates cause brittle response and fragmentation. The transition depends on melt viscosity. At the right side eruption styles are indicated, based on typical strain rates (Dingwell 1996; Gonnermann and Manga 2003; Cashman and Scheu 2015)

increase in the rate of deformation of the magma upon decompression (strain rate), the increase in the magma viscosity and amount of silica, as well as the decrease in temperature and amount of water, and magma crystallinity (Fig. 1.18; Dingwell 1996; Gonnermann and Manga 2003; Cashman and Scheu 2015).

The fragmentation efficiency increases with increasing eruption energy, developing particles with smaller grain size. Therefore, explosive eruptions are classified based on their fragmentation index and dispersal area. Both parameters use maps with **isopachs**, or contours with equal thickness of the erupted deposit. The fragmentation index is the percentage of ejecta in an air fall deposit (or tephra, see below) with grain size of less than 1 mm, measured at the 0.01 H_{max} isopach (where H_{max} is the maximum thickness of the deposit), whereas the dispersal area is the area enclosed within the 0.01 H_{max} isopach of a given eruption air fall. Using these parameters, explosive eruptions have been classified into Hawaiian, Strombolian, sub-Plinian and Plinian (with increasing dispersal area) and Violent Strombolian, Vulcanian and Surtseyan (with increasing fragmentation; Fig. 1.19; Walker 1973). These eruptions are briefly described below.

Hawaiian and Strombolian eruptions

These are characteristic of silica-poor magmas, often of basaltic to basaltic andesite composition, that erupt at temperatures of 1000–1200 °C. The low viscosity of these magmas (10–10^4 Pa s) allows for efficient segregation of gas bubbles and ductile melt deformation over short timescales relative to the ascent of the magma. These eruption styles represent the most frequent manifestations of subaerial explosive volcanism.

Hawaiian eruptions are characterized by the sustained jetting of gas and fluidal pyroclasts through fountaining, often associated with lava flows (Fig. 1.20). Hawaiian activity consists or explosive episodes that are sustained on timescales of hours to days, producing fountaining of molten magma clots and little ash, without wall-rock lithic debris. Lava fountains can reach elevations of hundreds of metres and are typically episodic, with highly variable duration of episodes and length of repose periods. **Strombolian** eruptions consist of the bursting of buoyant gas bubbles at the free surface of a vent. This results in weak, discrete, short-lived explosions that are the surface manifestation of the impulsive release of pressurized pockets of gas resulting from bubble coalescence. In fact, at a shallow depth the rapid expansion of the gas pockets fragments and ejects adjacent molten magma clots and ash, plus any minor lithic debris (Taddeucci et al. 2015, and references therein). Pyroclasts from Hawaiian and Strombolian activity include scoria (i.e., moderately vesicular, mostly glassy pyroclasts dominated by mm- to cm-sized vesicles, constituting the most common product), bombs and spatter (commonly found in proximal settings).

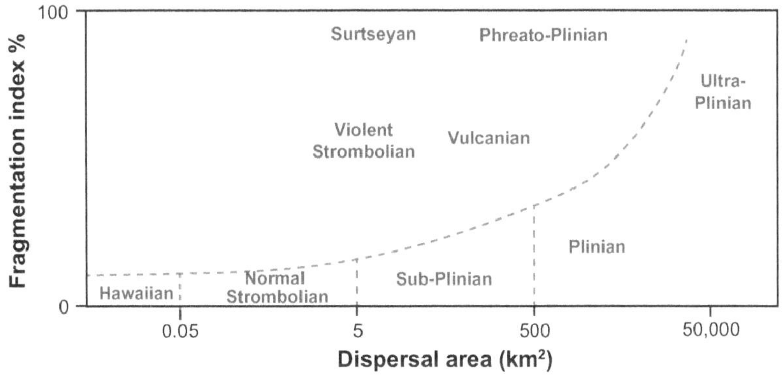

Fig. 1.19 The Walker eruption classification diagram, reporting the main types of eruptions as a function of the fragmentation index and the dispersal area

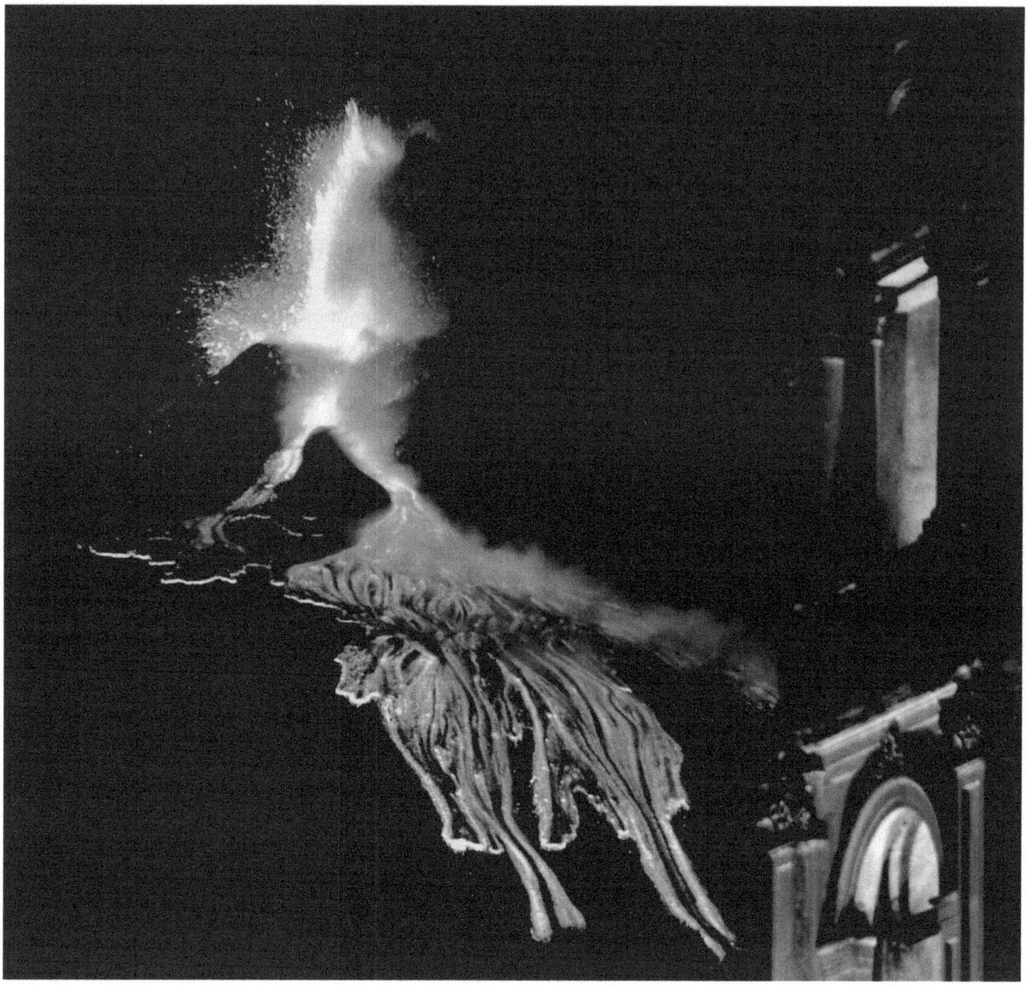

Fig. 1.20 Image of lava fountain and related flow occurred on the night of 24 April 2012 from the New SE crater at Mount Etna, Italy, as seen from the village of Zafferana Etnea (part of the village church is visible to the right). *Photo courtesy* Marco Neri

Vulcanian eruptions

A variety of explosive volcanic phenomena fall under this category. Their unpredictable (and hazardous) nature is related to the variability in magma composition, volatile content, degassing and crystallization patterns, and different pre-eruptive distribution of vesicularity and pressure, all of which control magma decompression in the conduit or feeding dike. Despite this, **Vulcanian eruptions** are well defined by their eruptive mechanism, emphasized by the critical role of a gas-impermeable plug composed of degassed and partly solidified magma that forms in the upper part of a slowly ascending magma column. This plug results from the sudden decompression of a conduit that contains highly pressurized crystallized bubbly magma of intermediate composition. In fact, as magma rises, water exsolves, promoting crystallization and increasing magma viscosity. This may form a viscous plug that leads to the stagnation of the underlying magma, which in turn increases its pressure. The high pressure gradient in the magma disrupts the conduit plug or dome, initiating the Vulcanian eruption (Fig. 1.21). Plug disruption occurs when pressure in the underlying conduit rises

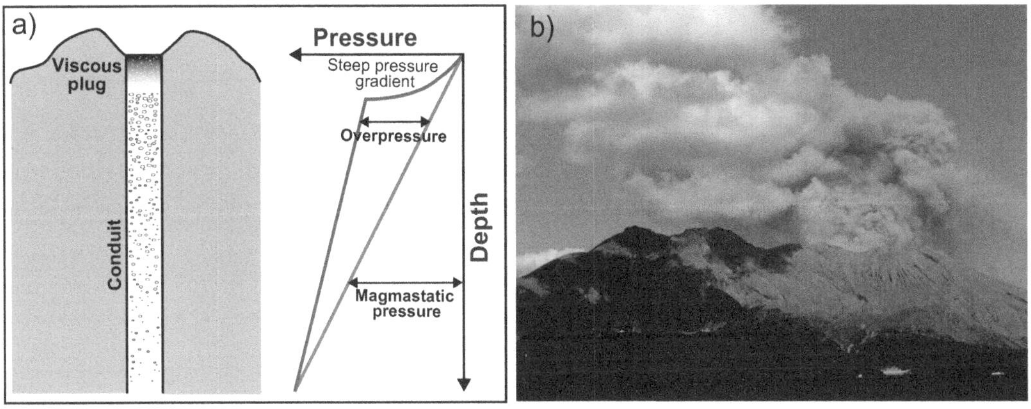

Fig. 1.21 **a** Conditions preparing a Vulcanian eruption. A viscous plug (grey) seals a conduit containing bubble-bearing magma (left), determining a depth-pressure profile with the magma passing from magmastatic (unplugged state, blue curve) to overpressured (plugged state, red curve) conditions (right; modified after Clarke et al. 2015). **b** Image of minor Vulcanian eruption (grey cloud to the left), from the Minamidake crater of Sakurajima volcano, Japan, as seen from northwest on July 22, 2013

sufficiently (up to 10–15 MPa, but usually less than 5 MPa, based on the strength of typical magmas) to exceed the mechanical strength of the plug or when the overlying dome collapses. This disruption results in the fragmentation of both the plug and underlying magma. At the fragmentation front, vesicular magma is disrupted into a gas-pyroclast mixture, propelled upward, and ejected from the vent into the atmosphere as a jet at sonic to supersonic velocities. Typically, only a portion of the magma in the conduit is fragmented and evacuated, such that Vulcanian eruptions are usually of relatively small volume and last only seconds to minutes: they may occur as single events or in sequence (Clarke et al. 2015, and references therein). Vulcanian explosions can precede large Plinian eruptions, as at Mount Pinatubo (Philippines) in 1991.

Plinian eruptions

Plinian eruptions are characterized by the sustained, high-speed discharge into the atmosphere of a high-temperature, multiphase mixture (gas, solid and liquid particles), forming a buoyant vertical column that reaches heights of tens of kilometres (Fig. 1.22). The composition of magma driving most Plinian and sub-Plinian eruptions is intermediate to silicic, while only few basaltic Plinian eruptions are known (as in 1886 at Tarawera, New Zealand and in 122 BC at Etna, Italy; e.g., Cioni et al. 2015, and references therein).

The Plinian regime encompasses sub-Plinian, Plinian, and ultra-Plinian styles. **Sub-Plinian** eruptions have lower intensity and magnitude (10 and 4, respectively), with fluctuations or temporary breaks in the discharge, explained by syn-eruptive degassing and viscosity increase of the magma column during its slow ascent. The partially degassed, high viscosity portion of the magma crystallizes rapidly, stagnating in the upper conduit and promoting a pressure build up that eventually reopens the system, as repeatedly occurred at Mount St. Helens after the 18 May 1980 climactic eruption. Plinian and **ultra-Plinian** eruptions share common eruption dynamics, but different values of intensity and magnitude (10–12, and 4–8, respectively); in fact, ultra-Plinian eruptions have higher mass flow rate, which reflects in higher columns and larger dispersal.

Fig. 1.22 Eruptive column from explosive episode piercing through the clouds developed during the 2009 Sarychev (Kuril Islands) eruption seen from the International Space Station. Pyroclastic flows developed from the volcano summit (image credit: NASA)

Magma fragmentation in Plinian eruptions is expected to occur when reaching a critical concentration of bubbles ($\sim$75 vol%). During the eruption a mixture of gas, magmatic and non-magmatic particles are discharged from a vent at high speed (typically 150–600 m/s). The conduit radius may experience significant enlargement (2–5 cm/min), due to its mechanical erosion, becoming several tens of metres wide; sub-Plinian eruptions would require conduit diameters on the order of ten metres. During the eruption a mixture of finely fragmented multiphase magma, solid fragments of pre-existing rocks and gas is injected in the atmosphere. After having attained its maximum height, the column eventually spreads laterally into an "umbrella" cloud (Fig. 1.23; Cioni et al. 2015, and references therein).

Magma fragments (juveniles) mainly consist of highly vesicular pumice clasts and ash fragments. In the gas-thrust region, the eruptive jet incorporates air and eventually gains positive buoyancy with respect to the atmosphere, shifting to a convective regime. Above the gas-thrust, the buoyant column rises in the atmosphere and, reached neutral buoyancy, propagates laterally as a radially spreading density current, advected by winds. Powerful eruption columns typically spread into the stratosphere. Sustained, buoyant plumes may rapidly shift to collapsing behaviour, generating pyroclastic density currents. The volume of ejected material produced by individual Plinian pulses ranges between 0.1 and 10 km^3, with pulses lasting between a few hours to days. The tapping of large amounts of gas-rich magma during Plinian eruptions may trigger the collapse of the roof of the magma reservoir, forming a caldera.

Due to the complexity of the eruptive regimes and occurrence of multiple eruptive pulses, Plinian eruption deposits exhibit wide variability: a general division may be done between fall and

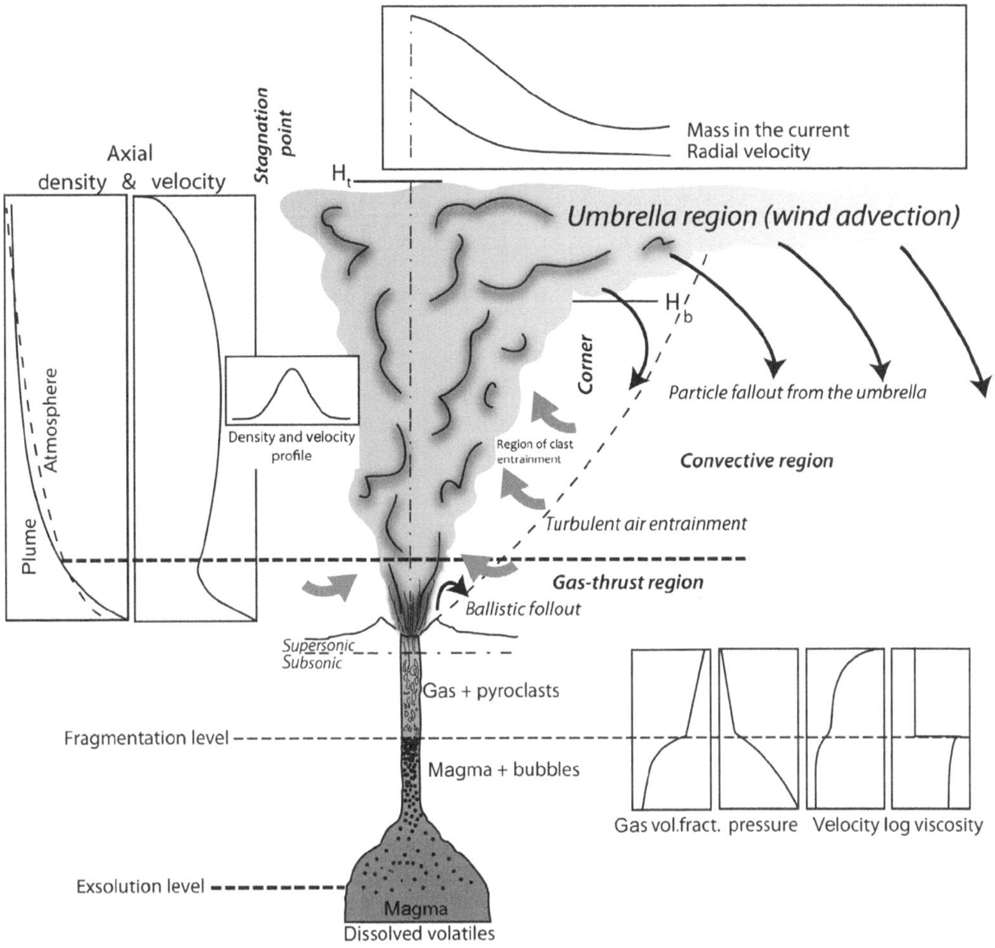

Fig. 1.23 General scheme of eruptive regimes for a strong plume and variation of physical parameters during Plinian eruptions; H_b is the level of neutral buoyancy and H_t is the height of the column (Cioni et al. 2015)

flow deposits. Fall deposits, also known as **tephra,** result from the settling on the ground of pyroclasts from the eruptive plume, forming sheet-like layers dispersed over large areas. Plume dynamics and wind field control tephra dispersal. Tephra consists of different components of both juvenile (fresh magma) and lithic (wall rock) particles and can vary from metre- to micron-sized particles, as ballistic blocks and bombs (>64 mm), lapilli (2–64 mm) and ash (<2 mm; Bonadonna et al. 2015, and references therein). Flow deposits are represented by **pyroclastic density currents** (PDC) produced by the collapse of Plinian columns. PDC are heterogeneous mixtures of volcanic particles and gas at high temperature (200–600 °C), which flow across the volcano flanks driven by gravity (Fig. 1.24). In particular, PDC include juvenile clasts derived from magma fragmentation, such as volcanic ash and pumice and scoria lapilli and pyroclasts sourced from the erosion of the volcanic conduit walls, and lithic and debris picked up by the current (Brown and Andrews 2015, and references therein). PDC encompass phenomena categorized in the past as pyroclastic surges (low particle concentration) and pyroclastic flows (dense particle mixtures). PDC are produced from eruptions of many scales, from small-volume events (dome collapse or small column collapse) to caldera-forming eruptions of

Fig. 1.24 Pyroclastic flow developed during the August 7, 1980 eruption at Mount St. Helens (Washington, USA). Credit: US Geological Survey, USGS

volumes of thousands of km^3. These currents can produce a range of deposits related to different depositional temperatures (Dufek et al. 2015). PDC generated during Plinian and sub-Plinian eruptions commonly sediment pumice- and ash-rich deposits, called **ignimbrites.** The juvenile material in ignimbrites covers the full range of magma compositions. Small volume ignimbrites can range from basaltic to rhyolitic composition, whereas large volume ignimbrites are typically of highly evolved compositions, ranging from dacitic to rhyolitic.

Phreatomagmatic eruptions

Ascending magma often encounters some type of water, as seawater, groundwater, rivers, lakes or glaciers. Where rising magma approaches and contacts water the result is an increase of thermal energy flux from the magma, which violently heats and pressurizes the water, manifested through a **phreatomagmatic eruption** (Zimanowski et al. 2015). The consequences of magma-water interaction in phreatomagmatic eruptions depend on internal factors (i.e., the thermal and hydro-mechanical properties of magma and water) and external factors (i.e., interaction geometry and dynamics). The outcome ranges from underwater effusion of lavas to explosive pulses so strong that up to 90% of the resulting deposits consist of powdered country rock that formerly hosted the conduit. In the latter case, approximately 30% of the available thermal energy is converted into mechanical energy within a fraction of a millisecond, and approximately 60% of this mechanical energy is emitted as shock waves able to pulverize even solid rocks within the conduit and vent region. Phreatomagmatic events are not only peculiar to the small eruptions of tuff rings, tuff cones and maar volcanoes, but occur across the full range of mass eruption rates and conditions. They occur not only with basaltic melts, but also with more evolved magmas. In an eruption, phreatomagmatic fragmentation may overlap or alternate with vesiculation and magmatic fragmentation, as in 2010 at Eyjafjallajokull (Iceland; Zimanowski et al. 2015, and references therein).

Eruptions where water invades the vent are named **Surtseyan**, after the 1963–1967 submarine and emergent eruptions that formed the island of Surtsey (Iceland). Surtseyan eruptions result in large-scale magma-water interaction, forming highly fragmented and fine-grained tephra (e.g., ash to fine ash) associated with mud, water, and steam. Base surges often also produce a ring-shaped cloud that travels horizontally above the water surface following collapse of the vertical column. This type of magma-water interaction is common during shallow submarine eruptions and at crater lakes, as for example at Ruapehu (New Zealand; Siebert et al. 2015).

Other types of volcanic activity

Volcanic heat may be also the source of **phreatic** (or steam-blast) explosions, in which heated and pressurized water fragments pre-existing volcanic materials. These events can be quite violent, and are still considered eruptions, even without the direct involvement of magma. Gas is an unequivocal volcanic product, but the gentle, continuous emission of fumarolic gases common to most volcanoes is not considered an eruption. Occasionally, though, large quantities of gas are suddenly vented, causing several victims, as in 1979 at Dieng volcano complex (Java, Indonesia, 149 victims), in 1984 at Lake Monoun and 1986 at Lake Nyos (Cameroon, 37 and 1746 victims, respectively) and in 2014 at Ontake volcano (Japan, 64 victims). Different processes explain these phreatic eruptions, including the sudden overturn of crater lake water with catastrophic exsolution of volcanic gas previously accumulated in the lake bottom (in Cameroon) and the pressurization of gases by the heat supplied by intruded magma (at Ontake; Siebert et al. 2015; Miyagi et al. 2020).

Lahars occur during (primary lahars) or after (secondary) volcanic eruptions on steep volcanic terrain when large masses of water mixed with sediment sweep down and off volcano slopes and commonly incorporate additional sediment and water. Because lahars are water-saturated, both liquid and solid interactions influence their behaviour and distinguish them from other phenomena, such as debris avalanches and floods. Lahar genesis requires: (a) an adequate water source; (b) abundant unconsolidated debris, which typically includes pyroclastic flow and fall deposits, glacial drift, colluvium, and soil; (c) steep slopes (dip commonly >25°) and substantial relief at the source; (d) a triggering mechanism (ice/snow melting, rains, floods, flank collapse). Water sources may include groundwater or hydrothermal water, rapidly melted snow and ice, subglacially trapped water, crater or other lake water, and rainfall runoff (Vallance and Iverson 2015).

1.10 Volcanic Hazard and Risk

An **active** volcano is defined as a volcano which has erupted in the last ~10,000 years. A **quiescent** volcano has not been active in the last ~10,000 years, but it is expected to erupt in the future. An **extinct** volcano did not produce eruptions in the last ~10,000 years and is not expected to erupt again. Even though quiescent and, sometimes, extinct volcanoes may pose hazards to humans and their environment, most of the hazards derive from active volcanoes. These **volcanic hazards** may be of very different nature and impact, and their manifestation may depend upon many factors, including the type and location of the volcano, type of activity, time since the last eruption, local climate and time of the year. Volcanic hazards at active volcanoes may occur not necessarily during an eruption (eruptive hazards), but also just before an eruption (pre-eruptive hazards) and even during long inter-eruptive quiescence (inter-eruptive hazards).

Inter-eruptive hazards may occur at any time during the life of a volcano. These mainly include degassing and lahar formation. Volcanic gases falling as acid aerosols may affect crops, vegetation, and communities downwind and even cause thousands of fatalities (as in the above-mentioned case of Lake Nyos, Cameroon, in 1986). Lahars, although usually immediately following explosive eruptions, as in 1991 at Pinatubo and in 2007 at Ruapehu, may also manifest in the inter-eruptive period. In fact, severe rainfall may mobilize at any time ash deposits on the slopes of reliefs, even far from the volcano. This for example occurred in 1998 at Sarno and Quindici villages, a few tens of kilometres to the east of Vesuvio (Italy), where ash from historical eruptions emplaced on the Apennines mountains was suddenly mobilized as mud flow by severe rainfall, causing 160 victims.

Pre-eruptive hazards occur during the **unrest** phase, which is identified by a variation in the geodetic, geophysical and geochemical monitoring parameters of a volcano that usually testifies the shallow accumulation and/or rise of magma, possibly leading to eruption (see Chap. 9). Almost all eruptions are preceded by unrest, though with very variable features. However, not every unrest culminates into eruption, with more than one third of the unrest episodes at volcanoes not leading to any eruption (Newhall and Dzurisin 1988; Phillipson et al. 2013; Acocella et al. 2015). The most frequent hazards occurring during unrest are surface deformation, seismicity and increased degassing. Surface deformation affects the volcanic edifice and usually manifests with displacement rates of a few cm/year, although when magma reaches very shallow levels it may dramatically displace the Earth's surface of several metres, as observed in 2000 at Usu (Japan), on the southern rim of Toya caldera (Fig. 1.25). Surface deformation may also culminate in the catastrophic collapse of the flank of a volcanic edifice, developing a deep-seated landslide capable of depressurizing the shallow magmatic system and promoting eruption, as for example at Mount St. Helens in 1980. Volcanic seismicity is another common pre-eruptive hazard. Volcanic earthquakes are less devastating than their regional counterparts, showing limited magnitude M (usually $M < 5.5$; see Sect. 8.4.1). Nevertheless, they may be extremely shallow and cause severe damage within and around a volcano well before any eruption. In inhabited areas shallow and prolonged seismicity may damage buildings and infrastructures (roads, bridges, lifelines), delaying response during a pre-eruptive evacuation, as proposed in the case of any future unrest at Vesuvio (Zuccaro et al. 2008).

Fig. 1.25 Building damaged during the 2000 Usu (Japan) eruption. The damage is testified by the tilt of the building and the collapse of some of its parts, as well as by the impact of erupted ejecta on the ash-filled roof

Eruptive hazards may vary significantly, mainly as a function of the type and size of the eruption. Effusive eruptions affect proximal and limited areas, with slow-advancing lava flows which allow to save lives and prepare mitigation measures. Explosive eruptions are more impactful: these may inject large quantities (from 10^{-1} to 10^{2} km^{3}) of volcanic ash into the atmosphere, much of which falls out locally, while a part might be transported around the globe. Ash fallout is one of the most severe volcanic hazards, both in the near- or far-field, impacting health, buildings, lifelines and ground and air traffic. The 2010 eruption of Eyjafjallajokull is an example of moderate ash eruption causing havoc to European air traffic because of the prevailing eastward wind direction. The collapse of an explosive eruption column may produce pyroclastic flows, which may quickly impact wide areas on the volcano slopes and base, constituting the most deadly volcanic hazard. Eruptions may also suddenly melt glaciers on the top of volcanoes, as during the minor explosive eruption (VEI 3) of Nevado del Ruiz (Colombia) in 1985, when syn-eruptive lahars propagated for several tens of kilometres down river valleys, finally hitting the village of Armero, causing ~23,000 casualties. Effusive and explosive eruptions are also accompanied by SO_2 release. Tall eruptive columns inject SO_2 into the stratosphere, forming an aerosol that may reflect and adsorb sun radiation, heating the stratosphere and cooling the troposphere below, affecting climate: a global cooling of ~0.5 °C of the Earth's surface has been measured for a few years after the 1991 Pinatubo eruption, the largest in recent decades (Robock 2000). Finally, most of the world's volcanoes are below the sea, and many are located in coastal areas, or on islands. Here the explosion or collapse of a volcano can cause a large body of water to move suddenly, creating a tsunami (or tidal wave) propagating for hundreds of kilometres with devastating effects.

The interpretation of the volcano monitoring signals during unrest may allow issuing a warning for impending eruption. However, eruptions and, more in general, volcanic hazards cannot be currently predicted in a deterministic way, even at well-monitored volcanoes. Volcanoes are in fact highly complex systems, whose internal processes are understood in a limited way and controlling parameters are poorly quantified, and often show non-linear (erratic and unpredictable) pre-eruptive behaviour. These limitations lead to a more popular probabilistic estimate of volcanic hazards and forecast of eruptions, which takes also into account for uncertainties (see Chap. 9; Sparks 2003).

The impact of a given volcanic hazard may be extremely variable depending, in a first approximation, on the amount of population living nearby the volcano generating the hazard. For example, a volcano erupting in a densely inhabited area is much more impactful than a volcano erupting, with the same type of activity, in an uninhabited area. This introduces the concept of **volcanic risk**, which takes into account not only the hazard of the volcano, but also the exposure and vulnerability of its environment. In particular, volcanic hazard is defined as the probability that a certain volcanic event affects an area in a given period. The exposure refers to the elements (number of human lives, capital value, land, buildings, factories, power plants, highways, lifelines, etc.) exposed to volcanic activity. Vulnerability is the tendency of an element to experience damage during volcanic activity. Volcanic risk is expressed as the product of these three terms:

$$\text{Volcanic risk} = \text{hazard} \times \text{exposure} \times \text{vulnerability} \qquad (1.7)$$

In general, the higher the hazard the higher the risk, although, for a given hazard, risk increases with population and urbanization. It is estimated that ~800 million people live within 100 km of an active volcano in 86 countries, particularly in densely inhabited countries in east Asia (as Japan), southeast Asia (as Philippines and Indonesia) or Central America (as Mexico, Guatemala and Nicaragua). In the last centuries several disasters have resulted from the impact of volcanic activity on the population living nearby active volcanoes (Fig. 1.26; Loughlin et al. 2015).

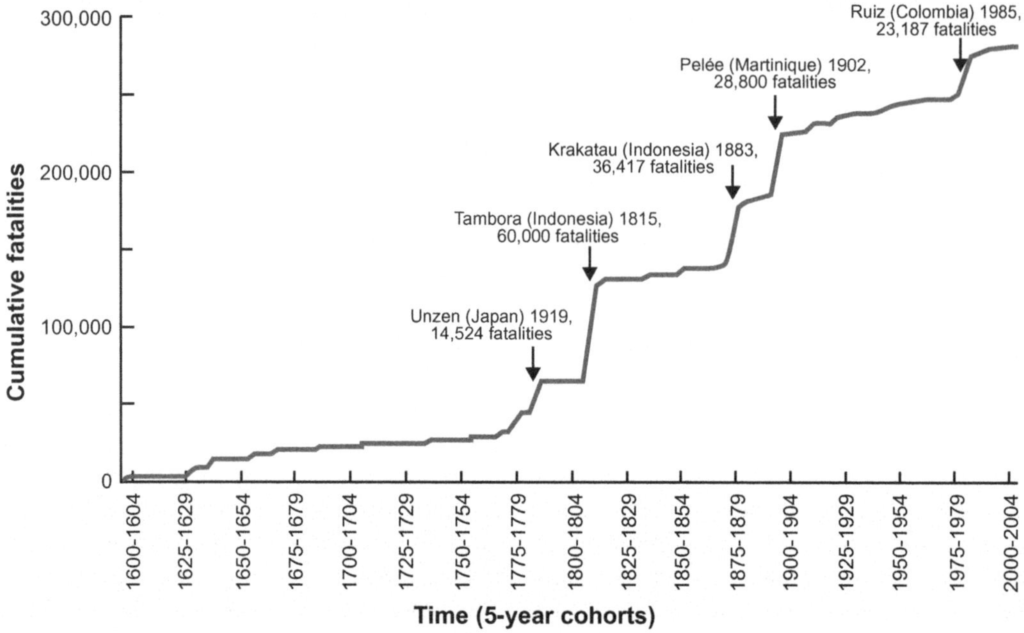

Fig. 1.26 Cumulative number of fatalities resulting from volcanic eruptions, shown using all 533 fatal volcanic incidents. The five largest disasters are: Tambora, Indonesia in 1815 (60,000 fatalities); Krakatau, Indonesia in 1883 (36,417 fatalities); Pelée, Martinique in 1902 (28,800 fatalities); Nevado del Ruiz, Colombia in 1985 (23,187 fatalities); Unzen, Japan in 1792 (14,524 fatalities; modified after Loughlin et al. 2015)

The most devastating recent eruptions occurred in the nineteenth century in Indonesia, at Tambora (in 1815, with ~60,000 fatalities) and Krakatau (in 1883, with ~36,500 fatalities), although in both cases most of the victims were caused by the indirect effects of the eruptions, as starvation and tsunami, respectively. In the twentieth century, the 1902 Pelée (Martinique) eruption caused ~30,000 victims, with pyroclastic flows devastating the village of St. Pierre, whereas the above-mentioned 1985 Nevado del Ruiz eruption caused ~23,000 victims. Human health, although definitely being the focus in evaluating the impact of volcanic hazards, is not the only concern in volcanic risk assessment. Indeed, many eruptions may not cause fatalities or direct health concern and, at the same time, may produce significant economic losses; for example, the estimated toll of the 2010 Eyjafjallajokull eruption, which did not cause any victim, was 5 billion US dollars.

In volcanic risk mitigation analysis, the risk should be properly quantified and, subsequently, reduced. The knowledge of the general volcanic processes and of the specific state of the monitored volcano, as well as of its hazards and the probabilistic forecast of these threats may allow properly defining the hazard component of volcanic risk (Eq. 1.7). However, for a given volcanic hazard, risk can be mitigated only decreasing the exposure and vulnerability components: this can be achieved through long-term activities of prevision and prevention. **Prevision** aims at identifying the risks, their causes and the affected area(s), whereas **prevention** is based on the operative activities aiming at minimizing any impact due to an event. Appropriate prevision and prevention measures may significantly reduce the impact of an adverse event. While hazards are inevitable, the worst of their consequences are not, as loss of life and infrastructure can be diminished pending adequate prevision

Fig. 1.27 Relationships between the frequency of natural disasters, the amount of exposed population and the related death toll (mean values) since 1900. The overall decrease in deaths, despite the significant increase in disasters, underlines the crucial importance of prevention in mitigating risk (from the OFDA/CRED Natural Disaster Database; www.emdat.be)

and prevention. In fact, while the (real or perceived) frequency of natural hazards and the related amount of exposed population have increased in the last century, the related death toll has significantly decreased (Fig. 1.27). This behaviour may be indeed explained by the increasing importance of prevention in mitigating risk, especially in the developed countries. The crucial role of prevention holds in any type of natural hazard. For example, a same chain of hazards (mega-earthquake followed by tsunami) may have a completely different outcome, in terms of victims, depending on the preparedness of the community, as suggested by the 2004 Sumatra earthquake (Indonesia, >200,000 victims) and the 2011 Tohoku earthquake (Japan, ~20,000 victims). While prevention in Sumatra was limited, decades-long prevention of natural disasters in Japan has likely decreased the death toll of one order of magnitude.

1.11 What is Volcano-Tectonics?

The overview on volcanoes and volcanic activity presented in the previous sections has underlined the importance of plate tectonics processes in explaining the location, distribution and composition of magma, as well as its main physical properties. Likewise, it has been mentioned that the coupling between the regional tectonic setting and these physical and chemical properties determines the mode of rise, emplacement and, eventually, eruption of magma, broadly affecting the architecture of the plumbing systems below volcanoes and the shape of the volcanic edifices. All these features are a main concern of a modern branch of volcanology that has rapidly grown in the last three decades, that is **volcano-tectonics**.

Volcano-tectonics investigates any stress-deformation process, occurring at any scale, related to the accumulation, transfer and eruption of shallow magma. Stress-deformation processes may mainly (a) be magma-induced or, conversely, (b) encourage or control the accumulation, transfer and eruption of shallow magma. In the first case, the magma-induced stress-deformation processes have local extent, usually occurring within the volcano. These processes include the formation of diapirs, dikes, sills, laccoliths, lopoliths, magma chambers, calderas, resurgences, pit craters, eruptive fissures, volcanic rift zones and any volcano flank instability: these are all discussed from Chaps. 3 to 9. In the second case, the stress-deformation processes encouraging or controlling magma accumulation, transfer and eruption commonly have a regional extent, as occurring also outside the volcano. These processes include the development of divergent and convergent plate boundaries and hot spots, with their regional fault systems, seismicity and magmatic activity, which are discussed from Chaps. 10 to 13.

In this frame, volcano-tectonics exploits the merging of knowledge from a wide range of disciplines. On the one side, these primarily focus on the stress-deformation relationships, as structural geology (focused on the geometry of the structures, usually at the local scale), tectonics (focused on a set of structures, usually at the regional scale), geodesy (including deformation and gravity monitoring), geophysics (including seismicity and seismic refraction), remote sensing (optical and thermal), and modelling (analytical, numerical and analogue models). On the other side, volcano-tectonics exploits the disciplines more closely related to volcanoes, including volcanology, stratigraphy, petrology, geochemistry and geochronology.

The general aim of volcano-tectonics is to capture the processes controlling the shallower and deeper structure of volcanoes, establishing the stress-deformation relationships between the magma and the country rock, to ultimately understand how volcanoes work, also in their regional context. This approach allows defining the dynamic behaviour of active volcanoes during unrest and eruptions, and thus being able to make reliable forecasts as to the likely scenarios. This concept is exemplified in Fig. 1.28, where a space image of the 2002 Etna eruption is compared to a structural sketch summarizing the shallow transfer of magma during the same eruption. The comparison allows relating our understanding of the shallow plumbing system of a volcano to the eruptive style and its environmental impact. As visible from the space image (Fig. 1.28a), three

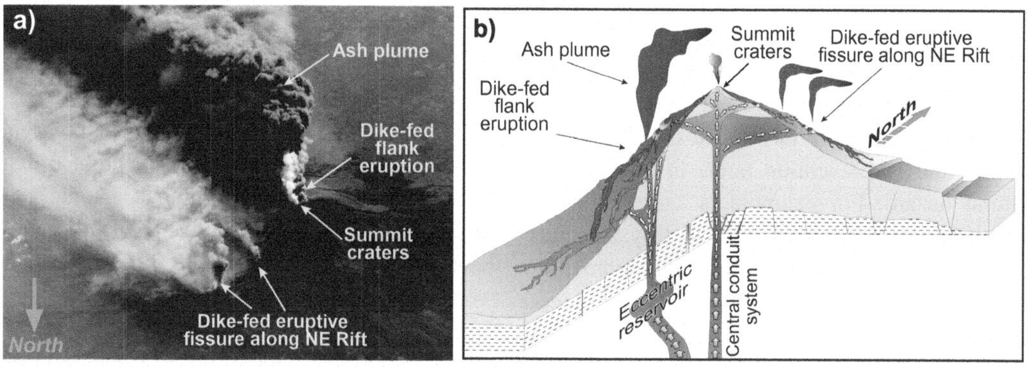

Fig. 1.28 Linking the eruptive style and environmental impact of an eruption to magma transfer within volcanoes (modified after Acocella 2014). **a** Space image of the 2002 eruption at Mt. Etna, Italy, activating the summit craters and the Northeast Rift (both erupting degassed magma, small white plumes), as well as a vent on the South Rift (erupting primitive undegassed magma, feeding the major and dark explosive ash plume). Image source: International Space Station (Credit: NASA). **b** Structural interpretation of the eruption, showing how the flank vent of the South Rift is fed by a dike from an eccentric reservoir independent of the summit conduit, responsible for the eruption of primitive magma and the generation of the ash plume. Conversely, the effusive summit and Northeast Rift vents are fed by the residual magma within the summit conduit, not generating any ash plume (Andronico et al. 2005). Note the opposite direction of the north in the two figures

groups of plumes were active during the eruption, with the largest and darkest ash-rich one of concern for population and aviation, as having also determined the temporary closure of the nearby Catania airport. The interpretation of the eruption (Fig. 1.28b; Andronico et al. 2005) shows that both the summit craters and the Northeast Rift of Etna erupt residual degassed magma migrating from the open summit conduit (the Northeast Rift is in fact fed by the summit conduit through a laterally propagating dike). Conversely, the vent on the South Rift is fed by an independent vertically propagating dike, filled with new, primitive and undegassed magma. These structural, volcanological and petrological differences in the shallow plumbing system are well reflected in the eruptive styles shown by the image: while the residual degassed magma feeds the low explosivity vapour plumes from the summit craters and the Northeast Rift, the primitive gas-rich magma feeds the more explosive higher and darker ash plume from the South Rift. This example summarizes the potential of volcano-tectonic studies in explaining eruptive styles and even apparently distant processes, as the generation of ash plumes impacting the environment, and in using this knowledge to forecast future events.

1.12 Summary

Volcanoes are unique and complex natural systems that provide impressive proof of the activity of our planet. Their study delivers challenging but exciting opportunities to better understand planetary processes, as well as to reduce the impact of their activity on human life and the environment.

This chapter has introduced the basic concepts of volcanology that provide the background to better understand the volcano-tectonic processes presented in the book. The tectonic setting of a volcano, as explained by the plate tectonics frame, allows understanding the variability in the composition and physical properties of its magma, as well as in the architecture of its plumbing system, the size and shape of its edifice and in its eruptive style. In particular, divergent plate boundaries and mantle plumes generate less viscous and denser mafic magma, which usually builds basaltic shields through effusive eruptions. Conversely, convergent plate boundaries generate more viscous and lighter felsic magma, which usually builds andesitic stratovolcanoes or large rhyolitic calderas with predominant explosive activity. The amount of explosivity of an eruption results from the efficiency in the fragmentation of the ascending magma by the gas bubbles exsolved during decompression, a feature promoted by magma viscosity.

The size of an eruption is empirically determined through the Volcanic Explosivity Index, although distinct concepts of eruption magnitude and intensity are also used. Active volcanoes generate different types of hazards, manifesting in the pre-eruptive phase (mainly related to seismicity, surface deformation and degassing), during the eruption (mostly related to the eruptive activity, but also including sector collapse) and after the eruption (as lahars remobilizing the erupted deposits). To properly define volcanic risk, the impact of these hazards should be considered in relation to the exposure and vulnerability of the area potentially affected by volcanic activity. Once defined, the risk may be mitigated through prevision and prevention.

Volcano-tectonics is a modern discipline of volcanology that considers any stress-deformation process, occurring at any scale, related to the accumulation, transfer and eruption of shallow magma, either as cause or consequence. This is useful not only to better understand how volcanoes work, but also to forecast hazards and their impact.

1.13 Main Symbols Used

C_p	Heat capacity
$\dot{e}$	Rate of strain
g	Acceleration due to gravity
H_{max}	Maximum thickness of the deposit
k	Thermal conductivity
L_M	Ellipse major axis
L_m	Ellipse minor axis
M	Magnitude
m	Mass

P	Pressure
T	Temperature
V	Volume
z	Depth of magma
α	Thermal diffusivity
ΔP_b	Magma buoyancy
η	dynamic viscosity
ρ_m	Magma density
ρ_r	Host rock density
σ	stress

References

Acocella V (2014) Great challenges in volcanology: how does the volcano factory work? Front Earth Sci 2:1–10. https://doi.org/10.3389/feart.2014.00004

Acocella V, Di Lorenzo R, Newhall C, Scandone R (2015) An overview of recent (1988 to 2014) caldera unrest: knowledge and perspectives. Rev Geophys 53. https://doi.org/10.1002/2015RG000492

Andronico D, Branca S, Calvari S, Burton M, Caltabiano T, Corsaro RA et al (2005) A multi-disciplinary study of the 2002–03 Etna eruption: insights into a complex plumbing system. Bull Volcanol 67:314–330. https://doi.org/10.1007/s00445

Bonadonna C, Costa A, Folch F, Koyaguchi T (2015) Tephra dispersal and sedimentation. In: Sigurdsson H, Houghton B, McNutt S, Rymer H, Stix J (eds) The encyclopedia of volcanoes, 2nd edn. Elsevier Academic Press, pp 587–597

Brown M (2013) Granite: from genesis to emplacement. Geol Soc Am Bull 125:1079–1113

Brown RJ, Andrews GDM (2015) Deposits of pyroclastic density currents In: Sigurdsson H, Houghton B, McNutt S, Rymer H, Stix J (eds) The encyclopedia of volcanoes, 2nd edn. Elsevier Academic Press, pp 631–647

Burchardt S (2018) Volcanic and igneous plumbing systems; understanding magma transport, storage, and evolution in the Earth's crust. Elsevier, 347 pp

Calder ES, Lavallee Y, Kendrick JE, Bernstein M (2015) Lava dome eruptions. In: Sigurdsson H, Houghton B, McNutt S, Rymer H, Stix J (eds) The encyclopedia of volcanoes, 2nd edn. Elsevier Academic Press, pp 343–361

Cashman KV, Scheu B (2015) Magmatic fragmentation. In: Sigurdsson H, Houghton B, McNutt S, Rymer H, Stix J (eds) The encyclopedia of volcanoes, 2nd edn. Elsevier Academic Press, pp 459–471

Cassidy M, Manga M, Cashman K, Bachmann O (2018) Controls on explosive-effusive volcanic eruption styles. Nature Commun 9:2839. https://doi.org/10.1038/s41467-018-05293-3

Cioni R, Pistolesi M, Rosi M (2015) Plinian and subplinian eruptions. In: Sigurdsson H, Houghton B, McNutt S, Rymer H, Stix J (eds) The encyclopedia of volcanoes, 2nd edn. Elsevier Academic Press, pp 519–535

Clarke AB, Esposti Ongaro T, Belousov A (2015) Vulcanian eruptions. In: Sigurdsson H, Houghton B, McNutt S, Rymer H, Stix J (eds) The encyclopedia of volcanoes, 2nd edn. Elsevier Academic Press, pp 505–517

Courtillot V, Davaille A, Besse J, Stock J (2003) Three distinct types of hotspots in the Earth's mantle. Earth Planet Sci Lett 205:295–308

Davies GF (1999) Dynamic earth. Plates, plumes and mantle convection. Cambridge Univ Press, 470 pp

de Silva S, Lindsay JM (2015) Primary volcanic landforms. In: Sigurdsson H, Houghton B, McNutt S, Rymer H, Stix J (eds) The encyclopedia of volcanoes, 2nd edn. Elsevier Academic Press, pp 273–297

Dingwell DB (1996) Volcanic dilemma: flow or blow? Science 273:1054–1055

Dufek J, Esposti Ongaro T, Roche O (2015) Pyroclastic density currents: processes and models. In: Sigurdsson H, Houghton B, McNutt S, Rymer H, Stix J (eds) The encyclopedia of volcanoes, 2nd edn. Elsevier Academic Press, pp 617–629

Fischer TP, Chiodini G (2015) Volcanic, magmatic and hydrothermal gases. In: Sigurdsson H, Houghton B, McNutt S, Rymer H, Stix J (eds) The encyclopedia of volcanoes, 2nd edn. Elsevier Academic Press, pp 779–797

Giordano D, Russell JK, Dingwell DB (2008) Viscosity of magmatic liquids: a model. Earth Planet Sci Lett 271:123–134

Gonnermann HM, Manga M (2003) Explosive volcanism may not be an inevitable consequence of magma fragmentation. Nature 426:432–435

Gonnermann HM, Manga M (2007) The fluid mechanics inside a volcano. Annu Rev Fluid Mech 39:321–356

Gregg P, Grosfils EB, de Silva S (2015) Catastrophic caldera-forming eruptions II: The subordinate role of magma buoyancy as an eruption trigger. J Volcanol Geoth Res 305:100–113

Harris AJL, Rowland SK (2015) Lava flows and rheology. In: Sigurdsson H, Houghton B, McNutt S, Rymer H, Stix J (eds) The encyclopedia of volcanoes, 2nd edn. Elsevier Academic Press, pp 321–341

Lagat J (2003) Geology and the geothermal systems of the southern segment of the Kenya Rift. In: International geothermal conference Reykjavík S04 Paper 107, pp 33–39

LaFemina PC (2015) Plate tectonics and volcanism. In: Sigurdsson H, Houghton B, McNutt S, Rymer H, Stix J (eds) The encyclopedia of volcanoes, 2nd edn. Elsevier Academic Press, pp 65–91

Le Maitre RW, Streckeisen A, Zanettin B, Le Bas MJ, Bonin B, Bateman P, Bellieni G, Dudek A et al (2002) Igneous rocks: a classification and glossary of terms, recommendations of the international union of geological sciences, subcommission of the systematics of igneous rocks. Cambridge University Press. ISBN 0-521-66215-X

Le Pichon X (1968) Sea-floor spreading and continental drift. J Geophys Res 73:3661–3697
Lesher CE, Spera FJ (2015) Thermodynamic and transport properties of silicate melts and magma. In: Sigurdsson H, Houghton B, McNutt S, Rymer H, Stix J (eds) The encyclopedia of volcanoes, 2nd edn. Elsevier Academic Press, pp 113–141
Lockwood JP, Hazlett RW (2010) Volcanoes. Global perspectives, Wiley-Blackwell, p 541
Loughlin SC, Vye-Brown C, Sparks RSJ, Brown SK, Barclay J, Calder E, Cottrell E, Jolly G at al (2015) An introduction to global volcanic hazard and risk. In: Loughlin SC, Sparks RSJ, Brown SK, Jenkins SF, Vye-Brown C (eds) Global volcanic hazards and risk. Cambridge University Press Cambridge, pp 1–79
Lucazeau F (2019) Analysis and mapping of an updated terrestrial heat flow data set. Geochem Geophys Geosyst 20:4001–4024. https://doi.org/10.1029/2019gc008389
Miyagi I, Geshi N, Hamasaki S, Oikawa T, Tomiya A (2020) Heat source of the 2014 phreatic eruption of Mount Ontake, Japan. Bull Volcanol 82:33
Mueller S, Scheu B, Spieler O, Dingwell DB (2008) Permeability control on magma fragmentation. Geology 36:399–402
Newhall CG, Self S (1982) The volcanic explosivity index (VEI) an estimate of explosive magnitude for historical volcanism. J Geophys Res 87:1231–1238
Newhall CG, Dzurisin D (1988) Historical unrest at large calderas of the world. US Geological Survey Professional Paper, 1109 pp
Okumura S, de Silva S, Nakamura M, Sasaki O (2020) Caldera-forming eruptions of mushy magma modulated by feedbacks between ascent rate, gas retention/loss and bubble/crystal framework interaction. Scientific Reports 9:15845
Papale P (2018) Global time-size distribution of volcanic eruptions on Earth. Sci Rep 8:6838
Phillipson G, Sobradelo R, Gottsmann J (2013) Global volcanic unrest in the 21st century: an analysis of the first decade. J Volcanol Geoth Res 264:183–196
Pyle DM (2015) Sizes of volcanic eruptions. In: Sigurdsson H, Houghton B, McNutt S, Rymer H, Stix J (eds) The encyclopedia of volcanoes, 2nd edn. Elsevier Academic Press, pp 257–263
Robock A (2000) Volcanic eruptions and climate. Rev Geophys 38:191–219
Rogers N, Hawkesworth C (2000) Composition of magmas. In: Sigurdsson H, Houghton B, McNutt S, Rymer H, Stix J (eds) The encyclopedia of volcanoes, 1st edn. Elsevier Academic Press, pp 115–131
Rogers N (2015) The composition and origin of magmas. In: Sigurdsson H, Houghton B, McNutt S, Rymer H, Stix J (eds) The encyclopedia of volcanoes, 2nd edn. Elsevier Academic Press, pp 93–111
Schmincke H-U (2004) Volcanism. Springer, Berlin, p 290
Siebert L, Cottrell E, Venzke E, Andrews B (2015) Earth's volcanoes and their eruptions: an overview. In: Sigurdsson H, Houghton B, McNutt S, Rymer H, Stix J (eds) The encyclopedia of volcanoes, 2nd edn. Elsevier Academic Press, pp 239–255
Simkin T, Siebert L (2000) Earth's volcanoes and eruptions: an overview. In: Sigurdsson H, Houghton B, McNutt S, Rymer H, Stix J (eds) The encyclopedia of volcanoes, 1st edn. Elsevier Academic Press, pp 249–261
Sigurdsson H, Houghton B, McNutt S, Rymer H, Stix J (2015) The encyclopedia of volcanoes, 2nd edn. Elsevier Academic Press. https://doi.org/10.1016/C2015-0-00175-7, 1456 pp
Sparks RSJ (2003) Forecasting volcanic eruptions. Earth Planet Sci Lett 210:1–15
Sparks S, Annen CJ, Blundy J, Cashman K, Rust A, Jackson MD (2019) Formation and dynamics of magma reservoirs. Philos Trans R Soc A 377:20190019
Stuwe K (2007) Geodynamics of the lithosphere: an introduction, 2nd edn. Springer, Berlin, 493 pp
Taddeucci J, Edmonds M, Houghton B, James MR, Vergniolle S (2015) Hawaiian and strombolian eruptions. In: Sigurdsson H, Houghton B, McNutt S, Rymer H, Stix J (eds) The encyclopedia of volcanoes, 2nd edn. Elsevier Academic Press, pp 485–503
Takada A (1994) The influence of regional stress and magmatic input on styles of monogenetic and polygenetic volcanism. J Geophys Res 99:13563–13573
Tibaldi A (1995) Morphology of pyroclastic cones and tectonics. J Geophys Res 100:24521–24535
Vallance JW, Iverson RM (2015) Lahars and their deposits. In: Sigurdsson H, Houghton B, McNutt S, Rymer H, Stix J (eds) The encyclopedia of volcanoes, 2nd edn. Elsevier Academic Press, pp 649–663
van der Pluijm BA, Marshak S (2004) Earth structure: an introduction to structural geology and tectonics. WW Norton & Company, 673 pp
Walker GP (1973) Explosive volcanic eruptions: a new classification scheme. Geol Rundsch 62:431–446
White JDL, McPhie J, Soule SA (2015) Submarine lavas and hyaloclastite. In: Sigurdsson H, Houghton B, McNutt S, Rymer H, Stix J (eds) The encyclopedia of volcanoes, 2nd edn. Elsevier Academic Press, pp 363–375
Zimanowski B, Buttner R, Dellino P, White JDL Wohletz KH (2015) Magma-water interaction and phreatomagmatic fragmentation. In: Sigurdsson H, Houghton B, McNutt S, Rymer H, Stix J (eds) The encyclopedia of volcanoes, 2nd edn. Elsevier Academic Press, pp 473–483
Zuccaro G, Cacace F, Baxter PJ, Spence R (2008) Impact of explosive eruption scenarios at Vesuvius. J Volcanol Geoth Res 178:416–453

2 Crustal Deformation

2.1 Introduction

As discussed in Chap. 1, volcano-tectonics focuses on the stress-deformation relationships related to volcanic activity, at the local and regional scale. While Chap. 1 has introduced volcanoes and their activity, this second chapter provides complementary introductory knowledge, summarizing the stress-deformation relations affecting the crustal rocks and completing the fundamentals section of the book. Crustal deformation is a wide term that includes numerous processes occurring from the micro- to the macro-scale, directly or indirectly affecting the motion and change in shape of the Earth's outermost shell. Crustal deformation is responsible for raising mountains and deepening oceanic trenches, shifting and shaping continents, and triggering earthquakes. Therefore, crustal deformation, together with volcanism, provides the most striking evidence of the extraordinary activity of our dynamic planet.

Defining how the crust responds to stresses necessitates knowledge of rock mechanics, rheology, structural geology, tectonics and geodynamics. A comprehensive description of each of these disciplines requires a dedicated monograph, with several texts available for an overview (e.g., Price and Cosgrove 1990; Ranalli 1995; Scholz 2002; Turcotte and Schubert 2002; van der Pluijm and Marshak 2004; Jaeger et al. 2007; Stuwe 2007; Fossen 2010; Gudmundsson 2011). Here this introductory chapter builds up on these texts to summarize the fundamental concepts of crustal deformation that are of main interest for volcano-tectonics, providing the background for the following chapters.

The main aims of this chapter are thus to introduce the:

- notions of stress and deformation;
- principles of rock rheology;
- main brittle deformation structures and their tectonic context.

2.2 Stress

According to Newton's Second Law of motion, a mass m experiencing acceleration a produces a force F, so that:

$$F = ma \tag{2.1}$$

A force is expressed in kg m/s^2, that is a Newton N. Force, like velocity or acceleration, is a vector, and can be defined by magnitude, direction and verse. **Stress** σ is defined as the force F acting on a unit area A

$$\sigma = F/A \tag{2.2}$$

thus providing a measure of the intensity of the force. Stress is expressed in N/m^2, or pascal (Pa) in the International System of Units (SI). In

V. Acocella, *Volcano-Tectonic Processes*, Advances in Volcanology,
https://doi.org/10.1007/978-3-030-65968-4_2

geology, however, the bar unit, which approximately equals one atmosphere, is still commonly used, so that:

$$1\,\text{bar} = 1\,\text{atmosphere} = 10^5\,\text{Pa}$$

or, considering the larger equivalents commonly found in geological problems:

$$1\,\text{kbar} = 10^8\,\text{Pa} = 100\,\text{MPa}$$

where the kbar is a kilobar and the MPa is a megapascal.

Stress that acts on a plane is a vector quantity, whereas stress acting on a body is described by a higher order entity, called stress tensor. Stress on an arbitrarily oriented plane can be resolved in two dimensions into a component normal and a component parallel to the plane (Fig. 2.1a; e.g., van der Pluijm and Marshak 2004, and references therein).

The vector component normal to the plane is the **normal stress** σ_n, where

$$\sigma_n = \sigma \cos^2 \theta \tag{2.3}$$

being θ the angle between the stress vector and the plane. The vector component along the plane is the **shear stress** τ, where

$$\tau = (\sigma \sin 2\theta)/2 \tag{2.4}$$

Both components vary with the orientation of the plane with regard to the stress vector.

In three dimensions, stress in a randomly oriented plane introduces some complication with regard to the two-dimensional case. This situation is best illustrated considering the state of stress at a given point in a rock, where planes in an infinite number of orientations may pass through the point. The stress state at this point has to describe the stresses acting on all planes in a body. Following Newton's Third Law of Motion, the stress on each plane must be balanced by one of opposite sign ($\sigma = -\sigma$). Because stress varies as a function of its orientation, the magnitude of the normal stress on each plane is different. A **stress ellipsoid** defines an envelope of the stress vectors for all the planes passing through the point, where the magnitude of a vector is represented by its length (Fig. 2.1b; e.g., van der Pluijm and Marshak 2004; Fossen 2010). Like all ellipsoids, the stress ellipsoid is defined by three mutually perpendicular axes, which are called the **principal stresses.** These principal stresses are: (1) orthogonal to each other and (2) perpendicular to three planes that do not contain shear stresses, or principal planes

Fig. 2.1 **a** Section through a cube showing face ABCD across which a vertical stress σ is applied. This force is resolved into orientations parallel (τ) and perpendicular (σ_n) to a plane (red line) that makes an angle θ with the horizontal. The magnitudes of the shear τ and normal σ_n stress vectors are a function of the angle θ. **b** The stress envelope is an ellipsoid defined by three mutually perpendicular principal stress axes ($\sigma_1 \geq \sigma_2 \geq \sigma_3$), normal to the principal planes of stress (modified after van der Pluijm and Marshak 2004)

of stress. Therefore, the stress state of a body is defined by specifying the orientation and magnitude of the three mutually perpendicular principal stresses. As mathematically the stress ellipsoid is described by a 3 × 3 matrix (called a second-rank tensor, or stress tensor), the stress state of a body in three dimensions is defined by nine components. Of these, three are normal stresses and six are shear stresses (with two orthogonal shear stresses on each plane). In the general condition that the principal stresses are unequal in magnitude, the stress is **anisotropic.** By convention, the maximum principal stress is given the symbol σ_1, the intermediate and minimum principal stresses acting along the other two axes are given the symbols σ_2 and σ_3, respectively: thus, $\sigma_1 \geq \sigma_2 \geq \sigma_3$. If the three principal stresses are equal in magnitude, so that $\sigma_1 = \sigma_2 = \sigma_3$, the stress is **isotropic**: this stress state is graphically represented by a sphere, rather than an ellipsoid.

A practical way to represent and deal with the crustal stresses is to use the **Mohr diagram** (Fig. 2.2a; e.g., van der Pluijm and Marshak 2004). The horizontal and vertical axes of the diagram represent the normal σ_n and shear τ stresses acting on a plane, respectively. On the horizontal axis, the values of the maximum σ_1 and minimum σ_3 principal stresses, as well as their difference (σ_1–σ_3, or **differential stress**), are also included. The distance between σ_1 and σ_3 defines the diameter of a circle (the Mohr circle) whose centre lies at σ_m, which represents the mean stress $\sigma_m = (\sigma_1 + \sigma_3)/2$. The Mohr circle describes the normal and shear stresses acting on planes of all orientations through a point in the rock. The orientation of the plane is expressed with regard to the principal stresses. As on a Mohr diagram angles are doubled, if θ is the angle between the normal to the plane and σ_1, then the angle between the radius to the point on the circle and the horizontal axis is 2θ (Fig. 2.2a). Usually, all principal stresses in the lithosphere are positive. However, in the presence of a tensile stress (negative σ_n), the Mohr circle shifts to the left of the origin, into the tensile field. The Mohr diagram also illustrates the relationships

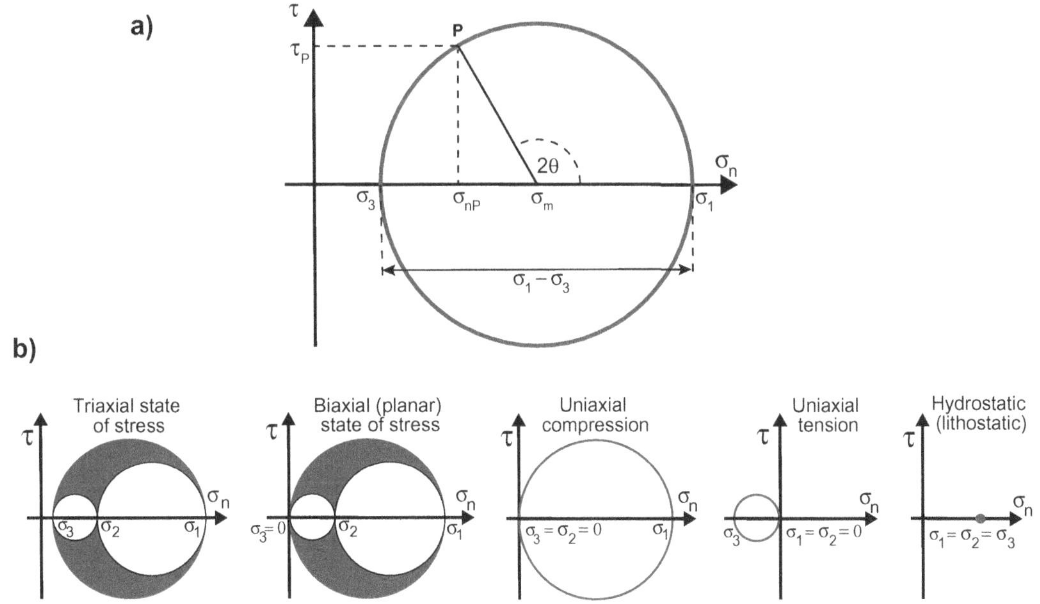

Fig. 2.2 **a** The Mohr circle: θ is the angle between the largest stress and the normal to a given plane; the state of stress on a plane P is given by the shear τ_P and normal σ_{nP} stresses on that plane; **b** Characteristic states of 3D stress illustrated in Mohr diagrams described by three circles connecting the three principal stresses. The largest circle contains σ_1 and σ_3. The three circles may reduce to two or one in case of specific states of stress

between the magnitude of the differential stress and the possibility to have rock failure, as well as the attitude of planes along which the shear stress is greatest for a given state of stress (see Sect. 2.5.2).

Several important **states of stress** can be represented in a Mohr diagram, as schematically illustrated in Fig. 2.2b (e.g., van der Pluijm and Marshak 2004; Gudmundsson 2011). In the most general case of triaxial (anisotropic) stress, all three principal stresses have nonzero values ($\sigma_1 > \sigma_2 > \sigma_3 \neq 0$). The biaxial (planar) stress implies a two-dimensional analysis, with one of the principal stresses assumed to be zero (e.g., $\sigma_3 = 0$). Uniaxial compression assumes that two of the principal stresses are zero, and the third is positive ($\sigma_2 = \sigma_3 = 0$; $\sigma_1 > 0$). Uniaxial tension assumes that two of the principal stresses are zero, and the third principal stress is negative ($\sigma_1 = \sigma_2 = 0$; $\sigma_3 < 0$), placing the Mohr circle to the left of the τ axis. Finally, isotropic stress, often called hydrostatic stress, is represented by a single point on the σ_n axis of the Mohr diagram (positive for compression, negative for tension), because all three principal stresses are equal in magnitude ($\sigma_1 = \sigma_2 = \sigma_3$). The isotropic stress state is also called lithostatic stress state. Note that the term lithostatic stress (or lithostatic pressure P_l) is commonly used also to indicate the normal stress due to the weight of the rocks at a certain depth, expressed as:

$$P_l = \rho g z \tag{2.5}$$

where ρ is the rock density (kg/m^3), g is the acceleration due to gravity (~9.8 m/s^2) and z is the depth at which the stress is considered (e.g., Turcotte and Schubert 2002; Zang and Stephansson 2010). The increase in lithostatic stress with depth depends on the density of the rocks: assuming an average rock density $\rho = 2750$ kg/m^3 for the crust, it occurs with a mean rate of ~27 MPa/km.

Tectonic stresses may introduce important deviations in these states of stress. **Tectonic stresses** develop as a consequence of regional tectonic processes, and are thus related to plate tectonics. However, tectonic stresses may also include the transient stresses induced by seismic activity. In fact, the Earth is not an inert body, as movements in the crust and mantle occur continuously, during the build up of stress caused by steady plate motion, and discontinuously, as during the local stress relief caused by earthquake ruptures. Anisotropic stresses that arise from these processes contribute to the resulting stress state at depth. The differential stresses of these anisotropic stress components, however, may be many orders of magnitude less than the lithostatic stress. In the crust, differential stresses may reach a few hundred megapascals, but in the mantle, where lithostatic pressure is high, they are only on the order of tens of megapascals or less; yet, such low differential stresses are responsible for the slow motion of the mantle. The traditional classification of tectonic stress states was proposed by Anderson (1951). This is based on the assumption that there is no shear stress at the Earth's surface, so that one of the principal stresses must be vertical and the other two horizontal. These Andersonian conditions are valid for an isotropic crust under coaxial (i.e., non-rotational) deformation. Depending on the nature of the vertical stress σ_v, Anderson defined three tectonic stress states, corresponding to three regimes of faulting (see Sect. 2.7):

(a) $\sigma_v = \sigma_1$: normal fault regime;
(b) $\sigma_v = \sigma_2$: strike-slip fault regime;
(c) $\sigma_v = \sigma_3$: thrust fault regime.

In addition to regional or far-field tectonic stresses, the emplacement, accumulation and transfer of magma may induce local and transient **magmatic stresses** in the crust. Similarly to tectonic stresses, magmatic stresses can be described in terms of maximum, intermediate and minimum principal components, although with a local or near-field extent. Their orientation and magnitude depend on the type (shape, size) of the magmatic intrusion and on its dynamic state (overpressure, neutral, underpressure). Magma-induced stresses are commonly larger than the lithostatic stress, thereby deforming the

Earth's crust. Magmatic stresses, particularly important in volcano-tectonics, are discussed in the following chapters.

2.3 Deformation

Deformation describes the collective displacements of points in a body, that is the complete transformation from its initial to final geometry. This transformation can include a translation (movement from one place to the other), a rotation (spin around an axis), and a distortion (change in shape, or strain): **strain** describes the changes of points in a body relative to each other, that is the distortion of a body produced by changes in lines lengths (Fig. 2.3; e.g., van der Pluijm and Marshak 2004; Fossen 2010).

If changes in lines lengths are observed, but the relative lengths of the lines remain the same, the notion of dilation should be also introduced to further characterize the deformation. Dilation refers to the area or volume change, and is positive or negative as the volume increases or decreases, respectively. Strain is homogeneous when any two portions of the body that were similar in form and orientation before are similar in form and orientation also after strain. Homogeneous strain is defined by its geometric consequences:

(a) originally straight lines remain straight;
(b) originally parallel lines remain parallel;
(c) circles become ellipses; in three dimensions, spheres become ellipsoids.

When one or more of these three restrictions does not apply, the strain becomes heterogeneous. In two-dimensional homogeneous strain, two orientations of material lines remain perpendicular before and after strain. These two lines form the axes of an ellipse, called the **strain ellipse.** In three dimensions, three material lines, called the **principal strain axes,** remain perpendicular to each other after strain and define the axes of the **strain ellipsoid.** The length of these lines defines the strain magnitude, with the symbols X, Y, and Z, and the convention that $X \geq Y \geq Z$.

Because heterogeneous strain is more complex to describe than homogeneous strain, heterogeneously strained bodies or regions are analysed by separating them into homogeneous portions. This implies that homogeneity of deformation is a matter of scale. The measure of strain that compares the initial and final configuration is called the **finite strain,** which is independent of the details of the steps toward the final configuration. When the intermediate strain steps are determined, they are called **incremental strains.** Therefore, the summation of all incremental strains is the finite strain.

If during strain the principal incremental strain axes coincide with the finite strain axes, the accumulation of the strain is coaxial (Fig. 2.4a):

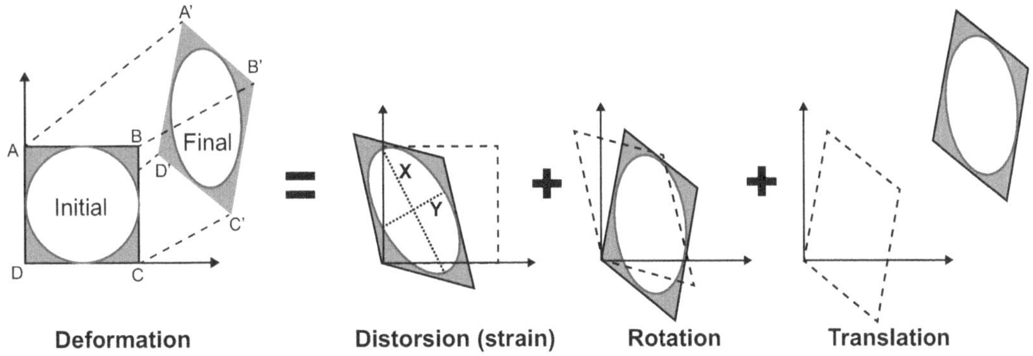

Fig. 2.3 The components of deformation. The generic deformation of a square ABCD (**a**) results from three independent components: (1) distortion, or strain (shown by the strain axes X and Y and the corresponding strain ellipse); (2) rotation; (3) translation. In each stage the previous state is shown by the dashed polygon

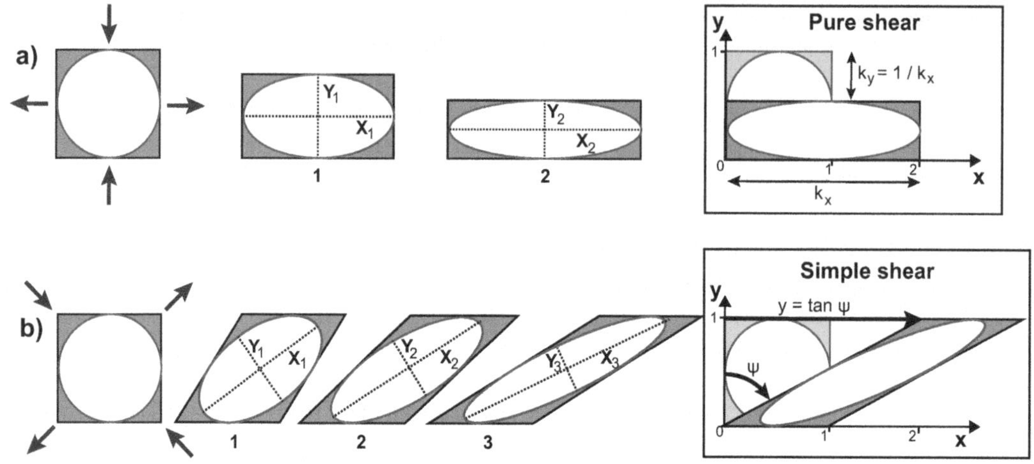

Fig. 2.4 Coaxial (**a**) and non-coaxial (**b**) strain. In coaxial strain the incremental strain axes (dashed lines) are parallel to the same material lines (pure shear conditions). In non-coaxial strain the incremental strain axes are different material lines (simple shear conditions)

in this case, the principal incremental strain axes are parallel to the finite strain axes. Conversely, in the more general case that the principal incremental strain axes are not the same throughout the strain history, or the principal incremental strain axes rotate relative to the finite strain axes, the strain accumulation is non-coaxial (Fig. 2.4b). Coaxial strain accumulation is often called **pure shear**, whereas the non-coaxial strain accumulation is also known as **simple shear** (e.g., Fossen 2010, and references therein).

Strain quantification is made using three dimensionless parameters: length change or longitudinal strain, volume change or volumetric strain, and angular change or angular strain (e.g., Fossen 2010). **Longitudinal strain**, defined as a change in length divided by the original length, is expressed by the **elongation**, e, which is defined as:

$$e = \left(L_f - L_0\right)/L_0 \qquad (2.6)$$

where L_f is the final length and L_0 is the original length. Negative values of e reflect shortening or contraction, whereas positive values represent extension (the related terms compression and tension are reserved for stress). Stretching of a line is designated as $\beta = 1 + e$; therefore, $\beta = L_f/L_0$. The **stretching factor** β is commonly used in the structural analysis of crustal sections (as across rift zones), with positive values indicating extension and negative values contraction (Fig. 2.5).

For **volumetric strain**, V_s, the relationship is:

$$V_s = (V - V_o)/V_o \qquad (2.7)$$

where V is the final volume and V_o is the original volume. Positive values of V_s represent volume gain, whereas negative values represent volume loss.

Angular strain measures the change in angle between two lines that were initially perpendicular. The change in angle is called the angular shear, ψ, but more commonly the tangent of this angle, called the **shear strain** γ, is used:

$$\gamma = \tan\psi \qquad (2.8)$$

The time interval it takes to accumulate a certain amount of strain is described by the **strain rate** $e^{\cdot}$, which is defined as elongation e per time t:

$$e^{\cdot} = e/t \qquad (2.9)$$

As elongation is dimensionless, the dimension of strain rate is t^{-1} and the unit is s^{-1}. Therefore, strain rate is related to variations of velocity,

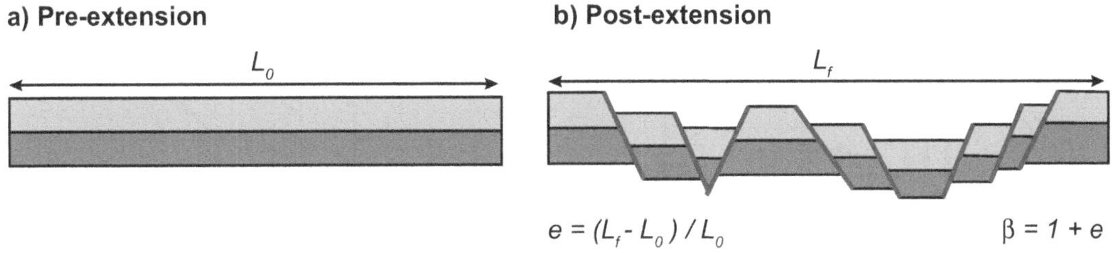

Fig. 2.5 Extension of a crustal portion (in section view) by faulting (red lines). The pre-extension crust (**a**) has original length L_0, whereas after extension the length of the crust becomes L_f. The amount of extension e is found by comparing the two lengths, whereas the stretching factor β quantifies the extension (modified after Fossen 2010)

rather than variations of displacement, as strain. Geologic processes, including tectonic processes, have typical strain rates of 10^{-13}–10^{-16} s^{-1}. Larger strain rates, on the order of 10^{-10} to 10^{-12} s^{-1}, are usually related to the shallow emplacement or transfer of magma. Extremely large strain rates, on the order of 10^{-2} to 10^{-4} s^{-1}, refer to rapid events, like meteorite impacts and explosive volcanism.

2.4 Rheology of Rocks

The stress-strain relationships describing the deformation behaviour of the crustal rocks mainly depend on the conditions under which the rocks are deformed, most notably the applied stress, strain rate, temperature and pressure. This section summarizes the main deformation behaviours of the rocks as a function of these conditions: the behaviours describe elastic, plastic and viscous deformation, as well as their combinations. Other terms are commonly used to describe the rheology of rocks, as "brittle" and "ductile" deformation, which are often used in opposition. However, both terms are not really described by a deformation law (i.e., relationship between stress and strain). **Brittle deformation** is localized and involves the formation of fractures; as based on the Mohr-Coulomb criterion of fracturing (see Sect. 2.5), brittle deformation only describes the stress state at which failure occurs. **Ductile deformation** is characterized by the presence of a ductile material, which accumulates permanent strain without visible fracturing. Ductile structures are well represented in metamorphic rocks that have been deformed in the lower crust: here viscous deformation is the law that is most commonly used to describe ductile structures. However, ductile structures may also occur in poorly unconsolidated sediments or in rocks experiencing low strain rates, where distributed deformation replaces discrete fracturing. Ductile behaviour may even occur by faulting, if thousands of small cracks take up the strain. Therefore, ductile deformation is a scale-dependent structural style, rather than a micro-scale deformation process, which preserves the continuity of originally continuous structures and layers and that can be achieved through a range of deformation mechanisms, including brittle and plastic deformation (e.g., Ranalli 1995; Stuwe 2007; Fossen 2010, and references therein).

2.4.1 Elastic Deformation

The distinctive characteristic of elastic deformation is its reversibility: once the stress is removed, an elastic material returns to its original shape. Reversibility implies that the (potential) energy introduced remains available for returning the system to its original state. Thus, elastic behaviour is characterized by recoverable strain. A further feature of elastic behaviour is the instantaneous response to stress, so that finite strain is achieved immediately. Releasing the

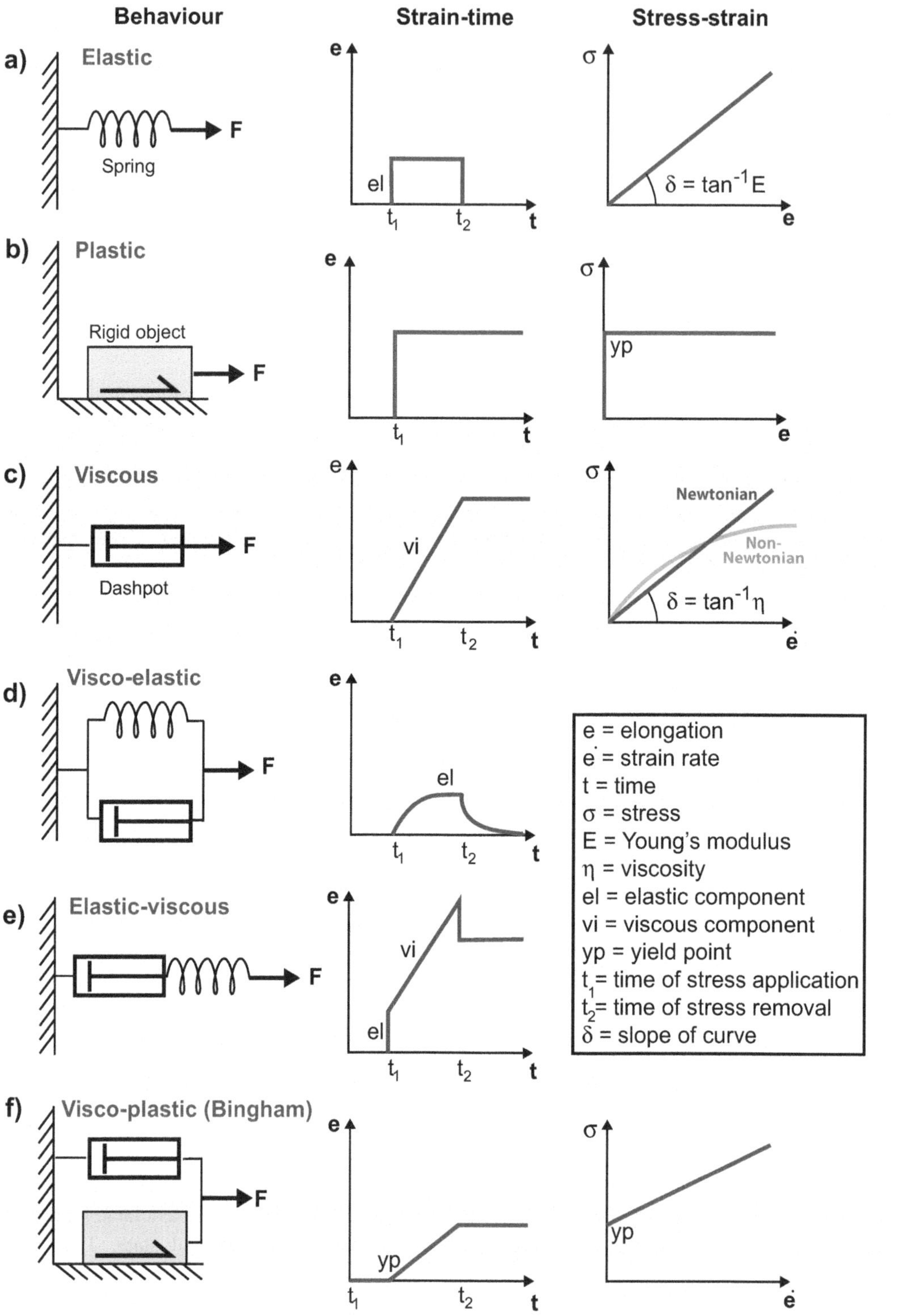

Fig. 2.6 Models of rheologies. Elastic (**a**), plastic (**b**), viscous (**c**), visco-elastic (**d**), elastic-viscous (**e**) and visco-plastic (**f**; Bingham) behaviours, illustrated by mechanical analogues (pure shear conditions under force F; left column), strain history curves (central column) and representative stress-strain curves (right column)

stress also results in an instantaneous return to a no-strain state. The mechanical analogue is a spring attached to a load (Fig. 2.6a). The general linear and proportional relationship describing the elastic behaviour between stress and strain, also known as **Hooke's Law**, is expressed by:

$$\sigma = Ee \tag{2.10}$$

where e is the elongation (i.e., one-dimensional strain case) and E is a constant of proportionality, called **Young's modulus** or, less formally, stiffness, that describes the slope of the line in the σ–e diagram (E is the tangent of the angle δ in the right diagram of Fig. 2.6a; e.g., van der Pluijm and Marshak 2004; Jaeger et al. 2007; Fossen 2010). As strain is dimensionless, the Young's modulus has the dimensions of stress and is expressed in Pascal. Typical values of E for crustal rocks are of 10^{10}–10^{11} Pa: a rock with a low value of E is considered mechanically weak (Table 2.1). The Young's modulus is closely related to the shear (or **rigidity**) **modulus** G. In uniaxial strain the relationship is:

$$E = 2G \tag{2.11}$$

G quantifies how hard it is to deform elastically a rock under simple shear conditions (Table 2.1). The corresponding constant of proportionality in volume change (dilation) is the **bulk modulus,** K_b, which quantifies the resistance of a material to compression, relating the pressure change $\Delta\sigma$ to the volume change (volumetric strain V_s):

$$K_b = \frac{\Delta\sigma}{\left[\frac{(V-V_0)}{V_0}\right]} \tag{2.12}$$

Perhaps more intuitive than the bulk modulus is its inverse, $1/K_b$, which is the bulk compressibility of a material. It is also common to use an alternative to the bulk modulus that expresses the relationship between volume change and stress, that is the **Poisson's ratio** v. This elastic constant is defined as the ratio of the elongation perpendicular to the compressive stress $e_\perp$ and the elongation parallel to the compressive stress $e_{||}$:

$$v = e_\perp / e_{||} \tag{2.13}$$

The Poisson's ratio describes the ability of a material to shorten parallel to the compression direction without corresponding thickening in a perpendicular direction. The ratio ranges from 0 to 0.5, from fully compressible to fully incompressible materials, respectively. Values for Poisson's ratio in natural rocks typically lie in the range 0.25–0.35 (Table 2.1; e.g., Jaeger et al. 2007, and references therein).

Elastic deformation occurs up to strains of a few percent. Above a threshold stress (or elastic limit), the behaviour of the material diverges from elasticity, entering the plastic deformation domain. The elastic limit is a function not only of the material, but also of temperature and pressure. In particular, rocks are approximately elastic at low pressure and temperature (i.e., in the upper lithosphere), and also in the bulk of the Earth for stresses at seismic frequencies. Several processes in volcano-tectonics, including the inflation and deflation of the crust due to pressure variations in magmatic or hydrothermal sources (see Chap. 8), or the bending of the lithosphere due to the load of large volcanic islands (see Chap. 13), are commonly described assuming elastic behaviours. At the lithospheric scale, the elastic behaviour of the lithosphere can be quantitatively expressed by its **elastic thickness**, that is the thickness of the lithospheric plate that best fits the observed flexure.

2.4.2 Plastic Deformation

For rocks and other solids beyond the elastic limit, permanent strain is added to the elastic strain. If this permanent strain keeps on accumulating under a constant stress the material undergoes **plastic** deformation (Fig. 2.6b; e.g., Ranalli 1995; Fossen 2010). Plastic deformation is characterized by its continuity, so that the material must not fracture at the scale of observation: in fact, plasticity is associated with microscale deformation mechanisms. Plastic strain thus consists of a permanent change in shape or size beyond the elastic

Table 2.1 Physical properties of rocks

Rock type	Density ρ (kg/m^3)	Young's modulus E (10^{11} Pa)	Shear Modulus G (10^{11} Pa)	Poisson's ratio v	Tensile strength T_0 (MPa)	Coefficient of friction μ
Igneous						
Basalt	2950	0.6–0.8	0.25–0.35	0.2–0.25	6–29	1.11–1.19
Granite	2650	0.4–0.7	0.2–0.3	0.2–0.25	4–25	0.60–1.73
Diabase	2900	0.8–1.1	0.3–0.45	0.25	5–13	1.19–1.43
Gabbro	2950	0.6–1.0	0.2–0.35	0.15–0.2	5–29	0.18–0.66
Diorite	2800	0.6–0.8	0.3–0.35	0.25–0.3	10–14	
Pyroxenite	3250	1.0	0.4			
Anorthosite	2640–2920	0.83	0.35	0.25		
Granodiorite	2700	0.7	0.3	0.25		
Syenite	2550–2700	0.3–0.5		0.2–0.3	10–13	
Tuff	1600	0.005–0.01		0.25–0.3	0.1–0.8	0.21–0.36
Rhyolite	2150–2500	0.2–0.3		0.14–0.19	16–21	
Sedimentary						
Shale	2100–2700	0.1–0.7	0.1–0.3	0.1–0.2	2–10	0.27–0.58
Sandstone	1900–2500	0.1–0.6	0.04–0.2	0.1–0.3	0.4–25	0.50–0.70
Limestone	1600–2700	0.5–0.8	0.2–0.3	0.15–0.3	1–25	0.70–1.20
Dolomite	2700–2850	0.5–0.9	0.2–6.4	0.1–0.4	3–25	0.40
Metamorphic						
Gneiss	2600–2850	0.4–0.6	0.2–0.3	0.15–0.25	4–20	0.60–0.71
Amphibole	2800–3150		0.5–1.0	0.4		
Marble	2670–2750	0.3–0.8	0.2–0.35	0.2–0.3	5–20	0.62–1.20
Mantle						
Peridotite	3250					
Dunite	3000–3700	1.4–1.6	0.6–0.7			

Density ρ, Young's modulus E, shear or rigidity modulus G, Poisson's ratio v, tensile strength T_0 and coefficient of internal friction μ (modified after Turcotte and Schubert 2002; Gudmundsson 2011). A more recent study suggests slightly lower values for the Young's modulus and the shear modulus (Heap et al. 2020)

limit of a solid body, but without developing fractures: deforming sand is a good example of plastic deformation. The mechanical analogue for plasticity is a rigid object resting on a friction surface, indicating that a yield stress (schematically represented by the friction along the surface) must be reached before the plastic deformation (Fig. 2.6b). Unlike elastic deformation, plastic deformation is not characterized by a predominant deformation law, and different criteria can be used to describe plasticity, including the Tresca and the Von Mises criteria. Both criteria are based on the assumption that plastic deformation starts when the maximum shear stress in a rock reaches a threshold, or critical **yield stress** (or yield strength, that is the maximum stress a rock can take before non-recoverable deformation; e.g., Jaeger et al. 2007).

2.4.3 Viscous Deformation

While elasticity may work well for small strains in the upper crust, rocks at higher temperatures and pressures tend to flow, storing permanent and irreversible deformation, a process described by viscous behaviour. Viscous deformation is the law that is most commonly used to describe ductile deformation at the crustal scale, where the rocks accumulate permanent strain without visible fracturing (e.g., Stuwe 2007, and references therein). Viscous deformation is characterized by the accumulation of strain as a function of time, that is by a strain rate, in a fluid. The general relationship between stress and strain rate is described (see also Eq. 1.3) as:

$$\sigma = \eta e^{\cdot} \tag{2.14}$$

where η is a constant of proportionality, called **viscosity** ($\eta = tan\ \delta$; see Fig. 2.6c), and $e^{\cdot}$ is the strain rate; the SI unit of viscosity is the unit of stress multiplied by time, which is Pa s. Equation 2.14 indicates that, the higher the stress, the faster the flow. Viscous deformation is time-dependent, so that the strain also depends on the duration of the applied stress (e.g., Ranalli 1995; van der Pluijm and Marshak 2004; Karato 2008, and references therein).

Viscous flow is irreversible and produces permanent or non-recoverable strain when the force is removed. The physical model for this type of behaviour is the dashpot (Fig. 2.6c), which is a leaky piston that moves inside a fluid-filled cylinder. The resistance encountered by the moving piston reflects the viscosity of the fluid. The viscosities of some common materials and rocks are reported in Table 2.2. Temperature is a fundamental parameter governing viscosity: the temperature dependence of viscosity is described by the Arrhenius relationship:

$$\eta = C\ e^{(-Q/RT)} \tag{2.15}$$

where C is a material-specific constant (or pre-exponent constant), R is the universal gas constant and T and Q are the absolute temperature and the activation energy (in J mol^{-1}), respectively (Stuwe 2007; Fossen 2010). The Arrhenius relationship states that the viscosity of any material will trend towards infinity at absolute zero and will decrease exponentially from there to approach the value C asymptotically at high temperatures.

The concept of viscosity should be applied only to fluids. In geology only a few rock types (including molten rock and salt) may be modelled as truly viscous media. However, the behaviour of large portions of the Earth, as for example the upper mantle, may be also approximated as viscous over the long time available to geologic processes.

The viscous behaviour represented by Eq. 2.14 is commonly referred to as linear, or Newtonian, with $e^{\cdot} \propto \sigma$. The term linear emphasizes the distinction from non-linear viscous (or non-Newtonian) behaviour, which has been observed experimentally in deforming hot

Table 2.2 Viscosities of fluids in nature

Fluid	Viscosity η (Pa s)
Air	10^{-5}
Water	10^{-3}
Ice	10^{12}
Basaltic lava, 1050–1200 °C	10^{2}–10^{3}
Basaltic magma, 1200 °C	10^{2}
Andesitic lava, 950–1170 °C	10^{4}–10^{7}
Rhyolite lava, 700–900 °C	10^{9}–10^{13}
Rhyolitic magma, 900 °C	10^{5}
Rock salt	10^{15}–10^{17}
Asthenosphere	10^{20}–10^{21}

Modified after van der Pluijm and Marshak (2004), Gudmundsson (2011)

rocks and is probably more widespread than linear viscosity. Non-linear viscous behaviour implies a non-linear relationship between stress and strain rate, in the form of $e^{\cdot} \propto \sigma^n$, with the strain rate being proportional to the nth power of stress ($n > 1$). The non-linear proportionality between stress and strain rate results in a variation of viscosity. For example, in Fig. 2.6c the viscosity (the tangent to the slope) decreases with increasing stress and strain rate, which implies a faster flow. Thus, in non-Newtonian materials one can only define a certain **effective viscosity** for a given stress or strain rate, conversely to Newtonian viscosity, which is a property of the material. A same material (including the silicate polycrystals constituting the lithosphere and the mantle) may show a linear or non-linear viscous behaviour as a function of its pressure, temperature and level of stress.

2.4.4 Combined Deformation Models

Rocks and other natural materials are rheologically complex and generally do not behave as perfect elastic, plastic or viscous materials. It is therefore useful to combine these three main types of deformation in order to describe more realistic natural rock behaviours (e.g., Fossen 2010 and references therein).

The cases where a deformation process is reversible, but the strain accumulation and strain recovery are delayed, define a **visco-elastic** behaviour. This can be exemplified by a spring (elastic behaviour) and a dashpot (viscous behaviour) in parallel (Fig. 2.6d). When stress is applied, both the spring and the dashpot move simultaneously. However, the dashpot delays the extension of the spring. When the stress is released, the spring will try to return to its original configuration, but again this movement is delayed by the dashpot.

A material behaving elastically at the first application of stress, but then behaving in a viscous manner, has an **elastic-viscous** behaviour (Fig. 2.6e). Here, when the stress is removed, the elastic portion of the strain is recovered, but the viscous component remains. This behaviour can be modelled by placing a spring and a dashpot in series.

A **visco-plastic** or Bingham material is one that flows as a perfectly viscous material, but only above a certain yield stress (characteristic of plastic behaviour). Below this yield stress there is no deformation. The mechanical analogue is a combination of a dashpot and a rigid object resting on a friction surface (Fig. 2.6f). Rheological experiments on lava and the morphology of lava flows suggest that, over a significant range of temperatures, lava behaves like a visco-plastic fluid, with yield stress; this follows from the crystal content of lava.

2.4.5 The Role of the Boundary Conditions

A same rock may experience most of the above-mentioned stress-strain relationships, depending upon different boundary conditions, which are mainly related to, in addition to the applied stress, its temperature, pressure, the presence of pore pressure and the strain rate. Laboratory experiments allow defining the role of these conditions on the general deformation behaviour of a rock, and in particular on its **strength**, which is the ability of a rock to support differential stress, or the maximum differential stress before a rock fails (e.g., van der Pluijm and Marshak 2004; Jaeger et al. 2007; Fossen 2010).

Increasing the temperature, the elastic portion of the strain decreases, fracturing becomes inhibited and the ductility increases, which is also most noticeable at elevated confining pressure. The maximum stress that a rock can support until it flows (its yield strength) decreases with increasing temperature. Therefore, rocks have lower strength and are more ductile with increasing depth (and temperatures) in the Earth.

Increasing the confining pressure results in greater strain accumulation before failure, thus increasing the strength of a rock by hindering fracturing. In addition, increasing the pressure results in an increase of the viscous behaviour, and therefore of the rock's ability to flow: in fact, higher confining pressures resist the opening of rock fractures, so any shape change that occurs is mainly viscous.

Experiments also show that increasing any pore-fluid pressure decreases the elastic component and produces a drop in the sample's strength, promoting fracturing and reducing the ductility. Therefore, when the pore-fluid pressure is high, rocks are weaker. The pore-fluid pressure P_f acts equally in all directions and counteracts the confining pressure, resulting in an effective pressure σ' that is less than the confining pressure σ_n, as:

$$\sigma' = \sigma_n - P_f \tag{2.16}$$

Thus, increasing the pore-fluid pressure has a similar effect as decreasing the confining pressure. Fluid content explains why many minerals and rocks deform relatively easily even under moderate stress conditions. In addition, experiments show that the same material with low fluid content ("dry" conditions) would resist deformation, but at high fluid content ("wet" conditions) deformation may occur readily.

Finally, strain rate variations also play an important role. In fact, decreasing the strain rate results in decreased elastic component and rock strength, and in increased ductility. In general, strain rate decrease in rock experiments produces effects similar to temperature increase.

The overall relationship between strain and time may vary in a same experiment: in fact, the rate of strain may decrease, increase, or remain constant under constant stress. The increase of the stresses necessary to continue the deformation suggests that the rock becomes stronger (**strain hardening**) with increasing strain. Conversely, the decrease of the stresses suggests that the rock becomes weaker (**strain softening**) with increasing strain. Strain hardening and softening also depend upon temperature, pressure and strain rate variations.

2.4.6 Rheology of the Crust and Magma

After having introduced the basic principles of rock rheology, this section summarizes the rheological properties of the crust and the magma, which are of most interest in volcano-tectonics.

Combining present-day stress and paleostress data with experimental data on flow properties of rocks and minerals gives general strength profiles for the crust and lithosphere. These strength curves represent the depth variation of the differential stress magnitude σ_1–σ_3 required to deform the rocks, given assumptions on their composition and temperature. A typical strength profile, as that in Fig. 2.7a, consists of two types of curves (e.g., Ranalli 1995; Stuwe 2007; Karato 2008).

The straight blue lines show increasing rock strength with increasing depth as the normal stresses in the crust increase with depth, as based on empirical observations: these curves are consistent with brittle fracture (for details see Sect. 2.5). The curved red lines describe viscous flow deformation: the strength they describe decreases exponentially downwards, because temperature increases with depth roughly linearly. This strength profile thus divides the crust into an upper part dominated by brittle deformation, where seismicity is usually confined, and a lower part where viscous flow dominates. A rock at a given depth will deform according to the deformation mechanism that requires less stress. The transition between brittle fracture and viscous flow is referred to as the **brittle–ductile transition**, which is characterized by marked strength contrast. For an ideal uniform crust with gradual depth increase in temperature, a single brittle–ductile transitional zone at the depth where temperature-activated flow becomes important is expected. At higher depths, below the Moho, the properties of the olivine-rich lithospheric mantle are quite different from the quartz-dominated crust, and this in turn produces

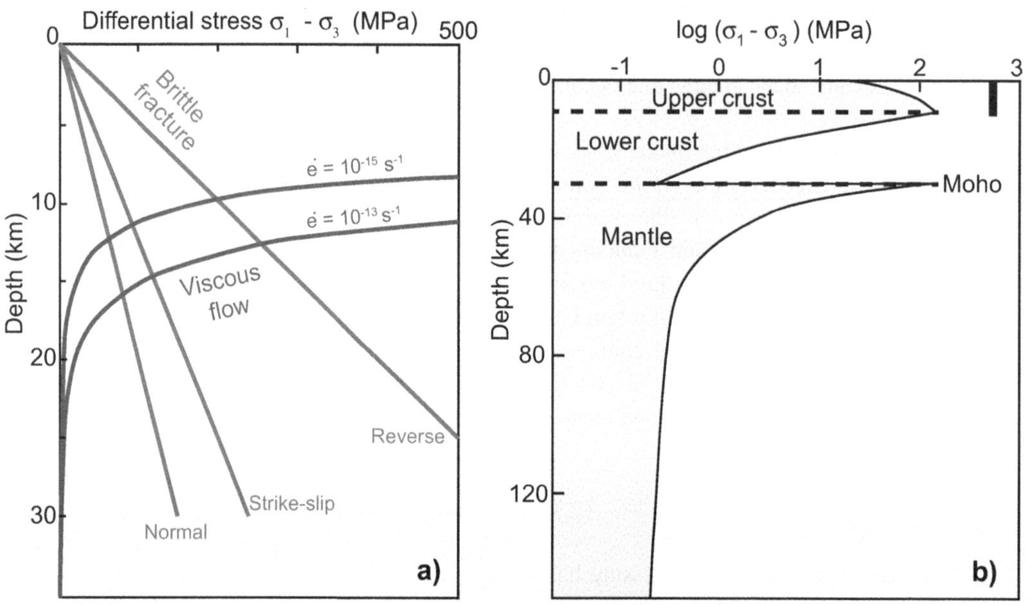

Fig. 2.7 **a** Strength (differential stress) increases downwards through the brittle crust (blue lines, each indicating a stress regime), until the temperature is high enough to activate viscous flow (red lines). Brittle and viscous deformations have two different strength profiles, and the intersection between the two defines the brittle–ductile transition. For a given geothermal gradient, the location of the brittle–ductile transition depends upon the tectonic regime and the strain rate $e^{\cdot}$ (modified after Fossen 2010). **b** Strength curve showing the variation in differential stress magnitude with depth in the lithosphere for a region with high geothermal gradient (e.g., area of continental extension). The black vertical bar at the right side indicates where seismic activity may occur

a sharp increase in strength and, therefore, a return to brittle behaviour. Following the increase in temperature with depth in the upper mantle, strength decreases as viscous behaviour replaces the brittle regime of mantle rocks, similarly to the crustal profile. The resulting "Christmas tree"-like profile in Fig. 2.7b emphasizes the role of the compositional stratification of the lithosphere and provides reasonable, first-order predictions of rheology for use in models.

The shape of the strength profile, including the depth to the brittle–ductile transition, depends upon several factors (Karato 2008; Fossen 2010). The geothermal gradient influences the location of the brittle and ductile domains. In particular, a higher geothermal gradient encourages viscous flow, thinning the brittle domain(s), and reduces the differential strength up to one order of magnitude. The tectonic regime also plays an important role. In fact, the stress required for thrusting is significantly larger than that required for normal faulting, so that each of the three stress regimes has an associated trend (Fig. 2.7a). The strength profile also depends on the pore fluid pressure, because "dry" rocks are more resistant to deformation, whereas pore fluid pressure reduces friction. In addition, varying mineralogical composition through the lithosphere can lead to several layers of alternating brittle and plastic rheologies. As quartz is one of the softer minerals and most crustal rocks contain quartz, the viscous deformation of the crust may in general be described with the rheological data for quartz. Rocks in the lithospheric mantle are quartz absent and dominated by olivine, and thus the rheology of olivine may describe the general behaviour of the lithospheric mantle. Variations from these simple general compositions may introduce deviations in the behaviour of portions of the lithosphere. Finally, each curve is for a given strain rate that is assumed to be constant over the entire lithosphere. Note that strain rate shifts the viscous flow laws (and thus the brittle–

ductile transition) vertically, so that a higher strain rate will yield a curve that has a higher strength at a given depth (Fig. 2.7a).

Defining the rheology of magma is less straightforward, as magma consists of three phases (melt, crystals, bubbles), which have each a distinct influence on rheology (e.g., Burgisser and Degruyter 2015, and references therein). Melt viscosity is a function of composition, which includes the amount of dissolved volatiles (mainly water). As the amount of dissolved water decreases during magma ascent, melt viscosity increases by several orders of magnitude (see Sect. 1.5). This increase is so large that it dominates the rheological changes during decompression. At low shear rates, melt viscosity is Newtonian, that is, it deforms linearly with the applied stress. As shear rate increases, however, the melt becomes visco-elastic and finally breaks: this implies that pure silicate melt fragments under sufficient shear rate (Fig. 1.18). Experiments have shown that these rates too depend on melt composition through a linear dependence on viscosity. At magmatic temperatures, silica-rich melts, such as rhyolites, break under shear rates on the order of 10^{-2}–10^{4} s^{-1}, whereas silica-poor melts, such as basalt, break under rates on the order of 10^{4}–10^{8} s^{-1}.

When crystals are present in the melt, they influence the rheology in two ways: first, the larger the crystal volume fraction, the harder it is to deform magma, indicating that magma rheology is a strong function of crystal volume fraction. Second, the more elongated the crystals, the more the suspension resists deformation. The details of the relationship between crystal shape and rheology are complex. One popular way to approach these complexities is to treat melt and crystals as a perfectly homogeneous mixture with a bulk rheology. The simplest case is that in which the bulk behaviour of the mixture is assumed to be Newtonian, although non-Newtonian behaviours have been also recently considered (e.g., Burgisser and Degruyter 2015, and references therein). Bubbles can play a role similar to that of crystals if they remain spherical at all times, which tends to occur at low shear rates. This increases bulk viscosity to a lesser degree than in the case of crystals because of the internal motion of the gas within the bubbles. At high shear rates, however, bubbles deform and ease flowage. The situation becomes quickly complex as high shear and deformation promote bubble coalescence and gas channelling. A common simplification is to calculate two extreme bulk viscosities, one for fully deformed bubbles and one for non-deforming spheres. When the three phases are together, experiments have shown that shear localization and complex rheological response occur. For example, bubbly magma may compact by foam collapse, and the resultant rheology may exhibit strain hardening. On the other hand, as strained crystals and bubbles organize themselves into flow bands, and crystals plastically deform or fail, the viscosity may decrease, promoting strain weakening (Calder et al. 2015, and references therein).

Therefore, while the viscous response of the magma is likely to involve a purely Newtonian role for the melt phase, the presence of bubbles and crystals nears the behaviour of magma to a non-Newtonian rheology, at times similar to that of a Bingham fluid, where the shear stress must exceed a critical value before flow occurs. As anticipated, lavas also tend to behave as Bingham fluids. Basaltic mixtures of fluid, crystals and bubbles typically have yield strength of $\sim 10^{3}$ Pa, that of silicic lavas being of $\sim 10^{5}$ Pa.

2.5 Brittle Deformation

A brittle material deforms by fracturing when subjected to stress beyond the yield point, usually at finite strains of $\sim 5\%$ or less, after the material has already undergone some elastic and/or plastic behaviour. During brittle fracturing grains are crushed and reorganized, and strain becomes localized. **Brittle deformation** is thus the permanent change that occurs in a solid material due to the growth of fractures. In contrast to elastic strain, brittle deformation is non-recoverable, meaning that the distortion remains when the stress is removed.

Fractures are the most common product of brittle deformation. A **fracture** is any surface of discontinuity across which the material is no longer bonded. There are many types of fractures, mainly depending on their kinematics and any filling (Fig. 2.8; e.g., Pollard and Aydin 1988; Price and Cosgrove 1990; Pollard and Fletcher 2005; Gudmundsson 2011).

A fracture without any measurable shear displacement is called **joint,** extension fracture or tensile fracture (for details see Sect. 2.5.1). If a fracture fills with minerals precipitated out of a hydrous solution, it is a **vein**. If it fills with magmatic rock, it forms a magma-filled fracture which, depending on its orientation with regard to the layering, may be a **dike** (discordant and perpendicular to the layering), an inclined **sheet** (discordant and at moderate angle) or a **sill** (concordant and parallel). A fracture characterized by contraction, often filled with immobile residue from the country rock, is called **stylolite**. A fracture characterized by shear displacement or sliding is called shear fracture, or **fault**.

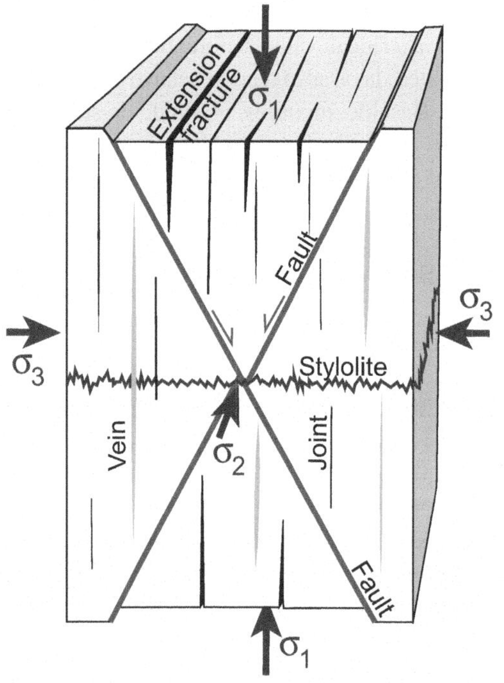

Fig. 2.8 Geometric relationships between different types of fractures and the principal stresses (modified after Fossen 2010)

Fractures may form at stresses of less than ~10 MPa, that are hundreds of times less than the theoretical strength of rock. This is possible because the far-field or **remote stress** (stress applied at a distance from a region of interest) concentrates at the sides of flaws (e.g., holes) inside a material. For example, in the case of a circular hole within an elastic sheet subjected to tensile stress, the near-field or **local stress** (i.e., stress at the point of interest) tangent to the sides of the hole is three times the remote stress magnitude (Fig. 2.9a; Jaeger et al. 2007; Fossen 2010). If the hole has the shape of an ellipse, the concentration of the local stress increases further (Fig. 2.9b). In fact, values for stress concentration at the ends of an elliptical hole depend on the hole eccentricity. The higher the eccentricity, the greater the stress concentration: pre-existing fractures, displaying highest eccentricity, promote the highest local stress. This concept of stress concentration at the ends of elliptical holes to explain fracture development at lower stresses than expected is at the base of the **Griffith theory** (Griffith 1924).

In rock mechanics it is common to classify fractures into three modes, accordingly to their displacement (Fig. 2.10; e.g., Pollard and Aydin 1988; Pollard and Fletcher 2005). Mode I is the opening mode, where displacement is perpendicular to the walls of the crack; Mode II represents the sliding (shear) perpendicular to the edge and Mode III involves vertical slip parallel (tearing) to the edge of the crack.

Mode I cracks include magma-filled fractures, and are thus of particular importance in volcano-tectonics. The following equation defines the conditions at which Mode I cracks propagate:

$$K = \sigma_t J(\pi L)^{1/2} \tag{2.17}$$

where K is the **stress intensity factor,** σ_t is the remote tensile stress, J is a dimensionless number that takes into account for the shape of the crack and L is half of the crack's length (see also Sect. 3.4.1; e.g., van der Pluijm and Marshak 2004; Gudmundsson 2011). In this equation the cracks are assumed to have very high eccentricity, and cracks with lower eccentricity require

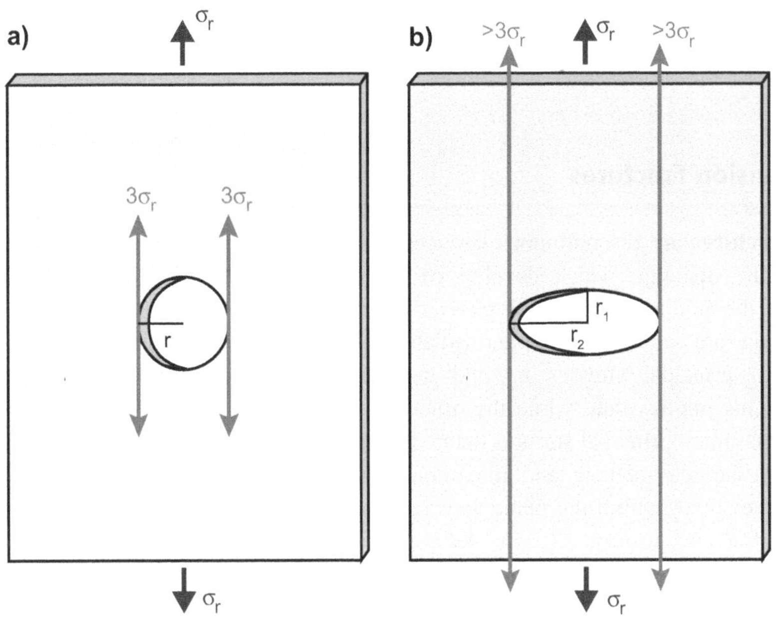

Fig. 2.9 Stress concentration adjacent to a hole in an elastic sheet. If the sheet is subjected to a remote tensile stress at its ends (purple arrows, σ_r), then stress magnitudes at the sides of the holes (blue double arrows) are equal to $n\sigma_r$, where n, the stress concentration factor, is $(2r_2/r_1)$ +1. **a** For a circular hole, n = 3. **b** For an elliptical hole, n > 3 (modified after van der Pluijm and Marshak 2004)

higher stresses to propagate. Equation (2.17) shows that the stress intensity factor increases when the remote stress increases and, other factors being equal, a longer crack generally propagates before a shorter crack. Also, a longer crack is less sensitive to the background stress (Maccaferri et al. 2019). A crack begins to propagate when K attains a critical value K_c, known as critical stress intensity factor, or fracture toughness. The **fracture toughness** can be considered as a material's resistance against continued growth of an existing fracture, and is constant for a given material. High fracture toughness implies high resistance against fracture propagation and therefore low propagation rates. Both the stress intensity factor and fracture toughness are expressed in MPa $m^{1/2}$.

The types of fractures not filled by magma most relevant for volcano-tectonics, that are extension and shear fractures, are described in

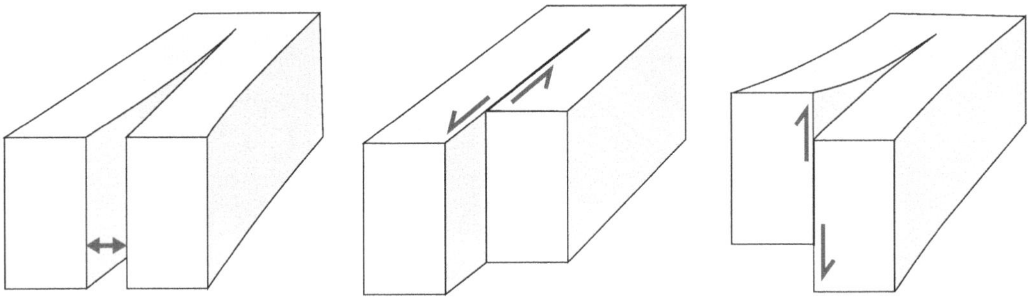

Fig. 2.10 Schematic diagrams of Mode I (opening), Mode II (sliding) and Mode III (tearing) fractures

Sects. 2.5.1 and 2.5.2, respectively, whereas magma-filled fractures are considered in the following chapters.

2.5.1 Extension Fractures

Extension fractures are discontinuous surfaces characterized by opening, which develop perpendicular to the minimum principal stress σ_3 and thus contain (are parallel to) the intermediate and maximum principal stresses σ_2 and σ_1, respectively. This implies that, while the orientation of the minimum principal stress is defined, the plunge of the intermediate and maximum principal stresses lying within the plane approximating the fracture surface is not defined (Fig. 2.11a). Extension fractures are most common approaching the Earth's surface, as they develop with very small strains, at low or no confining pressure and low differential stress. In fact, any increase in the confining pressure and/or differential stress promotes shear fractures, or faults, unless any high fluid pressure reduces the effective stress. In the latter case, the resulting fractures are called **hydrofractures** (see Sect. 2.5.2; e.g., Fossen 2010; Gudmundsson 2011).

Tensile fractures are extension fractures that form where the least principal stress σ_3 is tensile, that is negative. Tensile fractures commonly develop along the axis of divergent plate boundaries, where the σ_3 oriented perpendicularly or at high angle to the plate boundary may be negative at the surface. Here metres wide and hundreds of metres long tensile fractures reach a few hundred of metres of depth (Fig. 2.11b).

Ideally, an extension fracture grows radially from a nucleation point, so that the propagation front (fracture tip) has elliptical shape. During propagation, the stress readjustments or oscillations due to local heterogeneities at the crack tip cause the latter to bend and bifurcate, forming off-plane smaller joints. This results in long and narrow peripheral planes slightly oblique to the main fracture trend, in some cases forming a **plumose** (feather-like) structure (e.g., Fossen 2010). This is mechanically explained as, while entering an area with different stress orientation, the twisted joints forming this fringe zone

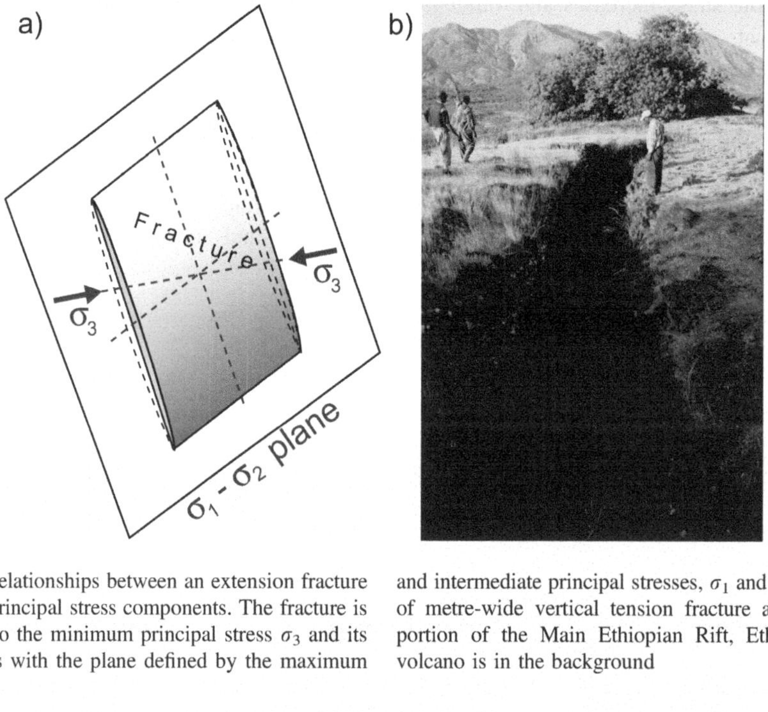

Fig. 2.11 **a** Relationships between an extension fracture and the three principal stress components. The fracture is perpendicular to the minimum principal stress σ_3 and its plane coincides with the plane defined by the maximum and intermediate principal stresses, σ_1 and σ_2. **b** Example of metre-wide vertical tension fracture along the axial portion of the Main Ethiopian Rift, Ethiopia; Fantale volcano is in the background

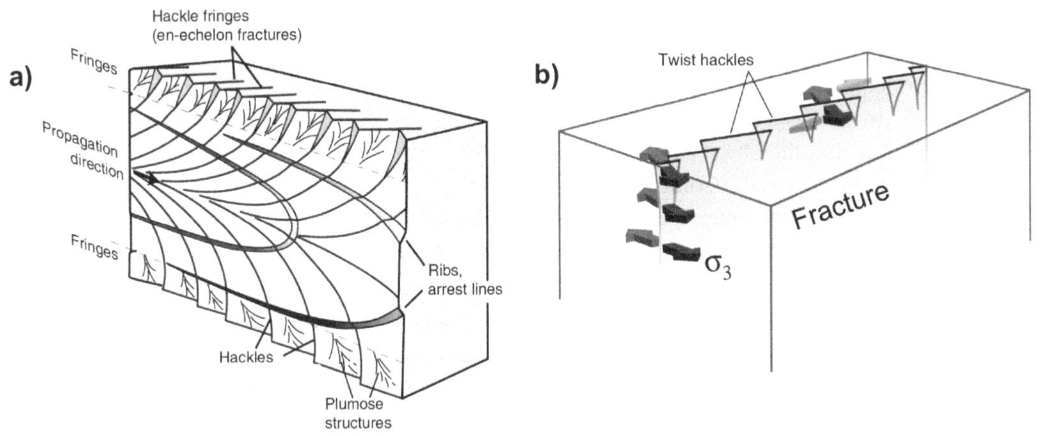

Fig. 2.12 Plumose joints. **a** Schematic illustration of the structures characteristic of a plumose joint. **b** Detail showing the twisting of the extension fractures as they reach an interface with a mechanically different rock layer. Note the parallel twisting of σ_3 and the fractures (hackles) (modified after Fossen 2010)

reorient perpendicular to the local σ_3, approaching an **en-echelon** configuration: the latter is an arrangement of subparallel fractures of similar length in a zone of fairly constant width, which trends slightly obliquely to the fractures (Fig. 2.12). The stress reorganization while the fracture is propagating occurs for two main reasons. First, any rock is never perfectly isotropic and homogeneous, and its material properties may vary from point to point. This inhomogeneity distorts the local stress field at the tip of a growing joint, so that here the local principal stresses are not necessarily parallel to the remote σ_3. Second, being the stress intensity proportional to the joint length (Eq. 2.17), the local stress field at the tip of a joint changes as the joint tip propagates. If the stress magnitude at the tip exceeds a critical value, the energy available for cracking rock exceeds the energy needed to create a single surface. This develops microscopic cracks that splay off the main joint. If the energy becomes very large, the crack may even split into two separate, parallel surfaces. Plumose structures are of interest in volcano-tectonics, as they represent smaller-scale mechanical analogues of the peripheral en-echelon segments observed in some dikes while these approach the surface. These peripheral dike segments feed vents with a direction oblique to their parent dike at depth, as for example observed at the Inyo dike, California (USA; see also Fig. 3.28; Reches and Fink 1988). **En-echelon joint arrays**, in addition to forming plumose structures, may be a common expression of the shear motion along a

Fig. 2.13 **a** Formation of an en-echelon system of veins (orange ellipses) within a shear zone. **b** Growth of the fractures as sigmoidal veins, due to the clockwise rotation of their older, central part

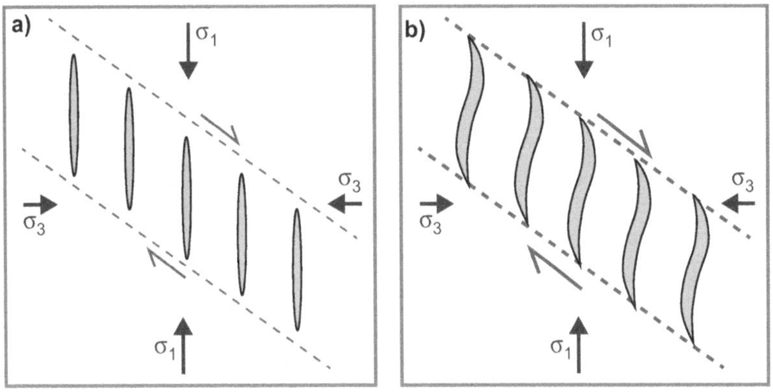

fault zone (Fig. 2.13a). Here the fractures comprising an en-echelon array initiate parallel to the maximum principal stress σ_1, typically at an angle of ~45° to the shear zone. Fractures open as displacement across the shear zone develops, and fill with vein material. Continued shear rotates the veins, which may become sigmoidal (Fig. 2.13b; van der Pluijm and Marshak 2004).

Intrusive rocks without strong anisotropy (such as granites) commonly contain a set of joints that roughly parallels the boundary of the intrusion, whose spacing decreases progressively towards the boundary: such joints are called **exfoliation joints**. The stresses causing exfoliation joints are residual. A residual stress remains in a rock even if it is no longer loaded externally and may develop in several ways. In the case of a pluton, residual stresses develop because its thermal properties (e.g., coefficient of thermal expansion) differ from those of the surrounding country rock, and because during cooling the pluton cools by a greater amount than the country rock. The pluton and the country rock tend to undergo different elastic strains as a result of thermal changes during cooling and later unroofing. If the pluton shrinks more than the country rock, tensile stresses develop perpendicular to its boundary (Fig. 2.14a; e.g., van der Pluijm and Marshak 2004). At depth, compressive stress due to the overburden counters these tensile stresses, but near the surface residual tensile stress perpendicular to the boundary of the pluton may exceed the weight of the overburden and produce exfoliation joints parallel to the boundary. An example of exfoliation joints is found on the top of the granitic Half Dome of Yosemite (California; Fig. 2.14b). The joints frequently observed along the sides of dikes, parallel to the contact with the country rock and decreasing in distance while approaching the contact, also share a similar origin. Here the dike and the country rock undergo different elastic strains because of the thermal changes developed during cooling, resulting in a shrinking of the dike and the development of tensile stresses along its margins.

Columnar joints are joints that break magmatic rocks into approximately hexagonal columns. They form during cooling and contraction in shallow intrusions (dikes, sills, laccoliths) or lava flows (Fig. 2.15). In the case of magma-filled fractures (as dikes and sills), the major axes of the columns tend to be perpendicular to the boundaries of the intrusion (usually subhorizontal in dikes and subvertical in sills). However, in some intrusions the columns may curve. Notable examples of columnar joints include the Giant's Causeway (Northern Ireland) and the Devil's Postpile (California).

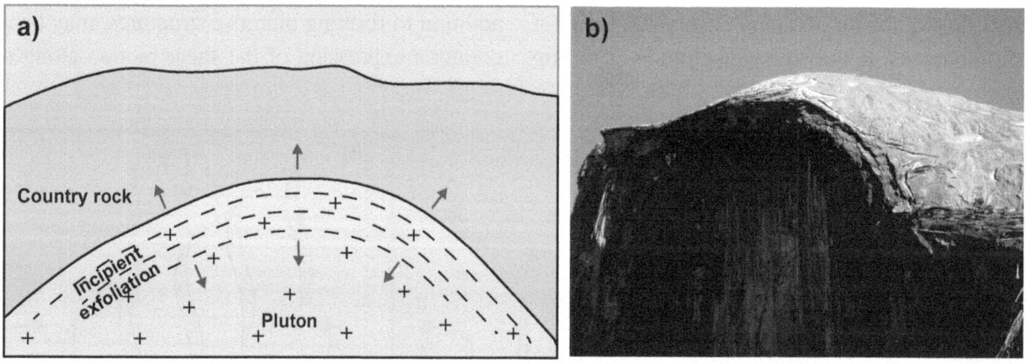

Fig. 2.14 **a** The residual tensile stresses (red arrows) developed at the contact between a solidifying pluton and the country rock generate a joint pattern parallel to the contact within the pluton (incipient exfoliation); **b** example of exfoliated granite dome at Yosemite (California; USA): note the subhorizontal joints parallel to the upper surface of the dome in the right portion of the image

Fig. 2.15 Top (left) and side (right) view of columnar joints developed within basalts at Asbyrgi, Iceland

2.5.2 Shear Fractures

Shear fractures are characterized by a slip, or displacement, parallel to the fracture surface and typically develop at ~30° to the direction of the maximum principal stress σ_1. When developing, shear fractures usually form conjugate pairs, at an angle of ~60° from each other, bisected by the maximum principal stress σ_1 (Fig. 2.8). Shear fractures develop under low to moderate temperature and pressure conditions, as those found in the upper crust. Here shear fractures form when the differential stress (σ_1–σ_3) exceeds the rock strength. The larger the confining pressure, the larger the rock strength and thus the larger the differential stress to fracture the rock. Without differential stress, as in isotropic stress state, there is no force pulling or pushing the rock volume, thus inhibiting the formation of shear fractures.

The stress conditions required to develop a shear fracture can be conveniently described by the **Mohr-Coulomb fracture criterion**. This considers the critical shear τ and normal σ_n stresses acting on a potential failure and relates these to two other parameters. The first is the coefficient of internal friction μ, which is expressed as $\mu = \tan\phi$ (where ϕ is the angle of internal friction of the material); for rocks, assuming ϕ as ranging between ~25° and ~35°, μ corresponds to a dimensionless number that varies between ~0.45 and ~0.7. The second parameter is defined by the cohesive strength, or cohesion, c of the material; for most rocks the cohesion is on the order of 1–10 MPa (e.g., Ranalli 1995). The Mohr-Coulomb fracture criterion relates these parameters in the following equation:

$$\tau = \sigma_n \tan\phi + c \qquad (2.18)$$

Therefore, the higher the normal stress acting on a potential failure plane, the higher must be the shear stress to initiate failure. Even in the case of σ_n=0, a minimum critical shear stress, corresponding to the cohesion of the material, must be reached for failure to initiate. The Mohr diagram provides an appropriate way to further illustrate the Mohr-Coulomb criterion, which has the general formula for a straight line (Fig. 2.16a; e.g., van der Pluijm and Marshak 2004; Jaeger et al. 2007; Fossen 2010). Many different critical states of differential stress, represented by the circles in Fig. 2.16a, may induce failure; the deeper is the rock, the stronger it is and the larger is the differential stress required to fracture it. The line tangent to all these circles, called **Coulomb failure envelope**, represents the Mohr-Coulomb fracture criterion of the rock expressed by Eq. 2.18. Any Mohr circle that does not reach the envelope represents a stable condition of the rock, that does not undergo failure. Therefore, the Coulomb failure envelope separates the stable field where the rock does not fracture (below the envelope) from the unstable field above the envelope, which is in principle unachievable

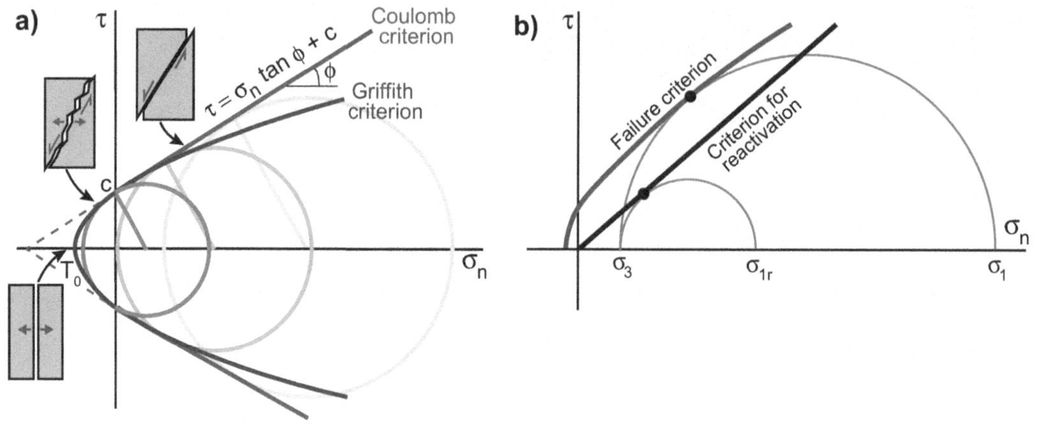

Fig. 2.16 **a** The Coulomb fracture criterion occurs as two straight lines (red) in the Mohr diagram. The blue line represents the Griffith criterion, mostly used in the tensile regime. The grey circles represent examples of critical states of stress. c = cohesive strength of the rock; T_0= tensile strength of the rock. The diagram includes the conditions for tensile, hybrid and shear failure, as exemplified by the related brown specimens in section view. **b** The effect of a pre-existing fracture (plane of weakness) illustrated in the Mohr diagram. The criterion for reactivation (purple line) is different from that of the same type of unfractured rock (red line), as the cohesion c =0 and the differential stress required to reactivate the fracture is considerably smaller $(\sigma_{1r}-\sigma_3)$ than that required to generate a new fracture in the rock $(\sigma_1-\sigma_3)$ (modified after Fossen 2010)

because fracture prevents such states of stress occurring. Ideally, the point at which a Mohr circle touches the failure envelope represents the orientation of the plane of failure, as well as the shear and normal stresses acting on that plane at failure (see Sect. 2.2). Each rock has its own failure envelope.

The point where the Mohr envelope intersects the x axis represents the critical stress at which tensile fractures start to grow, or critical **tensile stress**. The **tensile strength** T_0 of a rock is the maximum amount of tensile stress that it can take before failure. Experimental data show that the Coulomb criterion, while predicting the critical state of stress needed to create shear fractures, does not successfully predict tensile fractures. In fact, approaching the tensile regime, where at least one of the principal stresses is negative, the envelope defined by experimental tests becomes non-linear, with the tensile stress experimentally found to be approximately half the cohesive strength c. Also, the fact that the Coulomb criterion relies on the angle of internal friction, which is physically meaningless for tensile normal stress, indicates that the Coulomb criterion is inappropriate in the tensional regime.

In contrast to the Coulomb criterion, the **Griffith fracture criterion** proposes a non-linear relationship between the principal stresses for a rock on the verge of fracture (Jaeger et al. 2007; Fossen 2010). Based on the notion that macroscopic cracks grow from pre-existing microscopic flaws, as postulated by the Griffith theory, this criterion provides a much more realistic estimate of the tensile strength. The Griffith criterion relationship has the form of:

$$\tau^2 + 4T_0\sigma_n - 4T_0^2 = 0 \tag{2.19}$$

This equation defines a parabola, which can be represented in the Mohr diagram with the tensile strength T_0 intersecting the x axis (Fig. 2.16). Here the cohesive strength of a rock is twice its tensile strength (c =2 T_0), in agreement with experimental data. Using Eq. (2.19), it is possible to reformulate the Coulomb fracture criterion as:

$$\tau = 2T_0 + \sigma_n\mu \tag{2.20}$$

This formulation allows combining the Coulomb criterion for the compressional stress regime with the Griffith criterion for the tensile regime.

The Griffith criterion has important implications in several volcano-tectonic studies. For example, some fractures, as those commonly found along divergent plate boundaries or those generated by the shallow emplacement of dikes, show both a tensile and shear component and thus represent a hybrid between tensile cracks and shear ruptures. The conditions for initiating these hybrid fractures can be represented by the Griffith criterion on a Mohr diagram in the steeply sloping portion of the parabolic failure envelope (Fig. 2.16a; e.g., van der Pluijm and Marshak 2004).

Most rocks carry some form of planar anisotropy, due to sedimentary or tectonic fabrics, including bedding, lamination, foliation and pre-existing tectonic fractures. In these cases, the difference in the critical differential stress leading to failure may vary by several hundred percent, depending on the orientation of the anisotropy. In fact, when an anisotropy approaches an orientation similar to that typical of shear fractures in isotropic rocks, shear fractures develop along the anisotropy at gradually lower differential stress. Therefore, whether a rock fails or not at lower differential stress along a weak pre-existing fracture depends on the orientation of the fabric or fracture relative to the stress field. This implies that renewed stress build up may reactivate pre-existing fractures at a lower level of stress, rather than creating a new fracture through the energetically more demanding process of growth and linkage of minor flaws in the rock (Fig. 2.16b; e.g., Fossen 2010). Reactivation of smaller fractures is a pre-requisite for major faults to develop: had reactivation not happened, the crust would have been packed with short fractures with small displacements.

Pore pressure may also promote the formation or reactivation of fractures. If the fluid is unable to escape poorly permeable rocks, then the pore pressure within the rock increases. An increase in the fluid pressure can reactivate faults and fractures. In fact, the fluid pressure counteracts the normal stress on the fracture, so that the related shear stress may be sufficient for reactivation. Injection of pressurized magma provides similar conditions, where the normal stress is balanced by magma pressure. Therefore, fluid or magma pressure may weaken the rock, so that deformation can occur at a differential stress that would otherwise be insufficient for failure. This implies that whether a new fracture forms or an existing one is reactivated depends also on the effective stress σ', that is the difference between the applied stress and the fluid pressure (Eq. 2.16). Fluid pressure has the following effect on the Coulomb fracture criterion:

$$\tau = \left(\sigma_n - P_f\right) \tan \phi + c \qquad (2.21)$$

An increase in pore pressure decreases the mean stress σ_m to $(\sigma_m - P_f)$, while the differential stress $(\sigma_1 - \sigma_3)$ remains constant (Fig. 2.17; e.g., Fossen 2010, and references therein). On a Mohr diagram, this is represented by a shift of Mohr's circle to the left. As anticipated, if the effective stress is tensile (negative σ_3), tensile fractures can form. While in dry rocks tensile fractures can be expected only at very shallow depths (less than a few hundred metres of depth), fluid overpressure makes it possible to have local tensile stress even at several kilometres of depth: this process is known as **hydraulic fracturing.**

Shear fractures, as well as extension fractures, are usually approximated by surfaces with elliptical shape, with the inner part usually carrying most of the displacement, which gradually decreases approaching the tip.

2.6 Faults

A **fault** is a shear fracture dominated by brittle deformation mechanisms (Fig. 2.18). A fault usually consists of a zone of deformation (fault zone) with variable thickness, which includes several smaller subparallel slip surfaces close to each other and crushed rock within. A fault zone may be also characterized by several smaller faults (**splays**) diverging from the main fault. In a fault zone there is usually a central core, or slip surface, which can consist of breccia, fault gouge, or cataclasite, each characterized by a

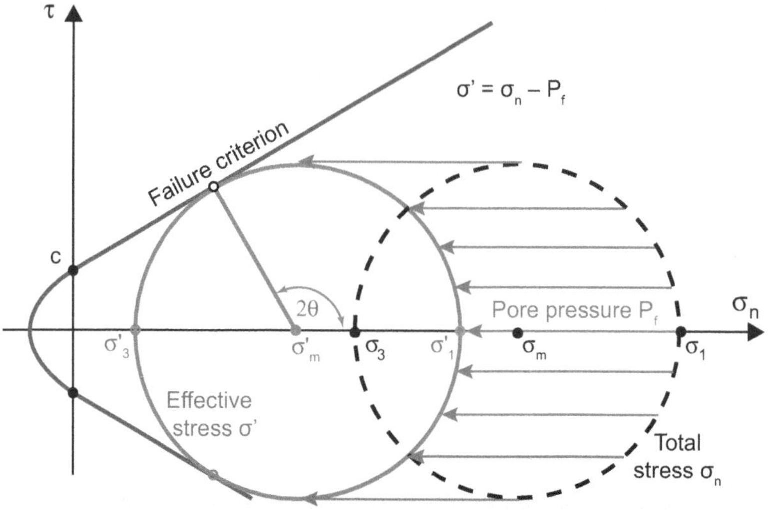

Fig. 2.17 Effect of pore pressure P_f (in blue) on enhancing faulting in a Mohr diagram through the reduction of the total stress components σ_1 and σ_3 (dashed circle) and mean stress σ_m to the effective stress components σ'_1 and σ'_3 (full circle) and mean stress σ'_m

progressive increase in the amount of the matrix of unconsolidated fault rock, or in the amount of deformation. The core is surrounded by brittlely deformed country rock, that is the fault damage zone. The thickness of a fault zone is usually much smaller than the offset and several orders of magnitude less than the fault length.

The vector connecting two points on opposite sides of a fault that were next to each other before faulting indicates the displacement, or net **slip** direction (Fig. 2.19). The net slip along a fault may be decomposed in the three dimensions (vertical, horizontal along-strike and horizontal across-strike), with each component providing an apparent and smaller portion of the total displacement. In ideal Andersonian conditions, faults are expected to exhibit a vertical (dip-slip faults) or an along-strike horizontal slip (strike-slip faults), with any across-strike horizontal component inversely proportional to the fault plane dip. Under these conditions, faulting takes place in a pair of conjugate planes which pass through the direction of the intermediate principal stress σ_2 and are equally inclined at angles less than 45° (usually ~30°) to the direction of the maximum principal stress σ_1 (Fig. 2.20; e.g., Price and Cosgrove 1990; van der Pluijm and Marshak 2004; Fossen 2010). In particular, **normal faults** form with vertical σ_1: the direction of the conjugate set of normal faults is parallel to the intermediate principal stress σ_2 and perpendicular to the horizontal least principal

Fig. 2.18 Front view of an active normal fault in northern Iceland: the hanging wall has been downthrown of a few metres with regard to the foot wall

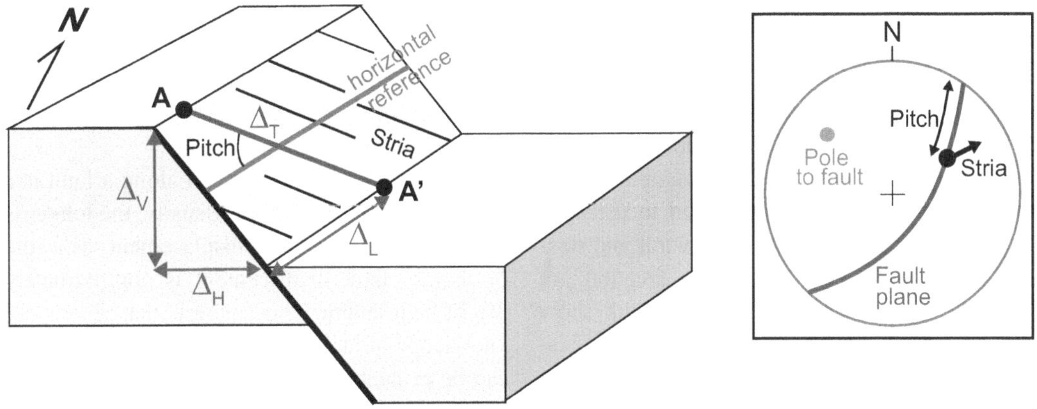

Fig. 2.19 The total displacement Δ_T along points A-A' on opposite sides of a fault can be decomposed in a vertical displacement Δ_V, horizontal (across-strike) displacement Δ_H and lateral (along-strike) displacement Δ_L. The slip direction can be expressed through the pitch, that is the angle between a horizontal line (in blue) and the slip direction (as for example indicated by the parallel black striations) on the fault plane. The pitch ranges from 0° to 180°: at these values it corresponds to pure strike-slip motions. Conversely, a pitch of 90° corresponds to pure dip-slip (normal or reverse) motions. The diagram to the right represents the fault and its kinematics on a lower-hemisphere equal-area projection

stress σ_3. In normal faulting the crustal portion above the fault (hanging wall) moves downward with regard to that below the fault (foot wall), and the faults usually have dip of ~60°, that is at ~30° with regard to the σ_1 direction. **Thrust faults** form with vertical σ_3: the direction of the conjugate set of normal faults is parallel to the intermediate principal stress σ_2 and perpendicular to the horizontal maximum principal stress σ_1. In thrust faulting the hanging wall moves upward with regard to the foot wall, and thrust faults usually have dip of ~30°; however, high angle **reverse faults** (dip of ~60°) may also form. Normal and thrust faults, having displacement vectors in the dip direction, are dip-slip faults. **Strike-slip faults** form with vertical intermediate

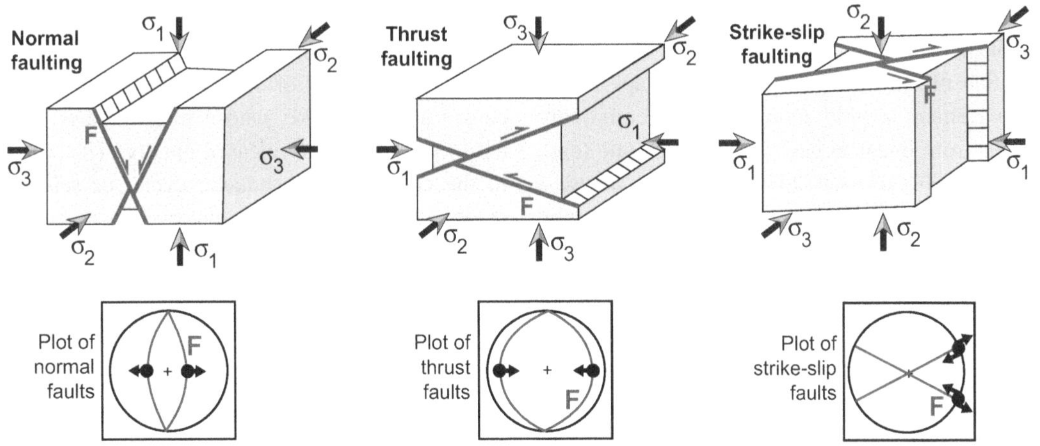

Fig. 2.20 Top: relationships between the three main types of faulting (parallel black lines indicate slip vectors on fault planes) and the configuration of the three principal components of the stress tensor, σ_1, σ_2 and σ_3. Bottom: for each case, the faults and slip vectors referring to the pair of faults above are plotted on a lower-hemisphere equal-area projection, as common in structural geology

principal stress σ_2. Here a conjugate pair of vertical faults develops, one with sinistral (or left-lateral) and the other with dextral (or right-lateral) motion (the sinistral or dextral motion is defined by the movement of the block across the fault); the direction of each conjugate fault is oblique (usually $\sim 30°$) with regard to the horizontal σ_1 direction. Therefore, sinistral and dextral faults, having a horizontal slip direction, are strike-slip faults. In reality, many faults show some deviation from the Andersonian idealization of dip- or strike-slip displacement, being their net slip vector oblique. These are thus called **oblique-slip** faults. The proportion between the dip- and strike-slip components of a fault, which defines its overall kinematics, is summarized by the **pitch** (or rake), which is the angle between a horizontal line and the slip vector on the fault plane. In the field, the slip vector on the fault plane is commonly identified by striations (Fig. 2.19).

Faults develop at all scales, from microscopic faults offsetting the boundaries of grains to megascopic faults that cut thousands of kilometres of crust. But even the biggest faults do not extend infinitely. Faults may terminate in several ways. Where a fault does not terminate against another structure (fault, intrusion, layer), it must die out decreasing its displacement. The boundary between the slipped and unslipped region at the end of a fault, that is where the amount of displacement decreases and becomes zero, is the **tip** line of the fault. The resulting shape of the displacement profile along a fault commonly varies from linear to bell-shaped or elliptic (e.g., Scholz 2002; Fossen 2010). In some cases, a fault ends splitting into numerous splays, thereby creating a fan of small fractures with decreasing displacement, called a **horsetail**. Alternatively, the deformation associated with the fault may die out in a zone of ductile deformation (e.g., folding). In fact, layers around faults are commonly deflected, or folded, showing a characteristic drag, which consists of a systematic change in the orientation of layers or markers, genetically related to the activity and growth of the fault. More in general, a **fold** is an originally planar layer which becomes bent or curved as a result of permanent deformation. A fold that forms ahead of a propagating fault tip is called **fault propagation fold**. Thus, many drag folds are fault propagation folds.

The maximum displacement along a fault is a function of the fault length. Usually, the longer is the fault, the larger is its displacement: as a rule of thumb, fault displacement is approximately 3% of fault length. This suggests that, given any knowledge of fault length, the fault displacement can be evaluated, or vice versa. Observations on faults with displacements up to 100 m also show that the fault displacement is usually similar or larger than the damage zone thickness, and ~ 100 times the fault core.

Faults grow from microfractures or deformation zones and accumulate displacement over time as deformation proceeds. Therefore, the slip observed on most faults is the sum of several increments, each with its own individual displacement or slip vector, as individual slip events may have had different slip directions. A fault is thus a discontinuity that may fail again, growing further, during renewed stress build up. Under a constant stress, faults may grow by two mechanisms (e.g., Scholz 2002, and references therein; Fossen 2010). The most common is the **stick-slip** mechanism, where slip accumulates during sudden seismic events, separated by periods of no slip (Fig. 2.21a). Stress builds up between slip events until it exceeds the frictional resistance of the fault. Each slip event causes an earthquake, whose magnitude is related to the amount of released energy. This leads to the concept of earthquake cycle, or **seismic cycle.** The idea of the seismic cycle was developed by H.F. Reid to explain the observations of the 1906 San Francisco (California) earthquake. The earthquake related deformation cycle consists of three phases (Fig. 2.21b): (a) the inter-seismic phase (period between two consecutive earthquakes, during which stress and elastic strain are accumulated); (b) the co-seismic phase (short period in which the accumulated stress and strain are suddenly released during the earthquake); (c) the post-seismic phase (a longer

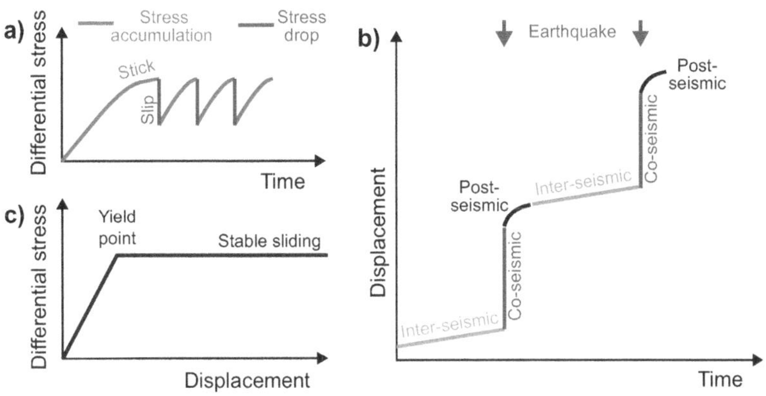

Fig. 2.21 Diagrams illustrating an ideal stick-slip process of faulting (**a**), a portion of the related seismic cycle (**b**) and an aseismic slip process (**c**)

period immediately after the earthquake, which exhibits relatively higher rates of deformation wherein the material deforms in response to the co-seismic release of strain). The concept of seismic cycle is thus based on the repeated build up of the stress on a fault over a long period, which is then rapidly released during the earthquake. Any single earthquake is unlikely to add more than a few meters of displacement to a fault, and usually affects only a portion of the fault. For example, an earthquake of magnitude 6.5–6.9 activating a 15–20 km long fault may add up to one metre of displacement to the fault; only the largest earthquakes may generate displacements up to ~20 m. Depending on the depth of nucleation and magnitude, an earthquake may create a surface rupture, which manifests through faults, fractures and folds. Displacement on an emergent fault with dip-slip component creates a step in the ground surface, called fault scarp. The other mechanism of fault growth is by continuous sliding, or **aseismic slip**, during which displacement accumulates at constant rate (Fig. 2.21c). Aseismic slip mainly occurs with small normal stress across the fault, such as at very shallow crustal levels, especially with weak and porous rocks. A same fault may show evidence of both aseismic and stick-slip. This occurs also in active volcanic areas, as for example along the Pernicana Fault on the eastern flank of Mt. Etna or along the Casamicciola Fault at Ischia Island (Italy), where the seismic events, responsible for most of the faults displacement, represent accelerations of a steady aseismic slip (Neri et al. 2004; Trasatti et al. 2019).

Faults are unlikely to grow as individual and isolated structures. As a fault grows, it commonly interacts and links with nearby faults, forming a single and larger structure (e.g., Fossen 2010, and references therein). Let us consider two growing faults whose tips approach each other. Before the tips have reached each other, but after their strain fields have started to interact, the faults **underlap** (Fig. 2.22a). Once the tips have passed each other, the faults **overlap** (Fig. 2.22b). Underlapping and overlapping faults are soft-linked, as they are not in direct physical contact. The layers in the overlap zone become gently folded, with the fold itself called **relay ramp**. Eventually the ramp breaks, forming a breached relay ramp, or hard-link, physically connecting the two faults (Fig. 2.22c). Overlap zones and relay ramps are formed and destroyed continuously during the growth of nearby faults. The increase in displacement associated with the two interacting faults follows a recurrent behaviour, summarized as follows. The propagation rate of the fault tips in the underlap area is reduced, increasing the local displacement gradient. This results in asymmetric displacement profiles, where the maximum displacement of each fault lies toward the overlapping tip (Fig. 2.22d). This asymmetric displacement becomes more pronounced as the faults overlap (Fig. 2.22e). As the deformation

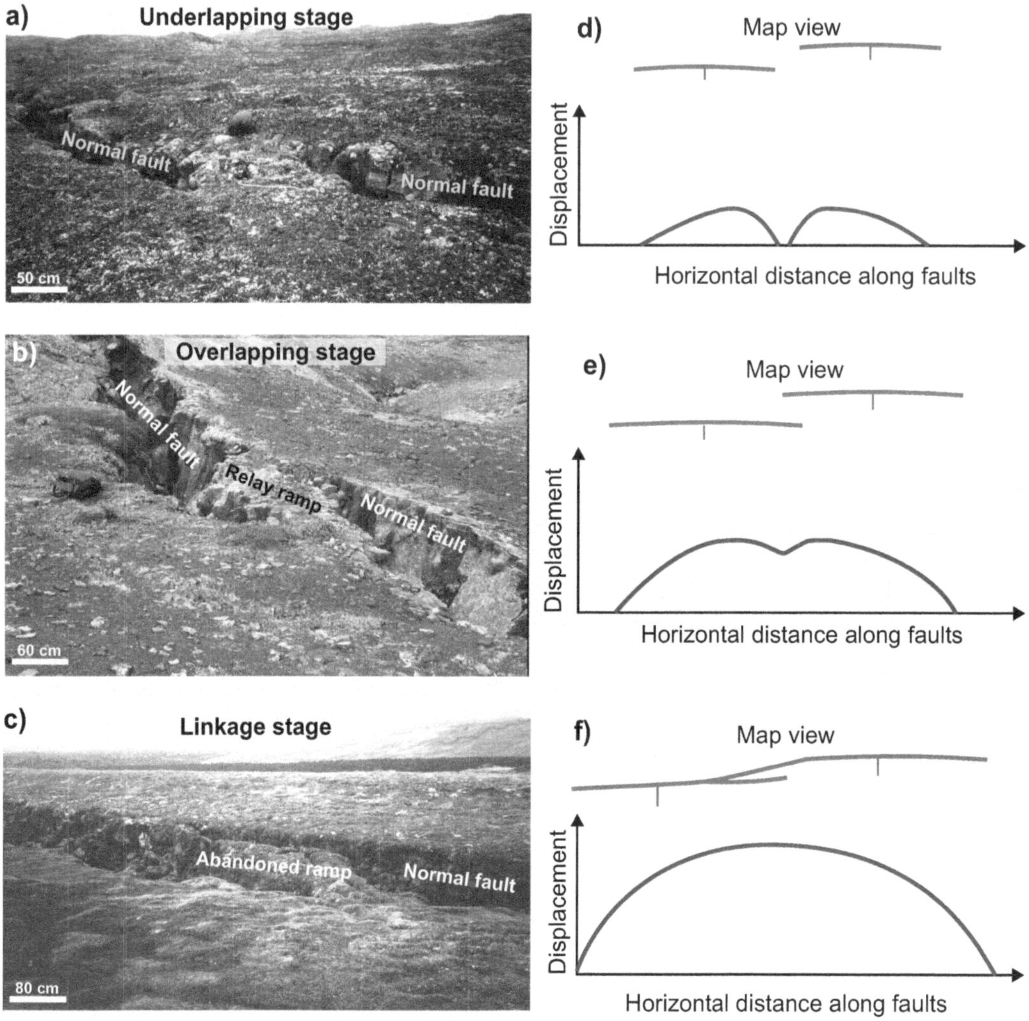

Fig. 2.22 Stages of linkage exemplified by three different pairs of normal faults in Iceland. **a** Underlapping stage; **b** overlapping stage; c) linkage stage, developing a single continuous structure. **d–f** Related (underlapping, overlapping and linkage, respectively) schematic profiles of displacement along the faults (in red) and, above each case, the corresponding fault configuration in map view (in blue)

proceeds and the two faults link, the displacement profile approaches that of a single fault, with a central maximum (Fig. 2.22f).

2.7 Tectonic Regimes

Faults are widespread on the surface of our planet. Active faults, that are likely to slip in the future, may be found in intraplate settings, due to near- or far-field stresses of different nature, although their frequency dramatically increases in correspondence of plate boundaries. As illustrated in Fig. 2.20, the predominant type of faulting in a certain crustal volume indicates if the latter experiences extension, contraction or strike-slip deformation, determining the **tectonic regime.** In particular, in extensional regimes the vertical stress σ_v coincides with the maximum principal stress σ_1 and the horizontal stress perpendicular to the extensional zone coincides with the minimum principal stress σ_3; this regime

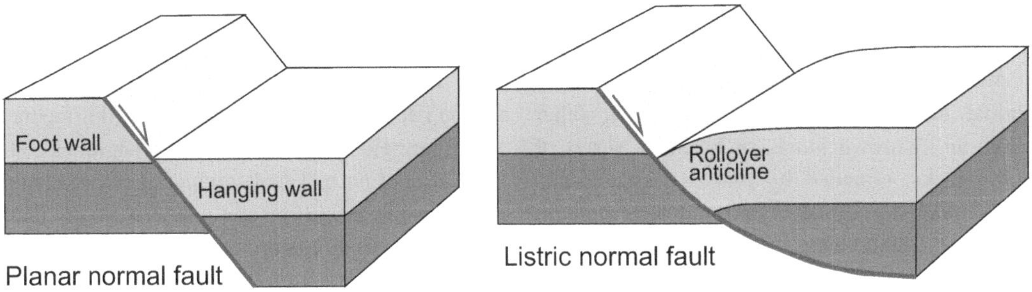

Fig. 2.23 Extensional structures. Diagram of planar (left) and curved (listric, right) normal fault; the curvature of the listric fault may generate a characteristic fold (rollover anticline) in the hanging wall

promotes normal faults. Contractional regimes occur where the vertical stress σ_v coincides with the minimum principal stress σ_3 and the horizontal stress perpendicular to the contractional zone coincides with the maximum principal stress σ_1; this regime promotes thrust faults. Strike-slip regimes occur where the vertical stress σ_v coincides with the intermediate principal stress σ_2 and the horizontal stresses are oblique to the fault zone; this condition promotes strike-slip faults. The predominance of a fault type in a specific tectonic regime develops representative structural styles whose general features can be found indiscriminately in intraplate settings and along plate boundaries. The main structural features resulting from these tectonic regimes, producing extensional, contractional and strike-slip tectonics, are summarized below, whereas their relationships with volcanic activity are illustrated in the third part of the book, in Chaps. 10–12.

2.7.1 Extensional Tectonics

Extensional tectonics is mainly found as a result of lithospheric extension, as along continental and oceanic divergent plate boundaries, in extending arcs and back-arcs in convergent plate boundaries and in intraplate rifts. Extensional tectonics may also result from gravitational processes, as in collapsing orogens and passive margins.

Normal faults are the basic structural feature of extensional tectonics. The largest faults, called master faults, are commonly associated with minor faults that may be synthetic (dipping in the same direction as the master fault) or antithetic (dipping in the opposite direction). Normal faults may be **planar** (with the fault surface approaching a plane) or **listric** (with a concave-up shape decreasing the fault dip at depth). Movement on a listric normal fault generally results in the rotation of hanging wall blocks around a horizontal axis, developing a fold called **rollover anticline** (Fig. 2.23; e.g., Price and Cosgrove 1990; Peacock et al. 2000).

Multiple normal faults with similar size may be arranged in different ways, generating scale-independent structural styles. Two parallel normal faults dipping toward each other create a downthrown block or depression named **graben**, whereas an asymmetric depression with a normal

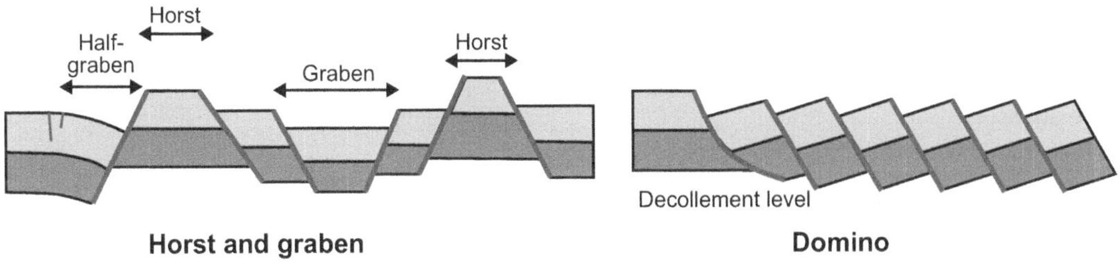

Fig. 2.24 Horst, graben and half-graben structures (left) and domino structure (right) typical of extensional regimes

fault on one side and minor faulting, or a broad flexure, on the opposite side defines a **half-graben** (Fig. 2.24). Conversely, two parallel normal faults dipping away from each other create an upthrown block, or **horst**. A series of rigid blocks, bounded by planar normal faults with similar dip, dip direction and displacement that rotate simultaneously in a uniform sense form a **domino** structure. A main feature promoting a domino is a weak basal layer, whose plastic flow allows solving the problem of the space to be accommodated at the bottom of the tilted blocks. The weak layer can consist of an overpressured formation, as clays or salt. The absence of such a weak layer favours horst and graben systems.

The portion of lithosphere experiencing extension and characterized by multiple and closely-spaced active normal faults identifies a **rift**. Rifts may occur along divergent plate boundaries, as the continental East African Rift System and the oceanic Mid-Atlantic Ridge, or in intraplate settings, as the Baikal Rift (central Asia), the Rhine Graben (western Europe) and the Rio Grande Rift (southern USA). Even though the direction of extension is often perpendicular to the rift axis, oblique rifting, characterized by extension at an acute angle to the rift axis, is common (see Sect. 10.3). Oblique rifting may occur where the geometry of the rift faults is controlled by a pre-existing fabric, with the reactivated rift faults oblique to the regional extension direction. Oblique rifting may also result from variations in the extension direction after a rift has initiated, so that the rift axis becomes no longer perpendicular to the later extension direction.

A rift system begins as a series of unconnected rift segments a few hundred of kilometres long (e.g., Rosendahl 1987). Each segment contains a set of normal faults, which dip dominantly in the same direction and die out along their length. At this immature stage, the section view of the rift resembles a half-graben. As the amount of extension increases, the rift segments grow along strike until adjacent segments overlap and interact, forming a more continuous zone of extension. Regions where two distinct rift portions or segments interact are called **accommodation zones,** characterized by complex kinematics, varying from dip- to strike-slip. Any significant difference in the amount of extension between nearby rift segments may be accommodated by **transfer faults** oriented at high angle to the rift axis and characterized by predominant strike-slip motion. A higher amount of extension associated with a continuous rift promotes a more symmetric configuration, resembling a graben structure. The sides of the rift may reach elevations up to 1–2 km above the rift floor, resulting from a combination of subsidence within the rift and elastic rebound of its margins.

The symmetry of a rift is determined not only by its maturity but, for a mature rift, also by the mechanism of extension, as associated with a "**pure shear**" or "**simple shear**" process, or a combination thereof (Fig. 2.25; McKenzie 1978; Wernicke 1985). In the pure shear model the zone of upper crustal extension coincides with the zone of lithospheric mantle attenuation,

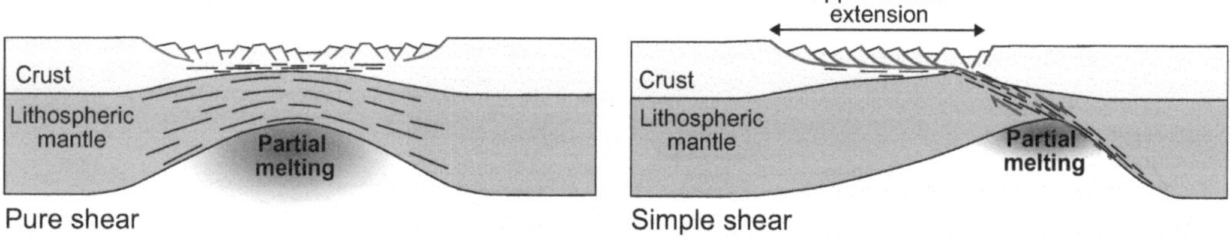

Fig. 2.25 Two idealized models for lithospheric extension and rifting. The pure shear model (left) is symmetric, with maximum melting underneath the middle of the rift. The simple shear model (right) is dominated by a low angle shear zone that produces an asymmetry on the rift (modified after Fossen 2010)

developing a symmetric structure bordered by oppositely-dipping high angle normal faults and magmatic activity along the rift axis. Conversely, in the simple shear model the zone of upper crustal extension is laterally offset from the zone of lithospheric mantle attenuation and develops an asymmetric structure consisting of a main low angle fault zone, or detachment. Any magmatic activity is asymmetrically distributed with respect to the rift axis, on the dipping side of the detachment.

Several factors may affect rifting, as the pre-rift tectonic history, the rheologic, thermal and stress conditions of the lithosphere and the rift history. These factors determine three modes of continental lithospheric extension, which may represent different stages of a spectrum or result from the potential of the lithosphere to localize or distribute strain during extension (Buck 1991; Corti et al. 2003). (a) Narrow rift, with concentrated (10^1 km wide) crustal and mantle lithospheric extension, responsible for large gradients in crustal thickness and topography. High heat flow corresponds to localized lithospheric thinning, suggesting hot upper mantle beneath, also highlighted by volcanism. Crustal extension, mainly achieved through high angle normal faults, is usually less than 30%. (b) Wide rift, with uniform crustal and mantle lithospheric thinning wider than the lithospheric thickness. Wide rifts develop a large number of separated basins over 10^2–10^3 km wide regions, with uneven strain. Underplating, or the emplacement of magma at the base of the crust, may occur, although volcanism is limited. (c) Core complex, with concentrated upper crustal extension and lower crustal thinning over a broad area, associated with a main low angle detachment. Any volcanism is limited. These modes of extension show that volcanism is not a straightforward consequence of any mode of rifting. It is difficult to generate melt in the asthenosphere by decompression through extension distributed over a wide lithospheric portion, either in normal or elevated asthenospheric temperatures. Therefore, wide rifting and core-complexes associated with simple shear fail to account for significant magma production. Conversely, the focusing of extension in a limited lithospheric portion, as in pure shear conditions, promotes melt generation (Fig. 2.26; Latin and White 1990). These different behaviours account for the more significant volcanism along narrow rifts with respect to wide rifts or core complexes. As a result, volcanic activity in extensional settings focuses along the continental narrow rifts and, especially, oceanic ridges of the divergent plate boundaries. To a minor extent, volcanic activity is also found along limited portions of intra-continental narrow rifts.

2.7.2 Contractional Tectonics

Contractional tectonics focuses along convergent plate boundaries, where contractional structures may be found also at considerable distance from the plate boundary. Nevertheless, contractional structures may be also locally found in intra-plate settings, as resulting from near- or far-field stresses.

Contractional structures form when rocks are shortened by tectonic or gravitational forces. The basic structural features developed in contractional settings are **thrust** faults (dip of $\sim 30°$) and, less commonly, **reverse** faults which, dipping $\sim 60°$, do not usually accommodate large shortening (Fig. 2.27; e.g., Boyer and Elliott 1982; Butler 1982; Price and Cosgrove 1990).

A low angle thrust fault may show variable dip portions. The portions running parallel to the bedding are called **flats,** while the portions cutting across the bedding are called **ramps**. If the fault has not been folded subsequent to its formation, flats are subhorizontal, whereas ramps have dips of 30°–45°. Ramps may also develop thrusts or reverse faults with opposite dip direction: such back-thrusts result from geometric complications in ramp locations, favoured by steep ramps. As with normal fault systems, faults in thrust fault systems tend to comprise parallel arrays. An **imbricate fan** consists of thrust faults that either intersect the ground surface or die out up dip, whereas a **duplex** consists of thrusts that span the interval of rock between a higher-level detachment, called roof thrust, and a lower level

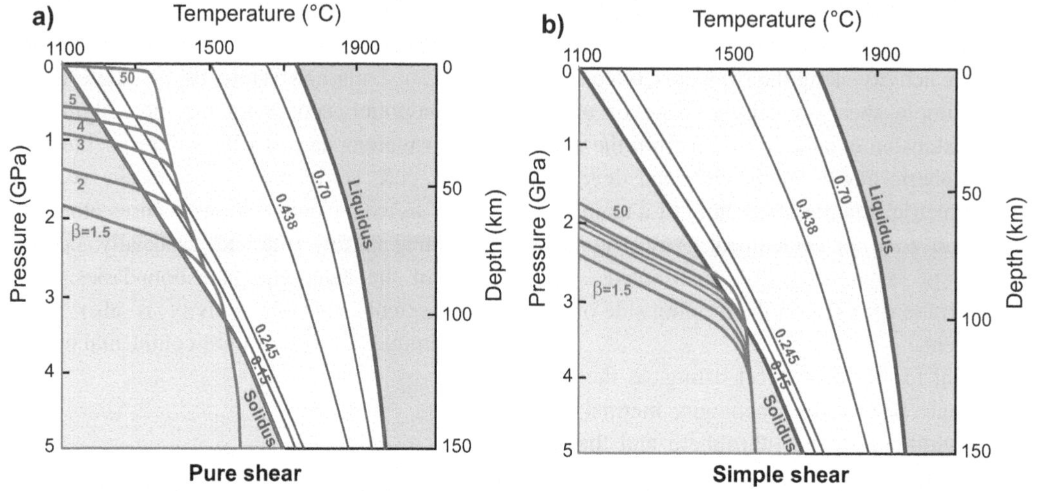

Fig. 2.26 Relationships between lithospheric amount of extension (blue curves) and produced amount of melt (red curves): the adiabatic upwelling due to stretching of convective geotherm by different values of β (blue curves) is shown for the pure shear (**a**) and simple shear (**b**) cases. Mechanical boundary layer thickness = 100 km; interior potential temperature at 1480 °C (modified after Latin and White 1990)

detachment, called floor thrust (Fig. 2.27). The along-strike differential shortening of one part of a thrust system relative to another may be accommodated by **tear faults**, or subvertical faults parallel to the regional transport direction, with dominant strike-slip motion. Within a collision zone, the hinterland is the area in the central portion and the foreland is the marginal part and thus farthest into the continent. The foreland-directed progression of duplexes and imbricate zones is called in-sequence thrusting. Thrust faults that do not follow this pattern of activation are out of sequence. Folds accompany the contractional deformation. Indeed, regional shortening of the upper crust yields a distinctive suite of thrust faults, folds and associated structures, called **fold-and-thrust belts**. Among the various types of folds, are the detachment, or **decollement,** folds. These form where layers fold above a detachment and shorten more than their

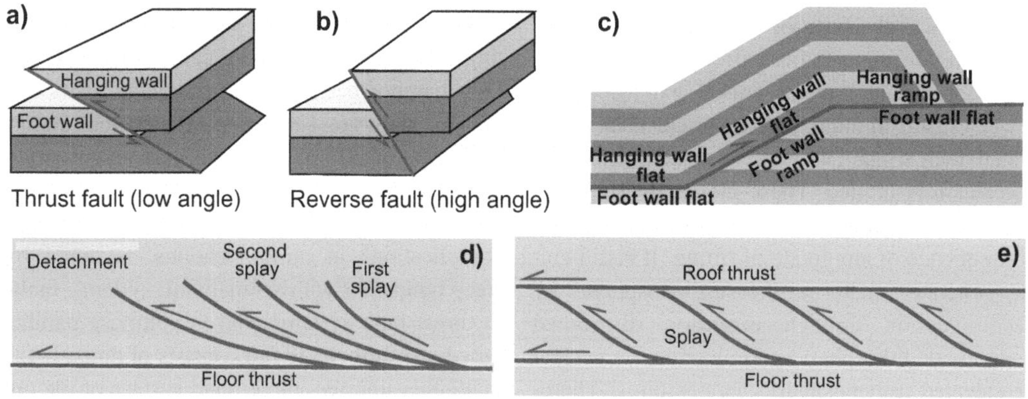

Fig. 2.27 Contractional structures. Schematic oblique views of a low angle thrust fault (**a**) and a high angle reverse fault (**b**). Schematic section views of a ramp-flat structure associated with fold (**c**), of an imbricate fan (**d**) and of a thrust duplex (**e**)

substrate, which usually remains undeformed. Detachment folds tend to develop above weak layers, such as overpressured clays or evaporates. Some volcanic edifices with flank instability, as Kilauea (Hawaii, USA), show detachment folds.

Convergent plate boundaries associated with magmatic activity are related to the bending and sinking, or **subduction**, of the denser oceanic lithospheric plate (or slab) within the mantle, beneath the overriding plate. Several distinctive tectonic features can be found across a subducting convergent plate boundary (Fig. 2.28; van der Pluijm and Marshak 2004). A deep submarine trough, the trench, marks the shallowest portion of the contact between the subducting and overriding plates. The edge of the overriding plate adjacent to the trench, or accretionary wedge, consists of intensely deformed sediments and oceanic basalts. A forearc basin with poorly-deformed strata lies between the accretionary wedge and the chain of volcanoes, which is called **volcanic arc**. The portion behind the arc is the back-arc region. Slabs and volcanic arcs are closely related. In fact, when the slab reaches a depth of 80–150 km, it heats up and releases volatiles (mainly H_2O and CO_2) into the overlying asthenosphere, promoting partial melting and yielding mafic magma. The melt rises, intruding and transferring heat within the overlying crust, causing further partial melting and yielding intermediate to felsic magma. Eventually the magma erupts along the volcanic arc (for details see Chap. 10). Island arcs form where one oceanic plate subducts beneath another oceanic plate, or where the volcanic arc grows on a sliver of continental crust that has separated from a continent. Continental arcs grow where an oceanic plate subducts beneath continental lithosphere. Volcanism at island arcs formed on oceanic crust mostly produces mafic and intermediate rocks, whereas volcanism at continental arcs also produces intermediate and felsic rocks, including granites.

he overriding plate may thus consist of either oceanic or continental lithosphere, but the subducting plate consists of oceanic lithosphere, as continental lithosphere is too buoyant to subduct. In the case of convergence between two buoyant continental plates, **continental collision** replaces subduction. Continental collision usually lacks the conditions to generate magma related to subduction, although local volcanism may result from other processes (see Chap. 10).

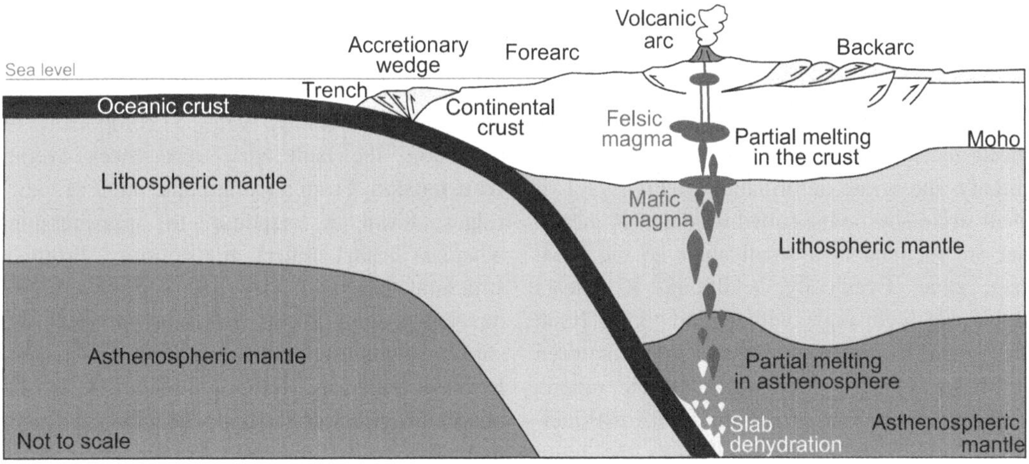

Fig. 2.28 Convergent plate boundaries with subduction: anatomy of an Andean-type continental arc experiencing subduction, with a contractional back-arc region. Here the volcanic arc grows on continental crust, above a zone of generation and rise of magma overlying the slab (modified after van der Pluijm and Marshak 2004)

2.7.3 Strike-Slip Tectonics

Dominant strike-slip tectonics focuses along oceanic and continental transform plate boundaries, although it may be important also in obliquely convergent plate boundaries. Strike-slip tectonics may be also found in intraplate settings, as for example associated with escape tectonics, that is the lateral motion of a crustal block following plate convergence. Finally, strike-slip structures may locally dominate also within portions of fold-and thrust-belts and rift zones.

The basic structural feature is the strike-slip fault, with the displacement vector parallel to its strike. Strike-slip faults are steeper than other faults, and in map view appear as fairly straight structures, up to thousands of kilometres long. Regional-scale strike-slip deformation does not produce a single, simple fault plane. Rather, it results in a broader zone of deformation, containing numerous subparallel strike-slip faults of varying lengths, as well as other structures with different direction and kinematics, such as normal and reverse faults and folds. Locally, one fault dies out where another one, parallel but usually not collinear, initiates: the region in between is a **stepover** (e.g., Price and Cosgrove 1990). To understand how strike-slip fault zones develop, consider an experiment where a layer of clay is placed over two wooden blocks, and then one block is horizontally sheared past the other (Fig. 2.29). As the blocks move relative to one another, the clay begins to deform, partly by plastic mechanisms and partly by brittle failure. Initially, the brittle deformation yields arrays of small strike-slip faults, called Riedel or R shears, that are inclined at a small angle to the basal shear zone. Eventually, additional R' shears develop at high angle with regard to the basal shear zone. Increasing the displacement between the basal blocks, the Riedel shears merge, forming a continuous strike-slip fault (or Principal Displacement Zone) parallel to the basal shear zone. Extension fractures may also develop as a series of parallel fractures that form within the fault zone and define an en-echelon array that tends to dilate. The fact that most of the subsidiary faults in a strike-slip zone initiate as Riedel shears indicates that strike-slip zones develop from several minor subparallel fault segments, or subsidiary fault splays. Very similar structures and evolution are found in strike-slip zones developed on the flanks of active volcanoes, as at Mt. Etna (Italy; Fig. 2.29).

Local bends along a strike-slip fault result in either local contraction or extension, depending on the configuration of the bend with regard to the kinematics of the fault. Locations where the bend is oriented such that blocks on opposite sides of the fault are squeezed together are **restraining bends,** whereas locations where the bend is oriented such that blocks on opposite sides of the fault pull away from each other are **releasing bends** (Fig. 2.30; e.g., Sylvester 1988; Woodcock and Schubert 1994; Westaway 1995). In releasing bends normal faults are promoted and the block of crust adjacent to the bend subsides. Displacement at a releasing bend may yield a rhomboid-shaped depression, called **pull-apart** basin. Its dimension and amount of subsidence depend on the size of the bend and the amount of extension. Both restraining and releasing bends can exist simultaneously along the same fault zone.

Regional strike-slip faults typically splay into many structures, which in cross section resemble a flower and are thus called **flower structures**. Where a more general movement across a strike-slip fault zone results in a component of shortening, the fault zone experiences overall **transpression**; conversely, with a component of extension, the fault zone experiences overall **transtension.** From a topographic point of view, ridges form in response to transpression, whereas basins reflect transtension. From a structural point of view, transpressive zones develop positive flower structures, in which the slip on subsidiary faults has a thrust component. Reverse faults and folds at high angle to the maximum principal stress σ_1 are associated with transpression. Conversely, transtensive zones develop a negative flower structure, in which the slip has a normal component. Extensional fractures and normal faults subparallel to the maximum principal stress σ_1 are associated with transtension.

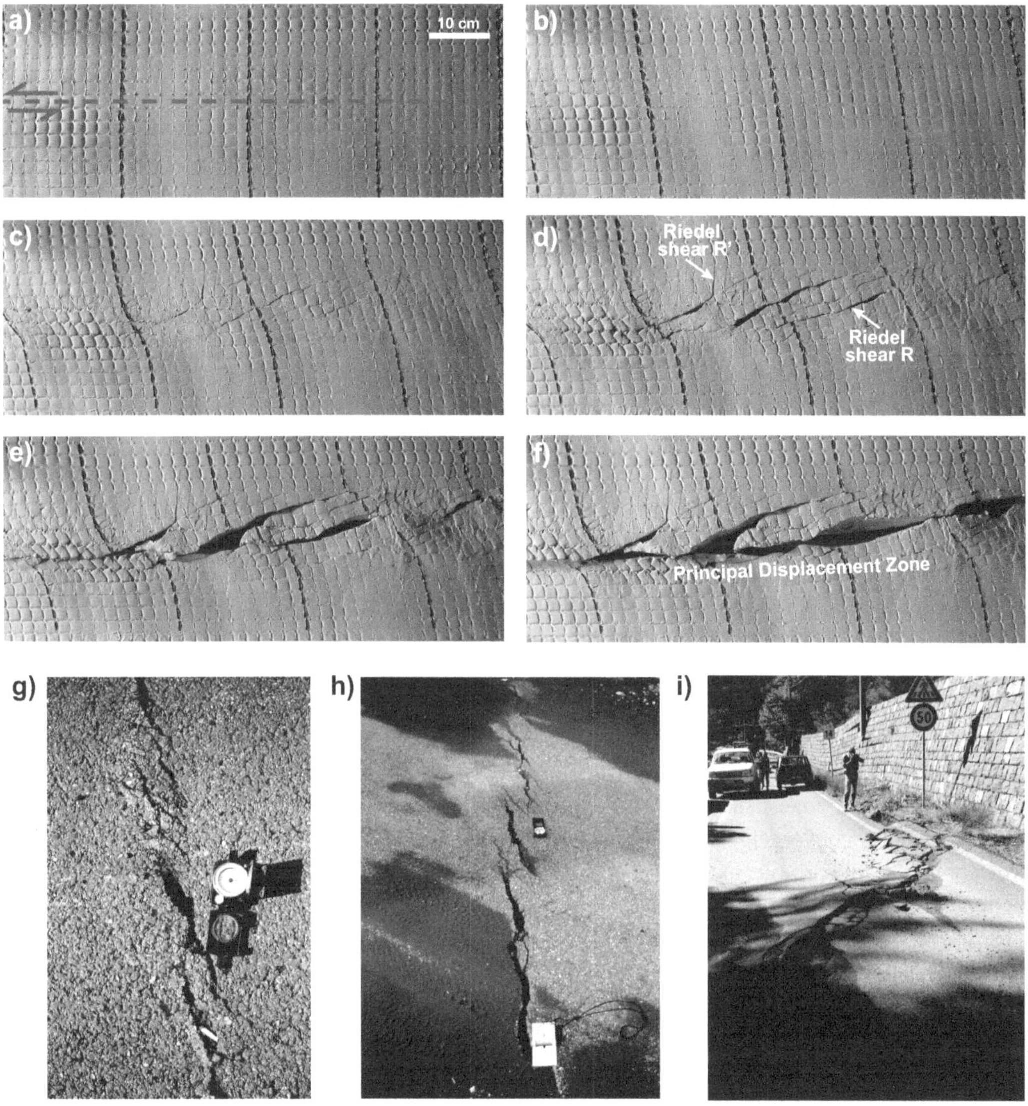

Fig. 2.29 Strike-slip structures. **a–f** Clay experiment depicting the evolution of a strike-slip (sinistral, in this case) fault zone in map view as a function of the progressive increase in the displacement at the base of the clay layer. **a** Undeformed stage of model, with trace of the discontinuity imposed at the base of the clay (red dashed line). **b–f** Progressive increase in the displacement, with a broad elastic deformation (**b**), followed by the development of Riedel shears R (sinistral, at a low angle to basal discontinuity) and R' (dextral, high angle to basal discontinuity) (**c**, **d**, **e**), and then by the coalescence of the R shears in a principal displacement zone (**f**). **g–i** Different stages of growth of the sinistral Pernicana Fault, on the eastern flank of Mt. Etna (Italy) during the 2002 eruption of the Northeast Rift. The fault is characterized by an increase in displacement from the coastline to the Northeast Rift, showing how the sinistral Riedel shears R (**g**) coalesce and form longer structures (**h**) and then, towards the Northeast Rift, develop a single continuous fault (**i**). All faults are on paved roads

Transform faults constitute a widespread type of strike-slip fault, occurring in a variety of settings and at a variety of scales. At the widest scale, transform faults can form plate boundaries up to thousands of kilometres long, on both continental and oceanic lithosphere. Transform faults have distinctive features: a) their active portion terminates at discrete endpoints, where

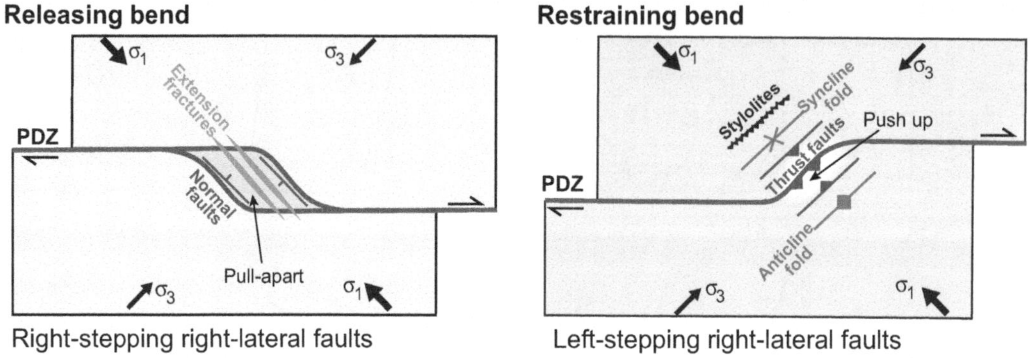

Fig. 2.30 Strike-slip structures: diagrams of the structures associated with a releasing bend (right-stepping right-lateral faults; left) and a restraining bend (left-stepping right-lateral faults, right) along a Principal Displacement Zone (PDZ)

the transform intersects other shortening structures (e.g., convergent boundary, thrust fault, or stylolite) or extensional structures (e.g., divergent boundary, normal fault, or vein); b) the length of a transform fault can be constant, or it can increase or decrease over time; c) the displacement across a transform fault can be constant and greater than the length of the fault (e.g., van der Pluijm and Marshak 2004).

Strike-slip fault zones may be also found along obliquely convergent plate boundaries, most commonly along volcanic arcs above subduction zones, and along collisional orogens (e.g., Tikoff and Teyssier 1994). In the first case, the strike-slip component along the arc results from the oblique amount of convergence, with a component parallel to the plate boundary. This obliquity generates **strain partitioning** within the overriding plate, developing contraction towards the trench and strike-slip structures along the volcanic arc. In the second case, strike-slip structures in the core of the orogen are associated with overall oblique collision between two plates.

In general, a strike-slip plate boundary or fault zone does not provide alone the conditions to generate magma; as a result, strike-slip plate boundaries and structures usually lack volcanism. However, in some cases additional conditions may allow magma generation along strike-slip fault zones. For example, mantle plumes may rise below transform plate boundaries, feeding volcanism partly controlled by the strike-slip structures. This is the case of the Azores Archipelago, partly located along the western portion of the transform boundary between the Eurasia and Africa plates (see Sect. 13.7).

2.8 Summary

This introductory chapter begins with the stress-deformation relationships affecting crustal rocks, illustrating the main deformation mechanisms and the role of the boundary conditions. For very low strains, and at low temperature and pressure, a linear behaviour between stress and strain defines a recoverable elastic deformation. Increasing the strain under a constant stress leads to the elastic limit, beyond which permanent plasticity, or continuous microscale deformation, is added. At higher temperatures and pressures, rocks store permanent and irreversible deformation flowing, thus describing a viscous behaviour: here the strain rate may show a linear relationship with stress (for Newtonian fluids) or not (non-Newtonian fluids). Rather than approximating perfectly elastic, viscous or plastic materials, crustal rocks often deform in more complex ways, combining different deformation laws. Characteristic strength profiles determine the deformation behaviour of crustal rocks as a function of the depth, applied stress, temperature, pressure, strain rate and pore pressures conditions. Magma also has a complex rheology: while the melt phase may approach Newtonian behaviour, the presence of bubbles and crystals nears magma behaviour to non-Newtonian or Bingham rheologies.

Brittle deformation, according to the Mohr-Coulomb criterion, is described only by the stress state at which failure occurs, and not by a real deformation law. Brittle deformation as associated with extension and shear fractures, characterized by opening and slip, respectively. On a Mohr diagram the extension fractures fall in the tensile regime and the shear fractures in the compressive regime. Faults are shear fractures characterized by dip-slip (normal and thrust faults) and/or strike-slip (dextral and sinistral faults) components of motion. Faults usually grow incrementally in size and displacement, with each step usually associated with seismicity, defining a seismic cycle consisting of inter-, co- and post-seismic phases.

Active faults focus in extensional, contractional and strike-slip tectonic regimes. Extensional regimes show dominant normal faults arranged in different structural styles, depending on the mechanism and maturity of extension: these are mostly represented along divergent plate boundaries. Contractional regimes show dominant thrust faults arranged in different structural styles and mostly represented in convergent plate boundaries experiencing subduction (with volcanic arc) or collision. Strike-slip regimes show dominant strike-slip faults: these are mostly represented in transform plate boundaries (which are usually non-magmatic) and to a lesser extent along volcanic arcs experiencing strain partitioning during oblique convergence.

2.9 Main Symbols Used

a	Acceleration
A	Area
C	Constant
c	Cohesion
e	Elongation
E	Young's modulus
$\dot{e}$	Strain rate
$e_{\parallel}$	Elongation parallel to the compressive stress
$e_{\perp}$	Elongation perpendicular to the compressive stress
F	Force
g	Acceleration due to gravity
G	Rigidity or shear modulus
J	Dimensionless number
K	Stress intensity factor
K_b	Bulk modulus
K_c	Fracture toughness
L	Half of the crack's length
L_0	Original length
L_f	Final length
m	Mass
n	Power law exponent
P_f	Pore pressure
P_l	Lithostatic pressure
Q	Activation energy
R	Gas constant
T	Absolute temperature
T_0	Tensile strength
V	Final volume
V_o	Original volume
V_s	Volumetric strain
X	Major axis of strain ellipsoid
Y	Intermediate axis of strain ellipsoid
Z	Minor axis of strain ellipsoid
z	Depth
β	Stretching factor
δ	Angle
γ	Shear strain
$\Delta\sigma$	Pressure change
η	Viscosity
θ	Angle between stress vector and plane
μ	Coefficient of internal friction
v	Poisson's ratio
ρ	Density
σ	Stress
σ'	Effective pressure
σ_1	Maximum principal stress
σ_2	Intermediate principal stress
σ_3	Minimum principal stress
σ_m	Mean stress
σ_n	Normal stress
σ_t	Remote tensile stress
σ_v	Vertical stress
τ	Shear stress
ϕ	Angle of internal friction
ψ	Angular shear

References

Anderson EM (1951) The dynamics of faulting. Oliver & Boyd Edinburgh, 206 pp

Buck WR (1991) Modes of continental extension. J Geophys Res 96:20161–20178

Boyer SE, Elliott D (1982) Thrust systems. Am Asso Petrol Geol Bull 66:1196–1230

Burgisser A, Degruyter W (2015) Magma ascent and degassing at shallow levels. In: Sigurdsson H, Houghton B, McNutt S, Rymer H, Stix J (eds) The encyclopedia of volcanoes, 2nd edn. Elsevier Academic Press, Cambridge, pp 225–237

Butler RW (1982) The terminology of structures in thrust belts. J Struct Geol 4:239–245

Calder E, Lavallee Y, Kendrick JE, Bernstein M (2015) Lava Dome Eruptions. In: Sigurdsson H, Houghton B, McNutt S, Rymer H, Stix J (eds) The encyclopedia of volcanoes, 2nd edn. Elsevier Academic Press, Cambridge, pp 343–361

Corti G, Bonini M, Conticelli S, Innocenti F, Manetti P, Sokoutis D (2003) Analogue modelling of continental extension: a review focused on the relations between the patterns of deformation and the presence of magma. Earth Sci Rev 63:169–247

Fossen H (2010) Structural geology. Cambridge University Press, Cambridge, 481 pp

Griffith AA (1924) The theory of rupture. In: Biezeno CB, Burgers JM (eds) First international congress on applied mechanics Delft J. Waltman, pp 55–63

Gudmundsson A (2011) Rock fractures in geological processes. Cambridge University Press, Cambridge, 569 pp

Heap MJ, Villeneuve M, Albino F, Farquharson JI, Brothelande E, Amelung F et al (2020) Towards more realistic values of elastic moduli for volcano modelling. J Volcanol Geoth Res 390:106684

Jaeger JC, Cook NGW, Zimmermann RW (2007) Fundamentals of rock mechanics. Blackwell 475 pp

Karato SI (2008) Deformation of Earth materials: an introduction to the rheology of solid earth. Cambridge University Press, Cambridge, 482 pp

Latin D, White N (1990) Generating melt during lithospheric extension: pure shear vs. simple shear. Geology 18:327–331

Maccaferri F, Smittarello D, Pinel V, Cayol V (2019) On the propagation path of magma-filled dikes and hydrofractures: the competition between external stress, internal pressure, and crack length. Geochem Geophys Geosyst 20:2064–2081

McKenzie D (1978) Some remarks on the development of sedimentary basins. Earth Planet Sci Lett 40:25–32

Neri M, Acocella V, Behncke B (2004) The role of the Pernicana Fault System in the spreading of Mt. Etna (Italy) during the 2002–2003 eruption. Bull Volc 66:417–430

Peacock DCP, Knipe RJ, Sanderson DJ (2000) Glossary of normal faults. J Struct Geol 22:291–305

Pollard DD, Aydin A (1988) Progress in understanding jointing over the past century. Geol Soc Am Bull 100:1181–1204

Pollard DD, Fletcher RC (2005) Fundamentals of structural geology. Cambridge University Press, Cambridge, 500 pp

Price NJ, Cosgrove JW (1990) Analysis of geological structures. Cambridge University Press, Cambridge, 502 pp

Ranalli G (1995) Rheology of the Earth. Chapman and Hall, London, 400 pp

Reches Z, Fink J (1988) The mechanism of intrusion of the Inyo Dike, Long Valley Caldera, California. J Geophys Res 93:4321–4334

Rosendahl BR (1987) Architecture of continental rifts with respect to East Africa. Ann Rev Earth Planet Sci 15:445–503

Scholz CH (2002) The mechanics of earthquakes and faulting. Cambridge University Press, Cambridge, 523 pp

Stuwe K (2007) Geodynamics of the lithosphere. Springer, Berlin, 493 pp

Sylvester AG (1988) Strike-slip faults. Geol Soc Am Bull 100:1666–1703

Tikoff B, Teyssier CT (1994) Strain modeling of displacement-field partitioning in transpressional orogens. J Struct Geol 16:1575–1588

Trasatti E, Acocella V, Di Vito MA, Del Gaudio C, Weber G, Aquino I et al (2019) Magma degassing as a source of long-term seismicity at volcanoes: the Ischia island (Italy) case. Geophys Res Lett. https://doi.org/10.1029/2019GL085371

Turcotte DL, Schubert G (2002) Geodynamics. Cambridge University Press, Cambridge, 863 pp

van der Pluijm BA, Marshak S (2004) Earth Structure: an introduction to structural geology and tectonics. WW Norton & Company, 673 pp

Wernicke B (1985) Uniform-sense normal simple shear of the continental lithosphere. Canadian J Earth Sci 22:108–125

Westaway R (1995) Deformation around stepovers in strike-slip fault zones. J Struct Geol 17:831–846

Woodcock NH, Schubert C (1994) Continental strike-slip tectonics. In: Hancock PL (ed) Continental deformation. Pergamon Press, Oxford, pp 251–263

Zang A, Stephansson O (2010) Stress field in the Earth's crust. Springer, Berlin 322 pp

The Rise of Magma Through the Crust

3

3.1 Introduction

After the fundamentals summarized in Chaps. 1 and 2, this chapter introduces the second and main part of the book. This part, from this Chapter to Chap. 9, focuses on the general path followed by magma in passing through the crust, ideally consisting of its rise, shallow accumulation, shallow transfer and eruption. In particular, the path begins with the rise of magma through the crust (described in this chapter). The rise of magma is a requisite for promoting shallow magma accumulation, and thus developing magma chambers (Chap. 4). A significant withdrawal of magma from a chamber may determine a vertical collapse, or caldera, at the surface (Chap. 5). A withdrawal of magma rising towards the surface may promote the instability of the flank of a volcanic edifice, triggering lateral, or sector, collapse (Chap. 6). In both types of collapse, the modifications in the shape of the volcanic edifice introduce deviations from the expected patterns of shallow transfer of magma feeding eruptions (Chap. 7). Both the shallow accumulation and transfer of magma can be detected and tracked through a multi-parametric volcano monitoring system (Chap. 8), which allows understanding the state of an active volcano, ultimately providing the basis for eruption forecasting (Chap. 9).

In more detail, this chapter considers the general processes controlling the rise of magma within the crust, progressing from classic studies on magma rise through diapirs to modern perspectives, which challenge or anyway limit diapiric ascent in favour of dike propagation. As regards dikes, this chapter considers the general mechanisms of dike propagation and arrest, independently of the level of nucleation of the dikes, may this be in the lower or upper crust. The description of the more specific conditions controlling dike nucleation from magma chambers and dike propagation approaching the surface and within volcanic edifices are considered in Chaps. 4 and 7, respectively.

The main aims of this chapter are thus to define the:

- mechanisms of development of magmatic diapirs;
- mechanisms of propagation and arrest of dikes;
- general conditions determining the rise of magma through diapirs or dikes in the crust.

V. Acocella, *Volcano-Tectonic Processes*, Advances in Volcanology,
https://doi.org/10.1007/978-3-030-65968-4_3

3.2 Magmatic Diapirs

The rise of magma through the crust has been ascribed for decades to the development of diapirs. A **diapir** is a massive magmatic intrusion piercing through the surrounding country rock mainly through density contrast, or buoyancy, between the lighter magma and the denser crust. As felsic magma is lighter and more buoyant, diapirs commonly have a felsic, preferably granitoid, composition. Diapirs, consisting of massive and relatively viscous magma, commonly interrupt their ascent and emplace at depth, where they slowly crystallize and form intrusions. The emplacement of diapirs has been also often invoked to explain the formation of **plutons**, which are massive and solidified intrusions of magma accumulated in the crust (see Chap. 4). A rising magmatic diapir is expected to have an inverted teardrop shape, with domed roof, evolving towards a mushroom-like shape during its final emplacement, that is when the intrusion starts spreading laterally (Fig. 3.1).

The rise of a diapir occurs through the viscous flow of the massive magma body coupled with that of the surrounding host rock. In this context, a rising diapir is accompanied by marked lateral gradients within the country rock, with the temperature and viscosity rapidly decreasing and increasing away from the intrusion contacts, respectively. Conversely, ascent velocity is largest (and positive) at the intrusion centre, to become negative (highlighting downward flow) within a thermally softened zone of the country rock next to the intrusion. The time scale of granitoid diapirism over crustal distances is estimated on the order of 10^4–10^5 years (Mahon et al. 1988).

Several studies have analysed the generation and development of diapirs in theory and experiments. These are consistent in describing an overall variation in the shape of the ascending diapir from its mobilization to its arrest, identifying three main stages: initiation, rise and emplacement (Fig. 3.2; Berner et al. 1972; Dixon 1975; Ramberg 1981; Bittner and Schmeling 1995). These are described in the sections below.

3.2.1 Diapir Initiation

Diapirs initiate as **Rayleigh–Taylor instabilities**, that are instabilities which develop along the interface between immiscible fluids, with a lighter fluid below a heavier one. In the case of magmatic diapirs, the lighter magma flows and rises within the denser country rock of the viscous lower crust (e.g., Turcotte and Schubert 1982, and references therein). The interface between the magma and the country rock is in a state of unstable equilibrium, as any perturbation or disturbance is amplified by the Earth's gravity, with the result that the heavier fluid tends to be displaced downward and an equal volume of lighter fluid upward. These disturbances grow, so that an increasing amount of denser material is displaced downward and lighter material upwards, leading to an overall decrease in potential energy of the system.

The size and rate of growth of an instability depend upon several boundary conditions, including the density and viscosity contrasts between the two fluids, the thickness of the upper layer and, in the case of magma, its degree of partial melting (Biot and Odè 1965). Theoretical calculations suggest that the wavelength associated with a dominant disturbance is $\sim 2.56\ T_u$, where T_u is the thickness of the upper layer, indicating that the thinner is the upper layer the more closely spaced are the instabilities. This theoretical relationship is in agreement with the distribution of salt diapirs in nature (Jackson and Talbot 1986). The wavelength of the disturbance also depends upon the viscosity contrast: as the viscosity of the upper layer increases, the dominant wavelength increases and its rate of amplification (i.e., vertical growth of wavelength) decreases. Assuming any viscosity contrast null, the time of growth of the dominant disturbance is given by:

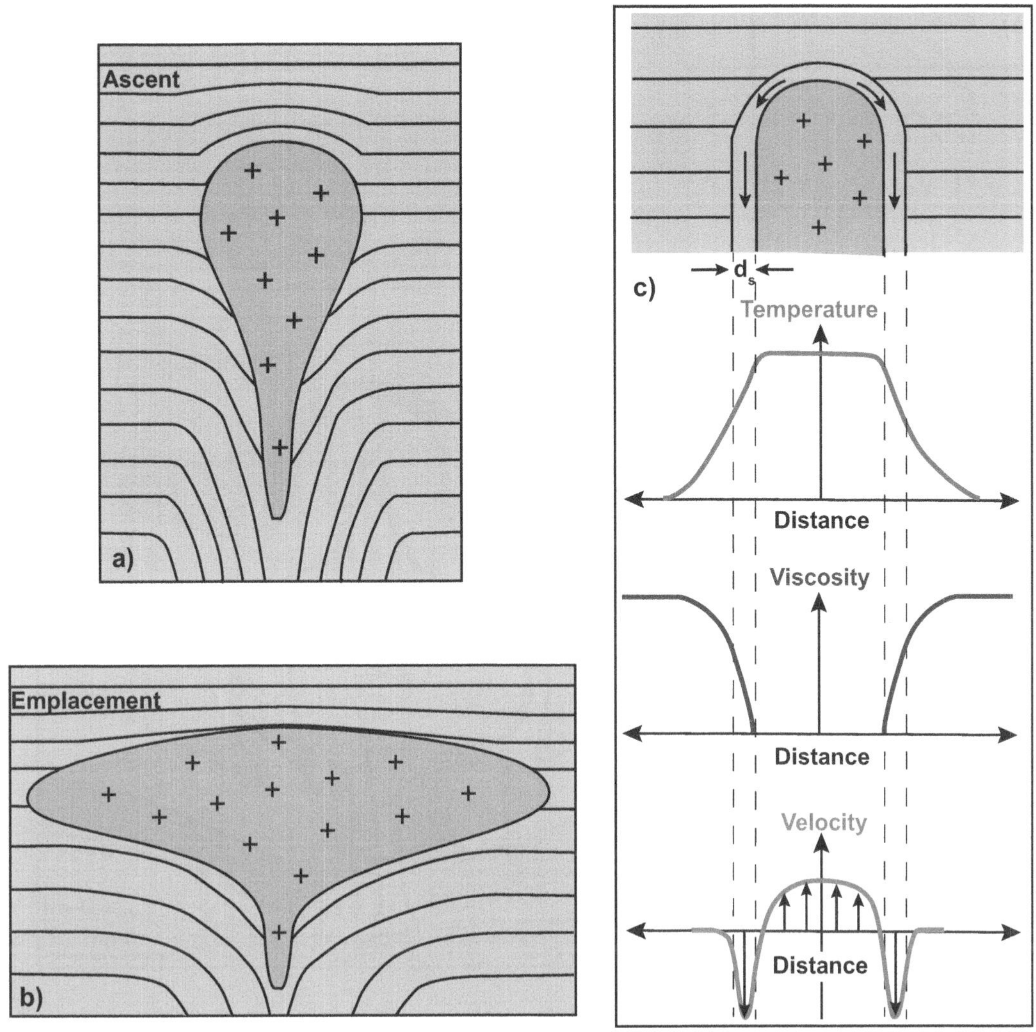

Fig. 3.1 Section view scheme of a **a** rising magmatic diapir, resembling an inverted teardrop, and an **b** emplaced diapir, attaining a mushroom-like shape, **c** temperature, viscosity and velocity variations within the top part of a rising diapir and the country rock; d_s = thickness of softened region (modified after Marsh 1982)

$$t = \frac{13.04\eta}{(\rho_1 - \rho_2)gT_u} \quad (3.1)$$

where η is the viscosity of both fluids, ρ_1 and ρ_2 are the densities of the upper and lower layer, respectively, and g is the acceleration due to gravity (e.g., Turcotte and Schubert 1982). Equation 3.1 indicates that the more viscous the fluids and the smaller their density difference, the longer it takes to grow an instability.

The initiation of an instability develops a diapir flowing upwards, which becomes progressively taller and narrower. The diapir rises within the country rock, eventually becoming detached from its root and acquiring an inverted tear-drop shape, with wider domed head and narrower tail.

3.2.2 Diapir Rise

The ascent rate of a magmatic diapir may be derived from the Stokes' law, although the common assumption of the isoviscous

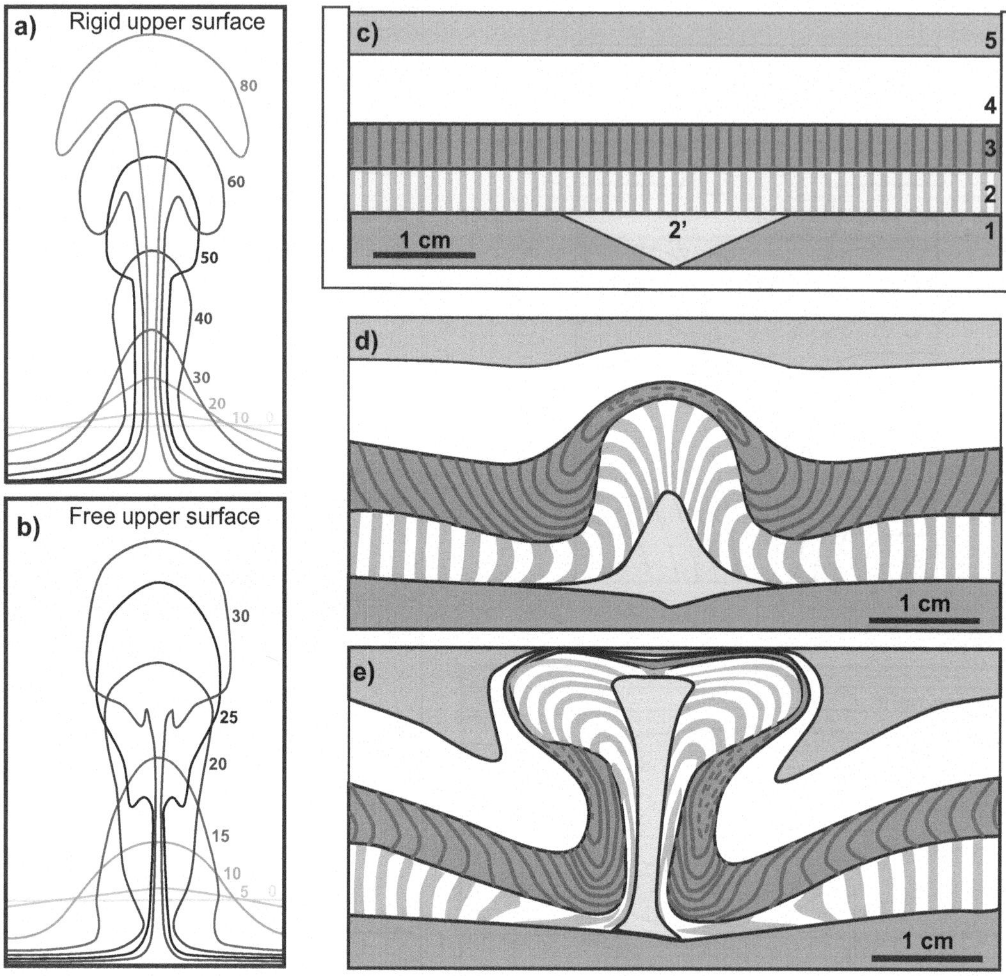

Fig. 3.2 **a**, **b** Numerical experiments of rising magmatic diapirs, showing the section view shape of diapirs at different times, in seconds, with different boundary conditions: rigid upper surface (**a**) and free upper surface (**b**), the latter with rise about twice fast. In both cases, the width of the diapir first decreases and then increases (Berner et al. 1972). **c–e** Centrifuge experiments (at 2000 g) of rise of lighter viscous material (1400 kg/m^3; layers 2 and 2′) into denser viscous material (1480–1560 kg/m^3; layers 3, 4 and 5); **c** schematic section view of the model before deformation; **d** section of model after 310 s in the centrifuge; **e** section of model after 505 s (Ramberg 1981)

surrounding medium of the latter may not be appropriate for a hot diapir inducing a thermal gradient within the country rock. In fact, the ascent rate and extent of the diapir are strongly dependent on the width of the deformation layer (with thickness d_s in Fig. 3.1) of variable viscosity that is formed by the transfer of heat to the country rock and on the buoyancy force which propels the diapir through it. In particular, the ascent velocity U_a is described by the expression:

$$U_a = \frac{2\Delta\rho g d_s^3 Z_v}{\eta r} \quad (3.2)$$

where $\Delta\rho$ is the density contrast between the diapir and the country rock, g is the acceleration due to gravity, d_s is the deformation zone width,

Z_v is a parameter which describes the increase in viscosity across the zone, η is the country rock viscosity at the contact, and r is the diapir radius (Marsh 1982; Mahon et al. 1988). Equation 3.2, which leads to the definition of the "hot Stokes model", shows that the ascent velocity varies with the diapir size, with larger bodies moving more slowly and ascending for longer periods relative to smaller bodies. Also, the ascent velocity of a diapir is inversely proportional to the viscosity of the surrounding medium, which should be less than 10^{15}–10^{16} Pa s to ensure diapir rise. Although a single diapir would have difficulty penetrating through a highly viscous region in the crust, once a conduit has been established by the previously hot rising magma, the resulting viscosity decrease of the country rock may promote diapiric transport through an otherwise less permeable medium. The ascent distance of the magma depends on the rheology and ambient temperature of the encompassing rock and on the initial diapir temperature, which determines the amount of released thermal energy. Because of the proportionately larger amount of available thermal energy, larger bodies will rise greater distances.

The relationship between growth rate of amplitude and time highlights a sigmoidal pattern, characterized by an exponential slow-growth phase, a linear rapid phase and an exponential stationary phase, defining an overall S-like shape. The final volume of the diapir is acquired at the beginning of the linear rapid phase of growth, implying reduced alimentation from the trunk in the later phase. The growth rate depends on the degree of melting and the volume of the magma, as well as on its viscosity (Berner et al. 1972; Bittner and Schmeling 1995).

In the case of isoviscous surrounding medium (or even of a thin envelope characterized by thermal softening outside the diapir), the rise of the diapir is accompanied by the progressive flattening and rotation of material above and the progressive stretching below, identifying the domains represented in Fig. 3.3 (Cruden 1988).

In particular, the diapir roof is gently dipping and unfaulted, overlaid by flat lying roof strata. Towards the diapir margin these strata surrounding the diapir roll over into a steeply dipping **rim synform**, or annular downward closing fold, resulting from the viscous drag forces related to gravitationally driven roof collapse and

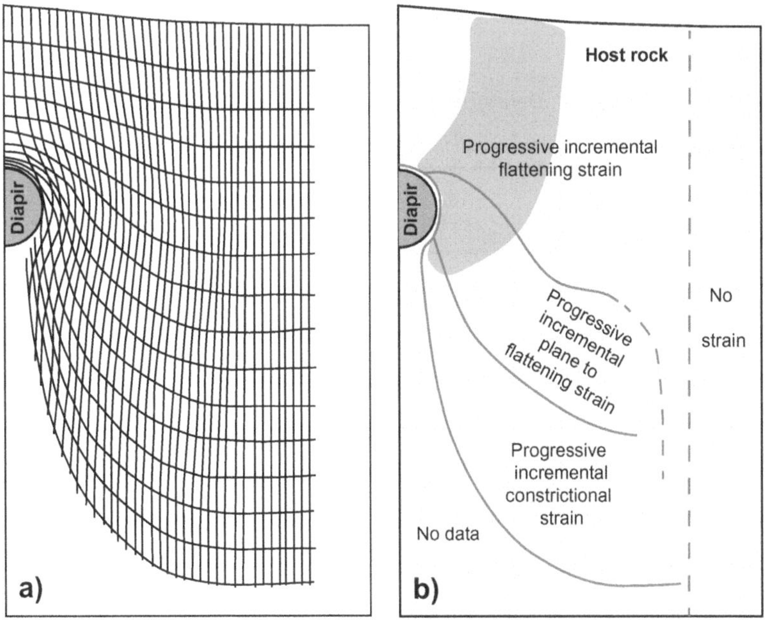

Fig. 3.3 **a** Reconstruction of the deformation within markers in a fluid to the side (half image) of a sphere simulating a rising diapir within ductile crust. **b** Map showing regions of progressive incremental strain in the fluid passing the sphere; grey shaded area highlights the region where the rotational component of deformation is significant (>70% of rotation; modified after Cruden 1988)

channel flow along the diapir margins (see the velocity profile in Fig. 3.1; England 1992; Paterson and Farris 2006). In the field, rim synforms are commonly observed at mid-crustal levels, to the sides of eroded Cordilleran plutons, where the strain due to the rim synform on the diapir margin is expected to increase with depth. This suggests widespread downward transport of country rock during pluton emplacement, involving important volumes of material and ultimately contributing to crustal thickening (Paterson and Farris 2006).

3.2.3 Diapir Emplacement

The emplacement of a diapir occurs when the diapir stops ascending. In a first approximation, the emplacement depth depends on the buoyancy, size and temperature of the diapir. The buoyancy depends on the **Level of Neutral Buoyancy** (LNB), that is when the intrusion density ρ_m equals the country rock density ρ_r (see also Eq. 1.2). A magma lighter than the country rock ($\rho_m < \rho_r$) has positive buoyancy and will tend to rise; conversely, a denser magma ($\rho_m > \rho_r$) has negative buoyancy and will tend to sink. As the density of the Earth generally increases with depth, a same magma may be positively buoyant at the base of the crust and negatively buoyant near the surface (Fig. 3.4). As for the diapir size, small diapirs (<5 km in diameter) with minor density contrast with regard to the country rock ($\sim$100 kg/m^3) are expected to stall in the lower crust. Large diapirs (>5 km in diameter) and with significant density contrast ($\sim$400 kg/m^3) may rise to shallower levels, possibly even faulting the overlying rocks (Burov et al. 2003). As for the temperature, the hotter the magma, the lower its viscosity and the easier is the flow of the diapir within the crust; a hotter magma is also lighter, increasing its buoyancy; finally, the hotter the magma, the softer is the surrounding deformation layer, further enhancing the rise of the diapir through the country rock.

The level of emplacement of a diapir also depends on the physical state of the country rock. In fact, if the strength of the country rock becomes significantly larger than the force promoted by buoyancy, a diapir may stop its ascent and start to emplace even before reaching any level of neutral buoyancy. This condition is commonly met in correspondence of the brittle-ductile transition, whose depth mainly depends on the geothermal gradient, the strain rate and the

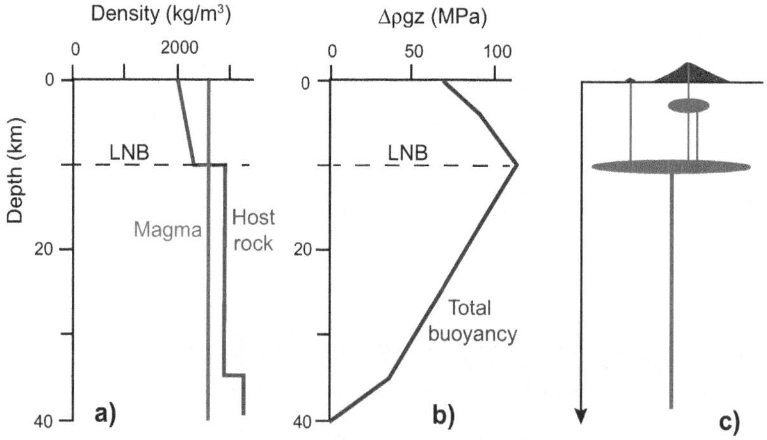

Fig. 3.4 Buoyancy force acting on a magma column in a stratified lithosphere. **a** Variation of density with depth; **b** variation of total magmastatic head with depth; **c** at the Level of Neutral buoyancy (LNB) magma may accumulate and differentiate in "hot zones," from where shallow crustal reservoirs are supplied (modified after Gonnermann and Taisne 2015)

tectonic setting: here the strength of the crustal rocks dramatically increases, hindering the further rise of diapirs (see Sect. 3.2.5; Bateman 1984; Bittner and Schmeling 1995). The depth of emplacement of a diapir also depends on the rheological behaviour of the ductile crust around it. In fact, diapirs are inferred to rise faster and emplace at shallower levels through a power law (non-Newtonian) crust than through a Newtonian crust, and such faster rise is not accompanied by the large heat loss expected with Newtonian crust. This behaviour results from the decrease in the effective viscosity, and thus resistance, of the non-Newtonian crust, which promotes strain rate softening (Weinberg and Podladchikov 1994).

When the diapir emplaces, its rate of growth decreases significantly, while its rate of crystallization increases. During emplacement, the diapir head spreads laterally, attaining a mushroom-like shape. This growth is propelled by the rise of the still positively buoyant main body and tail of the diapir. Eventually, the $\sim 10\%$ increase in the density of an emplaced diapir during final crystallization may promote its foundering within the country rock, particularly if the latter is rich in low density minerals (quartz, calcite). This possibility is supported by field evidence showing rim anticlines (i.e., annular upward closing folds) surrounding several plutons, as in the Great Basin of western North America (Glazner and Miller 1997).

Recognizing emplaced diapirs in the field is not straightforward, as other types of intrusions may be commonly mistaken for diapirs. This is the case of magmatic "balloons" produced by **ballooning**, which is the in-situ (at the level of emplacement), radial expansion (without predominant direction of movement) of an intrusion or magma chamber. In the common ballooning model, magma ascends until loss of heat or buoyancy causes the outermost magma to crystallize and cease ascent, while the hotter magma tail continues to rise and expand the already crystallized outer margin. This expansion forms a concentric, flattening-type strain in the pluton periphery and pushes aside the surrounding country rock, a process which may take hundreds of thousands of years (Fig. 3.5). This process may be encouraged by regional tectonic activity. For example, shortening at the base of the pluton may squeeze the magma upwards, causing expansion at higher crustal levels and promoting ballooning (Bateman 1985; Ramsay 1989; Rosenberg et al. 1995).

There are many similarities between balloons and diapirs. Both are subcircular in map view,

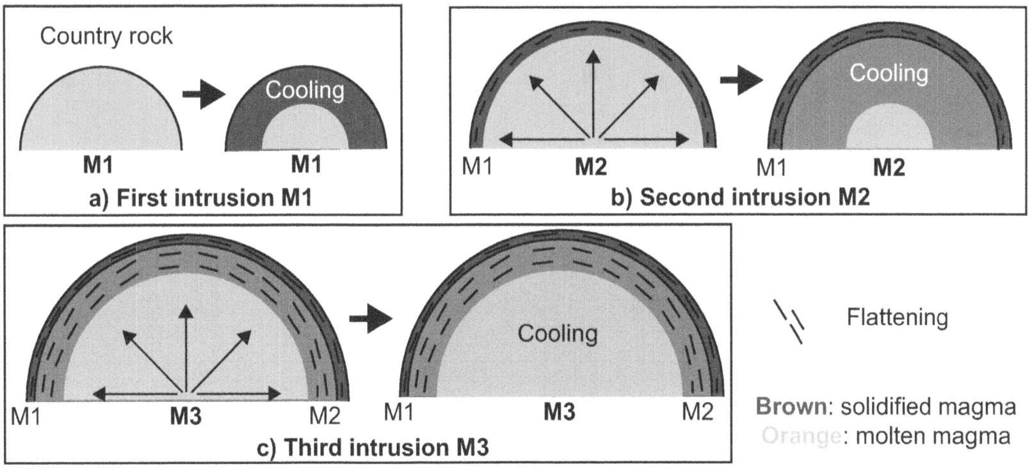

Fig. 3.5 Possible development of a composite intrusion through ballooning, section view. After a first intrusion M1, partially solidified (brown part, **a**), a second inner intrusion of magma M2 pushes away and flattens the outer shell of the first intrusion (**b**); when the second intrusion is solidified, a third inner intrusion M3 emplaces and flattens the shell of the first and second injections, eventually solidifying (**c**; modified after Ramsay 1989)

with concentric fabric and concentrically deformed wall rocks, with the strain increasing toward the pluton contact. However, a few differences do exist (e.g., Paterson and Vernon 1995). In diapirs, magma ascent may occur by rise of large magma batches, instead of transport in dikes followed by emplacement and ballooning. More importantly, the evaluation of the predominant direction of movement of the intrusion relative to the country rock may allow distinguishing balloons from diapirs. Ballooning is the radially directed expansion of a body in situ with no predominant movement: hence, while balloons should show no pluton-up kinematics, diapirs should be associated with pluton-up kinematics, and steep, well-developed lineations. Therefore, a balloon has not moved unidirectionally from its point of inflation, whereas a diapir can only be satisfactorily recognized if it has risen more than one body diameter. In other words, diapirism can be proven beyond doubt where it is possible to demonstrate uplift of country rock in excess of the vertical thickness of the intrusion. However, this requires accurate knowledge of the three dimensional shape of the intrusion and its country rock, a condition which is rarely met, resulting in an overall difficulty to demonstrate any diapiric ascent (England 1990; Molyneux and Hutton 2000). The difficulty in distinguishing diapirs from balloons can be also overcome identifying marker deflections in the field. These are useful for evaluating country rock displacements during diapirism or ballooning, provided that either the pre-emplacement geometry of the markers is known, or that rigid rotations and stretches are known (Paterson and Fowler 1996).

Using these approaches, several field studies recognized eroded portions of magmatic diapirs, usually with granitoid compositions (Ramsay 1989; Galadi Enriquez et al. 2003; He et al. 2009). These include the Chindamora granitic batholith (Zimbabwe), built by magmatic pulses at more than 15 km depth and also showing important ballooning, or the La Bazana granite (Spain), with a teardrop–pipe shape suggesting lateral expansion and vertical flattening, emplaced at 7–10 km of depth. The Cretaceous Fangshan granodiorite pluton (North China) has a bulk wall-rock shortening by ductile flow of ~4 km, similar to the pluton radius, and a rim synform formed during the late-stage rise of the diapir, emplaced at a depth of 3–6 km (Fig. 3.6). A rim synform is also found around the Northern Arran Coarse Granite (Scotland; England 1992). These cases suggest that diapirism may be a viable mechanism for the migration and emplacement of magma up to 5–10 km in depth, depending on the local conditions.

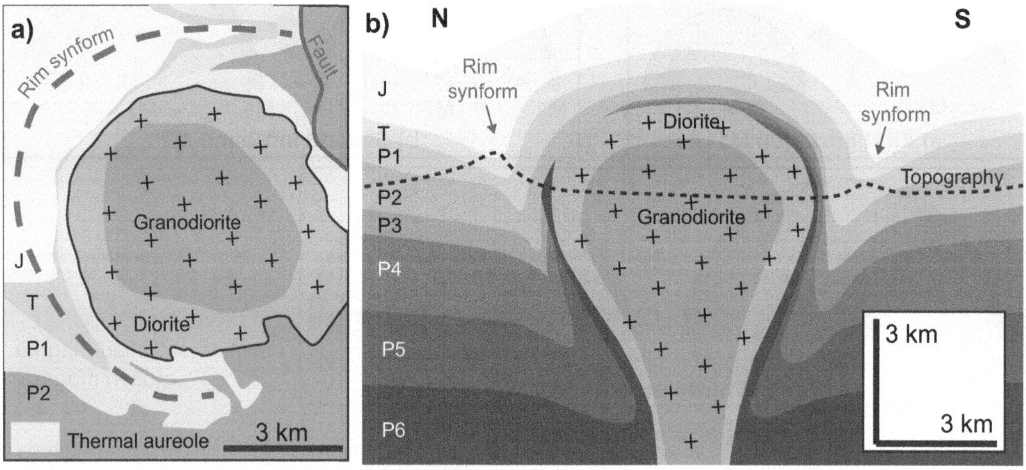

Fig. 3.6 Map **a** and schematic section view (**b**; approximately N–S) of the eroded Fengshan diapir (China), associated with a rim synform to the side (modified after He et al. 2009)

There is also indirect evidence of ongoing diapir emplacement, as suggested by the tens of kilometres wide uplifted area surrounded by a broader zone of subsidence in the Altiplano-Puna region of the Central Andes. This specific deformation pattern is inferred to result from the emplacement of a mid-crustal diapiric magma body at ~20 km of depth, although this interpretation is debated (Fialko and Pearse 2012; Henderson and Pritchard 2013).

3.2.4 Diapirism and Stoping

The rise of diapirs is often accompanied by magmatic stoping. **Stoping** consists of disconnecting and surrounding a piece of country rock (namely, the thermal halo) by magma, moving blocks relative to their position prior to emplacement (Fig. 3.7).

Stoped blocks may be displaced downward (sinking) or, less commonly, upward (floating), depending on their density contrast with the magma. Stoping is a process through which the magma essentially mines its way upward, and it is not limited to diapirs. Stoping may in fact occur also along magma chamber roofs, sides, and floors, at any crustal level, and blocks may form through a number of different processes, as thermal cracking, diking, tectonic stresses and focused porous flow (Marsh 1982; Paterson et al. 2008). Thermal cracking is possibly the easiest means to promote stoping at the top of voluminous magmatic intrusions. Here the heat of the intrusion causes expansion of the country rock, producing tensional stresses and developing cracks. The stresses required for fracturing necessitate only moderate changes in temperature, on the order of 100 °C, implying limited energy transfer. A model alternative to thermal shattering has been proposed for the development of stoped blocks in the Cordilleran batholith of Oregon (USA). Here stoping is seen as a multi-stage process involving the emplacement of sills along bedding planes and formation of short connecting dikes cutting across bedding. This results in a sill–dike network enclosing rectangular roof blocks along the batholith–roof contact (Zak et al. 2012). A further model has related stoping to short-lived roof collapse events in active magma chambers: accordingly, stoped blocks may record collapse of the roof during large caldera-forming eruptions (Hawkins and Wiebe 2004).

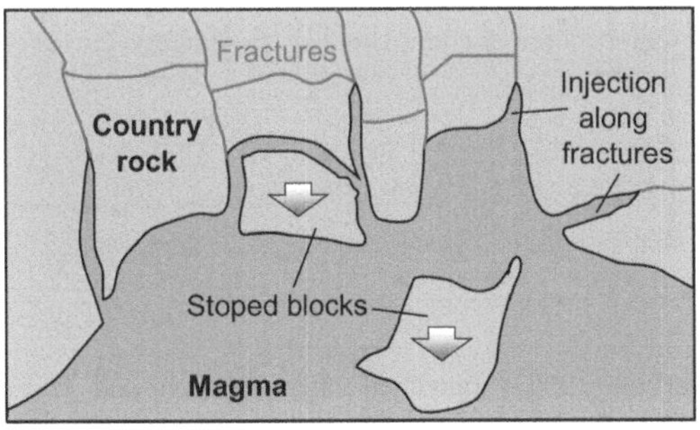

Fig. 3.7 Scheme illustrating the mechanism of stoping, where fractures at the top of a magmatic intrusion promote injection of magma and subsequently the foundering of the blocks surrounded by the fractures within the magma

Stoping requires the rate of advance of the diapir be determined by the rate of removal of blocks from its roof, while the ascent distance is probably limited to a few body heights, because stoped blocks rapidly congest the magma. Although stoping does not make space, it can facilitate magma ascent by vertical mass exchange. Field evidence from plutons cited in favour of stoping includes the following: (a) sharp, discordant contacts between plutons and wall rocks; (b) a lack of pluton-related ductile deformation of wall rocks; (c) xenoliths in plutons, also from diverse sources; (d) geochemical evidence for magma contamination (Glazner and Bartley 2006).

The importance of magmatic stoping has been debated, with studies considering it unlikely to be a volumetrically significant process and others highlighting it to be widespread in all types of plutons, where it may account up to 20% or the observed magma mass. Several field studies support this latter view, considering magmatic stoping as a dominant process of downward transport of country rock in upper crustal plutons, or of magma evolution also in the middle and lower crust (Dumond et al. 2005; Zak and

Paterson 2006; Glazner and Bartley 2008; Paterson et al. 2008; Yoshinobu and Barnes 2008). Indeed, there is much to be gained by studying stoping with regard to high-temperature rheological behaviour and magma emplacement, as stoped blocks may provide information about paleohorizontal during emplacement of plutons and about the timing, kinematics and magma viscosities during fabric formation (Paterson and Miller 1998a).

3.2.5 Limitations of Diapirism as a Mechanism for Magma Transport

Until the 1990s, diapirism was a very popular mechanism to explain the rise of magma, with the main limitation that a diapir, unless being very large (with a small surface area to volume ratio) would soon run out of heat and hence rise only short distances in the crust (Marsh 1982). This limitation has been partly relaxed considering a ductile crust with power law behaviour, encouraging the faster and shallower rise of diapirs and limiting the large heat loss inferred to occur with a Newtonian crust. A further limitation inferred for the rise of diapirs has been due to the crystallinity of magma. While the rise of crystal-rich magma is generally enhanced by diapirism, with significant crystallization (>70%) diapiric ascent is prevented and any further magma injection may inflate the diapir through ballooning (Bateman 1985).

Studies in the 1990s highlighted a further and crucial limitation for diapirism, underlining the inadequacy of the diapiric ascent of magma through buoyancy in the upper brittle crust. In particular, magma buoyancy increases with the density contrast with the country rock and with the height of the magma column (Fig. 3.8a; Paterson and Farris 2006). For example, assuming a density contrast between the magma (~2400 kg/m^3) and surrounding crust (~2750 kg/m^3) of ~350 kg/m^3 leads to a buoyancy force of ~3.5 MPa per 1 km of height of magma. Such a value is two orders of magnitude less than the stress value required to

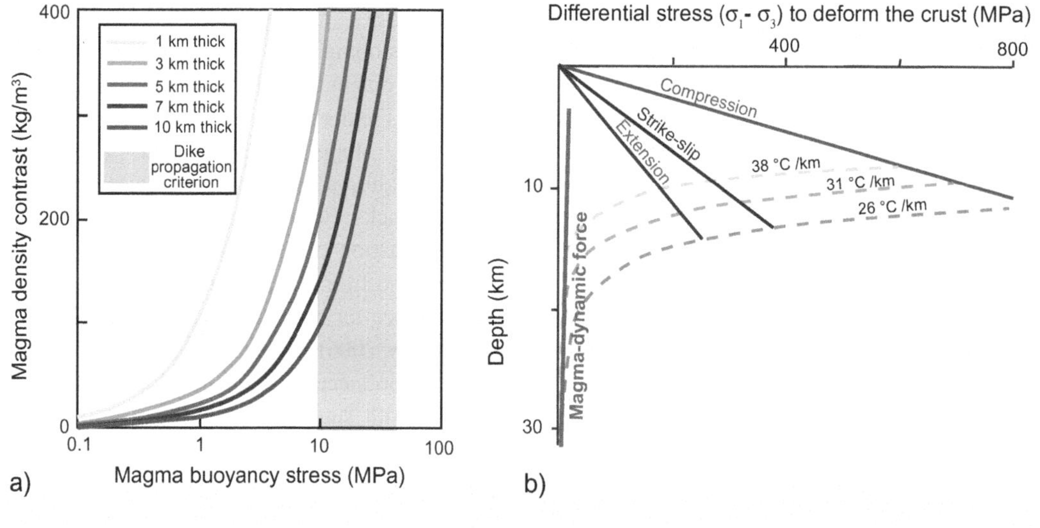

Fig. 3.8 **a** Plot of magma/country rock density contrast and buoyancy stress for magma columns of varying height. Thin (1 km thick) plutons generate small buoyancy stresses. Magma columns of more than 5–10 km high and density contrasts larger than 100–200 kg/m^3 are needed to propagate dikes and initiate wall rock fractures from buoyancy stresses alone. The grey area indicating the dike propagation criterion is from Jellinek and DePaolo (2003) (modified after Paterson and Farris 2006). **b** Schematic rheological profile showing the variation in the crustal strength as a function of the tectonic setting and geothermal gradient. The dynamic force of a magma column starting at the Moho (red line) shows that this does not match the differential strain required to break the brittle crust (modified after Vigneresse 1995)

penetrate the brittle-ductile zone at a depth of approximately 10 km, that is on the order of a few hundred of MPa (Fig. 3.8b; Vigneresse 1995).

The discrepancy between the force caused by the buoyancy of the magma and the differential stress required to fracture the brittle crust remains high even assuming a lighter magma incorporating an important percent of fluid phase. Only an unrealistically tall column of diapiric magma (hundreds of km) would overcome the shear resistance of the crust by buoyancy. This implies that, at the brittle-ductile transition zone, the buoyancy force of a magmatic diapir becomes insufficient to fracture the upper brittle crust and the magma cannot rise further only by buoyancy. The brittle-ductile transition should be thus considered as the upper limit of penetration of magmatic diapirs. However, this transition should not be perceived as a fixed barrier, as its depth is affected by the thermal and mechanical state (i.e., pre-existing fractures) of the crust. For example, repeated magma emplacement, also through diapirs, may progressively heat the crust, shallowing the brittle-ductile transition and the emplacement level of magmatic diapirs (Cao et al. 2016).

Finally, the diagram in Fig. 3.8b also shows that the strength required to break the brittle crust is about four times larger under compression than under extension. This implies that the easiest condition for magma to intrude the brittle crust is within extensional tectonic settings, where the stress level is lowest. Nevertheless, extensional volumes may be found also in localized zones within strike-slip or contractional settings (see Sect. 4.5.2 and Chap. 10; Vigneresse 1995).

3.3 Magma-Filled Fractures

Diapirs may be a viable means for the rise of magma in the crust although, because of the limitations discussed in the previous section, they are likely not the predominant mode, especially at shallow levels. In fact, dikes provide the main mode of rise of magma within the crust, at any level. Dikes, with inclined sheets and sills, are **magma-filled fractures**, or subplanar intrusions whose thickness is much smaller than their vertical or lateral extent. In particular, **dikes** are magma-filled fractures discordantly cutting at high angle across layers, **sheets** are magma-filled fractures discordantly cutting at moderate angle across layers, and **sills** are magma-filled fractures concordant with the country rock layers (Fig. 3.9). As a consequence, in presence of horizontal or poorly-tilted layers, as in active volcanic areas, dikes are usually subvertical, sheets are inclined and sills subhorizontal. These terms are used for both the actively propagating magma-filled fractures and the solidified intrusions. In this section the emphasis is on magma-filled fractures, whereas the following sections of the chapter focus on the general principles of dike propagation and arrest within the crust. Sills, as mainly responsible for magma emplacement and accumulation, and often anticipating the development of magma chambers, are considered in Chap. 4; dike nucleation from magma chambers is also considered in Chap. 4. The shallower propagation of dikes and, to a lesser extent, sheets and sills within volcanoes and the implications for eruptive activity are described in Chap. 7.

Magma-filled fractures are much thinner than their length and width: their usual thickness is on the order of metres, while the typical length and width is on the order of kilometres, or more. This leads to aspect ratios on the order of 10^{-3} to 10^{-4}. Even though magma-filled fractures are often assumed to be planar or subplanar, they may exhibit deviations from planarity, both along strike and dip.

Dikes ensure the transfer of magma at any crustal level, also feeding eruptions from shallow magma chambers. Such a transfer may be characterized by predominant vertical propagation, responsible for the rise of magma, or predominant lateral propagation, transferring magma away for tens of kilometres and eventually feeding distal eruptions outside volcanoes. Conversely, sills are usually responsible for the emplacement and accumulation of magma in the crust, especially at shallow levels (few km of depth), although in several cases sill systems

Fig. 3.9 Field examples of magma-filled fractures. **a** Series of inclined sheets (highlighted by the yellow arrows) intruded in a succession of sub-horizontal Tertiary lava flows; the intrusions belong to the Midhyrna-Lysuskard cone sheet system, Snaefellsnes Peninsula, northwestern Iceland (Photo courtesy: Alessandro Tibaldi). **b** Three en-echelon dike segments observed in an eroded portion of a rift zone in eastern Iceland. **c** A feeder dike outcropping in the Capelinhos Peninsula, site of the last eruptive activity (1957–1958) in Faial (Azores). A portion of a sill **d** and a set of stacked sills **e** observed at Piton des Neiges, La Reunion Island; note the blunt left termination of the lighter reddish sill in the lower part of **e**

have been responsible for relevant lateral magma propagation. Also, while dikes can propagate independently of pre-existing fractures (see Sect. 3.5), sill formation is often enhanced in correspondence of pre-existing subhorizontal anisotropies, as layering or fractures (see Sect. 4.3). In between, sheets may transfer and/or emplace magma in the crust, mainly depending upon their dip and the characteristics of the country rock (as far-field stress, anisotropies).

An important feature common to all magma-filled fractures is that, to ensure propagation and thickening, these tend to develop perpendicular to the least principal stress σ_3. Therefore, the plane approximating the magma-filled fracture contains the maximum and intermediate principal stresses σ_1 and σ_2, although their exact plunge cannot be determined from the orientation of the magma-filled fracture alone. Because of this, from a dynamic point of view magma-filled fractures behave as extension fractures. When the emplacement of magma-filled fractures results from far-field (regional, or tectonic) stresses, subvertical dikes are expected to form in extensional or strike-slip settings, where the least principal stress σ_3 is horizontal, and subhorizontal sills are expected to form in compressional settings, where the least principal stress σ_3 is vertical (Fig. 3.10).

Significant deviations from this simple behaviour may occur if far-field stresses are overprinted by any near-field (local) stress, as induced by important topographic variations, magmatic sources, layering, fractures or other features. For example, the development of inclined magmatic sheets requires an inclined least compressive stress σ_3 and thus non-Andersonian conditions. The latter are usually inhibited by far-field stresses and promoted by near-field stresses, as those developed by magma chambers. Therefore, the presence and orientation of dikes, sheets or sills at a given place within the crust ultimately depends on the combination of both the far-field and near-field acting stresses. The stress conditions for the development of dikes and sills, with the respective vertical σ_v and horizontal σ_h stress components, which may result from any far- and/or near-field stress, are reported in Fig. 3.11 (Jaeger 1969; Gressier et al. 2010). For both dikes and sills the magmatic pressure reduces the normal stresses, but not the deviatoric stress. The Mohr circle thus shifts leftward and, if it touches the failure envelope, fracturing occurs. In an anisotropic medium, characterized for example by layering, sill formation depends on the deviatoric stress and the tensile strength parallel and perpendicular to the layering.

Magma-filled fractures are expected to form and grow with low viscosity magmas, as with mafic and low viscosity felsic compositions. In fact, high viscosity felsic sheets will be likely halted by freezing soon after the magma encounters rock at temperatures below the magma solidus (Emerman and Marrett 1990; Rubin 1995a). Magma composition may also affect the geometry of magma-filled fractures, as revealed by estimates of the viscosity of several dikes in different tectonic settings, which show that dike thickness increases with magma viscosity (Wada 1994). In particular, mafic magmas with viscosities of 10^1–10^2 Pa s tend to form metres-thick dikes, while felsic magmas with viscosities of 10^6–10^7 Pa s tend to form thicker dikes. Therefore, magma viscosity, as determined by its composition, appears an important factor to explain the thickness variations in dikes.

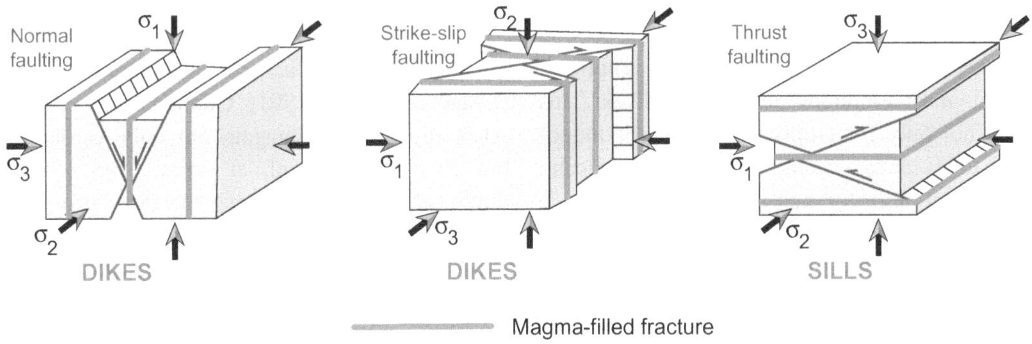

Fig. 3.10 Types of magma-filled fractures (dikes and sills) expected to develop as a function of the regional tectonic regime. Inclined sheets, not included here, are commonly induced by the local stress field induced by a magma chamber

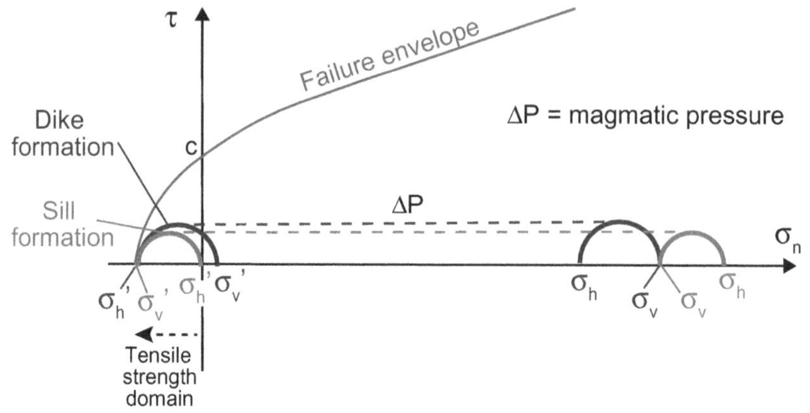

Fig. 3.11 Mohr Coulomb diagram illustrating the stress conditions for the development of dikes (purple) and sills (blue). These conditions refer to a zone in the country rock at the tip of a propagating dike or sill. The magmatic pressure ΔP shifts the Mohr circles to the left, passing from a vertical σ_v and horizontal σ_h to σ_v' and σ_h', allowing the propagation of sills and dikes (modified after Gressier et al. 2010)

Independently of their orientation, magma-filled fractures, developing perpendicular to the least principal stress σ_3, share overall similar mechanisms and conditions of formation and propagation. These mechanisms and conditions are described below in the case of dikes, although with minor modifications they may be extended to sills and sheets.

3.4 Dike Propagation

As this chapter discusses the rise of magma, this section focuses on the general case of dike propagation within the crust, neglecting the mechanism of dike nucleation from magma chambers, which is considered in Sect. 4.6.3.

Dikes are responsible for the transfer of magma at any crustal level, as well as within volcanic edifices, where they feed eruptions. Indeed, most eruptions are fed by dikes. This holds not only for eruptive fissures, for monogenic fields, or for eccentric vents on the volcano slopes, but also for summit craters fed by a central conduit in polygenic volcanoes. The importance of dikes in building central conduits is underlined by observations at several volcanoes, including drilling data at Unzen felsic stratovolcano (Japan), which indicate that the same central conduit consists of a dike swarm (Sakuma et al. 2008). For these reasons, understanding dike propagation is crucial to define shallow magma transfer and to ultimately forecast eruptions (see also Chap. 9).

Dikes can be approximated as magma-filled cracks with overall ellipsoidal shape. Their reported thickness to length aspect ratios are usually between 10^{-2} and 10^{-4} (Fig. 3.12). Dikes can propagate anywhere within the crust, with velocity on the order of m/s, without suffering the limitations mentioned in Sect. 3.2.5 for magma rising through diapirs. Dikes in fact focus the forces required for magma ascent at their tips, where these become much larger than the fracture resistance of the country rock. Because in situ tensile strengths of typical country rocks of dikes and sheets are approximately 0.5–6 MPa, the theoretical crack-tip values of dikes are from 100 to 10,000 times greater than the tensile strength of country rocks (Lister and Kerr 1991; Gudmundsson 2002). Also, dikes can propagate not only vertically, but also laterally and, at times, even with a downward component along the slope of a volcano or in the crust. The substantial independence of the dike propagation direction from the direction of gravity is a further factor that distinguishes dike ascent from diapirism and from porous flow in the source region of magma (Rubin 1995b).

Fig. 3.12 Subvertical dikes with variable thickness observed along the south coast of Tenerife (Canary Islands); the white boats provide a scale

Dike formation and propagation have been commonly investigated focusing on the deformation processes within the country rock (the "solid mechanics" approach) or on the viscous flow of magma within ("fluid mechanics" approach). Both approaches are considered below.

3.4.1 Solid Mechanics

The growth of a dike involves both elastic (associated with the thickening of the dike along its length) and inelastic (of limited extent, mainly fracturing at the dike tip) processes within the country rock. Therefore, excepting the tip region, deformation of the country rock is assumed to be largely elastic, and this motivates treating dikes as cracks in a linear elastic body. Let us first consider the elastic behaviour of the country rock. For uniform loading stresses, the thickness profile of a dike is elliptical and its half-thickness w at the centre is assumed to be (Fig. 3.13; Rubin 1995b, and references therein):

$$w = \frac{P_e}{E_s} L \tag{3.3}$$

with

$$P_e = P - \sigma_n \tag{3.4}$$

and

$$E_s = \frac{G}{1 - \nu} \tag{3.5}$$

being L the half-length of the dike, P_e the **magmatic excess pressure**, given by the difference between the internal magma pressure P and the ambient compressive stress perpendicular to the

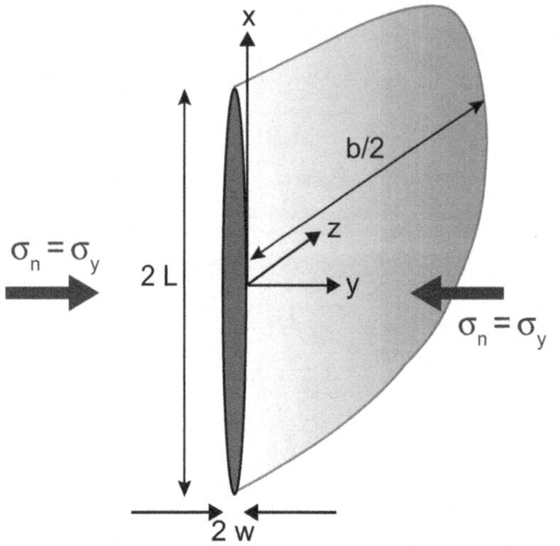

Fig. 3.13 Schematic view of dike of length *2L* (along the direction of propagation), width *b* and thickness *2w* experiencing a normal pressure σ_n on its walls

dike plane σ_n, and being E_s the elastic stiffness of the country rock, determined by its elastic shear modulus *G* and Poisson's ratio *ν*. Considering that the in situ elastic stiffness E_s is on the order of a few tens of GPa or less, and observing in the field the length and thickness of outcropping dikes, Eq. 3.3 can be used to estimate the related excess pressures: these are on the order of a few to ~ 10 MPa, although considerable variation is expected.

Equation 3.3, despite of its wide applicability, provides a simplified approach, as non-elastic processes may also control dike thickness. Recent field studies on thin dikes (thickness < 1 m) show that these are relatively thick at their ends and narrow in the middle, with shapes poorly matching the elastic theory. These shapes can be explained by country rock inelastic deformation and magma chilling at the dike tips, which prevent closure as magma pressure declines (Fig. 3.14; Daniels et al. 2012). While the application of these results to thicker dikes is not straightforward and requires further investigation, other studies confirm the importance of non-elastic processes in constraining dike thickness. These studies, based on thousands measurements from dikes in different tectonic settings, suggest that the country rock lithology, in particular the strength as defined by material defects and fracture toughness, is a dominant factor in controlling dike thickness (Krumbholz et al. 2014).

As for the inelastic behaviour occurring at the dike tips, it is useful to distinguish three regions: the crack, where the country rock is completely separated; the intact country material, where deformation is essentially elastic; the region in between surrounding the crack tip, or "process zone", where the strength of the material has been exceeded, but where inelastic deformation has not proceeded to develop a crack. The criteria commonly used to analyse the propagation of the fractures at the dike tip consider the local stress or the energy balance (Rubin 1995b, and references therein). In the first case, if there is no macroscopic crack and the material experiences uniaxial tension, the appropriate fracture criterion is that at the dike tip the tension due to the magmatic pressure exceeds the tensile strength of the material. If the material is under compression, the tensile failure criterion postulates that the

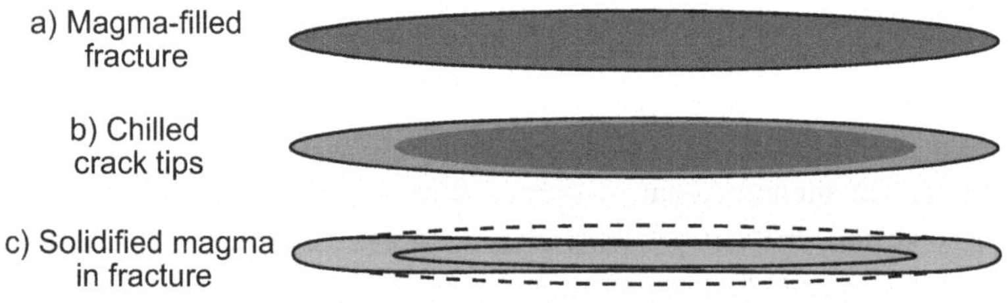

Fig. 3.14 Scheme showing the impact of cooling on dike geometry. **a** Magma creates a dike (section view); **b** cooling ensues at the dike margins (brown portions); **c** complete solidification creates the observed dike profile (outer solid line), with thinner centre and thicker tips with regard to an expected profile of a pressurized magma-filled fracture in elastic medium (the latter is exemplified by the dashed line; modified after Daniels et al. 2012)

magmatic pressure at the dike tip exceeds the least compressive stress by some tensile strength (Fig. 3.11). If a macroscopic crack exists, the stress concentration at the tip allows propagation to occur at a lower applied tension or magmatic pressure. The second case considers dike propagation to occur when the associated release of potential energy is sufficient to provide the energy required for fracturing at the tip. This potential energy consists of elastic strain energy, plus any work done on the country rock. In particular, dike propagation occurs when the potential energy release per unit increase in dike length equals the critical value required for propagation. Linear Elastic Fracture Mechanics (LEFM) provides a useful framework for this approach, quantifying the stress required to propagate a fracture and using the notions of crack-tip stress intensity factor K and rock fracture toughness K_c.

The **stress intensity factor** is used to describe the stress intensity near the tip of a crack: all else being equal, the longest crack has the highest stress intensity factor (see Sect. 2.5). **Fracture toughness** is the property that describes and quantifies the ability of a material to resist fracture. The fracture toughness of a material is determined from the stress intensity factor at which a thin crack in the material begins to grow. Thus, at the onset of crack propagation, the fracture toughness becomes the critical stress intensity factor, so that crack propagation occurs when $K = K_c$ (Lister and Kerr 1991; Rubin 1995b). The stress intensity factor K is a fundamental component of the stress field near the crack or dike tip. While K depends upon the crack size and applied forces, being

$$K = P_e\sqrt{L} \tag{3.6}$$

for a uniformly pressurized bladelike fracture of size 2 L under pure tensile failure (Mode I crack), K_c is a material property that can be experimentally defined, usually being on the order of 1 MPa $m^{1/2}$ and increasing with pressure. If K slightly exceeds K_c, the dike (or a crack) will propagate and magma will flow toward the tip. The propagation of magma within the fracture is thus described by Eq. 3.6, substituting K_c to K. However, magma cannot flow too fast, or the associated drop in pressure (see Sect. 3.4.2) toward the tip would drop K below K_c. Therefore, the propagation speed of a dike is limited by the fracture toughness. If $K < K_c$, the dike may propagate very slowly, only by chemical corrosion at its tip.

3.4.2 Fluid Mechanics

The fluid mechanics approach focuses on the flow of magma within the dike and at its tip. In this approach it is usually sufficient to consider the magma flux averaged over the channel thickness, assuming laminar flow. For laminar flow, the velocity is zero at the dike walls and varies parabolically across the channel thickness. Laboratory experiments also indicate that crystal- and bubble-free magmas may be treated as Newtonian viscous fluids.

Near the dike tip the average flow velocity in the propagation direction approaches the tip velocity. Here the magma decreases its pressure, attaining a value P_t before reaching the narrow crack tip, and a gap, or **tip cavity**, forms between the tip and the magma front (Fig. 3.15; Lister and Kerr 1991; Rubin 1995b, and references therein). This low pressure region results from the dynamical difficulty of forcing a viscous fluid to flow into a thin gap. The tip cavity is mainly filled by the volatiles exsolving from the magma, as its low pressure promotes exsolution even when the dike is well below the level at which the magma would normally become saturated. Examples of relict tip cavities include cracks extending beyond the magma front in a dike and pegmatite zones at the tips of granitic dikes. The dike tip experiences underpressure, or tip suction, defined as $P_s = P_t - \sigma_n$. The greater the tip suction, the shorter the tip cavity, while the larger the dike, the larger the tip cavity. In a first approximation, as the tip cavity has limited extent with regard to the dike size, it also has limited control on its thickness and velocity of propagation. Therefore, the existence of the tip cavity can be ignored in most calculations

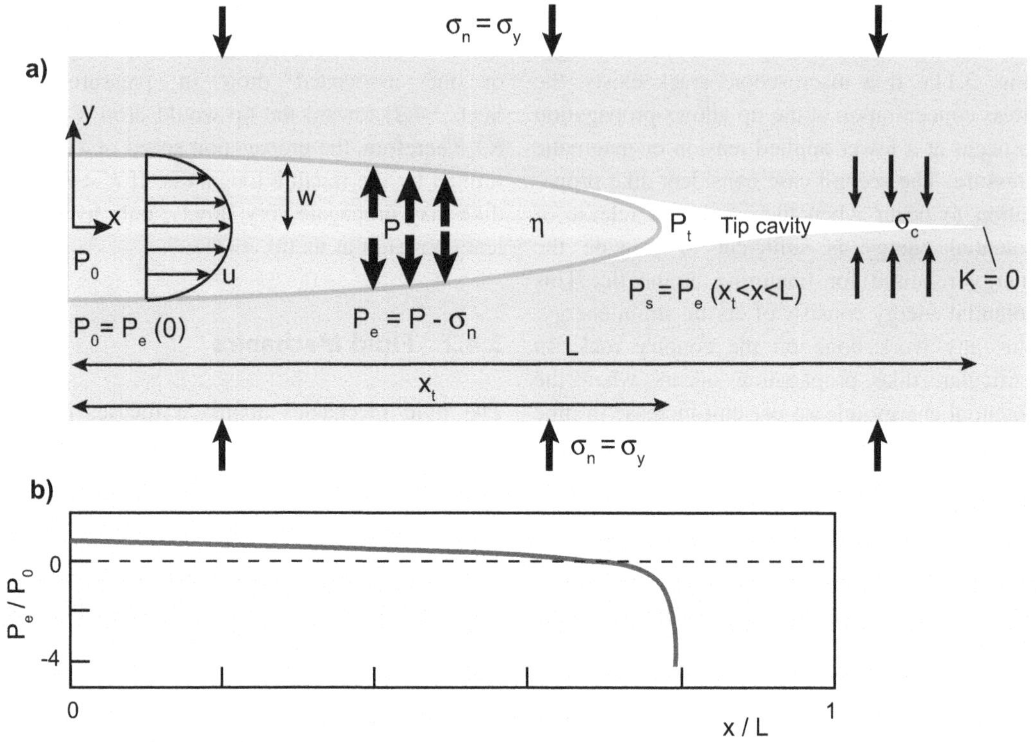

Fig. 3.15 **a** Two-dimensional not to scale scheme of propagating dike, with emphasis on the tip region. Material properties: magma viscosity η, cohesive stress σ_c acting over the critical crack-wall separation. Boundary conditions: ambient dike perpendicular compression σ_n (or σ_y), magma pressure at dike entrance P_0, magma pressure P, magma excess pressure P_e, pressure at tip cavity P_t, tip suction P_s, dike half-thickness w, position of magma front x_t, average flow velocity u. Propagation criterion is $K = 0$ at tip of cohesive zone. **b** Computed excess magma pressure P_e within the dike, normalized by P_0, as referring to **a** (modified after Rubin 1995b)

dealing with dike propagation. Nevertheless, the tip cavity may play a fundamental role in promoting inelastic deformation, and thus seismicity, at the dike front. Behind the tip cavity lies the bulbous nose of the dike, with maximum thickness some 25% greater than $2w$ (Fig. 3.16a; Lister and Kerr 1991; Rubin 1995b). This swollen zone results from the impossibility to have magma underpressure along the main walls of the dikes, as suggested by calculations. In fact, if the dike narrowed monotonically towards the tip, the pressure gradient for flow would also increase monotonically, generating the underpressure along the dike; conversely, an increased thickness at the head of the dike implies a reduced pressure drop along the dike.

Generally three sources of pressure may promote magma flow in dikes and contrast the viscous resistance to flow: the excess magma pressure at the source, the magma buoyancy and the differential stress acting on the dike. These three contributions define the driving pressure, or **magmatic overpressure** ΔP_m, responsible for magma flow in a dike, which is expressed as (Gudmundsson 1990; Rubin 1995b; Gudmundsson et al. 1999):

$$\Delta P_m = P_e + (\rho_r - \rho_m)\, g\, h + \Delta\sigma \qquad (3.7)$$

where P_e is the magmatic excess pressure above the lithostatic pressure (resulting for example from magma addition, gas exsolution), h is the vertical distance from the magma source chamber and $\Delta\sigma = \sigma_v - \sigma_h$ is the difference between the vertical σ_v and the horizontal σ_h far-field principal stresses. Usually, for dike propagation

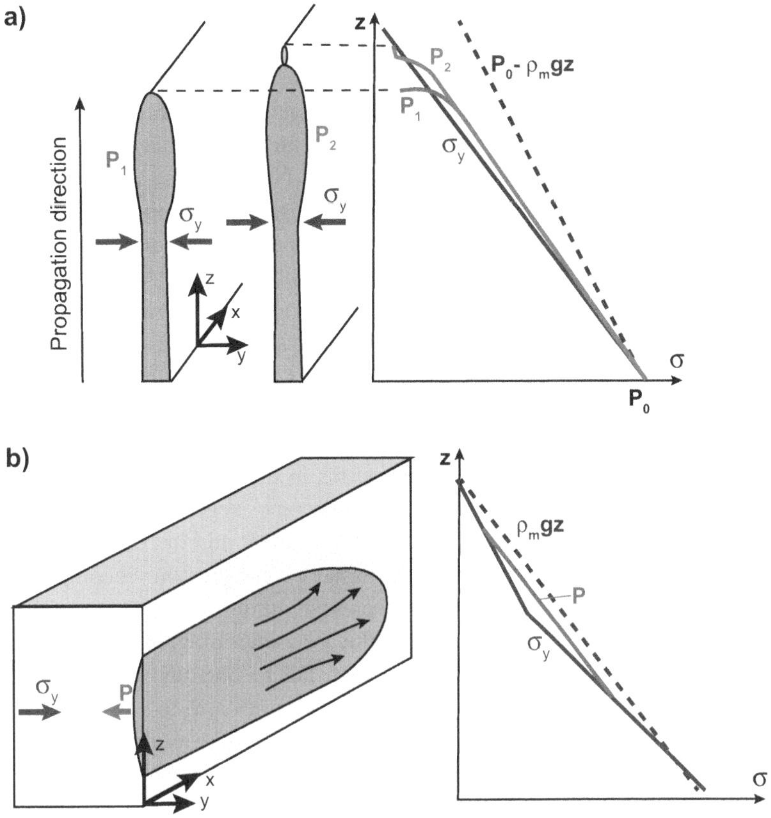

Fig. 3.16 **a** Scheme showing two stages of vertical propagation of a dike. Magma pressures P_1 and P_2 deviate from ambient dike-perpendicular stress σ_y only over a certain length below tip. Dashed green line represents the static magma pressure curve, with P_0 being the magma pressure at dike entrance. At stage 1 (left cartoon), tip suction is large and tip cavity is small; at stage 2 (right cartoon), magma front is closer to volatile saturation depth and tip cavity may become unstable. **b** Scheme showing the lateral propagation of a blade-like dike from a shallow source. The related stress state required for lateral propagation is shown in the right diagram. The kink in σ_y may be due to a density step or to variations in tectonic stress (modified after Rubin 1995b)

the horizontal stress is the minimum principal stress σ_3, while the vertical stress is the maximum principal stress σ_1: therefore, $\Delta\sigma$ may be commonly approximated by the deviatoric stress. Equation 3.7 thus indicates that the magmatic overpressure increases with the following conditions. (a) "Internal" loading (magma-related), as increasing the magmatic excess pressure, for example by adding magma or exsolving gases. (b) "External" loading (country rock-related), as increasing the deviatoric stress by decreasing the σ_3 component, as for example under extension. (c) The interaction between internal and external factors (the density difference, or buoyancy), as injecting lighter magma in denser crustal rocks.

As the overpressure ΔP_m in a vertically propagating dike also increases with the dike length L, once the dike grows sufficiently that $(\rho_r - \rho_m) gh > P_e$, magma buoyancy begins to provide the dominant pressure gradient for flow. In addition to increasing its length, during the early stages of propagation the dike also widens along the direction parallel to its strike, or width. Once the dike has reached a critical width driven by its elastic and buoyancy pressures, it does not widen further, due to the finite fracture toughness of the country rock. The dike will then predominantly lengthen at a rate governed by a balance between its viscosity and buoyancy (Fig. 3.16a; Gonnermann and Taisne 2015).

As at intermediate to great crustal depths the country rock density usually exceeds that of the magma, buoyancy is an important contribution in Eq. 3.7. At shallow crustal depths, however, rocks have densities similar to that of the magma. The density of the country rock immediately above the magma chamber may be even less than that of the magma, making it more difficult for the magma to rise through buoyancy alone. When a dike reaches the Level of Neutral Buoyancy (LNB) of the magma, it may stop ascending: if this occurs, magma may propagate laterally through dikes or sills, depending on the stress field. The requirement for the lateral propagation of a dike is that the excess pressure at the dike centre exceeds that at the dike top and bottom and, at the same time, shows a lateral gradient (Fig. 3.16b; Lister and Kerr 1991; Rubin 1995b). The condition for lateral magma propagation has been commonly referred to in the literature as related to the LNB. Nevertheless, recent experimental studies significantly relax the importance of buoyancy in the lateral propagation of dikes, rather highlighting the importance of rigidity layering (a stiffer layer overlying a weaker one) and topography (see Chap. 7; Urbani et al. 2018). Also, the filter role of the LNB on magma ascent is relaxed by three possible conditions promoting an additional rise of a dike above the LNB. (a) Degassing, which may further decrease magma density in the dike, providing some additional buoyancy (Gudmundsson et al. 1999). (b) The fact that the local **magmastatic pressure** gradient $(\rho_r - \rho_m)g$ does not need to be positive everywhere in the dike: it needs to be positive only when averaged over the dike length. Therefore, magma could be erupted in a negatively buoyant upper crust, provided that there is a sufficiently large underlying region in which the magma is positively buoyant (Fig. 3.4). (c) The elastic stress build-up due to magma accumulation at the LNB may also cause an overshoot beyond the LNB through a vertical distance depending upon the dike width and the supply rate from below. The same overshoot may also drive the magma laterally along the LNB (Lister and Kerr 1991). These results are generally confirmed by numerical results, which show that the velocity of a vertically propagating dike initially increases with the dike length, reaches a peak when the dike tip is well above the LNB and then rapidly decreases to zero after the dike propagates further into the region above (Chen et al. 2011). Nevertheless, numerical calculations of dike propagation through layered rocks emphasize that penetration through low density rocks is determined by the local buoyancy only in the swollen nose region, rather than the total buoyancy of the magma column integrated from the source to the tip (Taisne and Jaupart 2009). This feature may strengthen the filtering role of the LNB.

If both the density and the far-field stress terms in the right side of Eq. 3.7 approach zero or become negative (for example at shallow crustal levels and in compressional settings), the initial excess pressure becomes the only driving pressure of the dike. Because of the pressure drop due to viscous drag, the excess pressure may also soon fall to zero and the vertically propagating dike becomes arrested. In fact, while the flow of magma in a dike is driven by a gradient in the fluid pressure, a main force contrasting dike propagation is the pressure drop in the magma due to its viscous resistance to flow. The **viscous pressure drop** ΔP_d in laminar flow along the length $2L$ of a dike with magma viscosity η in a time t since the initiation of the dike is estimated as (Lister and Kerr 1991):

$$\Delta P_d \approx \frac{\eta 4L^2}{w^2 t} \tag{3.8}$$

If the flow is turbulent, Eq. 3.8 underestimates the viscous pressure drop. However, magma flow in thin dikes is likely to be laminar, and most treatments of magma flow within dikes approximate the flow as one-dimensional, laminar, and incompressible. As concerns magma flow, the mean velocity of magma U_d with pressure P within the dike propagating along the x direction (see Fig. 3.13) averaged across the dike thickness is (Rubin 1995b):

$$U_d = -\frac{1}{3\eta} w^2 \frac{dP}{dx} \tag{3.9}$$

This allows defining the **volumetric flow rate** Q_f of a vertically oriented dike as:

$$Q_f = -\frac{b}{3\eta} w^3 \frac{dP}{dx} \qquad (3.10)$$

where b is the dike breadth or width, which is the dimension perpendicular to the direction of propagation (x) and to the half-thickness w of the dike (Fig. 3.13). Thus, the volumetric flow rate of magma into a dike is inversely proportional to magma viscosity and proportional to the dike thickness cubed. This relationship implies an expected time delay, on the order of hours to days for basalts and years for granites, before an incipient dike tip becomes sufficiently pressurized to propagate (Gonnermann and Taisne 2015). The focusing (i.e., channeling) of magma from a linear dike to a central eruptive vent fed by the dike just below the surface is also a consequence of the strong cubic dependence of flow rate on dike thickness expressed by Eq. 3.10. Because most of the flow becomes channelled along the thicker sections of the dike, the rate of magma cooling during ascent will be lowest there, perhaps even promoting melting or erosion of the dike walls. In contrast, thinner sections of the dike will have lower flow rates and faster rates of magma cooling. This will increase the viscosity of the magma, as well as its solidification along the dike walls. The overall result is a progressive reduction of the portion of the dike actively feeding flow, as also discussed in Sect. 7.7.

As the propagation velocity of dikes driven by excess source pressure is inversely proportional to the magma viscosity, dikes filled by viscous magma could also produce significant viscous deformation of the country rock. Under common conditions of excess magma pressure and elastic stiffness of the country rock, the magma-country rock viscosity contrast must exceed 11–14 orders of magnitude for the viscous contribution to dike thickness to be negligible. These large contrasts are met by all basalts and most granites (Rubin 1993). However, very viscous granites (viscosity $\eta \geq 10^8$ Pa s) might have lower viscosity contrasts with regard to hot crustal portions, so that the viscous displacements along the dike wall away from the tip can exceed the elastic displacements by 1–2 orders of magnitude. In this case, a viscous deformation of the country rock occurs. Nonetheless, the resulting intrusions are still narrow, with thickness to length ratios expected to be no larger than 10^{-2}: in fact, even in this case elastic strains continue to dominate near the tip. These theoretical predictions are confirmed by experimental studies and field observations. In particular, experiments suggest that the aspect ratios of dikes around the brittle–ductile transition approach diapir–dike hybrids deviating from dikes in an elastic medium (Fig. 3.17; Sumita and Ota 2011). In the field, an exhumed dike complex in the Scandinavian Caledonides records increasing ductile features (dikes with rounded shapes and mingling between the soft country rock and the intruding mafic magma) approaching the paleo brittle-ductile transition (Kjoll et al. 2019). Similarly, intrusions in the Cascades (northwest North America) and in the North American Cordillera emplaced at depths between 10 and 30 km show a vertically elongated shape, with length to width

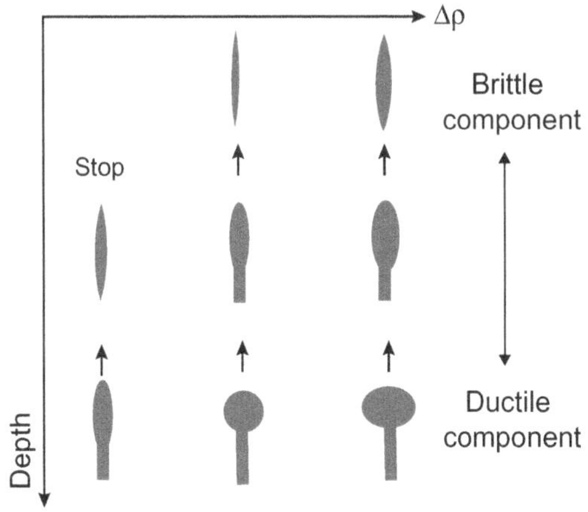

Fig. 3.17 Schematic diagram showing how magma may change its shape as it ascends through the brittle-ductile transition zone. Here three cases for a same magma volume, but different density difference $\Delta\rho$ are shown (modified after Sumita and Ota 2011). Diapir–dike hybrid intrusions may occur at intermediate crustal depths, consistently with what observed in the field

ratios <100, surrounded by country rock deforming by visco-elastic processes, and for which ascent is driven by buoyancy plus regional stress. These features suggest that hybrid mechanisms of ascent, involving both a diapiric and a diking component, are common at mid-crustal levels (Paterson and Miller 1998b; Miller and Paterson 1999).

3.4.3 Thermal Constrains

The above discussion neglects the influence of the temperature on the rheology of the magma and country rock. However, because dikes carry hot magma into colder rock, heat flow must be an important aspect of magma intrusion: while freezing of magma might prevent dikes from erupting, melting of the country rock might alter the composition and properties of the magma. The understanding of the thermal aspects of dike propagation is probably less advanced than that of the mechanical aspects, and only an overview is given here.

Magma cooling and/or solidification and melt against the dike walls are a consequence of the conduction of heat from the magma into the dike walls. The degree to which magma cools depends on the balance between along-dike heat transport by the flowing magma (which depends on the magma propagation velocity) and the across-dike diffusion of heat to the conduit walls (which depends on the gradient in temperature across the magma within the thermal boundary layer close to the dike walls). The ratio between these two competing heat transfer mechanisms leads to the dimensionless parameter Π (Bruce and Huppert 1989; Gonnermann and Taisne 2015):

$$\Pi \approx \frac{w^4 \Delta P}{\alpha \eta 4 L^2} \tag{3.11}$$

where $\Delta P/2L$ is the average pressure gradient driving magma flow within the dike and α is the thermal diffusivity. For small half-thickness of the dike w, magma cooling dominates. Cooler magma will increase the viscosity, which will reduce magma velocity and pressure, implying a further decrease in heat transfer through the dike relative to conductive heat loss to the dike walls. The result will be flow localization, that is, sections of the dike with small Π tend to close. Conversely, sections with large Π remain open (Fig. 3.18; Gonnermann and Taisne 2015, and references therein). In particular, during dike propagation a thermal boundary layer develops closest to the dike walls, where heat transfer is dominated by conduction, as opposed to along-dike convection by magma flow. When the thermal boundary layer thickness is much less than the dike thickness, the frozen margin thickness w_f grows with time as:

$$w_f = 2\lambda(\alpha t)^{1/2} \tag{3.12}$$

where λ is a numerical coefficient which depends upon the initial magma and country rock temperatures, the freezing temperature, the latent heat of crystallization and the heat capacity C_p (Carslaw and Jaegar 1959; Rubin 1995b, and references therein). The time taken for a dike to

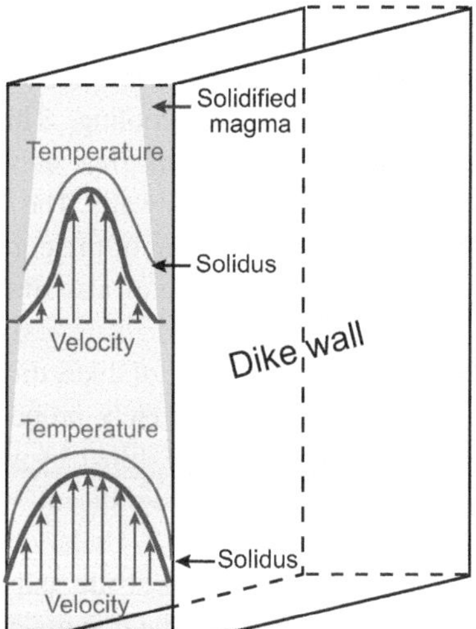

Fig. 3.18 Scheme illustrating the evolution of the velocity and temperature profiles for magma ascending within a dike. As the magma cools against the conduit walls (purple wedges, upper part), viscosity increases and velocity decreases, reducing the heat transport, resulting in a positive feedback between cooling, increasing viscosity and decreasing velocity (modified after Gonnermann and Taisne 2015)

completely solidify is given by (Lister and Kerr 1991):

$$t = \frac{w^2}{\lambda\alpha} \quad (3.13)$$

The short duration of heat transfer (on the order of hours) and small temperature increase (well below that of the magma) in the country rock during the emplacement of a dike under ordinary conditions explain the lack of contact metamorphic (i.e., heating) effects next to many dikes (Delaney and Pollard 1982). An estimate of the distance that dikes may propagate before freezing is given by multiplying the solidification time (Eq. 3.13) by the flow velocity (Eq. 3.9). The resulting propagation distance L_p can be expressed as (Rubin 1995b):

$$L_p = \frac{1}{12\lambda^2}\frac{w^4}{\alpha\eta}\frac{dP}{dx} \quad (3.14)$$

where η is the magma viscosity and dP/dx is the variation in magma pressure along the direction of propagation of the dike. These results show that a dike cannot propagate beyond a critical length, which may be estimated as order of magnitude by Eq. 3.14. More precise estimates are mainly limited by the difficulty in calculating the numerical coefficient λ deriving from models that incorporate both the mechanics of propagation and the thermodynamics of solidification. Considering these limitations, applying this equation to a rhyolite four orders of magnitude more viscous than a basalt suggests that the rhyolite dike would have to be ten times thicker in order to propagate as far. This implies that dikes filled with mafic magma commonly have critical thickness of a very few metres, one order of magnitude lower than that of dikes filled with granitic magma (Lister 1995).

3.4.4 Seismicity Constraints

The propagation of dikes and, more generally of any magma-filled fracture, is usually responsible for seismicity. Indeed, the generated earthquakes often represent the strongest evidence for dike propagation, allowing to constrain dike geometry, kinematics and dynamics, and carrying a wealth of information on the physics of diking. Dike seismicity is briefly considered here to further define the mechanism of propagation of dikes in the upper crust.

Magma emplacement commonly stresses the nearby country rock; these stresses create the fractures and faults which are ultimately responsible for the generation of earthquakes. In particular, most dike-induced seismicity, with frequency typical of tectonic earthquakes, should be interpreted as resulting from slip along suitably aligned existing fractures to the sides of the dike. Dike-induced earthquakes typically have minor to moderate magnitude, not exceeding 6. As a result, the seismicity associated with dike propagation lacks the higher magnitude spectrum typical of tectonic earthquakes. Observations from the 1975–1984 rifting episode at Krafla (Iceland) first highlighted that propagating dikes induce migrating seismicity. However, the migration of seismicity does not necessarily reflect the migration of the dike. Rather, the seismicity is linked to the evolution of the stress field associated with the opening of the dike and to the stress previously stored in the crust. Therefore, the distribution of the recorded dike-induced seismicity reflects the distribution of ambient stresses that are near to failure and does not necessarily reflect the true extent of the dike. This implies that seismicity can also occur on a different time scale to the migration of the dike tip. While the data often hint at a broad equivalence between tip migration and seismicity, there is evidence that the tip is not the only and at times not even the primary source of seismicity (Rubin and Gillard 1988; Rivalta et al. 2015; Amhed et al. 2016). Despite such a possible independence between the extent of a dike and the distribution of its seismicity, studies on several intrusions also highlight a proportional relationship between the average thickness of a dike and the mechanical energy released by its propagation. This feature may help predicting the total seismic moment that will be released by a propagating dike once its thickness is constrained by geodetic data (Bonaccorso et al. 2017).

In the case of a laterally propagating dike, models and observations show that when seismicity is at intermediate depth between the dike top and bottom, earthquakes focus at the leading edge of the dike. Then, as the dike tip advances and the seismogenic volume falls into a stress shadow, seismicity stops. Conversely, when seismicity is above or below the dike, it continues at a decaying rate after the dike tip passes (Segall et al. 2013). While both space–time patterns have been observed in nature, seismic observations from recent diking events in different tectonic settings reveal some additional variability in the earthquake distribution during dike propagation, where even similar events may show important changes. For example, there is a marked difference between the seismicity associated with the lateral propagation of tens of kilometres long dikes emplaced between 2007 and 2009 at Dabbahu (Afar) and that in 2014 at Bardarbunga (Iceland). At Dabbahu the seismicity focused at the lateral tip and at the top of the propagating dikes, with the largest magnitude earthquakes generated by normal faulting above the dikes (Grandin et al. 2011; Belachew et al. 2013). Conversely, at Bardarbunga seismicity focused at the bottom of the propagating dike tip, with earthquakes showing evidence of both left- and right-lateral faulting behind the leading edge of the dike (Fig. 3.19; Agustdottir et al. 2016; Woods et al. 2019).

The kinematics of the faults responsible for seismicity can be detected through the focal mechanisms of the earthquakes, which describe the deformation in the source region (for details see Sect. 8.4.1). Both Dabbahu and Bardarbunga diking events show a consistent pattern of focal mechanisms along the tip of the dike, generally coherent with a local maximum principal stress σ_1 directed outwards along the propagating dike tip and a local minimum principal stress σ_3 perpendicular to the propagating tip of the dike. This stress distribution broadly results in seismicity characterized by normal focal mechanisms, ideally forming conjugate normal faults, on the top of the dike, and by dextral and sinistral mechanisms, ideally forming conjugate strike-slip faults, to the sides of the laterally propagating dike tip (Fig. 3.20).

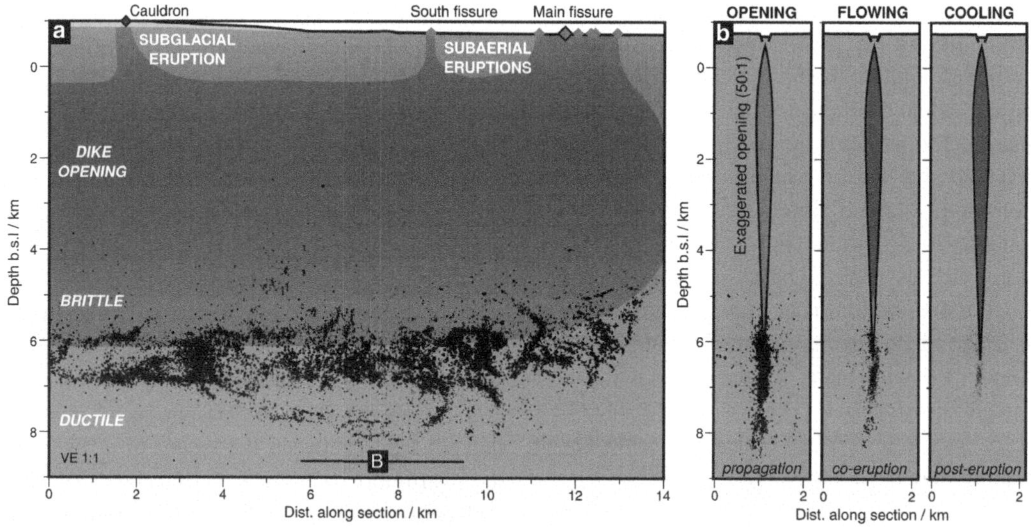

Fig. 3.19 Seismicity of the 2014 Bardarbunga (Iceland) fissure eruption. **a** Dike-parallel cross-section with schematic dike opening shaded and **b** dike-perpendicular cross-section with interpretative cartoon for propagation, eruption and post-eruption phases. Black dots show relocated dike-induced earthquakes. Ice cauldron indicated by dark purple diamond; eruption fissures shown as orange diamonds. (Courtesy of Jennifer Woods; Woods et al. 2019)

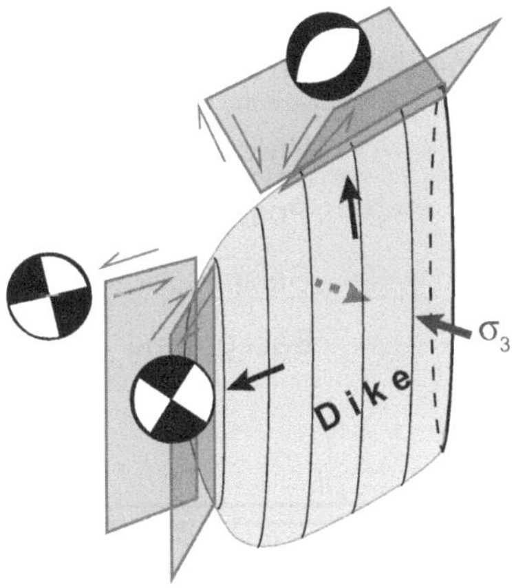

Fig. 3.20 Scheme not to scale summarizing the main fault systems (red rectangles) and seismicity (focal mechanisms highlighted by the beach balls seen in map view; see Sect. 8.4.1 for details) induced at the upper and lateral tips of a propagating dike (orange; local sense of propagation shown by black arrows). Conjugate normal faults predominate on the top of the dike, as observed at Dabbahu (Afar, Ethiopia), whereas conjugate strike slip-faults predominate to the sides, as observed at Bardarbunga (Iceland)

Other diking events presented common features to Dabbahu and Bardarbunga: these include the laterally propagating dike in Miyakejima (Japan) in 2000, with strike-slip events at the front of the shorter dike edge, normal fault events along the upper dike edge and the whole range of intermediate kinematics between these types. Similarly, the dikes propagating at Aden in 2010–2011 and the Southern Red Sea in 2011 induced seismicity with focal mechanisms indicating predominant normal faulting to the front and above the dikes (Passarelli et al. 2015; Amhed et al. 2016; Eyles et al. 2018). In addition to fault-related (double-couple) mechanisms, the seismicity associated with dike emplacement may also involve a fluid related (non-double-couple) component, indicating faulting mechanisms deviating from pure shearing and including a tensile component (see Sect. 8.4.1).

Focal mechanisms may provide information not only on the orientation, shape and pressurization level of the dikes, but also on any variation of these parameters. For example, the study of focal mechanisms associated with diking has allowed obtaining the stress variations before, during and after eruptive events. In the case of the 1992 eruption at Mount Spurr (Alaska, USA), the seismicity highlighted a rotation of the stress field related to the pressurization of a network of dikes, with stress changes corresponding to changes in the state of activity at the volcano (Fig. 3.21a; Roman et al. 2004). In particular, the NE–SW direction of the maximum principal stress locally revealed by the earthquakes prior to the eruption was perpendicular to the NW–SE regional maximum principal stress. After the eruption, the former rotated of $\sim 90°$, becoming parallel to the regional maximum principal stress. A few months later, during a seismic swarm induced by a magmatic intrusion, the local maximum principal stress rotated, assuming again a NE–SW orientation. Finally, in the years after the eruption, the stress direction was again parallel to the NW–SE regional stress. This horizontal rotation of the maximum principal stress likely resulted from the pressurization of a system of interconnected dikes, promoting stress increase in the country rock to the sides of the dikes responsible for seismicity. This feature is consistent with theoretical predictions, as in absence of mechanisms for relaxing the stresses due to intrusions, repeated dike emplacement can raise the dike-perpendicular stress until it is no longer the least principal stress, causing future sheet intrusions to take the form of sills or dikes of opposite orientation (Fig. 3.21b; Rubin 1995b; Vigneresse et al. 1999; Xu and Jonsson 2014).

Some aspects of the patterns of dike-induced seismicity are still poorly understood, as the temporal distribution of seismicity at the inferred advancing and retreating fronts of migrating epicentres of a laterally propagating dike. In fact, an advancing front, usually shown by a convex-upward trend in the case of lateral migration, is commonly trailed by a retreating front delimiting a frame of active seismicity (Fig. 3.22; Rivalta et al. 2015). While the advancing front has generally a simple shape, the retreating front may show a more complex pattern, with variable distance with regard to the advancing front.

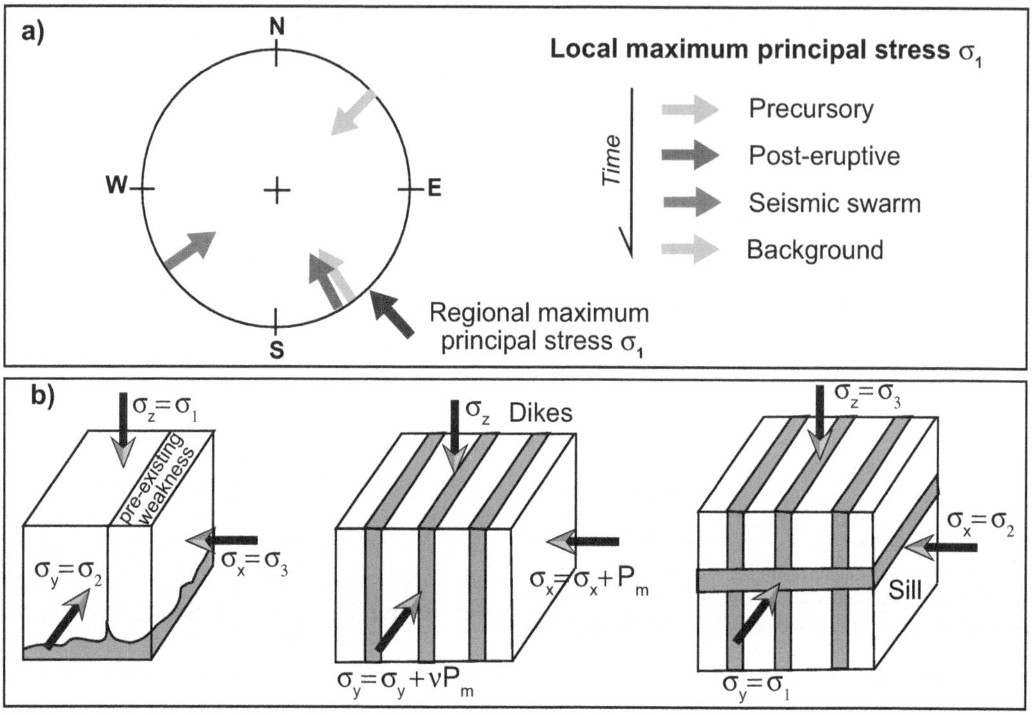

Fig. 3.21 **a** Stereograph showing the orientation of the local maximum principal stress at Mount Spurr (Alaska, USA) during four defined periods with regard to the regional maximum principal stress (modified after Roman et al. 2004). **b** Scheme of stress axes exchange under the influence of magma intrusion in extensional conditions: (left) initially, the opening plane is vertical (*yz*; forming a dike) and perpendicular to the minor stress component (σ_3); (centre) due to the pressure of magma intrusion (P_m), the minor stress increases to $\sigma_3 + P_m$, whereas the intermediate axis increases proportionately to the Poisson coefficient (ν); (right) provided the two horizontal components overcome the lithostatic component, the opening plane becomes horizontal, forming a sill (Vigneresse et al. 1999)

3.5 Dike Arrest

Eroded portions of extinct volcanic areas, as in eastern and western Iceland, show that most dikes become arrested in the crust and do not reach the surface to feed eruptions. Thus, some important mechanism determining dike arrest must exist. Since the possibility to have eruption is determined by the factors either encouraging dike propagation or determining dike arrest, understanding the conditions that lead to the stall of dikes may also allow better forecasting for eruptions.

Several factors may contribute in arresting a propagating dike. Among these is a poor or no contribution from any of the three factors affecting magmatic overpressure and thus dike propagation in Eq. 3.7: magmatic excess pressure, buoyancy and deviatoric stress. Decreasing magmatic excess pressure may be related to a reduced or chocked supply of magma from the source. The release of a limited amount of magma may induce the arrest of a dike or a magma-filled fracture at a finite distance from the source, consistently with observations at volcanoes (Taisne et al. 2011; Galetto et al. 2019). In fact, the smaller the volume of available magma, the more likely is that a dike stalls. There is indeed a minimum magma volume required for eruption as a function of magma buoyancy and source depth. The equation for this minimum volume V_m of magma that allows eruption can be written as:

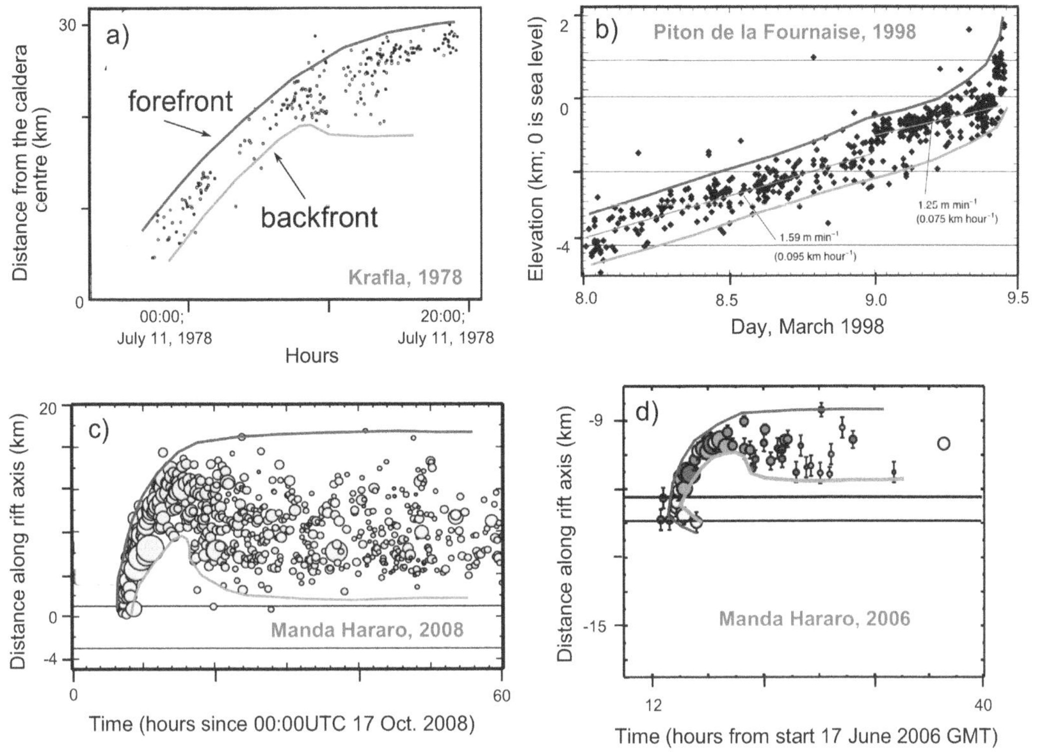

Fig. 3.22 Seismicity patterns during different dike intrusions. **a** Seismicity induced by the 1978 lateral dike propagation at Krafla (Iceland); **b** seismicity induced by the 1998 vertical dike intrusion at Piton de la Fournaise (Reunion Island); **c** seismicity induced by dike 11 of the rifting episode in the Manda Hararo rift segment (Afar, Ethiopia); **d** seismicity induced by the 17 June 2006 dike of the rifting episode in the Manda Hararo rift segment (Afar, Ethiopia; Courtesy of Eleonora Rivalta; after Rivalta et al. 2015)

$$V_m = \left(\frac{z}{\zeta}\right)^4 \frac{(1-\nu)\Delta\rho g}{G} \tag{3.15}$$

where z is the source depth, $\zeta = 2L/b$ (where $2L$ is the dike length and b is the dike width) and G is the shear modulus of the country rock (Taisne et al. 2011). The resulting minimum magma volume that is required for eruption as a function of the depth of the magma source is shown in Fig. 3.23. In addition, any stratification of density will encourage the arrest of magma, so that, even with large volumes of magma available, eruption may be hindered by a shallow and thick low density layer, annulling magma buoyancy (Taisne et al. 2011). In this case, magma may accumulate in a swollen nose region at the base of the low density layer, where magma overpressure is largest and may become large

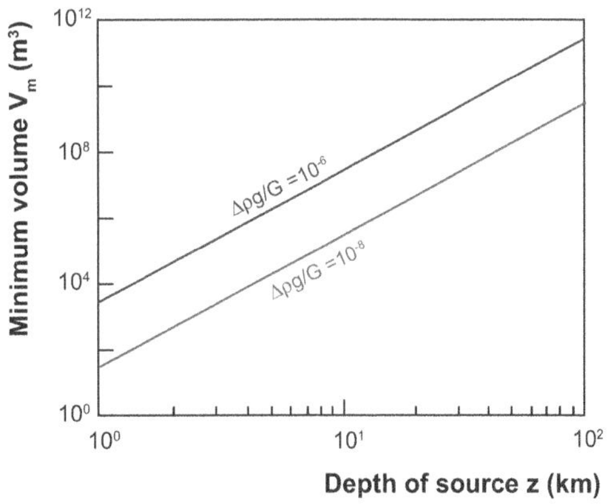

Fig. 3.23 Diagram showing the minimum magma volume V_m that is required for an eruption as a function of the depth of the magma source z. The two lines are drawn for two different values of $\Delta\rho g/G$ (where $\Delta\rho$ varies between 10 and 100 kg/m^3 and G between 10^9 and 10^{10} Pa; modified after Taisne et al. 2011)

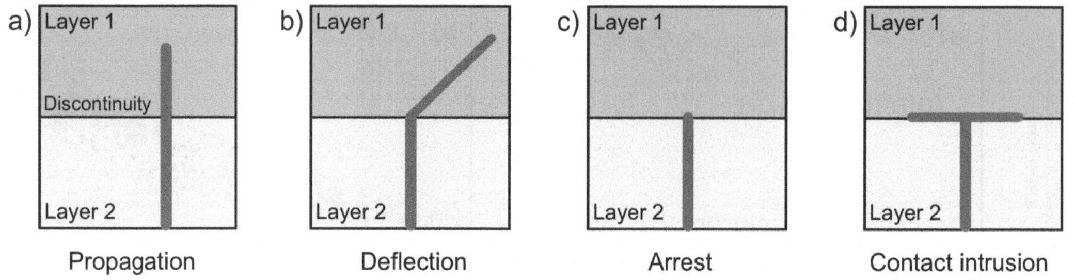

Fig. 3.24 A dike (red line) meeting a discontinuity between two layers with different mechanical properties may propagate in the upper layer (**a**), be deflected in the upper layer (**b**), arrest at the contact (**c**) or intrude (symmetrically or not) along the contact, forming a sill (**d**)

enough to induce horizontal fractures in the dike walls and lateral propagation of a dike or a sill, preventing eruption. Also, magma ascent may be hindered by a compressive stress field in the upper crust, due to either large-scale tectonic forcing or loading by a tall volcanic edifice: this may decrease or nullify the deviatoric stress contribution in Eq. 3.7.

In addition to the factors included in Eq. 3.7, other conditions may promote the stalling of dikes. Among these are the cooling and solidification of the magma within the dike (which has been considered in Sect. 3.4.3) and the presence of pre-existing anisotropies (layering, fractures, faults) within the country rock. In particular, a significant amount of studies have been recently focusing on the role of subhorizontal crustal heterogeneities in controlling the propagation of dikes, promoting their deflection or arrest (Fig. 3.24). As for the deflection of dikes, mathematical simulations suggest that, when an inclined crack crosses a boundary between layers with different rigidity, a sudden change in the dip of the propagating crack occurs. If an inclined crack enters a softer medium, it becomes steeper, while if it enters a harder medium it becomes shallower. If the rigidity contrast is large, the crack may even become arrested forming a horizontal discontinuity along

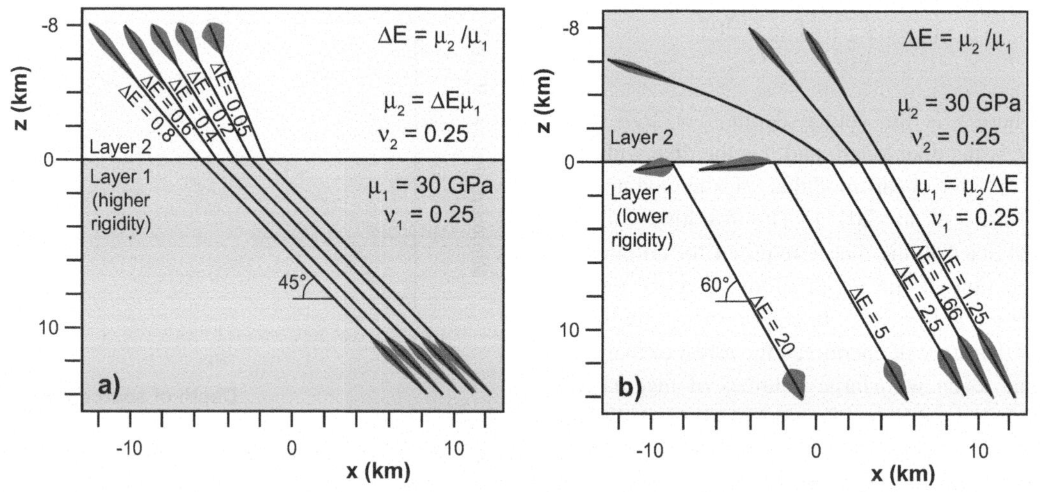

Fig. 3.25 Modelling results showing the deflection of dikes (black lines) propagating through a layered medium with different rigidity contrasts given by $\Delta E = \mu_2/\mu_1$, where μ_2 and μ_1 are the rigidities of the upper and lower layers, respectively. **a** Deflection of dikes propagating at 45° passing from higher to lower rigidity medium; **b** Deflection of dikes propagating at 60° passing from lower to higher rigidity medium. The initial and final shapes of the dikes (in red) are exaggerated by a factor of 500 (modified after Maccaferri et al. 2010)

the interface (Fig. 3.25; Maccaferri et al. 2010). The arrest of dikes due to subhorizontal pre-existing structures can be achieved through two main conditions. A first is related to the presence of a mechanical layering within the country rock, also with layers sharing similar mechanical properties. In this case, any weak contact may open up as a result of dike-induced tensile stresses. This **Cook–Gordon mechanism** is likely to operate primarily at shallow depths, as supported by field observations and numerical models (Fig. 3.26a; Gudmundsson 2011). In particular, when a vertically propagating dike approaches a contact, especially in the uppermost 1–3 km of the crust, the dike-induced tensile stress may open up the contact before the dike-tip reaches it. In this mechanism the tensile strength of the contact determines if the debonding takes place. On meeting such an open contact, a dike may change into sill or become offset. This process may be more effective at a certain distance from the pressure source (as from a magma chamber), as the intensity of the tensile stress associated with a pressured magma chamber falls off rapidly with distance. A second condition for dike arrest due to pre-existing structures is related to the presence of a mechanical layering within the country rock with markedly different mechanical properties, as the stiffness of nearby layers (Gudmundsson 2002, 2006, 2011). Here **stress barriers** may develop along the contact between layers with relevant rigidity contrast, mainly through the rotation of the principal stresses. In fact, along a stress barrier the maximum principal stress σ_1 may become horizontal and the minimum principal stress σ_3 vertical (Fig. 3.26b). Since a dike is an extension fracture, it must propagate parallel to σ_1 and perpendicular to σ_3, thus potentially becoming a sill in correspondence with a stress barrier with vertical σ_3. In addition, abrupt changes in tensile stresses occur between layers of contrasting Young's moduli (**material-toughness** mechanism). These may develop both in passing from stiffer (higher Young's modulus) to softer (lower Young's modulus) layers and from softer to stiffer layers. However, a dike propagating through a softer layer

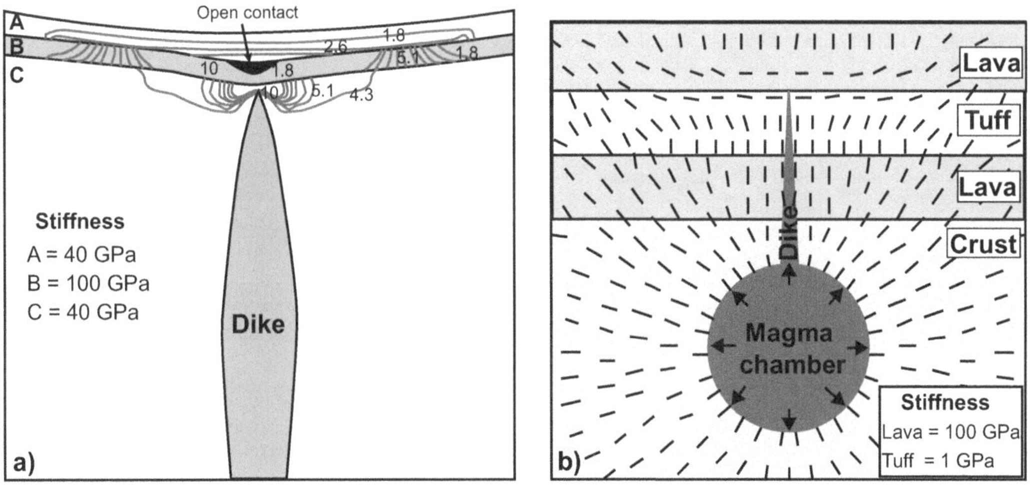

Fig. 3.26 a Result of boundary-element model of the dike-induced tensile stresses (in MPa) and opening of a weak, shallow contact between mechanically dissimilar layers. Only a part of the stresses (in red) propagate through the upper layer, developing an open contact between layers. This model highlights the potential effects of the Cook–Gordon debonding mechanism for propagating dikes (modified after Gudmundsson 2011). **b** Finite-element model of the rotation of the maximum principal stress σ_1 (black lines) at the upper contact between a soft pyroclastic layer and a stiff lava flow. The only loading is the magmatic excess pressure in the chamber. This model suggests that, depending on the magmatic overpressure, the dike would either become deflected into the contact to form a sill or become arrested (modified after Gudmundsson 2011)

towards a stiffer layer would be more likely to deflect than a dike propagating from a stiffer layer towards a softer layer. Therefore, where softer layers, as for example clays or evaporites, underlie stiffer layers, as lava flows or intrusions, sheet and dike deflection or arrest are encouraged. Abrupt changes in stiffness between layers are also commonly associated with weak and partly open contacts and other discontinuities, especially at shallower crustal levels. In these conditions, a soft layer may not receive any significant tensile stress associated with the tip of a propagating dike, thus arresting it. Conversely, when the upper layer has similar stiffness to the lower layer (hosting the dike), there is little tendency for the dike to become deflected into the contact. These behaviours are supported by experiments on fracture propagation (Kavanagh et al. 2006; Menand 2008) and field evidence (Fig. 3.27; Klausen 2006; Gudmundsson 2011; Drymoni et al. 2020).

A further classical problem in dike propagation and arrest regards the role of high angle pre-existing structures, such as fractures and faults, in controlling the path of a dike. While most studies agree on the possibility that dikes penetrate pre-existing structures, the debate is largely on the real importance of this process for dike propagation (e.g., Ruch et al. 2016). A necessary condition for a dike to penetrate a pre-existing fracture (at the length scale of the dike) is that the magma pressure exceeds the ambient compression perpendicular to the fracture. If the magma pressure is only slightly greater than the least principal stress acting at the time of intrusion, then only fractures nearly perpendicular to the least principal stress can be dilated. If the magma

Fig. 3.27 Field examples of terminations and captures of magma-filled fractures. **a** Dike arrested in correspondence of a change in lithology constituted by alternating lava flows and scoria layers, Tenerife (Canary Islands). **b** Lateral sharp termination of a sill, Piton des Neiges (La Reunion Island). **c** Dike partially deflected horizontally in correspondence of the lower margin of a sill, Piton des Neiges (La Reunion Island). Approximate scales are given

pressure exceeds the maximum principal stress, then fractures of any orientation can be dilated (Delaney et al. 1986; Rubin 1995b). However, in this case a dike may not penetrate such a fracture for long. If the fracture is not aligned with the principal stresses and the ratio of shear to opening of the dike walls is not small, the shear stress concentration at the tip fractures intact rock, determining the new direction of propagation. In addition, for a dike to intrude a pre-existing fracture, the effective ambient dike-normal stress should be small compared to the tensile strength of the country rock. This condition is not easily met at mid- to lower-crustal levels, where the

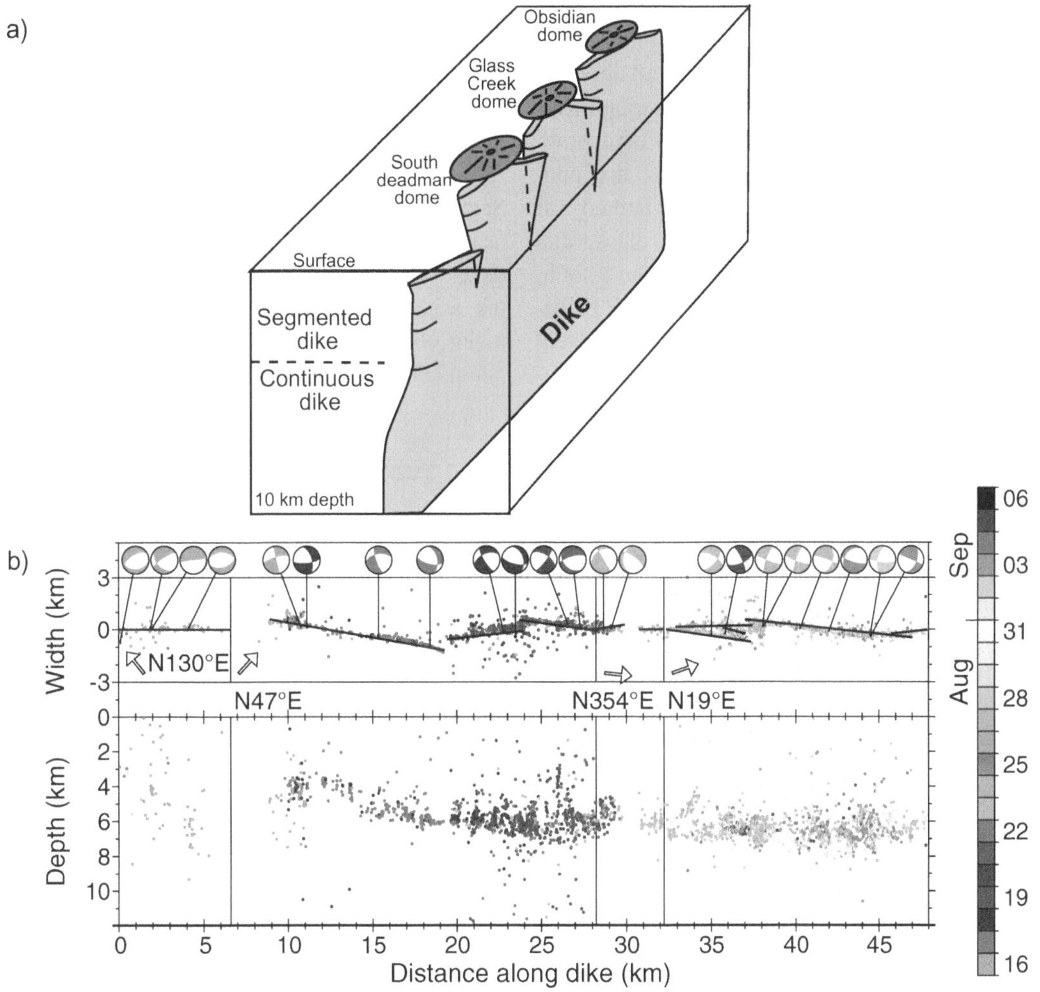

Fig. 3.28 **a** Three-dimensional view of the reconstructed segmented Inyo dike nearby Long Valley caldera (California; USA; modified after Reches and Fink 1988). **b** Seismicity along the Bardarbunga (Iceland) dike, 16 August–6 September 2014. Top: Plan-view of four rotated areas along the dike; arrows indicate geographic north. Coloured dots (for timing see reference bar to the right) denote epicentres and black lines dike segments; focal mechanisms for selected earthquakes are also shown. Bottom: earthquake depths referenced to sea level (modified after Sigmundsson et al. 2015. Image courtesy: Freysteinn Sigmundsson)

high ambient dike normal stress becomes significant, unless there is an abnormally high pore pressure. Studies studies also highlight the importance of the dip of a fault in the fault being captured by a dike, a process mainly occurring for steeply-dipping faults (Ziv et al. 2000; Gaffney et al. 2007). According to other studies, since the fracture strength of the country rock provides negligible resistance to dike propagation, there is little mechanical advantage in intruding along pre-existing fractures. As a consequence, the orientation of a dike will mainly depend upon the direction of the regional stresses (Lister and Kerr 1991). The relative independence of dike propagation from pre-existing fractures is also consistent with the evidence that, in faulted rift zones, most monogenic cones tend to cluster at a small distance (few hundred of m) from a fault, on its foot wall, and not along the fault (Maccaferri et al. 2015). In some cases the presence of faults may even hinder or arrest dike propagation. This has been proposed for Miyakejima (Japan) in 2000, when a pre-existing strike-slip fault, pre-loaded with significant stress, released gradually its energy during the interplay with a dike propagating laterally and perpendicularly to it, confining dike propagation. Quite similarly, the slip along the graben faults developed above a vertically propagating dike in 2009 at Harrat Lunayyir (Saudi Arabia) induced compression above the upper tip of the dike, holding back its further propagation (Maccaferri et al. 2016; Xu et al. 2016). In other cases, the interaction between a dike and nearby parallel pre-existing fractures may induce a change in the trajectory or the decrease in velocity and propagation of the dike, as suggested by analogue models (Le Corvec et al. 2013).

Finally, many dikes are often segmented in section (Fig. 3.9b) or, more commonly, in map view, especially at shallow crustal levels, with the segmentation highlighted by the distribution of seismicity and/or the distribution and orientation of vents forming fissure eruptions. Segmentation commonly develops dikes with en-echelon configuration, where each dike segment is offset, also with slightly different orientation, with regard to the previous one. This is interpreted to result from the rotation of the stresses or, equivalently, from a mixed mode of fracture, including opening (Mode I) and tearing (Mode III) of the dike tip. The angle of the segments relative to the overall strike of the array of en-echelon dikes is related to the rotation angle of the stress and to the extensional to shear stress ratio (Pollard et al. 1982; Rivalta et al. 2015). Segmentation is commonly observed at the tip of extension fractures, as for example when forming plumose structures. Similarly to extension fractures, each segment of a dike may be considered as branching from a single common parent structure at depth, as postulated for the Inyo dike nearby Long Valley caldera (California, USA; Fig. 3.28a; Reches and Fink 1988). Segmentation has been also observed along the nearly 50 km long Bardarbunga (Iceland) dike developed in 2014. In this case, the distribution of the seismicity allowed highlighting at least 6 major offset segments, each with different orientation (Fig. 3.28b; Sigmundsson et al. 2015).

3.6 Summary

This chapter has shown how magma may rise in the crust through different mechanisms, mainly depending upon its composition, temperature, crystallization and strain rate, and the physical (mechanical and thermal) state of the crust. In a lower crust experiencing ordinary thermal conditions, diking and diapirism are both viable processes for the rise of magma. As for the role of magma composition, while the less viscous mafic dikes have little difficulty leaving the source region, the more viscous felsic dikes freeze soon after the magma encounters rock at temperatures below the magma solidus. As a result, the rise through larger diapirs, cooling less rapidly, is more common in felsic magma. Magma temperature has a similar effect, with hotter magma expected to rise through dikes and relatively colder magma rising mainly through diapirs. Also, magma too crystalline (>25%) may not continue the ascent through dikes, but only through diapirs. However, with significant crystallization (>70%) diapiric ascent is prevented and any further magma injection may inflate the

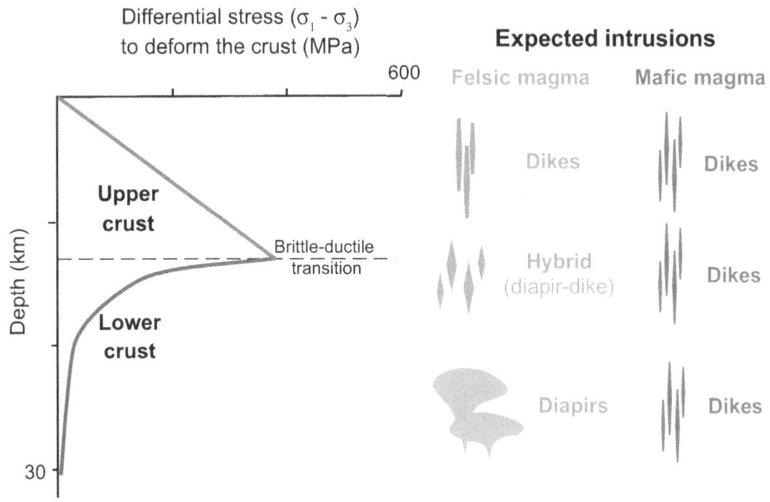

Fig. 3.29 Summary of the expected dominant intrusions accompanying the rise of magma in the upper brittle crust (dikes), in the lower viscous crust (dikes and diapirs) and along the brittle-ductile transition (hybrid and dikes)

diapir through ballooning (Bateman 1985). High strain rates induced by magmatic overpressures (in the range of 10^{-10} to 10^{-12} s^{-1}) promote elastic-brittle deformation, thus enhancing diking (DeSaint Blanquat et al. 2001).

These conditions imply that in the lower crust dikes are commonly associated with the less viscous mafic magmas, whereas diapirs are usually associated with the more viscous felsic magmas. However, hybrid mechanisms of ascent involving thicker dikes (aspect ratio on the order of 10^{-1} to 10^{-2}) may be also found, especially approaching the brittle-ductile transition (Fig. 3.29). In fact, at the brittle-ductile transition there is a rapid increase in the strength of the crustal rocks, so that in the upper brittle crust the ascent of buoyant diapirs is inhibited and the further rise of magma occurs through dikes, independently of the composition. Felsic and/or colder dikes must be thicker to survive freezing and usually propagate to shorter distances. Mafic and/or hotter magmas can propagate through thinner dikes to farther distances. Magma transfer through dike propagation, being widespread in the brittle upper crust, is thus of particular interest for volcano-tectonic processes, especially those leading to eruption. Dike propagation can be conveniently described considering both a solid and a fluid mechanics approach. The former focuses on the elastic behaviour of the country rock and the non-elastic behaviour at the dike tip, where the stress intensity factor competes with the fracture toughness of the country rock. The fluid mechanics approach considers the competition between the factors that allow the propagation of magma (summarised in the concept of magma overpressure) and the resisting factors (as the viscous pressure drop of the magma). For a complete analysis, the thermal interactions between the magma and the country rock, as deriving from conduction of heat, should be also considered. In addition to the factors controlling dike propagation, those controlling dike arrest should be also taken into account: these may be related to an insufficient magmatic overpressure and/or to the presence of mechanical anisotropies in the country rock. Indeed, it is the knowledge of the conditions determining both the propagation and arrest of dikes that allows understanding shallow magma transfer and ultimately help in forecasting eruptions.

3.7 Main Symbols Used

b	Dike width or breadth
C_p	Heat capacity
d_s	Thickness of softened region
E_s	Elastic stiffness of the host rock
g	Acceleration due to gravity
G	Rigidity or shear modulus of the host rock

h	Dike height
K	Stress intensity factor
K_c	Fracture toughness
L	Half-length of the dike
L_p	Propagation distance
P	Magma pressure
P_e	Excess magma pressure
P_s	Tip suction
P_t	Pressure at tip cavity
Q_f	Volumetric flow rate
r	Diapir radius
t	Time
T_u	Thickness of upper layer
U_a	Ascent velocity
U_d	Velocity of magma within the dike
V_m	Minimum volume of magma that allows eruption
w	Half-thickness of dike
w_f	Thickness of frozen margin of dike
z	Source depth
Z_v	Parameter describing the increase in viscosity
α	Thermal diffusivity
$\Delta P/2L$	Pressure gradient
ΔP_d	Viscous pressure drop
ΔP_m	Magmatic overpressure
$\Delta\rho$	Density contrast between magma and host rock
$\Delta\sigma$	Difference between far-field vertical and horizontal principal stresses
ζ	Dimensionless parameter
η	Viscosity
λ	Numerical coefficient
v	Poisson's ratio
Π	Dimensionless parameter
ρ_1	Density of the upper layer
ρ_2	Density of the lower layer
ρ_m	Intrusion density
ρ_r	Country rock density
σ_1	Maximum principal stress
σ_2	Intermediate principal stress
σ_3	Minimum principal stress
σ_h	Far-field horizontal principal stress
σ_n	Ambient compressive stress perpendicular to the dike
σ_v	Far-field vertical principal stress

References

Ágústsdóttir T, Woods J, Greenfield T, Green RG, White RS, Winder T et al (2016) Strike-slip faulting during the 2014 Bardarbunga-Holuhraun dike intrusion, central Iceland. Geophys Res Lett 43:1495–1503

Ahmed A, Doubre C, Leroy S, Kassim M, Keir D, Abayazid A et al (2016) Seafloor spreading event in western Gulf of Aden during the November 2010–March 2011 period captured by regional seismic networks: evidence for diking events and interactions with a nascent transform zone. Geophys J Int 205:1244–1266

Bateman R (1984) On the role of diapirism in the segregation, ascent and final emplacement of granitoid magmas. Tectonophysics 110:211–231

Bateman R (1985) Progressive crystallization of a granitoid diapir and its relationship to stages of emplacement. J Geol 93:645–662

Belachew M, Ebinger C, Cote D (2013) Source mechanisms of dike-induced earthquakes in the Dabbahu-Manda Hararo rift segment in Afar, Ethiopia: implications for faulting above dikes. Geophys J Int 192:907–917

Berner H, Ramberg H, Stephansson O (1972) Diapirism in theory and experiment. Tectonophysics 15:197–218

Biot MA, Odè H (1965) Theory of gravity instability with variable overburden and compaction. Geophysics 30:213–227

Bittner D, Schmeling H (1995) Numerical modelling of melting processes and induced diapirism in the lower crust. Geophys J Int 123:59–70

Bonaccorso A, Aoki Y, Rivalta E (2017) Dike propagation energy balance from deformation modelling and seismic release. Geophys Res Lett 44:5486–5494

Bruce PM, Huppert HE (1989) Thermal control of basaltic fissure eruptions. Nature 342:665–667

Burov E, Jaupart C, Guillou-Frottier L (2003) Ascent and emplacement of buoyant magma bodies in brittle-ductile upper crust. J Geophys Res 108:2177. https://doi.org/10.1029/2002JB001904

Cao W, Kaus BJP, Paterson S (2016) Intrusion of granitic magma into the continental crust facilitated by magma pulsing and dike-diapir interactions: numerical simulations. Tectonics 35:1575–1594

Carslaw HS, Jaeger JC (1959) Conduction: heat in solids. Oxford Clarendon, 2nd edn, 510 pp

Chen Z, Jin ZH, Johnson SE (2011) Transient dike propagation and arrest near the level of neutral buoyancy. J Volcanol Geoth Res 203:81–86

Cruden AR (1988) Deformation around a rising diapir modeled by creeping flow past a sphere. Tectonics 7:1091–1101

Daniels KA, Kavanagh JL, Menand T, Sparks RSJ (2012) The shapes of dikes: evidence for the influence of cooling and inelastic deformation. Geol Soc Am Bull 124:1102–1112

Delaney PT, Pollard DD (1982) Solidification of basaltic magma during flow in a dike. Am Journ Science 282:856:885

Delaney PT, Pollard DD, Ziony JI, McKee EH (1986) Field relations between dikes and joints' emplacement processes and paleostress analysis. J Geophys Res 91:4920–4938

de Saint Blanquat M, Law RD, Bouchez JL, Morgan SS (2001) Internal structure and emplacement of the Papoose Flat pluton: an integrated structural, petrographic and magnetic susceptibility study. Geol Soc Am Bull 113:976–995

Dixon JM (1975) Finite strain and progressive deformation in models of diapiric structures. Tectonophysics 28:89–124

Drymoni K, Browning J, Gudmundsson A (2020) Dyke-arrest scenarios in extensional regimes: insights from field observations and numerical models, Santorini, Greece. J Volcanol Geoth Res 396:106854

Dumond G, Yoshinobu AS, Barnes CG (2005) Midcrustal emplacement of the Sausfjellet pluton, central Norway: ductile flow, stoping, and in situ assimilation. Geol Soc Am Bull 117:383–395

Emerman SH, Marrett R (1990) Why dikes? Geology 18:231–233

England RW (1990) The identification of granitic diapirs. J Geol Soc London 147:931–933

England RW (1992) The genesis, ascent and emplacement of the Northern Arran Granite, Scotland: implications for granitic diapirism. Geol Soc Am Bull 104:606–614

Eyles JHW, Illsley-Kemp F, Keir D, Ruch J, Jónsson S (2018) Seismicity associated with the formation of a New Island in the Southern Red Sea. Front Earth Sci 6:141. https://doi.org/10.3389/feart.2018.00141

Fialko Y, Pearse J (2012) Sombrero uplift above the Altiplano-Puna magma body: evidence of a ballooning mid-crustal diapir. Science 338:250–252

Gaffney EF, Damjanac B, Valentine GA (2007) Localization of volcanic activity: 2. Effects of pre-existing structure. Earth Planet Sci Lett 263:323–338

Galadi Enriquez E, Galindo Zaldivar J, Simancas F, Exposito I (2003) Diapiric emplacement in the upper crust of a granitic body: the La Bazana granite (SW Spain). Tectonophysics 361:83–96

Galetto F, Bagnardi M, Acocella V, Hooper A (2019) Noneruptive unrest at the caldera of Alcedo Volcano (Galápagos Islands) revealed by InSAR data and geodetic modeling. J Geophys Res 124. https://doi.org/10.1029/2018JB017103

Glazner AF, Miller DM (1997) Late-stage sinking of plutons. Geology 25:1099–1102

Glazner AF, Bartley JM (2006) Is stoping a volumetrically significant pluton emplacement process? Geol Soc Am Bull 118:1185–1195

Glazner AF, Bartley JM (2008) Reply to comments on "is stoping a volumetrically significant pluton emplacement process?" Geol Soc Am Bull 120:1082–1087

Gonnermann H, Taisne B (2015) Magma transport in dikes. In: Sigurdsson H, Houghton B, McNutt S, Rymer H, Stix J (eds) The encyclopedia of volcanoes, 2nd edn. Elsevier Academic Press, pp 215–224

Grandin R, Jacques E, Nercessian A, Ayele A, Doubre C, Socquet A et al (2011) Seismicity during lateral dike propagation: insights from new data in the recent Manda Hararo–Dabbahu rifting episode (Afar, Ethiopia). Geochem Geophys Geosyst 12: Q0AB08. https://doi.org/10.1029/2010GC003434

Gressier JB, Mourgues R, Bodet L, Matthieu JY, Galland O, Cobbold P (2010) Control of pore fluid pressure on depth of emplacement of magmatic sills: an experimental approach. Tectonophysics 489:1–13

Gudmundsson A (1990) Emplacement of dikes, sills and crustal magma chambers at divergent plate boundaries. Tectonophysics 176:257–275

Gudmundsson A, Marinoni LB, Marti J (1999) Injection and arrest of dykes: implications for volcanic hazards. J Volcanol Geoth Res 88:1–13

Gudmundsson A (2002) Emplacement and arrest of sheets and dykes in central volcanoes. J Volcanol Geoth Res 116:279–298

Gudmundsson A (2006) How local stresses control magma-chamber ruptures, dyke injections, and eruptions in composite volcanoes. Earth Sci Rev 79:1–31

Gudmundsson A (2011) Rock fractures in geological processes. Cambridge University Press 569 pp

Hawkins DP, Wiebe RA (2004) Discrete stoping events in granite plutons: a signature of eruptions from silicic magma chambers? Geology 32:1021–1024

He B, Xu Y, Paterson S (2009) Magmatic diapirism of the Fangshan pluton, southwest of Beijing, China. J Struct Geol 31:615–626

Henderson ST, Pritchard ME (2013) Decadal volcanic deformation in the Central Andes Volcanic Zone revealed by InSAR time series. Geochem Geophys Geosyst 14:1358–1374

Jackson MPA, Talbot CJ (1986) External shapes, strain rates, and dynamics of salt structures. Geol Soc Am Bull 97:305–323

Jaeger JC (1969) Elasticity, Fracture and Flow, with engineering and geological applications. Springer Netherlands 268 pp, https://doi.org/10.1007/978-94-011-6024-7

Jellinek AM, De Paolo DJ (2003) A model for the origin of large silicic magma chambers: precursors of caldera-forming eruptions. Bull Volcanol 65:363–381

Kavanagh JL, Menand T, Sparks RSJ (2006) An experimental investigation of sill formation and propagation in layered elastic media. Earth Planet Sci Lett 245:799–813

Kjoll HJ, Galland O, Labrousse L, Andersen TB (2019) Emplacement mechanisms of a dyke swarm across the brittle-ductile transition and the geodynamic implications for magma-rich margins. Earth Planet Sci Lett 518:223–235

Klausen MB (2006) Geometry and mode of emplacement of dike swarms around the Birnudalstindur igneous centre, SE Iceland. J Volcanol Geoth Res 151:340–356

Krumbholz M, Hieronymous CF, Burchardt S, Troll VR, Tanner DC, Friese N (2014) Weibull-distributed dyke thickness reflects probabilistic character of host-rock strength. Nat Commun 5:3272. https://doi.org/10.1038/ncomms4272

Le Corvec N, Menand T, Lindsay J (2013) Interaction of ascending magma with pre-existing crustal fractures in monogenetic basaltic volcanism: an experimental approach. J Geophys Res 118:968–984

Lister JR, Kerr RC (1991) Fluid-mechanical models of crack propagation and their application to magma transport in dikes. J Geophys Res 96:10049–10077

Lister JR (1995) Fluid-mechanical models of the interaction between solidification and flow in dykes. In: Baer, Heimann (eds) Physics and chemistry of dykes. Balkema Rotterdam, pp 115–124. ISBN 9054105518

Maccaferri F, Bonafede M, Rivalta E (2010) A numerical model of dyke propagation in layered elastic media. Geophys J Int 180:1107–1123

Maccaferri F, Acocella V, Rivalta E (2015) How the differential load induced by normal fault scarps controls the distribution of monogenic volcanism. Geophys Res Lett 42. https://doi.org/10.1002/2015GL065638

Maccaferri F, Rivalta E, Passarelli L, Aoki Y (2016) On the mechanisms governing dike arrest: insight from the 2000 Miyakejima dike injection. Earth Planet Sci Lett 434:64–74

Mahon KI, Harrison TM, Drew DA (1988) Ascent of a granitoid diapir in a temperature varying medium. J Geophys Res 93:1174–1188

Marsh BD (1982) On the mechanics of igneous diapirism, stoping, and zone melting. Am J Sci 282:808–855

Menand T (2008) The mechanics and dynamics of sills in elastic layered media and their implications for the growth of laccoliths. Earth Planet Sci Lett 267:93–99

Miller RB, Paterson SR (1999) In defense of magmatic diapirs. J Struct Geol 21:1161–1173

Molyneux SJ, Hutton DHW (2000) Evidence for significant granite space creation by the ballooning mechanism: The example of the Ardara pluton, Ireland. Geol Soc Am Bull 112:1543–1558

Passarelli L, Rivalta E, Cesca S, Aoki Y (2015) Stress changes, focal mechanisms, and earthquake scaling laws for the 2000 dike at Miyakejima (Japan). J Geophys Res 120:4130–4145

Paterson SR, Vernon RH (1995) Bursting the bubble of ballooning plutons: a return to nested diapirs emplaced by multiple processes. Geol Soc Am Bull 107:1356–1380

Paterson SR, Fowler TK (1996) Local displacement of diapir contacts and its importance to pluton emplacement study: discussion. J Struct Geol 18:711–712

Paterson SR, Miller RB (1998a) Stoped blocks in plutons: paleo-plumb bobs, viscometers, or chronometers? J Struct Geol 20:1261–1272

Paterson SR, Miller RB (1998b) Mid-crustal magmatic sheets in the Cascades Mountains, Washington: implications for magma ascent. J Struct Geol 20:1345–1363

Paterson SR, Farris DW (2006) Downward host rock transport and the formation of rim monoclines during the emplacement of Cordilleran batholiths. Trans Roy Soc Edinburgh Earth Sci 97:397–413

Paterson SR, Pignotta GS, Farris D, Memeti V, Miller RB, Vernon RH et al (2008) Is stoping a volumetrically significant pluton emplacement process? Discussion. Geol Soc Am Bull 120:1075–1079

Pollard DD, Segall P, Delaney PT (1982) Formation and interpretation of dilatant echelon cracks. Geol Soc Am Bull 93:1291–1303

Ramberg H (1981) Gravity, deformation and the Earth's crust in theory, experiments and geological applications. Academic Press, London

Ramsay JG (1989) Emplacement kinematics of a granite diapir: the Chindamora batholith, Zimbabwe. J Struct Geol 11:191–209

Reches Z, Fink J (1988) The mechanism of intrusion of the Inyo Dike, Long Valley Caldera, California. J Geophys Res 93:4321–4334

Rivalta E, Taisne B, Bunger AP, Katz RF (2015) A review of mechanical models of dike propagation: schools of thought, results and future directions. Tectonophysics 638:1–42

Roman DC, Moran SC, Power JA, Cashman KV (2004) Temporal and spatial variation of local stress fields before and after the 1992 Eruptions of Crater Peak Vent, Mount Spurr Volcano, Alaska. Bull Seismol Soc Am 94:2366–2379

Rosenberg CL, Berger A, Schmid SM (1995) Observations from the floor of a granitoid pluton: inferences on the driving force of final emplacement. Geology 23:443–446

Rubin AM, Gillard D (1988) Dike-induced earthquakes: theoretical considerations. J Geophys Res 103:10017–10030

Rubin AM (1993) Dikes versus diapirs in viscoelastic rock. Earth Planet Sci Lett 119:641–659

Rubin AM (1995a) Getting granite dikes out of the source region. J Geophys Res 100:5911–5929

Rubin AM (1995b) Propagation of magma-filled cracks. Annual Rev Earth Planet Sci 23:287–336

Ruch J, Wang T, Xu W, Hensch M, Jonsson S (2016) Oblique rift opening revealed by reoccurring magma injection in central Iceland. Nat Commun 7:12352. https://doi.org/10.1038/ncomms12352

Sakuma S, Kajiwara T, Nakada S, Uto K, Shimizu H (2008) Drilling and logging results of USDP-4—penetration into the volcanic conduit of Unzen Volcano, Japan. J Volcanol Geoth Res 175:1–12

Segall P, Llenos AL, Yun S-H, Bradley AM, Syracuse EM (2013) Time-dependent dike propagation from joint inversion of seismicity and deformation data. J Geophys Res 118:5785–5804

Sigmundsson F, Hooper A, Hreinsdóttir S, Vogfjörd KS, Ófeigsson BG, Heimisson ER et al (2015) Segmented lateral dyke growth in a rifting event at Bardarbunga volcanic system, Iceland. Nature 517:191–195

Sumita I, Ota Y (2011) Experiments on buoyancy-driven crack around the brittle–ductile transition. Earth Planet Sci Lett 304:337–355

Taisne B, Jaupart C (2009) Dike propagation through layered rocks. J Geophys Res 114:B09203

Taisne B, Tait S, Jaupart C (2011) Conditions for the arrest of a vertical propagating dyke. Bull Volcanol 73:191–204

Turcotte DL, Schubert G (1982) Geodynamics: application of continuum physics to geological problems. John Wiley, NY, 450 pp

Urbani S, Acocella V, Rivalta E (2018) What drives the lateral versus vertical propagation of dikes? Insights from analogue models. J Geophys Res 123. https://doi.org/10.1029/2017JB015376

Vigneresse JL (1995) Crustal regime of deformation and ascent of granitic magma. Tectonophysics 249:187–202

Vigneresse JL, Tikoff B, Améglio L (1999) Modification of the regional stress field by magma intrusion and formation of tabular granitic plutons. Tectonophysics 302:203–224

Wada Y (1994) On the relationship between dike width and magma viscosity. J Geophys Res 99:17743–17755

Weinberg RF, Podladchikov Y (1994) Diapiric ascent of magmas through power law crust and mantle. J Geophys Res 99:9543–9559

Woods J, Winder T, White RS, Bransdottir B (2019) Evolution of a lateral dike intrusion revealed by relatively-relocated dike-induced earthquakes: the 2014–15 Bárdarbunga-Holuhraun rifting event, Iceland. Earth Planet Sci Lett 506:53–63

Yoshinobu AS, Barnes CG (2008) Is stoping a volumetrically significant pluton emplacement process? Discussion. Geol Soc Am Bull 120:1080–1081

Xu W, Jónsson S (2014) The 2007–8 volcanic eruption on Jebel at Tair island (Red Sea) observed by satellite radar and optical images. Bull Volcanol 76:795

Xu W, Jónsson S, Ruch J, Aoki Y (2016) The 2015 Wolf volcano (Galápagos) eruption studied using Sentinel-1 and ALOS-2 data. Geophys Res Lett 43:9573–9580

Zak J, Paterson SR (2006) Roof and walls of the Red Mountain Creek pluton, eastern Sierra Nevada, California (USA): implications for process zones during pluton emplacement. J Struct Geol 28:575–587

Zak J, Vemer K, Johnson K, Schwartz JJ (2012) Magma emplacement process zone preserved in the roof of a large Cordilleran batholith, Wallowa Mountains, northeastern Oregon. J Volcanol Geoth Res 227–228:61–75

Ziv A, Rubin AM, Agnon A (2000) Stability of dike intrusion along preexisting fractures. J Geophys Res 105:5947–5961

Magma Emplacement and Accumulation: From Sills to Magma Chambers

4

4.1 Introduction

The previous chapter has discussed the rise of magma throughout the crust. This fourth chapter focuses on the arrest, emplacement and accumulation of magma. These are indeed primary and widespread processes at volcanoes, as most of the rising magma remains stalled in the crust, with only a fraction being erupted (approximately one tenth; e.g., Shaw 1985; White et al. 2006). Moreover, the accumulated magma may develop magma chambers, whose dynamics can be detected to define the state of active volcanoes, also providing a warning for forecasting any impending eruption.

This chapter describes the processes associated with the shallow emplacement and accumulation of magma, progressing from elementary intrusions (sills) to larger and more complex intrusions (laccoliths, bysmaliths, lopoliths, plutons). Finally, the specific conditions leading to the development of magma reservoirs and magma chambers, considering also dike nucleation herein, are discussed.

The main aims of this chapter are thus to define the conditions for the:

- Arrest, emplacement and accumulation of magma in the crust;
- Development of the intrusions characterizing magma emplacement and accumulation;
- Formation of magma chambers and the related nucleation of dikes.

4.2 General Features

As anticipated in Sect. 3.5, several factors may hinder the rise or propagation of magma in the crust. These factors may cause magma **arrest** (i.e., magma stops rising, maintaining the same shape as during ascent, as a terminating dike; Fig. 3.27a), eventually leading to its **emplacement** (magma stops rising, generating an intrusion with different shape, as dike-fed sills or laterally expanding diapirs) and **accumulation** (prolonged, usually incremental, emplacement of magma within the same crustal volume).

In particular, magma may stop ascending:

- with limited or discontinuous supply from the source;
- with low-density crustal layers, which may decrease or nullify magma buoyancy;
- with low far-field (regional) deviatoric stress, which hinders dike propagation;
- if highly viscous, a condition determined by its temperature, composition and crystal content;

V. Acocella, *Volcano-Tectonic Processes*, Advances in Volcanology,
https://doi.org/10.1007/978-3-030-65968-4_4

- with pre-existing crustal anisotropies generating debonding contacts, as layering and fractures;
- with mechanical heterogeneities within the crust promoting stress barriers and rotations, as layers with different stiffness.

These factors may not necessarily have the same weight in interrupting magma ascent, Parametric modelling studies allow defining the relative importance of the factors controlling magma arrest, suggesting a possible hierarchy (Kavanagh et al. 2006; Menand et al. 2010; Chanceaux and Menand 2014; Urbani et al. 2018; Sili et al. 2019). The most important factor determining the arrest of magma is probably related to its internal resistance to flow, which includes the viscosity, thermal state, composition, crystal fraction and flux rate of the magma. Another important condition to arrest dikes is a strong rigidity contrast (i.e., a stiffer layer overlying a less stiff one) within the country or host rock. Other important, through probably subordinate, factors are the density variations within the host rock and the far-field, or regional, stresses.

These factors do not only determine whether magma may rise or arrest in the crust but, in case of arrest, also contribute in controlling the shape and size of the stalled magma. In fact, different types of intrusions (including sills, laccoliths, bysmaliths, lopoliths, plutons, batholiths) may accompany magma emplacement and accumulation. Despite the variability in shape and size, some features common to these intrusions can be observed. The building unit from which the intrusions accompanying magma emplacement and accumulation commonly originate is a sill. Sills, the focus of Sect. 4.3, are a simple and common way to emplace magma with extremely variable size at any crustal level. An isolated pulse of magma may form a single sill, whereas incremental pulses may form multiple sills, stacked one above the other. If stacked sills inflate and/or amalgamate, thicker and larger intrusions may develop, as laccoliths, bysmaliths or lopoliths, which are discussed in Sect. 4.4. These may also have extremely variable dimensions, with the largest size commonly found in plutons, which are often with subhorizontal or tabular shape, resembling laccoliths (Petford et al. 2000; Cruden and McCaffrey 2001). Plutons, the focus of Sect. 4.5, store significant volumes of mafic and, mostly, felsic magma in the crust, especially when assembled to form larger batholiths. Many, if not most, plutons have not fed any eruption, even if emplaced at shallow depths, as for example the ~4 km deep Larderello granite (Italy), active for more than 4 Ma. However, several plutons have also fed eruptions, thus acting as magma chambers nucleating dikes, which are the focus of Sect. 4.6. These magma chambers and magma reservoirs form in the upper bound of the inferred magma supply rates for plutons, as discussed in Sect. 4.7.

4.3 Sills

4.3.1 Overview

Sills are magma-filled fractures concordant with the layers and bedding of the host rock. Therefore, in young sedimentary basins or active volcanic areas sills are subhorizontal or gently dipping. Sills are usually subcircular to elliptical in map view, with the ellipticity mainly controlled by the direction of the feeder dike. Sills are much thinner than long and wide, with their thickness ranging from less than a metre to a few kilometres and their length ranging from a few metres to several hundreds of kilometres (Airoldi et al. 2012): accordingly, their thickness to length aspect ratio clusters between 10^{-1} and 10^{-3}. Large sills may store a significant amount of magma, contributing to crustal thickening. Large sills may also promote lateral magma transfer below volcanoes, even explaining the location of distal eccentric vents. Larger (and thicker) sills, if shallow, can also uplift the overburden and dome the Earth's surface.

Most evidence for sills derives from eroded crustal portions, as below extinct volcanoes, in the roots of magmatic arcs or in sedimentary basins, from field data and geophysical imaging (mainly high resolution seismic profiles). For

Fig. 4.1 **a** Series of basaltic sills, partly stacked, intruded in a succession of horizontal lavas and volcano-sedimentary deposits, southeast Iceland (photo courtesy Alessandro Tibaldi); **b** The Little Minch dolerite sills of Trotternish peninsula, northern Skye Island (Scotland). The thicker, subhorizontal sill crops out in the upper part of the rock cliff and is characterised by widespread columnar joints; the sill intruded Jurassic sediments. Photo courtesy Alessandro Tibaldi

example, large sills, ~100 km long, ~60 km wide and ~200 m thick, have been seismically imaged in the lower crust of the northern North Sea (Wrona et al. 2019). In the field, the distinction between a sill and a lava flow may not be straightforward, especially within a lava pile (Fig. 4.1; Pasquarè and Tibaldi 2007; Famin and Michon 2010). Diagnostic features to distinguish a sill include heated contacts (or contact metamorphism) along the bottom of the upper unit and top of the lower unit (conversely to a lava flow, heating only the top of the lower unit), the lack of vesicles, as the higher pressure of emplacement prevents the escape of gases (conversely, lava flows show widespread vesicles, especially towards their top) and the lack of any weathering on the upper portion (lava flows often show some weathering).

Sills frequently consist of simple subplanar structures (concordant sills) splaying towards the periphery as multiple, slightly offset segments connected to the feeder. Any sill can be accompanied by stacked systems, with multiple intrusions emplaced one above the other, without necessarily involving distinct episodes of injection (Chanceaux and Menand 2016). Sometimes sills consist of complex structures, partly concordant and partly discordant, as staircase and saucer-shaped sills (Fig. 4.2). **Staircase sills** consist of multiple sills emplaced at different levels within a succession, each connected by discordant magmatic sheets. **Saucer-shaped sills** are symmetric intrusions with a central concordant sill whose periphery turns into an inward dipping discordant sheet, which may further flatten outwards. Saucer-shaped sills likely develop from the growth of a sill which spreads laterally and domes the overburden, promoting stresses along the dome edge. As a result, the sill tips propagate initiating the outer inclined portion

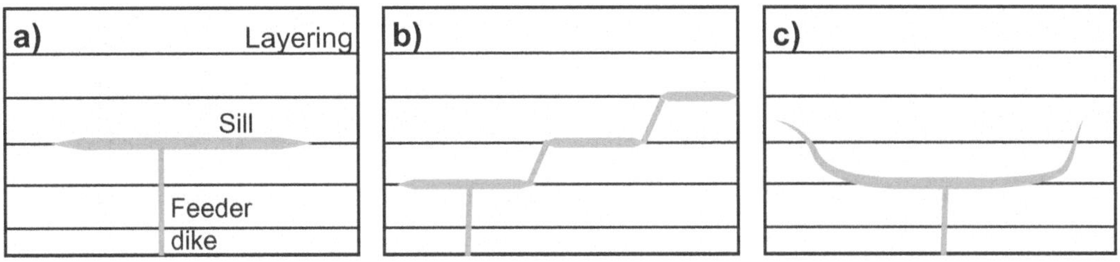

Fig. 4.2 Types of sills: concordant sill **a**, and partly discordant sills: staircase-like sill **b** and saucer-shaped sill **c**

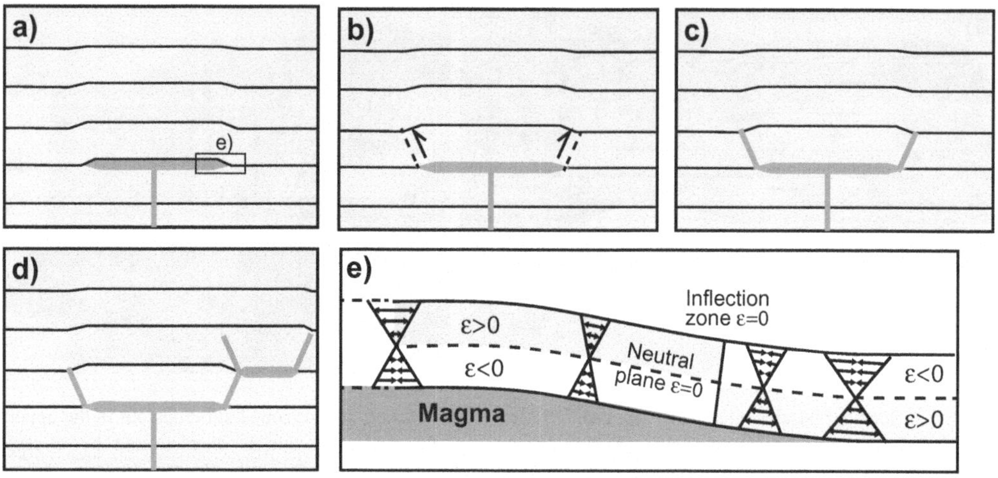

Fig. 4.3 Formation of a saucer-shaped sill: stresses focus at the tip of the early emplaced sill **a**, which may experience shear **b** and promote the propagation of magmatic sheets **c**, eventually feeding a successive shallower saucer-shaped sill **d** Radial strain ε along a plate bending above a sill **e**: along the neutral plane and at the inflection point, $\varepsilon = 0$; white areas correspond to compression ($\varepsilon < 0$), whereas grey areas correspond to stretching ($\varepsilon > 0$). Modified after Pollard and Johnson (1973)

of the sheet with a combination of tensile and shear failure, although pre-existing structures may be also reactivated (Fig. 4.3; Haug et al. 2017; Coetzee and Kisters 2018; Galland et al. 2018; Siregar et al. 2019). Field observations and modelling results show linear relationships

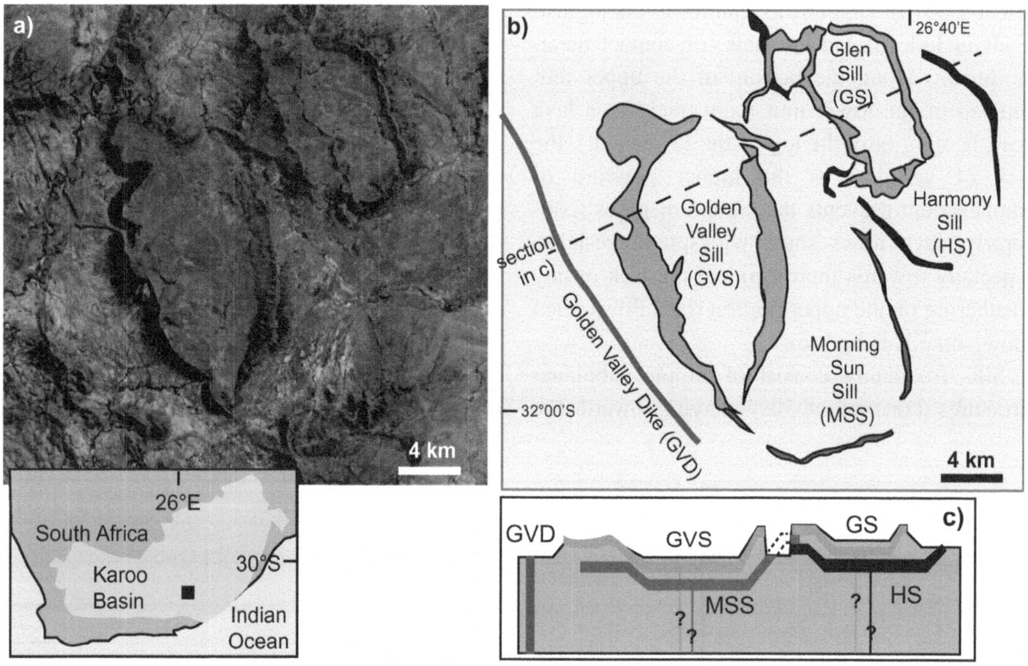

Fig. 4.4 Saucer-shaped sills in the Karoo Basin (South Africa): **a** satellite image (from Google Earth) and related location (black square) in the bottom inset; **b** associated simplified geological map; **c** schematic section view of the sills along the line in **b**. Modified after Galland et al. (2018)

between the depth of emplacement and size of the saucer-shaped sills. Saucer-shaped sills are common in sedimentary basins, as offshore in the Norwegian, North Sea and Faeroe-Shetland (North Atlantic) basins and onshore in the Karoo Basin (South Africa; Fig. 4.4; Thomson 2007; Polteau et al. 2008). Saucer-shaped sills in the Faeroe-Shetland Basin show radial magma flow propagation patterns, consistently with field observations in the Karoo Basin. Also, these sills do not need to be fed by dikes, with the sudden upward flux of magma from the inclined periphery of a sill forming a new saucer-shaped intrusion. Detailed 3-D seismic imaging of the plumbing system of Kora volcano in the Taranaki Basin, offshore New Zealand and the Ceduna Sub-basin, offshore southern Australia, reveals the possible relationships between saucer-shaped sills and central volcanoes (Magee et al. 2013; Morley 2018). At Kora, the shallow (<8 km depth) plumbing system is dominated by stacked saucer-shaped sills and laccoliths distributed in an oval shape around the volcano. The deeper sills are usually wider, and the shallower ones, closer to the volcano, form a concentric complex that dips away from the volcano. At Ceduna, shield volcanoes usually overlie the sills tips, suggesting that the eruption phases correspond to the incremental emplacement of magma pulses within the sills. Volcanogenic hydrothermal vents are also associated with the sill tips, promoted by fractures within the host rock. These cases confirm that pre-eruptive deformation patterns may be offset from the eruption site, as at active calderas (Rivalta et al. 2019).

Growing sills are often inferred to inflate active volcanoes, as at Campi Flegrei (Italy), where repeated sill emplacement has been recently promoting several episodes of uplift and seismicity (Amoruso et al. 2014; Macedonio et al. 2014; D'Auria et al. 2015). The recurrent emplacement of shallow sills has been also detected in the years preceding the 2010 eruption of Eyjafjallajokull (Iceland), which caused disruption of the air traffic in the northern hemisphere, whereas sill contraction has been observed during the 2008 eruption of Alu-Dalafilla, Afar (Fig. 4.5; Sigmundsson et al. 2010; Pagli et al. 2012). Larger sill complexes of compositionally evolved magma constitute wide reservoirs at 5 to 18 km depths beneath the calderas of Yellowstone (Wyoming, USA) and Long Valley (California, USA; Jiang et al. 2018).

Sills are studied also for their potential economic relevance, as providing the ideal conditions for hydrocarbons production in sedimentary basins, including maturation, reservoir, trap and seal. In fact, the heat of the intruded magma maturates the organic matter in the surrounding sediments. Once the intrusion cools, it becomes highly fractured, providing a reservoir. The intruded magma, especially if forming stacked sills, may dome and fracture the nearby overburden, forming a trap in a medium with locally increased permeability. The less deformed upper portion of the overburden, if consisting of sediments with limited permeability, then acts as seal.

4.3.2 Emplacement Conditions

Analogue models, intruding air or water within homogeneous or layered gelatine, have highlighted the main factors controlling the formation of sills from upward propagating dikes, as well as their relative importance, or hierarchy (Kavanagh et al. 2006; Chanceaux and Menand 2014; Sili et al. 2019). The most important factors controlling sill formation are likely related to the flux and thermal state of the intruding magma, which appear more important than the presence of any interface in the host rock, that would favour but not necessarily promote sill formation. The magma flux and thermal state can be expressed in terms of injection rate and injection temperature of an upward propagating dike in layered medium, both defining distinct experimental behaviours (Fig. 4.6a; Chanceaux and Menand 2014). In particular, domains with highest injection temperatures and rates contrast magma solidification, promoting the piercing of an upward propagating dike through the interface between layers, whereas domains with lowest injection temperatures and rates do not promote dike

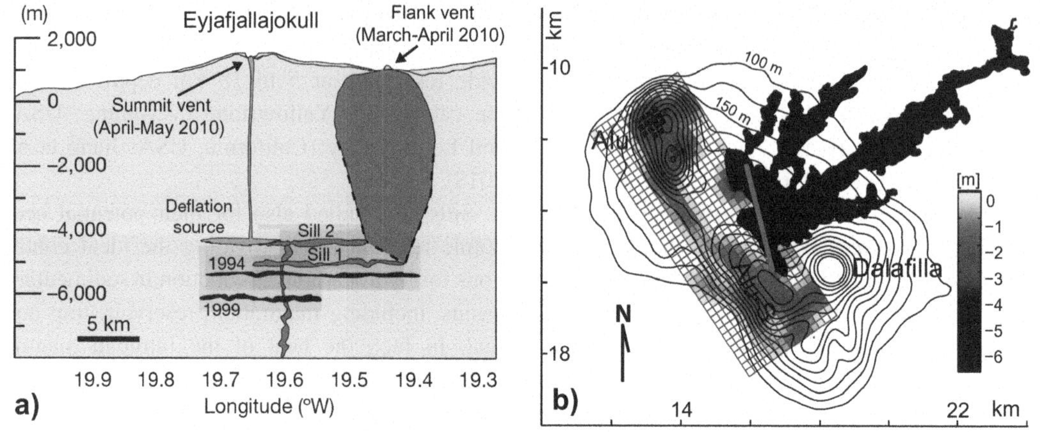

Fig. 4.5 Examples of active sills beneath volcanoes. **a** Schematic east–west cross-section across the summit area of Eyjafjallajokull volcano (Iceland), showing the inferred location of the sill-like magmatic sources responsible for the 2010 eruption plotted at their best-fit depth (vertical exaggeration by a factor of 2); grey shaded background indicates source depth uncertainties (95% confidence), which overlap (modified after Sigmundsson et al. 2010). **b** Plan view of the distributed sill contraction during the 2008 fissure eruption (red line) at Alu-Dalafilla volcanoes (Afar, Ethiopia), as derived from the inversion of InSAR data, overlaid by topographic contours and the lava flow. (In black; modified after Pagli et al. 2012; image courtesy of Carolina Pagli)

propagation. Sills form only within a restricted domain of intermediate injection temperatures and rates, responsible for moderate solidification.

Once a sill has formed, the thermal state of the intrusion plays an important role also on its propagation and growth. High injection temperatures and fluxes produce limited solidification effects, and the sill propagates with regular and smooth surface. Conversely, low injection temperatures and fluxes promote solidification, with the sill propagating discontinuously, displaying folds, ropy structures and lobes (Chanceaux and Menand 2016). The thermal state of the host rock is also important. Theoretical models suggest that magma can delay freezing and develop a sill if the country rock temperature is close to the magma solidus temperature and the vertical thermal gradient is very small, although the existence of a thermal lag between the sill and the host rock, filled with magmatic gases, delays magma solidification and encourages sill spreading. Therefore, depending on the intrusion and host rock temperatures and the radial thermal gradient, sill propagation appears largely thermally controlled (Rubin 1993; Amoruso et al. 2017; Amoruso and Crescentini 2019). The general thermal requirements for the solidification of magma within dikes, discussed in Sect. 3.4.3, are valid for any magma-filled fracture, including sills. These theoretical requirements are also supported by field observations. For example, for what concerns the minimum thickness of a sill to sustain magma flow, field and petrology studies at the Isle of Mull (Scotland) confirm that sustained flow creating channels active for up to 5 months occurred only in sill portions thicker than 3.5 m. Conversely, narrower portions experienced a single and shorter injection of magma (Holness and Humphreys 2003). Indirectly related to the thermal state of the sill is also its capacity to propagate due to heated pore fluids. In fact, field evidence at the Golden Valley saucer-shaped sill in South Africa shows that this consists of several lobes constructed from coalesced magma fingers. These fingers formed through the fluidization of the host rock, initiated by the boiling of pore fluids caused by heating from the magma and tensional failure of the roof strata. Once fluidization occurred, the mechanical

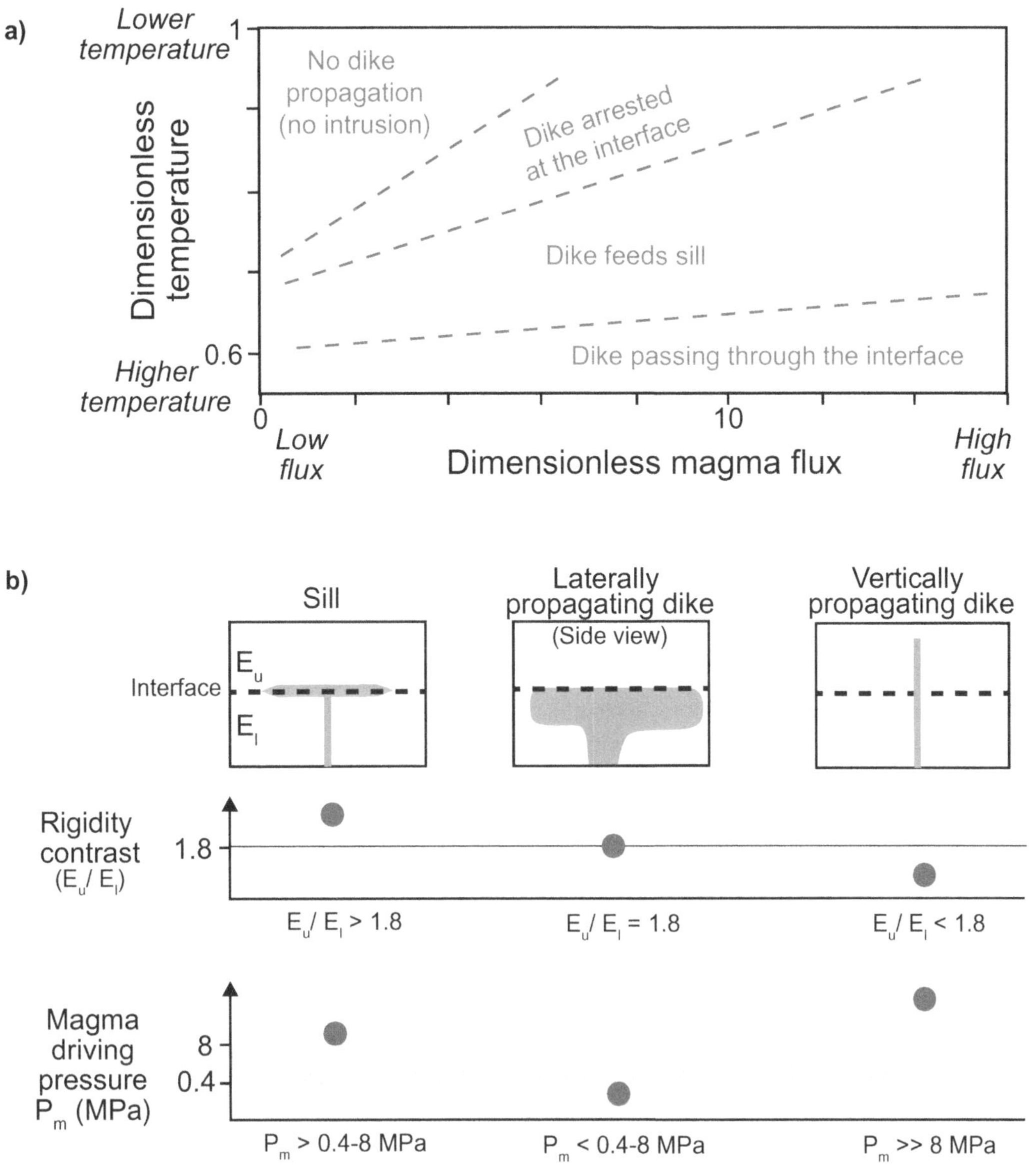

Fig. 4.6 Sill formation in experiments. **a** Fate of a dike approaching an interface as a function of its dimensionless temperature (temperature of intrusion increasing downward) and dimensionless magma flux. Depending on the relative values of these parameters the dike may not propagate, become stalled at the interface, form a sill or pass the interface (modified after Chanceaux and Menand 2014). **b** Fate of a dike approaching an interface between layers as a function of the rigidity contrast between the upper and lower layer E_u/E_l and the magma driving pressure P_m (red dots). A higher rigidity contrast promotes sills. A higher driving pressure promotes vertically propagating dikes, whereas sills develop only with relatively high pressures; if pressures are lower, the dike propagates laterally beneath the interface. Modified after Sili et al. (2019)

heterogeneity controlling the propagation of the concordant sill was broken and the intrusion propagated discordantly, forming the inclined portion of the intrusion (Schofield et al. 2010).

In addition to the thermal state of the magma and host rock, two experimental conditions appear both necessary and sufficient to develop sills from an upward propagating dike (Fig. 4.6b; Kavanagh et al. 2006; Sili et al. 2019). The first requires a significant stiffness contrast between adjacent crustal layers, with the upper layer being at least ~ 1.8 times stiffer than the lower layer to inhibit dike propagation and promote sill formation. Slightly lower rigidity contrasts require other second-order contributions to develop sills, as weak interface and compressive stress (see below). However, without any rigidity contrast sill formation is always inhibited. The second condition requires a relatively large driving pressure of the intruded magma (of a few MPa), providing the force to intrude the contact between layers. Driving pressures below a certain threshold would arrest the dike or promote its lateral propagation within the lower layer. Conversely, if the driving pressure is much higher than that required to form sills, as for example in the case of extremely buoyant (gas-rich) magmas, the dike may pierce the stiffer upper layer and propagate. Therefore, sills form with an appropriate balance between the stiffness contrast between adjacent layers and the driving pressure of the magma. Experiments also show that the onset of sill propagation is accompanied by a simultaneous contraction within the underlying feeder dike, suggesting a magmatic pressure drop of up to $\sim 60\%$ induced by the fluid from the feeder dike extracted to help grow the sill (Kavanagh et al. 2015).

As sills are commonly subhorizontal, associated with a subvertical least principal stress σ_3, it has been often underlined that regional compression promotes sill formation (Gill and Walker 2020, and references therein). Experiments have been investigating the role of lateral compression on an upward propagating dike, showing that this may deflect the dike (Menand, et al. 2010; Sili et al. 2019). However, dike-to-sill rotation occurs only for large compressive stress or small buoyancy and, even in this case, a dike-to-sill transition in response to lateral compression would take several hundreds of metres to a few kilometres. This is not consistent with geological observations, which show that sills form within much thinner lithological units, thus suggesting a subordinate role in sill formation for regional stresses, if these are the only operating mechanism. In a similar fashion, a weak interface, or sharp mechanical discontinuity, between the layers with different rigidity, appears as second-order governing parameter. Both regional compression and weak interface may enhance sill formation only if supported by relatively high stiffness contrasts (between 1 and 1.8) and driving pressures (very few MPa). Modelling data suggest that density variations within the host rock have even minor impact on sill formation. The limited role of density variations is also supported by the evidence of sills with analogous composition exposed at different stratigraphic levels below eroded volcanoes (Thomson 2007; Menand 2011). Therefore, any level of neutral buoyancy is unlikely to play an important role, although it could assist sill formation.

Even though these different controls on sill formation are not mutually exclusive, they do not necessarily operate on the same length scale. For example, the length scale associated with layers with different stiffness determines the depth at which rigidity-controlled sills will form. This may differ from the depth at which a cluster of weak interfaces may control sill emplacement. Therefore, the depth at which sill form ultimately depends on the local dominance of a specific controlling mechanism (Menand 2011).

As sills are magma-filled fractures similar to dikes, their propagation involves both elastic (along-length thickening) and inelastic (tip fracturing) processes within the host rock. The mainly elastic deformation of the host rock, except for the tip regions, motivates considering sills as cracks in a linear elastic body, whose formation and propagation follow the general principles of the Linear Elastic Fracture Mechanics (LEFM) described for dikes in Sect. 3.4.1. The tip zone, being the site of large principal stress differences and large stress gradients, may be characterized by significant inelastic deformation, similar to dikes (Fig. 4.7).

The sill tip propagates by tensile opening perpendicular to the local least principal stress σ_3. The opening at the tip front determines a cavity between the magma front and the intrusion tip, filled with volatiles exsolved from the intrusion. At the tip, three possible mechanisms for sill

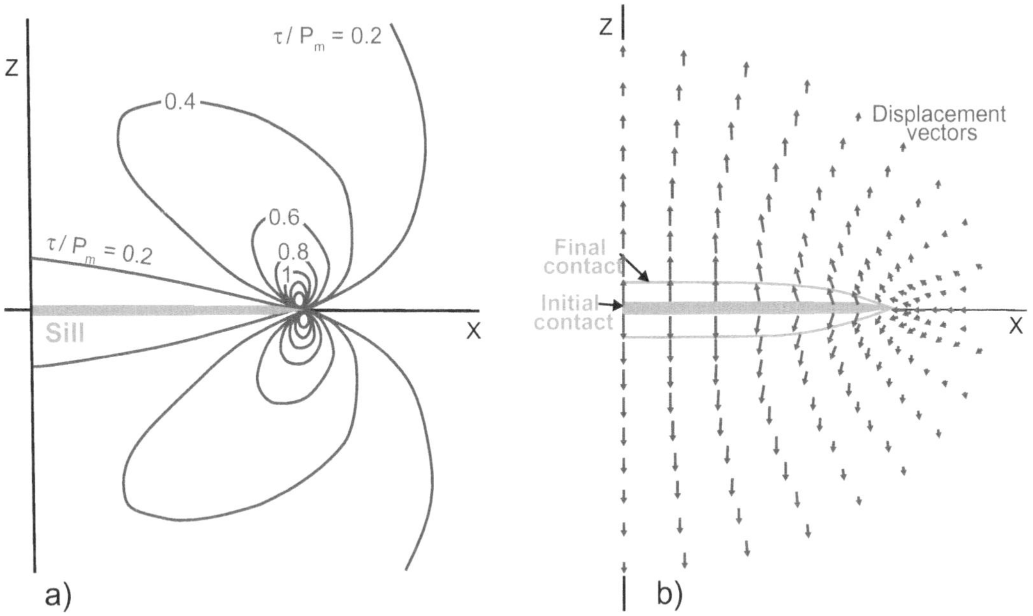

Fig. 4.7 **a** Shear stress τ distribution at the tip of a sill, normalized to the driving pressure P_m of the magma within the sill. Blue contours are loci of equal maximum shear stress and numbers refer to the stress magnitude; the symmetric distribution reveals maximum shear stresses at the sill tip. **b** Displacement distribution at the sill tip; the maximum displacement occurs towards the sill centre and decreases towards the sill tip. The solutions in both diagrams are generally valid for the tip of any linear magma-filled fracture. Modified after Pollard and Johnson (1973)

propagation are extension fracturing, brittle faulting, and ductile faulting. In particular, in a brittle crust at low confining pressures and temperatures, a sill grows by splitting the host rock along a single extension fracture, presumably facilitated by the existence of the discontinuity or the weak interface the sill is potentially intruding. At relatively high confining pressures, the host rock near the front of the sill tips deforms by brittle faulting. At higher confining pressures and temperatures, ductile faulting occurs (Fig. 4.8; Pollard 1973). Analytical models of shallow sill formation show that, for geologically realistic values, the horizontal extent of the plastic zone at the sill tip is much smaller than the radius of the intrusion, decreasing during propagation and with the intrusion's depth. Conversely, the extent of the plastic zone increases if the host rock is weaker. Plasticity effects might be thus small for large sills, but relevant in early sill propagation, also doubling the overpressure to propagate the sill (Scheibert et al. 2017).

4.4 Laccoliths, Bysmaliths and Lopoliths

4.4.1 Overview

Laccoliths, bysmaliths and lopoliths are thick tabular intrusions with slightly different shape, but sharing important features, as the origin from inflating sills and the possibility to accommodate the space required to emplace the magma independently of regional tectonics (Fig. 4.9). **Laccoliths** are concordant tabular intrusions with domed upper surface and flat lower surface, developed through roof uplift (Fig. 4.9a). **Bysmaliths** are thicker intrusions, with conical or sub-cylindrical shape, whose emplacement requires faulting at the lateral contact with the overburden, resulting partly concordant (along their floor) and partly discordant (along the fault-bounded portion of their roof; Fig. 4.9b). Bysmaliths develop from thickened laccoliths,

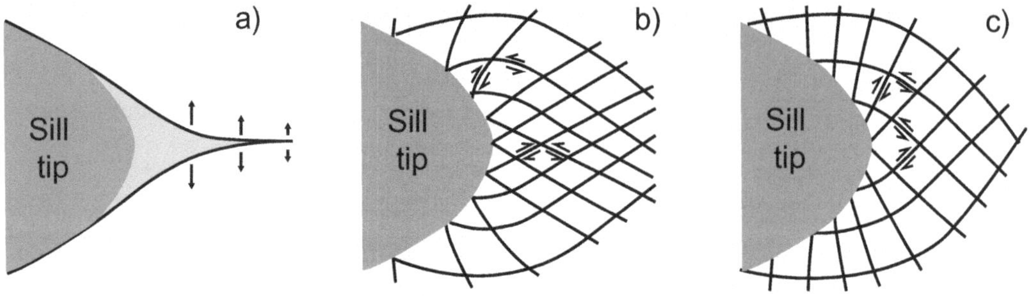

Fig. 4.8 Mechanisms of propagation of a sill through: **a** a single extension fracture in a brittle host rock (grey), **b** multiple faults in a brittle host rock; c) multiple faults in a ductile host rock. These mechanisms are generally valid for the tip of any magma-filled fracture. Modified after Pollard (1973)

usually when these are very shallow and/or with highly viscous magma. **Lopoliths** are concordant tabular intrusions with depressed lower surface and flat upper surface, developed through floor depression (Fig. 4.9c).

It has been long postulated that these intrusions develop from inflating sills (e.g., Pollard and Johnson 1973). In particular, sills must have a magma pressure that is greater than the lithostatic pressure, otherwise they could not exist. This pressure results in the elastic and inelastic deformation of the country rock at both the top and bottom of the sill, although at shallow crustal levels the bending of a thinner upper plate is much easier than the deformation of a much thicker country rock below. This feature explains the preferred development of laccoliths at shallow crustal levels, up to 7–8 km of depth (Pollard and Johnson 1973; Acocella 2000). In the case of highly viscous, particularly shallow and/or voluminous intrusions, the magma can pierce through the domed overburden along a fault-bounded side with overall trapdoor uplift, forming a bysmalith. Conversely, in the deeper crust the bending of a thicker upper plate is hindered and intrusion thickening starts to occur by depression of the floor through elastic deformation and, especially, viscous flow of the host rock. When the bending of the floor of the intrusion becomes the dominant mode of sill thickening, a lopolith develops. Therefore, lopoliths are more common at mid- to lower-crustal levels, also explaining why they are observed much less frequently than laccoliths.

Laccoliths, bysmaliths and lopoliths accommodate shallow magma emplacement regardless of any contribution from regional tectonics. In fact, laccoliths and bysmaliths create the space for their growth simply displacing the Earth's surface upwards, while lopoliths mainly create

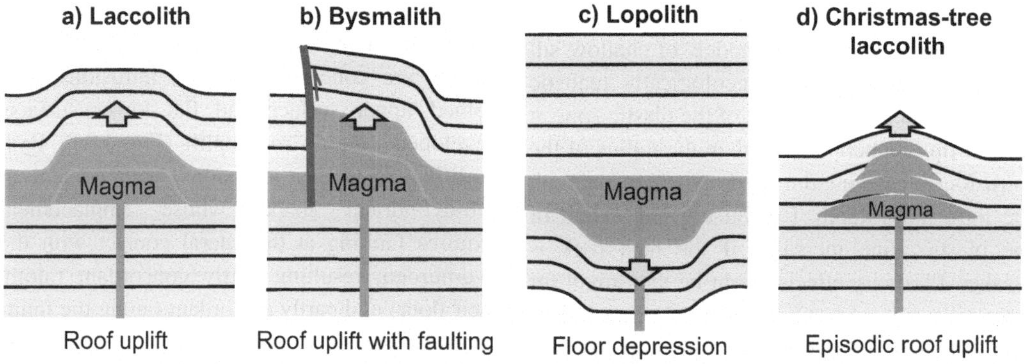

Fig. 4.9 Schematic illustrations of a laccolith **a**, a bysmalith **b**, a lopolith **c** and a Christmas-tree laccolith consisting of stacked sills **d**; the associated mechanisms of emplacement are indicated below

the space through viscous flow of the lower crust. These tabular intrusions are often stacked one above the other and connected by feeder dikes, forming "Christmas-tree"-like intrusions (Fig. 4.9d; Corry 1988). These stacked systems can store a significant amount of magma, irrespective of the regional tectonic setting, as supported by their rapid emplacement strain rates, which are much faster than those due to regional tectonics. For example, in the Neogene extensional setting of Tuscany (Italy) several mid- to upper-crustal granitic laccoliths emplaced with magmatic strain rates, estimated from theoretical models or petrological studies, of 10^{-10} to 10^{-12} s^{-1}, much higher than the tectonic rates associated with regional extension, on the order of 10^{-15} s^{-1} (Acocella and Rossetti 2002). This and other cases highlight a subordinate role of the regional tectonic processes in assisting the emplacement of tabular intrusions, suggesting that the growth of laccoliths, bysmaliths and, likely, lopoliths is the easiest means to accumulate shallow magma.

While elasticity explains most of the deformation related to sill formation, the assumption of elastic deformation associated with the development of laccoliths, bysmaliths and lopoliths may carry significant limitations. In the case of laccoliths, elasticity holds for a relatively small thickness of the intrusion compared to its length. Under these conditions, theoretical calculations show that the roof of the laccolith is the site of maximum bending and that the maximum differential stresses focus at the laccolith tips (Fig. 4.10). With thicker laccoliths, elasticity is restricted to the early stages of development; analysis for later stages should consider elasto-plastic constitutive equations. In the case of bysmaliths and lopoliths, fracture mechanics and visco-elastic approaches should be also considered, respectively.

Laccoliths, being the most common type of thick tabular intrusion, with several thousands of cases recognized worldwide, have provided the opportunity for detailed investigations. Their ellipticity, expressed as the map view ratio of their long to short axes, ranges from 1 to 8, although clustering towards a subcircular shape (Corry 1988). This shape provides ideal thermal conditions, as circular sills maintain the least amount of surface area per volume of fluid, thus limiting solidification. The deviation from the circular shape is explained taking into account for the strike of the feeder dike, magma viscosity and solidification, crustal heterogeneities and incremental assembly (Currier et al. 2017). Estimated times for the emplacement of laccoliths vary between a few weeks to a few years: for example, a laccolith thicker than 200 m developed in only one month below Cordon Caulle volcano (Chile), between 20 and 200 m of depth (Castro et al. 2016).

Superb laccolith exposures are found in the Henry Mountains (Utah, USA), where the term laccolith was indeed defined, with a population of intruded Oligocene diorites ranging from 10 to 200 m in thickness and 1 to 3 km^2 in areal extent (Gilbert 1877; Johnson and Pollard 1973). Here the deformation in the host rock is concentrated above the intrusions, around the laccoliths peripheries and, mostly, at their lateral terminations, consistently with theoretical calculations (Fig. 4.10). These lateral terminations show blunt geometries, or dikes cutting upwards across the host rock, or abrupt terminations against subvertical faults (as bysmaliths). Among the intrusions is the subcircular Black Mesa diorite, 1.7 km in diameter and between 150 and 250 m thick, resembling a sheeted laccolith on its western side and a bysmalith (fault bounded) on its eastern side. It emplaced in less than 100 years at depth between 2.5 and 2.7 km, growing through asymmetric vertical stacking of sills (Fig. 4.11; De Saint Blanquat et al. 2006).

A spectacularly exposed laccolith is the 12.5 Ma old Torres del Paine (Chilean Patagonia), consisting of a composite intrusion with peripheral subvertical feeding system (Fig. 4.12; Michel et al. 2008; Leuthold et al. 2012). The laccolith grew first by under-accretion of granitic sills over $\sim$90 ka, followed by underplating of mafic sills after $\sim$20 ka of quiescence, with individual sills injected by over-accretion during $\sim$41 ka. The average growth rate of the laccolith was 0.0005 km^3 yr^{-1}, with total volume of $\sim$88 km^3. Construction rates were variable during

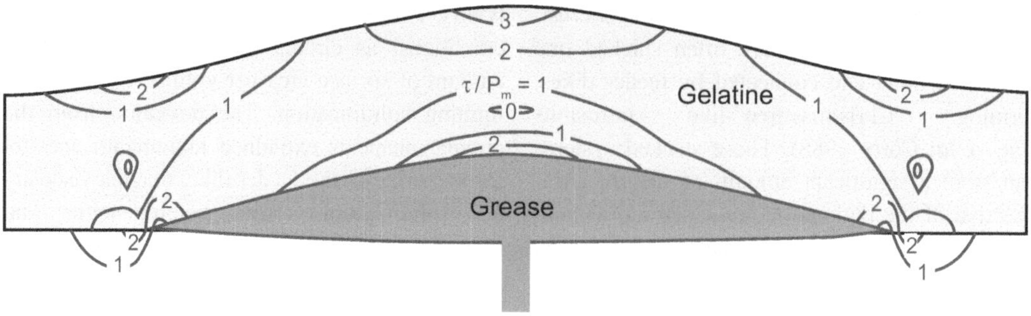

Fig. 4.10 Distribution of the maximum shear stress τ normalized to the driving pressure P_m within a gelatine layer intruded by grease with laccolith shape obtained through photoelastic models. Contours are loci of equal maximum shear stress and numbers refer to the stress magnitude. Modified from Pollard and Johnson (1973)

growth, with felsic magmas accumulated at a faster rate than the mafic magmas. The thermal effect between individual units was not sufficient to allow large amounts of partial melting of previously emplaced, solidified, granitic rocks. In contrast, the ductile interactions between individual mafic sills suggest that the emplacement rate of the mafic magmas was faster than the time

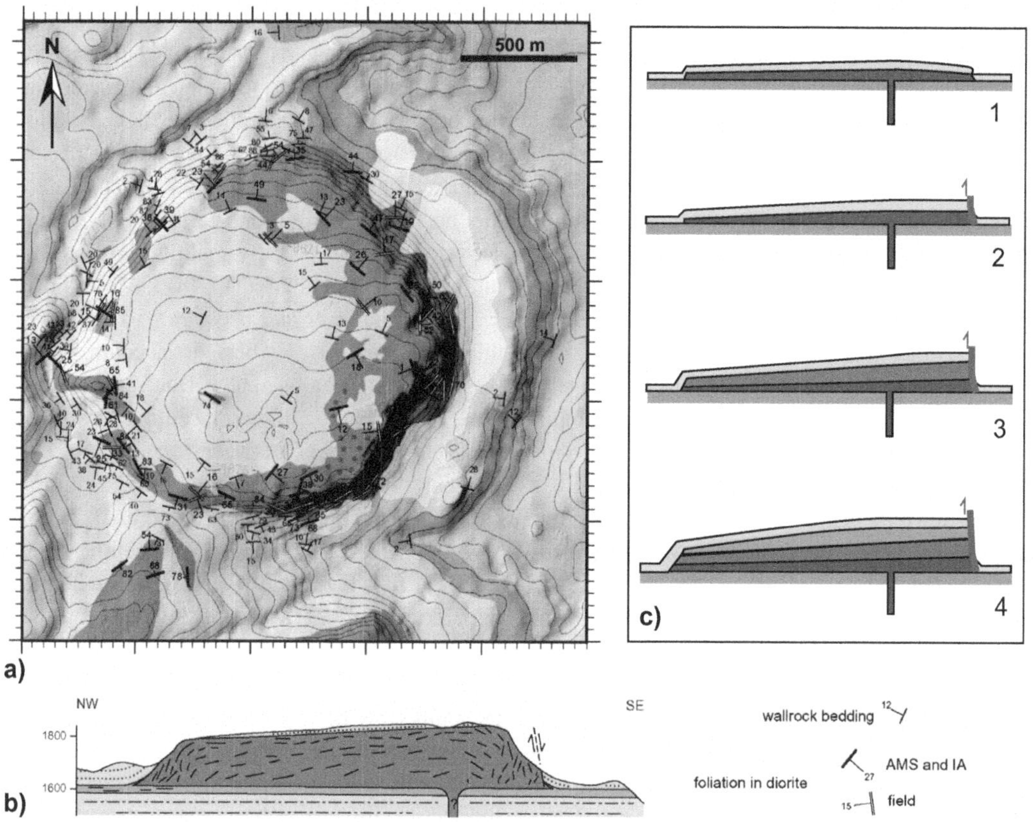

Fig. 4.11 Map **a** and NW–SE section **b** views of the Black Mesa bysmalith (Utah, USA). The inferred mechanism of formation, involving roof uplift and faulting on the southeastern edge, is summarized in the progressive stages in **c**. Modified after de Saint-Blanquat et al. (2006)

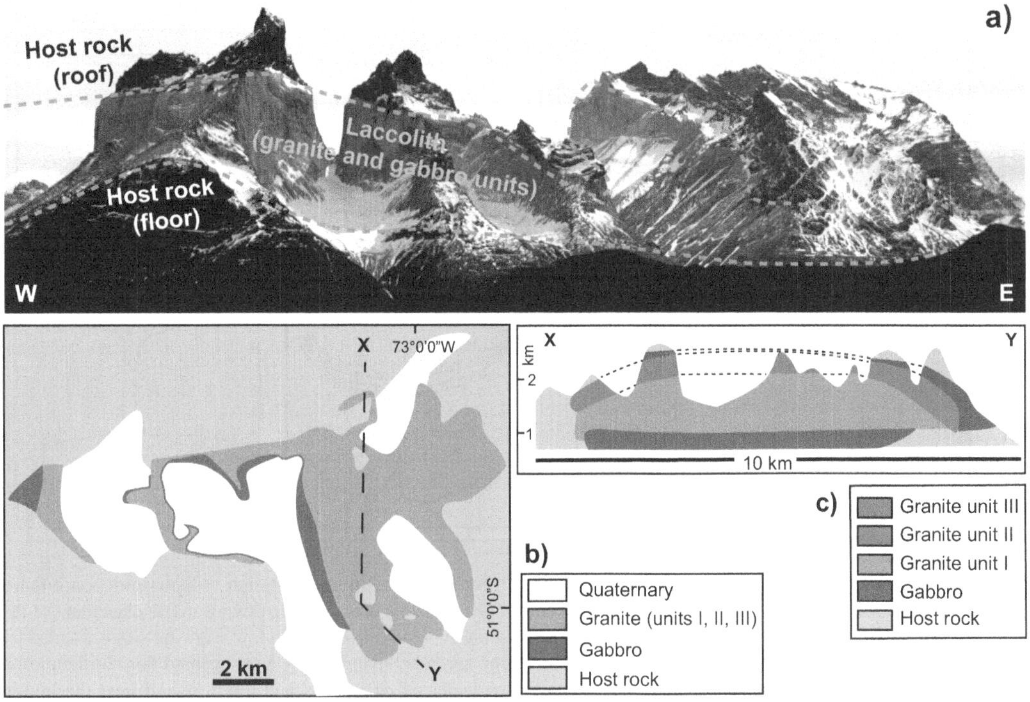

Fig. 4.12 Torres del Paine Intrusive Complex laccolith (Chile). **a** View of the laccolith from the south. **b** Simplified geological map. **c** North–south section; stippled lines are correlation between outcrops of individual granite units (not specified in **b**) and their contacts with host rock; granite unit I is the oldest and granite unit III is the youngest. Modified after Michel et al. (2008)

required to solidify successive mafic pulses. The growth of the laccolith was thus incremental, with individual magmatic pulses in a 20 ka range separated by quiescence.

Deformation during laccolith emplacement primarily occurs in the host rock. However, shallowly emplaced laccoliths may show extensive internal deformation, as for example the rhyolitic Sandfell laccolith (Iceland), which formed in a single intrusive event at a depth of 500 m. In this laccolith a large volume of magma became viscously stalled and was unable to flow further, with the magma solidifying along the rim deformed by intrusive stresses (Mattsson et al. 2018).

Lopoliths, because of their usually higher emplacement depth (mid- to lower-crust) are less frequent. Among the known lopoliths is the Jurassic Qitianling pluton (South China), emplaced at a depth of 10–15 km. The Qitianling pluton forms a subcircular ~25 km long and ~5 km thick granite lopolith that was incrementally emplaced by under-accretion of multiple magma sheets deflected from magma pulses that ascended through pipe-like channels. Other examples of lopoliths include the Scuzzy and Urquhart tonalitic plutons, British Columbia (Canada; Fig. 4.13; Brown and McLelland 2000; Liu et al. 2018).

4.4.2 Emplacement Conditions

Most of the knowledge on the mechanism of emplacement of thick tabular intrusions derives from studies on laccoliths. Therefore, the focus of this section is on laccoliths, although some notion may regard also bysmaliths and lopoliths.

Laccoliths develop uplifting the overburden. As anticipated, with relatively thin laccoliths, this process can be described using the elastic theory, where the overburden of thickness z is elastically

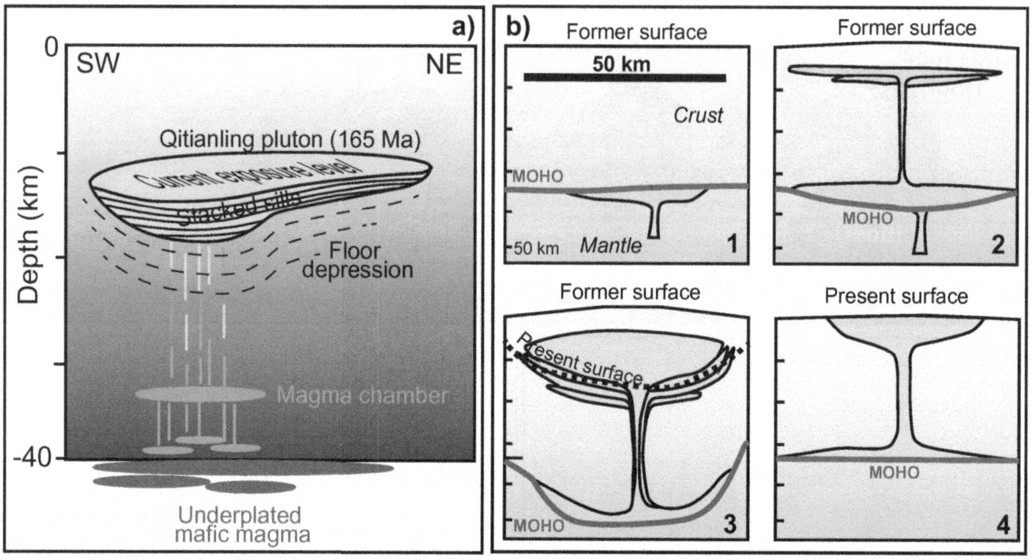

Fig. 4.13 Examples of lopoliths. **a** Scheme of the Qitianling (China) lopolith, consisting of multiple stacked sills (not to scale; modified after Liu et al. 2018). **b** Emplacement model for the Scuzzy and Urquhart plutons, British Columbia, Canada. Magma ponds in and thickens the lower crust (1). The Moho sags and the Earth's surface becomes domed (2). First intrusions in the upper crust are subhorizontal sheets fed by magma rising through dikes. Subsequent high-volume intrusion inflates one or more sheets, causing downward displacement of the underlying column into the lower crustal space vacated by the ascended magma. The early-formed floor of the pluton is displaced downward and tilted inward near the pluton margins (3). Erosion and consequent isostatic uplift exposes deep levels of the pluton and country rock (4). Modified after Brown and McLelland (2000)

bent upward by the pressure p of the magma that forms the laccolith. Considering an upward pressure p applied uniformly and instantaneously at the bottom of a bi-dimensional elastic plate, the maximum deflection, or laccolith thickness, w at the centre of the laccolith in cross section is (Fig. 4.14; Turcotte and Schubert 2002):

$$w = -\frac{(p - \rho g z)L^4}{384D} \quad (4.1)$$

where ρ is the average density of the plate, g is the acceleration due to gravity, L is the length of the laccolith and D is the flexural rigidity of the overburden (see below). Here the plate is assumed to be thin compared with its length ($z \ll L$) and its maximum vertical deflection w is taken to be small ($w \ll L$). Also, the opening induced by the intrusion is not uniform; the surface deformation induced by an intrusion with uniform opening (sill-like) is considered in Sect. 8.3.2 (see Eq. 8.7). A very similar relationship is obtained also for a circular elastic plate (three-dimensional case; Pollard and Johnson 1973):

$$w = -\frac{(p - \rho g z)L^4}{1024D} \quad (4.2)$$

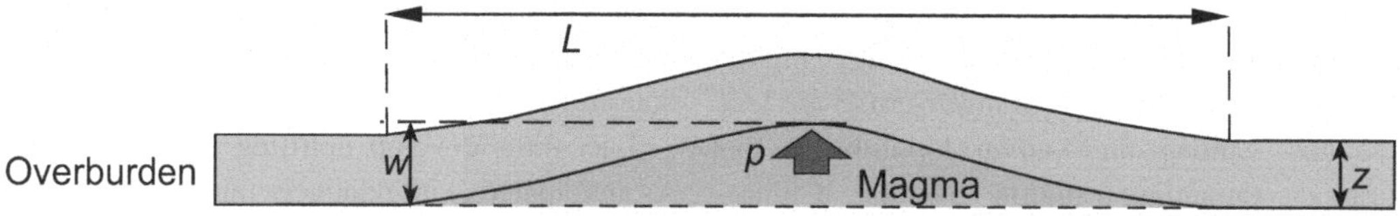

Fig. 4.14 Flexuring of a plate (overburden) with thickness z induced by a tabular intrusion with length L, thickness w and upward magma pressure p. Modified after Turcotte and Schubert (2002)

Equations 4.1 and 4.2 show that the longer the laccolith, the larger its uplift. However, this relationship is not linear, as the amount of upward bending increases as the fourth power of the distance the magma spreads: this implies that, all else being equal, doubling the length of a laccolith increases sixteen times the uplift of its roof.

The **flexural rigidity** D of the overburden represents the resistance to bending offered by a layer of thickness z and is given by

$$D = \frac{Ez^3}{12(1 - \nu^2)} \quad (4.3)$$

where E is the Young's modulus and ν the Poisson's ratio of the overburden. Equation 4.3 shows that the thicker the overburden, the more difficult is to bend it. This relation is also not linear, with the resistance to bending relating to the cube of the overburden thickness: this implies that a layer of certain thickness bends eight times with regard to a layer having twice that thickness. While it is common to approximate the thickness of the overburden to a single layer (especially in the case of shallow laccoliths), in practice the overburden usually consists of several layers. These, because of the discontinuities between them, may slide one above the other during bending, decreasing the flexural resistance of the overburden. Therefore, to properly express the flexural rigidity of multiple layers, the **effective** (or **equivalent**) **thickness** is commonly considered in Eq. 4.3. This is defined as the thickness of a single layer which has the same flexural rigidity as the stack of free-sliding layers. Therefore, the effective thickness is smaller than the true thickness of the stacked layers (e.g., Pollard and Johnson 1973).

The geometric behaviours described by Eqs. 4.1 to 4.3 are well supported by field data in the Henry Mountains and by experiments (Fig. 4.15; Pollard and Johnson 1973; Roman Berdiel et al. 1995). The latter show that laccoliths change from lenses (lower w and higher L) to bell-shaped (higher w and lower L) with decreasing overburden thickness z (and, in turn, D). A similar analysis may be applied also to the early stage development of bysmaliths, which share their initial growth with laccoliths.

More generally, an overview of the dimensions of geologically and geophysically known laccoliths shows relatively consistent length to thickness ratios. These can be approximated by an overall scale invariant relationship between the laccolith length L and thickness w, expressed by a simple empirical power law equation of the form:

$$w = jL^n \quad (4.4)$$

where j is a constant (for laccoliths $j \sim 0.02$ m) and n is the power law exponent (for

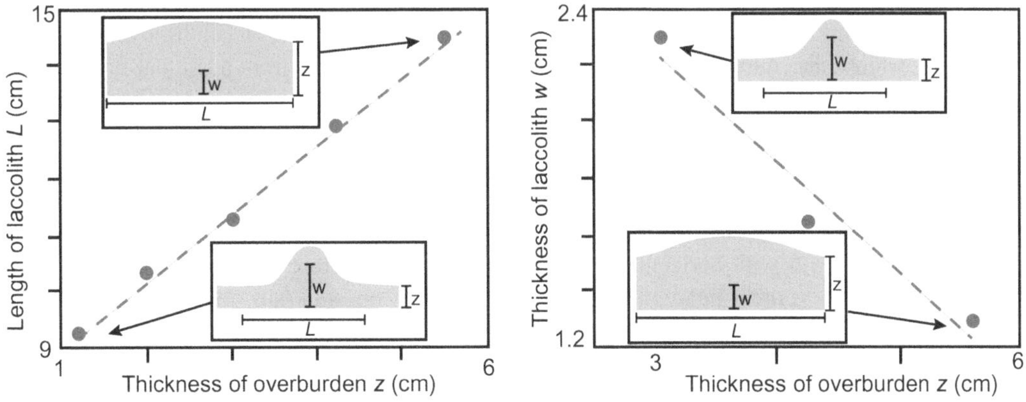

Fig. 4.15 Experimental models of laccolith formation, showing the proportion between the overburden thickness z and the laccolith length L (left) and the inverse proportion between the overburden thickness z and the laccolith thickness w (right; modified after Roman Berdiel et al. 1995)

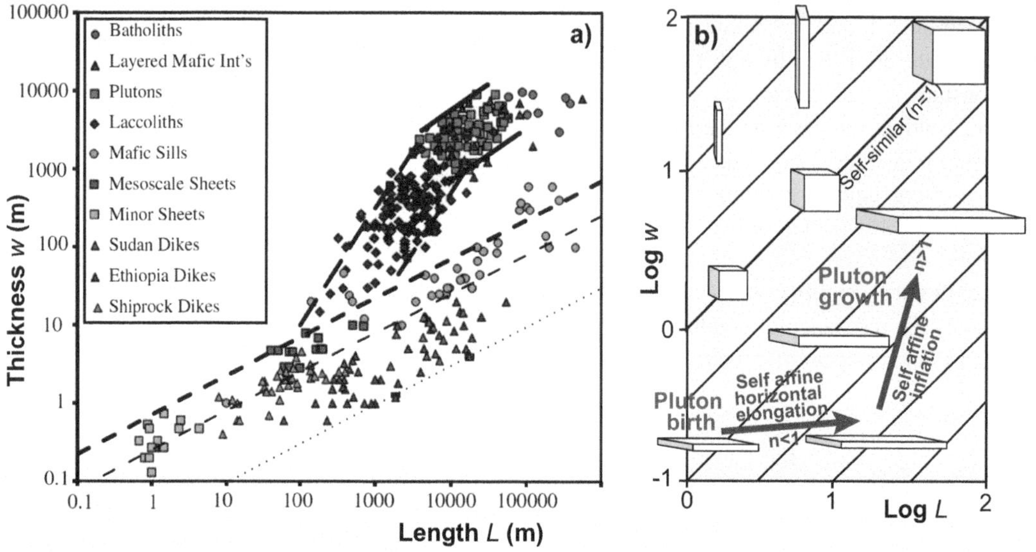

Fig. 4.16 **a** Log thickness (w) versus log length (L) plot for different types of intrusions, including dikes, sills, laccoliths, plutons and batholiths; solid and dashed lines are bounding curves for different classes of intrusion (after Cruden et al., 2017). **b** Possible theoretical modes of growth of intrusions, highlighting an example of two-stage growth model for sheet intrusions: lateral spreading or self-affine horizontal elongation ($n < 1$) is followed by vertical thickening or self affine inflation ($n > 1$). The net result is a self-affine (n slightly less than 1) scaling of sheet intrusions. Inclined parallel black lines indicate self-similar growth. Modified after McCaffrey and Petford (1997)

laccoliths $n \sim 1.5$; McCaffrey and Petford 1997; Cruden and McCaffrey 2001; Cruden et al. 2017). The diagram of Fig. 4.16a illustrating such a L-w relationship, which also includes the geometrical features of other types of intrusions (see Sect. 4.5.3), confirms that the laccoliths are consistently much longer than thick, displaying a tabular shape. Two growth modes may be in principle proposed for laccoliths (Fig. 4.16b; McCaffrey and Petford 1997). (a) Self-similar growth, in which the intrusion horizontal to vertical dimensional ratios are the same during growth *($n = 1$)*. (b) Self-affine growth, where the ratio of horizontal to vertical dimension varies during growth, with changes in the exponent n; this includes a possible two-stage growth, passing from $n < 1$ (growth with horizontal elongation dominating over vertical inflation) to $n > 1$ (vertical inflation dominating over horizontal elongation), or also passing from $n > 1$ to $n < 1$. The observed power law relationship with $n \sim 1.5$ for laccoliths excludes self-similar growth, but cannot discriminate between any of the two self-affine growth modes. Also, a value of $n \sim 1.5$ suggests that laccoliths grow differently from mafic and felsic sills, which are associated with $n \sim 0.5$, indicating that during their growth horizontal lengthening dominates over vertical thickening (Fig. 4.16a; Cruden et al. 2017).

Mechanical models and field data suggest that a two-stage growth for laccoliths, bysmaliths and lopoliths may occur from a sill that first intrudes laterally and then inflates as some critical length is reached (Fig. 4.16b). There is evidence of this growth path for laccoliths and to some extent also for bysmaliths, where the confinement by a fault indicates that their lateral growth must have also occurred before inflation and faulting. Less is known on any sill to lopolith transition, although a growth path similar to that of laccoliths may be inferred. In particular, the sill to laccolith transition may occur when a sill reaches a length of approximately 3.3 times the effective overburden thickness: at this stage, the sill starts inflating. This approach, or "mechanical model", finds analytical justification combining Eqs. 4.1 and 4.3 in a form that describes the growth of a laccolith with thickness w as:

$$w = \frac{P_e L^4 (1 - \nu^2)}{2 E z_e^3} \quad (4.5)$$

(where the magma excess pressure $P_e = p - \rho g z$, that is the difference between the magmatic and the lithostatic pressure, and z_e is the effective thickness of the overburden), and then combining Eq. 4.5 with Eq. 3.3, which describes the thickening of a magma-filled fracture:

$$w = \frac{P_e}{E_s} L \quad (4.6)$$

where the elastic stiffness $E_s = G / (1 - \nu)$, being G the elastic shear modulus and L the sill length. Equating these two relations, it is found that $L = z_e \sqrt[3]{4}$, which indeed suggests that a laccolith starts forming when the length of a sill reaches ~3.3 times the effective overburden thickness (Pollard and Johnson 1973). The concept of minimum length required for a sill to start to inflate finds physical justification in the Pascal's principle, whereby relatively small magma pressures may produce large vertical forces when surface areas are increased (e.g., Morgan 2018). As changes in fluid pressure are transmitted through the fluid everywhere equally, pressure is always equal as long as the fluid is connected. The pressure within the sill (P_2) is the same as that of the feeder dike (P_1), so that:

$$P_1 = P_2 = \frac{F_1}{A_1} = \frac{F_2}{A_2} \quad (4.7)$$

being F_1 and F_2 the forces acting on areas A_1 (the feeder dike) and A_2 (the sill). Therefore, the force applied by the fluid is a function of the area exposed to the fluid pressure, so that the larger is the area, the larger the force: in this way the force can be multiplied. In the case of a sill with limited length, the magma does not have enough area to increase the upward force to overcome the lithostatic load until the sill lengthens. This lengthening increases the force on the overburden and, when the force equals the lithostatic load, the sill can inflate and develop a laccolith.

The intrusions outcropping at Elba Island (Italy) support such a two-stage mechanical model of growth of laccoliths (Rocchi et al. 2002). Here nine Miocene granites connected by feeder dikes built up three nested Christmas-tree laccoliths, emplaced at depths between 1.9 and 3.7 km exploiting crustal discontinuities. These intrusions represent a population of thinner-than-average laccoliths, probably resulting from an abundance of lithologic and/or tectonic discontinuities in the host rock that acted as magma traps and hindered the coalescence of the multilayer laccoliths. These relatively thin laccoliths suggest a two-stage growth, characterized by horizontal spread and subsequent incomplete thickening, also by amalgamation of smaller sheet-like bodies.

Not all available evidence supports such a simple relation between laccolith length and emplacement depth to justify a two-stage sill to laccolith transition. More recent studies focus on the importance of the solidification of the lateral portions of the sill to initiate a laccolith. Accordingly, laccolith growth follows three stages, implying a pulsatile development ("thermal model", Fig. 4.17; Corry 1988; Currier and Marsh 2015).

Stage I. The sill propagates radially from a feeder zone. The lateral growth of the sill is at first rapid, but then decays as the finite flux from the feeder must fill an ever-increasing growing volume (Delaney and Pollard 1982; Currier and Marsh 2015). The resulting decrease in the propagation rates promotes solidification along the sill periphery.

Stage II. This stage shows a slowed or stalled flow at the sill periphery, inhibiting lateral growth. As magma continues to flow from the feeder dike, sill growth is accomplished by thickening. At this stage, a laccolith or lopolith (depending upon the bending of the intrusion roof or floor) forms; the laccolith may generate a bysmalith.

Stage III. This stage is characterized by lateral propagation via breakouts, with the initiation of a breakout reinvigorating the lateral growth of the

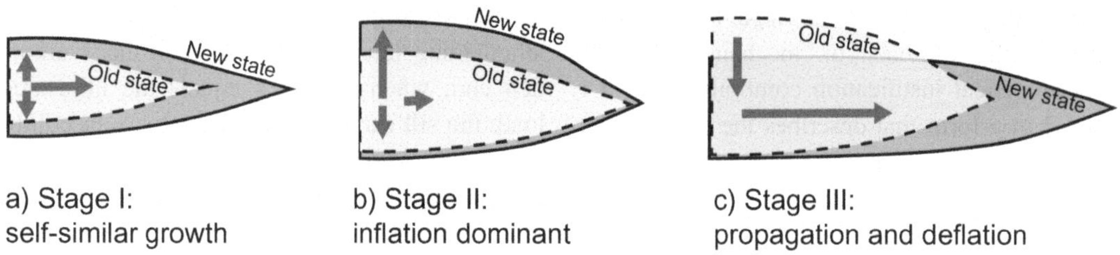

Fig. 4.17 Three main stages of laccolith development according to the thermal model: intrusions begin as stage I **a**, while stages II **b** and III **c** are the effect of solidification. Arrows show the directions of propagation of the intrusion. See text for details; modified after Currier and Marsh (2015)

intrusion. If the flux of magma through this burst section is faster than that at the conduit, deflation of the intrusion occurs.

The viscosity of magma, as also determined by its composition, is a main feature in controlling the sill to laccolith transition in the thermal model of laccolith formation. While low-viscosity magma can propagate to long distances promoting sill formation, high-viscosity magma cannot flow over large distances, accumulating and inflating by substantial doming the overburden, developing laccoliths. Indeed, most sills consist of low-viscosity magma with mafic to intermediate composition, whereas laccoliths commonly consist of more viscous felsic magma (Bunger and Cruden 2011).

The mechanical and thermal models of sill to laccolith transition are not mutually exclusive, and any contribution from each may not be easily recognized. A distinctive feature is that in the mechanical model propagation always predates inflation, whereas in the thermal model propagation predates and postdates inflation. Often the growth of a laccolith may be more complex than that described by the mechanical and thermal models. In fact, a laccolith may result from the stacking of multiple sills, rather than from the inflation of a single sill, as for example observed for the Torre del Paine intrusion (Leuthold et al. 2012). In this case, the development of the laccolith results directly from the conditions promoting repeated sill emplacement, which may differ from those promoting the emplacement of a single sill. In fact, once a sill has emplaced, its presence may create a favourable rigidity and viscosity anisotropy to develop successive sills by over-accretion, under-accretion or mid-accretion, offering ideal conditions to form a stacked system, which may eventually generate not only laccoliths, but also bysmaliths and lopoliths (Menand 2008; Galetto et al. 2017).

4.5 Plutons

4.5.1 Overview

A **pluton** is a massive and solidified intrusion of magma accumulated in the crust. Plutons are usually felsic in composition (i.e., mainly granitic) and may represent extinct magma chambers, thus feeding volcanic activity (see Sects. 4.6 and 4.7). A **batholith** is a larger intrusion deriving from the assemblage of several plutons and usually forming the core of a magmatic arc or mountain range at mid to deep crustal levels.

Most plutons are composite, built from several magma injections, each of small size and with different composition. Field and geochronologic evidence indicate that large and relatively homogeneous plutons can accumulate incrementally over millions of years, mainly through magmatic sheets, across different scales and tectonic environments. For example, the felsic Tenpeal and Mount Stuart plutons (Washington, USA) must have grown by the emplacement of distinct pulses of magma over a time longer than the solidification times of magma chambers with the same dimensions as the plutons (Matzel et al. 2006). Plutons may commonly form incrementally without ever existing as a large magma body, possibly in several increments analogously to the growth of crack-seal veins, which contain a superficially uniform fill that accumulates incrementally as the crack opens in several discrete events. Similarly, pluton growth may occur through the incremental assemblage of sheeted intrusions (Fig. 4.18a; Glazner et al. 2004; Bartley et al. 2006; de Saint Blanquat et al. 2011).

The earliest emplaced intrusions have the greatest temperature contrast with the host rock, suggesting quick cooling and preservation of the contacts. As in an incrementally assembled pluton early intruding magma becomes the host rock for later intruding magma, the lack of non-magmatic host rocks screens in the youngest portions of the plutons may be common. As a result, incrementally assembled plutons spend significant intervals at or near magmatic temperatures, contributing to erase contacts between increments (Coleman et al. 2016). Because of incremental growth, many plutons consist of assembled sills and/or dikes: a pluton composed of sills is tabular with oblate (laccolithic or lopolithic) shape, whereas a pluton composed of dikes is more prolate (Fig. 4.18b; Glazner et al. 2004). In some cases the correct evaluation of the size and volume of a pluton may be significantly affected by thousands of "erosion events" occurred during emplacement. In the Tuolumne Intrusive Complex (California, USA) these may have removed 35–55% of the original plutonic material from the presently exposed surface, with some (possibly ~25%) being recycled into

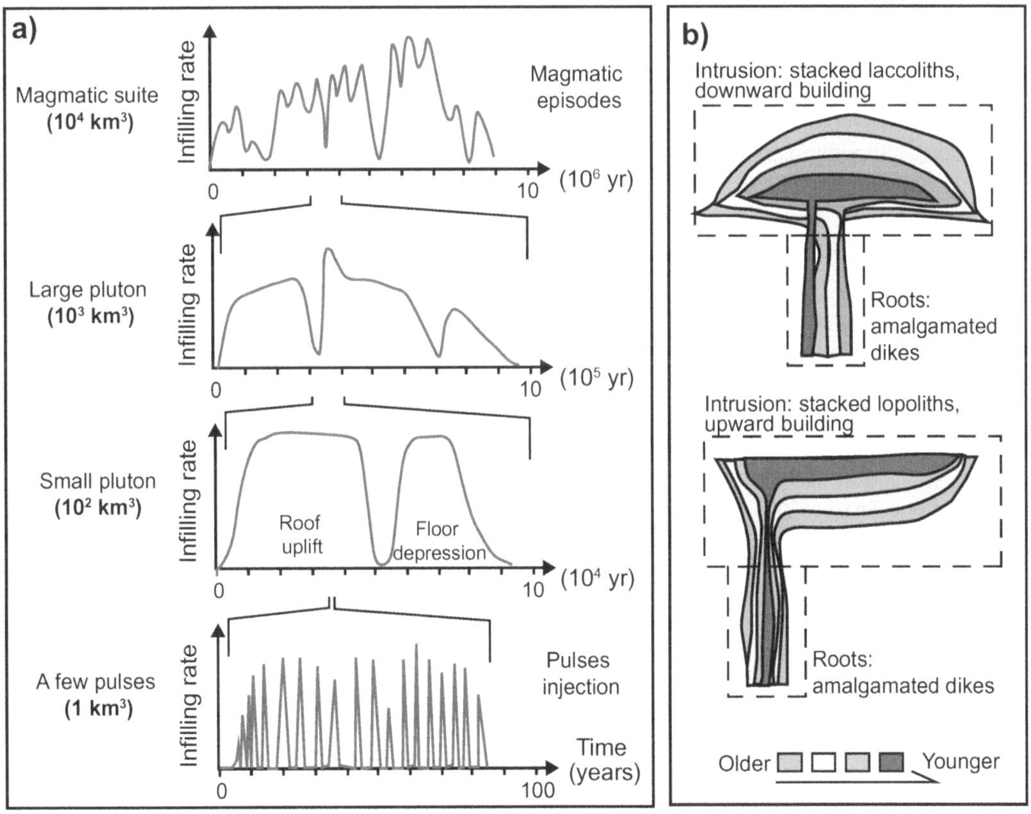

Fig. 4.18 **a** Pluton construction with time, from geochronological, petro-structural and thermal modelling data. The multiscale cyclicity of plutonic activity results from the incremental pulses in the smallest plutons (bottom) to the longer magmatic episodes of the largest suites of plutons (top; modified after de Saint Blanquat et al. 2011). **b** Mode of growth of plutons in a manner analogous to the growth of crack-seal veins, that is through the incremental assemblage of sheeted intrusions, as dikes (root zone) and sills (developing laccoliths or lopoliths): colours indicate the relative age of the intrusions. Modified after Bartley et al. (2006)

younger magmas and the remainder either erupted or displaced downward (Paterson et al. 2016).

Plutons indicate important accumulation of magma in the crust, even in the brittle portion. Therefore, a first-order problem regards the definition of the conditions allowing pluton development, including the resolution of the "**space problem**", that is how space is created to host the pluton. For decades, local or regional extension has been invoked as the solution to create the space required to accumulate the magma of the pluton. This should not be interpreted as magma intruding within empty spaces or voids, as rock strength and crustal stresses prevent these from existing at depth. Rather, regional or local extension would reduce the least principal stress σ_3, promoting the growth of magma-filled fractures (as dikes or sills) perpendicularly to its direction and, ultimately, pluton growth. This behaviour marks a tectonically-controlled, or passive, or **permitted** intrusion mechanism. Permitted intrusion mechanisms have become less popular in the last decades, although magma accumulation rates in some plutonic environments were recently discovered to be far slower than previously considered and in line with tectonic rates (Coleman et al. 2016). As alternative, several studies have been emphasizing the magma-controlled (active, or **forceful**) emplacement and growth processes, where magma plays an active role in creating the space for its accumulation, displacing itself the host rock without contribution from regional or local extension (Paterson and Fowler 1993; Cruden 1998; Tikoff et al. 1999; Petford et al. 2000). In addition to the permitted and forceful processes, the space required for the incremental growth of plutons may be obtained by the crystallization of the previously-emplaced solidifying magma batches, as proposed for the Tertiary Adamello pluton (Italy; John and Blundy 1993). Here pulses of emplaced magma cooled and solidified, contracting by 15%-20% and creating an unpressured volume, which allowed subsequent magma to emplace.

Currently, the solution to the room problem does not have a single answer of general applicability, and both the tectonically-assisted and magma-induced processes should be considered. Even though the permitted and forceful processes are presented separately below (Sects. 4.5.2 and 4.5.3, respectively), each mechanism alone may be an oversimplification of the real pluton

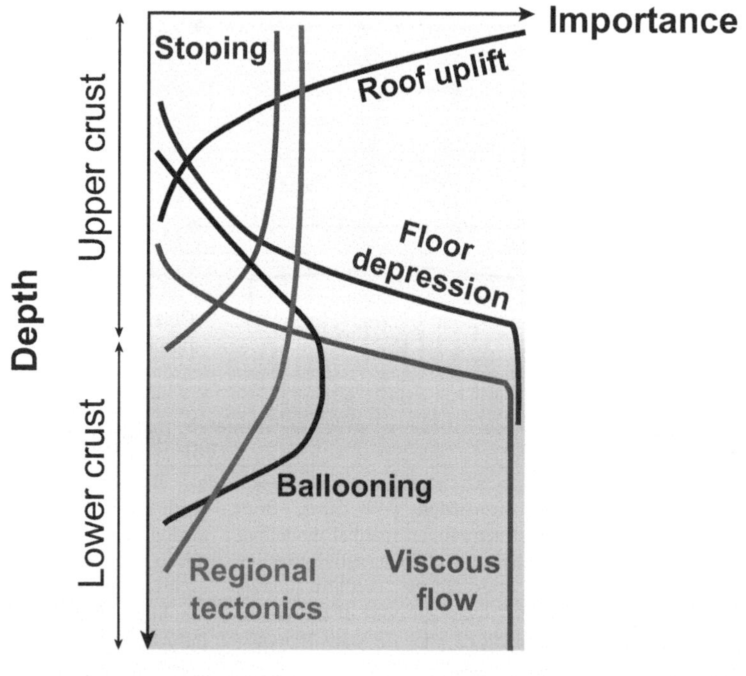

Fig. 4.19 Schematic summary of the importance of the main regional (tectonically-assisted) and local (magma-induced) processes, identified by the different coloured lines, encouraging pluton growth as a function of the depth in the brittle (yellow) and viscous (orange) crust

assembly process, as the growth of many plutons benefits from both permitted and forceful conditions (e.g. McNulty et al. 2000). This feature is supported by the scheme summarizing the importance of the main processes assisting pluton growth as a function of depth in Fig. 4.19, which highlights the coexistence of permitted and forceful contributions at different crustal levels.

4.5.2 Tectonically-Assisted Plutons

Regional or local tectonics may control not just the accumulation, but also the segregation and ascent of the magma forming the plutons. Tectonic shearing helps to segregate magma and feed conduits, creating local extension promoting magma injection. Tectonically-assisted plutons often derive from multiple granite sheeting parallel to the shear zone walls, obliquely to the principal stress directions, reflecting the exploitation of shear and faults zones active during emplacement. This is a common mechanism of magma ascent in the crust, which also promotes pluton growth (Hutton 1992; Vigneresse 1995).

Tectonically-assisted plutons grow under favourable tectonic conditions, which depend on the tectonic regime (extensional, strike-slip, contractional). The most favourable growth conditions result from local extension created by the activity of strike-slip structures. This commonly occurs along convergent plate boundaries, where any obliquity of convergence induces strain partitioning in the overriding plate, developing strike-slip structures along the magmatic arc, with plutons forming below the volcanoes (see Chaps. 10 and 12). Indeed, most convergent plate boundaries currently exhibit an oblique displacement vector, suggesting that oblique convergence must have been important also in the past. For example, Mesozoic plutonism along intrabatholithic faults in California correlates with oblique subduction and trench-parallel transport (Glazner 1991; Tikoff and De Saint Blanquat 1997; De Saint Blanquat et al. 1998). In arcs with strike-slip component, magma ascends in dikes to feed plutons constructed from hundreds of magma pulses emplaced in transient dilational sites along strike-slip fault systems. In particular, these plutons commonly grow within volumes of localized extension induced by releasing bends, pull-aparts, subvertical tension fractures, en-echelon shear arrays, tensional bridges, horsetail structures, branches of flower structures, and tectonic escape of crustal wedges bounded by conjugate strike-slip zones (Fig. 4.20). These structures may develop plutons with different shape, as also supported by analogue experiments of pluton emplacement along strike-slip zones (Hutton 1988; Tikoff and Teyssier 1992; Tobisch and Cruden 1995; Aranguren et al. 1996; Fernandez and Castro 1999; Corti et al. 2005).

Plutons may also grow in extensional settings, according to several mechanisms. These include pluton development within subhorizontal extensional shear zones, in growth wedges of listric normal faults, at releasing bends of transfer-faults, along extensional jogs of normal faults and through coalescing dikes oriented perpendicular to the regional extension (Fig. 4.20; Hutton et al. 1990; Walker et al. 1995; Richards and Collins 2004; Ciancaleoni and Marquer 2006). Plutons may even grow in arcs experiencing contraction, according to different mechanisms. Plutons may develop within fold-and-thrust belt structures, where the top of frontal thrust ramps may constitute feeder zones and the releasing steps at the ramp tops may serve as initial loci of magma emplacement, assisted by back-thrusts. Plutons may also form as subvertically sheeted complexes into steep reverse shear zones or at the core of growing thrust anticlines. In addition, shallow pluton growth may be driven by deeper regional compression, rather than buoyancy (Fig. 4.20; Rosenberg et al. 1995; Paterson and Miller 1998; Kalakay et al. 2001; Mahan et al. 2003; Musumeci et al. 2005).

In synthesis, while most plutons are found in extensional zones created by the activity of strike-slip structures, in principle each tectonic regime may provide various opportunities to emplace and accumulate magma. Therefore, provided that magma is available, plutons may locally grow supported by any tectonic regime.

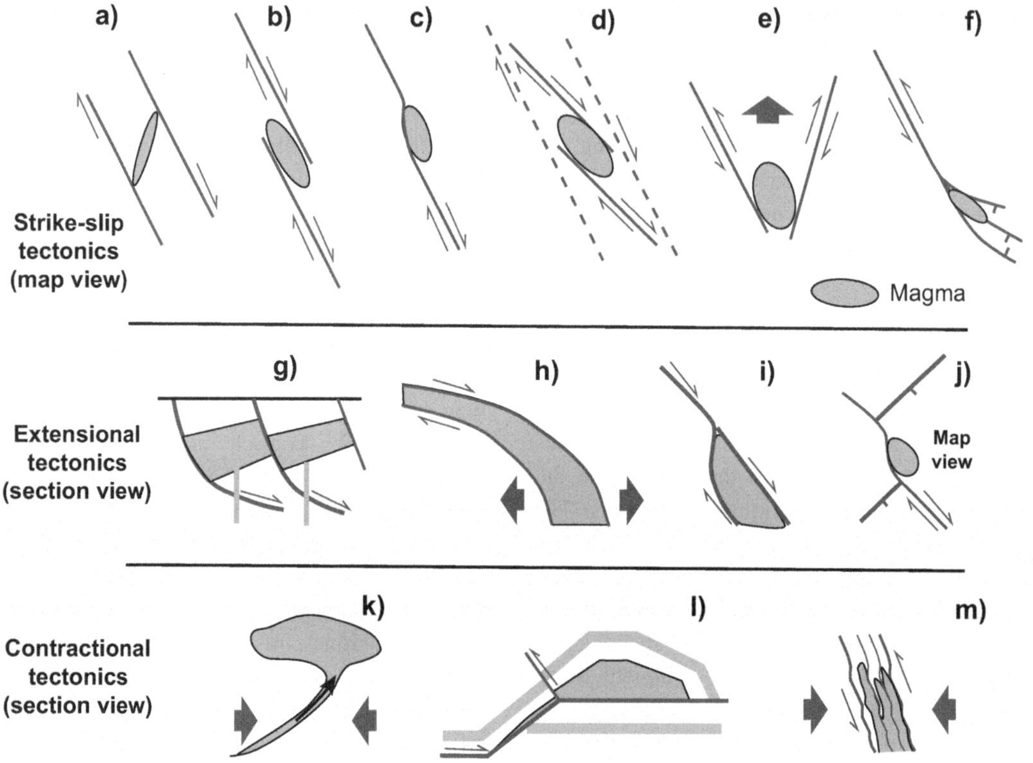

Fig. 4.20 Possible processes controlling pluton growth as a function of the regional tectonic context. In strike-slip magmatic arcs in oblique convergent settings pluton growth may occur, as shown in map view, along: **a** extension fractures oblique to shear zones, **b** pull-apart structures, **c** releasing bends, **d** extensional zones controlled by second-order en-echelon shear zones, **e** escaping blocks bounded by strike-slip faults, **f** horsetail structures. In extensional settings pluton growth may occur, as shown in section view, in: **g** growth wedges related to the activity of listric normal faults; **h** extensional shear zones; **i** extensional jogs; **j** releasing bends along transfer faults (this is shown in map view). In contractional settings pluton growth may occur, as shown in section view: **k** at shallow crustal levels, following stronger shortening at deeper crustal levels; **l** in hanging wall ramps due to thrust activity, with the feeder along the hanging wall flat; **m** in high angle reverse shear zones

4.5.3 Magma-Induced Plutons

In principle, the growth of the largest plutons, built over longer periods, may reflect the influence of a regional tectonic setting and be tectonically-assisted to a higher extent than the smallest plutons, built over shorter periods. In practice, available data show that the size of a pluton does not necessarily imply a predominant tectonically-assisted growth. In fact, the development of several plutons, as Amiata and Larderello (Italy), Central Extremadura (Spain), Mount Stuart (Washington, USA) and the Tuolumne Intrusive Suite (California), yields magmatic strain rates of 10^{-2} to 10^{-13} s^{-1}, peaking at 10^{-11} s^{-1} (Fernandez and Castro 1999; Acocella and Rossetti 2002; Albertz et al. 2005). These plutons are characterized by different size (the Central Estremadura is so large to approach a batholith), activity (Larderello has been active for ~4 Ma), shape and tectonic setting. Despite this variability in the features of the plutons, their magmatic strain rates remain consistently higher than the regional tectonic ones (which are on the order of 10^{-13} to 10^{-15} s^{-1}), and in line with modelled host rock strain rates of dike-fed plutons, which are up to six orders of magnitude faster than the fastest accepted regional tectonic strain rates (Johnson et al. 2001). Therefore, plutons may grow extremely rapidly, imposing high strain rates in the host rock and creating the space required to accommodate their growth regardless of their size, longevity, shape and independent of, even though not insensitive to, the tectonic setting (de Saint Blanquat et al. 2001, 2011). This important feature may explain the significant amount of recent studies highlighting magma-induced processes during pluton growth.

To define the magma-induced processes creating the space for pluton development let us first consider the shape of the known plutons. These are mostly tabular granites, which tend to be circular to elliptical in map view, ~10 to 100 km long and usually ~3 km thick. Data on their length (*L*) and thickness (*w*) define an empirical power law relationship similar to that obtained for laccoliths in Eq. 4.4, but with slightly different values for *j* and *n*, as:

$$w = 1.08\,L^{0.8} \tag{4.8}$$

with the exponent value $n < 1$ indicating that plutons have a tendency for horizontal lengthening over vertical thickening. Continuing on the same trend, batholiths are longer than 100 km, with thickness w ~10 km and overall n~0 (Fig. 4.16a; Cruden et al. 2017). Laccoliths, plutons and batholiths define a continuous curve that bifurcates from the magmatic sheets curve at L ~500 m towards higher thickness values. This thickness increase is expected to result from the incremental growth of stacked sheets in laccoliths and plutons. Considering the depth of emplacement *z* of all these intrusions, two first-order emplacement regimes may be defined (Cruden et al. 2017). Intrusions with length *L* much smaller than their emplacement depth ($L \ll z$) will propagate in a regime under which the dynamics of emplacement, explained by LEFM, does not involve interaction with Earth's free surface. Conversely, intrusions with $L \gg z$ will be emplaced under conditions, explained by elastic plate models, where their growth is strongly coupled to the deformable free surface and space is made predominantly by roof lifting. The transition between these two regimes occurs when $L \sim z$, which in Fig. 4.16a roughly corresponds to the bifurcation between the sheets and sills array and the laccoliths array at $L \sim 1000$ m. The distribution of the plutons in Fig. 4.16a indicates an increased thickness and length with regard to laccoliths. As there is an overall tendency for laccoliths, plutons and batholiths to become deeper as their length, thickness and volume increase, the transition between laccoliths and plutons corresponds to a change in emplacement mechanism from one dominated by roof lifting to one dominated by floor depression. As such, the transition represents a "lifting limit", which is likely controlled by the tendency for plutons to be emplaced at deeper crustal levels than laccoliths (Cruden et al. 2017). This point can be further appreciated considering the distribution of the thickness (*w*) versus length (*L*) and horizontal area (*A*) of all the tabular intrusions (Fig. 4.21a; McCaffrey and Petford 1997; Cruden 1998; Brown 2013). The distribution defines an overall S-shaped path when plotted in log *w* versus log *A* or log *L*. The vertical limits for these tabular cracks and intrusions are of $w < 1$ m (when $L < 10$ m) and $w > 10$ km (when $L > 500$ km). Between these, *w* increases with increasing *L* and *A*, with maximum slope at $L \sim 1$ km, and L/w decreases because vertical thickening dominates over horizontal lengthening. However, as *w* approaches 10 km, horizontal lengthening begins to dominate over vertical thickening, which is limited to ~15 km. The interpretation for this behaviour is that the space for the growth of these plutons is provided by roof-lifting (generating laccoliths) and/or floor-depression (generating lopoliths) following repeated sill emplacement. In particular, shallower sills may more easily bend a thinner overburden, developing laccoliths, whereas deeper sills may more easily deform viscously the lower crust, developing lopoliths. The sill to laccolith transition occurs with $L/z > 2$–3: this reflects shallow emplacement (z = 0.5–5 km) for intrusions where $L = 1$–10 km, with the change of slope at $L = 0.1$–1 km recording the increasing ability of shallow intrusions to lift their roofs as their area increases (Fig. 4.21b). The steepest

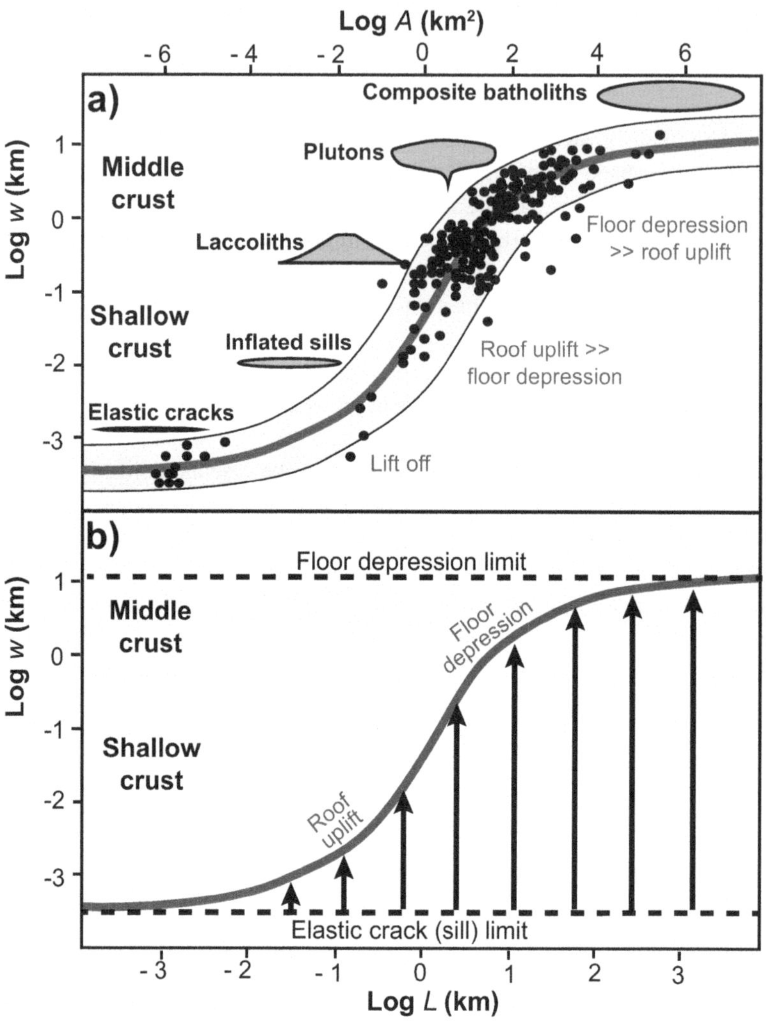

Fig. 4.21 **a** Compilation of intrusion thickness *w*, length *L* and horizontal area *A* data from sills, laccoliths, plutons, and composite batholiths. The solid S-curve is drawn though the data by eye, and the shaded area represents the approximate limits of the data about the curve. Representative intrusion styles are shown adjacent to appropriate parts of the curve. **b** Interpretation of the S-curve in terms of minimum and maximum growth limit, dominant emplacement mechanism, and depth of emplacement in the crust. Arrows show vertical growth trajectories along a possible end-member growth curve. Modified after Brown (2013), and references therein

slope likely represents a limiting curve for laccoliths to expand vertically from a sill by roof lifting along a growth line with slope $n > 1$. After this point, vertical growth of moderate to large plutons occurs dominantly by floor depression through downbending of the floor if L/z is small (deep emplacement) and the material beneath the pluton is weaker than that above it, as expected approaching the brittle-ductile transition and below it (Brown 2013).

Therefore, magma-induced plutons grow through the thickening of sill-like intrusions responsible for the uplift of their roof and/or the depression of their floor, similarly to laccoliths and lopoliths, respectively. This mechanism of thickening and coalescence of intrusions may also explain the development of batholiths up to 10–20 km thick as deriving from the stacking of thinner plutons. Equation 4.1 defines the linear relationship between magma pressure and intrusion thickness, as reported in the diagram of Fig. 4.22. As the total available magma pressure in the crust is unlikely to exceed 100 MPa, this limits the maximum thickness of individual intrusions to ~2.5 km, suggesting that thicker batholiths result from the coalescence of thinner plutons (McCaffrey and Petford 1997).

Magma-induced plutons with laccolithic or lopolithic shape have been recognized in different tectonic settings. The above-mentioned Black

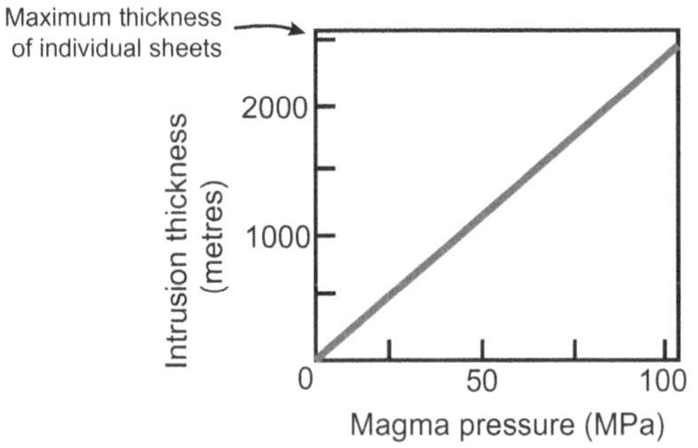

Fig. 4.22 Variation of intrusion thickness as a function of magma pressure, derived from Eq. 4.1 assuming a Young's modulus $E = 3 \times 10^{10}$ Pa and Poisson's ratio $\nu = 0.2$. Modified after McCaffrey and Petford (1997)

Mesa pluton grew through asymmetric vertical stacking of sill-like horizontal magma sheets within ~100 years, requiring a minimum vertical displacement rate of the pluton roof larger than 2 m/yr. These uplift rates, similar to those found at a few restless and resurgent calderas, confirm that plutonic and volcanic processes could be similar (De Saint Blanquat et al. 2006). The granitic Papoose Flat pluton (California) initially emplaced as inclined sill and subsequent inflation developed a laccolith within 30 ka. Cooling from the roof of the pluton downward resulted in cessation of vertical inflation on its west side and promoted lateral expansion and floor depression (Morgan et al. 1998; De Saint Blanquat et al. 2001). La Tojiza (Spain) and Amiata (Italy) granitic intrusions share common features, as both developed in an extensional regime, with gravimetry data suggesting a laccolith shape with maximum preserved thickness of 1 (La Tojiza) or 2 km (Amiata) and a root zone at the border (Fig. 4.23; Acocella 2000; Aranguren et al. 2003). Lopolithic plutons have been recognized in the Great Eucrite gabbroic intrusion (Scotland), with overall funnel-shape geometry, or in the felsic Coastal Plutonic Complex (British Columbia, Canada). Here, in response to ballooning due to multiple injections of magma, space was primarily made by downward displacement of country rock and concomitant depression of the Moho (Fig. 4.13b; Brown and McClelland 2000; O'Driscoll et al. 2006). Finally, present-day stacked sills actively assembling a tabular pluton have been imaged in southern Tibet as bright spots of high conductivity at 15–20 km depth, corresponding to zones of seismic attenuation and correlated to high heat flow at the surface (Gaillard et al. 2004).

Therefore, an effective solution to the space problem during pluton growth is the translation of the roof and/or the floor of the intrusion, which can be achieved irrespective of the regional tectonic setting. However, the growth and shape of these tabular plutons may still not be insensitive to the tectonic setting. In contractional settings, the tabular shape of plutons is determined by the horizontal maximum principal stress σ_1. In extensional settings, subhorizontal tabular intrusions may be promoted by the transient stress changes induced by dike emplacement, with the horizontal regional σ_3 being replaced by the magma-induced σ_1 (Fig. 3.21b; Vigneresse et al. 1999).

4.6 Magma chambers and magma reservoirs

4.6.1 Overview

Magma chambers and magma reservoirs are the sites where shallow magma accumulates to feed eruptions, although most of the accumulated magma does not reach the surface and remains permanently stored in the crust, eventually forming solidified plutons. The terms magma chamber and magma reservoir have been often considered synonymous and commonly approximated by some well-defined sphere permanently filled with molten magma. Recent studies, mainly based on geophysical and petrological data, have challenged this classic and popular approach, underlining a spatial and temporal more complex distribution of the magma stored below a volcano.

Geophysical studies have repeatedly aimed at constraining the depth, size and shape of magma reservoirs below active volcanoes, especially

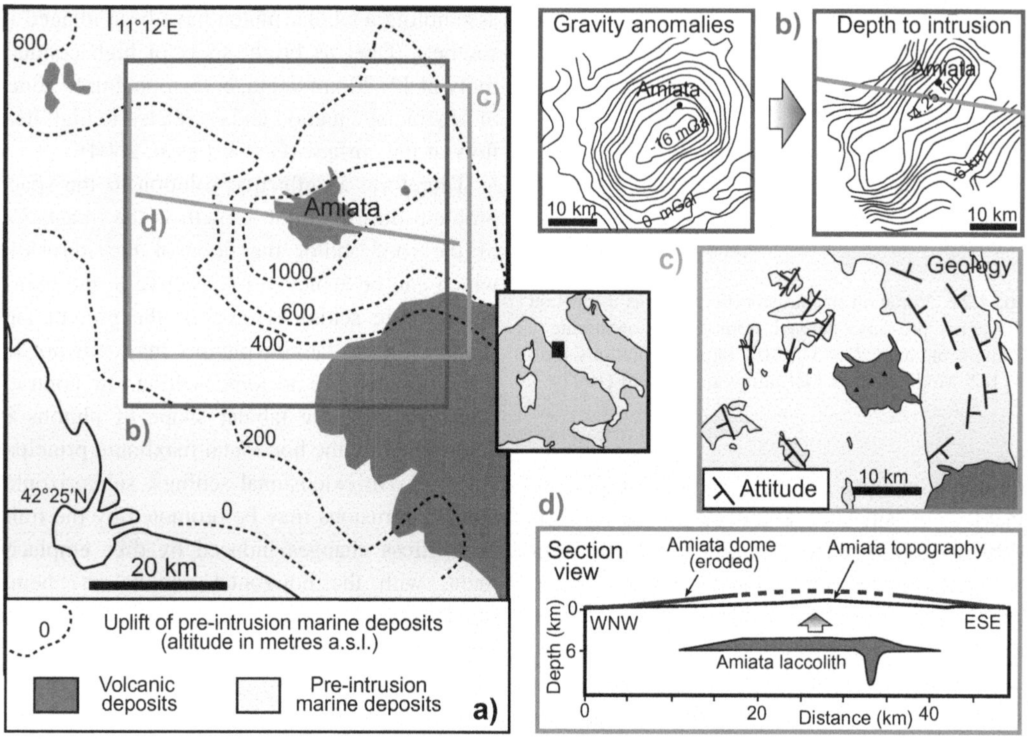

Fig. 4.23 The buried granite laccolith at Amiata (Italy) responsible for the uplift of the Lower Pliocene (pre-intrusion) marine deposits **a**, as highlighted by gravity data and their inversion, which suggests the pluton's depth **b**, by the domed attitude of the pre-intrusive deposits **c** and by the resulting section view **d**. Modified after Acocella (2000)

through tomography produced by the passage of seismic waves. In fact, the 3D variations in the velocity of the seismic waves are often used to try to identify magma bodies, assuming that high velocity anomalies may highlight denser solidified plumbing systems and low velocity anomalies may highlight lighter melt zones (for details see Sect. 8.4.1"). Tomographic studies have often struggled to detect melt-dominated crustal portions, with most recognized anomalies in the range of ± 10% perturbation and seldom higher than 30% (Lees 2007; e.g., McVey et al. 2019). Yellowstone caldera, with its record of and potential for **super-eruptions** (which are eruptions producing > 450 km^3 of magma during a single episode), lies above a partially molten rock with only 2–15% melt present. Similarly, Rabaul caldera (Papua New Guinea), which has experienced prolonged unrest since the 1970s culminating in repeated eruptions since the early 1990s, overlies low velocity zones suggesting only ~10% of melt fraction (Bai and Greenhalgh 2005; Coleman et al. 2016). These low estimates of melt fraction, even below large and erupting calderas, highlight a significantly lower amount of eruptible magma than expected.

Petrological data show that phenocrysts in volcanic rocks commonly became suspended in their host magma only for very limited periods prior to eruption, from years to centuries. For example, recharge for the ~3.6 ka old Minoan eruption at Santorini (Greece), despite the gestation of 18 ka with the previous major eruption, occurred only during the century before the eruption (Druitt et al. 2012). Similarly, at Taupo caldera (New Zealand) the onset of rapid heating and priming of the host silicic mush occurred less than 120 years prior to each of the youngest three eruptions (2.15–1.7 ka), with subsequent melt accumulation occurring in only decades. In some cases, as for the 27 ka Oruanui super-eruption, the rapidity of melt extraction and accumulation suggests an external driver, possibly a rift-related regional event promoting important stress changes (Allan et al. 2012; Barker et al. 2016). Concerning more recent events, the final recharge event of the 1912 Novarupta-Katmai (Alaska, USA) VEI 6 eruption occurred just weeks prior to the eruption itself (Singer et al. 2016). These short recharge times do not seem to be affected by the eruption size, with similar estimates (a few tens of years) obtained for final magma recharge events in eruptions of 10^1 km^3 or 10^3 km^3 of magma at Taupo, Okataina (New Zealand) and Toba (Sumatra; Matthews et al. 2012). The evidence that magma resides in reservoirs for only a brief period is paired by studies on plutons, as for example at the 7 Ma old and larger than ~ 80 km^3 Monte Capanne intrusion in Elba Island (Italy): here the transition from eruptible magma to immobile granitic mush and pluton formation occurred in only 10–40 ka (Barboni and Schoene 2014). These studies show that before eruption magma can accumulate on timescales as short as decades, geologically negligible in comparison to the preceding repose period, and whole magma systems can solidify and be rebuilt in millennia. The short pre-eruption storage times imply that the detection of voluminous reservoirs of eruptible magma, even beneath active super-volcanoes, may be possible only prior to impending eruption, also raising important implications for the monitoring of long-dormant, but potentially active, large magmatic systems (see Chap. 8; Reid 2008; Wotzlaw et al. 2015; Wilson and Charlier 2016). At the same time, the large spread in radiometric ages of suspended crystals (notably zircon) from erupted products is explained by very long residence times in magma reservoirs. For example, the Black Peak intrusive complex (Cascades, Washington, USA) may have remained "mushy" for long periods (10^5 years; Shea et al. 2016). All these observations have been reconciled proposing that magmatic systems consist of crustal portions wherein melt is dispersed within a dominating framework of crystals responsible for an overall mushy state, except for transient periods, where lens with higher proportion of melt, the magma chambers themselves, may form (Kennedy et al. 2018; Sparks et al. 2019).

Therefore, geophysical and petrological data suggest that magma chambers usually host a low amount of melt, with most of their lifetime spent in a crystalline state, except for very limited periods in the magmatic history of the volcano, when the amount of melt suddenly increases, eventually feeding eruptions. This condition also applies to super-volcanoes, where semisolid crystal mushes, rather than liquid-rich magma chambers, may be the prevalent condition. According to this modern view, a **magma reservoir** is the deep portion of the volcano which includes both crystal- and melt-dominated region(s), as well as pockets of exsolved magmatic fluids. Portions of crust from the surface to the Moho may be characterized by this condition, thereby identifying a transcrustal magma reservoir (Fig. 4.24a). The crystal-dominated region is the **mush**, a melt within a crystalline framework where melt segregation and heat loss alternate with faster destabilisation and magma transport. The melt-dominated region is the **magma chamber**, an ephemeral magma accumulation which may form by incremental magma intrusion into sub-solidus rocks or the segregation and merging of melt-rich layers within mush regions. Therefore, a magma reservoir may show considerable heterogeneity, with different parts separated by sub-solidus rock or connected and continuous, and may consist of one or several transient magma chambers. The modern concept of magma reservoir implies a strongly transient behaviour, with the timescales of magmatic processes highlighting prolonged storage (10^3–10^6 years), punctuated by rapid magma transfer, remobilization and ascent during unrest episodes and, eventually, eruptions. In the latter case, large magma volumes may evolve as a single enlarging reservoir or through the rapid amalgamation of small previously independent magma pockets (Fig. 4.24b; Sparks and Cashman 2017; Cashman et al. 2017; Edmonds et al. 2019). As a

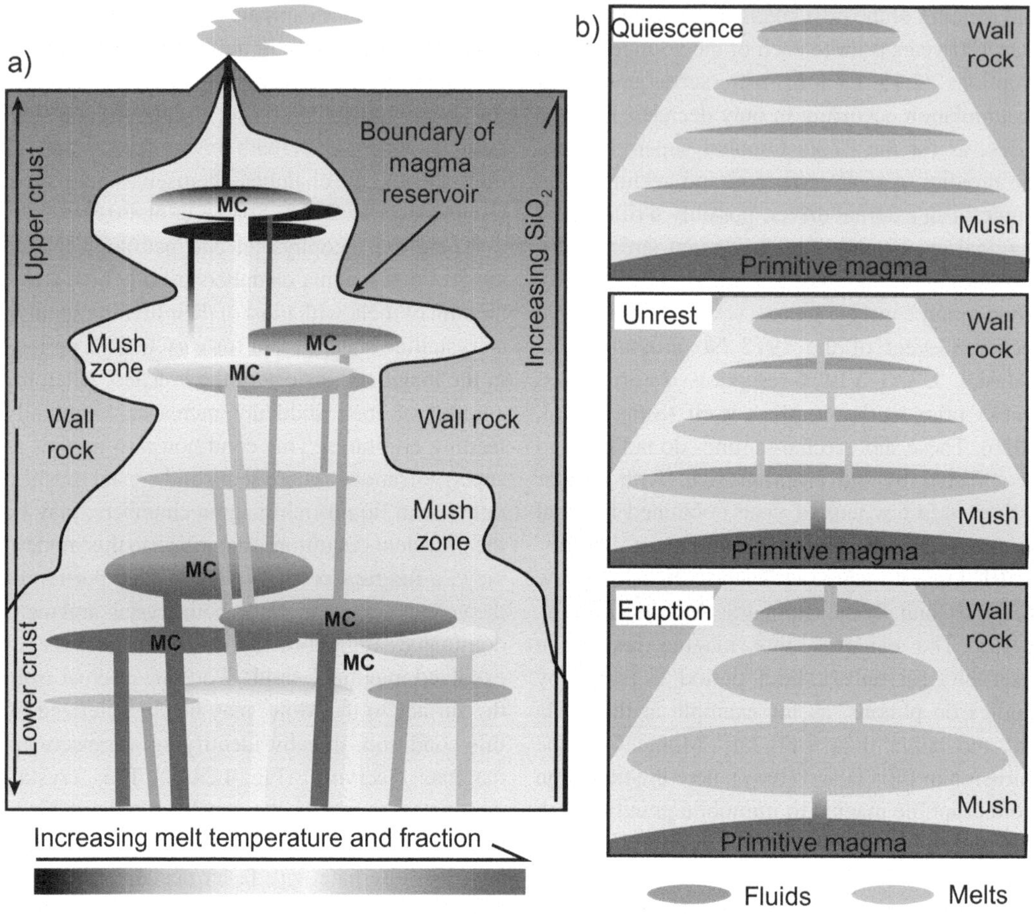

Fig. 4.24 **a** Scheme of mature transcrustal magma reservoir, characterized by a mush zone (light orange) hosting several magma chambers (MC) at different temperature, usually decreasing towards the surface (modified after Sparks et al., 2019). **b** Variations in the state of a transcrustal magma reservoir. During quiescence isolated melt pockets slowly segregate and exsolve fluids; during unrest the pockets connect and move upward, eventually forming transient larger magma chambers

result of these transients, a transcrustal magmatic reservoir may be affected, especially at shallow levels, by major physical variations in temperature, viscosity and strength. Magma addition also modifies the crustal gradients, shifting the brittle-ductile transition upward, promoting a hotter and weaker crust (Fig. 4.25).

Despite these modern perspectives, the abstraction of a well-defined magma chamber with simple and smooth geometry still provides an enduring and useful reference for thermal calculations and for the magmatic sources obtained through the inversion of tomographic and geodetic data at active volcanoes. This simplification is somehow supported by the fact that thermal considerations suggest that a long-lived magma chamber cannot be very irregular in shape: any irregularity would expose a relatively larger area to the contact with the colder host rock, rapidly transferring heat, cooling and solidifying the magma (e.g., Jaeger 1964; Gudmundsson 1990). Therefore, any irregularity tends to be smoothed and erased during the lifetime of the chamber, which should ideally approach a sub-spherical shape. Tomographic and geodetic data often lack the resolution (<1 km) required to constrain the precise melt distribution within magma reservoirs, providing

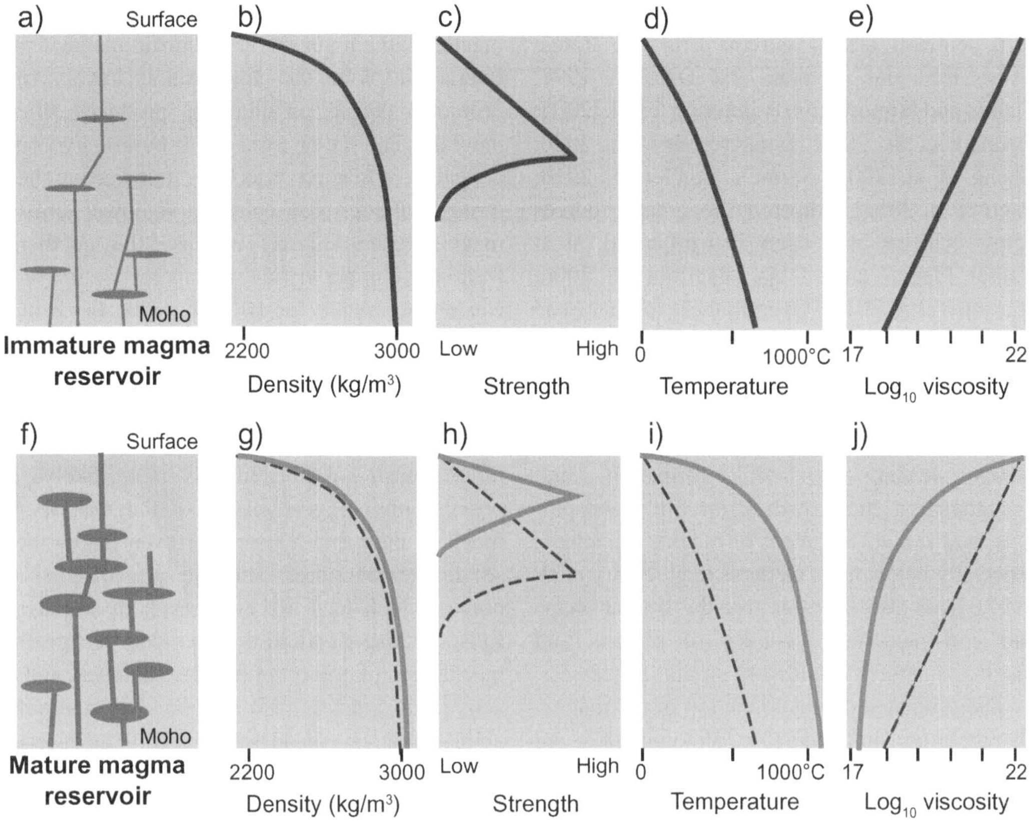

Fig. 4.25 Schematic depth profiles of density, strength, temperature and viscosity variations of transcrustal magma reservoirs; parameter values are indicative. Two end-members are shown. **a–e** An unperturbed state of immature transcrustal magma reservoir which might characterize either the early stages of magmatism or low magma flux conditions. **f–j** A mature transcrustal magma reservoir where igneous activity has strongly perturbed the physical structure of the crust; the unperturbed profiles (corresponding to states a-e) are here shown as dashed dark violet lines. The background colours underlie the upward shift of hot, weak, viscous lower crust (predominant light orange), as opposed to cold, strong, brittle upper crust (predominant light brown), as the transcrustal magma reservoir reaches maturity. Modified after Sparks et al. (2019)

only schematic and simplified perspectives. With these limitations, tomography has commonly identified shallow (<10 km deep) and sill-like magma reservoirs below volcanoes, including Vesuvio and Campi Flegrei (Italy), Rabaul (Papua New Guinea), Okmok (Aleutians, USA), Newberry (Oregon, USA), Soufrière Hills (Montserrat) and Katla and Askja (Iceland) (Gudmundsson et al. 1994; Auger et al. 2001; Bai and Greenhalgh 2005; Zollo et al. 2008; Beachly et al. 2012; Paulatto et al. 2012; Mitchell et al. 2013; Lu and Dzurisin 2014). Elsewhere, tomography has highlighted less distinctive chamber shapes, as at Mount St. Helens (Oregon), Valles (New Mexico, USA), Deception Island (Antarctica) and Axial Seamount (Juan de Fuca Ridge), or more complex melt networks, as at Klyuchevskoy (Kamchatka, Russia) or Long Valley (California; Roberts et al. 1991; West et al. 2001; Waite and Moran 2009; Zandomeneghi et al. 2009; Koulakov et al. 2011; Seccia et al. 2011). The inversion of geodetic data also suggests magmatic sources with different shapes. Among these, sill-like sources are particularly frequent below calderas, including Kilauea (Hawaii, USA), Yellowstone, Sierra Negra (Galapagos) and Uzon (Kamchatka, Russia), and elsewhere, as Eyjafjallajokull (Iceland),

Three Sisters (Oregon), Socorro (Rio Grande Rift, southern USA), and the Erta Ale Range (Afar, Ethiopia) (Dvorak and Dzurisin 1997; Fialko and Simons 2001; Jonsson et al. 2005; Dzurisin et al. 2006; Lundgren and Lu 2006; Chang et al. 2007; Sigmundsson et al. 2010; Pagli et al. 2012). Ellipsoidal magmatic sources below calderas have been also proposed, as at Campi Flegrei and Askja (Pagli et al. 2006; Amoruso et al. 2014). Overall, these tomographic and geodetic data highlight a recurrent sill-like or tabular shape for magma chambers, broadly consistent with that commonly inferred for plutons. This strengthens the continuum in the processes leading to sill formation and amalgamation, the growth of laccoliths and plutons, and the development of magma chambers, especially below large calderas (e.g., Henry et al. 1997). This also suggests that the similar shape and constructional dynamics of plutons and magma chambers result from similar processes.

The estimated depths to the top of the magma chambers are quite variable, although they are shallower than 10 km. Available geodetic, tomographic, geological and petrological data suggest that these depths primarily depend on the tectonic setting of the volcano: the higher the contribution from regional or local tectonic extension, the shallower the chamber. Therefore, the shallower magma chambers are found along divergent plate boundaries and, subordinately, in zones of local extension in oblique arcs. Conversely, the deeper magma chambers are found in arcs experiencing dominant contraction (see Chap. 12; Purdy et al. 1991; Chaussard and Amelung 2012).

4.6.2 Development Conditions

Petrological studies allow understanding the formation and development of magma chambers, partly clarifying also their eruptive activity (e.g., Bachmann et al. 2007; Collins et al. 2016; Edmonds et al. 2019). The formation of magma chambers is commonly explained as due to episodic magma addition from depth. Hydrous basaltic magma, produced by melting of the mantle, is the main driver of crustal magmatism. Interaction with the non-basaltic lower crust promotes partial melting and produces silicic magma. The latter rises and builds hydrous, crystal-rich magma reservoirs transferring heat through the crustal column. A lower crustal magma reservoir may increase the geotherm forming an upper crustal magma reservoir which can remain stable for 10^5–10^6 years. Once in a mush state, the silicic magma may solidify without erupting or be periodically reactivated by the direct input of hotter mafic or felsic magma, eventually leading to eruptions. The first possibility (mush solidification) is more likely, as observations suggest that most magmas spend much of their time close to or even below their solidus temperatures, and are heated in short bursts, which may not even result in eruptions. Episodic magma addition is a widely recognized mechanism of growth of magma chambers and at active volcanoes is also supported by the variability of unrest episodes alternated to quiescence (see Chap. 9). Similarly to plutons, the growth of magma chambers is inferred to be mainly promoted by the incremental assemblage of multiple dike-fed sills, eventually developing shallow laccoliths (Fig. 4.26; e.g., Gudmundsson 2012). The geophysical and geodetic evidence that most magma chambers have tabular shape supports this view.

In addition to the widely recognized episodic addition from depth, petrological models suggest that magma chamber formation may occur through the rapid amalgamation of small, initially independent shallow magma pockets. In this "mush model", magma chambers are created by extracting interstitial liquid from mushes and collecting it in lenses, not necessarily requiring an external supply (Bachmann and Bergantz 2008). The mush model assumes a crystal fraction window (between $\sim$45 and $\sim$65 vol% crystals) in which the separation of crystals and interstitial liquid is particularly efficient. This window hinders the convection currents that efficiently stir the magma chamber (and develop below 45 vol% crystals), allowing the separation

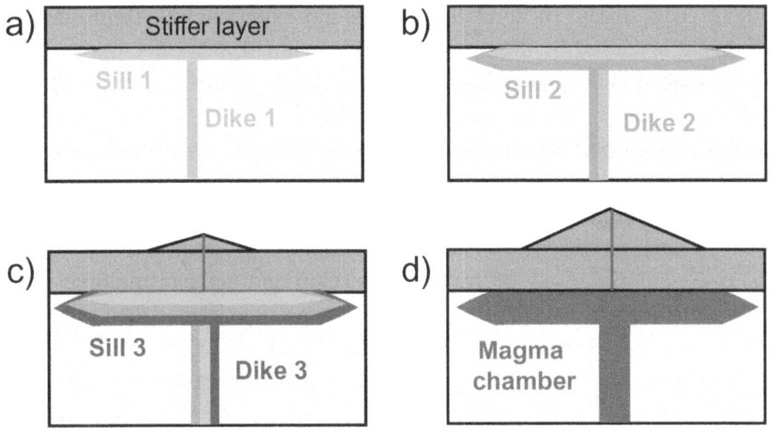

Fig. 4.26 Scheme of magma-chamber formation through repeated sill injection. A sill forms from a feeder dike at the contact between mechanically different layers **a**. Subsequent feeder dikes become arrested, developing further sills, whose magmas become partly absorbed by the original sill **b** and **c**. The sill cluster expands and forms a sill-like magma chamber that feeds a volcano **d**; here it is assumed that all the sills remain liquid during the formation of the chamber. Modified after Gudmundsson (2012)

of crystals from liquid through compaction (see also Sect. 7.4).

The stability of a formed magma chamber depends upon several features, including its depth and the interactions between its thermal and mechanical processes. As for the depth, this is in general related to the thermal state of the crust: deeper emplacement levels in hotter environments can both extend magma survival and increase melt availability (Gelman et al. 2013a). However, from a petrological perspective the situation may be more complex. In fact, simulations suggest that, at pressures lower than 1.5 kbar, the presence of an exsolved magmatic volatile phase hinders chamber growth because eruptive volumes are usually larger than the recharges feeding the system. Also, at pressures higher than 2.5 kbar, the viscosity of the crust may be sufficiently low to inhibit most eruptions. Therefore, sustainable eruptible magma chambers may develop mostly within a relatively narrow range of pressures around 2 $\pm$ 0.5 kbar, where the amount of exsolved volatiles fosters growth, while the high viscosity of the crust promotes the necessary overpressurization for eruption (Huber et al. 2019). As for the interactions between the thermal and mechanical processes, the former involve the growth and state (molten, mushy, solid) of the chamber, whereas the latter refer to the possibility to rupture a chamber due to a dike injection. For a stable and growing chamber there should be a balance between the thermal processes avoiding solidification, where more heat is supplied by convection than lost by conduction, and mechanical processes leading to eruption. In fact, magma supply rates should not be as high as to trigger eruptions and thereby prevent magma accumulation (Karlstrom et al. 2010; Edmonds et al. 2019). Modelling suggests that a stable equilibrium of a chamber results when its volume changes are negligible compared to its total volume, with the volume of total melt approximately constant in time. Temporary stability may be achieved also through the slow growth of the chamber, as a progressive and steady increase in flux does not result in relevant magma accumulation. A sudden and significant step-like flux increase of at least one order of magnitude above the baseline may produce large magma chambers, although soon leading to instability deriving from chamber expansion and ultimately rupture. Eruption thus occurs for relatively high fluxes, even though this depends strongly on the depth and size of the chamber, with larger and deeper chambers being the most stable (Schopa and Annen 2013). The decay of a chamber deriving from freezing is caused by the opposite process, when the supplied melt volume is only a small fraction of the chamber volume, not providing enough enthalpy to prevent progressive solidification. This leads to a decrease in the size of the magma chamber. Even though smaller and shallower chambers are more susceptible to freezing, below a certain threshold of magma flux, any chamber will freeze.

Therefore, **magma flux** appears a crucial factor controlling chamber stability. This is usually calculated as the amount of magma provided to the system in a certain period (km^3 yr^{-1}), although recent studies suggest that an area-normalized flux (km^3 km^{-2} yr^{-1}) provides a more useful estimate (Silwinski et al. 2019). As the growth of magma bodies is episodic, magma fluxes may vary by orders of magnitude with time. Integration of geochronological and field data with numerical simulations suggests that magmas accumulate to form melt-rich chambers only during episodes of high magma flux. In particular, thermal simulations of the formation of large magma chambers within plutons that grow by vertical stacking of sills have estimated that only if magma fluxes exceed a threshold between 10^{-3} and 10^{-2} km^3 yr^{-1} are the volumes of eruptible magma large enough to feed voluminous silicic explosive eruptions, and yet the fluxes must not be too high or an eruption would occur and prevent magma accumulation (Annen 2009; Gelman et al. 2013a). These estimates are consistent with available calculations of eruptive rates from the largest silicic provinces, which range between 10^{-3} and 10^{-2} km^3 yr^{-1}, although the latter are minimum estimates as lacking the intrusive component, which is usually one order of magnitude larger (e.g., White et al. 2006, and references therein). Unrest episodes at active volcanoes also allow recognizing transient high fluxes in magma chambers. For example, the Laguna del Maule volcanic field (Chile) experiences a decade-long, surface inflation at more than 20 cm/yr, representative of high magma fluxes and comparable with those that have produced moderate to large volume caldera-forming eruptions elsewhere. Similarly, Campi Flegrei experienced an episode of high magma flux in the 1982–1984 unrest episode (Kilburn et al. 2017; Andersen et al. 2017; Singer et al. 2018).

In addition to the highest magma fluxes, another requisite for the accumulation and mobilization of thousands of km^3 of magma during large eruptions is the possibility to develop and activate multiple magma bodies at the same time, rather than a single large magma chamber. For example, the current architecture of the super-magmatic system beneath Yellowstone may consist of several isolated magma batches, although each with limited amount of melt (Wotzlaw et al. 2014; Edmonds et al. 2019). In addition to a more frequent and significant magma supply, the coalescence of multiple chambers, as deriving from large-scale extraction of interstitial liquid (the above mentioned mush model), may be a means to accumulate the magma to fuel large eruptions. Numerical models show that multiple chambers coalesce only under the highest melt supply rates and in the deepest (hotter) systems, when mechanical ruptures and mush mobilization occur. The models also suggest that large eruptions are expected to occur long after the main intrusive activity, as the highest temperatures and melt fractions are reached during a period of thermal relaxation after prolonged melt injection has ceased. Following large eruptions, the mush is expected to crystallize significantly (up to 70–80 vol% crystals) due to syn-eruptive decompression (Gelman et al. 2013b; Biggs and Annen 2019).

At the other end of the spectrum, the long-term average pluton emplacement rates usually inferred from geochronological data (10^{-4} km^3 yr^{-1}) are too low to allow for the formation of large magma chambers. This implies that many intrusive suites making up the roots of eroded magmatic arcs may not represent cumulates left behind following large eruptions and thus may not be genetically linked to the magma chambers responsible for these eruptions (Annen 2009; Davis et al. 2012). Indeed, there is a missing record of granitoid plutonic complexes representing large fossil upper crustal magma chambers, with only one reported case (the Eocene

Golden Horn batholith, Washington) bearing intrusive rates of $\sim$0.0125 km^3 yr^{-1}, that are high enough to build silicic magma chambers: this case may thus represent the first fossil large-volume silicic magma chamber ever recognized (Eddy et al. 2016). Therefore, the common assumption that magma chambers represent melt-rich regions expelled from mushy portions of reservoirs which later crystallize to form plutons should be significantly relaxed, at least for large silicic systems (see Sect. 4.7).

The magma flux supplying a chamber also affects the geometry of the plumbing system and ultimately, the size of a volcano. Stratovolcanoes producing frequent minor to moderate eruptions are associated with smaller magma chambers deriving from lower fluxes, consisting of magma pulses with limited areal extent. Conversely, calderas producing less frequent and larger eruptions are associated with wider and tabular magma chambers deriving from higher fluxes of magma. The latter would consist of large pulses, likely in the form of sills, that spread laterally over a significant distance (Annen et al. 2015).

A more general long-standing question regarding the evolution of magma chambers is how much magma is erupted versus how much is retained in the form of solidified intrusions. This may be estimated through the Volcanic to Plutonic Ratio (VPR), expressed as the ratio of eruptible magma (having a melt fraction > 0.4) to non-eruptible magma (melt fraction $\leq$ 0.4). Theoretical models show that the VPR varies as a function of the magma flux, or supply rate to the system. In high-flux small magmatic systems, VPR $\sim$0.5 for most of the lifetime, dropping when supply ceases: here VPR transiently exceeds 1 as eruptible volumes exceed non-eruptible volumes. In low-flux large systems, VPR remains low, usually lower than 0.2. The VPR thus varies not only for different magmatic systems, but also during the lifetime of the same system, with estimates varying from 0.03 to 1, 0.2 being a typical value, suggesting that low-flux systems are dominant (White et al. 2006; Blundy and Annen 2016).

4.6.3 Magma Chambers Attracting and Nucleating Dikes

Once formed, a magma chamber becomes able to generate a local stress field influencing the path of dikes propagating in the nearby crust and, most importantly, to nucleate magma-filled fractures, which may eventually feed eruptions.

Analogue models provide insights on how dikes propagating through the crust respond to the stress field generated by a nearby magma chamber (Pansino and Taisne 2019). Generally the dike propagation path depends on the sign of the chamber pressure (i.e., inflated/overpressurized or deflated/underpressurized) and on the dike's initial orientation (i.e., if it strikes radially, circumferentially, or obliquely to the chamber). Dikes that are initially radial respond by propagating toward overpressurized chambers and away from underpressurized chambers. Otherwise-oriented dikes respond by changing orientation, twisting and curling following the principal stresses. With overpressurized chambers, dikes approach the chamber propagating radially to it; with underpressurized chambers, dikes avoid the chamber propagating tangentially. These interactions are explained by the orientation of the minimum principal stress σ_3 induced by a chamber with a given pressure, to which a dike tends to be perpendicular. In particular, around overpressurized reservoirs σ_3 becomes tangential, attracting radial dikes, whereas around underpressurized reservoirs it becomes radial, deflecting dikes tangentially. In both underpressure and overpressure conditions, the higher the chamber-induced stress, the more pronounced the reaction. The extent of this influence, or capture radius, is commonly 10–20 times the chamber size, implying that magma chambers attracting dikes may maintain a stable spacing between volcanoes. This may also explain the spacing between adjacent but distinct magmatic systems, each feeding from its capture zone (Karlstrom et al. 2009).

Despite the capacity to influence the propagation path of crustal dikes, the most distinctive feature of a magma chamber remains the capacity to nucleate magma-filled fractures, including the dikes that may feed eruptions. Polygenic volcanoes and many monogenic volcanoes would not form without active and overpressurized magma chambers nucleating feeder dikes. Here the conditions controlling dike nucleation from a magma chamber are considered, complementing Sect. 3.4, which discusses the more general problem of dike propagation at any crustal level. The conditions to initiate a dike, or more generally any magma-filled fracture, from a magma chamber are determined by the total pressure P_{tot} existing within the chamber, the in situ tensile strength of the host rock T_0 and the minimum principal stress σ_3 acting in the host rock (Gudmundsson 2012). The total pressure includes the lithostatic pressure P_l at that depth and the excess pressure of the magma P_e above the lithostatic pressure, so that:

$$P_{tot} = P_l + P_e \qquad (4.9)$$

To nucleate a dike, the total pressure at the chamber contact must balance the sum of the pressures σ_3 and T_0, which resist dilation, as follows:

$$P_l + P_e = \sigma_3 + T_0 \qquad (4.10)$$

If the magma chamber is completely molten and behaves as a fluid, then the stress within becomes isotropic, so that $\sigma_1 = \sigma_2 = \sigma_3 \sim P_l$; Eq. 4.10 thus can be approximated as:

$$P_e \sim T_0 \qquad (4.11)$$

which implies that, to nucleate a dike from a magma chamber, the excess pressure should reach the tensile strength of the host rock. The fact that the latter is experimentally determined to focus in the range of 0.5–9 MPa suggests that relatively low excess pressures are needed to nucleate dikes (Gudmundsson 2012). Nevertheless, these low magma excess pressures may be required only in the case of lithostatic pore pressure conditions. In fact, numerical models suggest that under zero pore pressure conditions (i.e., in dry rock) the overpressure required to nucleate dikes increases linearly with depth, potentially reaching a few hundred of MPa at 5 km depth (Albino et al. 2018).

More generally, Eq. 4.10 shows that the stress conditions for magma-chamber rupture and the nucleation of magma-filled fractures can be achieved through internal and external triggers. Internal triggers promote an increase of the magma excess pressure, as for example injecting magma or exsolving volatiles: this increase encourages the nucleation of dikes, sheets and sills (e.g., Bachmann and Bergantz 2008). Magma viscosity may affect the rate at which the excess pressure required to nucleate a dike increases within the chamber. If a less viscous mafic magma chamber is suddenly pressurized, the delay before dike nucleation is of hours to days; for more viscous felsic magma, the delay may last many years. Therefore, more viscous magma requires greater chamber overpressures and/or longer time delays to promote dike nucleation. As a result, repose periods are longer and dikes may propagate further before freezing, eventually resulting in more vigorous eruption (McLeod and Tait 1999).

Assuming a host rock with constant tensile strength (see Table 2.1), the external trigger of Eq. 4.10 consists of a reduction of the minimum principal stress σ_3. This may occur (a) along divergent plate boundaries or (b) with the generation of transient stresses, and usually encourages the nucleation of dikes (Gudmundsson 2012). As for (a), divergent plate boundaries are characterized by quite stable conditions of low magnitude of the minimum principal stress σ_3, which may even become tensile at very shallow crustal levels (less than 1 km). Therefore, this tectonic setting provides permanent conditions for a reduced σ_3. As for (b), transient stresses nearby magma chambers may result from other pressurized magma bodies, promoting stress interactions which are strongly dependent on the distance and shape of the bodies or, more commonly, from seismic activity, particularly if affecting the rupture of crystalline, solid-like silicic reservoirs (reaching crystallinities beyond 50 vol.%). In the latter case, if experiencing rapid strain rates (>10^{-3} or 10^{-4} s^{-1}) during near-field seismicity, these reservoirs may more easily fail (see also Sect. 9.3.3; e.g., Gottsmann et al. 2009). Pre-existing magma-filled cracks or other perturbations in the chamber surface may also promote dike nucleation at much lower magma pressures than those commonly required for tensional fracturing.

In addition to the internal and external triggers included in Eq. 4.10, other conditions may influence the nucleation of magma-filled fractures from magma chambers, as the shape and size of the chamber. The shape of a pressurized magma chamber, also when coupled with a remote stress field, may affect its rupture conditions, as suggested by the distribution and intensity of the induced tensile stresses, which encourage the nucleation of magmatic sheets (Gudmundsson 2006, and references therein). In the case of a spherical magma chamber with excess pressure of 5 MPa, the highest tensile stresses, reaching 7 MPa, focus in two areas (in section view) on the upper part of the chamber, as indicated by the contours of the minimum principal stress σ_3. The focusing of the stresses on the upper part of the chamber results from the influence of the nearby free surface (Fig. 4.27a). If a spherical magma chamber is subject only to a remote horizontal tensile stress, the highest computed tensile stresses reaches 15 MPa at the top of the chamber, close to the free surface (Fig. 4.27b). In the case of a sill-like oblate magma chamber with excess pressure of 5 MPa, the tensile stresses peak, reaching 50 MPa, at the upper parts of the lateral ends of the chamber, which are the sites of maximum curvature (Fig. 4.27c). Therefore, both the magnitude and location of the tensile stresses around magma chambers depend on the chamber shape and the application of any remote stress field. The presence of a volcanic edifice may further affect the tensile stresses around the magma chamber, ultimately controlling the location and path of the nucleating dikes (Pinel and Jaupart 2003). Without any edifice, the stresses around a spherical chamber reach a maximum in two areas (in section view) away from the top of the chamber, similarly to the case of Fig. 4.27a. Conversely, with a volcanic edifice the maximum stress concentration and thus failure occurs at the top of the chamber, beneath the edifice summit, similarly to the case with remote tensile stress of Fig. 4.27b. This feature may explain the higher frequency of eruptions from the central summit crater of a volcanic edifice. The size of a magma chamber is a further factor controlling the nucleation of magma-filled fractures and the possibility that these reach the surface. Recent models suggest a minimum volume of the magma chamber, estimated from ~0.01 to 10 km^3 depending on several factors, required to nucleate feeder dikes. Below this threshold eruptions are suppressed or of limited volume (less than the dike volume; Townsend and Huber 2020). However, larger chambers (with volume > 10^2 km^3) cannot be sufficiently overpressurized to form dikes through a warmer and viscous host rock and are more prone to store

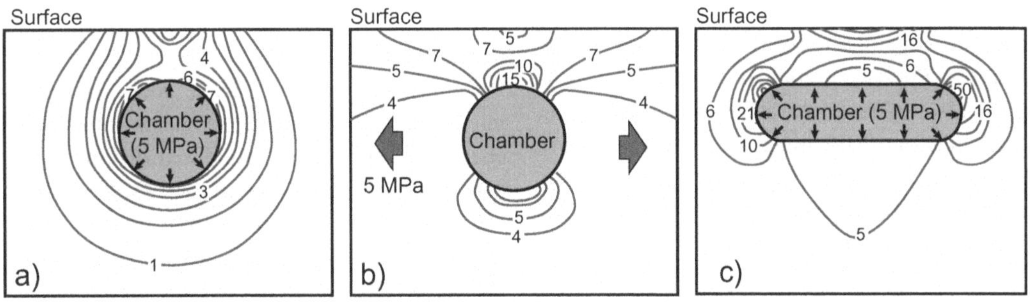

Fig. 4.27 Modelling results showing the maximum principal tensile stress concentration contours (blue lines; black numbers refer to values in MPa) in section view around magma chambers with **a** spherical shape with internal pressure of 5 MPa, **b** spherical shape experiencing only a remote tensional stress of 5 MPa, and **c** elongated shape with internal pressure of 5 MPa. The computed related stress concentrations reach 7 (in a), 15 (in b) and 50 MPa (in c) at different locations around the chamber. Here a homogeneous, isotropic crust with stiffness of 40 GPa and a Poisson's ratio of 0.25 are assumed. Modified after Gudmundsson (2006)

magma, possibly requiring additional triggering eruptive mechanisms (Jellinek and DePaolo 2003; Gregg et al. 2013).

The lengthening of a dike nucleated and propagating from a magma chamber as a function of time may be described in relatively simple terms by a logarithmic law, which may be also used to forecast dike growth (Grossman-Ponemon et al. 2020). If the nucleated dike meets the propagation conditions described in Sect. 3.4, it may reach the surface and feed an eruption. In this case, assuming a magma chamber partly filled with crystals and thus with poroelastic behaviour, the average volume of magma V that flows out of the chamber through the feeder-dike (including the erupted and intruded volumes) may be estimated as (Gudmundsson 2006):

$$V = fP_e(\beta_c + \beta_m)V_c \tag{4.12}$$

where f is porosity of the chamber (corresponding to the melt volumetric fraction; if the shallow magma chamber is totally molten $f = 1$), P_e is the magma excess pressure in the chamber before rupture, β_c is the compressibility of the host rock, β_m is the magma compressibility and V_c is the total volume of the chamber.

The overall volumetric flow rate Q_f of a dike nucleated from a magma chamber varies through time t as:

$$Q_f = Q_e e^{-Ct} \tag{4.13}$$

where Q_e is the initial volumetric flow rate and C a constant related to the excess pressure and the compressibility and volume of the reservoir. For an eruption, the volumetric flow rate is also referred to as the effusion rate (e.g., Gudmundsson 2012, and references therein). A similar exponential decrease describes the variation of the magma excess pressure in the chamber P_c with time:

$$P_c = P_e e^{-(Q_f t/V)} \tag{4.14}$$

where P_e is the excess pressure at the time of magma chamber rupture ($t = 0$) and the exponent is dimensionless. This similarity implies that both the magmatic excess pressure in the chamber and the volumetric flow rate through the feeder dike decrease exponentially during the eruption and, when the excess pressure approaches zero, the eruption comes to an end. The exponential decrease in the volumetric flow rate is consistent with observations at several erupting volcanoes. These exponential behaviours rely on the assumptions that the contribution of the exsolved volatiles and any replenishment into the magma chamber have negligible effects on the excess pressure evolution in the chamber during the eruption.

4.7 The Plutonic-Volcanic Link

Despite a few exceptions, plutons and volcanoes have been studied for decades by distinct scientific communities, which resulting in limited interactions. In the last two decades, important efforts have been made to move beyond any evidence of a long-established first-order connection between large felsic plutons and their generated volcanic ash-flows at a given volcano; these efforts are also based on the awareness that many of the uncertainties in understanding volcanic activity, including unrest, may be reduced by better knowledge of the intrusive processes. As a result, most studies relating volcanic activity to the processes occurring in magma chambers, investigating the "plutonic-volcanic link", are relatively recent and still accompanied by major questions.

In the past, plutons have been commonly interpreted as solidified remnants of magma chambers. Although all solidified magma chambers become plutons, not all plutons derive from extinct chambers, with most intrusions probably not having directly fed any eruption. Therefore, the more recent studies considering the plutonic-volcanic link have focused on the sites where both the intruded and erupted components of a magmatic system are accessible. This condition is usually met at partly eroded calderas, which allow comparing plutonic and volcanic rocks from a same magmatic system. The Tertiary magmatic provinces in the Rocky Mountains and in the Central Andes provide ideal windows into the general plutonic-volcanic connections, even sharing common boundary conditions, as having similar areal extent, composition, eruptive volumes and duration of volcanism (de Silva and Gosnold 2007; Lipman 2007; Lipman and Bachmann 2015). Indeed, both provinces record multistage histories of magma accumulation, fractionation and solidification in upper parts of large subvolcanic plutons that were sufficiently liquid to erupt, with intrusive to extrusive ratios of ~10:1 or more, consistent with estimates elsewhere (Edmonds et al. 2019). Age and compositional similarities suggest that ignimbrites and plutons are closely related, with more than 10 km thick plutons recording non-erupted parts of long-lived cogenetic magmatic systems beneath calderas. Ignimbrite eruptions occurred at times of peak power input, following vertically extensive focused and proximal intrusions mixing and variably homogenizing successively emplaced incremental inputs.

Modelling studies indicate that, to sustain such a volumetrically significant ignimbritic activity, magma reservoirs require supply rates of at least ~0.01 $km^3\ yr^{-1}$, which are consistent with geodetic estimates of magma flux only beneath a very few currently inflating volcanoes. These rates are much higher than the long-term growth rate of plutons, which is orders of magnitude slower (Fig. 4.28; Menand et al. 2015; Coleman et al. 2016). This discrepancy in rates is supported by the general missing record of granitoid plutonic complexes that would represent large fossil upper crustal magma chambers, implying that most intrusive suites are not genetically linked to the magma chambers responsible for eruptions. This is consistent with the broader evidence that magma chambers are short-lived and erupt rapidly after being recharged, representing extremely productive but transient conditions (Glazner et al. 2004; Miller 2008; Annen 2009; Davis et al. 2012; Hooper 2012; Menand et al. 2015).

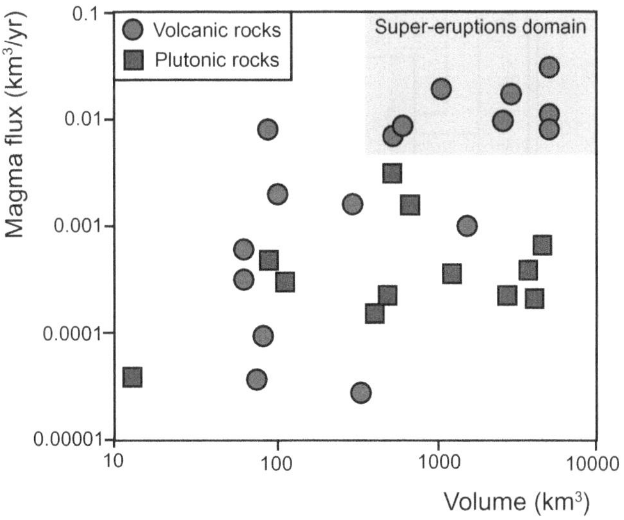

Fig. 4.28 Magma volume and flux (in log–log scale) estimates for granitic plutons (blue squares) and volcanic rocks (red circles). Super-eruption magmas (grey area to upper right) appear to accumulate orders of magnitude faster than comparable volumes of plutonic rocks. Modified after Coleman et al. (2016)

In synthesis, the available information on the plutonic-volcanic link provides complementary insights. On the one side, studies of representative cases from the Rocky Mountains and Central Andes highlight close relationships, in terms of rates and genesis, between volcanic and plutonic processes. On the other side, more general observations challenge a wider applicability of these relationships, highlighting that plutons feed eruptive activity only under particularly sustained conditions of magma flux. The merging of these two perspectives highlights one of the currently most important questions on the volcanic-plutonic link, that is under which conditions an intrusive suite may fuel ordinary or large volcanism. Recognizing that volcanism requires on average higher magma fluxes than plutonism, thus constituting an end-member condition in magmatic processes, provides only a preliminary working hypothesis.

4.8 Summary

This chapter summarizes the conditions and mechanisms leading to the emplacement and accumulation of magma in the crust. The essential unit for this is the sill, which may eventually inflate and/or amalgamate with other sills to form thicker intrusions, as laccoliths, bysmaliths and lopoliths. These intrusions are scale independent, but their largest equivalents, resulting from incremental assemblage, commonly constitute the tabular plutons which are widespread in the crust and at times merge to form larger crustal-scale batholiths. Pluton growth may be magma-induced and/or tectonically-assisted. In the likely case that the strain rates due to magma emplacement are larger than the tectonic strain rates, the progression from sill to pluton is largely independent of, but not completely insensitive to, regional tectonics.

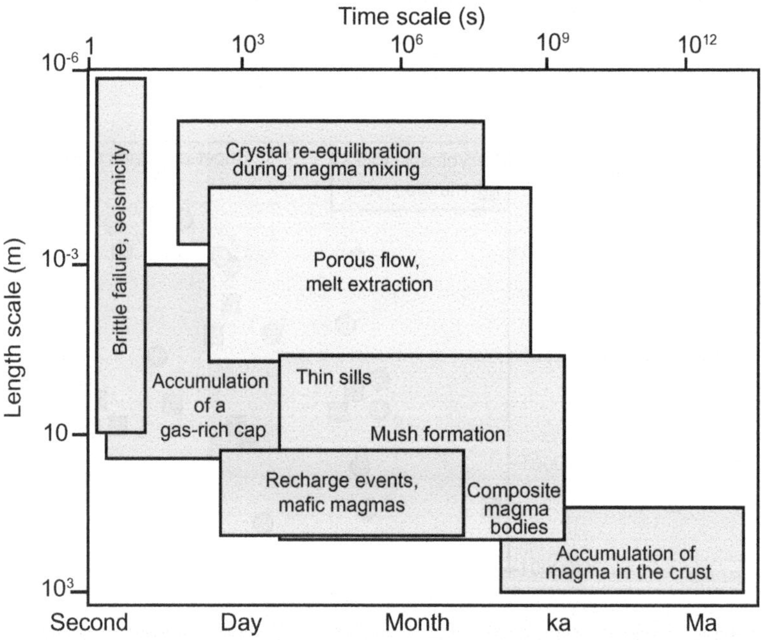

Fig. 4.29 Summary of the time and length scales associated with different magmatic processes in the crust. Modified after Edmonds et al. (2019)

Magma reservoirs consist of both crystal- and melt-dominated regions. Complex and diverse time- and length-scales control different aspects in their formation and evolution (Fig. 4.29; Edmonds et al. 2019). The shorter diffusional re-equilibration timescales, related to mixing and magma ascent prior to eruption, are usually found in the mafic systems, consistent with their lower viscosity and higher flow rates. Older ages derived from radiometric dating, in contrast, provide information about the longevity of the reservoir and the timescales of crystallization and the modality of accumulation of larger magma bodies.

Magma chambers represent ephemeral components of magma reservoirs, resulting from the rapid and transient accumulation of melt. They may affect the propagation path of nearby dikes, as well as nucleate the shallow magma-filled fractures potentially feeding eruptions, in the latter case mainly promoted by increasing the magma excess pressure or decreasing the minimum principal stress σ_3. Magma chambers may eventually become plutons, although only the few plutons associated with the highest magma flux derive from extinct chambers, raising important questions on the plutonic-volcanic connection.

4.9 Main Symbols Used

A	Intrusion area
A_1	Area at base of feeder dike
A_2	Area at top of sill
C	Constant
D	Flexural rigidity of the overburden
E	Young's modulus
E_s	Elastic stiffness of the host rock
f	Porosity
F_1	Force acting on area A_1
F_2	Force acting on area A_2
g	Acceleration due to gravity
G	Shear modulus
j	Constant
L	Laccolith or sill length
n	Power law exponent
p	Magmatic pressure
P_1	Pressure at the base of feeder dike
P_2	Pressure at the top of sill
P_c	Magma excess pressure in the chamber
P_e	Excess pressure of the magma
P_l	Lithostatic pressure
P_{tot}	Total pressure
Q_f	volumetric flow rate
Q_e	Initial volumetric flow rate
t	Time
T_0	Tensile strength of host rock
V	volume of magma
V_c	Total volume of the chamber
w	Maximum thickness of laccolith
z	Thickness of the overburden or magma emplacement depth
z_e	Effective thickness of overburden
β_c	Compressibility of host rock
β_m	Magma compressibility
υ	Poisson's ratio
ρ	Density
σ_1	Maximum principal stress
σ_2	Intermediate principal stress
σ_3	Minimum principal stress

References

Acocella V (2000) Space accommodation by roof lifting during pluton emplacement at Amiata (Italy). Terra Nova 12:149–155

Acocella V, Rossetti F (2002) The role of extensional tectonics at different crustal levels on granite ascent and emplacement: an example from Tuscany (Italy). Tectonophysics 354:71–83

Airoldi G, Muirhead JD, Zanella E, White JDL (2012) Emplacement process of Ferrar Dolerite sheets at Allan Hills (South Victoria Land, Antarctica) inferred from magnetic fabric. Geophys J Int 188:1046–1060

Albertz M, Paterson SR, Okaya D (2005) Fast strain rates during pluton emplacement: Magmatically folded leucocratic dikes in aureoles of the Mount Stuart Batholith, Washington, and the Tuolumne Intrusive Suite, California. Geol Soc Am Bull 117:450–465

Allan ASR, Wilson CJN, Millet M-A, Wysoczanski RJ (2012) The invisible hand: tectonic triggering and modulation of a rhyolitic supereruption. Geology 40:563–566

Albino F, Amelung F, Gregg P (2018) The role of pore fluid pressure on the failure of magma reservoirs: Insights from Indonesian and Aleutian arc volcanoes. J Geophys Res 123:1328–1349

Amoruso A, Crescentini L, Sabbetta I (2014) Paired deformation sources of the Campi Flegrei caldera (Italy) required by recent (1980–2010) deformation history. J Geophys Res 119:3276–3289

Amoruso A, Crescentini L, D'Antonio M, Acocella V (2017) Thermally-assisted magma emplacement explains restless calderas. Scientific Reports 7:7948. https://doi.org/10.1038/s41598-017-08638-y

Amoruso A, Crescentini L (2019) An approximate approach to nonisothermal emplacement of kilometer-sized kilometer-deep sills at calderas. J Geophys Res 124:1236–1253

Andersen NL, Singer BS, Jicha BR, Beard BL, Johnson CM, Licciardi JM (2017) Pleistocene to Holocene Growth of a Large Upper Crustal Rhyolitic Magma Reservoir beneath the Active Laguna del Maule Volcanic Field, Central Chile. J Petrol 58:85–114

Annen C (2009) From plutons to magma chambers: Thermal constraints on the accumulation of eruptible silicic magma in the upper crust. Earth Planet Sci Lett 294:409–416

Annen C, Blundy JD, Leuthold J, Sparks RSJ (2015) Construction and evolution of igneous bodies: Towards an integrated perspective of crustal magmatism. Lithos 230:206–221

Aranguren A, Tubia JM, Bouchez JL, Vigneresse JL (1996) The Guitiriz granite, Variscan belt of Northern Spain: extension-controlled emplacement of magma during tectonic escape. Earth Planet Sci Lett 139:165–176

Aranguren A, Cuevas J, Tubia JM, Roman-Berdiel T, Casas-Sainz A, Casas-Ponsati A (2003) Granite laccolith emplacement in the Iberian arc: AMS and gravity study of the La Tojiza pluton (NW Spain). J Geol Soc 160:435–445

Auger E, Gasparini P, Virieax J, Zollo A (2001) Seismic Evidence of an Extended Magmatic Sill Under Mt Vesuvius. Science 294:1510–1512

Bachmann O, Miller CF, de Silva S (2007) The volcanic–plutonic connection as a stage for understanding crustal magmatism. J Volcanol Geoth Res 167:1–23

Bachmann O, Bergantz J (2008) The magma reservoirs that feed supereruptions. Elements 4:17–21

Bai C, Greenlangh S (2005) 3D multi-step travel time tomography: imaging the local, deep velocity structure of Rabaul volcano, Papua New Guinea. Phys Earth Planet Inter 151:259–275

Barboni M, Schoene B (2014) Short eruption window revealed by absolute crystal growth rates in a granitic magma. Nat Geosci 7:524–528

Barker SJ, Wilson CJN, Morgan DJ, Rowland JV (2016) Rapid priming, accumulation, and recharge of magma driving recent eruptions at a hyperactive caldera volcano. Geology 44:323–326

Bartley JM, Coleman DS, Glazner AF (2006) Incremental pluton emplacement by magmatic crack-seal. Transactions of the Royal Society of Edinburgh. Earth Sciences 97:383–396

Beachly MW, Hooft EEE, Toomey DR, Waite GP (2012) Upper crustal structure of Newberry Volcano from P-wave tomography and finite difference waveform modeling. J Geophys Res 117. https://doi.org/10.1029/2012JB009458

Biggs J, Annen C (2019) The lateral growth and coalescence of magma systems. Philosophical Transactions Royal Society a 377:20180005

Blundy JD, Annen CJ (2016) Crustal magmatic systems from the perspective of heat transfer. Elements 12:115–120

Brown EH, McClelland WC (2000) Pluton emplacement by sheeting and vertical ballooning in part of the southeast Coast Plutonic Complex, British Columbia. Geol Soc Am Bull 112:708–719

Brown M (2013) Granite: From genesis to emplacement. Geol Soc Am Bull 125:1079–1113

Bunger AP, Cruden AR (2011) Modeling the growth of laccoliths and large mafic sills: Role of magma body forces. J Geophys Res 116:B02203. https://doi.org/10.1029/2010JB007648

Cashman KV, Sparks RSJ, Blundy JD (2017) Vertically extensive and unstable magmatic systems: a unified view of igneous processes. Science 355:1280

Castro JM, Cordonnier B, Schipper CI, Tuffen H, Baumann TS, Feisel Y (2016) Rapid laccolith intrusion driven by explosive volcanic eruption. Nature Communications 7:13585. https://doi.org/10.1038/ncomms13585

Chanceaux L, Menand T (2014) Solidification effects on sill formation: an experimental approach. Earth Planet Sci Lett 403:79–88

Chanceaux L, Menand T (2016) The effects of solidification on sill propagation dynamics and morphology. Earth Planet Sci Lett 442:39–50

Chang WL, Smith RB, Wicks C, Farrell JM, Puskas CM (2007) Accelerated uplift and magmatic intrusion of the Yellowstone Caldera, 2004 to 2006. Science 318:952–956

Chaussard E, Amelung F (2012) Precursory inflation of shallow magma reservoirs at west Sunda volcanoes detected by InSAR. Geophys Res Lett 39:L21311. https://doi.org/10.1029/2012GL053817

Ciancaleoni L, Marquer D (2006) Syn-extension leucogranite deformation during convergence in the Eastern Central Alps: example of the Novate intrusion. Terra Nova 18:170–180

Coetzee A, Kisters AFM (2018) The elusive feeders of the Karoo Large Igneous Province and their structural controls. Tectonophysics 747–748:146–162

Coleman DS, Mills RD, Zimmerer RJ (2016) The Pace of Plutonism. Elements 12:97–102

Collins WJ, Wuang H-Q, Jiang X (2016) Water-fluxed crustal melting produces Cordilleran batholiths. Geology 44:143–146

Corry CE (1988) Laccoliths: mechanics sof emplacement and growth. Geological Society of America Special Publications 220:110

Corti G, Moratti G, Sani F (2005) Relations between surface faulting and granite intrusions in analogue models of strike-slip deformation. J Struct Geol 27:1547–1562

Cruden AR (1998) On the emplacement of tabular granites. J Geol Soc London 155:853–862

Cruden AR, McCaffrey KJW (2001) Growth of Plutons by floor subsidence: implications for rates of emplacement, intrusion spacing and melt-extraction mechanisms. Phys Chem Earth 26:303–315

Cruden AR, McCaffrey KJW, Bunger AP (2017) Geometric scaling of tabular igneous intrusions: implications for emplacement and growth. In: Bretkreuz C, Rocchi S Physical geology of shallow magmatic systems; dykes, sills and laccoliths. advances in volcanology Springer, vol 2, pp 11–38

Currier RM, Marsh BD (2015) Mapping real time growth of experimental laccoliths: the effect of solidification on the mechanics of magmatic intrusion. J Volcanol Geoth Res 302:211–224

Currier RM, Forsythe P, Grossmeier C, Laliberte M, Yagle B (2017) Experiments on the evolution of laccolith morphology in plan-view. J Volcanol Geoth Res 336:155–167

Davis JW, Coleman DE, Gracely JT, Gaschnig R, Stearns M (2012) Magma accumulation rates and thermal histories of plutons of the Sierra Nevada batholith, CA. Contrib Miner Petrol 163:449–465

D'Auria L, Pepe S, Castaldo R, Giudicepietro F, Macedonio G, Ricciolino P et al (2015) Magma injection beneath the urban area of Naples: a new mechanism for the 2012–2013 volcanic unrest at Campi Flegrei caldera. Scientific Reports 5. https://doi.org/10.1038/srep13100

de Saint Blanquat M, Tikoff B, Teyssier C, Vigneresse JL (1998) Transpressional kinematics and magmatic arcs. In: Holdsworth RE, Strachan RA, Dewey JF (eds) 1998. Continental transpressional and transtensional tectonics. Geological Society of London Special Publications, vol 135, pp 327–340

de Saint Blanquat M, Law RD, Bouchez JL, Morgan SS (2001) Internal structure and emplacement of the Papoose Flat pluton: an integrated structural, petrographic and magnetic susceptibility study. Geol Soc Am Bull 113:976–995

de Saint Blanquat M, Habert G, Horsman E, Morgan SS, Tikoff B, Pauneau P et al (2006) Mechanisms and duration of non-tectonically assisted magma emplacement in the upper crust: The Black Mesa pluton, Henry Mountains, Utah. Tectonophysics 428:1–31

de Saint Blanquat M, Horsman E, Habert G, Morgan S, Vanderhaeghe O, Law R et al (2011) Multiscale magmatic cyclicity, duration of pluton construction, and the paradoxical relationship between tectonism and plutonism in continental arcs. Tectonophysics 500:20–33

de Silva SL, Gosnold WD (2007) Episodic construction of batholiths: insights from the spatiotemporal development of an ignimbrite flare-up. J Volcanol Geoth Res 167:320–335

Delaney PT, Pollard DD (1982) Solidification of basaltic magma during flow in a dike. Am J Sci 282:856–885

Druitt TH, Costa F, Deloule E, Dungan M, Scaillet B (2012) Decadal to monthly timescales of magma transfer and reservoir growth at a caldera volcano. Nature 482:77–83

Dvorak JJ, Dzurisin DD (1997) Volcano geodesy: the search for magma reservoirs and the formation of eruptive vents. Rev Geophys 35:343–384

Dzurisin DD, Lisowski M, Wicks CW, Poland MP, Endo ET (2006) Geodetic observations and modeling of magmatic inflation at the Three Sisters volcanic center, central Oregon Cascade Range, USA. J Volcanol Geoth Res 150:35–54

Eddy MP, Bowring SA, Miller RB, Tepper JH (2016) Rapid assembly and crystallization of a fossil large-volume silicic magma chamber. Geology 44:331–334

Edmonds M, Cashman KV, Holness M, Jackson M (2019) Architecture and dynamics of magma reservoirs. Philosophical Trans Royal Soc 377:20180298

Famin V, Michon L (2010) Volcano destabilization by magma injections in a detachment. Geology 38:219–222

Fernandez C, Castro A (1999) Pluton accommodation at high strain rates in the upper continental crust. The example of the Central Extremadura batholith Spain. J Struct Geol 21:1143–1149

Fialko Y, Simons M (2001) Evidence for on-going inflation of the Socorro magma body, New Mexico, from Interferometric Synthetic Aperture Radar imaging. Geophys Res Lett 28:3549–3552

Gaillard F, Scaillet B, Pichavant M (2004) Evidence for present-day leucogranite pluton growth in Tibet. Geology 32:801–804

Galetto F, Acocella V, Caricchi L (2017) Caldera resurgence driven by magma viscosity contrasts. Nature Communications 8:1750. https://doi.org/10.1038/s41467-017-01632-y

Galland O, Bertelsen HS, Eide CH, Guldstrand F, Haug OT, Leanza HA (2018) Storage and transport of magma in the layered crust—formation of sills and related flat-lying intrusions. In: Burchardt S (ed) Volcanic and igneous plumbing systems; understanding magma transport, storage, and evolution in the earth's crust. Elsevier, pp 113–138

Gelman SE, Gutierrez F, Bachmann O (2013a) On the longevity of large upper crustal silicic magma reservoirs. Geology 41:759–762

Gelman SE, Deering CD, Gutierrez FJ, Bachmann O (2013b) Evolution of the Taupo Volcanic Center, New Zealand: petrological and thermal constraints from the Omega dacite. Contributions in Mineralogy and Petrology 166:1355–1374

Gilbert GK (1877) Report on the geology of the Henry Mountains. U.S, Government Printing Office Washington DC

Gill SPA, Walker RJ (2020) The roles of elastic properties, magmatic pressure, and tectonic stress in saucer-shaped sill growth. J Geophy Res 124: e2019JB019041

Glazner A (1991) Plutonism, oblique subduction and continental growth: an example from the Mesozoic of California. Geology 19:784–786

Glazner AF, Bartley JM, Coleman D, Gray W, Taylor RZ (2004) Are plutons assembled over millions of years by amalgamation from small magma chambers? GSA Today 14. https://doi.org/10.1130/1052-5173

Gottsmann J, Lavallee Y, Martì J, Aguirre-Diaz G (2009) Magma–tectonic interaction and the eruption of silicic batholiths. Earth Planet Sci Lett 284:426–434
Gregg PM, de Silva SL, Grosfils EB (2013) Thermomechanics of shallow magma chamber pressurization: Implications for the assessment of ground deformation data at active volcanoes. Earth Planet Sci Lett 384:100–108
Grossman-Ponemon BE, Eimisson ER, Lew AJ, Segall P (2020) Logarithmic growth of dikes from a depressurizing magma chamber. Geophysical Research Letters 47:e2019GL086230
Gudmundsson A (1990) Emplacement of dikes, sills and crustal magma chambers at divergent plate boundaries. Tectonophysics 176:257–275
Gudmundsson A (2006) How local stresses control magma-chamber ruptures, dyke injections, and eruptions in composite volcanoes. Earth Sci Rev 79:1–31
Gudmundsson A (2012) Magma chambers: Formation, local stresses, excess pressures, and compartments. J Volcanol Geoth Res 237–238:19–41
Gudmundsson O, Bransdottir B, Menke W, Sigvaldson GE (1994) The crustal magma chamber of the Katla volcano in South Iceland revealed by 2-D seismic undershooting. Geophys J Int 110:277–296
Haug OT, Galland O, Souloumiac P, Souche A, Guldstrand F, Schmiedel T (2017) Inelastic damage as a mechanical precursor for the emplacement of saucer-shaped intrusions. Geology 45:1–4
Henry CD, Kunk MJ, Muehlberger WR, McIntosh WC (1997) Igneous evolution of a complex laccolith-caldera, the Solitario, Trans-Pecos, Texas: Implications for calderas and subjacent plutons. Geol Soc Am Bull 109:1036–1054
Holness MB, Humphreys MCS (2003) The Traigh Bhan na Sgurra Sill, Isle of Mull: Flow Localization in a Major Magma Conduit. J Petrol 44:1961–1976
Hooper A (2012) A volcano's sharp intake of breath. Nat Geosci 5:686–687
Huber C, Townsend M, Degruyter W, Bachmann O (2019) Optimal depth of subvolcanic magma chamber growth controlled by volatiles and crust rheology. Nat Geosci 12:762–768
Hutton DHW (1988) Igneous emplacement in a shear zone termination: the biotite granite at Strontian, Scotland. Geol Soc Am Bull 100:1392–1399
Hutton DHW, Dempster TJ, Brown PE, Becker SD (1990) A new mechanism of granite emplacement: intrusion in active extensional shear zones. Nature 343:452–455
Hutton DHW (1992) Granite sheeted complexes: evidence for the dyking ascent mechanism. Trans Royal Soc Edinburgh Earth Sci 83:377–382
Jaeger JC (1964) Thermal effects of intrusions. Rev Geophys 2:443–466
Jellinek AM, DePaolo DJ (2003) A model for the origin of large silicic magma chambers: precursors of caldera-forming eruptions. Bull Volcanol 65:363–381
Jiang C, Schmandt B, Farrell J, Lin F-C, Ward KM (2018) Seismically anisotropic magma reservoirs underlying silicic calderas. Geology 46:727–730
John BE, Blundy JD (1993) Emplacement-related deformation of granitoid magmas, southern Adamello Massif, Italy. Geol Soc Am Bull 105:1517–1541
Johnson AM, Pollard DD (1973) Mechanics of growth of some laccolithic intrusions in the Henry Mountains, Utah, I. Field observations, Gilbert's model, physical properties and flow of the magma. Tectonophysics 18:261–309
Johnson SE, Albertz M, Paterson SR (2001) Growth rates of dike-fed plutons: Are they compatible with observations in the middle and upper crust? Geology 29:727–730
Jonsson S, Zebker H, Amelung F (2005) On trapdoor faulting at Sierra Negra volcano, Galapagos. J Volcanol Geoth Res 144:59–71
Kalakay TJ, John BE, Lageson DR (2001) Fault-controllled pluton emplacement in the Sevier fold and thrust belt of southwest Montana, USA. J Struct Geol 23:1151–1168
Karlstrom L, Dufek J, Manga M (2009) Organization of volcanic plumbing through magmatic lensing by magma chambers and volcanic loads. J Geophys Res 114:B10204. https://doi.org/10.1029/2009JB006339
Karlstrom L, Dufek J, Manga M (2010) Magma chamber stability in arc and continental crust. J Volcanol Geoth Res 190:249–270
Kavanagh JL, Menand T, Sparks RSJ (2006) An experimental investigation of sill formation and propagation in layered elastic media. Earth Planet Sci Lett 245:799–813
Kavanagh JL, Boutelier D, Cruden AR (2015) The mechanics of sill inception, propagation and growth: experimental evidence for rapid reduction in magmatic overpressure. Earth Planet Sci Lett 421:117–128
Kennedy BM, Holohan EP, Stix J, Gravley DM, Davidson JRJ, Cole JW (2018) Magma plumbing beneath collapse caldera volcanic systems. Earth Sci Rev 177:404–424
Kilburn CRJ, De Natale G, Carlino S (2017) Progressive approach to eruption at Campi Flegrei caldera in southern Italy. Nature Communications 8:15312. https://doi.org/10.1038/ncomms15312
Koulakov I, Gordeev EI, Dobretsov NL, Vernikovsky VA, Senyukov S, Jakovlev A (2011) Feeding volcanoes of the Kluchevskoy group from the results of local earthquake tomography. Geophys Res Lett 38:L09305. https://doi.org/10.1029/2011GL046957
Lees JM (2007) Seismic tomography of magmatic systems. J Volcanol Geoth Res 167:37–56
Leuthold J, Muntener O, Baumgartner L, Putlitz B, Ovtcharova M, Schaltegger U (2012) Time resolved construction of a bimodal laccolith (Torres del Paine, Patagonia). Earth Planet Sci Lett 325:85–92

Lipman PW (2007) Incremental assembly and prolonged consolidation of Cordilleran magma chambers: Evidence from the Southern Rocky Mountain volcanic field. Geosphere 3:42–70

Lipman PW, Bachmann O (2015) Ignimbrites to batholiths: Integrating perspectives from geological, geophysical, and geochronological data. Geosphere 11:1–39

Liu H, Martelet G, Wang B, Erdmann S, Chen Y, Faure M et al (2018) Incremental emplacement of the Late Jurassic midcrustal, lopolith-like Qitianling pluton, South China, revealed by AMS and Bouguer gravity data. J Geophys Res 123:9249–9268

Lu Z, Dzurisin D (2014) InSAR Imaging of Aleutian volcanoes. Springer, Monitoring a volcanic arc from Space, p 390

Lundgren P, Lu Z (2006) Inflation model of Uzon caldera, Kamchatka, constrained by satellite radar interferometry observations. Geophys Res Lett 33:L06301. https://doi.org/10.1029/2005GL02

Macedonio G, Giudicepietro F, D'Auria L, Martini M (2014) Sill intrusion as a source mechanism of unrest at volcanic calderas. J Geophys Res 119:3986–4000

Magee C, Stewart EH, Jackson CAL (2013) Volcano growth mechanisms and the role of sub-volcanic intrusions: Insights from 2D seismic reflection data. Earth Planet Sci Lett 373:41–53

Mahan KH, Bartley JM, Coleman DS, Glazner AF, Carl BS (2003) Sheeted intrusion of the synkinematic McDoogle pluton, Sierra Nevada, California. Geol Soc Am Bull 115:1570–1582

Matthews NE, Huber C, Pyle DM, Smith VC (2012) Timescales of Magma Recharge and Reactivation of Large Silicic Systems from Ti Diffusion in Quartz. J Petrol 53:1385–1416

Mattsson T, Burchardt S, Ronchin ABSG, E, (2018) Syn-Emplacement Fracturing in the Sandfell Laccolith, Eastern Iceland—Implications for Rhyolite Intrusion Growth and Volcanic Hazards. Front Earth Sci 6:5. https://doi.org/10.3389/feart.2018.00005

Matzel JEP, Bowring SA, Miller RB (2006) Time scales of pluton construction at differing crustal levels: Examples from the Mount Stuart and Tenpeak intrusions, North Cascades, Washington. Geol Soc Am Bull 118:1412–1430

McCaffrey KJW, Petford N (1997) Are granitic intrusions scale invariant? J Geol Soc London 154:1–4

McLeod P, Tait S (1999) The growth of dykes from magma chambers. J Volcanol Geoth Res 92:231–245

McNulty B, Tobisch OT, Cruden AR, Gilder S (2000) Multistage emplacement of the Mount Givens pluton, central Sierra Nevada batholith, California. Geol Soc Am Bull 112:119–135

McVey BG, Hooft EEE, Heath BA, Toomey DR, Paulatto M, Morgan JV et al (2019) Magma accumulation beneath Santorini volcano, Greece, from P-wave tomography. Geology 48:231–235

Menand T (2008) The mechanics and dynamics of sills in layered elastic rocks and their implications for the growth of laccoliths and other igneous complexes. Earth Planet Sci Lett 267:93–99

Menand T, Daniels KA, Benghiat P (2010) Dyke propagation and sill formation in a compressive tectonic environment. J Geophys Res 115:B08201. https://doi.org/10.1029/2009JB006791

Menand T (2011) Physical controls and depth of emplacement of igneous bodies: A review. Tectonophysics 500:11–19

Menand T, Annen C, de Saint Blanquat M (2015) Rates of magma transfer in the crust: Insights into magma reservoir recharge and pluton growth. Geology 43:199–202

Michel J, Baumgartner L, Putlitz B, Schaltegger U, Ovtcharova M (2008) Incremental growth of the Patagonian Torres del Paine laccolith over 90 k.y. Geology 36:459–462

Miller JS (2008) Assembling a pluton…one increment at a time. Geology 36:511–512

Mitchell MA, White RS, Roecker S, Greenfield T (2013) Tomographic image of melt storage beneath Askja Volcano, Iceland using local microseismicity. Geophys Res Lett 40:5040–5046

Morgan SS, Law RD, Byman MW (1998) Laccolith-like emplacement model for the Papoose Flat pluton based on porphyroblast-matrix analysis. Geol Soc Am Bull 110:96–110

Morgan S (2018) Pascal's Principle, a Simple Model to Explain the Emplacement of Laccoliths and Some Mid-Crustal Plutons. In: Burchardt S (ed) Volcanic and igneous plumbing systems; Understanding Magma Transport, Storage, and Evolution in the Earth's Crust. Elsevier, pp 139–162

Morley CK (2018) 3-D seismic imaging of the plumbing system of the Kora Volcano, Taranaki Basin, New Zealand: The influence of syn-rift structure on shallow igneous intrusion architecture. Geosphere 14:2533–2584

Musumeci G, Mazzarini F, Corti G, Barsella M, Montanari D (2005) Magma emplacement in a thrust ramp anticline: The Gavorrano Granite (northern Apennines, Italy). Tectonics 24:TC6009. https://doi.org/10.1029/2005TC001801

O'Driscoll B, Troll V, Reavy RJ, Turner P (2006) The Great Eucrite intrusion of Ardnamurchan, Scotland: Reevaluating the ring-dike concept. Geology 34:189–192

Pagli C, Sigmundsson F, Arnadottir T, Einarsson P, Sturkell E (2006) Deflation of the Askja volcanic system: constraints on the deformation source from combined inversion of satellite radar interferograms and GPS measurements. J Volcanol Geoth Res 152:97–108

Pagli C, Wright TJ, Ebinger CJ, Yun SH, Cann JR, Barnie T et al (2012) Shallow axial magma chamber at the slow spreading Erta Ale Ridge. Nat Geosci. https://doi.org/10.1038/NGEO1414

Pansino S, Taisne B (2019) How magmatic storage regions attract and repel propagating dikes. J Geophys Res 124:274–290

Pasquarè F, Tibaldi A (2007) Structure of a sheet-laccolith system revealing the interplay between tectonic and magma stresses at Stardalur Volcano, Iceland. J Volcanol Geoth Res 161:131–150

Paterson SR, Fowler TK (1993) Extensional pluton-emplacement models: Do they work for large plutonic complexes? Geology 21:781–784

Paterson SR, Miller RB (1998) Magma emplacement during arc-perpendicular shortening: an example from the Cascades crystalline core, Washington. Tectonics 17:571–586

Paterson SR, Memett V, Mundil R, Zak J (2016) Repeated, multiscale, magmatic erosion and recycling in an upper-crustal pluton: Implications for magma chamber dynamics and magma volume estimates. Am Miner 101:2176–2198

Paulatto M, Annen C, Henstock TJ, Kiddle E, Minshull TA, Sparks RSJ et al (2012) Magma chamber properties from integrated seismic tomography and thermal modeling at Montserrat. Geochem Geophys Geosyst 13:Q01014. https://doi.org/10.1029/2011GC003892

Petford N, Cruden AR, McCaffrey KJW, Vigneresse JL (2000) Granite magma formation, transport and emplacement in the Earth's crust. Nature 408:669–673

Pinel V, Jaupart C (2003) Magma chamber behavior beneath a volcanic Edifice. J Geophys Res 108:2072. https://doi.org/10.1029/2002JB001751

Pollard DD (1973) Derivation and evaluation of a mechanical model for sheet intrusions. Tectonophysics 19:233–269

Pollard DD, Johnson AM (1973) Mechanics of growth of some laccolith intrusions in the Henry Mountains, Utah, II; bending and failure of overburden layers and sill formation. Tectonophysics 18:311–354

Polteau S, Ferré EC, Planke S, Neumann E-R, Chevallier L (2008) How are saucer-shaped sills emplaced? Constraints from the Golden Valley Sill, South Africa. J Geophys Res 113:B12104. https://doi.org/10.1029/2008JB005620

Purdy GM, Kong LSL, Christenson GL, Solomon SC (1991) Relationship between spreading rate and the seismic structure of mid-ocean ridges. Nature 355:815–817

Reid MR (2008) How Long Does It Take to Supersize an Eruption? Elements 4:23–28

Richards SW, Collins WJ (2004) Growth of wedge-shaped plutons at the base of active half-grabens. Trans Royal Soc Edinburgh Earth Sci 95:309–317

Rivalta E, Corbi F, Passarelli L, Acocella V, Davis T, Di Vito MA et al (2019) Stress inversions to forecast magma pathways and eruptive vent location. Science Advances 5:eaau9784

Roberts PM, Aki K, Fehler MC (1991) A low velocity zone in the basement beneath the Valles caldera, New Mexico. J Geophys Res 96:21583–21596

Rocchi S, Westerman DS, Dini A, Innocenti F, Tonarini S (2002) Two-stage growth of laccoliths at Elba Island, Italy. Geology 30:983–986

Roman Berdiel T, Gapais D, Brun JP (1995) Analogue models of laccolith formation. J Struct Geol 17:1337–1346

Rosenberg CL, Berger A, Schmid SM (1995) Observations from the floor of a granitoid pluton: Inferences on the driving force of final emplacement. Geology 23:443–446

Rubin AM (1993) On the thermal viability of dikes leaving magma chambers. Geophys Res Lett 20:257–260

Schofield N, Stevenson C, Reston T (2010) Magma fingers and host rock fluidization in the emplacement of sills. Geology 38:63–66

Scheibert J, Galland O, Hafver A (2017) Inelastic deformation during sill and laccolith emplacement: insights from an analytic elasto-plastic model. J Geophys Res 122:923–945

Schopa A, Annen C (2013) The effects of magma flux variations on the formation and lifetime of large silicic magma chambers. J Geophys Res 118:926–942

Seccia D, Chiarabba C, De Gori P, Bianchi I, Hill DP (2011) Evidence for the contemporary magmatic system beneath Long Valley Caldera from local earthquake tomography and receiver function analysis. J Geophys Res 116:B12314. https://doi.org/10.1029/2011JB008471

Shaw HR (1985) Links between magma-tectonic rate balances, plutonism, and volcanism. J Geophys Res 90:11275–11288

Shea EK, Miller JS, Miller RB, Bowring SA, Sullivan KM (2016) Growth and maturation of a mid- to shallow-crustal intrusive complex, North Cascades, Washington. Geosphere 12:1489–1516

Sigmundsson F, Hreinsdottir S, Hooper A, Arnadottir T, Pedersen R, Roberts MJ et al (2010) Intrusion triggering of the 2010 Eyjafjallajokull explosive eruption. Nature 468:426–430

Sili G, Urbani S, Acocella V (2019) What controls sill formation: An overview from analogue models. J Geophys Res 124. https://doi.org/10.1029/2018JB017005

Sliwinski J, Farsky D, Lipman PW, Guillong M, Bachmann O (2019) Rapid Magma Generation or Shared Magmatic Reservoir? Petrology and Geochronology of the Rat Creek and Nelson Mountain Tuffs, CO, USA. Front Earth Sci 7:271. https://doi.org/10.3389/feart.2019.00271

Singer BS, Costa F, Herrin JS, Hildreth W, Fierstein J (2016) The timing of compositionally zoned magma reservoirs and mafic 'priming' weeks before the 1912 Novarupta-Katmai rhyolite eruption. Earth Planet Sci Lett 451:125–137

Singer BS, Le Melvel H, Licciardi JM, Cordova L, Tikoff B, Garibaldi N et al (2018) Geomorphic expression of rapid Holocene silicic magma reservoir growth beneath Laguna del Maule, Chile. Sci Adv 4:eaat1513

Siregar E, Omosanya KO, Magee C, Johansen SE (2019) Impacts of fault-sill interactions on sill emplacement in the Vøring Basin, Norwegian North Sea. J Struct Geol 126:156–174

Sparks RSJ, Cashman KV (2017) Dynamic magma systems: implications for forecasting volcanic activity. Elements 13:35–40

Sparks RSJ, Annen CJ, Blundy J, Cashman K, Rust A, Jackson MD (2019) Formation and dynamics of magma reservoirs. Philosophical Trans Royal Soc Math Phys Eng Sci 377:20190019

Thomson K (2007) Determining magma flow in sills, dykes and laccoliths and their implications for sill emplacement mechanisms. Bull Volcanol 70:183–201
Tikoff B, Teyssier C (1992) Crustal-scale, en echelon "P shear" tensional bridges: a possible solution to the batholithic room problem. Geology 20:927–930
Tikoff B, de Saint Blanquat M (1997) Transpressional shearing and strike-slip partitioning in the Late Cretaceous Sierra Nevada magmatic arc, California. Tectonics 16:442–459
Tikoff B, de Saint Blanquat M, Teyssier C (1999) Translation and the resolution of the pluton space problem. J Struct Geol 21:1109–1117
Tobisch O, Cruden AR (1995) Fracture controlled magma conduits in an obliquely convergent continental magmatic arc. Geology 23:941–944
Townsend M, Huber G (2020) A critical magma chamber size for volcanic eruptions. Geology 48:431–435
Turcotte DL, Schubert G (2002) Geodynamics: application of continuum physics to geological problems. John Wiley NY pp 450
Urbani S, Acocella V, Rivalta E (2018) What drives the lateral versus vertical propagation of dikes? Insights from analogue models. J Geophys Res 123. https://doi.org/10.1029/2017JB015376
Vigneresse JL (1995) Crustal regime of deformation and ascent of granitic magma. Tectonophysics 249:187–202
Vigneresse JL, Tikoff B, Améglio L (1999) Modification of the regional stress field by magma intrusion and formation of tabular granitic plutons. Tectonophysics 302:203–224
Waite GP, Moran SC (2009) VP Structure of Mount St. Helens, Washington, USA, imaged with local earthquake tomography. J Volcanol Geoth Res 182:113–122
Walker JD, Fletcher JM, Fillmore RP, Martin MW, Taylor WJ, Glazner AF et al (1995) Connection between igneous activity and extension in the central Mojave metamorphic core complex, California. J Geophys Res 100:10477–10494
West M, Menke W, Tolstoy M, Sebb S, Sohn R (2001) Magma storage beneath Axial volcano on the Juan de Fuca mid-ocean ridge. Nature 413:833–836
White SM, Crisp JA, Spera FJ (2006) Long-term volumetric eruption rates and magma budgets. Geochem Geophys Geosyst 7. https://doi.org/10.1029/2005GC001002
Wilson CJN, Charlier BLA (2016) The Life and Times of Silicic Volcanic Systems. Elements 12:103–108
Wotzlaw J-F, Bindeman IN, Watts KE, Schmitt AK, Caricchi L, Schaltegger U (2014) Linking rapid magma reservoir assembly and eruption trigger mechanisms at evolved Yellowstone-type supervolcanoes. Geology 42:807–810
Wotzlaw J-F, Bindeman IN, Stern RA, D'Abzac F-X, Schaltegger U (2015) Rapid heterogeneous assembly of multiple magma reservoirs prior to Yellowstone supereruptions. Scientific Reports 5:14026. https://doi.org/10.1038/srep14026
Wrona T, Magee C, Fossen H, Gawthorpe RL, Jackson CAL, Faleide JI (2019) 3-D seismic images of an extensive igneous sill in the lower crust. Geology 47:729–733
Zandomeneghi D, Barclay A, Almendros J, Godoy JMI, Wilcock WSD, Ben-Zvi T (2009) Crustal structure of Deception Island volcano from P wave seismic tomography: Tectonic and volcanic implications. J Geophys Res 114:B06310. https://doi.org/10.1029/2008JB006119
Zollo A, Maercklin N, Vassallo M, Dello Iacono D, Virieux J, Gasparini P (2008) Seismic reflections reveal a massive melt layer feeding Campi Flegrei caldera. Geophys Res Lett 35:L12306. https://doi.org/10.1029/2008GL034242

5 Calderas

5.1 Introduction

After having described magma chamber formation and growth in Chap. 4, this fifth chapter considers the surface processes resulting from the dynamics of magma chambers, including magma accumulation on the long- and short-term, and rapid magma withdrawal and eruption. The rapid withdrawal of magma may develop a caldera. Calderas often represent the surface expression of long-lived and large magmatic reservoirs, which may be responsible for the most destructive eruptions. Calderas are active on different time scales as, in addition to experiencing eruptions and collapse (lasting from days to weeks), also experience degassing, seismicity and surface deformation testifying unrest (from months to decades), eventually culminating in long-term uplift, or resurgence (from hundreds to thousands of years). Commonly lacking a central conduit, calderas may be associated with eruptive vents scattered over a wide area, posing a higher risk than central volcanoes. Finally, calderas may also provide resources, through geothermal and ore deposits exploitation.

Calderas thus constitute an end-member type of volcano, providing the most challenging yet exciting ground for volcanologists.

The original version of this chapter was revised: Fig. 5.13 has been corrected. The correction to this chapter is available at https://doi.org/10.1007/978-3-030-65968-4_14

The main aims of this chapter are to:

- describe the structure and evolution of calderas, including the nature of their ring faults;
- discuss resurgence, resulting from long-term magma accumulation, and its mechanisms;
- introduce caldera unrest, resulting from short-term magma accumulation, and discuss its relations to resurgence;
- discuss how the caldera structure controls the transfer of the magma accumulated below, eventually feeding eruptions at different locations.

5.2 General Features of Calderas

Calderas are subcircular depressions in volcanic areas, with diameter between approximately one and several tens of kilometres (Fig. 5.1). Caldera formation is associated with the withdrawal of magma from an underlying chamber; the withdrawal induces the foundering of the chamber roof, determining the depression at the surface. The withdrawal may result from the rise and eruption, at times with explosive behaviour and significant size, of magma from the chamber. Alternatively, calderas may result from the lateral intrusion of magma through a dike or sill, eventually feeding a minor to moderate distal effusive eruption. Therefore, calderas may be associated with effusive or explosive eruptions of any size (Druitt and Sparks 1984; Branney and Acocella 2015).

V. Acocella, *Volcano-Tectonic Processes*, Advances in Volcanology,
https://doi.org/10.1007/978-3-030-65968-4_5

Fig. 5.1 Views of different calderas. **a** Erta Ale caldera, Afar, Ethiopia, hosting two pit craters. **b** Las Canadas caldera, Tenerife (Canary Islands), as seen from the International Space Station on June 6, 2006; the stratovolcano on the caldera rim in the foreground is Pico de Teide, 3715 m high; image property NASA. **c** Dolomieu caldera (Piton de la Fournaise, La Reunion Island), formed in April 2007. **d** Western rim of the 1875 Öskjuvatn caldera (Iceland), filled by the lake, nested within the larger and older Askja caldera, whose rim is visible in the background (see also Fig. 5.4b). **e** Caldera of the Caldeira stratovolcano on Faial, Azores

The most distinctive feature of a caldera consists of one or more ring faults, usually arcuate in map view, along which the roof of the chamber collapses. The **ring fault** separates an outer non-collapsed zone from an inner collapsed zone, thus representing the caldera **structural boundary** (Fig. 5.2). The ring fault forms during caldera collapse, but it may be soon intruded by the magma, generating a ring-dike ultimately feeding eruptive vents (see Sect. 7.5.3). In particular, the ring fault propagates upward from the chamber upon magma withdrawal. If the amount

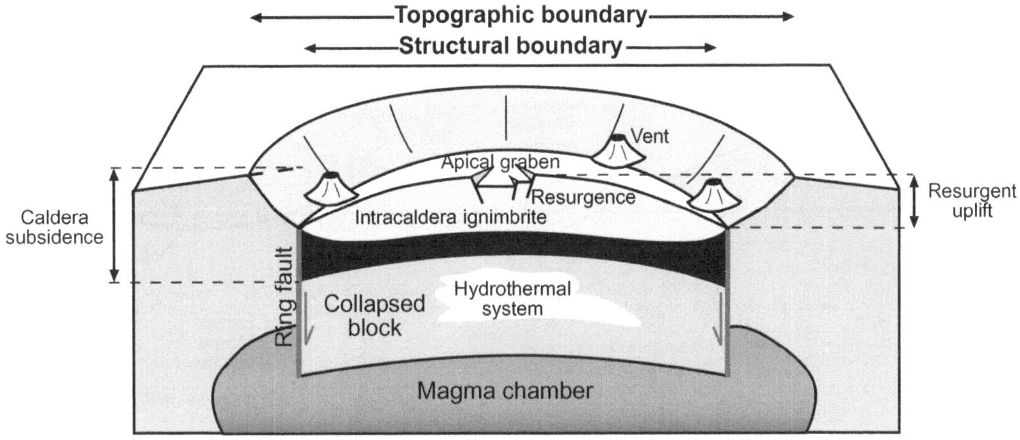

Fig. 5.2 Main morphological, structural and magmatic features of a caldera; not to scale (modified after Cole et al. 2005)

of subsidence of the caldera is significant, the ring fault may carry an important displacement and reach the surface. However, massive erosion during and after caldera collapse usually prevents observing a pristine ring fault, leaving an erosional scarp highlighted by a steep slope: this is the caldera **topographic boundary**, which may be as high as one kilometre (Lipman 1997; Cole et al. 2005). Because of retrogressive erosion, the topographic boundary lies externally to the structural boundary. Conversely, if the amount of subsidence is not significant, the ring fault may not reach the surface and a broad flexure, or monocline, forms above it; in this case, the downsag-like caldera boundary lacks a topographic scarp.

The caldera depression may be filled with volcanic deposits (including any intra-caldera ignimbrite), landslide breccia from the structural and topographic boundaries, and deposits deriving from any water infill (sea, lake, groundwater). The amount of subsidence of the caldera results from the difference in elevation of a known pre- or syn-collapse marker outside and within the caldera. However, any post-collapse vertical motion may significantly displace the caldera floor. This is the case of **resurgence**, consisting of a significant (tens of metres at least) and prolonged (hundreds of years at least) uplift of the caldera floor, highlighted by the anomalous height of its infill, accompanied by doming and/or faulting (see Sect. 5.6). A **hydrothermal system** commonly lies between the caldera floor and the magma chamber, focusing at depths between $\sim$1 and $\sim$4 km; here meteoric and magmatic fluids are pressurized within a fracture network that increases permeability (e.g., Garden et al. 2020).

These basic ingredients may be complicated by several factors, leading to elongated, overlapping, nested and asymmetric collapses. Elongated collapses are elliptical in map view and usually result from the activity of a regional stress field (see Sect. 5.5), as at Long Valley caldera (California, USA). In a few cases, elongated collapses may result from aligned overlapping calderas due to reservoir withdrawal along a same direction, as Las Canadas (Tenerife, Canary Islands; Fig. 5.1b). Nested calderas consist of one collapse structure within the other and may be related to a pair of eruptions or the same eruptive event (see Sect. 5.3). Asymmetric (or trapdoor) collapses are common, with one side subsided much more than the opposite. These result from asymmetries in the magma chamber shape, in the withdrawal of magma from the chamber, or from the drag effect reducing resistance along the foundering block within a molten chamber (Marti and Gudmundsson 2000; Acocella 2007; Kennedy et al. 2008).

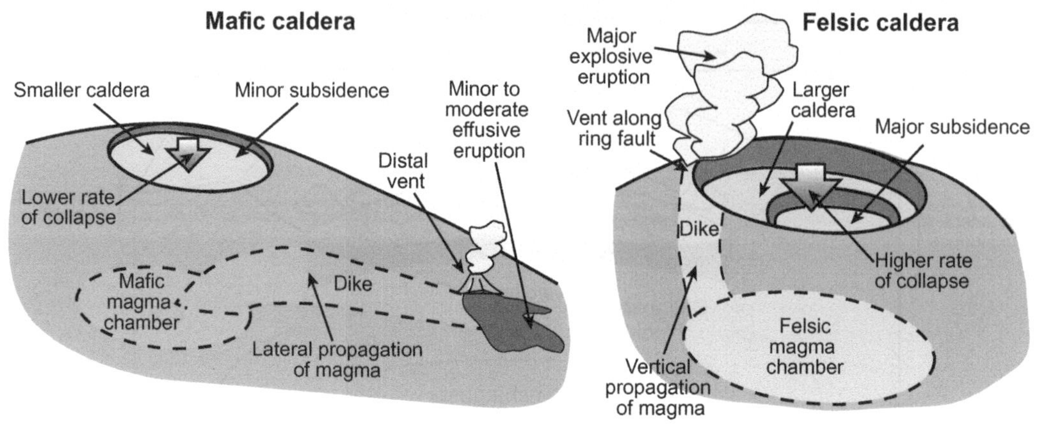

Fig. 5.3 Recurrent features commonly associated with mafic (left) and felsic (right) calderas; not to scale

Generally there is a broad relationship among the amount (from a few tens of metres to a very few kilometres) and rate (from cm/day to km/day) of collapse of calderas, the mechanism of magma withdrawal (lateral intrusion with effusive eruption, or vertical intrusion with explosive eruption), the size (from 10^{-1} to 10^3 km^3) and the composition (mafic or felsic) of the involved magma (Fig. 5.3; Newhall and Dzurisin 1988; Branney and Acocella 2015). Minor effusive eruptions often occur in the distal portion of smaller mafic calderas subsiding moderately (tens to hundreds of m) over longer periods (days to months); these calderas and eruptions are usually associated with the lateral intrusion of mafic magma. Conversely, major explosive summit eruptions are often related to felsic calderas formed in shorter time spans (hours to days) with significant subsidence (up to a very few kilometres); these calderas and eruptions are related to the vertical propagation and eruption of felsic magma. These general, but not unique, features suggest that ultimately the composition of the involved magma, related to the size of the system, controls the type and location of the eruption, as well as the rate of magma withdrawal and the amount of subsidence. Based on this, it is useful to further distinguish between the behaviour of mafic and felsic calderas.

Mafic calderas usually occur on shield volcanoes and erupt predominantly basaltic magma, although more evolved compositions (up to rhyolites) may be also erupted. These calderas are commonly up to several kilometres wide, with total amount of subsidence of a few hundred of metres, achieved with rates from cm/day to m/day. These mafic calderas may consist of smaller and structurally simpler systems, as Kilauea and Mauna Loa (Hawaii, USA), Piton de la Fournaise (La Réunion) and Erta Ale (Afar, Ethiopia), or larger and structurally more complex systems, as Askja (Iceland), Fernandina and Sierra Negra (Galapagos) (Fig. 5.4). The smaller and structurally simpler calderas usually show a moderate amount of collapse along a single subcircular ring fault, whose topographic expression consists of a hundred metres high subvertical scarp on a pile of lava flows. Extension fractures, up to a few metres wide, form in the upper portion (head) of the subvertical scarp zone and parallel to this, as induced by the lack of confinement created by the scarp. The fractures bound unstable blocks which eventually topple in the caldera, promoting the erosional retreat of the scarp. These calderas are commonly associated with distal effusive eruptions resulting from the lateral intrusion of magma through dikes (see Sect. 5.10.3). Mafic calderas may be also wider and structurally more complex (for example showing nested depressions), with higher subsidence associated with larger and more explosive eruptions, occurring also within the caldera itself. This type of calderas are

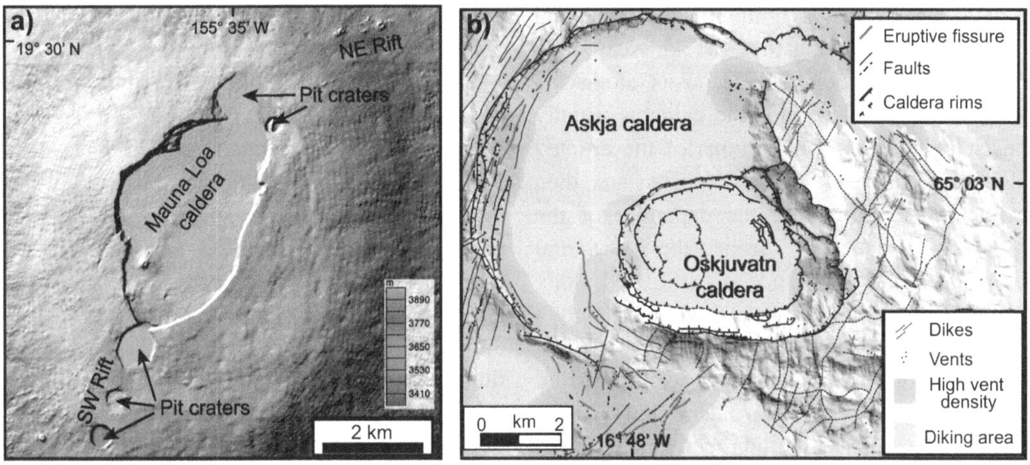

Fig. 5.4 Examples of mafic calderas. **a** Digital elevation model of Mauna Loa caldera (Hawaii, USA), showing the main caldera and the pit craters along the volcanic rift zones departing from the volcano summit (base from GeoMapApp). **b** Map view structure of the more complex nested caldera of Askja, Iceland, showing the larger Askja and the younger (formed in 1875) Oskjuvatn calderas (modified after Trippanera et al. 2018)

relatively common in the western Galapagos Islands, where the steeper summit of the shield-like volcanic edifice, resembling an "overturned soup plate", also hosts distinctive circumferential eruptive fissures (see Sect. 7.5; Chadwick and Howard 1991).

Smaller-scale analogues of mafic calderas, as sharing magma withdrawal at depth, are **pit craters,** which are metres to hundreds of metres wide subcircular depressions found along volcanic rift zones, whose formation is commonly related to the lateral flow and local stoping of magma along dikes. Pit craters are often associated with calderas, occurring within (as at Erta Ale or Kilauea) or to the side of the caldera (as at Mauna Loa and Piton de la Fournaise). The eroded Nindiri and Santiago pit craters at Masaya (Nicaragua), a few hundred of metres wide and several tens of metres deep, offer the rare opportunity to observe their deeper structure, consisting of concentric reverse (inner) and normal (outer) faults (Okubo and Martel 1998; Rymer et al. 1998; Harris 2009)

Felsic calderas are frequently associated with the explosive eruption of evolved magma and show greater variability in size, structure and activity, and often experience resurgence. Their width can reach several tens of kilometres, whereas the amount of subsidence a few kilometres, with subsidence rates of km/day. Being able to accumulate magma and generate major eruptions, felsic magma reservoirs below calderas are often large and long-lived. The true spatial extent of the felsic magma chambers may not be reflected by the size of their calderas. In fact, caldera size may be affected by the rheology of crystal-rich felsic magma, behaving elastically along the periphery of the chamber and developing calderas before complete remobilization of the entire reservoir. This may for example explain the magma chamber wider (90 km wide) than the caldera (60 km wide) at Yellowstone (Wyoming, USA; Karlstrom et al. 2012; Farrell et al. 2014).

The architecture of felsic calderas may consist of a single, major ring fault system accommodating subsidence (as Gariboldi, Ethiopia) or, more often, of pairs of nested concentric collapses, as Campi Flegrei and Latera (Italy), Guayabo (Costa Rica), La Pacana (Chile) or Taupo (New Zealand; Acocella 2007; Delgado and Pavez 2015). Many of these pairs of nested collapses are activated during the same eruptive event, not necessarily requiring two distinct eruptions, as at Campi Flegrei, where both the outer and inner collapses formed during the ~39 ka Campania Ignimbrite eruption and were reactivated during the ~15 ka Neapolitan Yellow Tuff eruption. The deposits along the topographic boundary of felsic calderas usually consist of the proximal portions of the erupted caldera ignimbrites. These deposits are then easily eroded, promoting the departure of the topographic boundary from the structural boundary. If the subsidence is partitioned along several parallel ring faults, each carrying only a part of the displacement, these may better resist erosion; this is observed along the well-preserved eastern margin of Bolsena caldera (Italy), characterized by staircase faulting, conversely to the poorly-preserved northern margin, where the single ring fault carrying the subsidence experienced stronger erosion (Acocella et al. 2012).

Felsic systems also form the largest calderas (super-calderas) on Earth, with diameter of several tens of kilometres, such as Yellowstone and Toba (Sumatra, Indonesia). The Yellowstone volcanic field was formed during three major caldera-forming eruptions at ~2.05, ~1.3, and ~0.64 Ma, erupting ~2500 km^3, ~280 km^3 and ~1000 km^3 of magma, respectively (see Sect. 13.8). The youngest caldera-forming eruption generated the current caldera, followed by ~50 rhyolitic and basaltic events, the youngest at ~70 ka. The 40 × 60 km wide caldera hosts two fractured resurgent domes. Monitoring data indicate a restless behaviour, attributed to repeated intrusions of magma and/or the pressurization of the hydrothermal system by fluids released from more than 15,000 km^3 of rhyolitic magma at depths from 8 to 18 km, largely emplaced as sill complexes (Smith and Braile 1994; Christiansen 2001; Lowenstern et al. 2006; Chang et al. 2007; Jiang et al. 2018). Toba has been active in the past 1.3 Ma, producing the Earth's largest Quaternary eruption at ~75 ka, ejecting ~5300 km^3 DRE of magma. The 100 × 45 km caldera is NW–SE elongated, parallel to the nearby Great Sumatra Fault (Fig. 5.5; see Sect. 12.3.2). The caldera

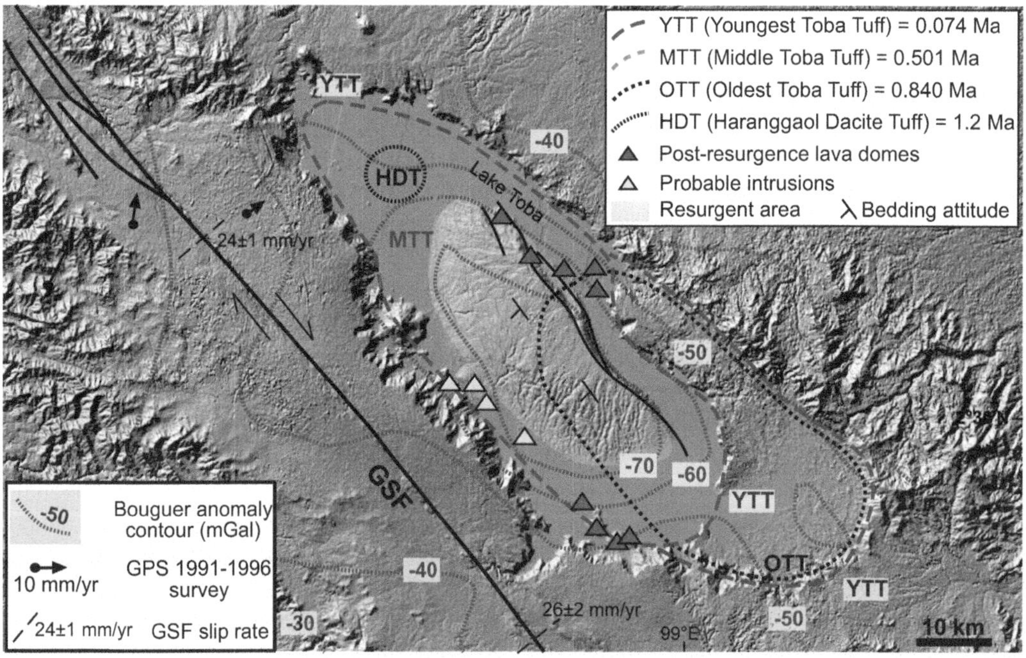

Fig. 5.5 Simplified structure of the large felsic caldera of Toba, Sumatra (Indonesia), including the collapses from the major eruptions, the resurgence area, the post-resurgence vents, the Bouguer anomalies and the slip rate along the Great Sumatra Fault (GSF; after Genrich et al. 2000; Masturyono et al. 2001; Chesner 2012; de Silva et al. 2015)

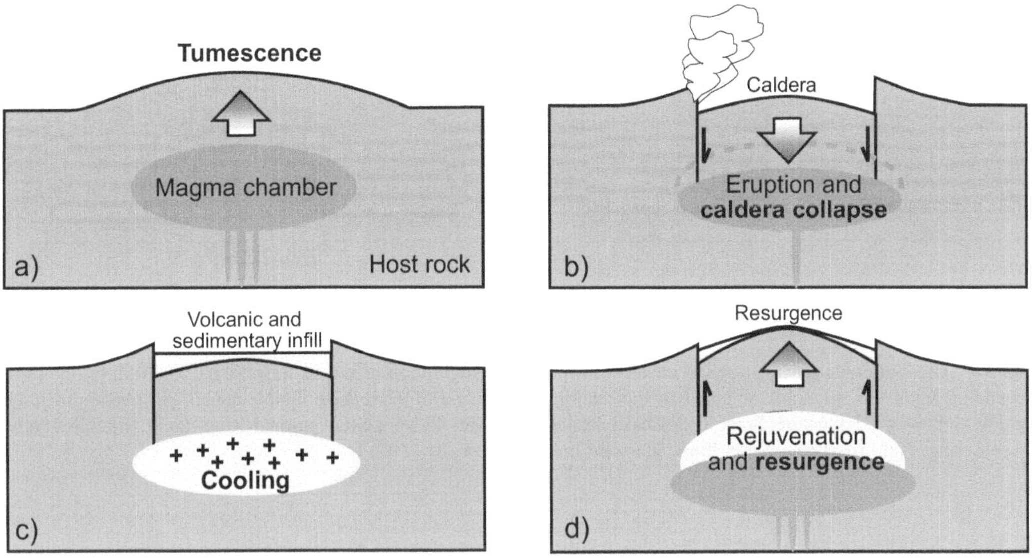

Fig. 5.6 Possible main evolutionary stages of a large felsic volcano with caldera: **a** initial tumescence, due to magma accumulation; **b** major eruption(s) and caldera formation; **c** evanescence and cooling of the shallow chamber; **d** rejuvenation, due to the input of new magma, promoting resurgence (modified after Smith and Bailey 1968)

hosts the NW–SE elongated Samosir resurgent block, uplifted asymmetrically of ~1100 m in less than 75 ka. Two nearby reservoirs, from 8 to 14 km depth, lie below the caldera (Vazquez and Reid 2004; Koulakov et al. 2009; Chesner 2012; de Silva et al. 2015; Solada et al. 2020).

An ideal evolution of large felsic magmatic complexes has been summarized in a few representative stages (Fig. 5.6; Smith and Bailey 1968; Lipman 1984). (a) Initial tumescence, due to prolonged magma emplacement inducing broad surface uplift. (b) One or more major eruptions, promoting caldera formation. (c) Decaying activity, when the mature magma chamber experiences discontinuous supply and faster cooling. (d) Possible resurgence, when new magma rejuvenates the shallow reservoir, uplifting the caldera floor.

5.3 Structure and Evolution

Despite recent episodes of caldera collapse, field and geophysical data provide limited access to understand the structure of calderas, in particular the resolution of the "space problem", that is how to accommodate the subsidence of the collapsed portion, which also determines the geometry and kinematics of the caldera's ring faults. To this aim, analogue, analytical and numerical models have been performed, providing insights on the architecture and development of calderas (e.g., Geyer and Martì 2014).

All analogue models have shown that, independently of the different boundary conditions, the structure of a caldera is mainly a function of its amount of subsidence (Fig. 5.7; Acocella 2007). For minor subsidence, the upward propagating ring fault does not reach the surface and thus a flexure forms above the blind ring fault (Stage 1). Increasing the subsidence, the ring fault reaches the surface, replacing the flexure with a fault scarp (Stage 2). The fault is outward dipping with reverse kinematics, which results from the differential vertical motion between the two blocks, responsible for a deflection of the maximum principal stress σ_1 from the vertical path. Therefore, these reverse faults are not related to any lateral compression (see detail of σ_1 trajectories in stage 2, Fig. 5.7; Mandl 1988). An increase in subsidence produces a peripheral flexure, resulting from the upward propagation of

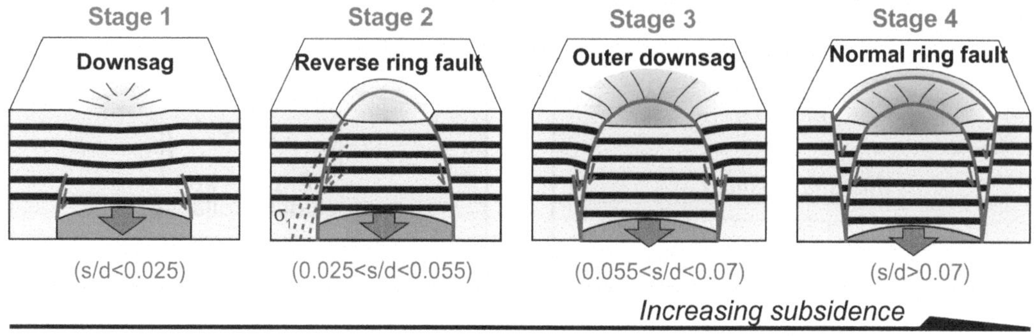

Fig. 5.7 The four evolutionary stages of caldera collapse, depending upon the *s*/*d* ratio, obtained in the analogue models and observed in nature. Stage 1: downsag; stage 2: reverse ring fault (dashed purple lines illustrate local configuration of the σ_1 trajectories controlling the development of the reverse fault); stage 3: peripheral downsag; stage 4: peripheral normal ring fault (modified after Acocella 2007)

an outer blind ring fault, driven by the gravitational collapse of the unstable wedge of the hanging wall of the reverse ring fault (Stage 3). In the final stage, increasing again the subsidence, the outer ring fault reaches the surface, producing a second peripheral and concentric scarp, or nested collapse (Stage 4). This outer fault is inward dipping and has a normal kinematics due to the collapse of the wedge above the reverse fault. Any further increase in subsidence does not change this deformation pattern and both ring faults continue to be active.

These results, being obtained by independent studies under different boundary conditions, provide a robust platform to define the structure of calderas as a function of the amount of subsidence. However, calderas with different diameter may show different structures (corresponding to the above-mentioned evolutionary stages) even with similar amount of subsidence. For example, the same amount of subsidence may produce immature (stage 1) collapse in wider calderas or mature (stage 4) collapse in narrower calderas. This results from the fact that a same amount of subsidence in wider calderas develops a wider flexure accumulating larger vertical displacement before faulting. Therefore, to compare the structure and evolutionary stage of calderas, it is important to relate the amount of subsidence *s* to the caldera diameter *d*, that is considering the *s*/*d* ratio.

A second-order parameter controlling caldera structure is the depth to the magma chamber with regard to its width. For a given magma chamber width, shallower chambers produce the simple pattern described in Fig. 5.7, whereas deeper chambers promote multiple sets (whose number increases with the chamber depth) of overlapping outward dipping reverse ring faults, and only an uppermost normal ring fault (Roche et al. 2000). Often the entire collapse structure above these deep and narrow magma chambers is slightly inclined, forming a "sliding-trapdoor", or ring-fault architecture that consists of outward-inclined reverse portions and inward-inclined normal portions on opposite sides of the ring fault. This feature has been recognized at recent caldera collapses at Miyakejima (Japan), Dolomieu (Reunion) and Bardarbunga (Iceland; Bathke et al. 2015). However, in case of deep magma chamber with limited subsidence, the caldera faults may not reach the surface, delaying the expected evolutionary stages. Therefore, the deeper the magma chamber, the least developed is the collapse at the surface, as a larger amount of withdrawn magma is required to have collapse (Roche and Druitt 2001; Geyer et al. 2006). This feature has important implications in terms of magma transfer and eruption (see Sect. 5.9).

Analytical and numerical models of calderas provide less consistent results than analogue models, probably for the difficulty to simulate the

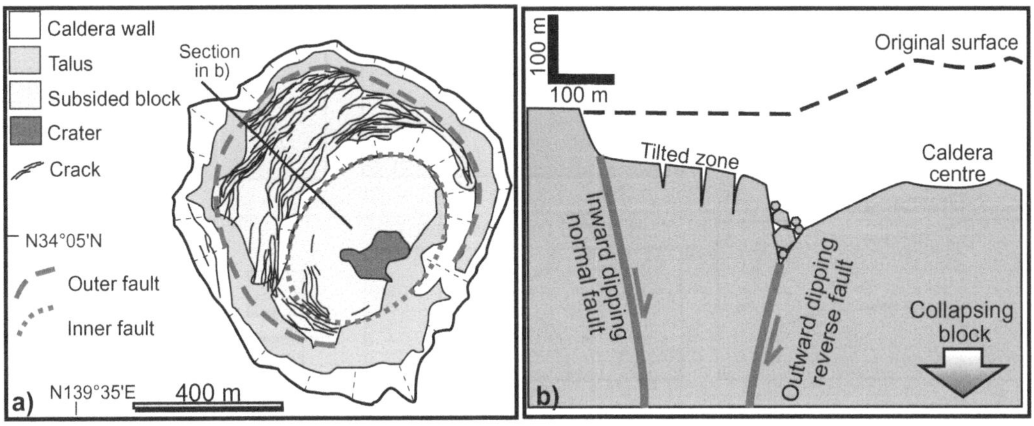

Fig. 5.8 Map (left) and section (right) views of Miyakejima caldera (Japan), as formed in 2000 (Japan; modified after Geshi et al. 2002). Compare with Fig. 5.7, stage 4

activity of shear fractures. In fact, only discrete element models in a frictional medium simulated caldera formation consistently with analogue models, developing upward propagating outward dipping reverse ring faults and peripheral inward dipping normal ring faults (Gudmundsson 2007; Hardy 2008; Holohan et al. 2011; Gregg et al. 2012; Kabele et al. 2017).

More importantly, the analogue modelling results are supported by geological and geophysical evidence from several tens of active mafic and felsic calderas, and direct observation of recent caldera collapses, in particular at Miyakejima, Axial Seamount and Kilauea (Fig. 5.8; Geshi et al. 2002; Acocella 2007; Wilcox et al. 2016; Levy et al. 2018; Baillard et al. 2019; Neal et al. 2019). This consistency indicates a robust frame to understand caldera structure and evolution, with the four stages allowing to define the overall structure of a caldera simply knowing its *s*/*d* ratio or, conversely, to define the amount of subsidence from the surface deformation pattern. Determining the *s*/*d* ratio of a caldera also allows classifying calderas (see Sect. 5.4) and better defining magma transfer below calderas, possibly forecasting the location of eruptive vents (see Sect. 5.9; Acocella and Rivalta 2019).

Kinematically, analogue models also show that collapse may occur with several modes and velocities (continuous, incremental or sudden), mainly depending on the aspect ratio (thickness/width) of the crust overlying the chamber and the reactivation of pre-existing ring faults. This feature is consistent with geophysical (mainly seismicity) and geodetic (mainly tilt variations) data from collapsing calderas (Michon et al. 2011; Ruch et al. 2012; Munekane et al. 2016). Also, models show that the propagation rates for the reverse faults during collapse increase with magma evacuation rates, with magma chamber dynamics and fault propagation being kinematically coupled (Seropian and Stix 2018). However, this relation may be influenced by the friction along the ring fault. For example, direct access to the ring fault of the extinct Jangsan caldera (Korea) reveals extremely low dynamic frictions, suggesting that an extraordinarily large fault slip (>100 m), causing a large earthquake, was associated with caldera formation. This slip generated pseudotachylytes, indicating that frictional melting was an important fault zone process during caldera collapse (Han et al. 2019; Kim et al. 2019).

5.4 Classification

A first popular approach to a systematic definition of caldera types was based on field data from active and extinct eroded calderas, defining five end-member geometries: piston, piecemeal,

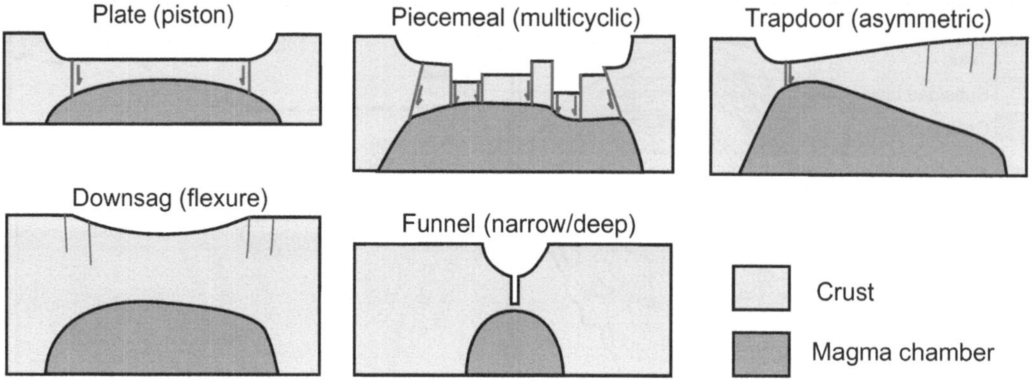

Fig. 5.9 The five established geometric caldera types (modified after Lipman 1997; Cole et al. 2005)

trapdoor, downsag and funnel (Fig. 5.9; Walker 1984; Lipman 1997, 2000; Cole et al. 2005). **Piston**-type calderas are bordered by a ring fault, delimiting a sinking central block. **Piecemeal** calderas result from the differential subsidence of multiple blocks, usually along several pre-existing faults. **Trapdoor** calderas show asymmetric subsidence accommodated by different types of structures along the rim, as a flexure on one side and a fault on the opposite. **Downsag** calderas are depressions with unfaulted and inward tilted, or flexured, margins. **Funnel** calderas are narrow and deep cone-shaped depressions.

While this geometric classification is based on easily detectable field features, justifying its popularity, it also carries several limitations concerning the structure and evolution of calderas. In particular: (a) this classification does not take into account for the resolution of the "space problem"; (b) this classification is based on a static perspective of calderas, not considering their development, maturity and possible evolutionary relationships; (c) several of the five proposed caldera types may be found within a same caldera, making it difficult to unequivocally capture its distinctive structure and evolutionary stage. For example, Bolsena caldera (Italy) contains a central piston collapsed asymmetrically, with one margin interpreted as a downsag and the other as a piecemeal (Acocella et al. 2012): accordingly, Bolsena should be awkwardly defined as a caldera with "trapdoor piston and downsagged and piecemeal margins".

The analogue models and their comparison to nature allow passing these limitations and proposing, using the *s/d* ratio of a caldera, a revised and more practical classification. As the shallow structure of calderas corresponds to a precise architecture at depth, recognizing diagnostic surface features allows a caldera to be categorized within a precise structural and evolutionary context. The general relationship between the evolutionary stages of a caldera and their range of *s/d* ratios explains the architecture and maturity of calderas along a continuum mainly controlled by progressive subsidence. In this context, stage 1 (downsag) calderas are characterized by $s/d < 0.025$, stage 2 (reverse fault) by $0.025 < s/d < 0.055$, stage 3 (outer downsag) by $0.055 < s/d < 0.07$ and stage 4 (normal fault) by $s/d > 0.07$ (Fig. 5.7; Acocella 2007). Using such a classification it is also possible to define the deeper structure of a caldera from the available surface evidence: the apparently complex structure of Bolsena would be simply explained by a stage 3 asymmetric collapse.

A more recent classification of calderas relies on the pressure variation inside the magma chamber with regard to eruption, considering underpressure and overpressure calderas (Martì et al. 2009). Underpressure calderas result from subsidence after decompression of the magma chamber following a pre-caldera eruptive episode, according to the established model of caldera formation. Here the initiation of collapse is a consequence of the eruption, and this is expected to constitute the condition most frequently occurring in nature. However, in some cases there may be evidence that the stress conditions developing the caldera ring faults are achieved before the eruption. This is the theoretical rationale behind overpressure calderas, which may form when, for example, an overpressurized sill-like magma chamber is loaded by regional doming or undergoes regional extension: here the initiation of collapse is coeval to the eruption onset (Gudmundsson 1998, 2007).

5.5 Relationships to Regional Tectonics

Modelling suggests that the structure and development of calderas are largely independent of any regional tectonics. Despite this, regional tectonics may still affect some features, as the caldera shape in map view, generating an elliptical collapse with major axis parallel to the minimum horizontal principal stress σ_m.

These elliptical calderas may develop from elliptical or even circular (in map view) chambers at depth, in any case due to a regional tectonic contribution (Acocella 2007). Most commonly elliptical calderas reflect the shape of anisotropic (elongated) magma chambers at depth. Elongated chambers may form similarly to what observed at borehole breakouts, where the cavity is narrowed along the direction of the maximum horizontal principal stress σ_M and elongated parallel to minimum horizontal principal stress σ_m. This differential horizontal stress may explain the elongation, parallel to the direction of regional extension σ_3, of several elliptical calderas, as along the oceanic ridge of Iceland (Askja, Krafla), the East African Rift System (Suswa, Silali, Paka) and the Taupo Volcanic Zone of New Zealand (Okataina, Taupo; e.g., Bosworth et al. 2003). Elongated magma chambers leading to elliptical calderas may also form through the coalescence of multiple dikes aligned perpendicular to the direction of regional extension σ_3 or intruding pre-existing regional structures. In both cases, the result is an elongated magma chamber growing along the direction of the dikes. Examples of calderas elongated parallel to the rift axis (and thus to the direction of dikes) and perpendicular to the direction of regional extension include Erta Ale and Pantelleria (Italy).

In principle, elliptical calderas may also derive from the collapse of circular (in map view) magma chambers, provided a regional stress field is active. This may occur in three situations (Acocella 2007). (a) When an outward dipping ring fault locally reactivates a pre-existing regional normal fault dipping towards the caldera centre: this shifts the upper part of the ring fault outwards, locally widening the caldera. The reactivation may occur only if the ring fault is tangent to the regional fault. The field relationships between regional normal faults and the western rim of Askja caldera suggest that even a difference in strike of very few tens of degrees hinders reactivation. (b) In presence of an active regional stress field, which affects the dip of the portion of the outward dipping ring fault striking perpendicular to the minimum horizontal principal stress σ_m or to the maximum horizontal principal stress σ_{M} (Holohan et al. 2005). In particular, the portion of the outward dipping ring fault striking perpendicular to the σ_m direction (extensional condition) becomes steeper, widening the caldera in that direction, whereas the portion of outward dipping ring fault striking perpendicular to the σ_M direction (compressional condition) becomes gentler, narrowing the caldera in that direction. As a result, in both cases a) and b) the major axis of the elliptical ring fault is parallel to the minimum horizontal principal stress σ_m. In any case, the models suggest that with pre-existing regional structures or regional stress field the expected maximum eccentricity is ~ 1.2. Therefore, only a part of the eccentricity of more elongated calderas may be ascribed to pre-existing regional extensional structures or a regional stress field. (c) A further process possibly generating elliptical calderas from originally circular magma chambers is post-collapse faulting. Regional normal faults may in fact dissect and elongate a caldera along the minimum principal stress σ_3, as proposed at Fieale, Afar (De Chabalier and Avouac 1994). However, the efficiency of this process appears limited, as highly dipping normal faults do not usually increase caldera eccentricity of more than $\sim 10\%$.

At the extreme, significant magma chamber withdrawal may reactivate regional normal faults, if present, forming graben calderas. **Graben calderas** are elongated rift segments bounded by normal faults collapsing during large eruptions under regional extension and whose width approximates that of the erupting reservoir below. Graben calderas, inferred on continental (southern Basin and Range, western USA), transitional (central Afar) and oceanic rifts (East Pacific Rise), thus represent a strongly elongated end-member type of caldera (Lagabrielle and Cormier 1999; Aguirre Diaz et al. 2008; Acocella 2010). Because of the "space problem", a block bounded only by inward dipping regional faults may not sink: therefore, the reactivation of outward-dipping regional normal faults with inverted (reverse) kinematics is expected.

Finally, regional tectonic stresses may not only affect the caldera shape, but also its anticipated or delayed formation. In fact, numerical results indicate that regional extension decreases the stability of the roof rock overlying a magma chamber, thereby promoting early-onset caldera collapse, or even suppressing any resurgence. Alternatively, moderate amounts of regional compression (≤ 10 mm/year) on relatively short timescales ($<10^4$ years) increase roof rock stability, delaying collapse (Cabaniss et al. 2018).

5.6 Resurgence

Some calderas experience notable uplift of their floor, providing evidence for resurgence. **Resurgence** is a spectacular, albeit still relatively poorly-known, feature of many calderas, so that it is commonly defined on the evident uplift of the caldera floor, rather than on the responsible process(es). The uplift may be recognized from unusually raised deposits within the caldera, including the caldera-forming eruption or marine and lacustrine sediments. The uplifted area is usually circular to elliptical, from a few kilometres (Pantelleria, Italy) to a few tens of kilometres wide (La Pacana, Chile), somehow proportional to the size of the caldera; in some cases, resurgence may show a polygonal shape, due to the reactivation of pre-existing regional structures (Ischia, Italy; Fig. 5.10).

The amount of uplift ranges from a few tens of metres, as at Alcedo (Galapagos), to more than a thousand of metres, as at Valles (New Mexico), Toba (Sumatra) and Cerro Galan (Argentina). The duration and uplift rate of resurgence are usually poorly constrained, being at best inferred through the cumulated uplift achieved within a geologically-defined period (often tens of thousands of years), which is usually longer than that of effective uplift. So far, the best-constrained uplift paths for resurgence derive from Campi Flegrei, Toba, Siwi (Vanuatu) and Iwo-Jima (Japan) (Fig. 5.11; Newhall and Dzurisin 1988, and references therein; Chen et al. 1995; de Silva et al. 2015; Isaia et al. 2019). These cases show very different duration of resurgence, which lasts from ~ 1 to 70 ka, and different uplift rates, which range from centimetres to tens of centimetres per year, that is similar to or one order of magnitude faster than the regional tectonic rates. However, these rates are usually not constant. For example, the resurgence at Campi Flegrei shows uplift episodes alternated with subsidence. While net uplift dominates, the temporary subsidence may be still significant at times, on the order of several tens of metres over a few thousands of years. Also, the eruptive periods (epochs) broadly correlate with periods of uplift during resurgence. At Siwi and Iwo-Jima the available data describe incremental uplift phases interrupted by stases and minor subsidence. The resurgence at Toba shows an apparently continuous, albeit not constant, uplift. Therefore, available evidence indicates that most resurgent uplift occurs episodically, within a very few thousands of years, interrupted by stasis or even subsidence, although some calderas show a more continuous behaviour (Acocella 2019). In some

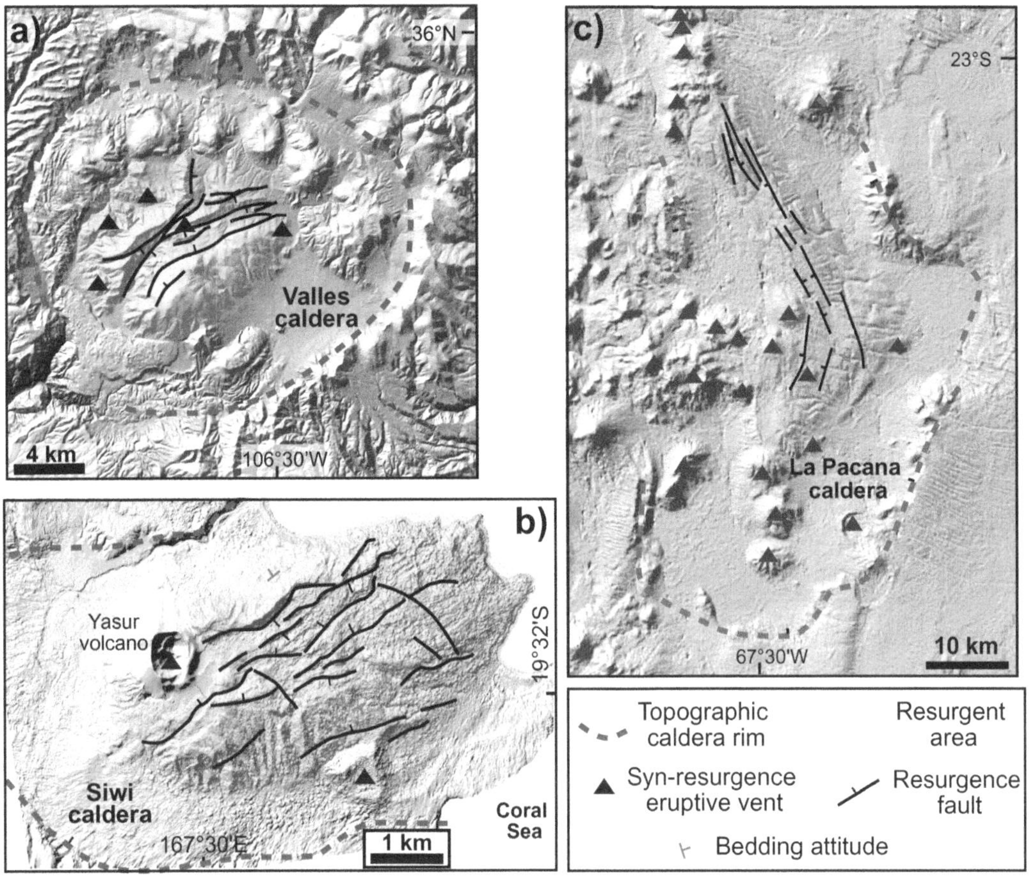

Fig. 5.10 Examples of resurgent calderas; note the different scales. **a** Valles caldera, New Mexico (USA); **b** Siwi caldera, Vanuatu; **c** La Pacana caldera, Chile (Siwi DEM courtesy: Elodie Brothelande; modified after Lindsay et al. 2001; Galetto et al. 2017)

cases, as at Pantelleria and Ischia, geodetic data indicate that the resurgent portion of the caldera has been experiencing subsidence for decades, suggesting an interruption of the longer-term resurgence. At Ischia, the subsidence is accompanied by recurrent, very shallow (~ 1 km depth) and destructive seismicity along the faults bordering the most uplifted part of the resurgent area. This behaviour may result from accelerations of the ongoing subsidence, which is interpreted to derive from the degassing of the previously intruded and non-erupted magma partly responsible for resurgence (Trasatti et al. 2019). Resurgence is commonly associated with felsic volcanism (Table 5.1). In fact, most of the resurgent calderas, including those with strong uplift reaching a thousand of metres, as at Valles, Long Valley, Toba and Ischia, are consistently felsic. Resurgence in mafic calderas is found only at Loihi (Hawaii) and Alcedo and Sierra Negra (Galapagos), with moderate uplift between ~ 30 and ~ 100 m (Acocella 2019; Clague et al. 2019).

In many resurgent calderas the uplifted deposits define a broad subcircular or elliptical **resurgent dome**, associated with an apical graben with inward-dipping normal faults, as observed at Valles, Long Valley or Kutcharo, Japan. The apical graben, often elongated perpendicular to the regional extension direction, results from the crestal extension created by the doming, which may also be responsible for a

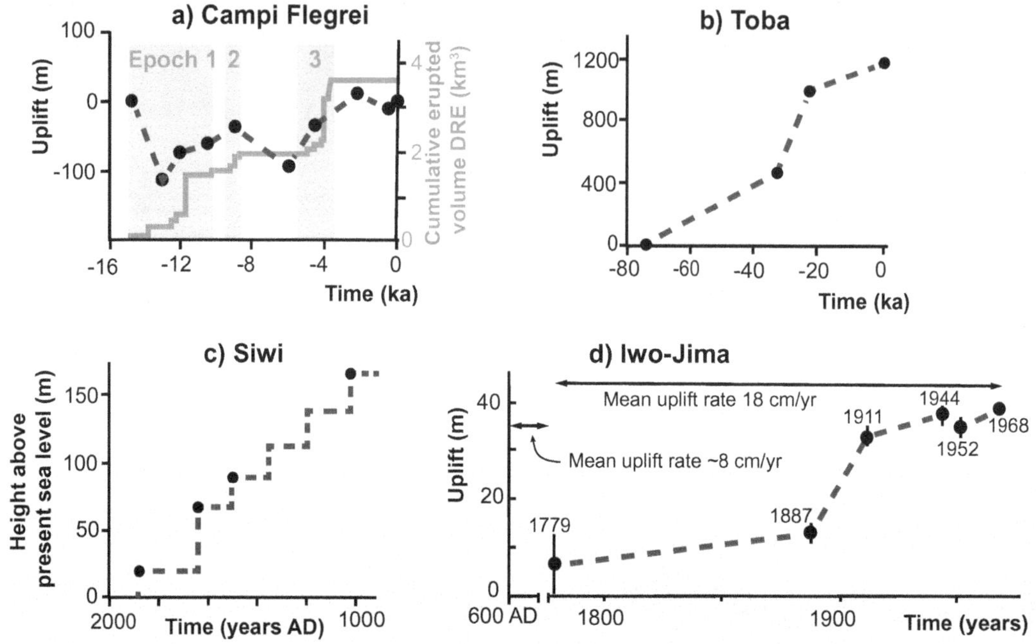

Fig. 5.11 Uplift histories of felsic resurgent calderas at Campi Flegrei (Italy), Toba (Sumatra), Siwi (Vanuatu) and Iwo-Jima (Japan). Eruptive periods (orange bands) and cumulative erupted products (orange curve) are also included for Campi Flegrei (modified after Acocella 2019)

localized decompression promoting the apical volcanism. These domes in fact also show syn-resurgence eruptive vents within the dome and along the boundary, likely resulting from the pervasive intrusion of magma within the dome faults and the resurgence ring fault (Self et al. 1986; Goto and McPhie 2018).

In other calderas the uplifted deposits may define a **resurgent block**, where the non-bended deposits have a consistent attitude. These may be tilted of a very few tens of degrees about a horizontal axis, showing a marked asymmetry in the section view of the block, as at Ischia, Pantelleria, Alcedo, Sierra Negra and Suswa (Orsi et al. 1991; Skilling 1993). Resurgent blocks show syn-resurgence eruptive vents along their sides, in correspondence with the border faults, or outside. The development of a resurgent dome or block does not depend on the amount of uplift, as resurgent blocks may be uplifted as much as domes (Table 5.1). The possibility to develop a resurgent dome or a block may depend on the aspect ratio A_r (thickness z/width L) of the crust overlying the magma chamber (Fig. 5.12; Acocella et al. 2001). Aspect ratios $A_r \sim 1$, where the depth to the magma chamber is similar to the width of the uplifted area, develop resurgent blocks with consistent attitude of the layers and peripheral volcanic activity (Ischia and Pantelleria type). Aspect ratios $A_r \sim 0.4$, where the depth to the magma chamber is smaller than the width of the uplifted area, develop resurgent domes with apical depression, domed layers and peripheral and internal volcanic activity (Valles and Long Valley type). These features may result from the different flexural response of the thicker and thinner uplifted crusts. The large and apparently resurgent block of Toba may provide an exception to this model.

The boundary of the resurgent portion may be sharp and with evident morphological expression, such as a scarp, suggesting the activity of faults confining the uplift. This is well exemplified along the most uplifted part of resurgent blocks, as at Ischia and Pantelleria. Modelling

Table 5.1 Main features of active resurgent calderas

Name	AC (km^2)	AR (km^2)	Uplift (m)	Vr (km^3)	Vv (km^3)	Type	t (ka)	tr (ka)	Composition	Supply	Reg. ext. (mm/year)
Ischia	46	~20	~1000	9 ± 2	<1	B	22 ± 5	<33	T	Yes	< 2
Pantelleria	42	7	~350	1.5 ± 0.3	0.5 ± 0.2	B	27	<18	T, R	Yes	< 2
Campi Flegrei	~200	~30	~250	3 ± 0.5	4 ± 0.5	D	<4	<15	T	Yes	< 2
Long Valley	450	~80	>400	~15	(≪100)	D, G	(100)	(80)	R, Da	Yes	0.6
Valles	283	~65	≥1000	32 ± 5	<5	D, G	(<40)	27 ± 27	R, RD	Yes	< 4
Yellowstone SC	2500	214	?	>27	0	D, G	(<84)	(<123 ± 9)	R	Yes	< 4.3
Yellowstone ML	2500	164	?	20 ± 7	(<600)	D, G	~440		R	Yes	< 4.3
Toba	2350	950	1100	320 ± 30	20 ± 5	B	(<40)	<70	R, Da	Yes	0
Siwi	36	18	>250	4.5 ± 1	0.5 ± 0.3	D, G	(<18)	(2)	Ba, TA	Yes	≥ 0
Iwo-Jima	63	17	>120	>1.5	0	D	(>2)	<0.8	TA	Inferred	< 2

AC Area of caldera; *AR* Area of resurgence; *Vr* Volume of resurgence; *Vv* DRE volume of syn-resurgence volcanism; *B* Block, *D* Dome, *G* Graben; *t* Time between caldera collapse and resurgence; *tr* Duration of resurgence. Composition: *T* Trachyte, *R* Rhyolite, *Da* Dacite, *RD* Rhyodacite, *Ba* Basalt, *TA* Trachyandesite; supply: evidence of new magma supply; reg. ext.: rates of regional extension. Yellowstone SC: Yellowstone Sour Creek dome, Yellowstone ML: Yellowstone Mallard Lake dome. Values in () are poorly constrained and may not be representative

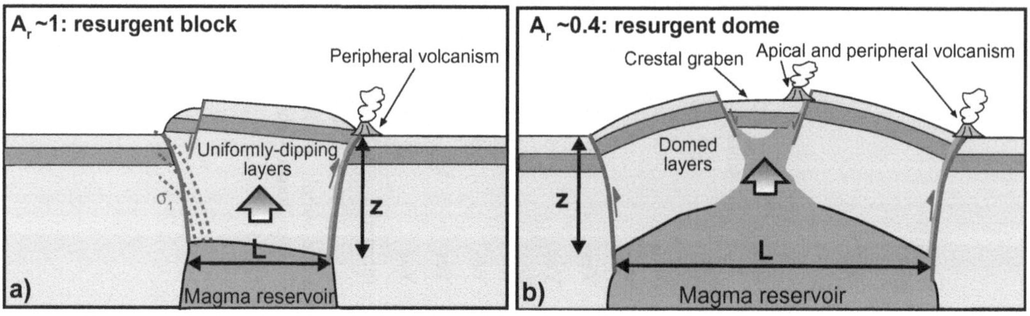

Fig. 5.12 Summary of the two main resurgence modes observed in experiments and nature mainly as a function of the aspect ratio A_r (where A_r = thickness z/width L) of the crust overlying the magma chamber. **a** A higher A_r develops a resurgent block with consistent layers within and peripheral volcanic activity (Ischia and Pantelleria type; dashed purple lines illustrate local configuration of the σ_1 trajectories controlling the development of the reverse ring fault). **b** A lower A_r develops a resurgent dome with domed layers, apical depression and volcanic activity at the periphery and within the dome (Valles and Long Valley type; modified after Acocella et al. 2001)

and field data (as at Valles) suggest that the boundary of resurgent domes may also experience non-elastic deformation, thus being faulted. Therefore, the presence of resurgence bounding faults raises a "space problem" similar to caldera collapse. As a consequence, the nature of the resurgence's bounding faults has been object of field and modelling studies (Tibaldi and Vezzoli 1998; Acocella et al. 2001). Resurgent blocks and domes cannot be bordered by major outward dipping normal faults, as the uplift of a horst-like structure would require unrealistic syn-resurgence extension outside the uplifted area. Therefore, most resurgent areas are expected to be bordered by subvertical to high angle inward dipping, reverse structures. High angle inward dipping reverse faults, with displacement of several hundreds of metres, have been observed to border resurgent blocks, as at Ischia. These features are also common in resurgence experiments, resulting from the upward rotation of the maximum principal stress σ_1 during differential vertical motion, similarly to what occurs for the caldera reverse faults (see detail of stress distribution in Fig. 5.12a). In addition to the reverse faults, marking the structural boundary of the resurgence, internal normal faults may result from the gravitational instability of the periphery of the uplifted area, similarly to what observed for the caldera normal faults.

Resurgence is a notable feature of many calderas. However, while many studies have investigated and explained the causes of caldera collapse, the processes controlling the opposite motion, that is resurgent uplift, have been more rarely investigated and, as such, remain less understood. In fact, while it has been generally accepted for decades that resurgence results from a build up in pressure in the magma chamber, or from shallow intrusions, a few models have been proposed to explain the details of this process. Possible mechanisms related to the dynamics of a magma chamber and triggering resurgence are regional detumescence, magmatic rebound and general magma pressurization (Marsh 1984; Kennedy et al. 2012). **Regional detumescence** is one of the first models proposed for resurgence (Smith and Bailey 1968; Marsh 1984). It implies that, upon caldera formation, an originally inflated crust relaxes, pushing downward on the magma chamber and squeezing magma upward against, around, and possibly through the fallen caldera block, triggering resurgence. This process does not really require any addition of new magma within the chamber and, as such, provides a self-sustaining explanation for resurgence after caldera formation. However, it would require a magma chamber much wider than the collapsed area and relaxation of a large previously uplifted area, features which are not really observed. Therefore, this model appears unsupported by observations. **Magmatic rebound**, and/or decompression following the formation of a caldera, has been proposed as an attractive and elegant explanation for resurgence (Marsh 1984; Kennedy et al. 2012). This is related to the sluggish response of the viscous magma to the sudden drawdown caused by the eruptions during caldera formation. In particular, the caldera-forming eruption may empty more than enough magma to bring the system to magma-static equilibrium, determining a pressure drop which may be balanced by the flow of new magma injected in the system. In this model, resurgence requires the input of new magma closely following caldera formation. A limitation to the widespread application of this model is that the time between caldera development and the onset of resurgence in natural cases often ranges from a few tens of thousands of years (Ischia, Pantelleria, Valles) to several hundreds of thousands of years (Yellowstone), suggesting the lack of a direct relationship between caldera collapse and resurgence (Table 5.1). **Magma pressurization** is a broad term, which includes an increase in the pressure within a magma chamber as resulting from magmatic convection, vesiculation or the influx of new magma at the base of the caldera block (Smith and Bailey 1968; Marsh 1984). This process usually requires the injection of new magma within the magma chamber, although it may also occur emplacing shallower laccolith-like intrusions, even at the base of the caldera infill. Evidence for such a shallower magma mainly derives from eroded calderas, as at Grizzly Peak and Lake City (Colorado, USA) and Kumano (Japan), and borehole stratigraphy, as at Long Valley (Fridrich et al. 1991; Kawakami et al. 2007; Kennedy et al. 2015; Hildreth et al. 2017). Magma pressure increase within the chamber, or at shallower levels, is a feasible possibility, although alone it still does not explain a crucial feature of resurgence, that is the anomalous accumulation of magma required to trigger dramatic uplift instead of feeding eruptions. This latter feature is underlined by the usually much higher volumes of the resurgent portion with regard to the volumes of magma erupted during resurgence. This imbalance suggests some process hindering the extrusion of the pressurized magma. This problem has been tackled postulating that the hindered eruption of the magma responsible for resurgence is related to its relatively low viscosity contrast (less than approximately 9 orders of magnitude) with the residual magma below a caldera (Galetto et al. 2017). Such a viscosity contrast provides a rheological barrier, hindering the upward propagation through dikes of the new injected magma, which accumulates in sill- or laccolith-like intrusions, stagnates and promotes resurgence (Fig. 5.13). This model considers a spectrum of combinations in the thickness and viscosity of the residual magma, which determine the permeability of the rheological barrier. Thicker and medium viscosity residual magma provides a less permeable barrier, leading to resurgence without eruptions within the resurgence area, as observed at Ischia, Yellowstone or Iwo-Jima. Other conditions of thickness and viscosity provide a more permeable barrier, leading to resurgence accompanied by diking and thus eruptions within the resurgent area, as observed at Campi Flegrei. This mechanism may explain resurgence through the input of new magma at different crustal levels, as within the magma chamber or at the base of the caldera infill. A neutral or compressive regional stress field would further hinder dike propagation and eruptions, thus encouraging the storage of the magma at depth and resurgence, consistently with the evidence that resurgent calderas are rare or absent in rifts experiencing relevant extension (several mm/yr or more, as in the Taupo Volcanic Zone of New Zealand). Similar low-viscosity contrasts in magma reservoirs may also provide the conditions to pass from magma eruption to accumulation, which is essential to develop large magma chambers before major eruptions.

5.7 Caldera Unrest

A distinctive shorter-term behaviour of calderas is their frequent state of unrest, as revealed by surface deformation, seismicity and degassing.

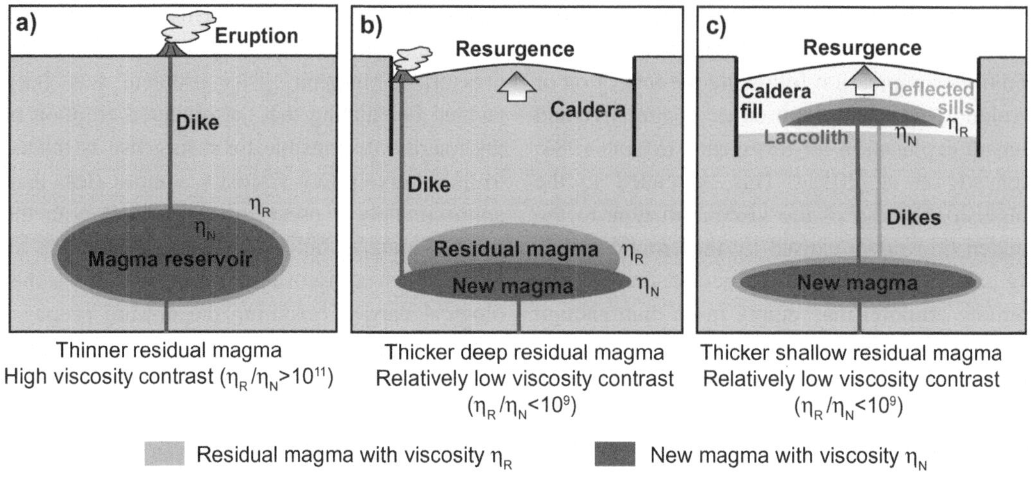

Fig. 5.13 Conditions for resurgence. **a** A thinner layer of viscous residual magma (high viscosity contrasts with the new magma) promotes dike propagation and eruption, without resurgence. **b** A thicker layer of medium viscosity residual magma (with relatively low viscosity contrast with the new magma) hinders dike propagation, promoting stagnation and resurgence; peripheral eruptions may occur. **c** Same as (**b**), but shifted at shallower levels: dikes (orange) may arrest within the altered intracaldera tuff, developing one or more sills (orange) constituting a rheological barrier for successive dikes (red), stagnating in laccoliths and promoting resurgence. Mechanism (**c**) may be alternative to, or combined with, (**b**) (modified after Galetto et al. 2017)

Unrest is a deviation, lasting from days to centuries, from the normal quiescent state towards a state which may lead to eruption. Unrest, although most common at calderas, may occur at any volcano, representing a widespread condition. Because of its general occurrence and potentially hazardous implications, unrest is discussed, together with eruption forecasting, in Chap. 9; here only an overview of the specific caldera unrest features is given.

More than a thousand episodes of caldera unrest have been documented based on the appearance of phenomena related to the eruption preparation, or possible precursors. As in a typical year some form of unrest occurs at nearly 20 calderas on average, unrest should be considered a common state for active calderas, although in approximately half of the cases unrest does not culminate into eruption (Phillipson et al. 2013). The most frequently reported features characterizing caldera unrest include local earthquake swarms, volcanic tremor; uplift, subsidence, tilt, ground fissuring, changes in the temperature of soil, water, or gas, changes in fumarolic activity and, ultimately, eruptions. These seismic, geodetic, geochemical and thermal variations are mostly limited to parts of calderas, without being necessarily concentric. But seismicity, uplift, degassing, thermal activity and eruptions at a given caldera do not always share a common centre. Furthermore, centres of each unrest may shift from one episode to the next, or even within a single episode. At depth, unrest may reach the lowermost crust (Newhall and Dzurisin 1988; Acocella et al. 2015; Hotovec-Ellis et al. 2018).

In particular, as regards surface deformation, all types and scales of ground deformation may occur in any part of a caldera during unrest. Occasionally, ground deformation above a conduit or a dike is so dramatic that it can be directly witnessed without any instrument. Normally, however, ground deformation is subtle and within the elastic domain (with rates of millimetres or centimetres per year), so that it can be detected only through appropriate equipment (see Chap. 8). Broad, subtle, years to decades long uplift has occurred within several large calderas. A common deformation pattern consists of a dome-like inflation, with the maximum uplift centred on an internal portion of the caldera.

Surface deformation can be usually modelled with sources shallower than 10 km depth, consistent with the depth to the top of large geophysically imaged magmatic reservoirs (Newhall and Dzurisin 1988).

As regards seismicity, this may be found around caldera margins and elsewhere within the caldera. Most seismic swarms within or near calderas consist of relatively small earthquakes, with magnitude lower than 3 and rarely reaching 6 and depth less than 15 km. Some of these result from the brittle failure of country rock under magma intrusion (volcano-tectonic earthquakes, see Sect. 8.4.1). Failure may be facilitated by high pore pressures within hydrothermal systems. Other events reflect release of regional tectonic stress, shear of viscous magma along the walls of a conduit, explosions in magma, or collapse as magma is erupted or drains laterally into a rift zone, producing low period events and volcanic tremor (see Sect. 8.4.1). Eruptions can occur at any stage in the waxing or waning of seismic energy release: patterns are too varied and too poorly understood to conclude that any specific behaviour will lead to eruption. Also, the intensity of precursory seismic activity is not always proportional to the size of any associated eruption (Newhall and Dzurisin 1988).

As regards degassing, this may occur in any part of the caldera, also outside. Within this variability, degassing usually focuses in defined portions of the caldera, as vents, craters or along fault zones, largely depending upon increased local permeability conditions. Changes are commonly noted in fluid composition, flux of fumaroles or hot springs, level and composition of ground water, or water level in crater lakes. The most common changes are increases in total discharge, in the discharge and proportion of acid gases (SO_2, H_2S, HCl and HF) and in CO_2 emission. Also, several calderas have major thermal areas on their floors or along their rims, as Yellowstone and Uzon (Kamchatka, Russia). Temperatures of fumaroles range from less than 100 °C to more than 950 °C, and temperature changes of several hundred degrees are not uncommon during unrest. Temperature changes have been reported for several caldera lakes (Newhall and Dzurisin 1988).

Caldera unrest usually persists for months to years, and sometimes even decades to centuries. Unrest is usually longer at silicic calderas, possibly resulting from the higher viscosity and larger size of silicic magma systems, which slow any response to disturbance. Some unrest at large calderas, such as Campi Flegrei, Rabaul (Papua New Guinea), Aira and Iwo-Jima (Japan), has been lasting for centuries (Newhall and Dzurisin 1988; Jellinek and DePaolo 2003; Dvorak and Dzurisin 1997). Unrest is likely to be intermittent, rather than to show a systematic increase until culmination. In fact, during unrest activity may wax or wane several times before culminating in any eruption or a shallow intrusion or before returning to quiescence (Fig. 5.14). Seismicity often occurs in repeated swarms, rather than in a continuous pattern and uplift often alternates with subsidence (as at Yellowstone and Campi Flegrei) and may be episodic, as suggested by the multiple terraces at Toba and Iwo-Jima. The intermittent nature of most unrest at calderas poses a challenge for eruption forecasting, because successful forecasts must distinguish between short lulls and long periods of quiescence in order to be socially useful.

Caldera unrest reflects tectonic, magmatic and hydrologic processes. However, where established, the root cause for unrest is magmatic. While external earthquakes or hydrothermal processes may definitely encourage unrest, no unrest is of purely hydrothermal or tectonic origin: magma appears as the main ingredient, although in many cases it may also induce important changes in the hydrothermal system, supplying this with fluids and energy (Newhall and Dzurisin 1988; Acocella et al. 2015). As a result, it may become difficult to define whether unrest has a dominant magmatic or hydrothermal involvement, although the outcome of one condition or the other may be different. If the hydrothermal component dominates, a phreatic eruption may be the expected outcome from the pressurization of the hydrothermal system. Conversely, the involvement of a dominant magmatic component may lead to a stronger and more

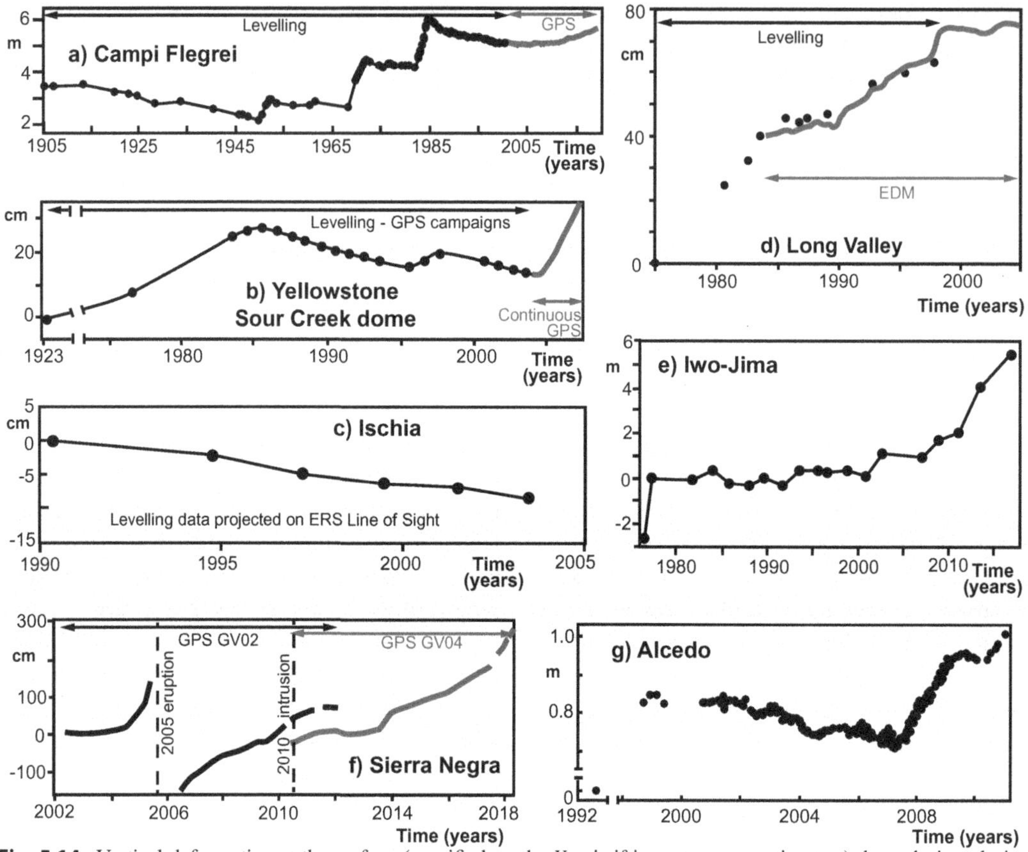

Fig. 5.14 Vertical deformation at the surface (specified on the *Y*-axis if in metres or centimetres) through time during unrest episodes at felsic (**a–e**) and mafic (**f**, **g**) resurgent calderas (modified after Acocella 2019)

impacting magmatic eruption. Therefore, while discriminating the nature of the dominant unrest component is challenging, it is also important for appropriate hazard assessment and eruption forecasting. The recommendation is to combine multiple types of observations, as no one type of data uniquely pinpoints the nature of unrest (Pritchard et al. 2019). Among the various approaches, the inversion of volcano deformation data contributes to distinguish pressurized magmatic from hydrothermal sources, constraining their location, volume change and shape (see Sect. 8.3.2). In fact, a first-order distinction between a dominant magmatic or hydrothermal involvement in unrest should consider the depth of the retrieved source: if this is larger than 3–4 km, then chances are that the hydrothermal system has minor involvement. Magmatic unrest at calderas is often consistent with pressurised sill-like intrusions or oblate (with minor vertical axis) ellipsoids (see Chap. 9). In the case of repeated and intermittent unrest, recurrent magma emplacement should be considered. This condition requires a persistently hot crust, where magma emplacement is assisted by the thermal anomaly created by the previously intruded magma (Amoruso et al. 2017). Sometimes more than one dominant source is needed to explain observations, highlighting complex distributions of melt in the shallow plumbing system, especially below silicic calderas. Caution should be also taken in interpreting the caldera deformation data for geodetic inversion. In fact, some calderas have shown patterns of surface deformation consisting of broad subsidence affecting the wider volcanic edifice and stronger localized subsidence focused inside the caldera ring faults. Physical models in elastic half-space (see Sect. 8.3.2) commonly explain these observations with the pressurization of two magma sources at different depths. However, taking into account for the deformation along the ring fault, both the broader and the more focused subsidence can be explained by the shrinking of a same deeper magma source, above which the activity of the ring-fault localizes the deformation at the surface. Therefore, omitting ring faulting and using multiple point/sill-like sources in models of subsiding (or inflating) calderas can result in erroneous estimates of magma reservoir depths and volume changes (De Natale and Pingue 1993; Liu et al. 2019).

5.8 Long- Versus Short-Term Deformation

Many calderas show the superimposition of longer-term geological uplift (hundreds to thousands of years; during resurgence) and shorter-term geodetic deformation often characterized by uplift (years to decades; during unrest), frequently insisting on the same area (Fig. 5.15). These uplifts refer to different timescales and can be detected only through different approaches: therefore, trying to understand the relationships between the longer- to the shorter-term uplift, as well as the responsible processes, is not straightforward. In fact, resurgence has been often assumed to result from a continuous, although not necessarily constant, uplift, related to a single or anyway distinct supply of magma. Conversely, unrest is accompanied with a more discontinuous behaviour, characterized by statis and deflation episodes. This apparently different behaviour has suggested that resurgence is a distinct process, independent of unrest in mechanism and scale (e.g., Marsh 1984; Kennedy et al. 2012; de Silva et al. 2015).

Comparing the shorter- and longer-term deformation histories at resurgent calderas provides an alternative understanding of these

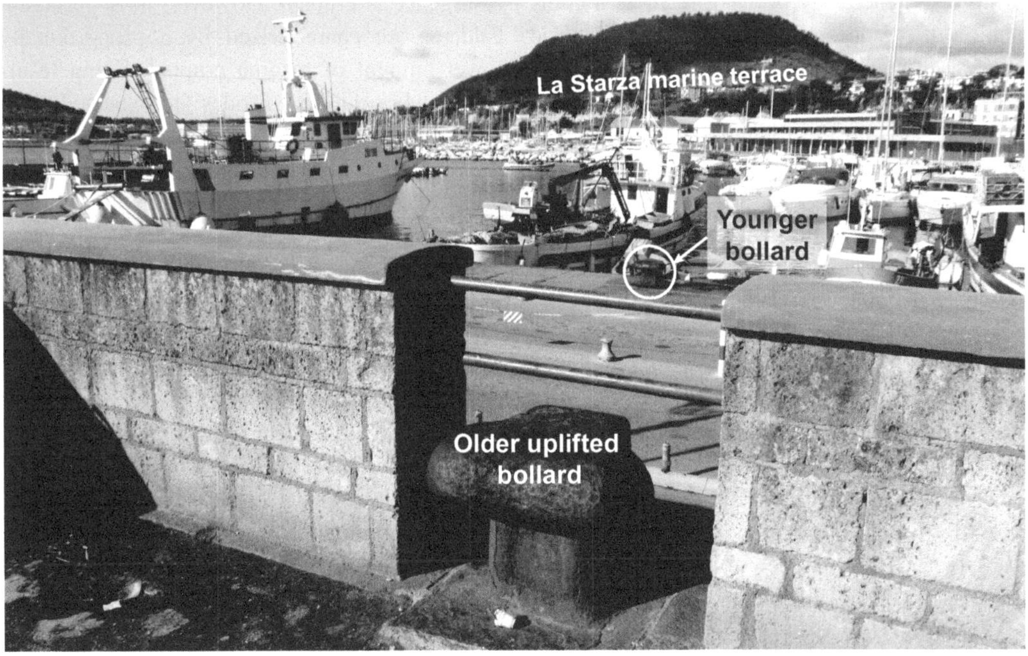

Fig. 5.15 Example of notable shorter- and longer-term deformation at Campi Flegrei caldera, Italy. Foreground: older dock level of the Pozzuoli harbour, in the caldera centre, uplifted of ~2 m during the 1982–1984 unrest, with the newer and currently active dock level, built after the unrest, lying approximately 2 m below; both docks are identified by the bollards. Background: La Starza marine terrace, uplifted of at least 60 m during resurgence of the caldera centre (photo courtesy Mauro Di Vito)

relationships. Available data show that, on the short-term, unrest at resurgent calderas (Long Valley, Yellowstone, Ischia, Campi Flegrei, Iwo-Jima, Sierra Negra, Alcedo) consists of uplift and subsidence, with the former dominating and producing net uplift (Fig. 5.14). On the long-term, the uplift history of the known resurgent cases (Campi Flegrei, Iwo-Jima, Toba and Siwi) includes, in addition to prevalent uplift, also stasis and subsidence episodes (Fig. 5.11). Therefore, both the short- and long-term deformation histories at resurgent calderas reveal distinct and repeated uplift episodes, highlighting an incremental behaviour. Also, in most resurgent calderas (Long Valley, Campi Flegrei, Iwo-Jima, Sierra Negra and Alcedo) the short- and long-term uplifts coincide in location and pattern. All these similarities suggest that unrest is an incremental episode of the much longer and cumulative resurgence (Acocella 2019). This would imply that resurgence is not the result of a constant, and not even continuous, uplift independent of unrest. Rather, the vertical deformation during resurgence may be interpreted as a combination of episodes of uplift, repose and subsidence, each with different duration, in which net uplift predominates on the longer-term. Also, consistently with the likelihood that caldera inflation results from magma emplacement, each uplift episode during resurgence may be interpreted as resulting from a pulse of magma intrusion. This view supports an incremental growth of magmatic reservoirs, where each episode of inflation may result from the emplacement of a sill-like intrusion, eventually developing a laccolith responsible for resurgence. This frame is consistent with observations from the growth of plutonic bodies, confirming that volcanic and plutonic processes are characterized by similar incremental growth modalities (see Chap. 4; Glazner et al. 2004; De Saint Blanquat et al. 2011; Coleman et al. 2016).

Therefore, resurgence seems the result of a complex longer-term evolution of a magmatic reservoir, in which repeated episodes of shallow magma emplacement responsible for net uplift predominate over possible episodes of stasis or subsidence. In this frame, during resurgence magma intrusion may be followed by quiescence (suggested by a stasis in the deformation), lateral magma propagation (suggested by short-lasting subsidence in the caldera centre), degassing and crystallization (both suggested by long-lasting subsidence), eruption (suggested by abrupt subsidence) or by a new episode of intrusion (suggested by a different uplift rate; Fig. 5.16).

This vision may also contribute in determining the capacity of volcanoes to store or erupt magma, forecasting their longer-term eruptive behaviour. In particular, the coupling or not between uplift episodes and eruptive periods during resurgence at calderas should allow understanding whether or not and, mostly, to which extent, a caldera is able to store its shallow magma. For example, Campi Flegrei highlights a coupling between uplift and eruptions during resurgence (Fig. 5.11a). This is in line with the minor resurgence with ~180 m of uplift and the relatively high amount of magma erupted during resurgence (compare the erupted and resurgent volumes in Table 5.1), suggesting that a relevant part of the magma intruded during resurgence is erupted. However, most resurgent calderas are characterized by a proportionally lower amount of magma erupted during resurgence, whose uplift episodes are expected to be less coupled, or even uncoupled, to eruptions. These diverse behaviours may be related to the different permeability of the residual magmatic reservoir to the new injected magma, as discussed in Sect. 5.6.

5.9 Magma Transfer and Eruptions

After having considered the long-term (resurgence) and short-term (unrest) accumulation of magma below calderas, this section discusses how the accumulated magma may be transferred, eventually feeding eruptions. More detailed aspects of shallow magma transfer below calderas, dealing with circumferential intrusions, are discussed in Sect. 7.5.

Calderas are associated with the largest eruptions. Indeed, they are capable of erupting the hundreds or thousands of km^3 of magma typical of a super-eruption, which is the most destructive

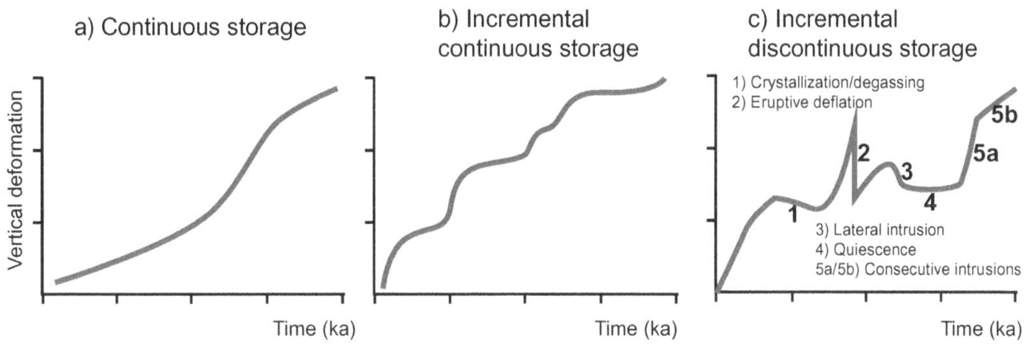

Fig. 5.16 Possible deformation models for resurgence. The incremental discontinuous storage model (case c) is more consistent with available deformation data: here net uplift is interrupted by possible stasis (quiescence) or subsidence, due to crystallization/degassing, eruptions or lateral intrusions, all included within resurgence (modified after Acocella 2019)

natural disaster associated with the activity of our planet. These cataclysmic events are relatively rare, occurring on average less than once every several tens of thousands of years. More frequent (occurring once every approximately 110 years) are the moderate-sized caldera-forming eruptions, as in 1991 at Pinatubo (Philippines; Newhall et al. 2018; Rougier et al. 2018). These moderate to large explosive eruptions are related to calderas because a significant ($\geq 10^1$ km^3 in volume) and rapid (days to weeks) removal of magma from a chamber often generates roof collapse.

The structural features of a caldera may not affect the size of an eruption, but may influence the transfer of magma feeding the eruption. These features are mainly related to the variations in the aspect ratio of the roof of the magma chamber A_r (where A_r is the ratio between the depth z and width L of the crust overlying the chamber; see for example Fig. 5.12) and the amount of subsidence s with regard to the caldera diameter d, expressed as the s/d ratio. While the roof aspect ratio A_r mainly regards larger caldera-forming eruptions, the subsidence/diameter ratio s/d is mainly applicable to minor eruptions, not necessarily related to caldera formation.

As far as the aspect ratio A_r of the roof of the magma chamber is concerned, when $A_r < 1$ the volume of the depression created at the surface is similar to that lost within the chamber during collapse. This implies a $\sim$ 1:1 ratio between the caldera depression volume and the chamber contraction. Conversely, when $A_r > 1$, the volume of the caldera at the surface becomes smaller than the volume lost in the chamber, because of the incoherent deformation developed in the thicker roof. In this case, the ratio between the caldera depression and the chamber volumes becomes less than 1, in line with the previously mentioned (Sect. 5.3) evidence that a larger volume of magma must be extruded at greater depth to form a caldera. For very high aspect ratios, a caldera may not even form at the surface (Fig. 5.17; Roche and Druitt 2001; Geyer et al. 2006). It follows that deeper and narrower magma chambers must evacuate a larger percentage of magma to reach a caldera structure similar to that of shallower and wider chambers.

Another factor related to the aspect ratio A_r regards the effective connectivity between the magma chamber and the surface during a caldera-forming eruption, potentially controlling syn-collapse eruptive activity (Acocella 2007). Lower aspect ratios ($A_r < 1.6$) form coherent collapse, ensuring connection from the chamber to the surface with newly formed gaping faults, which may be penetrated by the magma creating a continuous ring conduit and feeding vents capable of sustaining the eruption. Higher aspect ratios ($A_r > 1.6$) form incoherent collapse, which may hinder appropriate magma penetration, and thus connection from the chamber to the surface and, eventually, the continuation of the eruption, if the eruption is not of relevant size. Therefore,

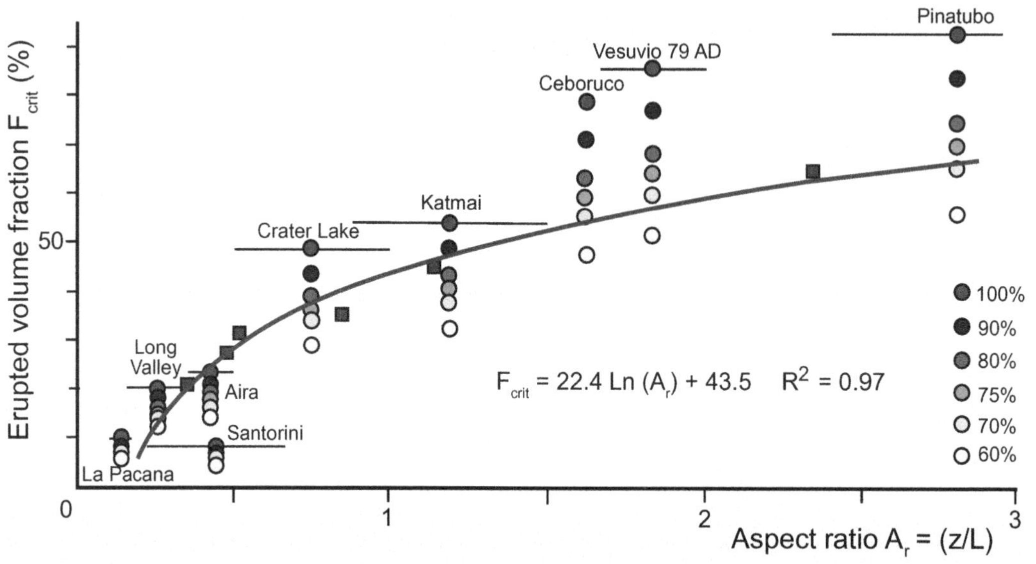

Fig. 5.17 Erupted volume fraction (F_{crit}) at the caldera onset as a function of the aspect ratio of the roof of the magma chamber A_r (where A_r = thickness z/width L of the roof). The erupted volume fraction is the portion of magma that needs to be erupted to trigger caldera collapse. Values of F_{crit} for natural examples are calculated considering different percentages (60–100%) of erupted magma; green squares indicate experimental values of F_{crit} at the collapse onset and green line indicates the related log-fit (modified after Geyer et al. 2006)

in addition to the availability of eruptible magma, the aspect ratio of the chamber roof should be considered as a factor capable of sustaining smaller eruptions.

The occurrence and location of syn- to post-collapse volcanism may also depend on the aspect ratio of the roof of the magma chamber coupled with the s/d ratio. A small s/d ratio ($s/d < 0.05$) generates downsag without ring fault. In this case, magma withdrawal is expected along the vent responsible for the eruption and/or pre-existing structures. With larger s/d ratio ($s/d > 0.05$), the roof aspect ratio A_r becomes important to control the eruption. In the case of shallow and wide magma chambers ($A_r < 1$), the ring faults easily reach the surface and the caldera structure may provide an effective path for the withdrawal for the magma. In this case, if the chamber is experiencing underpressure, the outward dipping ring faults may be preferably intruded during immediate post-collapse activity. This process may have occurred during the 700 km^3 Bishop Tuff super-eruption, at Long Valley. Here the strong ellipticity of the caldera may have favoured the reconstructed pattern of vent migration and ring fracture unzipping, with the maximum failure shear strain along the shorter axis of the elliptical caldera (Holohan et al. 2008). The intrusion of magma along the deeper part of the ring faults may depressurise the nearby portion of the chamber, enhancing the sinking of its roof and triggering a feedback between intrusion and collapse. Any intrusion along inward dipping ring faults requires a higher pressure, as a larger part of the caldera roof has to be uplifted against gravity. This may explain the common occurrence of post-caldera vents along the outward dipping ring fault of mature calderas. Conversely, in the case of larger roof aspect ratio A_r ($A_r > 1$), and still with large s/d ratio ($s/d > 0.05$), the condition for the establishment of an effective path for magma withdrawal may not be met. This occurrence depends, for a magma chamber with width L, on the amount of subsidence s with regard to the depth of the magma chamber z, so that a deeper chamber requires a larger subsidence to ensure connection with the surface, as previously explained.

As far as the *s/d* ratio is concerned, this also allows defining the distribution and intensity of the gravitational stresses beneath the caldera. These result from the mass excavation due to the caldera depression, which effectively unloads the crust and reorients the principal stresses, with a subvertical minimum principal stress σ_3 below the caldera centre gradually rotating and becoming subhorizontal below the caldera rim (Corbi et al. 2015; Gaete et al. 2019; Rivalta et al. 2019). The larger the subsidence *s*, the stronger the **unloading** (removal of mass due to the caldera depression), and the larger the caldera width *d*, the deeper the effect of such stress reorientation. However, subsidence depth should be corrected for the presence of any caldera infill (including any water from lakes, sea or ice), accounting for its density difference with the surrounding basement. Therefore, while in general a large subsidence *s* promotes a stronger stress reorientation, this behaviour may be less pronounced in filled calderas, depending upon the thickness and density of the infill. As magma propagates through magma-filled fractures oriented perpendicular to the least principal stress σ_3, this unloading affects the direction of magma propagation, so that any magma-filled fracture ascending below the caldera centre, as a dike, may be deflected towards the caldera rim: the higher the unloading, the stronger the deflection (Fig. 5.18; Gaete et al. 2019). This process is further described below considering two-end members, as Fernandina, a mafic caldera with evident topographic expression at the top of a volcanic edifice, and Campi Flegrei, a felsic caldera largely filled by post-collapse eruptive deposits, without evident edifice.

Fernandina hosts a $\sim$1 km deep and 6.5 $\times$ 4 km wide caldera, as well as circumferential eruptive fissures just outside the caldera rim and radial fissures on the outer slopes of the edifice (Simkin and Howard 1970; Chadwick and Howard 1991). The possible processes controlling this distinctive fissure pattern have been explored considering principal stress orientations. Studies have addressed the effect of magma chamber pressurization, the superposition of regional and magma chamber stresses, or the stress reorientation caused by gravitational loads of the volcanic edifice. An important contribution derives from the evidence that stresses due to the gravitational loading of a volcanic edifice are generally larger than magma chamber pressurisation stresses. For realistic edifice and magma chamber pressurisation levels, edifice stresses dominate everywhere except very close to the magma chamber walls (Roman and Jaupart 2014). In addition, recent geodetic data at Fernandina suggest the circumferential fissures have been fed by laterally propagating sills beneath the caldera, which progressively increased their dip becoming sheets and, subsequently, dikes (see Sect. 7.5; Bagnardi et al. 2013). These features underline the importance of the caldera depression of Fernandina in generating unloading with subvertical σ_3 at depth, promoting sill emplacement and lateral magma propagation. In particular, the rotation towards a subhorizontal σ_3 approaching the caldera rim (where the unloading effect vanishes) affects the lateral propagation of magma, passing from sills to inclined sheets and finally to subvertical circumferential dikes (Fig. 5.19; Corbi et al. 2015).

If the nucleating sill is slightly deeper, it rotates about its propagation direction to adjust to the circumferential σ_3 along the outer slope of the volcanic edifice, so that the resulting feeder dike becomes radial and vertical. Therefore, at Fernandina the superposition of an unloading stress within the caldera (responsible for a vertical σ_3 in the centre and a radial σ_3 at the rim) with a gravitational stress along the outer slope of the edifice (responsible for a circumferential σ_3; see Sect. 7.6.1) controls the distinctive pattern of circumferential fissures outside the caldera rim and radial fissures on the volcano flanks, respectively. A similar behaviour is observed at the other active Galapagos calderas, which share similar topographic features to Fernandina.

Campi Flegrei caldera, on the other hand, has a very different topographic signature. It is a $\sim$12 km wide depression hosting two nested calderas formed during eruptions at $\sim$39 and $\sim$15 ka. The subsidence associated with both depressions is on the order of $\sim$2 km. However, the sea, marine deposits and post-

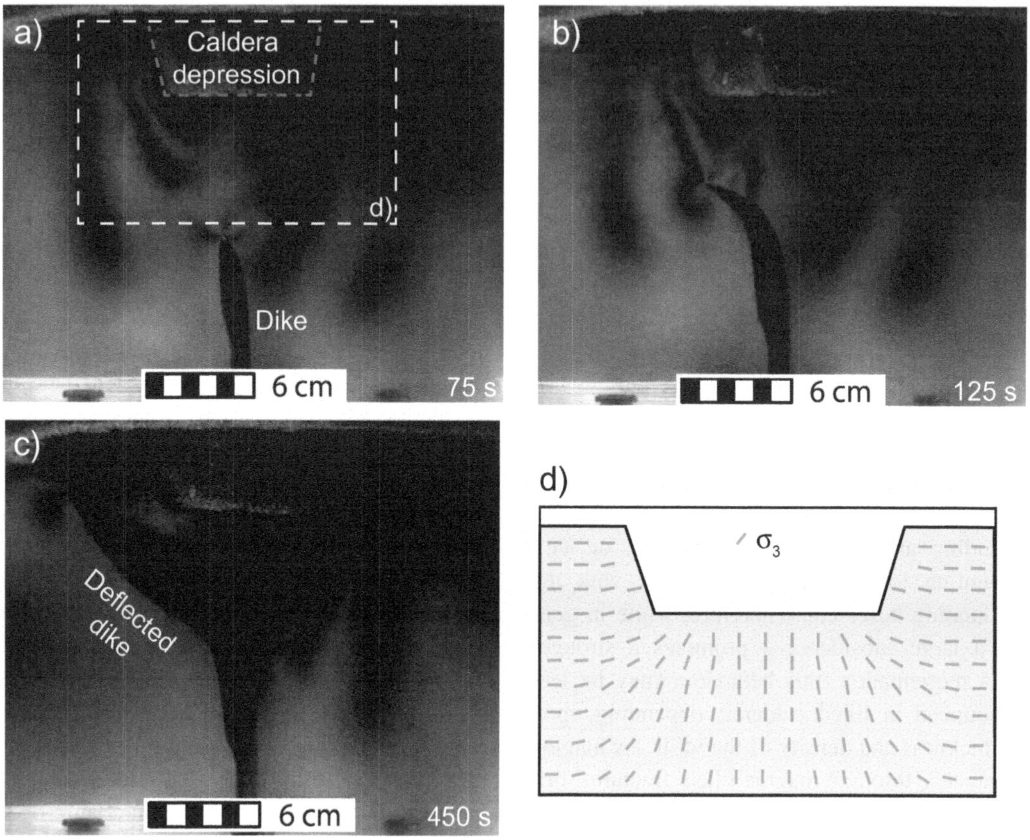

Fig. 5.18 **a–c** Section views in polarized light of three stages (running time indicated in seconds at bottom right) of an experiment where red water (simulating magma) is injected in gelatine (simulating the upper crust), highlighting the progressive deflection of the resulting dike below the depression imposed at the surface (red dashed line; simulating a caldera). **d** Simplified sketch of the trajectories of the minimum principal stress σ_3 acting below the depression, illustrating the induced unloading responsible for the deflection of the dike, which orients perpendicular to σ_3 (modified after Gaete et al. 2019)

collapse volcanism filled the caldera depression. In particular, post-caldera volcanism developed more than 70 monogenic vents focused predominantly in the northeast, presently onshore, sector of the caldera. Eruptive activity migrated progressively inward over Epochs 1 (15–9.5 ka), 2 (8.6–8.2 ka), and 3 (4.8–3.7 ka), filling most of the caldera (Orsi et al. 1996; Di Vito et al. 1999; Smith et al. 2011). The vent of the last Monte Nuovo eruption (occurred in 1538) lies just outside the central part of the caldera, but within its rim. Immediately before the last eruption, the intruded magma migrated laterally from an oblate chamber below the caldera centre to beneath Monte Nuovo and then propagated upward feeding a circumferential dike (Fig. 5.20; Di Vito et al. 2016). This limited lateral magma transfer, feeding an intra-caldera eruption, results from the minor difference in topography of the filled caldera, of a very few hundred of meters, and the slightly lower density of the caldera infill, which impose a relatively weak unloading. A limited lateral propagation is observed also for the vents of the last eruptive epoch, which cluster at a similar distance of ~3 km from the caldera centre (Di Vito et al. 2016).

Therefore, Fernandina and Campi Flegrei show how calderas with different topographic expression (deriving from different amounts of subsidence and caldera infill) determine different

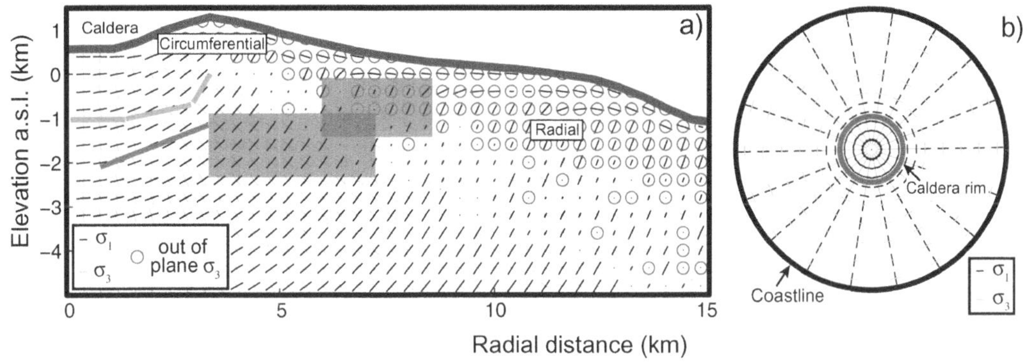

Fig. 5.19 Results of finite element models for an isotropic edifice subject to unloading applied to Fernandina caldera and edifice (Galapagos). **a** Axisymmetric plane projection; the orange lines and the purple line and rectangles represent the projections on the axisymmetric plane of the magmatic sheets feeding the 2005 and 2009 eruptions; note that the purple rectangles represent projections of shallow dipping sheets. The stress components controlling the propagation of the intrusions and resulting from caldera unloading and the load of the volcanic edifice are specified in the inset. **b** Map view distribution of the stress trajectories within the volcanic edifice calculated at sea level (modified after Corbi et al. 2015)

unloading, stronger at Fernandina and weaker at Campi Flegrei. A stronger unloading may drive the magma farther from the caldera centre (outside the caldera rim, Fernandina), whereas a weaker unloading drives magma nearer (intra-caldera, Campi Flegrei). The distal or proximal transfer of magma and, ultimately, the location of eruptive vents at calderas depends, in addition to the unloading stresses, also on the regional stresses, the depth to the magma chamber and magma buoyancy. While regional compression (horizontal σ_1) is expected to promote gentler dike trajectories leading to eruptions outside the caldera rim, regional extension (horizontal σ_3)

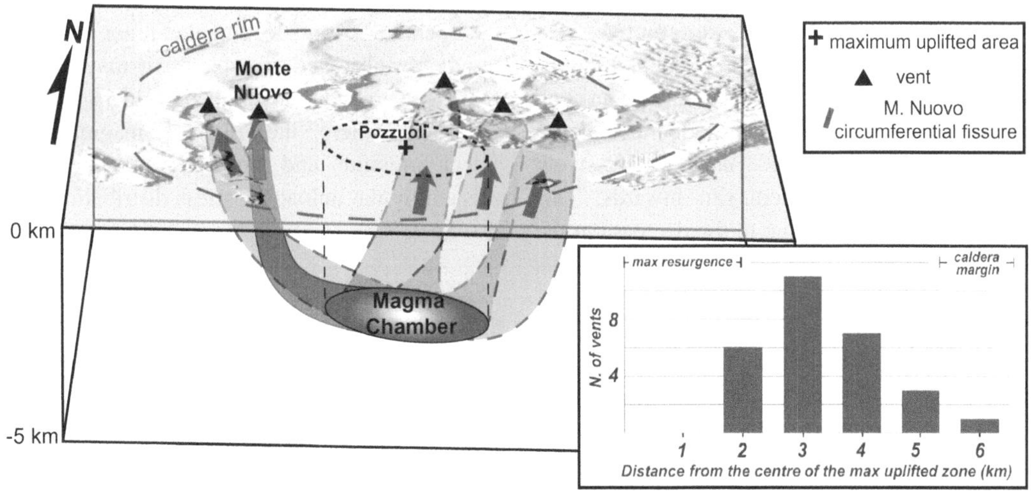

Fig. 5.20 Magma transfer below Campi Flegrei caldera (Italy) before the 1538 AD Monte Nuovo eruption: magma first propagated laterally from the oblate magma chamber at ~4.7 km depth below the caldera centre and then propagated vertically (red path) to feed the Monte Nuovo eruption. The inferred paths (in grey) for other representative eruptions of the last 5 ka are also reported. Inset at bottom right shows the distance from the caldera centre of the eruptive vents active in the last 5 ka: all vents cluster slightly outside the caldera centre, due to the moderate caldera unloading (modified after Di Vito et al. 2016)

has the opposite effect, steepening dike trajectories and promoting proximal eruptions. Also, a deeper magma chamber implies a longer feeder dike and, in turn, a higher magmatic pressure, which inhibits the lateral propagation of sills and promotes dike propagation. A similar effect is played by the buoyancy of the intruded magma, with more buoyant magma inhibiting the lateral propagation of sills and weakly buoyant magma feeding inclined sheets (see Sect. 7.5; Corbi et al. 2015; Gaete et al. 2019).

This rationale may be used to forecast the location of future eruptive vent at calderas. The trajectories of magma propagation are in fact controlled by the stress field within the caldera, which results from the following components: stresses due to the evolution of the volcano (including unloading), tectonic or regional stress, flank movements, magma chamber pressurization, previous intrusions and previous large earthquakes (see Sect. 7.2). The relative importance of these contributions may vary significantly, and determining their weight may be challenging. However, these contributions may be constrained comparing modelled magma trajectories with the location of the eruptive vents. The outcome is a probabilistic stress field which can be used to define magma propagation trajectories and thus probabilistic eruptive vent maps. This procedure has been applied to the post-caldera vents at Campi Flegrei (Rivalta et al. 2019). Here, after the last caldera collapse at $\sim$15 ka, the eruptive vents progressively migrated from the caldera rim inwards. This feature is explained as due to the decrease in the unloading stresses within the caldera, due to its progressive filling in the last 15 ka (Fig. 5.21). Progressive caldera filling has thus weakened unloading stresses and made them less dominant, steepening the dike trajectories and determining the inward migration of the vents in the three eruptive epochs. Moreover, the overall asymmetric topography of the caldera, with the highest relief along its northeast rim (200–300 m above sea level) and the lowest bathymetry in the southern part (50–100 m below sea level), explains also the asymmetric distribution of vents, which cluster in the currently onshore northeast part. In fact, any magma lens below the central portion of the caldera tends to be attracted by the zone of highest relief, where the deviatoric stress is highest, thus propagating towards that direction. The reconstructed evolution of the stress distribution at Campi Flegrei has allowed not only explaining the vent distribution in the last 15 ka, but also hindsight forecasting of the location of the last 1538 Monte Nuovo vent. This in fact lies in the onland annular portion, at $\sim$3 km from the caldera centre, which is also the most likely location for the opening of future vents (Rivalta et al. 2019). A consistent relationship between the location of vents and the stress distribution is observed also at other calderas with known eruptive history and overall stress state, suggesting a wider applicability of the approach, including the possibility to forecast the location of vents (Fig. 5.22). This method may provide an important advancement in building physics-based hazards maps of the probability of opening of a vent at a volcano. In fact, current hazards maps are typically description-based (considering only the location of the past vents, without explaining their distribution with physical models), resulting in limited forecast.

While the concept of stress unloading may generally explain the shallow transfer of magma and opening of vents feeding the minor eruptions commonly observed at calderas, it may be less applicable to the less frequent caldera-forming eruptions, where the rise of magma with higher pressure and buoyancy may be less affected by the unloading stress distribution (see also Sect. 7.5).

5.10 Lessons from Active Calderas

As introduced, calderas are the most active type of volcano, characterized by long-term resurgence, short-term unrest and eruptions of any size, with exclusive pertinence of the largest sizes. Active calderas have shown remarkable examples of these types of activity in the last decades, permitting to better understand their behaviour and processes. This section summarizes some relevant events, considering unrest

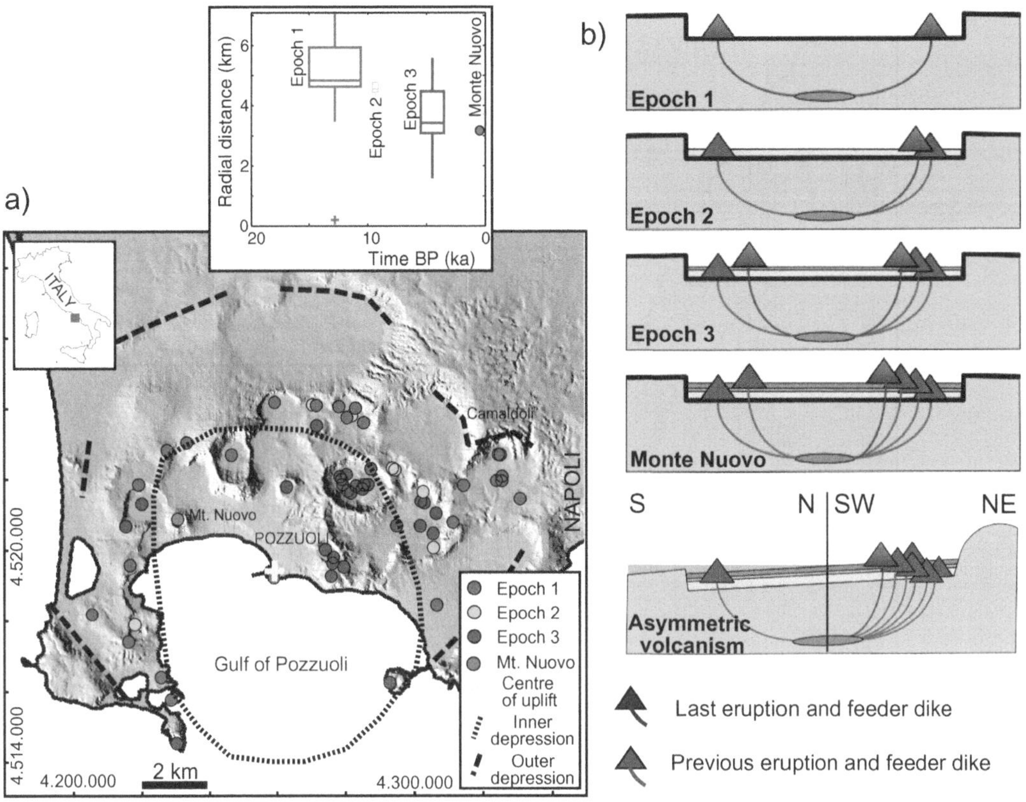

Fig. 5.21 **a** Distribution of the post-caldera vents (coloured circles) at Campi Flegrei as a function of its three epochs of eruptive activity; volcanism migrates inward with time (upper inset), clustering in the northeast part of the caldera. **b** This eruptive pattern results from the progressive thickening of the marine and volcanic infill within the caldera (yellow), which gradually shifts the location of the eruptive vents (triangles) inward. Also, the asymmetric distribution of the vents is explained by the highest relief to the northeast, which develops stronger gravitational loads and thus attracts magma (bottom diagram; modified after Acocella and Rivalta 2019)

culminating in eruption (eruptive unrest) from mafic and felsic calderas, as well as episodes of caldera collapse.

5.10.1 Sierra Negra

Eruptive unrest recently occurred at both mafic and felsic calderas. Among mafic calderas, the unrest predating the 2005 eruption at Sierra Negra shows several interesting features. Sierra Negra is the most voluminous volcano in the western Galapagos Archipelago. Above sea level, it is 60 × 40 km wide and 1140 m high, with a summit caldera larger (7 × 10 km) but shallower (110 m deep) than the other calderas in the western Galapagos. More than ten historical eruptions have occurred since 1813, producing lava flows on the north flank. InSAR results from three different intervals between 1992 and 1999 showed that the caldera floor inflated by 2.7 m. From 1992 to 1997, the inflation pattern was nearly axisymmetric to the caldera centre: this was modelled as due to the intrusion of magma into a sill beneath the caldera, at 2 km of depth. Between 1997 and 1998 the maximum uplift, consisting of ~1.2 m of slip along a steeply south-dipping normal fault, was centred on the southern limb of a pre-existing intracaldera fault system. The focus of inflation shifted back to the caldera centre between 1998 and 1999, again interpreted as magma filling a sill (Fig. 5.23;

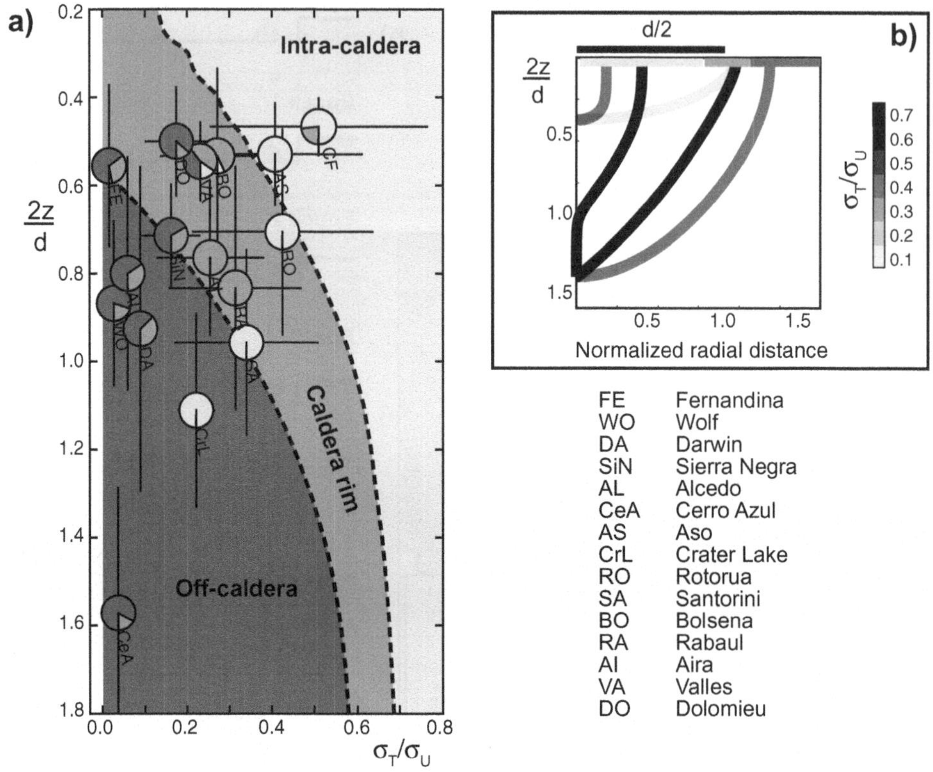

Fig. 5.22 **a** Application of the unloading model to several calderas worldwide (listed to the bottom right). The background colours represent the domain for the predicted location of the intra-caldera, caldera rim, and off-caldera vent locations as a function of the ratio between the depth to the magma chamber z and the caldera radius $d/2$, that is $2z/d$, and the ratio between the tectonic σ_T and unloading stress σ_U, that is σ_T/σ_U. Estimates for calderas in nature are represented by pie plots according to the proportion of observed intra-caldera, rim, or off-caldera vents, with the same colour coding as the background. Uncertainties are shown as black lines. The overall agreement (except for Crater Lake and Santorini) between the predicted and observed vent patterns suggests a broad control of unloading on shallow magma transfer at calderas. **b** Examples of streamlines of selected models with $2z/d = 0.4$ and $2z/d = 1.4$ and variable σ_T/σ_U, which controls trajectory concavity (modified after Rivalta et al. 2019)

Amelung et al. 2000; Jonsson et al. 2005). GPS data during 2000–2003 showed a deceleration in the uplift rate, followed by a change to subsidence of 9 cm/year. The deflationary source was modelled as a contracting sill at a depth of 2.1 km, similar to the inflationary source of the 1990s. In April 2003, deformation of the caldera floor reversed to inflation. The inflation rate gradually increased throughout 2004 and into 2005, for a total uplift of 89 cm, accompanied by 69 cm of horizontal extension across the caldera. Inflation was interrupted by an episode of inelastic trapdoor faulting, marked by a magnitude *M4.6* earthquake on 16 April 2005. This produced 84 cm of uplift on the southern part of the caldera and contraction of the cross-caldera line by 26 cm. Modelling results for this trapdoor faulting suggest slip on a high angle inward dipping reverse fault, extending from the surface down to the sill at 2.2 km depth (Fig. 5.23).

The inflation rate was not affected by this faulting event, approaching 1 cm/day. Parts of the caldera uplifted of 1.22 m between the faulting event and a successive *M5.4* earthquake on 22 October 2005, probably occurred again along the faults in the southern part of the

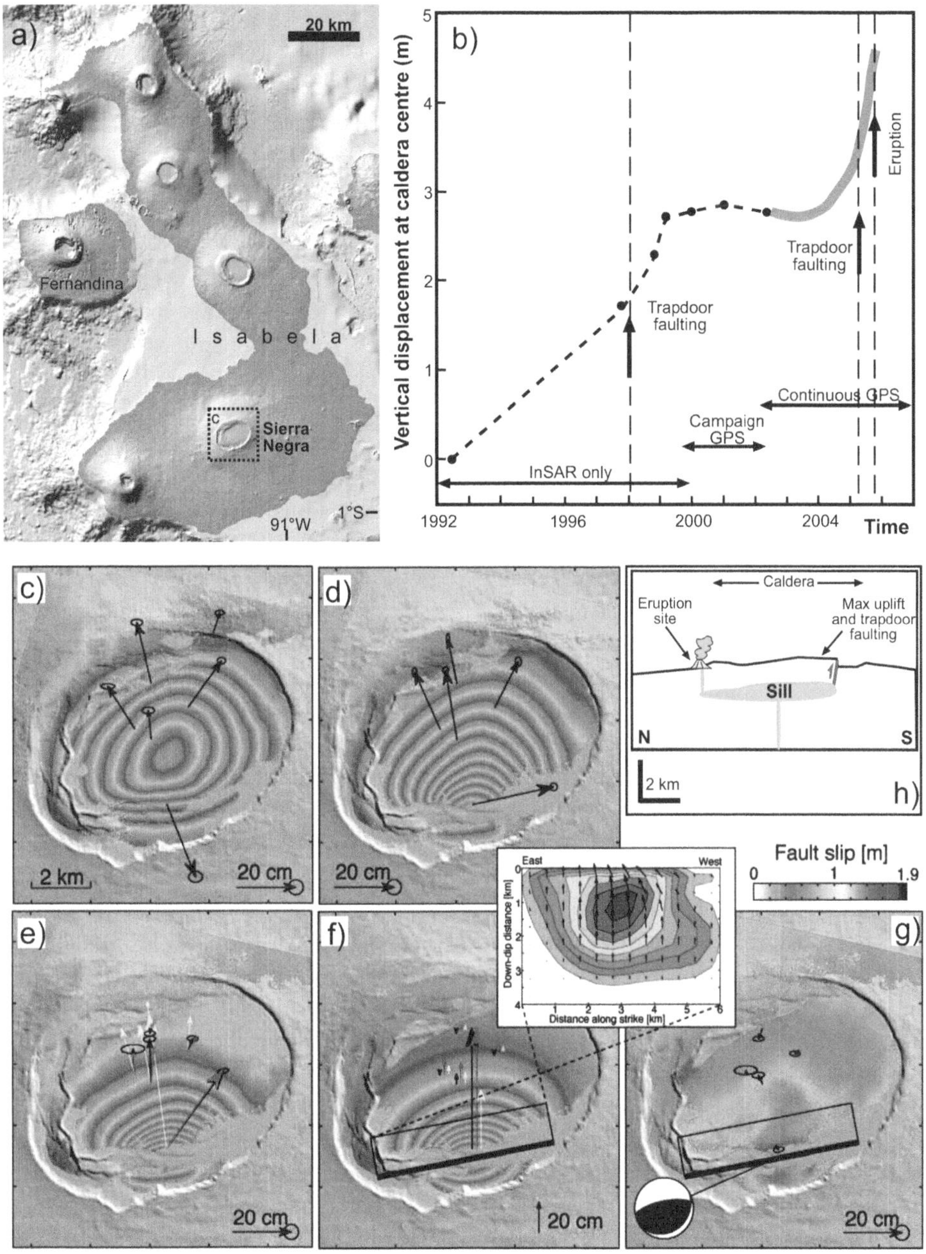

Fig. 5.23 Unrest and eruption at Sierra Negra caldera, Galapagos. **a** Location of Sierra Negra caldera. **b** Vertical displacement of caldera floor from 1992 to 2005, culminating in the 2005 eruption, detected through geodetic data. **c–g** InSAR and GPS data and comparison with trapdoor fault models: **c** uplift from February 12, 2004 to January 27, 2005; InSAR displacement is 10 cm per fringe; **d** uplift and trapdoor faulting from January 27, 2005 to May 12, 2005; **e** same as **d**, but corrected for deformation due to uplift and effectively only showing deformation due to trapdoor faulting on April 16, 2005; black and red vectors show observed and predicted horizontal displacements, respectively; yellow vectors in (**e**) and (**f**) show predicted displacements from previous models; **f** simulated interferogram from trapdoor fault model (inset); **g** residual between **e** and **f**; focal mechanism shows modelled sense of slip (after Chadwick et al. 2006). **h** Section view diagram summarizes the relationships between sill emplacement beneath the caldera, trapdoor faulting on the south side and eruptions on the north flank (after Jonsson et al. 2005)

caldera. The static stress transfer induced by the latter earthquake may have triggered tensile failure and catalysed the onset of the eruption three hours later (Gregg et al. 2018). The eruption, responsible for a 13 km high steam and ash plume, opened a fissure inside the north rim, on the opposite side with regard to the location of the *M4.6* and *M5.4* events, feeding a total eruptive volume of $\sim 150 \times 10^6$ m^3. From 1 April 2003 to 22 October 2005, at the eruption onset, the precursory maximum uplift reached $\sim$2.2 m, bringing the total amount of uplift since 1992 to nearly 5 m (Chadwick et al. 2006). Conversely, during the 8 days of the eruption, the caldera floor deflated $\sim$5 m and the volcano contracted horizontally $\sim$ 6 m (Geist et al. 2008).

This eruptive unrest has been characterized by the largest, although discontinuous, precursory inflation ($\sim$5 m) ever recorded at a mafic caldera. In addition, it shows two interesting features. First, the eruptive vent was located on the opposite side with regard to the most uplifted area, where the earthquakes were also detected. This is explained by the fact that the trapdoor faulting due to the *M4.6* and *M5.4* earthquakes relieved the pressure within the sill to the south, at the same time inducing compression south of the fault and inhibiting the lateral propagation of the sill. As a result, the sill thickened on this side, postponing any eruption. However, the faulting also caused a radially directed tension on the opposite side of the caldera, encouraging circumferential dike propagation at the northern periphery of the sill, finally feeding the eruption (Jonsson 2009). A second feature is the seismogenic reactivation of the reverse fault characterizing the trapdoor uplift, during the *M4.6* event. This fault borders the southern and most uplifted portion of the $\sim$100 m high tilted resurgent block of Sierra Negra. This pre-eruptive trapdoor uplift may thus constitute an episode of growth of the longer-term resurgence, similarly to what observed at the nearby Alcedo caldera during the 2007–2011 non-eruptive inflation, where a fault bordering the minor resurgent block was reactivated (Galetto et al. 2019). The consistency in the shorter- and longer-term deformation pattern at these mafic calderas supports the possibility that resurgence results from cumulated and distinct unrest episodes.

5.10.2 Rabaul

Rabaul caldera (Papua New Guinea) is one of the most active felsic calderas. It consists of a pair of elongated nested collapse structures, the inner being identified by focused seismicity along high-angle outward dipping faults (Fig. 5.24). Rabaul produced several caldera-forming eruptions: after the last, in AD 640, more than 8 intra-caldera eruptions have been occurring. The last three eruptions (1878, 1937–1941 and between 1994-Present) showed the simultaneous activity of the Tavurvur (andesitic-dacitic) and Vulcan (dacitic-andesitic) eruptive centres, lying on opposite sides along the inner, active caldera (Mori and McKee 1987; Nairn et al. 1995). Rabaul experienced several unrest episodes, culminating in an eruption sequence starting in 1994 (Fig. 5.24). Following two nearby *M8* tectonic earthquakes in 1971, Rabaul experienced progressive uplift and shallow seismicity (less than 3 km of depth) along the inner ring-fault. Between 1971 and 1983, thousands of earthquakes were accompanied by a cumulative uplift of 1 m. In 1983, both the seismicity and uplift rate dramatically increased, possibly following a *M7.6* regional earthquake 200 km to the east. This phase lasted until July 1985 and was characterized by several tens of thousands of earthquakes and uplift of $\sim$1 m. The 1983–1985 unrest phase has been interpreted as due to the injection of mafic magma. From August 1985 to April 1992 the seismicity and uplift rate significantly decreased. From May 1992 to September 1994, the uplift and seismicity rate increased again, although with a much lower rate than that observed between 1983 and 1985. A burst of seismicity occurred on Late August 1994, followed by quiescence until September 18, when a *M5.1* earthquake occurred at a depth of 1.2 km in

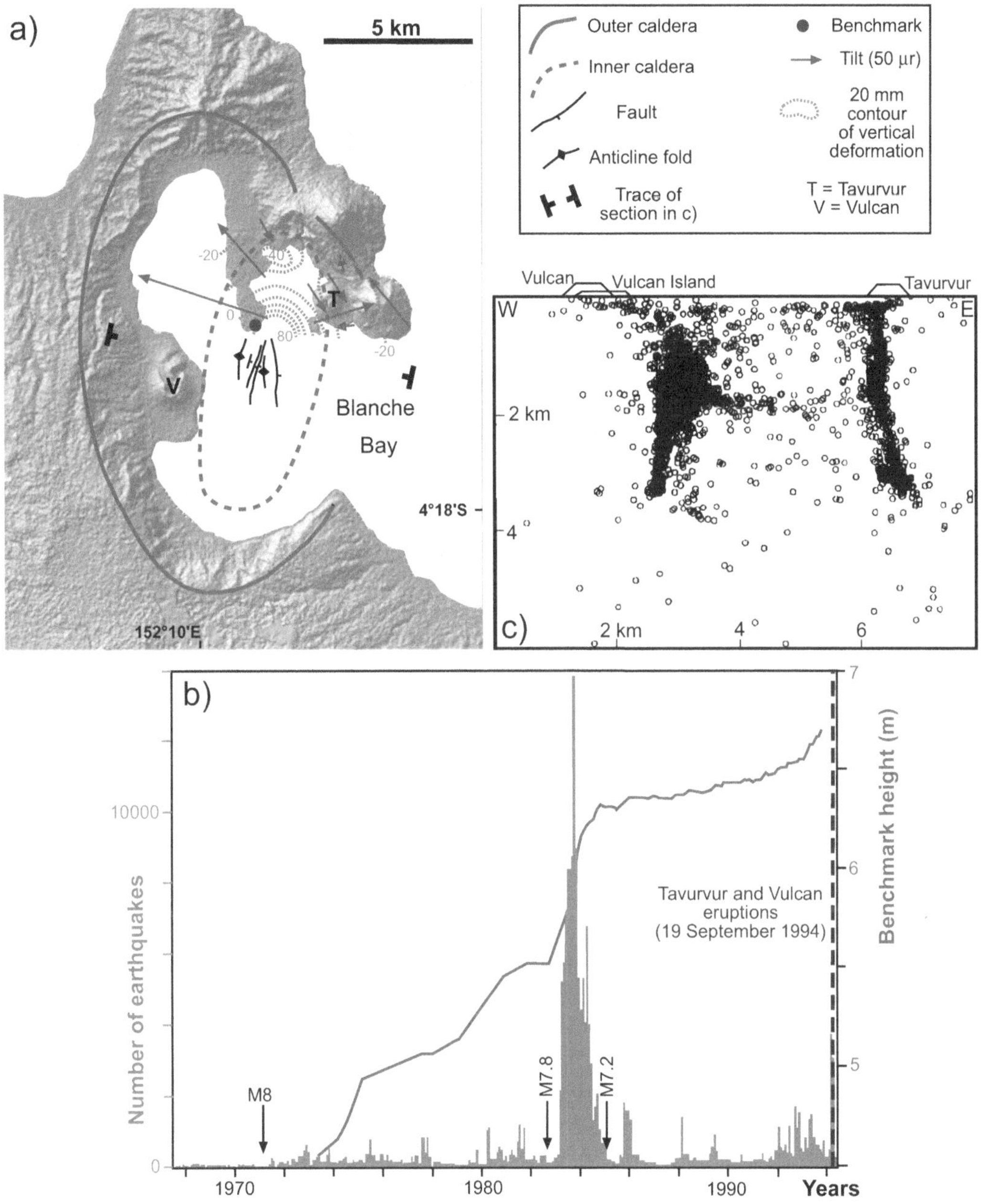

Fig. 5.24 Unrest and eruption at Rabaul caldera (Papua New Guinea). **a** Simplified structure, showing the two nested collapses and the deformation between 7/85 and 9/89, highlighted by the uplift contours, the tilt data and faults and folds (modified after Saunders 2001). **b** Monthly total earthquakes (blue histogram) and compounded uplift (red line) between 1968 and 1994; nearby major regional earthquakes are also indicated (modified after Johnson et al. 2010). **c** Approximately E-W-trending section (location in **a**) showing the projection of the seismicity in the central portion of the caldera between 1971 and 1992, with the seismicity clustering along outward dipping faults; the projection of Vulcan and Tavurvur vents is also included (modified after Jones and Stewart 1997)

the eastern part of the caldera. After only 27 h of sustained seismicity, 12 of which included low-frequency events, the eruption begun on September 19. Tavurvur, on the east rim, erupted first, immediately preceded by an uplift of 1–2 m. Approximately 1–1.5 h later Vulcan, following a localized uplift of 6 m also starting a few hours before the eruption, erupted from at least 4 vents. Several tsunamis were generated during the onset of the eruption. Soon after the eruption onset, the caldera deflated. The initial part of the VEI 4 eruption was the most violent, generating an ash cloud ~20 km high. Vulcan's eruption ended on 2 October, while the eruption at Tavurvur, after peaking during the first five days of activity, lasted until April. At the end of October, the subsidence of the caldera reached 1 m in the central part and 20–30 cm along the rim. Seismic activity along the caldera rim progressively decreased from the beginning of October to the end of November. At least 5 other explosive eruptive events, with VEI between 1 and 4, occurred from November 1995 to July 2010 (Johnson et al. 2010).

The 1971–1994 eruptive unrest at Rabaul shows several interesting features. First, higher seismicity and uplift between 1983 and 1985 were not immediately followed by any eruption, while moderate seismicity and uplift occurring between 1992 and 1994 culminated in a VEI 4 eruption. This behaviour may be understood only by considering the second phase as a continuation and dependent on the first: the first primed the system and the second set off the eruption (Acocella et al. 2015; Robertson and Kilburn 2016). Second, in at least two cases, in 1971 and 1983, the unrest of the caldera, or its intensification, occurred immediately after major regional earthquakes, suggesting interactions between tectonic events and a prepared magmatic reservoir. However, while Rabaul responded to a *M7* earthquake at a distance of 180 km with a pronounced earthquake swarm, it did not produce any detectable activity in response to a second *M7* earthquake two months later, at a distance of only 60 km (Johnson et al. 2010). Another remarkable feature is the simultaneous activation of two vents along opposite portions of the caldera ring fault. This feature has been also demonstrated for the 4.3 ka Solfatara and Averno eruptions at Campi Flegrei, fed by two vents within the caldera erupting products with different compositions (Pistolesi et al. 2016).

5.10.3 Recent Caldera Collapses

In addition to eruptions, calderas experience collapse episodes. Seven episodes of caldera collapse have been observed and/or monitored in the last decades: Fernandina (1968, Galapagos), Pinatubo (1991, Philippines), Miyakejima (2000, Japan), Dolomieu (2007, La Réunion), Bardarbunga (2014, Iceland), Axial Seamount (2015, Juan de Fuca Ridge) and Kilauea (2018, Hawaii). In addition, a more limited subsidence (2.5 m) has been detected at Ambrym caldera (2018; Vanuatu; Stix and Kobayashi 2008; Michon et al. 2011; Gudmundsson et al. 2016; Wilcox et al. 2016; Neal et al. 2019; Shreve et al. 2019). All these episodes were related to eruptions, although their size differed significantly, from minor eruptions, as the VEI 1 eruption at Dolomieu in 2007, to major eruptions, as the VEI 6 Pinatubo eruption in 1991. Also, with the exception of Pinatubo, all these collapses occurred with mafic magma. These episodes, especially the best-monitored later ones, have provided a wealth of valuable information to better understand the mechanism of collapse and its triggering factors.

As far as the mechanism of collapse is concerned, geodetic and seismicity observations at Miyakejima, Dolomieu, Bardarbunga and Kilauea have revealed a common step-by-step, incremental mode of collapse unfolding over days, weeks or months (e.g. Michon et al. 2011; Gudmundsson et al. 2016; Munekane et al. 2016; Neal et al. 2019). This can be mechanically described in simple terms by the vertical slip of a cylindrical piston driven by gravity and resisted by friction on the ring slip surface and by the rebound force exerted by the compressible magma onto the sliding piston (Kumagai et al. 2001). Both the 2007 Dolomieu and the 2018

Kilauea collapse showed a reduction of the time-interval between successive subsidence increments, which resulted from the acceleration of magma withdrawal and a progressive weakening of the edifice at the beginning of the sequence (Duputel and Rivera 2019). Where the caldera faults were not completely reactivated, because of the lower amount of collapse, the seismicity along the caldera ring faults was limited. This was the case of Bardarbunga in 2014, where caldera subsidence was largely aseismic, with seismicity accounting for 10% or less of the geodetic moment. Approximately 90% of the seismic moment release occurred on the most subsided northern rim, highlighting asymmetric collapse (Agustdottir et al. 2019).

As for the triggering factors, with the exception of the largest felsic eruption occurring at the summit of Pinatubo, all the remaining caldera collapse episodes, including the subsidence at Ambrym, were related to the withdrawal of the magma chamber induced by the lateral migration of dikes, eventually erupting outside the caldera. In some cases, as at Bardarbunga in 2014 and Kilauea in 2018, the eruption occurred downslope along the rift zone, 40–50 km away. This similarity among most collapse episodes highlights the importance of the lateral intrusion of magma in generating collapse at mafic calderas, even though the same process may show distinct features, in terms of amount and mode of collapse, distance of vents and eruptive activity (Sigmundsson 2019). At Bardarbunga, observations highlight an exponential decline of the magma flow rate at the eruption site at the tip of the laterally propagating dike, coupled with the exponential rate of volume change in the caldera, consistently with theoretical predictions (see also Sect. 4.6.3; Gudmundsson et al. 2016). The 2018 Kilauea collapse allowed clarifying in more detail the possible conditions that trigger the onset of collapse, as well as those controlling the eruptive flow (Fig. 5.25). At Kilauea failure began after less than 4% of magma was withdrawn from a shallow reservoir beneath the volcano's summit. Also, several cubic kilometres of magma were stored in the reservoir, and only a fraction was withdrawn before the end of the

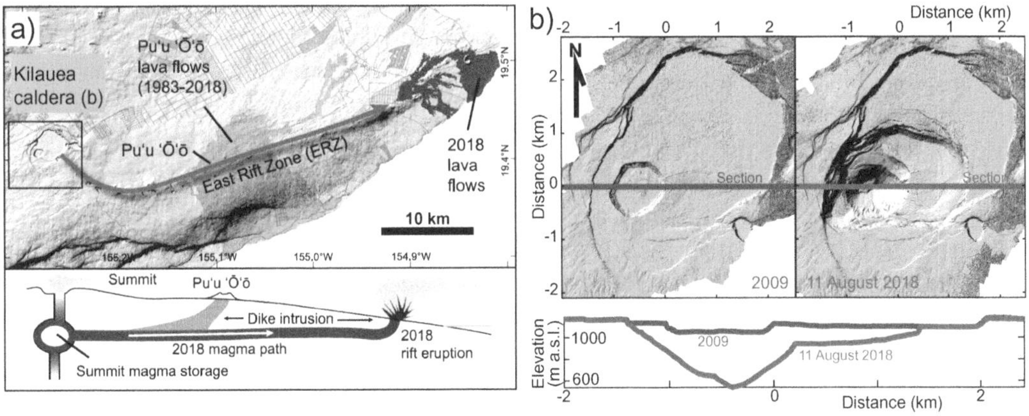

Fig. 5.25 Kilauea Volcano (Hawaii, USA) and the 2018 eruption. **a** During the eruption, magma flowed for more than 40 km underground subhorizontally from the summit (left) to the lower Eastern Rift Zone (ERZ) vents (above). Schematic cross section (not to scale) showing flow of magma from the summit to the lower ERZ (below; modified after Anderson et al. 2019). **b** Digital elevation models of Kilauea's summit from 2009 (left) and 11 August 2018 (right), highlighting the collapse of the caldera. The red and blue lines correspond to the locations of the cross sections shown at the bottom. The difference between the 2009 and 2018 profiles gives the amount of subsidence that occurred almost entirely since 1 May 2018 (images courtesy of Kyle Anderson and Christina Neal, USGS; modified after Neal et al. 2019)

eruption. This indicates that, in presence of a shallow reservoir, caldera formation may begin after withdrawal of only small amounts of magma, and may end well before source reservoirs are completely evacuated. In addition, episodic fault-bounded subsidence of the roof block above the reservoir increased magma pressure, sustaining the lateral flow of magma. Therefore, the collapsing block also drove out a large volume of magma as it subsided. This feature was supported by observations at the main vent along the rift zone, which exhibited cyclic eruption rates on long (tens of hours) time scales, due to pressure-driven surges in magma supply triggered by summit caldera collapse events 40 km upslope. Such a connection between the summit magma reservoir and the flank eruption allowed the episodic nature of summit collapses to be rapidly expressed as changes in eruption vigour on the flank (Anderson et al. 2019; Patrick et al. 2019).

5.11 Summary

Calderas are the most active yet challenging type of volcano. Caldera collapse may occur in very different conditions, most commonly during lateral intrusion of magma from mafic chambers and less often following large explosive eruptions from felsic chambers. Even though several differences exist in the collapse of mafic and felsic calderas, the mechanism, structure and evolutionary stages of collapse are the same and depend upon the amount of subsidence s with regard to the caldera diameter d, or the s/d ratio. This ratio allows defining four stages of collapse, with distinct deformation pattern, experienced by calderas.

Calderas sometimes experience longer-term uplift of their floor, or resurgence, which develops blocks or domes. Resurgence is the surface expression of the repeated shallow emplacement of new magma, unable to reach the surface and erupt because of the relatively low-viscosity contrast with the residual magma. On the shorter-term, caldera activity is often characterized by deformation, seismicity and degassing, highlighting unrest, which usually results from the shallow emplacement of magma. Comparison between the shorter-term unrest and longer-term resurgence suggests a connection, with resurgence likely resulting from the cumulative net uplift produced by multiple unrest episodes. Therefore, uplift during unrest at a resurgent caldera may be the short-term expression of resurgence.

The accumulated magma during resurgence and unrest may be finally transferred, eventually leading to eruption. For a given caldera diameter, the amount of subsidence of the caldera, albeit corrected for the presence of any infill, controls the shallow propagation path of magma. In fact, the mass removal due to the development of the caldera depression determines the local unloading stress. This, in addition to regional stresses and magma buoyancy, explains the propagation path of dikes and the distribution of eruptive vents at several calderas, and may also allow forecasting the location of future vents, as proposed for Campi Flegrei and other calderas. Nevertheless, the propagation of dikes feeding caldera-forming eruptions, as associated with magma with higher pressure and buoyancy, may be less affected by the unloading stresses.

5.12 Main Symbols Used

A_r	aspect ratio
d	caldera diameter
L	width of crust overlying the magma chamber
M	magnitude
s	caldera subsidence
z	depth of crust overlying the magma chamber
σ_1	maximum principal stress
σ_3	minimum principal stress
σ_M	maximum horizontal principal stress
σ_m	minimum horizontal principal stress

References

Acocella V, Cifelli F, Funiciello R (2001) The control of overburden thickness on resurgent domes: insights from analogue models. J Volcanol Geoth Res 111:137–153

Acocella V (2007) Understanding caldera structure and development: an overview of analogue models compared to natural calderas. Earth-Sci Rev 85:125–160

Acocella V (2010) Coupling volcanism and tectonics along divergent boundaries: collapsed rifts from Central Afar, Ethiopia. Geol Soc Am Bull 122:1717–1728

Acocella V, Palladino DM, Cioni R, Russo P, Simei S (2012) Caldera structure, amount of collapse and erupted volumes: the case of Bolsena Caldera, Italy. Geol Soc Am Bull 124:1562–1576

Acocella V, Di Lorenzo R, Newhall C, Scandone R (2015) An overview of recent (1988 to 2014) caldera unrest: knowledge and perspectives. Rev Geophys 53. https://doi.org/10.1002/2015RG000492

Acocella V (2019) Bridging the gap from caldera unrest to resurgence. Front Earth Sci 7:173. https://doi.org/10.3389/feart.2019.00173

Acocella V, Rivalta E (2019) Calderas: structure, unrest, magma transfer and eruptions. Reference module in earth systems and environmental sciences. Elsevier 15 p

Aguirre-Diaz GJ, Labarthe-Hernandez G, Tristan-Gonzalez M, Nieto-Obregon J, Gutierrez-Palomares I (2008) The ignimbrite flare-up and graben calderas of the Sierra Madre Occidental, Mexico. In: Marti J, Gottsmann J (eds) Caldera volcanism: analysis, modelling and response: developments in volcanology. Elsevier, vol 10, pp 285–311

Agustsdottir T, Winder T, Woods J, White RS, Greenfield T, Brandsdottir B (2019) Intense seismicity during the 2014–2015 Bardarbunga-Holuhraun rifting event, Iceland, reveals the nature of dike-induced earthquakes and caldera collapse mechanisms. J Geophys Res 124:8331–8357

Amelung F, Jonsson S, Zebker H, Segall P (2000) Widespread uplift and 'trapdoor' faulting on Galapagos volcanoes observed with radar interferometry. Nature 407:993–996

Amoruso A, Crescentini L, D'Antonio M, Acocella V (2017) Thermally-assisted magma emplacement explains restless calderas. Sci Rep 7:7948. https://doi.org/10.1038/s41598-017-08638-y

Anderson KR, Johanson IA, Patrick MR, Gu M, Segall P, Poland MP et al (2019) Magma reservoir failure and the onset of caldera collapse at Kīlauea Volcano in 2018. Sci 366:eaaz1822

Bagnardi M, Amelung F, Poland MP (2013) A new model for the growth of basaltic shields based on deformation of Fernandina volcano, Galápagos Islands. Earth Planet Sci Lett 377–378:358–366

Bathke H, Nikkhoo M, Holohan EP, Walter TR (2015) Insights into the 3D architecture of an active caldera ring-fault at Tendürek volcano through modeling of geodetic data. Earth Planet Sci Lett 422:157–168

Bosworth W, Burke K, Strecker M (2003) Effect of stress fields on magma chamber stability and the formation of collapse calderas. Tectonics 22:1042. https://doi.org/10.1029/2002TC001369

Branney M, Acocella V (2015) Calderas. In: Sigurdsson H, Houghton B, Rymer H, Stix J (eds) The encyclopaedia of volcanoes, 2nd edn. Academic Press, pp 299–315

Cabaniss HE, Gregg PM, Grosfils EB (2018) The role of tectonic stress in triggering large silicic caldera eruptions. Geophys Res Lett 45:3889–3895

Chadwick WW, Howard KA (1991) The pattern of circumferential and radial eruptive fissures on the volcanoes of Fernandina and Isabela islands, Galapagos. Bull Volcanol 53:259–275

Chadwick WW, Geist DJ, Jonsson S, Poland M, Johnson DJ, Meertens CM (2006) A volcano bursting at the seams: inflation, faulting, and eruption at Sierra Negra volcano, Galápagos. Geology 34:1025–1028

Chang W-L, Smith RB, Wicks C, Farrell JM, Puskas CM (2007) Accelerated uplift and magmatic intrusion of the yellowstone Caldera, 2004 to 2006. Science 318:952. https://doi.org/10.1126/science.1146842

Chen JK, Taylor FW, Edwards RL, Cheng H, Burr GS (1995) Recent emerged reef terraces of the Yenkahe resurgent block, Tanna, Vanuatu: implications for volcanic, landslide and tsunami hazards. J Geol 103:577–590

Chesner CA (2012) The Toba caldera complex. Quatern Int 258:5–18

Christiansen RL (2001) The quaternary and pliocene yellowstone plateau volcanic field of Wyoming, Idaho, and Montana. U.S. Geological Survey Professional Paper 729-G:120 pp

Clague DA, Paduan JB, Caress DW, Moyer CL, Glazer BT, Yoerger DR (2019) Structure of Loihi Seamount, Hawai'i and Lava Flow morphology from high-resolution mapping. Front Earth Sci 7:58. https://doi.org/10.3389/feart.2019.00058

Cole JW, Milner DM, Spinks KD (2005) Calderas and caldera structures: a review. Earth-Sci Rev 69:1–96

Coleman DS, Mills RD, Zimmerer RJ (2016) The pace of plutonism. Elements 12:97–102

Corbi F, Rivalta E, Pinel V, Maccaferri F, Bagnardi M, Acocella V (2015) How caldera collapse shapes the shallow emplacement and transfer of magma in active volcanoes. Earth Planet Sci Lett 431:287–293

De Chabalier JB, Avouac JP (1994) Kinematics of the Asal Rift (Djibouti) determined from the deformation of Fieale Volcano. Science 265:1677–1681

De Natale G, Pingue F (1993) Ground deformation in collapsed calderas structures. J Volcanol Geoth Res 57:19–38

De Saint Blanquat M, Horsman E, Habert G, Morgan S, Vanderhaeghe O, Law R et al (2011) Multiscale magmatic cyclicity, duration of pluton construction, and the paradoxical relationship between tectonism

and plutonism in continental arcs. Tectonophysics 500:20–33
de Silva SL, Mucek AE, Gregg PM, Pratomo I (2015) Resurgent Toba—field, chronologic, and model constraints on duration, time scales and mechanisms of resurgence at large calderas. Front Earth Sci 3. https://doi.org/10.3389/feart.2015.00025
Delgado F, Pavez A (2015) New insights into La Pacana caldera inner structure based on a gravimetric study (central Andes, Chile). Andean Geol 42:313–328
Di Vito MA, Isaia R, Orsi G, de Vita S, D'Antonio M, Pappalardo L et al (1999) Volcanism and deformation since 12000 years at the Campi Flegrei caldera (Italy). J Volcanol Geoth Res 91:221–246
Di Vito MA, Acocella V, Aiello G, Barra D, Battaglia M, Carandente A et al (2016) Magma transfer at Campi Flegrei caldera (Italy) before the last 1538 AD eruption. Sci Rep 6:32245. https://doi.org/10.1038/srep32245
Druitt T, Spark RSJ (1984) On the formation of calderas during ignimbrite eruptions. Nature 310:679–681
Duputel Z, Rivera L (2019) The 2007 caldera collapse of Piton de la Fournaise volcano: source process from very-long-period seismic signals. Earth Planet Sci Lett 527:115786
Dvorak JJ, Dzurisin DD (1997) Volcano geodesy: the search for magma reservoirs and the formation of eruptive vents. Rev Geophys 35:343–384
Farrell J, Smith RB, Husen DT (2014) Tomography from 26 years of seismicity revealing that the S. spatial extent of the Yellowstone crustal magma reservoir extends well beyond the Yellowstone caldera. Geophys Res Lett 41:3068–3073
Fridrich CJ, Smith RP, De Witt ED, McKee EH (1991) Structural, eruptive and intrusive evolution of the grizzly peak caldera, Sawatch Range, Colorado. Geol Soc Am Bull 103:1160–1177
Gaete A, Kavanagh JL, Rivalta E, Hazimb SH, Walter TR, Dennis DJC (2019) The impact of unloading stresses on post-caldera magma intrusions. Earth Planet Sci Lett 508:109–121
Galetto F, Acocella V, Caricchi L (2017) Caldera resurgence driven by magma viscosity contrasts. Nat Commun 8:1750. https://doi.org/10.1038/s41467-017-01632-y
Galetto F, Bagnardi M, Acocella V, Hooper A (2019) Noneruptive unrest at the caldera of Alcedo Volcano (Galápagos Islands) revealed by InSAR data and geodetic modelling. J Geophys Res 124. https://doi.org/10.1029/2018JB017103
Garden TO, Chambefort I, Gravley DM, Deering CD, Kennedy BM (2020) Reconstruction of the fossil hydrothermal system at Lake City caldera, Colorado, U.S.A.: constraints for caldera-hosted geothermal systems. J Volcanol Geotherm Res 393:106794
Genrich JF, Bock Y, McCaffrey R, Prawirodirdjo L, Stevens CW, Puntodewo SSO et al (2000) Distribution of slip at the northern Sumatran fault system. J Geophys Res 105:28327–28341
Geshi N, Shimano T, Chiba T, Nakada S (2002) Caldera collapse during the 2000 eruption of Miyakejima volcano, Japan. Bull Volcanol 64:55–68
Geyer A, Folch A, Martì J (2006) Relationship between caldera collapse and magma chamber withdrawal: an experimental approach. J Volcanol Geoth Res 157:375–386
Geyer A, Martì J (2014) A short review of our current understanding of the development of ring faults during collapse caldera formation. Front Earth Sci 2. https://doi.org/10.3389/feart.2014.00022
Glazner AF, Bartley JM, Coleman D, Gray W, Taylor RZ (2004) Are plutons assembled over millions of years by amalgamation from small magma chambers? GSA Today 14. https://doi.org/10.1130/1052-5173
Goto I, McPhie J (2018) Tectonics, structure, and resurgence of the largest Quaternary caldera in Japan: Kutcharo, Hokkaido. Geol Soc Am Bull 130:1307–1322
Gregg P, de Silva SL, Grosfils EB, Parmigiani JP (2012) Catastrophic caldera-forming eruptions: thermomechanics and implications for eruption triggering and maximum caldera dimensions on Earth. J Volcanol Geoth Res 241–242:1–12
Gregg PM, Le Mével H, Zhan Y, Dufek J, Geist D, Chadwick WW (2018) Stress triggering of the 2005 eruption of Sierra Negra volcano, Galápagos. Geophys Res Lett 45:13288–13297
Gudmundsson MT, Jónsdóttir K, Hooper A, Holohan EP, Halldórsson SA, Ófeigsson BG et al (2016) Gradual caldera collapse at Bárdarbunga volcano, Iceland, regulated by lateral magma outflow. Science 353 (6296): aaf8988
Gudmundsson A (1998) Formation and development of normal fault calderas and the initiation of large explosive eruptions. Bull Volcanol 60:160–170
Gudmundsson A (2007) Conceptual and numerical models of ring-fault formation. J Volcanol Geoth Res 164:142–160
Han R, Kim J-S, Kim C-M, Hirose T, Jeong JO, Jeon GY (2019) Dynamic weakening of ring faults and catastrophic caldera collapses. Geology 47:107–110
Hardy S (2008) Structural evolution of calderas: Insights from two-dimensional discrete element simulations. Geology 36:927–930
Harris AJL (2009) The pit-craters and pit-crater-filling lavas of Masaya volcano. Bull Volcanol 71:541–558
Hildreth W, Fiersetin J, Calvert A (2017) Early postcaldera rhyolite and structural resurgence at Long Valley Caldera, California. J Volcanol Geoth Res 335:1–34
Holohan EP, Troll VR, Walter TR, Munn S, McDonnell S, Shipton ZK (2005) Elliptical calderas in active tectonic settings: an experimental approach. J Volcanol Geoth Res 144:119–135
Holohan EP, Troll VR, van Wyk deVries B, Walsh JJ, Walter TR, (2008) Unzipping long valley: an explanation for vent migration patterns during an elliptical ring fracture eruption. Geology 36:323–326
Holohan EP, Schopfer MPJ, Walsh JJ (2011) Mechanical and geometric controls on the structural evolution of

pit crater and caldera subsidence. J Geophys Res 116: B07202. https://doi.org/10.1029/2010JB008032
Hotovec-Ellis AJ, Shelly DR, Hill DP, Pitt AM, Dawson PB, Chouet BA (2018) Deep fluid pathways beneath Mammoth Mountain, California, illuminated by migrating earthquake swarms. Sci Adv 4: eaat5258
Isaia R, Vitale S, Marturano A, Aiello G, Barra D, Ciarcia S et al (2019) High-resolution geological investigations to reconstruct the long-term ground movements in the last 15 kyr at Campi Flegrei caldera (southern Italy). J Volcanol Geoth Res 385:143–158
Jellinek AM, DePaolo DJ (2003) A model for the origin of large silicic magma chambers: precursors of caldera-forming eruptions. Bull Volcanol 65:363–381
Jiang C, Schmandt B, Farrell J, Lin F-C, Ward KM (2018) Seismically anisotropic magma reservoirs underlying silicic Calderas. Geology 46:727–730
Johnson R, Itikarai I, Patia H, McKee C (2010) Volcanic systems of the Northeastern Gazelle Peninsula, Papua New Guinea: Synopsis, evaluation, and a model for Rabaul Volcano, Papua New Guinea. Rabaul Volcano Work Report Papua New Guinea Dep Miner Policy Geohazards Manag Aust Agency Int Dev Port Moresby Papua New Guinea, 94 p
Jones RH, Stewart RC (1997) A method for determining significant structures in a cloud of earthquakes. J Geophys Res 102:8245–8254
Jonsson S, Zebker H, Amelung F (2005) On trapdoor faulting at Sierra Negra volcano, Galapagos. J Volcanol Geoth Res 144:59–71
Jonsson S (2009) Stress interaction between magma accumulation and trapdoor faulting on Sierra Negra volcano, Galápagos. Tectonophysics 471:36–44
Kabele P, Zak J, Somr M (2017) Finite-element modeling of magma chamber–host rock interactions prior to caldera collapse. Geophys J Int 209:1851–1865
Karlstrom L, Rudolph ML, Manga M (2012) Caldera size modulated by the yield stress within a crystal-rich magma reservoir. Nat Geosci 5:402–405
Kawakami Y, Hoshi H, Yamaguchi Y (2007) Mechanism of caldera collapse and resurgence: observations from the northern part of the Kumano Acidic Rocks, Kii peninsula, southwest Japan. J Volcanol Geoth Res 167:263–281
Kennedy B, Jellinek MA, Stix J (2008) Coupled caldera subsidence and stirring inferred from analogue models. Nat Geosci 1:385–389
Kennedy B, Wilcox J, Stix J (2012) Caldera resurgence during magma replenishment and rejuvenation at Valles and Lake City calderas. Bull Volcanol 74:1833–1847
Kennedy B, Stix J, Hon K, Deering C, Gelman S (2015) Magma storage, differentiation, and interaction at Lake City caldera, Colorado, USA. Geol Soc Am Bull 128:764–776
Kim C-M, Han R, Kim J-S, Sohn YK, Jeong JO, Jepng GY et al (2019) Fault zone processes during caldera collapse: Jangsan Caldera, Korea. J Struct Geol 124:197–210
Koulakov I, Yudistira T, Luehr BG, Wandono, (2009) P, S velocity and Vp/Vs ratio beneath the Toba caldera complex (northern Sumatra) from local earthquake tomography. Geophys J Int 177:1121–1139
Kumagai H, Ohminato T, Nakano M, Ooi M, Kubo A, Inoue H et al (2001) Very-long-period seismic signals and caldera formation at Miyake Island. Japan. Science 293:687. https://doi.org/10.1126/science.1062136
Lagabrielle I, Cormier MH (1999) Formation of large summit troughs along the East Pacific Rise as collapse calderas: an evolutionary model. J Geophys Res 104:12971–12988
Levy S, Bohnenstiehl DR, Sprinkle P, Boettcher MS, Wilcox WSD, Tolstoy M et al (2018) Mechanics of fault reactivation before, during, and after the 2015 eruption of Axial Seamount. Geology 46:447–450
Lindsay JM, de Silva S, Trumbull R, Emmermann R, Wemmer K (2001) La Pacana caldera, N Chile: a re-evaluation of the stratigraphy and volcanology of one of the world's largest resurgent calderas. J Volcanol Geoth Res 106:145–173
Lipman PW (1984) The roots of ash flow calderas in Western North America: windows into the tops of granitic batholiths. J Geophys Res 89:8801–8841
Lipman PW (1997) Subsidence of ash-flow calderas: relation to caldera size and magma-chamber geometry. Bull Volcanol 59:198–218
Lipman PW (2000) Calderas. In: Sigurdsson H, Houghton B, McNutt S, Rymer H, Stix J (eds) The encyclopedia of volcanoes, 1st edn. Elsevier Academic Press, pp 643–662
Liu Y-K, Ruch J, Bathke HV, Jonsson S (2019) Influence of ring faulting in localizing surface deformation at subsiding calderas. Earth Planet Sci Lett 526:115784
Lowenstern JB, Smith RB, Hill DP (2006) Monitoring super-volcanoes: geophysical and geochemical signals at Yellowstone and other large caldera systems. Philos Trans Roy Soc A 364:2055–2072
Mandl G (1988) Mechanics of tectonic faulting: models and basic concepts. Elsevier, Amsterdam, 401 p
Marsh BD (1984) On the mechanics of caldera resurgence. J Geophys Res 89:8245–8251
Martì J, Gudmundsson A (2000) The Las Canadas caldera (Tenerife, Canary Islands): an overlapping collapse caldera generated by magma-chamber migration. J Volcanol Geoth Res 103:161–173
Martì J, Geyer A, Folch A (2009) A genetic classification of collapse calderas based on field studies, and analogue and theoretical modelling. In: Thordarson T, Self S, Larsen G, Rowland SK, Hoskuldsson A (eds) Studies in volcanology: the legacy of George Walker. J Geol Soc Lond Spec Public IAVCEI 2:249–266
Masturyono, McCaffrey R, Wark DA, Roecker SW, Fauzi, Inbrahin G et al (2001) Distribution of magma beneath the Toba caldera complex, north Sumatra, Indonesia, constrained by three dimensional P wave velocities, seismicity, and gravity data. Geochem Geophys Geosyst 2:2000GC000096

Michon L, Massin F, Famin V, Ferrazzini V, Roult G (2011) Basaltic calderas: collapse dynamics, edifice deformation, and variations of magma withdrawal. J Geophys Res 116:B03209. https://doi.org/10.1029/2010JB007636
Mori J, McKee C (1987) Outward-dipping ring-fault structure at Rabaul Caldera as shown by earthquake locations. Science 235:193–195
Munekane H, Oikawa J, Kobayashi T (2016) Mechanisms of step-like tilt changes and very long period seismic signals during the 2000 Miyakejima eruption: insights from kinematic GPS. J Geophys Res 121:2932–2946
Nairn IA, McKee CO, Talai B, Wood CP (1995) Geology and eruptive history of the Rabaul caldera area, Papua New Guinea. J Volcanol Geotherm Res 69:255–284
Neal CA, Brantley SR, Antolik L, Babb JL, Burgess M, Calles K et al (2019) The 2018 rift eruption and summit collapse of Kīlauea Volcano. Science 363:367–374
Newhall CG, Dzurisin D (1988) Historical unrest at large calderas of the world. US Geological Survey Professional Paper 1109 p
Newhall C, Self S, Robock A (2018) Anticipating future Volcanic Explosivity Index (VEI) 7 eruptions and their chilling impacts. Geosphere 14:572–603
Okubo CH, Martel SJ (1998) Pit crater formation on Kilauea volcano, Hawaii. J Volcanol Geoth Res 86:1–18
Orsi G, Gallo G, Zanchi A (1991) Simple shearing block resurgence in caldera depressions. A model from Pantelleria and Ischia. J Volcanol Geoth Res 47:1–11
Orsi G, De Vita S, di Vito M (1996) The restless, resurgent Campi Flegrei nested caldera (Italy): constrains on its evolution and configuration. J Volcanol Geoth Res 74:179–214
Patrick MR, Dietterich HR, Lyons JJ, Diefenbach AK, Parcheta C, Anderson KR et al (2019) Cyclic lava effusion during the 2018 eruption of Kīlauea Volcano. Sci 366: eaay9070
Phillipson G, Sobradelo R, Gottsmann J (2013) Global volcanic unrest in the 21st century: an analysis of the first decade. J Volcanol Geoth Res 264:183–196
Pistolesi M, Isaia R, Marianelli P, Bertagnini A, Fourmentraux C, Albert PG et al (2016) Simultaneous eruptions from multiple vents at Campi Flegrei (Italy) highlight new eruption processes at calderas. Geology 44:487–490
Pritchard ME, Mather TA, McNutt SR, Delgado FJ, Reath K (2019) Thoughts on the criteria to determine the origin of volcanic unrest as magmatic or non-magmatic. Philos Trans Roy Soc 377:20180008
Rivalta E, Corbi F, Passarelli L, Acocella V, Davis T, Di Vito MA et al (2019) Stress inversions to forecast magma pathways and eruptive vent location. Sci Adv 5:eaau9784
Robertson RM, Kilburn CRJ (2016) Deformation regime and long-term precursors to eruption at large calderas: Rabaul, Papua New Guinea. Earth Planet Sci Lett 438:86–94
Roche O, Druitt TH, Merle O (2000) Experimental study of caldera formation. J Geophys Res 105:395–416
Roche O, Druitt TH (2001) Onset of caldera collapse during ignimbrite eruptions. Earth Planet Sci Lett 191:191–202
Roman A, Jaupart C (2014) The impact of a volcanic edifice on intrusive and eruptive activity. Earth Planet Sci Lett 408:1–8
Rougier J, Sparks RSJ, Cashman KV, Brown SK (2018) The global magnitude–frequency relationship for large explosive volcanic eruptions. Earth Planet Sci Lett 482:621–629
Ruch J, Acocella V, Geshi N, Nobile A, Corbi F (2012) Kinematic analysis of vertical collapse on volcanoes using experimental models time series. J Geophys Res 117:B07301. https://doi.org/10.1029/2012JB009229
Rymer H, van Wyk de Vries B, Stix J, Williams-Jones G (1998) Pit crater structure and processes governing persistent activity at Masaya Volcano, Nicaragua. Bull Volc 59:345–355
Self S, Goff F, Gardner JN, Wright JV, Kite WM (1986) Explosive rhyolitic volcanism in the Jemez Mountains: vent locations, caldera development and relation to regional structure. J Geophys Res 91:1779–1798
Seropian G, Stix J (2018) Monitoring and forecasting fault development at actively forming calderas: an experimental study. Geology 46:23–26
Shreve T, Grandin R, Boichu M, Garaebiti E, Moussalam Y, Ballu V et al (2019) From prodigious volcanic degassing to caldera subsidence and quiescence at Ambrym (Vanuatu): the influence of regional tectonics. Sci Rep 9:18868. https://doi.org/10.1038/s41598-019-55141-7
Sigmundsson F (2019) Calderas collapse as magma flows into rifts. Science 366:1200–1201
Simkin T, Howard KA (1970) Caldera Collapse in the Galapagos Islands, 1968. Science 169:429–437
Skilling IP (1993) Incremental caldera collapse of Suswa volcano, Gregory Rift Valley, Kenya. J Geol Soc Lond 150:885–896
Smith RL, Bailey RA (1968) Resurgent cauldrons. Geol Soc Ame Memor 116:613–662
Smith RB, Braile LW (1994) The Yellowstone hotspot. J Volcanol Geoth Res 61:121–187
Smith VC, Isaia R, Pearce NJG (2011) Tephrostratigraphy and glass compositions of post-15 kyr Campi Flegrei eruptions: Implications for eruption history and chronostratigraphic markers. Quatern Sci Rev 30:3638–3660
Solada KE, Reilly BT, Stoner JS, de Silva SL, Mucek AE, Hatfield RG et al (2020) Paleomagnetic observations from lake sediments on Samosir Island, Toba caldera, Indonesia, and its late Pleistocene resurgence. Quatern Res 95:97–112
Stix J, Kobayashi T (2008) Magma dynamics and collapse mechanisms during four historic caldera-forming events. J Geophys Res 113:B09205. https://doi.org/10.1029/2007JB005073
Tibaldi A, Vezzoli L (1998) The space problem of caldera resurgence: an example from Ischia Island, Italy. Geol Rundsch 87:53–66

Trasatti E, Acocella V, Di Vito MA, Del Gaudio C, Weber G, Aquino I et al (2019) Magma degassing as a source of long-term seismicity at volcanoes: the Ischia island (Italy) case. Geophys Res Lett 46. https://doi.org/10.1029/2019GL0853712

Trippanera D, Ruch J, Acocella V, Thordarson T, Urbani S (2018) Interaction between central volcanoes and regional tectonics along divergent plate boundaries: Askja. Iceland. Bull Volcanol 80:1. https://doi.org/10.1007/s00445-017-1179-8

Vazquez JA, Reid MR (2004) Probing the accumulation history of the voluminous Toba Magma. Science 305:991–994

Walker GPL (1984) Downsag calderas, ring faults, caldera sizes, and incremental caldera growth. J Geophys Res 89:8407–8416

Wilcock WSD, Tolstoy M, Waldhauser F, Garcia C, Tan YJ, Bohnenstiehl DR et al (2016) Seismic constraints on caldera dynamics from the 2015 Axial Seamount eruption. Science 354:1395–1399

Volcano Flank Instability and Collapse

6

6.1 Introduction

After having discussed magma chamber formation and dynamics in the previous two chapters, in this chapter the focus is on the shallower volcano-tectonic processes related to the instability and collapse of the flanks of volcanic edifices. Flank instability and collapse is in fact, together with caldera formation (considered in the previous chapter), a first-order process affecting the shape of a volcanic edifice and, in turn, controlling shallow magma propagation, which is the topic of the next chapter.

The flanks of many active or quiescent volcanic edifices are often unstable, because of magma emplacement, hydrothermal activity, pore pressure, weak basement and seismicity. Flank instability manifests in different ways, affecting minor or major portions of volcanic edifices, as well as their basement, with variable slip rates and frequency of occurrence. The instability may culminate in the catastrophic lateral collapse of a sector of the edifice, which may in turn affect the frequency and style of future eruptions. In addition, flank instability may be accompanied with surface deformation, seismicity, eruption, caldera collapse, landslide formation, flooding and tsunami generation, often posing a multi-hazard threat. Because of all these features, flank instability is a major and frequent destabilizing process affecting volcanic edifices.

The main aims of this chapter are to:

- describe the main features associated with flank instability;
- define the factors promoting volcano flank instability and collapse;
- illustrate the multi-hazard relationships associated with flank instability and, mostly, sector collapse.

6.2 General Features

Landslides are a ubiquitous feature of the Earth's surface, occurring on any slope, also underwater. In a similar fashion, the slopes of volcanic edifices are also subject to instability and failure. However, specific features promote, accompany and amplify slope instability in volcanoes. First, volcanic edifices result from the repeated emplacement and piling of erupted products in a geologically short period, that is commonly within 10^5 years. Because of this rapid construction, any edifice with significant height ($\sim 10^3$ m) may become unable to support its own load and thus unstable. Second, volcano flank instability is further promoted by the recurrent magmatic and shallow seismic activity, responsible for the internal deformation, steepening and shaking of the edifice. As a result, volcanic edifices are particularly prone to experience flank instability.

V. Acocella, *Volcano-Tectonic Processes*, Advances in Volcanology,
https://doi.org/10.1007/978-3-030-65968-4_6

Here the term **flank instability** includes any form of downslope (with relevant lateral component of slip) movement of a flank of a volcanic edifice, independent of its type, size, velocity and frequency. Flank instability may thus include different types of gravitational processes, as rockfalls, avalanches, shallow, deep-seated and spreading movements. Flank instability may develop from slow gravitational deformation, dominated by creeping, into a much faster flank or **sector collapse**, consisting of the sudden failure of a significant portion of a volcanic edifice, possibly also affecting the summit area. A sector collapse generates a **debris avalanche**, a flowing mixture of debris, rock, and moisture that moves downslope under the influence of gravity. The term sector collapse is used to emphasise the rapid, large and deep-seated extent of the instability in the context of the parent edifice (Fig. 6.1; Watt 2019). Flank instability may therefore manifest through a wide range of slope velocities, from a few mm/yr, as geodetically observed at Mount Damavand (Iran), to 10^2 m/yr, as on Monowai submarine volcano (Kermadec Arc), to sudden movements, as in 2002 at Stromboli (Italy; Calvari et al. 2005; Chadwick et al. 2008; Shirzaei et al. 2011). These velocities may not only differ from slope to slope and volcano to volcano, but may also vary in time within the same slope. In some cases, the gravitational instability may even affect the whole volcanic edifice, as at Mount Etna (Italy), where the entire volcano, ~3300 m high, is slowly sliding seawards (Murray et al. 2018). The conditions leading from slower flank instability to sudden sector collapse may be quite variable and often require an external trigger. In fact, while even significant slope movements of decimetres or metres per day may not produce catastrophic collapse, as at Stromboli in 2007 or Pacaya (Guatemala) in 2010, an earthquake may suddenly promote the collapse of an unstable flank, as in 1980 at Mount St. Helens (Washington, USA; Di Traglia et al. 2014; Schaefer et al. 2015; Karstens et al. 2019).

The frequency of the instability events may also vary significantly, from monthly for minor rock falls to 10^6 years for large submarine debris avalanches (Wright et al. 2008; Zernack et al. 2013). Considering only the major slope failures, more than 20 events occurred in the past 500 years, a rate similar to that of caldera collapse. A relationship between the frequency of volcanic mass movements and their volumes shows higher frequencies with volumes of 10^2 m^3 and lower frequencies with volumes of 10^{12} m^3 (Fig. 6.2a). This indicates that larger instability events are less frequent, similarly to what observed with the size of eruptions. The unstable volume is proportional to the size of the volcano, being the largest volcanoes associated with failure volumes of tens of cubic kilometres. In addition, the higher the edifice, the higher the percentage of collapsed volcanic edifices, with a threshold for significant instability above ~2500 m (Fig. 6.2b). The volcano size controls also the lateral extent reached by the failed deposits: in fact, the vertical drop H of the avalanches, from the volcano summit to the deposit toe (somehow proportional to the volcano height), correlates positively with the travel distance L_d, indicating that the products of the catastrophic collapse of higher volcanoes may reach several tens of kilometres (Fig. 6.2c). All these relationships also imply that the runout L_d/H of volcanic landslides correlates positively with their volume, with larger collapses travelling further, similar to non-volcanic landslides (Fig. 6.2d; McGuire 1996; Siebert et al. 1995; Dade and Huppert 1998; Voight 2000).

Flank instability may occur on any volcano, independently of the composition (mafic, felsic, intermediate), shape (shield volcanoes, composite volcanoes, stratovolcanoes, calderas, resurgent calderas) and tectonic setting (extensional, compressional, strike-slip). The dip of the slope of the edifice is important, as indicated by the higher frequency of major failures on steeper volcanoes compared to gentler ones, although the latter also show a minor peak (Fig. 6.2e). This double-peaked distribution suggests preferred but distinct conditions for flank instability. The major peak may be explained by the higher slope of the

Fig. 6.1 Subaerial and submarine volcanoes experiencing sector collapse. **a** Stromboli, Italy, highlighting the Sciara del Fuoco sector collapse, on August 18, 2014 (photo courtesy Marco Neri). **b** Near vertical space image of Bezymianny volcano (Kamchatka, Russia), acquired on 20 September 2012, affected by a horseshoe-shaped collapse scarp (photo credit: NASA). **c** Monowai Seamount, Kermadec Volcanic Arc (after Chadwick et al. 2008). **d** Near vertical view, as seen on 26 July 2006, of the lava dome-filled amphitheatre-shaped scar of Mount St. Helens (Washington, USA), formed during the 1980 flank collapse (photo credit USGS; US Geological Survey). **e** Space image of Fogo volcano (Cape Verde Islands), acquired on June 10, 2009, affected by a collapse scarp on its eastern portion (photo credit: NASA)

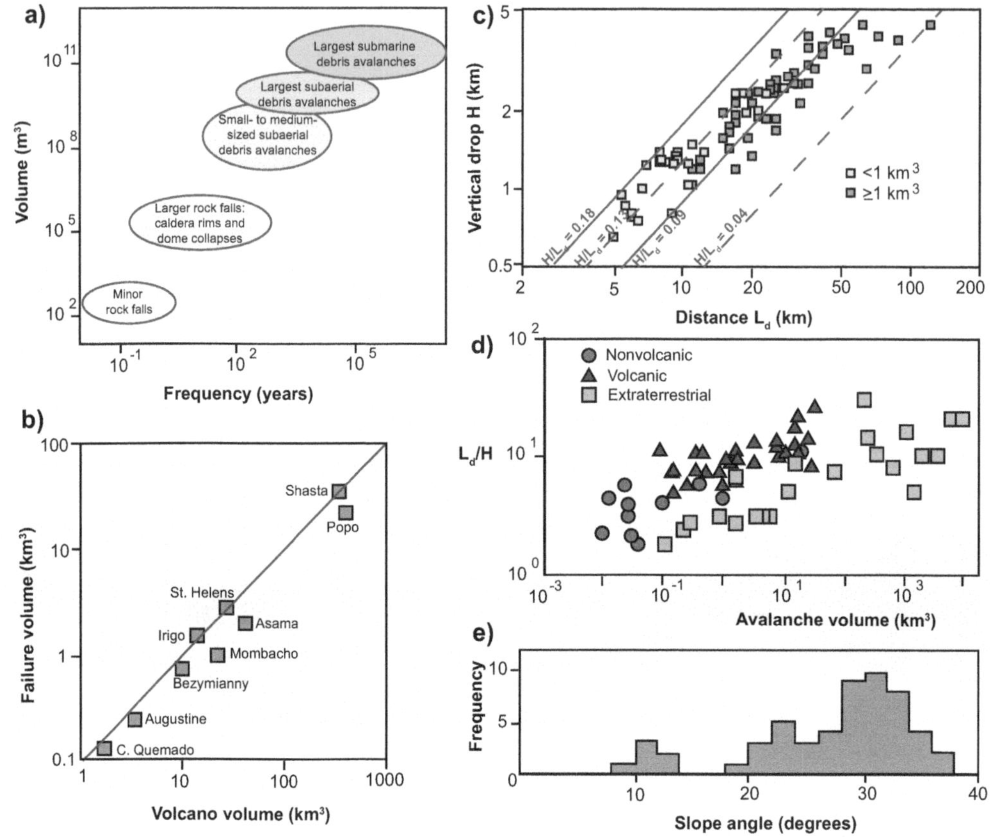

Fig. 6.2 Relationships between different parameters related to flank collapse. **a** Size versus frequency of collapse. **b** Size of collapse versus size of volcano. **c** Vertical drop H (log scale) of debris avalanche produced by flank collapse versus horizontal travel distance L_d; solid diagonal lines enclose data from deposits of less than 1 km^3 volume, whereas diagonal dashed lines enclose deposits larger than 1 km^3; **d** L_d/H ratio of debris avalanches versus collapsed volume. **e** Slope angle of volcano versus frequency of collapse (Siebert et al. 1995; McGuire 1996; Voight and Elsworth 1997; Dade and Huppert 1998)

steeper stratovolcanoes approaching the repose angle of the edifices, a condition common in volcanic arcs. The minor peak suggests that many gently-dipping shield volcanoes experience flank failure. This may be explained by the fact that many shield volcanoes lie in poorly-confined oceanic islands environments, are often dominated by volcanic rift zones responsible for flank destabilization and show quite homogeneous mechanical layering (mainly lava flows), which requires less energy for a failure to propagate (Masson et al. 2002; Gudmundsson 2009; Watt 2019). Therefore, stratovolcanoes in volcanic arcs and shield volcanoes in oceanic islands, although morphologically, structurally and compositionally distinct, provide the most favourable conditions for flank failure. Indeed, several studies have focused on the stability of these types of volcanoes in both settings. As regards volcanic arcs, in Kamchatka (Russia) 60% of the volcanoes have experienced sector collapse, some repetitively, and mostly related to major eruptions. On the Lesser Antilles arc (Central America), the frequency and direction of tens of sector collapses are affected by the configuration of the submarine slope, at times poorly-confining the volcanic edifices. Quite similarly, in parts of the eastern Mexican Volcanic Belt, the eastward sloping substrate controls the direction of the collapsing sectors in major volcanoes. In the western Aleutian Arc (USA), the orientation of submarine debris avalanche deposits appears influenced by regional faults, with the large collapsed volumes suggesting incorporation of submarine material. In the central Aleutian Arc, seven out of ten volcanoes have experienced edifice failure, with many deep-seated events not necessarily accompanied by large eruptions (Ponomareva et al. 2006; Carrasco-Nunez et al. 2006; Boudon et al. 2007; Coombs et al. 2007; Montanaro and Beget 2011). As regards oceanic islands, flank instability is common at basaltic shields. For example, the neighbouring active volcanoes of Mauna Loa and Kilauea (Hawaii, USA) are affected by widespread flank instability towards the ocean side (Fig. 6.3). On the Canary Islands, at least 14 sector collapses occurred on the flanks of El Hierro, La Palma and Tenerife, with collapsed volumes between 50 and 500 km^3, accounting for $\sim 10\%$ of the volume of the volcanic edifices and runout distances of up to 130 km: these collapses are primarily controlled by the activity of the volcanic rift zones. In a similar fashion, all the larger Cape Verde Islands show evidence of large sector collapse, although only Fogo and Santo Antao have failed in the last 400 ka (Carracedo et al. 1999; Masson et al. 2002, 2008; Lipman et al. 2006; Blahut et al. 2019).

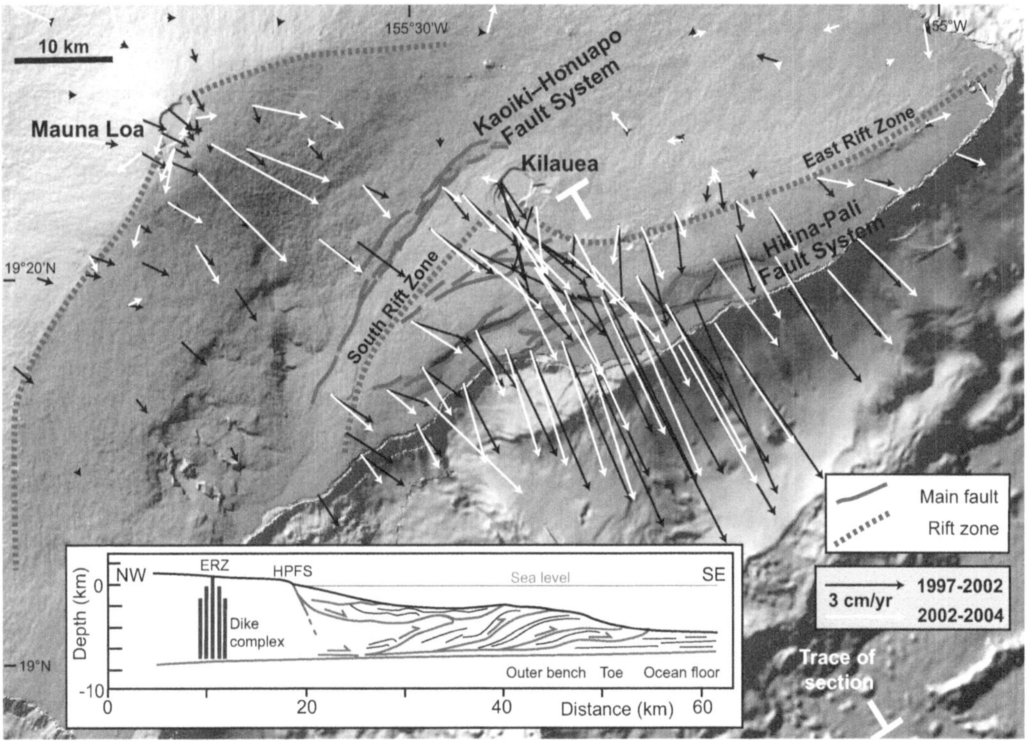

Fig. 6.3 Schematic structure of Mauna Loa and Kilauea volcanoes, Hawaii (USA), reporting the main rift zones, fault systems and the 1997–2002 (black) and 2002–2004 (white) GPS displacements mainly related to flank instability. The GPS data in the two periods highlight a consistent flank motion for Kilauea; conversely, Mauna Loa experienced flank motion between 1997 and 2002 and underwent summit dike intrusion between 2002 and 2005. Inset shows a NW–SE trending section view of the unstable flank, sliding along an inward-dipping detachment at the base of the edifice and an intensely contracted frontal portion of the unstable wedge; ERZ = East Rift Zone; HPFS = Hilina-Pali Fault System (Miklius et al. 2005; Phillips et al. 2008). DEM provided by GeoMapApp.

Sector collapse, because of the wide areas involved (up to ~100 km of distance from the volcano), its unpredictability and other potentially related hazards (see Sect. 6.7), may be particularly threatening; indeed, about 20,000 victims have resulted from historic volcano flank collapse (Siebert et al. 1987).

6.3 Anatomy of an Unstable Flank

Despite the variability in the conditions accompanying flank instability, the unstable or collapsing flanks of volcanoes usually show recurrent structural similarities. An unstable flank may be in fact approximated as a volume of the volcanic edifice bordered by structures with distinct geometry and kinematics (Fig. 6.4).

The uppermost part, or **head**, of the unstable flank is commonly characterized by tension, resulting from the separation of the unstable mass from the more stable rear portion of the volcano. This tension develops extensional structures, such as planar or curved normal faults and extension fractures, usually striking parallel to the direction of the slope. Tilted collapsing blocks may be also associated with these structures. A common but not ubiquitous feature at the rear of the collapse head is a volcanic rift zone, with focused volcanism and fracturing induced by the propagation of dikes feeding fissure eruptions (see Chap. 7). Repeated and frequent dike propagation along volcanic rift zones promotes lateral compression, progressively increasing the dip of the slopes to the sides, eventually destabilizing the flanks. In some cases, the head of the collapse scarp may coincide with the summit crater area, as for example at Stromboli (Italy), where the summit portion of the Sciara del Fuoco collapse coincides with the alignment of the active craters (Tibaldi 2001). In other cases, the head of the unstable portion may include the entire summit of the volcano, extending also to the upper part of the opposite slope.

The **sides** of the unstable flank may show transitional and/or sharp kinematic variations with regard to the nearby stable flanks. In the case of sharp variations, the sides experience downslope slip along faults with predominant strike-slip motion, as for example observed during the 1980 collapse of Mount St. Helens or in the northeast flank of Mount Etna (Solaro et al. 2010; Walter 2011). These structures to the sides

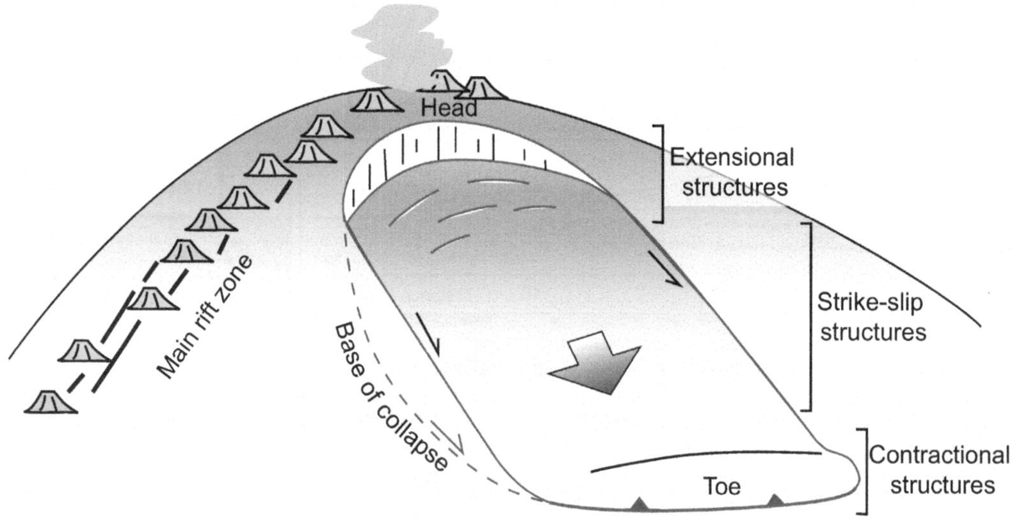

Fig. 6.4 Scheme of unstable flank of a volcano, highlighting the main structural features at its head, sides and toe. A volcanic rift zone is often present behind the head of the unstable flank, with flank instability related to dike emplacement along the rift zone

connect the head of the unstable portion to its base and may be characterized by constant or variable displacement. The variable displacement, decreasing or increasing outward (towards the periphery of the volcano), may result from different factors controlling the instability, as for example magma intrusion (associated with outward decrease in displacement) or unbuttressing at the base of the edifice (associated with inward decrease). Magma may intrude through dikes from the summit area along the sides of the unstable flank, as at Stromboli. In some cases, dike emplacement may be a precursor of faulting and collapse, as suggested by the obliquity of en-echelon dikes implying incipient strike-slip motion, as along the sides of the Guimar unstable flank at Tenerife (Canary Islands; Acocella and Tibaldi 2005; Del Camp et al. 2012).

The **toe** of the unstable flank, where the material accumulates, commonly lies at the break in slope at the base of the volcanic edifice. This zone consists of contractional structures (folds, thrust faults) responsible for shortening and at times experiences localized uplift. The exact structural style may vary significantly, being mainly a function of the amount of shortening and confinement: from the presence of gentle uplifting folds, as around part of the periphery of Mount Etna, to imbricate thrusts and folds, as offshore Kilauea (Fig. 6.3; Morgan et al. 2000; Solaro et al. 2010).

The base, or **detachment**, of the unstable flank constitutes the bottom surface connecting part or all of the above mentioned features. This may consist of, or be approximated by, a planar or a curved basal surface, with uniform translation or more complex rotational displacement, respectively. Alternatively, the base may consist of several subparallel shear splays at slightly different depth. The dip of the base of the unstable flank may be directed outward and/or inward. Some detachments are gently inward dipping, as below Mauna Loa and Kilauea, resulting from the sagging of the oceanic crust under the volcanoes load: here the oceanic clays below the volcanoes provide the basal detachment for the unstable flanks. However, most detachments are outward dipping. These may show a quite constant dip subparallel to the slope, as at La Palma (Canary Islands), possibly reactivating pre-existing discontinuities, as layering or unconformities, or a variable (decreasing) dip related to a rotational listric detachment, as at Stromboli (Tibaldi 2001; Morgan and McGovern 2005). The detachment at the base of the unstable flank is usually not directly accessible, but its geometric and kinematic features may be reconstructed using the distribution of seismicity and/or inverting geodetic data associated with the slip of the flank. In a few cases, the eroded portions of unstable or collapsed flanks may allow direct access to the detachment. For example, at La Gomera (Canary Islands) the rocks close to the sliding surface are strongly deformed, with folds, fault breccia and gauge and fracture cleavage, and with hydrothermal metamorphism (Casillas et al. 2010). At Piton des Neiges (La Reunion Island) a basal detachment concentrates brittle and ductile deformation, sill injections of basic magmas and hydrothermal metamorphic alteration, probably induced by the heat of the sills. The overlap of ductile and brittle deformation suggests that the basal slip was accommodated by slow, continuous creep and rapid co-intrusive sliding (Famin and Michon 2010).

Within the unstable flank, several structures (mainly fault zones) may accommodate any differential movement, forming kinematically distinct intra-blocks, with rigid or semi-rigid deformation. These blocks may come in a wide variety of configurations. In some cases, they may be vertically piled one on the other above a rotational detachment, as observed for Mount St. Helens or inferred for Stromboli (Walter 2011; Vezzoli and Corazzato 2016). In other cases, laterally juxtaposed blocks with distinct kinematics may lie above a detachment, as geodetically observed at Mount Etna (Solaro et al. 2010).

6.4 Causes of Flank Instability and Collapse

Flank instability may result from a single triggering factor or, more commonly, from a combination of circumstances and events acting together, although one specific action may set the mass in motion (Fig. 6.5; McGuire 1996; Carrasco-Nunez et al. 2006). For example, while the trigger of the 1980 sector collapse of Mount St. Helens was a shallow magnitude *M5.1* earthquake, the collapse had been prepared months before, with the emplacement of a viscous cryptodome within the northern portion of the volcanic edifice, which oversteepened the northern flank to be collapsed. This directionality may have in turn resulted from the fact that the intruding viscous magma deflected its originally subvertical propagation path approaching the northern flank because of the differential buttressing experienced by its non-centred rise (Voight et al. 1981; Donnadieu and Merle 2001). Therefore, flank instability and collapse may result from longer-term preparing and shorter-term triggering factors, which should be equally considered.

The above mentioned event at Mount St Helens is one of the few cases of monitored volcano collapse in the last decades for which the preparing and triggering factors could be determined. More generally, studies on past sector collapses rely on the geological record, from which the exact preparing and triggering factors of collapse are more difficult to be inferred. Therefore, while different factors may promote flank instability and collapse, in most cases establishing the precise succession of the preparing and triggering events remains challenging. Indeed, for most of the over 400 identified volcanic sector collapses worldwide the specific preconditions and trigger mechanisms are unknown (Siebert 1984).

In principle, several processes may play an important role on flank instability and collapse. In volcanoes, magma emplacement, in the form of both dikes and viscous intrusions, increases the stresses within the edifice and is probably the most common triggering factor for flank instability. Magma-induced hydrothermal alteration may also reduce the strength of the edifice, promoting flank instability. Excess pore pressures, especially when increased by thermal pressurization, similarly destabilize volcano flanks. Variations in mechanical properties within the volcanic edifice, or in its basement, may also locally decrease the strength of the rocks, inducing the collapse or spreading of the edifice. Finally, external factors, as fault activation and seismicity or local weather and climate, may further promote flank instability (Voight and Elsworth 1997). These processes are discussed in the sections below.

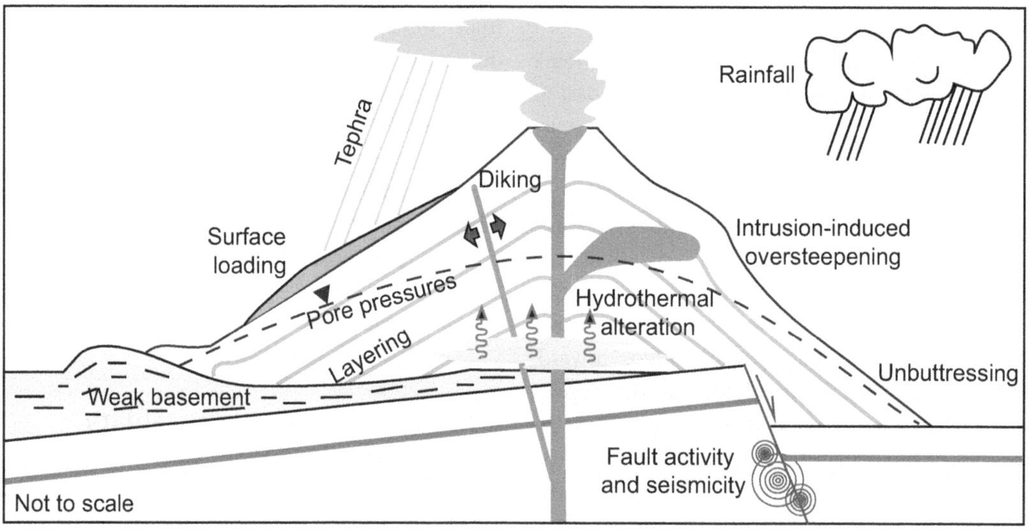

Fig. 6.5 Scheme showing the main factors capable of preparing and/or triggering flank instability on a volcano (modified after McGuire 1996)

6.4.1 Magma Emplacement (Dikes and Viscous Intrusions)

Magma emplacement within the volcanic edifice is probably the most common trigger of flank instability. The emplaced magma may have the shape of a dike, a sill, a plug, or a cryptodome, depending on its viscosity and structural context.

Dike emplacement is widespread in any type of volcanic edifice and, when occurring along the rift zones frequently active in mafic volcanoes, it may be especially effective in destabilizing flanks. As a dike emplaces along a volcanic rift, it compresses and uplifts the portions of the edifice to the sides of the rift zone. Repeated diking tilts and steepens the slopes to the sides of the rift, ultimately promoting flank instability: as a result, volcanic rift zones focus flank instability to their sides (Fig. 6.4). Episodes of acceleration of flank instability often coincide with eruptions along volcanic rift zones, a process that at active mafic volcanoes usually occurs with frequency of a few decades, as at Fogo (Cape Verde Islands) in 1951 and 1995, at Kilauea in 1975, 2007 and 2018, at Mount Etna in 2002 and 2018, and at Stromboli in 2002–2003 and 2007, after decades of quiescence (Fig. 6.6; Day et al. 1999; Amelung and Day 2002; Neri et al. 2009; Tibaldi

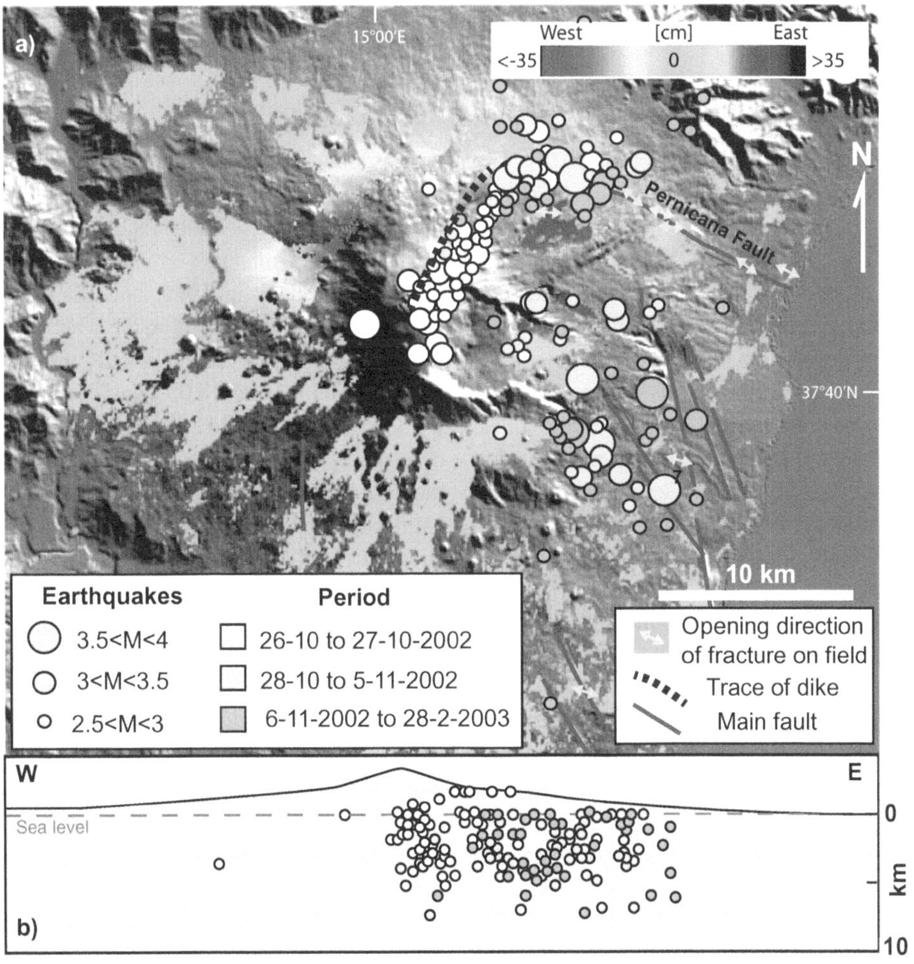

Fig. 6.6 **a** Dike-induced flank slip at Mount Etna (Italy) in 2002. The emplacement of the dike (purple dashed line) along the NE Rift on 27 October 2002 induced widespread flank instability on the eastern flank (shown by InSAR data: E-W component), accompanied by seismicity (circles proportional to magnitude M) and surface fracturing (opening direction shown by double yellow arrows) along main faults (red lines). **b** E-W section view of the volcano across the summit, highlighting the seismicity projected along the section focusing on the eastern unstable flank, shallower than 5 km below sea level

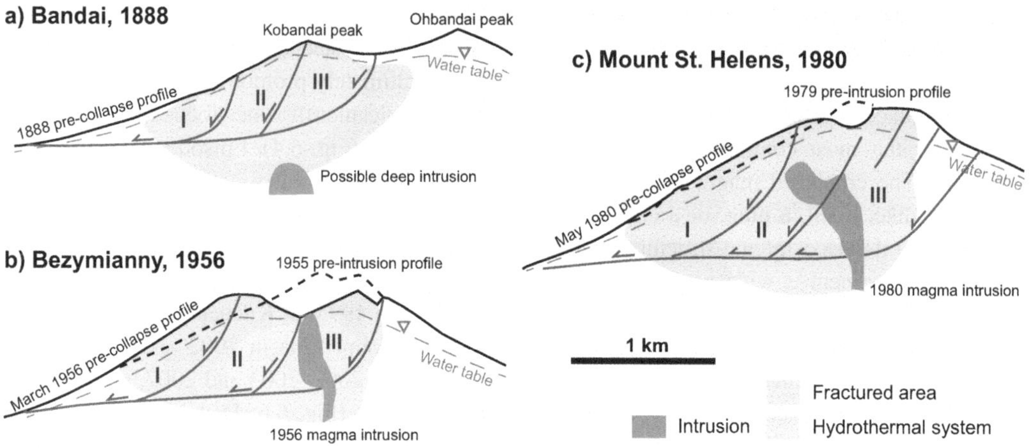

Fig. 6.7 Section view reconstructions of the sector collapses at Bandai (Japan, in 1888), Bezymianny (Kamchatka, Russia, in 1956) and Mount St. Helens (Washington, USA, in 1980) parallel to the direction of the collapse, highlighting the possible relationships between the pre- and post-collapse topography, the collapsed blocks (consecutively sliding from I to III) and the intrusions (after Voight and Elsworth 1997)

et al. 2009; Montgomery-Brown et al. 2011; Chen et al. 2019; De Novellis et al. 2019). The destabilizing role of dikes emplaced along volcanic rift zones, and more generally within the volcano flanks, includes mechanical and thermal fluid pressures along the basal detachment and magmastatic pressures at the dike interface. These may develop shallow-seated block instabilities for dike thicknesses of ~ 1 m and horizontal lengths higher than 1 km. For larger dikes, the destabilizing influence of the magmastatic column is significant and excess pore fluid pressures may not be necessary to initiate failure. The potentially destabilized block geometry changes from a shallow sliver, in the case of shorter dikes, to a deeper detachment, for longer dikes (Elsworth and Day 1999).

While the destabilizing nature of dikes is well acknowledged, there is also evidence of dike-induced stabilization of volcano flanks, as inferred at Kilauea in 1983, when a dike emplaced along the East Rift Zone dropped the flank earthquakes and slowed the downslope motion. In this case, the deflation of the reservoir accompanying lateral dike intrusion reduced the magma-reservoir pressure and the compressional load on the upper flank over a much larger area than intrusion of the dike added to it (Delaney and Denliger 1999).

Shallow viscous intrusions, or cryptodomes, usually andesitic to dacitic in composition, also propagate in volcanic edifices and destabilize their flanks, as at several felsic stratovolcanoes, including Bezymianny (Kamchatka) in 1956, Mount St. Helens in 1980, and Soufriere Hills (Montserrat) in 1997 (Fig. 6.7). The volume of these cryptodomes, accumulated within a few months, was between 0.03 (Soufriere Hills) and 0.2 km^3 (Mount St. Helens). The induced surface deformation, with uplift reaching 100–150 m, increased the instability of the edifices through oversteepening and destruction of rock mass strength, thus contributing to the collapse. In these three cases the resulting collapses suddenly depressurized the shallow magma reservoirs, developing lateral blasts and pyroclastic density currents. A deeper cryptodome, at the base of the volcanic edifice, may still produce a sector collapse without necessarily resulting in lateral eruptive blast, as for Bandai-san (Japan) in 1888, Harimkotan (Kuriles) in 1933 and Sheveluch (Kamchatka) in 1964 (Fig. 6.7; Belousov et al. 2007). The ascent path of a viscous intrusion within a volcano is controlled by any irregular topography of the edifice, as for example the presence of a previous collapse scarp, which promotes an asymmetric load deflecting the intrusion towards the least buttressed flank (see

Chap. 7). The deflected intrusion induces, in turn, a local stress field creating a curved major fault along its trajectory, responsible for an apical asymmetric graben and a lateral bulge (Donnadieu and Merle 1998; Acocella 2005).

A common effect of magma-induced sector collapse is the depressurization of the magmatic reservoir. If magma is very shallow (<1 km from the surface), the collapse-induced depressurization may increase magma degassing, triggering eruption. If magma is at greater depths, the depressurization may influence the magma plumbing system, erupting higher proportions of dense and less evolved magmas. Significant stress perturbations, with decompression up to a few megapascals, may act on shallow and large magma reservoirs underlying collapsed volcanoes. These decompression values are similar to the magmatic overpressures associated with eruptions, confirming that sector collapse may enhance or trigger volcanic activity (Fig. 6.8; Manconi et al. 2009; see also Sect. 6.6).

The possibility that volcanic activity is triggered by sector collapse, which in turn is generated by magma emplacement, highlights a feedback between magmatic and gravity processes. In this feedback, while magma emplacement prepares and promotes sector collapse, the collapse attracts successive magma intrusions, which in turn further destabilize the flank, attracting further magma. The result is the focusing of volcanic activity within or along the less stable areas of an edifice, as exemplified by the evolution of the Sciara del Fuoco collapse scarp at Stromboli, where magma is emplaced parallel to and driven by the collapse scarp, in turn promoting further collapse (see Sect. 7.6.2). This general self-sustaining mechanism between magmatic activity and gravity has been proposed to control the evolution also of other volcanoes, including Mount Etna, Kilauea, and Mauna Loa. In some cases, an originally linear volcanic rift zone responsible for sector collapse may curve and split into two distinct segments following the gravitational stress field to the rear of the collapse. This may lead to compression in the sector enclosed by the two rift zones and extension in the outer sector, developing a third volcanic rift zone, as proposed for the Anaga basaltic shield on Tenerife (Canary Islands; Fig. 6.9; Dieterich 1988; McGuire et al. 1990; Borgia 1994; Acocella 2005; Walter et al. 2005).

The emplacement of erupted products on the upper slopes of volcanic edifices may also affect

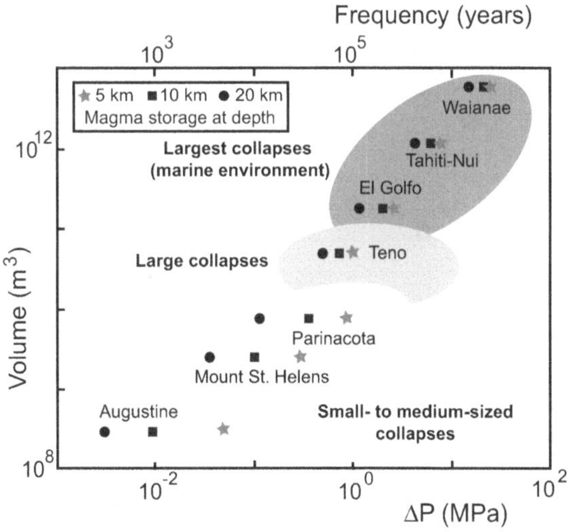

Fig. 6.8 Landslide-induced decompression calculated for various volcanic systems. Log-scale axes show decompression ΔP (bottom x-axis), recurrence time (top x-axis) and flank-collapse volume (y-axis). Decompression is calculated assuming a mechanically softer reservoir at three different depths beneath edifices: 5 km (star); 10 km (square); 20 km (circle; modified after Manconi et al. 2009)

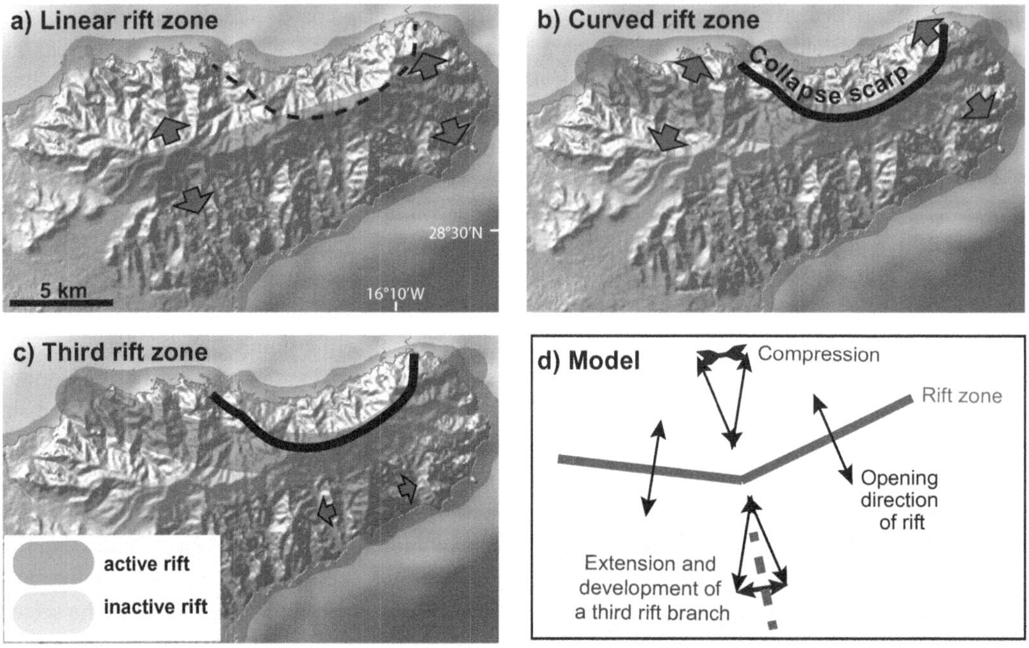

Fig. 6.9 Possible evolution of a volcanic rift zone following sector collapse, as inferred for the eroded Anaga basaltic shield on Tenerife (Canary Islands; Walter et al. 2005; DEM provided by GeoMapApp). **a** Incipient collapse (dashed line) develops to the side of a linear volcanic rift zone opening as shown by red arrows. **b** Sector collapse (black line) attracts dikes at its rear, curving the rift zone. **c** This process creates local extension in the outer zone between the curved branches, developing a third rift branch. **d** General model showing the opening of a third branch (red dashed line), indicated by the vector (purple double arrow), resulting from the joint opening (black double arrows) of two rift zones (red lines) at the rear of a collapse (not shown)

flank instability. In fact, these may oversteepen a volcano flank, increasing its instability; this process is most common at steep stratovolcanoes with open conduit, where loose deposits repeatedly accumulate, as for example Stromboli and Monowai.

6.4.2 Hydrothermal Alteration

Hydrothermal Alteration Consists of the chemical replacement of the original minerals in a rock by new minerals, where a hydrothermal fluid, usually weakly acidic, delivers the chemical reactants and removes the aqueous reaction products. This process requires an open system and may involve major textural, mineralogical and/or chemical changes. Hydrothermal alteration varies widely, also among individual deposits. Several ubiquitous types of alteration assemblages may create distinct groups of alteration effects, textures and mineral assemblages. The most important controlling parameters are the host rock composition, grain size, physical conditions and permeability, as well as the fluid chemistry, pH, Eh, and temperature and pressure conditions. Hydrothermal fluids are most effectively discharged along fracture zones within volcanoes, highlighting the control of the structure of a volcanic edifice on flank instability. In fact, continued hydrothermal activity reduces the strength of the rock, especially due to clay alteration and increased pore pressures, with time scales between a very few hundred of years to a few thousands of years (Lopez and Williams 1993; Day 1996). Effects may vary widely, depending upon the above mentioned boundary conditions. While at Mount Rainier and Mount Hood (Cascades Arc, western North America) rock strength differences greatly in excess of one order of magnitude were obtained from fresh and completely hydrothermally altered volcanic rock, at Teide (Tenerife, Canary Islands) the type of alteration appears more important than the degree of alteration in controlling the strength of altered soil (Watters et al. 2000; Del Potro and Hurlimann 2009). These cases suggest that generalizations are not advisable and each situation should be considered independently.

Flank instability has been related to hydrothermal alteration at several shields and stratovolcanoes. At La Gomera and Piton des Neiges shield volcanoes, collapse resulted from the weakness of clay-rich detachments under the emplacement of intrusions providing part of the heat required for alteration (Casillas et al. 2010; Famin and Michon 2010). At Nevado del Ruiz stratovolcano (Colombia), voluminous debris avalanches totalling 26 km^3 and travelling great distances repeatedly occurred in the last 3 ka, probably resulting from the alteration due to acid sulphate waters around fault zones. Large collapses of hydrothermally altered parts of Mount Rainier stratovolcano have similarly generated far-travelled debris deposits. Low temperature hydrothermal alteration also appears a trigger for catastrophic collapse of resurgent blocks, creating debris avalanche deposits, as at Ischia (Italy), where the Green Tuff trachytic ignimbrite uplifted of ~1 km during resurgence has been significantly weakened by repeated alteration events (Lopez and Williams 1993; Reid et al. 2001; Altaner et al. 2013).

6.4.3 Excess Pore Pressures

Excess pore pressures may develop within volcanic edifices, associated with magma emplacement, hydrothermal systems and/or confined aquifers. These pressures may significantly decrease the strength of a volcanic edifice, especially along basal detachments, preparing or triggering failure.

A hydrostatic pressure in pore water develops if the pores are interconnected and communicate with the Earth's surface. This pressure at any depth is equal in all directions and is related to the weight of the water column from this depth to the surface, with the pore fluid pressure increasing with depth. Considering the mean densities of the pore fluid (1000 kg/m^3) and of the rock column (2500–3000 kg/m^3), the hydrostatic pore pressure at any depth is 0.25–0.3 times the lithostatic pressure, which derives from the load of the rock column: this relation is expressed by the pore-fluid factor λ, or ratio between the fluid and lithostatic pressures (Zang and Stephansson 2010). The pore pressure can vary widely in the crust, so that for dry rocks $\lambda = 0$, whereas for hydrostatically pressured sediments $\lambda = 0.4$. Pore fluids not connected to the surface (confined) may generate overpressure, or pore pressure greater than hydrostatic (i.e., $\lambda > 0.4$). Excess pore pressures in volcanoes commonly result from the expansion of high temperatures fluids and from the generation of fluids from magmatic intrusions much faster than these can escape. Under these circumstances, fluids can considerably reduce the effective brittle strength of rocks. Fluid-triggered crack propagation likely assists emplacement of magmatic sheets, as dikes and sills.

In a hydrostatic stress state, the fluid pressure opposes the lithostatic stress, supporting some of the load that would otherwise be supported by the rock matrix. Recalling Eq. 2.16, the pore pressure P_f is a component of the normal stress σ_n which acts against the rock to give the effective normal stress σ' (Terzaghi 1936):

$$\sigma' = \sigma_n - P_f \tag{6.1}$$

Pore pressures allow a rock to behave as if the confining pressures were lowered by an amount equal to P_f. This is shown in the Mohr diagram of Fig. 2.17, where a "dry" circle represents the state of stress in a stable rock with zero pore pressure, where the effective normal stress σ'_1 equals the total normal stress σ_1. If the pore pressure is increased gradually the effective normal stresses σ'_1 and σ'_3 become smaller than the maximum and minimum principal stresses σ_1 and σ_3, respectively, but the differential stress remains constant, so that:

$$\sigma'_1 - \sigma'_3 = \sigma_1 - \sigma_3 \tag{6.2}$$

In other words, the "dry" circle keeps the same size, but translates horizontally towards

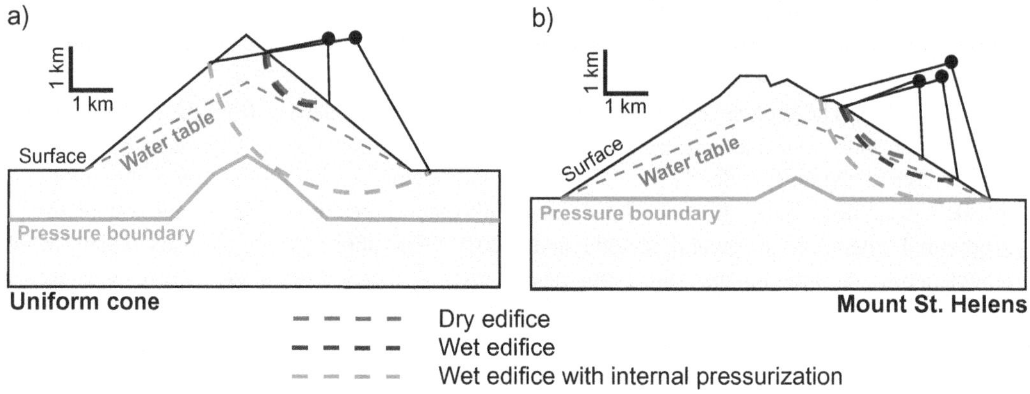

Fig. 6.10 Potential critical failure surfaces for a uniform cone (**a**) and an edifice with pre-1980 Mount St. Helens-like topography (**b**). Potential failure surfaces for a "dry" edifice, a "wet" edifice and a "wet" edifice plus internal gas pressurization are shown (modified after Thomas et al. 2004; pressure boundary refers to source area of internal fluid pressures)

lower compressive stresses, closer to the failure envelope. The amount of translation is determined by the fluid pressure P_f. If P_f is large enough, the circle will hit the Mohr envelope and failure is promoted. Pore pressures thus may allow failure even when the shear stresses are too small for faulting in the dry rock. Moreover, by simply increasing the fluid pressure under very low differential stresses, the outward push of the fluid creates a tensile stress sufficient to cause crack propagation, resulting in **hydraulic fracturing**, a process in which the rock is fractured by the pressurized fluid. In synthesis, increased pore pressure dampens down the effect of confining pressure, which strengthens a fault as it increases with depth. Faults with favourable orientation for slip or dilation constitute potential fluid flow pathways.

Excess pore pressures within volcanoes are often transiently triggered by magmatic intrusions, most commonly by dikes. The increase in pore pressure may occur through different processes. An intrusion may mechanically increase the pore pressures in the surrounding porous elastic media, a behaviour approximated by a moving volumetric dislocation, whose destabilizing effect depends upon the intrusion rate and width: the potential for slide initiation is not definitive. The intrusion may also induce thermal expansion of an aquifer, with long-distance lateral excess pore pressure transmission capable of developing large uplift forces beneath a potentially unstable block: in this case, the potential of failure is relevant when a dike acts as a feeder for several weeks. Finally, the intrusion may also promote, through its cooling or crystallization, retrograde boiling, or pressurized volatile separation from a residual liquid increasingly rich in gas, possibly leading to gas exsolution. If the intrusion is shallow, retrograde boiling may induce hydraulic fracturing, especially in weak porous layers, and/or steam drive pressure transmission in adjacent aquifers (Voight and Elsworth 1997).

Numerical models of heat and groundwater flow confirm that heating from magma intrusion may destabilize the core of an edifice, resulting in deep-seated collapse migrating to deeper levels with increasing internal pressure (Fig. 6.10; Reid 2004; Thomas et al. 2004).

There is widespread evidence that excess pore pressures play an important role in the destabilization and failure of volcanic edifices, although they may not necessarily be the only controlling factor, as at the Mount St Helens, Bezymianny and Bandai catastrophic sector collapses, or at the ~7.5 ka Valle del Bove scarp at Mount Etna (Voight and Elsworth 1997; Deeming et al. 2010). The sensitivity of pore pressures to perturbation by magmatic processes is highly dependent on the host rock permeability, which may itself be subject to rapid changes by fracturing, faulting and other processes. Once structural failure has initiated, water may also determine the behaviour of the collapsing mass, in particular whether the event is aborted or forms a debris avalanche. For example, the lack of pressurised fluids may have caused fault movement to cease, without debris avalanche generation at La Palma and El Hierro (Canary Islands). This possibility may allow reinterpreting large flank slumps, as at Kilauea, as aborted collapse events which might have generated debris avalanches in presence of high pore pressures either in the fault zones or within the slump blocks (McGuire 1996).

6.4.4 Weak Layers and Basement (Volcano Spreading)

The strength of a volcanic edifice and its capability to resist failure also depend on the presence of any weak layer, within the edifice or at its base.

Different types of eruptive products overlap within a volcanic edifice; these include thick solid lavas, loose pyroclastic deposits, ash layers and scoria deposits with variable welding. Intermittent volcanic activity may also allow sedimentation of non-volcanic deposits, as conglomerates, sands and clays with variable water content. As a result, any polygenic edifice includes layers of different nature, thickness and extent, whose strength may vary of several orders of magnitude. The strength of these layers may be also affected by different post-emplacement processes, as strain or thermal alteration and repeated cycles of inflation and deflation of the edifice. Flank failure on unstable volcanoes typically focuses along the weakest layers, consistently with their orientation with regard to the ideal failure plane. For example, weak paleosoils and pre-existing canyons created by erosion may have initiated large sector collapse on Tenerife. The weakening of the Mount St. Helens edifice in 1980, related to shear strain focused in pervasively shattered older dome rocks, could have provoked collapse even without transiently elevated pore-fluid pressures or earthquake ground shaking. Cyclic stressing by repeated inflation and deflation may also have affected the stability of the edifice, resulting in lower failure strength. A detailed knowledge of the mechanical stratigraphy of a volcano is therefore fundamental in assessing volcano stability and improving risk mitigation (Hurlimann et al. 1999; Watters et al. 2000; Reid et al. 2010; Kendrick et al. 2013).

A volcanic edifice may also experience flank instability and collapse as a result of a weak basement, a process often referred to as volcano spreading. **Volcano spreading** describes flank instability under a weak basement at the edifice base: this often consists of clays or evaporites showing ductile behaviour, but in some cases even loose pyroclastic deposits may play a role (Borgia et al. 1992; Goto et al. 2019). With a sufficiently large mass and magma influx, the volcanic edifice cannot be properly sustained by the flowing basal weak layer and spreads. A spreading volcano may grow through five main phases, which may overlap or repeat: building, compressing, thrusting, intruding and spreading (Borgia 1994). During the initial building phase, the volcano mass does not affect the stress field of the basement. However, as the mass increases, the volcano subsides, undergoing an overall contraction that inhibits further magma intrusion. When the stress acting at the volcano base overcomes the shear strength of the basal weak layer, thrusting begins on the detachment along the weak layer. Tangential thrust structures and folds develop at the outer base of the edifice, above and within the weak layer (Fig. 6.11). While radial compression dominates along the periphery of the edifice, the central part undergoes circumferential extension, allowing magma intrusion and developing a basal intrusive complex. This extension, with circumferential minimum principal stress σ_3, develops a set of radial grabens with arcuate normal faults in map view and a cusp termination towards the outer compressed zone. Such a spreading extends the upper flanks, contracting the lower slopes. The dimensionless parameters controlling this behaviour are the ratio of the height of the volcano with regard to its radius, or Π_1, and the ratio between the thickness of the brittle portion of the basement and that of the weaker viscous

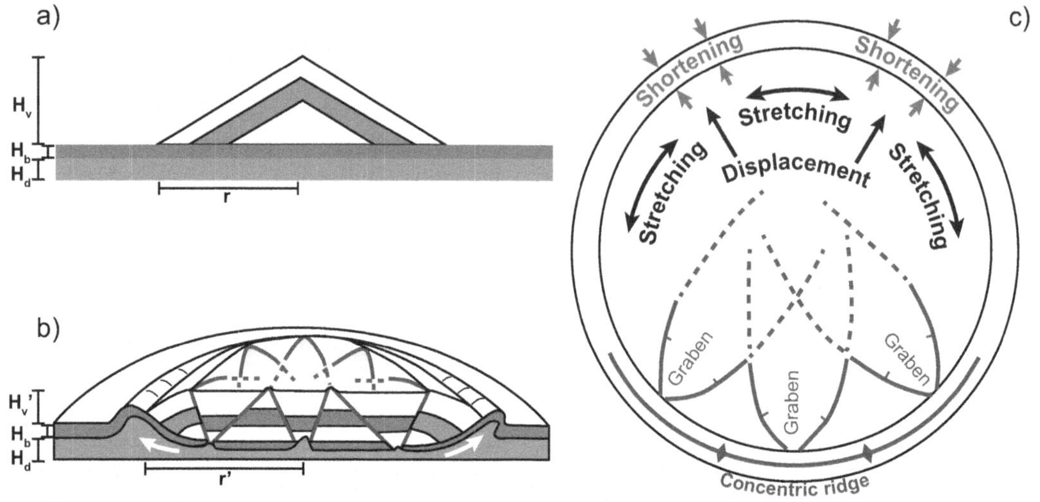

Fig. 6.11 **a** Section view of an undeformed cone of height H_v and radius r resting on a crustal rigid layer of thickness H_b (light brown) and an underlying weak viscous layer of thickness H_d (orange). **b** Peripheral section view of the cone after spreading, showing a decrease of the height H_v', increase of the radius r', outer radial grabens and, along the periphery, outer contraction above outward flow of the ductile layer. **c** Map view of the structure of the deformed cone (lower half) and associated structural conditions (upper half; modified after Merle and Borgia 1996)

substratum below it, or Π_2. These parameters define the possible non-spreading and spreading domains for volcanoes, the latter associated with higher volcanoes on a thicker viscous layer (Fig. 6.12a; Merle and Borgia 1996). Experiments suggest that the decrease in height of a volcano with time follows an exponential decay and is balanced by an exponential increase in the volcano radius. In this process, the onset of the development of the radial grabens precedes that of basal folding (Fig. 6.12b).

The viscosity of the weak layer also plays an important role, as lower viscosities lead to faster rates of deformation (e.g., van Wyk deVries and Matela 1998). For lower viscosities (10^{15} Pa s), the weak layer cannot sustain the erupted volcanics long enough (>10^3 years) for an edifice to grow. For higher viscosities (10^{19} Pa s), the time of deformation becomes larger (>10^6 years) than the lifespan of volcanoes. The characteristic time for the occurrence of spreading t is given by:

$$t = \frac{3c\eta}{\rho g H_v}\left(\frac{r}{H_d}\right)^2 \tag{6.3}$$

where c is a constant (with $c = 1$ or $c = 4$, depending on the motion of the brittle layer with regard to that of the viscous layer), η is the viscosity of the ductile layer, r is the radius of the volcano of density ρ and height H_v, g is the acceleration due to gravity and H_d is the thickness of the weak layer (Borgia et al. 2000, 2005). Estimated times to develop spreading vary between 10^3 and 10^5 years: within this time frame, spreading may occur with extremely variable rate, from continuous creep-like movement to catastrophic failure.

The mature stage of spreading allows magma accumulation within the edifice, promoting volcanism through rift zones. If the volcano is spreading radially, it enlarges its circumference and the resulting volcanic rift zones are radial. However, most spreading volcanoes are not uniformly confined to their sides, showing a preferred spreading direction. As a result, extension within the edifice may be uneven, focusing in one or two rift zones perpendicular to the main spreading direction. Therefore, volcano spreading may also manifest without radially oriented and equally developed graben zones,

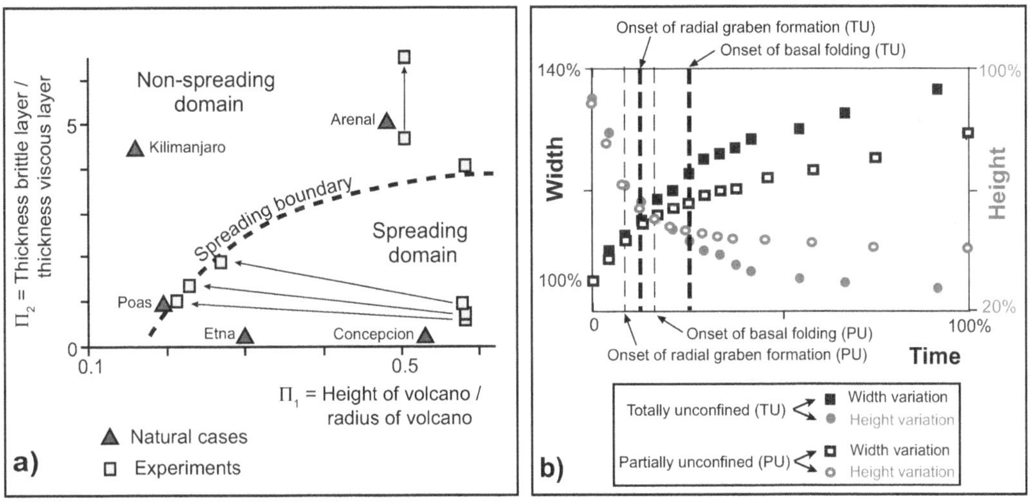

Fig. 6.12 a Boundary conditions between spreading and non-spreading volcanoes, based on Π_1 (height/radius of volcano) and Π_2 (thickness of brittle layer/thickness of viscous layer) in nature (red triangles) and experiments (yellow squares, evolving accordingly to the arrows; (Merle and Borgia 1996), b) Variation of dimensionless height and width of spreading cones as a function of time, in totally unconfined (TU, cone spreading radially in all directions) and partially unconfined experiments (PU, cone buttressed at three out of four sides)

becoming directional and being accommodated only by fewer and directional rift zones. This behaviour implies a causal relationship between the main spreading direction of a volcano and the perpendicular strike of its main rift zone(s). This relationship holds also for multiple juxtaposing spreading volcanic edifices. For example, partly overlapping volcanoes of different age and spreading at different rates may form a rift zone parallel to their boundary, as suggested for Mauna Loa and Kilauea, at Hawaii. Also, spreading edifices of similar age that partly overlap tend to develop a rift zone approximately perpendicular to their boundary, as suggested for Piton des Neiges and Piton de la Fournaise, at La Reunion Island (Walter et al. 2006).

The spreading model, initially proposed for large basaltic edifices, as Mount Etna and the Kilauea-Mauna Loa system, has been applied also to Piton de La Fournaise, several strato- and composite volcanoes, including Concepcion, Maderas and Mombacho (Nicaragua), Jacotitlan (Mexico), Socompa (Central Andes), Ardjuno, (Indonesia), Mt. Cameroon (Cameroon), Usu (Japan), and even to Martian volcanoes (Borgia et al. 1992; Borgia 1994; Merle and Borgia 1996; Borgia et al. 2000; van Wyk deVries et al. 2001; Borgia and van Wyk deVries 2003; Dufresne et al. 2010; Mathieu et al. 2011; Goto et al. 2019). For example, at Mount Etna spreading focuses on the least buttressed side of the volcano, that is the eastern flank facing the Ionian Sea, while peripheral basal compression dominates where the buttressing is higher. The 1.6 km high stratovolcano of Concepcion rises above one kilometre thick Quaternary lacustrine mudstones and is surrounded by a 20 km wide circular topographic rise, created by thrust-related folds and mud diapirs; its evolution well meets the above mentioned five evolutionary spreading stages. Socompa stratovolcano rises more than 2.1 km over its basement, with prominent sector collapse on its northwest flank resulting from the gravitational spreading of the basement; about 80% of the produced avalanche deposit consists of substrata formerly residing under the volcano (Froger et al. 2001; van Wyk deVries et al. 2001; Borgia and van Wyk deVries 2003; Solaro et al. 2010).

6.4.5 Fault Activation and Earthquakes

Active fault zones in the volcano basement (a) or within the volcanic edifice (b) may promote flank instability. As for case (a), the destabilization of a volcanic edifice usually results from the transient **dynamic stresses** imposed by the passage of the seismic waves generated by an earthquake activated by local or regional faults below the volcano or in the crust nearby. These stresses are responsible for the shaking of the edifice (see also Sect. 9.3.3). A few cases of this type of earthquake-induced collapse have been observed in the last decades. Among these is the above mentioned example of Mount St. Helens in May 1980, when a shallow volcanic earthquake triggered the collapse of the north flank of the edifice, which in turn suddenly decompressed the magma below and started the lateral eruptive blast (see Sect. 6.8.2). A similar condition was met at Krakatau (Indonesia) in 2018. Here two minutes after a small earthquake (magnitude between 2 and 3), seismic signals recorded the collapse of the volcano's flank. The collapse decapitated the volcanic edifice, increased its eruptive activity and triggered a tsunami that caused 430 fatalities. Less dramatic effects have occurred on the flanks of Arenal volcano (Costa Rica), which experienced repeated rockfalls and shallow (5–11 m thick) landslides between 2011 and 2013. Here the nearby 2012 Nicoya earthquake (*M7.4*) had no measurable impact on the velocities of the sliding units, but did result in an increase in the area affected by rockfall (Ebmeier et al. 2014; Walter et al. 2019).

As for case (b), the frequent spatial coincidence between fault traces and collapse scarps suggests that several collapses occurring in the past may have been induced or encouraged by the activity of faults within the volcanic edifice. Examples include the stratovolcanoes of Zampoala (Mexico), with several collapses partly coinciding with normal fault systems, and Ollague (Central Andes), where three main sector collapses developed along a normal fault system affecting the basement, with a migration of volcanism towards the collapses (Arce et al. 2008; Vezzoli et al. 2008). Quite similar cases include Mount Etna in 2002, when a shallow *M3.7* earthquake in September activated part of the Pernicana Fault, which confines the unstable flank to the northeast, promoting the October dike intrusion and fissure eruption along the nearby NE Rift and accelerated the slip of the eastern flank (Fig. 6.6). At a much smaller scale, in June 1997 a collapse removed the top of the dome growing at Soufriere Hills volcano, Montserrat. The collapse has been occurring on a three-days long culmination of earthquake swarms below the dome (Acocella et al. 2003; Green and Neuberg 2006).

In some cases, fault activity inducing or accompanying flank instability is associated with **slow slip events**, or long-lived slip along portions of faults with duration of days to years, geodetically observed at the surface and sometimes correlated with small bursts of coherent seismic energy. Eight slow slip events occurred at Kilauea between 1997–2007, with recurrence time of 0.9 years and equivalent magnitude between 5.3 and 6. These events focused along the basal detachment, between the area of unstable sliding near the rift zone and the distal portion with stable sliding, although some events also occurred within the Hilina-Pali normal fault system, accompanying the instability of the flank south of the East Rift. These episodes are not limited to secular fault creep movements of the southern flank, but also occur as a response of the emplacement of dikes along the rifts, as observed in June 2007 (Cervelli et al. 2002; Brooks et al. 2008; Montgomery-Brown et al. 2009).

6.4.6 Weather and Climate

Weather and climate may provide important preparing factors for flank instability and collapse at volcanoes, although detailed causal relationships require further investigation.

As far as the weather influence is concerned, periods of high rainfall usually correspond to increased sub-aerial slope failures, as suggested for the Eastern Mexican Volcanic Belt and the southeast coast of Hawaii (Iverson 1995;

Carrasco-Nunez et al. 2006). In particular, a recent study highlighted that before and during the 2018 eruption and collapse at Kilauea (see Sect. 6.7 for details on the event), infiltration of rainfall increased pore pressure at depths of 1–3 km by 0.1–1 kPa, to its highest pressure in almost 50 years. This suggests that weakening and mechanical failure of the edifice was driven by changes in pore pressure within the rift zone, prompting dike intrusion and ultimately facilitating the eruption. Such a precipitation-induced eruption trigger is consistent with the lack of precursory summit inflation. Moreover, statistical analysis of historic eruption occurrence suggests that rainfall patterns contribute substantially to the timing and frequency of Kilauea's activity, with more than 60% of intrusions from 1975 associated with periods of rainfall-induced pressure perturbation (Farquharson and Amelung 2020).

Climate may indirectly promote landslides in oceanic environments through precipitations and changes in sea level closely coupled with glaciation–deglaciation cycles. These regional and global effects may explain the general triggering of large landslides from oceanic islands and continental margins (Keating and McGuire 2004). Indeed, a link between variations in sea level and instability and collapse at coastal and island volcanoes is suggested by correlating the intensity of volcanic activity in marine settings with the Late Quaternary rate of sea level change. Late Quaternary sea level variations may cause significant stress changes along continental margins and island arcs and are also likely to have triggered increased levels of seismicity and perhaps volcanism (McGuire et al. 1997). For example, post-shield landslides on Tenerife correlate with major falls in sea level, which increase coastal erosion and reduce support of the volcanic flanks. However, there is still little evidence to suggest that slump activity is regularly related to episodes of sea level low stand. Also, while evidence from the Canary Islands and Madeira Abyssal Plain may support a relationship between climate change and ocean island and continental margin landslides, evidence from the Hawaiian archipelago is not supportive. Therefore, the existence and nature of a climate change link to volcano flank instability remains to be firmly established (Keating and McGuire 2004).

6.5 Analysis of Flank Instability

Studies of flank instability at volcanoes are conducted to define the conditions promoting failure in a hazard assessment perspective that should allow forecasting impending failures. To this aim, several actions should be considered (Voight 2000). (a) Examination of the geological and historical record of the site, which may have also experienced previous failures. (b) Definition of the distribution and properties of the rocks and of the discontinuities involved, accompanied by a geomechanical analysis of the related samples, and analytical or numerical models. (c) Establishment of a monitoring network (e.g. displacement, tilt, seismic strong motions, propagation and widening of visual cracks, etc.), which may allow detecting accelerations of the instability that may act as precursors of collapse.

As regards point (a), a detailed field survey is the pre-requisite to define the extent of the unstable area, its borders, the involved rocks and the kinematics observed at the surface, as given by displaced or tilted geological and anthropic (including buildings, roads, fences, walls) markers. In addition to the geometric and kinematic features, the field survey should also provide information on the possible causes (dynamics) of the instability, even if at this stage this evaluation may not be definitive. This knowledge may be integrated with any available historical information, including reports, chronicles or maps. Sampling of representative rocks may be useful for a successive geomechanical characterization. The goal of this first approach is to gain the fundamental field knowledge of the unstable flank, which may then provide the base for the successive mechanical analysis.

As regards point (b), it concerns the mechanical and modelling analysis of stability of the flank, partly based on the results from (a). In mechanical terms, flank instability can be analysed considering the forces acting along the sliding surface (Elsworth and Voight 1995; Voight 2000). The total force driving outward movement, F_D, is opposed by the resisting shear force, F_R, mobilized along a potential basal slide surface and lateral boundaries. F_R represents the integrated value of mobilized shear resistance, τ, over the area of failure. The shear resistance is commonly defined in terms of effective stress σ', cohesion c' and friction angle ϕ', whereby

$$\tau = c' + \sigma' \tan\phi' \tag{6.4}$$

and the primes signify parameter correspondence to effective stresses (see Eq. 6.1). In more detail, the total force driving outward movement, F_D, and that resisting it, F_R, result from the sum of several components associated with the specific conditions acting on the unstable wedge. For the case of an emplaced dike inducing flank instability, the forces acting on a wedge to the side of the dike are summarized in Fig. 6.13 (Elsworth and Voight 1995; Voight 1997). These are: the weight force M_p acting on the block, which takes into account for the porosity of the rock and the presence of water; the magmatic forces F_m due to the emplacement of the dike, made up of the magmastatic force (resulting from the lateral pressure of the magmatic column) and the overpressure (resulting from the magma overpressure driving the magma conduit flow); the total uplift force F_p, due to the mechanical F_{pm} and thermal F_{pt} pore fluid pressures triggered by the intrusion; the force applied by the seawater load F_s on any submerged part of the block; the uplift force F_{ps} due to static groundwater pressure; the friction along the basal failure plane and the force along the sides of the wedge F_l; the latter is due to the effective lithostatic pressure given by the ratio of vertical effective stress to horizontal effective stress. Considering δ as the inclination of basal failure surface and θ as the inclination of the failing block (Fig. 6.13), the destabilizing force components parallel to the shear surface F_D are:

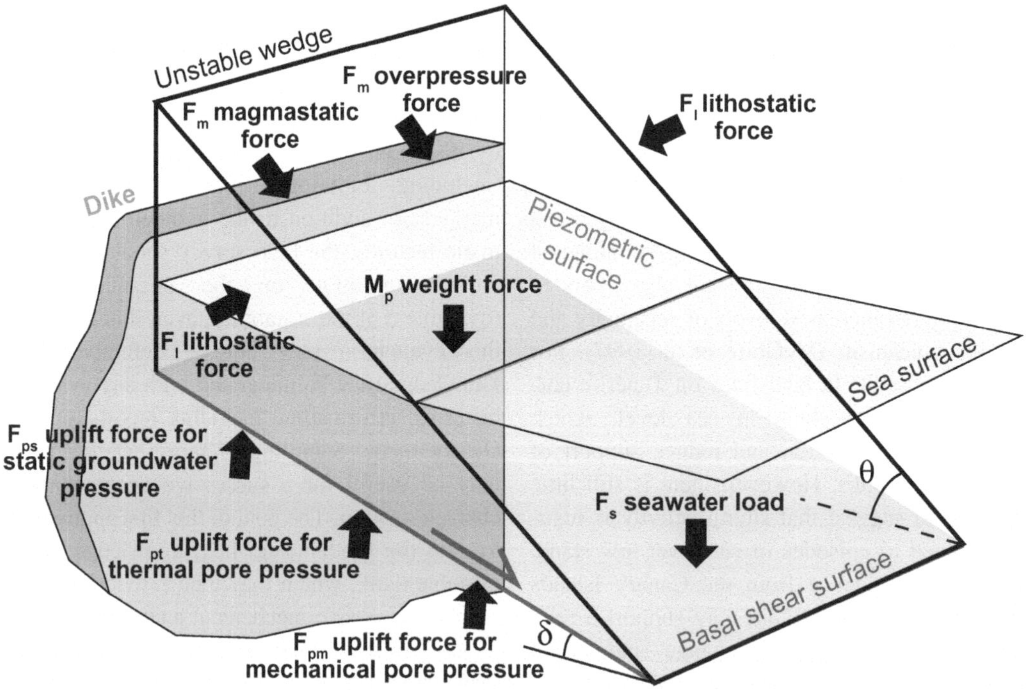

Fig. 6.13 Summary of the forces acting on an unstable wedge to the side of an emplaced dike (see text for details; modified after Elsworth and Voight 1995)

$$F_D = -M_p \sin\delta + F_m \cos\delta - F_s \sin(\theta + \delta) \quad (6.5)$$

These are countered by the resisting force F_R proportional to forces normal to the possible shear surface and to the sides of the wedge (stabilizing forces), given by:

$$F_R = M_p \cos\delta + F_m \sin\delta - F_{ps} - F_{pm} - F_{pt} + F_l + F_s \cos(\theta + \delta) \quad (6.6)$$

A normalized **factor of safety** $Fsa/\tan\phi$ represents the stability behaviour for a variety of undefined frictional resistance magnitudes (Voight and Elsworth 1997). This factor refers to the ratio of forces acting to resist failure to those destabilizing the block, as:

$$Fsa/\tan\phi = F_R/F_D \quad (6.7)$$

Normalized factors of safety in the range 1–2 represent the limiting equilibrium condition for angles of frictional resistance in the range 45°–27°, respectively. For a slope that has remained stable throughout a volcanic/seismic crisis, it is usually impossible to calculate a reliable value of *Fsa* other than to note that it exceeds unity.

Among the main computational procedures commonly used to evaluate slope stability in two- or three-dimensions is the Limit Equilibrium Analysis (LEA; Elsworth and Voight 1995; Voight 2000; Battaglia et al. 2011). This is a method that satisfies most or all conditions of equilibrium with a large number of trial circular or non-circular slip surfaces to be analysed, making it possible to identify critical slip surfaces with a relatively high degree of reliability. In a LEA a slip mechanism is postulated and the shearing resistance required to equilibrate the slip mechanism is calculated. This resistance is compared to the available shear strength in terms of factor of safety *Fsa*. The same procedure is then used for other postulated mechanisms, until the lowest *Fsa* is found by iteration.

The input data required for the mechanical analysis may be obtained through different approaches, including in situ and laboratory studies. In situ analysis includes the evaluation of the Geological Strength Index, for an overall characterization of the rock mass. This index is based upon the general visual impression of rock structures, in terms of blockiness and geological complexity, and of the surface conditions of the discontinuities expressed by roughness and weathering (Hoek et al. 1998, and references therein; Apuani et al. 2005). Laboratory studies are based on the collection of samples of intact rock and the related measurement of different physical properties, including bulk volume, void index and porosity, point load strength index, uniaxial compressive strength, elastic moduli and diametral compressive strength. Many components involved in these analyses may be easily estimated (e.g., the block weight), whereas others may be difficult or impossible to measure reliably (e.g. rock-mass strength, properties of magma). In this case, it is advisable to place some limiting bounds to provide a more careful examination through sensitivity analyses, which allow evaluating the weight of the uncertainties of the less defined parameters.

This mechanical analysis is adequate mainly to define the parameters controlling the instability and understand their possible relative importance. However, it is not considered adequate to enable the reliable and accurate computation of the factor of safety of a given edifice or slope, as complete information on distributions and properties of materials and discontinuities cannot be reliably obtained. In fact, while reasonable estimates may be given, these may not be reliable for a realistic hazard assessment and forecast. Therefore, in case of concern for edifice or flank stability, real-time observational monitoring, that is the point (c) mentioned above, becomes essential for prompt hazard assessment. Once a monitoring network is established, the early signs of instability or its acceleration, on which any hazard evaluation has to rely, may include cracking near the crest of the slope and outward bulging near the toe. These signs and the monitored evolution of the deformation field are more important than any theoretical analysis that is necessarily based on assumed or laboratory measured values (Voight and Elsworth 1997; Voight 2000; Morgan and McGovern 2005). For details on the approaches used to monitor volcanoes, including flank instability, see Chap. 8.

6.6 Sector Collapse and Eruptive Activity

Flank instability may not have direct connection with eruptions; however, sector collapse has a much stronger tie with eruptive activity. The possibility that volcanic activity is triggered by sector collapse, which in turn is generated by magma emplacement, results in the spatial coincidence and persistence of magmatism and flank instability, as previously mentioned. In addition, there is evidence that after large sector collapse eruptive activity undergoes long-lasting and important modifications. In fact, any collapse-driven unloading of the shallow plumbing system of a volcano may lead to changes in the erupted rates and compositions. These changes may manifest not only during or immediately after the collapse, as for example observed at Mount St. Helens in 1980, but may persist for much longer time spans, affecting the post-collapse eruptive activity and growth of the edifice for thousands of years, as for example suggested for Stromboli, El Hierro and Soufriere Hills (Tibaldi 2004; Manconi et al. 2009; Cassidy et al. 2015).

The control of sector collapse on the evolution of a magmatic reservoir, including its eruptive behaviour, is more evident after large collapse events, with volume exceeding $\sim$5 km^3, which are much less frequent in the history of individual volcanoes. The rapid nature of sector collapse induced stress changes is comparable to earthquake induced magmatic stresses, but the local static stress change is far larger (Watt 2019). The sudden removal of mass promotes disequilibrium between the magma and surface load, marking major changes in the development of the magmatic reservoir. These changes can be explained by theoretical models that suggest that mass redistribution following edifice collapse can influence pressurisation and failure conditions in the shallow magma plumbing system (Pinel and Jaupart 2005; Manconi et al. 2009; Pinel and Albino 2013). This condition can affect both arc stratovolcanoes and oceanic shield volcanoes. However, in the latter case it may be even more relevant, as oceanic volcanoes show some of the largest mass-movements (and removal) on Earth, on the order of hundreds of cubic kilometres.

The changes in the magmatic reservoir may manifest in several ways. The decompression due to sector collapse generally enhances volcanic activity, although in some cases enhanced volcanic activity may induce the partial emptying of the magma chamber and ultimately promote caldera collapse, as proposed for Tenerife (Hunt et al. 2018). In addition, the decrease in the edifice load following sector collapse may affect the local stress field, influencing magma chamber growth and dike nucleation. In particular, the collapse of a flank of a volcano may induce a significant deflection of deep magma pathways in the crust, similarly to what observed for the unloading induced by calderas, favouring the formation of new eruptive centres within the collapse embayment, as suggested for Fogo volcano (Cape Verde; Fig. 6.14; Maccaferri et al. 2017). The changes induced by sector collapse may also initiate the addition of heat or volatiles to the shallow plumbing system and be responsible for compositional changes in the magma. As surface loading usually inhibits the ascent of denser mafic magmas, more mafic magma, or magma with a greater influence of mafic input, may be erupted after collapse (Pinel and Jaupart 2004). Therefore, large sector collapse may determine permanent changes in eruption frequency, magnitude, location, composition and style. This suggests that long-term variations in the activity of a volcano do not necessarily reflect changes at its source, but can be modulated directly from the surface (Watt 2019). As this behaviour is evident in both arc and oceanic volcanoes, it is expected that the impact of large-volume sector collapse on volcanic activity is largely independent of the tectonic setting and magma composition.

More in detail, several changes may progressively occur in the magmatic reservoir of a volcano in different time frames after major sector

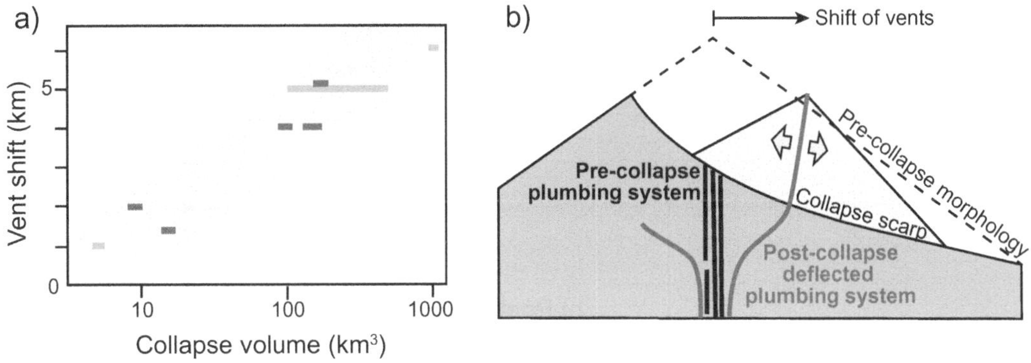

Fig. 6.14 Deflection of eruptive vents due to sector collapse. **a** Amount of shift of vent location from the centre of pre-collapse volcanism to the location of post-collapse volcanism as a function of the collapse volumes for several volcanoes. All shifts occurred towards the collapsed flank; red segments refer to volcanoes with no evidence of shallow reservoirs, while orange segments refer to volcanoes with shallow reservoirs. **b** Sketch illustrating the section view of a modified plumbing system (red) after a sector collapse, characterized by a deflection in the dike trajectories with regard to the pre-collapse plumbing system (purple). The new plumbing system is responsible for the growth of a new shifted cone. This model has been applied to Fogo, Cape Verde Islands (modified after Maccaferri et al. 2017)

collapse (Fig. 6.15; Watt 2019). The variations in the eruptive behaviour determined in the years immediately after the sector collapse may depend on the availability of eruptible magma destabilised by the surface mass redistribution. If magma is available, effusive activity is commonly observed, with more mafic composition and elevated eruption rates (Fig. 6.15b). This activity is not directly triggered by the collapse, or the magma responsible for it (as observed at Mount St. Helens). Rather, it is interpreted to result from the disruption of the shallow magma plumbing system which, following the mass redistribution induced by the collapse, allows the ascent of deeper, denser magma without being captured by more evolved upper-crustal reservoirs. Independently of any initial availability of magma, also the longer-term behaviour (up to 10^3–10^4 years after the collapse) is characterized by the ascent and eruption of mafic magma, responsible for volcano growth at elevated eruption rates (Fig. 6.15c). This condition may have occurred at several volcanoes, as for example Soufriere Hills. Here the basaltic volcanism for a few thousands of years after the collapse which occurred at ~ 130 ka was a sharp departure from the andesitic volcanism that characterized the pre-collapse activity. The basaltic magma's transit through the crust promoted by unloading was rapid (<100 years) and from mid-crustal depths (Cassidy et al. 2015). This compositional variation may be less evident in intraplate oceanic volcanoes, where the common absence of an evolved shallow reservoir results in less clear compositional shift, although here post-collapse changes in behaviour can be still explained via plumbing system disruption. On longer time spans (more than 10^3–10^4 years), the regrowth of the volcanic edifice promotes upper crustal storage and a return to more evolved compositions and pre-collapse eruptive rates, with a plumbing system spatially and temporally distinct from the pre-collapse one (Fig. 6.15d).

6.7 Multi-hazards Related to Flank Instability and Collapse

Flank instability is a primary source of hazard at volcanoes, and is also potentially responsible for a chain of consequential hazards, or **multi-hazards**, which may significantly amplify the impact of instability alone (Fig. 6.16). Flank instability may in fact be followed by sector collapse, eruption, shallow earthquakes, surface

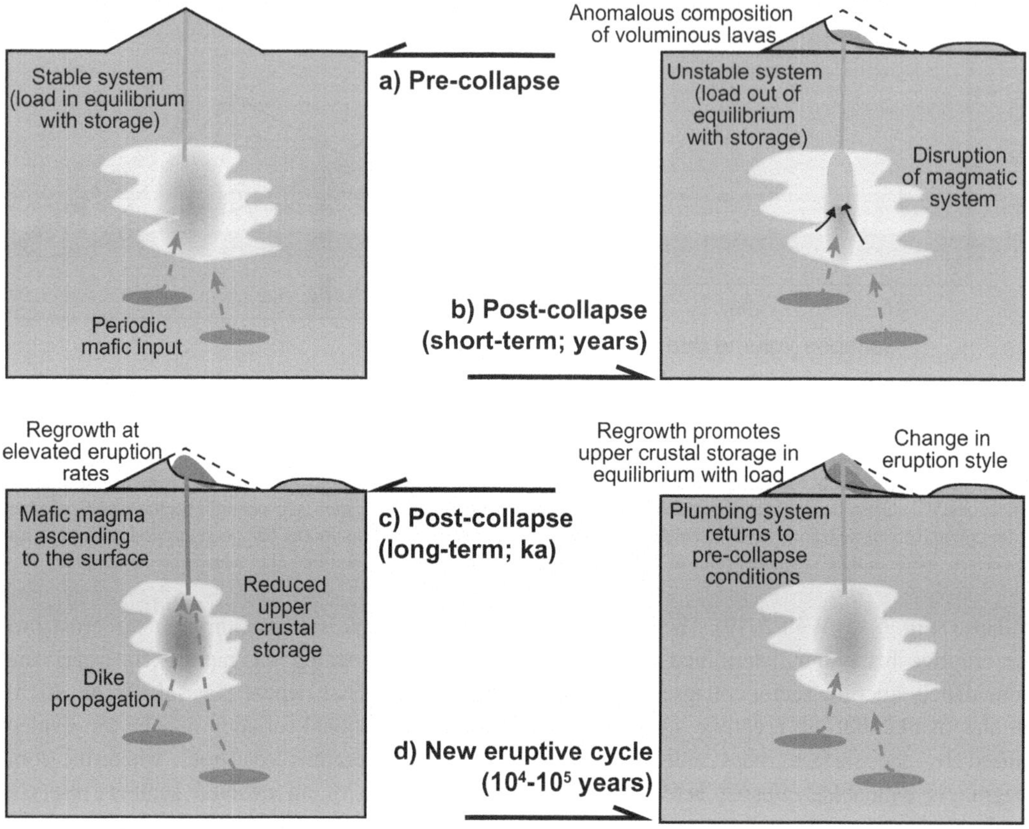

Fig. 6.15 Conceptual model summarizing the impact of large-volume sector collapse at an arc volcano. **a** Pre-collapse state, with magma reservoir in equilibrium with the surface load. **b** Short-term post-collapse behaviour depends on the presence of eruptible magma, although the magmatic reservoir becomes disrupted. **c** Cone rebuilding is promoted by subsequent ascent and eruption of deeper mafic magma. **d** Longer-term regrowth promotes upper crustal storage and a return to more evolved compositions, but with a plumbing system that is spatially and temporally distinct from the pre-collapse one (modified after Watt 2019)

fracturing, flooding (if the collapse deposits dam rivers), tsunami (in coastal areas) and even caldera formation (following voluminous eruption). These multiple and consequential hazards do not necessarily manifest all in a same event, although even a small segment of this hazardous chain may be extremely impacting. For example, the 1888 sector collapse of Ritter Island (Papua New Guinea) is the most voluminous volcanic island flank collapse in historic times. This collapse was

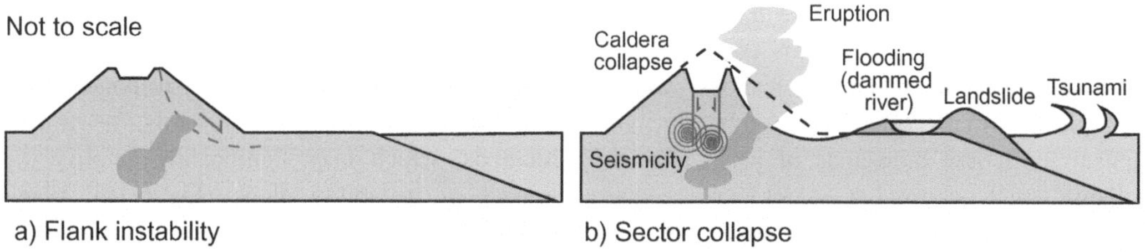

Fig. 6.16 Theoretical scheme of how sector collapse, as culmination of flank instability, may trigger different types of consequential hazards, creating a multi-hazard chain; magma shown in orange

probably induced by an earthquake or magma intrusion and, although without triggering any magmatic eruption, it generated a tsunami with run-up heights of more than 20 m on the neighbouring islands, reaching settlements 600 km away from its source (Karstens et al. 2019). In more recent times, the 1980 sector collapse at Mount St. Helens initiated a destructive eruption. This manifested as a lateral blast; however, would it not have been for the laterally directed blast, the eruptive column would have risen vertically, injecting a substantial amount of SO_2 in the stratosphere and potentially affecting global climate (Pallister et al. 1992). In addition, the magma intruded at the onset of the 2002–2003 Stromboli eruption was responsible for two contemporaneous sector collapses along the subaerial and submarine portions of the Sciara del Fuoco, which induced a local tsunami in December 2002 (Fig. 6.17; Verrucci et al. 2019). Two months earlier, the dike propagated along the NE Rift of Mount Etna promoted the slip of the eastern flank of the volcano, associated with shallow seismicity and surface fracturing and ultimately feeding a fissure eruption (Fig. 6.6; Acocella et al. 2003). In addition, as previously mentioned, the collapse of the flank of Krakatau in 2018 triggered a deadly tsunami and promoted subsequent eruptive activity (Walter et al. 2019). Another recent, well-monitored and inclusive multi-hazard case involving flank instability is offered by the 2018 eruption at Kilauea. Here from March to April 2018 a progressive increase in the level of the Halemaumau lava lake and pressure along the East Rift Zone (ERZ) were observed. On April 30, magma stored in the ERZ for decades intruded laterally along the lower ERZ, triggering eruptions in early May. On May 4, the largest *M6.9* earthquake on the island in 43 years occurred beneath Kilauea's south flank at $\sim$6 km depth, along the subhorizontal basal detachment between the volcanic pile and the seafloor: ground deformation models indicate up to $\sim$5 m of seaward fault slip. The motion induced by the flank slip generated a modest tsunami that arrived on the shores of the major Hawaiian Islands. Summit subsidence and lava lake withdrawal at Kilauea caldera began gradually on 1 May and accelerated in the days after the earthquake. By May 10 the summit lava lake level dropped more than 300 m and was soon accompanied by the $\sim$500 m of incremental collapse of Kilauea caldera, from May to early August (see Fig. 5.25). The summit collapse

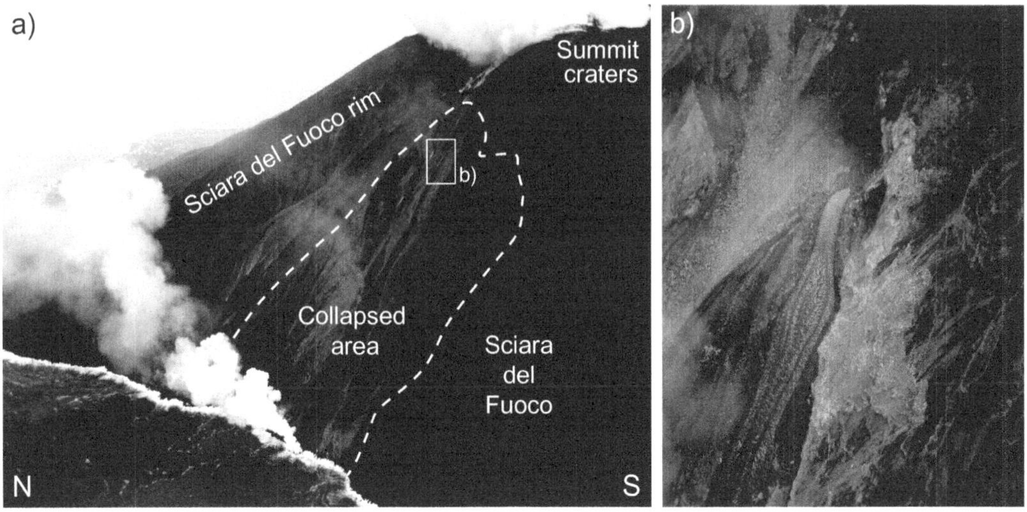

Fig. 6.17 Example of multi-hazards occurred at Stromboli (Italy) during the 2002–2003 eruption. **a** The magma intruded and, feeding the eruption, triggered a sector collapse within the unstable Sciara del Fuoco depression; the collapse generated a local tsunami with maximum run-up of 10–12 m. **b** Close-up of one of the vents fed by the dike responsible for part of the collapse (photo courtesy Marco Neri)

stopped by 4 August, at the end of the ERZ eruption. While the 4 May *M6.9* earthquake responsible for flank slip may have resulted from dike intrusion, it may have also aided magma transport, marking an important event in the magmatic cycle of Kilauea (Neal et al. 2019; Bai et al. 2018). Despite the uncertainties in the exact definition of the triggering factors in the multi-hazard chain, this episode highlights major flank slip within a conspicuous sequence of hazardous events, as dike emplacement, eruptions, caldera collapse and tsunami.

These behaviours highlight the importance of a multi-hazard analysis when considering flank instability. Indeed, no study on the hazards related to unstable flanks should be limited only to flank instability, as the outcome may be much more impactful than that deriving from the instability alone. In these cases, a multi-hazard evaluation, capturing the possible relationships also with volcanic, seismic, and tsunami hazards, should be considered.

Any approach for a multi-hazard analysis of volcanoes experiencing flank instability should rely on the: (a) knowledge of the recent activity of the volcano; (b) identification of the potential sources of hazard related, either as a cause or effect, to flank instability; (c) possibility to establish qualitative or quantitative causal relationships between different processes associated with flank instability. These conditions may be currently met at a few well-monitored and well-known volcanoes, including Mount Etna, Kilauea, Stromboli, Piton de la Fournaise and Mount St. Helens. For example, the evaluation of the potential occurrence of different types of hazard, as well as of their possible interactions, has been used at Mount Etna to define scenarios possibly manifesting in the case of unrest of its unstable flanks. These scenarios may provide a general reference and recommendation in case of multi-hazard processes related to flank instability. Even though these scenarios are still at a qualitative and descriptive level, they constitute a first integrated attempt to evaluate flank instability on active volcanoes and provide a semi-operative tool to assess the related hazards (e.g., Acocella and Puglisi 2013).

6.8 Lessons from Unstable Volcanoes

Flank instability is an important feature accompanying the evolution of many volcanic edifices. Its manifestation, particularly as sector collapse, may not only affect the short- and long-term eruptive behaviour of a volcano, but also determine a chain of consequential hazards with dramatic impact, even at hundreds of kilometres of distance. This section considers a few representative cases of flank instability and collapse which have manifested at a mafic (Mount Etna) and a felsic volcano (Mount St. Helens), or which have occurred less recently and may reoccur in the near future (Augustine, Alaska, USA). In particular, Mount Etna provides an ideal case of widespread and well-studied flank instability, also accompanied by accelerations involving multi-hazards, Mount St. Helens constitutes the most dramatic sector collapse witnessed within the last decades, and Augustine represents a fast growing volcanic edifice with recurrent flank instability, which may soon culminate in sector collapse.

6.8.1 Mount Etna

Mount Etna lies in eastern Sicily, between the Appennine-Maghrebian thrust Chain to the north, the Hyblean Foreland to the south, and on the foot wall of the Malta Escarpment, a lithospheric boundary separating the Ionian oceanic crust from the Sicilian continental one. The location of Mount Etna has been ascribed to the presence of an asthenospheric window in the western side of the retreating Ionian oceanic slab, below the Malta Escarpment. This condition induces localized mantle upwelling, eventually driving decompression melting and promoting volcanism (see Sect. 10.4).

The volcanic edifice has three radial rift zones (the NE, South and West rifts), usually fed by the central conduit through laterally propagating dikes. The transtensive dextral Timpe Fault System (TFS), on the east flank, is the most evident fault system and represents the onshore portion of the regional Malta Escarpment. The evolution of Mount Etna has been accompanied by flank instability in the last ~15 ka at least. This activity in the past has culminated also in sector collapses, as testified by the Valle del Bove (VdB) amphitheatre, formed at approximately 7.5 ka on the eastern slope of the volcano (Fig. 6.18a; Calvari et al. 2004).

Etna is currently experiencing diffuse flank instability, as shown by geodetic and geologic data. The entire volcanic edifice, ~3300 m high, is slowly sliding seawards (Murray et al. 2018), but flank instability is widespread at every scale on the volcano flanks, although the faster motion rates, between ~0.5 and ~5 cm/yr, are measured on the eastern and southern sides. This most unstable area is confined to the north by the sinistral transtensive Pernicana Fault and to the southwest by the dextral transtensive Ragalna Fault (Fig. 6.18b; Rust and Neri 1996). The head of the unstable portion, to the west, is bounded by the NE and the South rifts. Several faults, mostly coinciding with the above mentioned Timpe Fault System, are active between the Pernicana and Ragalna faults: these usually bound sectors, or rigid blocks, with differential kinematics, despite an overall similar eastward motion. The eastward motion of the eastern flank decreases eastward, from the summit area to the coastline. The western flank undergoes a more limited westward motion, with minor evidence for sectors with different kinematics. The periphery to the south, west and northwest of the volcano undergoes continuous and constant uplift, irrespective of any type of magmatic activity. This uplift probably results from the activity of blind thrusts accommodating the basal contraction induced by the load of the spreading of the edifice. Available data are not decisive in defining the deeper extent of any basal detachment. This may have different geometries and depths: it may dip westward, at 5–6 km below sea level (bsl), or at shallower levels; alternatively, it may dip eastward, between 0 and 3 km bsl; the best-constrained portion of the

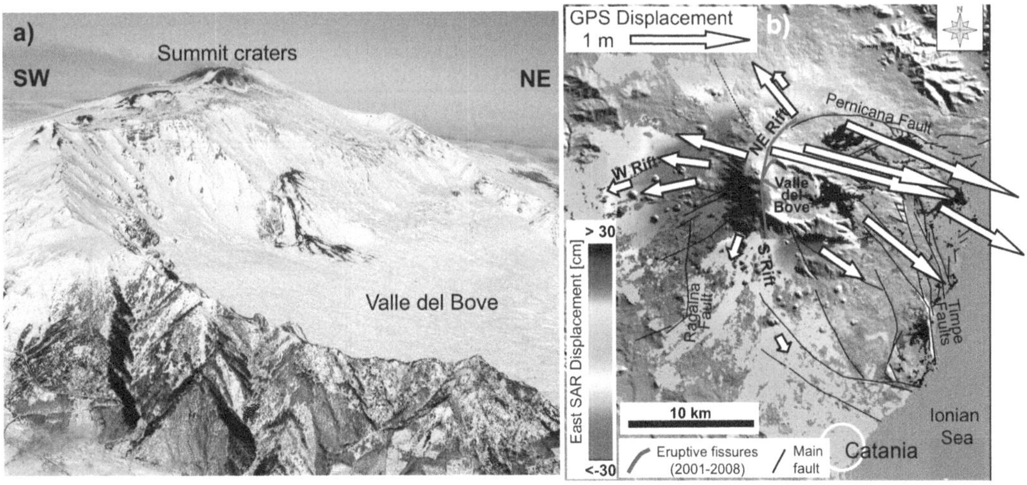

Fig. 6.18 Past sector collapse and current flank instability at Mount Etna (Italy): **a** overview of the Valle del Bove scarp, remnant of a sector collapse at ~7.5 ka (photo courtesy Marco Neri). **b** Geodetic data detecting widespread flank instability at Mount Etna during 1993–2006; colors represent east–west InSAR displacement, arrows show the GPS horizontal displacement (after Acocella and Puglisi 2010)

detachment is eastward dipping at ~2 km bsl between the NE Rift and the Pernicana Fault (Ruch et al. 2010; Siniscalchi et al. 2012). A close relationship between flank deformation, seismic and eruptive activity has been recognized during the 2002–2003 eruption. This highlighted the kinematic connection between the dike-induced opening of the NE Rift and the sinistral slip along the Pernicana Fault, both moving the northern portion of the eastern flank. This movement was accompanied with widespread seismicity, highlighting a connection between flank slip, eruptive and seismic activity. In 2018, magma intrusion along the South Rift induced major dextral slip along a fault on the southern portion of the eastern flank, again accompanied by eruptive and seismic activity. These slip events along the northern and southern portions of the eastern flank highlight a decades-long incremental and complementary activity (Acocella et al. 2003; De Novellis et al. 2019; Villani et al. 2020).

Therefore, in the last decades, the volcano flanks have experienced complex instability resulting from several factors acting on different timescales: (a) volcano load, as testified by the uplift of the periphery of the volcano over decades; (b) inflation due to magma accumulation, as testified by the uplift of the volcanic edifice over a few years; (c) dike emplacement, as testified by dikes propagating in a few days along the NE and S rifts. While the volcano load and inflation produce diffuse, minor and continuous flank instability, dike emplacement accelerates the instability in specific parts of the volcano. (Fig. 6.19; Borgia et al. 2000; Froger et al. 2001; Solaro et al. 2010; de Novellis et al. 2019).

Several factors prepare flank instability at Etna, including a weak, clayish basement, the differential buttressing conditions at the volcano base, the seaward growth of an intrusive complex at depth, and regional tectonic activity associated with the Malta Escarpment. The triggering factor for flank instability is commonly shallow magma emplacement, most effectively in the form of dikes. All these features reasonably explain the current deformation of the volcano's flanks (Norini and Acocella 2011).

In synthesis, Mount Etna represents a volcano with flank instability at different scales and variable rates in space and time, whose variety allows discriminating among the preparing and triggering factors. Here the common association with eruptions and shallow seismicity makes accelerated flank instability a first-order feature in controlling the evolution of the volcano and the manifestation of its main hazards.

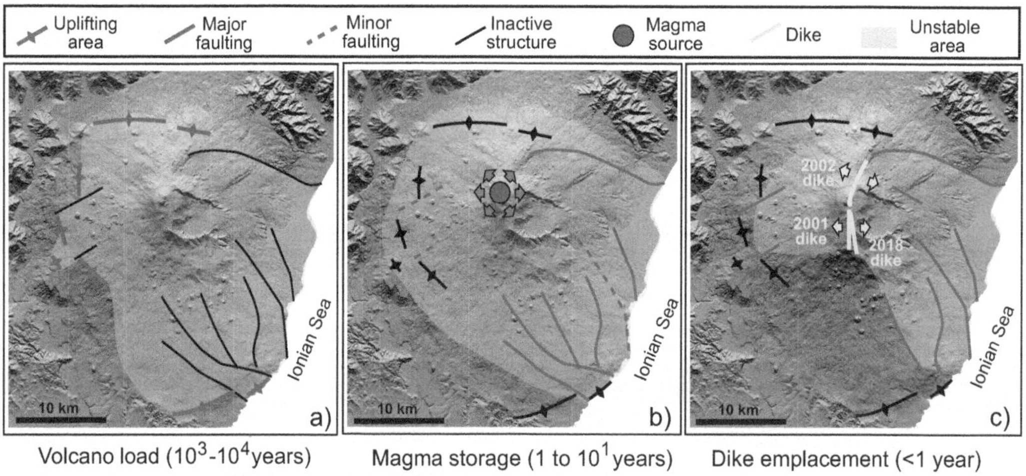

Fig. 6.19 Main processes responsible for flank instability (highlighted by the affected area and the main faults) observed in the last decades at Mount Etna. **a** Volcano load, promoting spreading; **b** magma accumulation beneath the summit, promoting inflation; **c** dike emplacement along the NE and South rifts, promoting flank slip (modified after Solaro et al. 2010)

6.8.2 Mount St. Helens

Mount St. Helens lies in the Cascade Arc, related to the subduction of the Juan de Fuca plate beneath the North American plate (see Sect. 12.2.1). The stratovolcano is on the margin of a mid-crustal zone of high electrical conductivity caused by partial melt and, more locally, on a prominent low-velocity zone north-northwest of the edifice, that corresponds with the St. Helens Seismic Zone (Fig. 6.20a; Hill et al. 2009). In particular, Mount St. Helens lies at a dextral offset along the right-lateral St. Helens Seismic Zone, characterized by localized extension and accompanied by earthquakes related to magmatic activity. Late Quaternary vents within this extensional volume strike northeast, similar to a set of pre-Quaternary faults and intrusive rocks north of the volcano and the deepest northeast aligned earthquakes. Since these NE-SW striking features are aligned approximately perpendicular to the regional minimum principal stress σ_3, the location of Mount St. Helens may be controlled by the junction of the right-stepping offset of the St. Helens Seismic Zone with an older set of NE-SW trending fractures, favourably aligned with respect to the regional tectonic stress directions for magma ascent. Beneath the edifice, a high-velocity body extends at least between depths of 0.5 and 2.1 km. Deeper, fine-grid models image a low-velocity zone between 2 and 3.5 km below sea level, which may correspond to a shallow magma storage zone. Low velocities also reveal a high melt fraction of 10–12% between 4 and 6 km depth, in the source region of the major eruptions over the past 4 ka (Weaver et al. 1987; Waite and Moran 2009; Kiser et al. 2018).

Mount St. Helens has explosively erupted dacitic magma discontinuously over the last 40 ka. Detailed stratigraphic data from the past 4000 years show that the volcano's edifice last collapsed ~2500 years ago; it then erupted

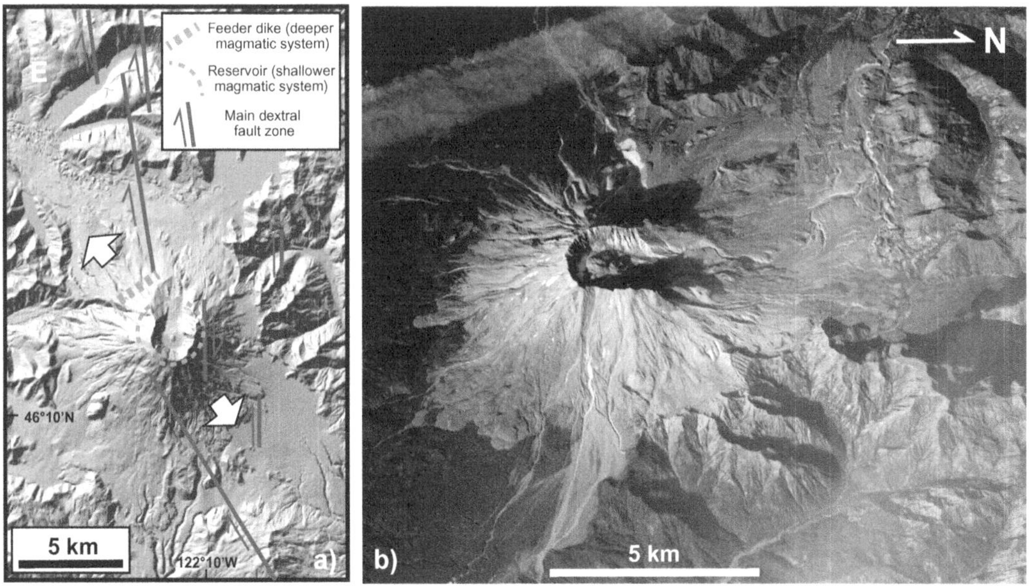

Fig. 6.20 **a** Tectonic setting of Mount St. Helens, Washington, lying along an extensional stepover created by dextral faults. Here a deep NE-SW trending magmatic reservoir feeds the volcano (Weaver et al. 1987); DEM provided by GeoMapApp. **b** Oblique view of Mount St. Helens taken from the International Space Station on October 28, 2008 (photo credit: NASA). Much of the rugged terrain in the amphitheatre is the new dome, but the 1980s dome can be recognized. The area impacted by the 1980 sector collapse is still largely visible to the north (https://earthobservatory.nasa.gov/IOTD/view.php?id=36298)

repeatedly rebuilding its edifice within several hundred years. There is a general correlation between the intensity of Plinian eruptions within eruptive cycles and the repose period prior to each cycle. Current data and models suggest relatively rapid transport from magma genesis to eruption, without evidence for protracted storage or recycling (Carey et al. 1995).

The 1980 eruption is a well-documented edifice failure (Fig. 6.21). The volcano became active in late March 1980, with substantial shallow seismicity on the north flank. Large deformations started to affect most of the north flank, with a graben crossing the summit area. The relation of this deformation to potentially hazardous slides was recognized by early April.

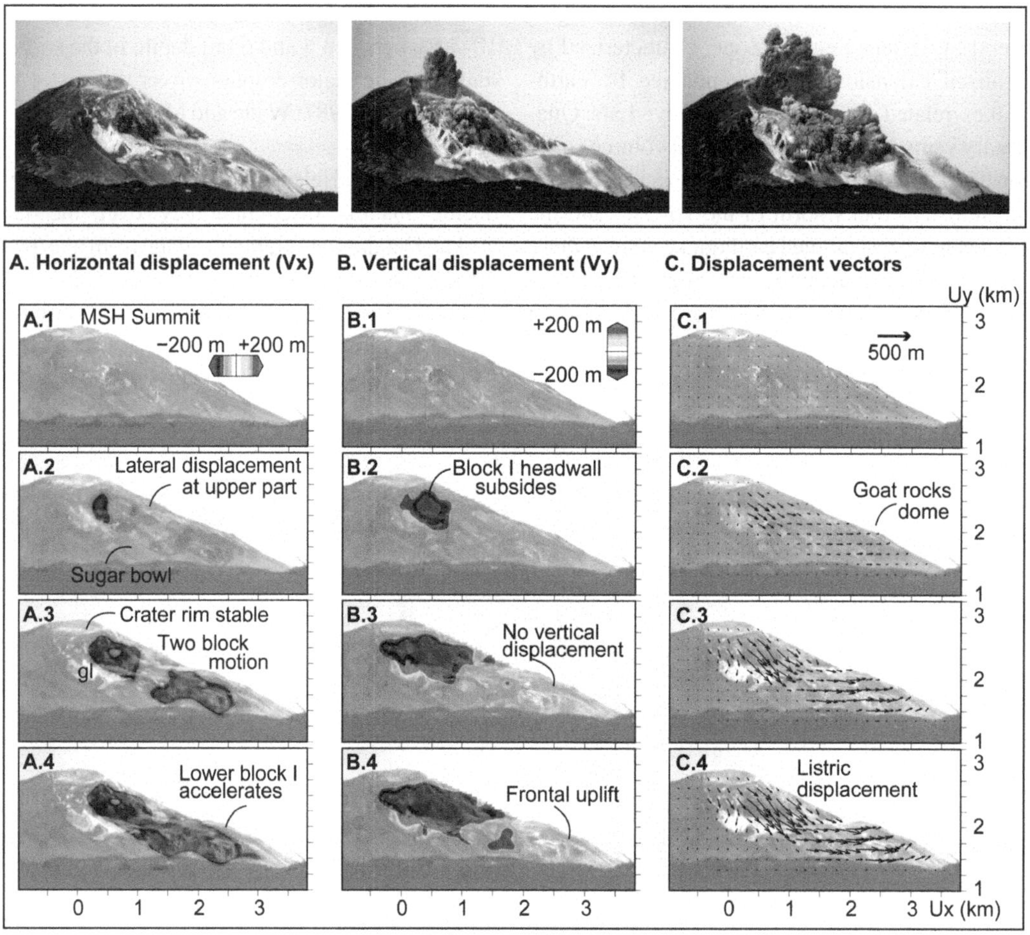

Fig. 6.21 Top images: general evolution of the May 18, 1980, Mount St. Helens flank collapse, showing the incipient failure, developing the two main sliding blocks (left), the onset (centre) and progression (right) of the eruption (Walter 2011). Bottom diagrams: flank displacement, as observed by digital image correlation, during the entire pre-eruptive incipient stage of collapse. MSH: Mount St. Helens. **A**: Horizontal displacement (Vx); gl: Forsyth Glacier. **B**: Vertical displacement (Vy). Colour scales are saturated: maximum Vx was 680 m and maximum Vy was 450 m. **C**: Displacement vectors. All illustrations display cumulative displacements that are perpendicular to field of view (for details see Walter 2011; image courtesy Tom Walter)

Ground displacement surveys initiated in late April confirmed that subhorizontal displacements were as much as 1.5–2.5 m/day. A report on stability hazards on May 1 suggested that a rockslide triggered by hydraulic pressures could be as much as 1 km thick and involve >1 km^3 of rock, with high emplacement avalanche velocities (Voight 2000). Explosive phreatic activity commenced on March 27 and continued through April 22. Indications that juvenile magma was approaching the surface included episodes of shallow (2.5–4 km) harmonic tremor, extreme deformation of the volcano's north flank and resumption of ash-producing eruptions between May 7 and 15. Without short-term warning or change in deformation rate, but triggered by a *M5.2* earthquake and enhanced by high pore-water pressures, the north flank collapsed in a sequence of three retrogressive slope failures on 18 May 1980 (Fig. 6.7c). These, labelled I, II, and III, respectively involved about 32, 25, and 43% of the total rockslide volume of 2.3 km^3; slide I failed first and was least affected by subsequent volcanic activity. Slides I, II, and III broke up into smaller fragments (debris-avalanche blocks) that moved together as the avalanche underwent a transition from a sliding to a flowing mode (Sousa and Voight 1995). Optical images analysis reveals the details of the deformation of the blocks, particularly the outermost block I. Strain localization commenced in the mid-upper flank and formed a landslide headwall that breached the surface. The same fault propagated downward to define block I, which began to slide both vertically and laterally. With progressive flank movement, additional smaller faults formed and merged in a main listric landslide detachment zone. The mid flank exhibited isolated points of high strain, but was also a zone of transition from vertical to horizontal displacement. The disintegration of block I occurred within the first 3–4 s of the rockslide avalanche. Therefore, block I resembled a disintegrated rock mass that sheared internally and fractured before moving away from the source volcano (Walter 2011). The slope failure evolved into a debris avalanche that filled adjacent lowlands and travelled as much as 28 km. The failed edifice mass had a volume of $\sim$2.3 km^3, and generated an avalanche deposit of 2.8 km^3 (Voight 2000).

The decompression associated with sequential slope movements resulted in multiple magmatic and hydrothermal explosions that generated a complex, northward-directed volcanic blast, that travelled $\sim$500 km/hour northward and blew down or scorched 625 km^2 of forest within 3 min. Within 15 min, a vertical plume of volcanic ash rose to an altitude of 25 km, discharging $\sim$0.5 km^3 of DRE dacitic magma (Pallister et al. 1992). The timing and location of the post-eruptive seismicity suggest that the earthquakes occurred in the country rock surrounding a magma body within the aseismic volume. Magma withdrawal caused a pressure decrease within the reservoir and the earthquakes occurred as a brittle response to the resulting stress change (Barker and Malone 1991).

Between 1980 and 1986, a lava dome was built in the crater of Mount St. Helens by intermittent intrusion and extrusion of dacite lava. Spectacular ground deformation was associated with the dome building events and included radial cracks and tangential thrust faults in the surrounding crater floor (see Sect. 7.7). These formed because the crater fill was displaced upward and radially outward from the feeder conduit, owing to rising magma. Radial cracks formed first and, as some evolved into tear faults, influenced the subsequent geometry of thrusting. Once faulting began, deformation was localized near the thrust scarps and their bounding tear faults. The magnitude of displacement systematically increased before extrusion, whereas the azimuth and inclination of displacements remained relatively constant. The eruptive period ended in October 1986. The seismicity in the late 1990s confirms a magma reservoir between 5.5 and 10 km depth, connected to the surface by a vertical conduit (Chadwick and Swanson 1989; Musumeci et al. 2002). A new eruptive phase started in late 2004 and continued until early 2008.

Overall, Mount St. Helens provides a well-studied and dramatic collapse event (despite timely evacuation, the eruption caused 57 victims), which, after four decades, still provides the reference to understand how arc stratovolcanoes may collapse.

6.8.3 Augustine

The 1220 m high Augustine stratovolcano lies on the eastern Aleutian Arc (Alaska), within the southwest part of the NE-SW trending Cook Inlet forearc basin, characterized by the activity of NE-SW trending thrust faults. The volcano is on continental crust ~35 km thick, which prevents the ascent of basaltic melts, which fractionate and assimilate at various depths to produce andesitic magmas (Fisher et al. 2009). Augustine has been historically the most active volcano in the eastern Aleutian Arc, with sustained lava effusion rates more than 10 times greater than is typical of plate-margin volcanoes (Fig. 6.22). As a result, the stratovolcano has repeatedly collapsed and regenerated, averaging 150–200 years per cycle, during the past 2000 years. Repeated failure of 5–10% of the edifice has produced mobile debris avalanches that reached the sea on all sides. High lava extrusion rates rapidly restore the volcano to its pre-failure configuration, highlighting an equilibrium between constructive and destructive processes (Beget and Kienle 1992; Siebert et al. 1995).

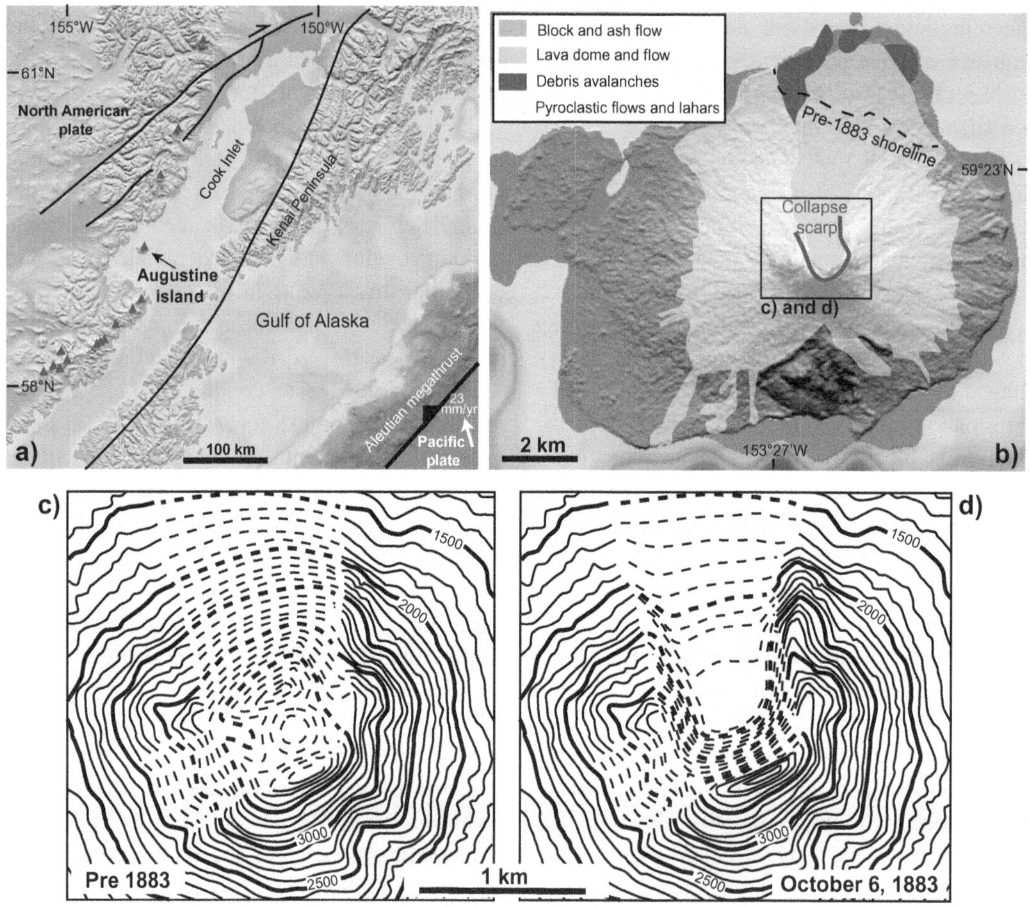

Fig. 6.22 **a** Tectonic setting of Augustine volcano, Alaska, USA; red triangles represent volcanoes; **b** Map of the distribution of the 1883–1884 deposits related to the last major sector collapse. DEM provided by GeoMapApp. **c**, **d** Comparison in the summit topography before (**c**) and after (**d**) the 1883 collapse (modified after Siebert et al. 1995)

A large volcanic tsunami was generated by a debris avalanche on the east side of Augustine ~400 A.D. and affected areas more than 7 m above high tide at 80 km from the volcano, on the Kenai Peninsula. The tsunami occurred about the same time as the collapse of the well-developed Kachemak culture in the southern Cook Inlet area. Another major collapse at ~1540 ± 110 A.D. produced the West Island debris avalanche. An accompanying lateral blast overtopped the avalanche, covering a 40° sector of the west flank. The debris avalanche produced a wave that affected areas 17 m above high tide on Augustine Island.

The most recent edifice collapse occurred at the onset of the October 1883 eruption (Fig. 6.22). Few precursors to the eruption were noted; smoke and severe earthquakes have been reported beginning in summer 1883. The paroxysmal phase of the eruption began on the morning of October 6, 1883, when residents on the Kenai Peninsula, 85 km across the Cook Inlet, heard a large explosion and observed dense volumes of smoke rolling from the summit of Augustine. Twenty-five minutes after the eruption a tsunami with a 9 m high wave struck nearby Port Graham, ~100 km away. The tsunami occurred at low tide, and there were no fatalities. The estimated height of the tsunami wave on Augustine Island was of 20 m. The dramatic events of October 6 apparently derive from collapse of the upper edifice of Augustine, which produced a 0.3 km^3 debris avalanche that reached the north coast with the destruction of much of the north side of the volcano. Three major slide blocks extended to the northeast shoreline up to 2 km. The duration of the 1883 eruption is uncertain. After its end, the breached crater formed by collapse was filled by a 0.09 km^3 lava dome and a 0.04 km^3 lava flow travelled down the north flank (Beget and Kienle 1992; Siebert et al. 1995).

After 1883, the volcano experienced five explosive eruptions, in 1908, 1935, 1963, 1976, 1986 and in January 2006. The 1986 eruption was fed by a vertically extensive system of interconnected dikes, rather than a single coherent magma chamber and/or conduit. InSAR data between 1992 and 2005 show no significant inflation before the explosive January 2006 eruption. However, a small signal was detected by GPS stations, suggesting a source of pressurization beneath the summit, at a depth of around sea level (Cervelli et al. 2006; Roman et al. 2006). Eight months before the onset of the 2006 eruption a significant increase in the number of rockfalls occurred at nearly the same time as precursory earthquake activity increased. The rockfalls originate from the steep summit dome and their high rate may constitute a precursory signal for the long-term monitoring strategies at the volcano (Deroin and McNutt 2012). The January 2006 eruption was fed by a NW-SE trending dike. During the next six months, the otherwise quiescent region 1 to 5 km below sea level centred beneath the summit became seismically active, with two groups of earthquakes. Deep longer-period earthquakes occurred during the eruption, possibly resulting from the ascent of magma, and the post-eruptive shorter-period earthquakes may be due to the relaxation of an emptied magma tube (Syracuse et al. 2011).

Today Augustine has rebuilt itself to a size similar to that which preceded the last edifice failure in 1883 (Fig. 6.23). The frequency of past collapses and the complete reconstruction of an oversteepened summit edifice by four major dome-building eruptions since the 1883 debris avalanche suggest that summit collapse is a possibility during any future eruption. A next major collapse is expected to involve 0.1–0.5 km^3 of the summit. The ensuing debris avalanche would likely reach the coast, producing a tsunami that could impact populated areas of the Kenai Peninsula. Tsunami magnitude is contingent on failure volume, direction and timing with respect to tides (Beget and Kienle 1992; Siebert et al. 1995). The predicted stability of all parts of the upper edifice is approximately the same, with an equal likelihood of slope failure and with the possible exception of a least stable sector on the east flank (Reid et al. 2006).

Fig. 6.23 Augustine volcano feeding a gas plume on January 24, 2006, during its 2005–2006 eruptive phase, as seen from the east (photo credit Cyrus Read, USGS; US Geological Survey)

6.9 Summary

Flank instability is a widespread feature of volcanic edifices. It may manifest with very different modalities, from slow creep-like movements to catastrophic sector collapse.

Despite the variability in size and rates of motion, the unstable portion of the flank of a volcano can be approximated as a sliding volume bordered by structures showing consistent geometric and kinematic features. These include an extensional zone at the head, a contractional zone at the toe, strike-slip structures along the sides and a detachment at the base. This approximation is also useful to evaluate the condition of stability of the slope, which considers the forces parallel to the sliding surface, with the total force driving the outward movement, F_D, being opposed by the resisting shear force, F_R, mobilized along a potential basal slide surface and lateral boundaries. These forces allow evaluating a theoretical factor of safety for the stability of the slope.

Flank instability may be triggered by a single process or, more commonly, by a combination of factors acting together, although one predominant action may set the mass in motion. Magma emplacement, through dikes or viscous intrusions, is the most common triggering factor. Indirectly related to magma, hydrothermal alteration due to volcano degassing and fluid circulation may also affect flank instability. Still related to fluids, excess pore pressures, especially when increased by thermal pressurization, can destabilize volcano flanks. Another destabilizing factor is provided by mechanically weak layers within the edifice or at its base, where a weak basement may induce volcano spreading. Additional external triggers include fault activation and seismicity and, subordinately, local weather (in particular rainfall) and climate.

Sector collapse may induce short- and long-term variations in the rate and composition of erupted magma, as observed at many volcanoes. More generally, sector collapse is a part of a potential chain involving volcanic, seismic, tsunami and climatic hazards, for which a multi-hazard analysis is required.

6.10 Main Symbols Used

c' cohesion under effective stress
F_D destabilizing force
F_l force along sides of the wedge
F_m magmatic forces
F_p total uplift force
F_{pm} mechanical pore fluid pressures
F_{ps} uplift force due to static groundwater pressure
F_{pt} thermal pore fluid pressures
F_R stabilizing force
F_s seawater load
Fsa factor of safety
g acceleration due to gravity
H vertical drop
H_b thickness of brittle layer
H_d thickness of weak layer
H_v volcano height
L_d travel distance
M magnitude
P_f pore pressure
r radius of volcano
t spreading time
δ inclination of basal failure surface
η viscosity of ductile layer
θ inclination of the failing block
λ pore-fluid factor
Π_1 ratio of the height of the volcano with regard to its radius
Π_2 ratio between thickness of brittle basement and ductile substratum
ρ density
σ' effective normal stress
σ'_1 effective maximum principal stress
σ'_3 effective minimum principal stress
σ_1 maximum principal stress
σ_3 minimum principal stress
σ_n normal stress
τ shear resistance
ϕ' angle of internal friction under effective stress

References

Acocella V, Behncke B, Neri M, D'Amico S (2003) Link between major flank slip and 2002–2003 eruption at Mt. Etna (Italy). Geophys Res Let 30:2286. https://doi.org/10.1029/2003GL018642

Acocella V (2005) Modes of sector collapse of volcanic cones: insights from analogue experiments. J Geophys Res 110:B02205. https://doi.org/10.1029/2004JB003166

Acocella V, Tibaldi A (2005) Dike propagation driven by volcano collapse: a general model tested at Stromboli, Italy. Geophys Res Lett 32:L08308. https://doi.org/10.1029/2004GL022248

Acocella V, Puglisi G (2010) Hazard mitigation of unstable volcanic edifices. EOS 91:357–358

Acocella V, Puglisi G (2013) How to cope with volcano flank instability? Recommending prototypal procedures for Mt. Etna. J Volcanol Geoth Res 251:137–148

Altaner S, Demosthenous C, Pozzuoli A, Rolandi G (2013) Alteration history of Mount Epomeo Green Tuff and a related polymictic breccia, Ischia Island, Italy: evidence for debris avalanche. Bull Volcanol 75:718

Amelung F, Day S (2002) InSAR observations of the 1995 Fogo, Cape Verde, eruption: implications for the effects of collapse events upon island volcanoes. Geophys Res Lett 29:1606. https://doi.org/10.1029/2001GL013760

Apuani T, Corazzato C, Cancelli A, Tibaldi A (2005) Stability of a collapsing volcano (Stromboli, Italy): limit equilibrium analysis and numerical modelling. J Volcanol Geoth Res 144:191–210

Arce JL, Macias R, Garcia Palomo A, Capra L, Marcias JL, Layer P et al (2008) Late Pleistocene flank collapse of Zempoala volcano (Central Mexico) and the role of fault reactivation. J Volcanol Geoth Res 177:944–958

Bai Y, Ye L, Yamazaki Y, Lay T, Cheung KF (2018) The 4 May 2018 Mw 6.9 Hawaii Island earthquake and implications for tsunami hazards. Geophys Res Lett 45:11040–11049

Barker SE, Malone SD (1991) Magmatic system geometry at Mount St. Helens modeled from the stress field associated with posteruptive earthquakes. J Geophys Res 96:11883–11894

Battaglia M, Di Bari M, Acocella V, Neri M (2011) Dike emplacement and flank instability at Mount Etna: constraints from a poro-elastic-model of flank collapse. J Volcanol Geoth Res 199:153–164

Beget JE, Kienle J (1992) Cyclic formation of debris avalanches at Mount St Augustine volcano. Nature 356:701–704

Belousov A, Voight B, Belousova M (2007) Directed blasts and blast-generated pyroclastic density currents: a comparison of the Bezymianny 1956, Mount St Helens 1980, and Soufrière Hills, Montserrat 1997 eruptions and deposits. Bull Volcanol 69:701–740

Blahut J, Balek J, Klimes J, Rowberry M, Kusak M, Kalina J (2019) A comprehensive global database of giant landslides on volcanic islands. Landslides 16:2045–2052

Borgia A, Ferrari L, Pasquarè G (1992) Importance of gravitational spreading in the tectonic and volcanic evolution of Mt. Etna. Nature 357:231–235

Borgia A (1994) Dynamic basis of volcanic spreading. J Geophys Res 99:17791–17804

Borgia A, Delaney PT, Denlinger RP (2000) Spreading volcanoes. Annu Rev Earth Planet Sci 28:539–570

Borgia A, van Wyk de Vries B (2003) The volcano-tectonic evolution of Concepcion, Nicaragua. Bull Volcanol 65:248–266

Borgia A, Tizzani P, Solaro G, Manzo M, Casu F, Luongo G et al (2005) Volcanic spreading of Vesuvius, a new paradigm for interpreting itsvolcanic activity. Geoph Res Lett 32:L03303, https://doi.org/10.1029/2004GL022155

Boudon G, Le Friant A, Komorowski J-C, Deplus C, Semet MP (2007) Volcano flank instability in the Lesser Antilles Arc: Diversity of scale, processes, and temporal recurrence. J Geophys Res 112:B08205. https://doi.org/10.1029/2006JB004674

Brooks BA, Foster J, Sandwell D, Wolfe CJ, Okubo P, Poland M et al (2008) Magmatically triggered slow slip at Kilauea Volcano, Hawaii. Science 321:1177

Calvari S, Tanner LH, Groppelli G, Norini G (2004) Valle del Bove, Eastern Flank of Etna Volcano: a comprehensive model for the opening of the depression and implications for future hazards. In: Bonaccorso A, Calvari S, Coltelli M, Del Negro C, Falsaperla S (eds) vol 143. Etna Volcano Laboratory. Geophysical Monograph Series American Geophysical Union, pp 65–75

Calvari S, Spampinato L, Lodato L, Harris AJL, Patrick MR, Dehn J et al (2005) Chronology and complex volcanic processes during the 2002–2003 flank eruption at Stromboli volcano (Italy) reconstructed from direct observations and surveys with a handheld thermal camera. J Geophys Res 110: B02201. https://doi.org/10.1029/2004JB003129

Carey S, Gardner J, Sigurdsson H (1995) The intensity and magnitude of Holocene plinian eruptions from Mount St. Helens volcano. J Volcanol Geoth Res 66:185–202

Carracedo JC, Day SJ, Guillou H, Perez Torrado FJ (1999) Giant Quaternary landslides in the evolution of La Palma and El Hierro; Canary Islands. J Volcanol Geoth Res 94:169–190

Carrasco-Nunez G, Diaz-Castellon R, Siebert L, Hubbard B, Sheridan MF, Rodriguez SR (2006) Multiple edifice-collapse events in the Eastern Mexican Volcanic Belt: the role of sloping substrate and implications for hazard assessment. J Volcanol Geoth Res 158:151–176

Casillas R, Fernandez C, Colmenero JR, de la Nuez G, Garcia-Navarro E, Martin MC (2010) Deformation structures associated with the Tazo landslide (La Gomera, Canary Islands). Bull Volcanol 72:945–960

Cassidy M, Watt SFL, Talling PJ, Palmer MR, Edmonds M, Jutzeler M et al (2015) Rapid onset of mafic magmatism facilitated by volcanic edifice collapse. Geophys Res Lett 42:4778–4785

Cervelli P, Segall P, Amelung F, Garbeil H, Meertens C, Owen S (2002) The 12 September 1999 Upper East Rift Zone dike intrusion at Kilauea Volcano, Hawaii. J Geophys Res 107. https://doi.org/10.1029/2001JB000602

Cervelli PF, Fournier T, Freymueller J, Power JA (2006) Ground deformation associated with the precursory unrest and early phases of the January 2006 eruption of Augustine Volcano, Alaska. Geophys Res Lett 33: L18304. https://doi.org/10.1029/2006GL027219

Chadwick WW, Swanson DA (1989) Thrust faults and related structures in the crater floor of Mount St. Helens volcano, Washington. Geol Soc Am Bull 101:1507–1519

Chadwick WW, Wright IC, Schwarz-Schampera US, Hyvernaud O, Reymond D, de Ronde CEJ (2008) Cyclic eruptions and sector collapses at Monowai submarine volcano, Kermadec arc: 1998–2007. Geochem Geophys Geosyst 9:Q10014. https://doi.org/10.1029/2008GC002113

Chen K, Smith JD, Avouac J-P, Liu Z, Song YT, Gualandi A (2019) Triggering of the Mw 7.2 Hawaii earthquake of 4 May 2018 by a dike intrusion. Geophys Res Lett 46:2503–2510

Coombs ML, White SM, Scholl DW (2007) Massive edifice failure at Aleutian arc volcanoes. Earth Planet Sci Lett 256:403–418

Dade WB, Huppert HE (1998) Long-runout rockfalls. Geology 26:803–806

Day SJ (1996) Hydrothermal pore fluid pressure and the stability of porous, permeable volcanoes. In: McGuire WJ, Jones AP, Neuberg J (eds) Volcano instability on the earth and other planets, vol 110. Journal of the Geological Society of London Special Publication, pp 77–93

Day SJ, Heleno da Silva SIN, Fonseca JFBD (1999) A past giant lateral collapse and present-day flank instability of Fogo, Cape Verde Islands. J Volcanol Geoth Res 94:191–218

De Novellis V, Atzori S, De Luca C, Manzo M, Valerio E, Bonano M et al (2019) DInSAR analysis and analytical modeling of Mount Etna displacements: The December 2018 volcano-tectonic crisis. Geophys Res Lett 46:5817–5827

Deeming KR, McGuire B, Harrop P (2010) Climate forcing of volcano lateral collapse: evidence from Mount Etna, Sicily. Philos Trans R Soc A 368:2559–2577

Del Camp A, Troll VR, van Wyk de Vries B, Carracedo JC, Petronis MS, Perez-Torrado FJ et al (2012) Dykes and structures of the NE rift of Tenerife, Canary Islands: a record of stabilisation and destabilisation of ocean island rift zones. Bull Volcano 74:963–980

Del Potro R, Hurlimann M (2009) The decrease in the shear strength of volcanic materials with argillic hydrothermal alteration, insights from the summit region of Teide stratovolcano, Tenerife. Eng Geol 104:135–143

Delaney PT, Denliger RP (1999) Stabilization of volcanic flanks by dike intrusion: an example from Kilauea. Bull Volcanol 61:356–362

Deroin N, McNutt S (2012) Rockfalls at Augustine Volcano, Alaska: the influence of eruption precursors and seasonal factors on occurrence patterns 1997–2009. J Volcanol Geoth Res 211–212:61–75

Di Traglia F, Nolesini T, Intrieri E, Mugnai F, Leva D, Rosi M et al (2014) Review of ten years of volcano deformations recorded by the ground-based InSAR monitoring system at Stromboli volcano: a tool to mitigate volcano flank dynamics and intense volcanic activity. Earth Sci Rev 139:317–335

Dieterich JH (1988) Growth and persistence of Hawaiian volcanic rift zones. J Geophys Res 93:4258–4270

Donnadieu F, Merle O (1998) Experiments on the indentation process during cryptodome intrusions: new insights into Mount St Helens Deformation. Geology 26:79–82

Donnadieu F, Merle O (2001) Geometrical constraints of the 1980 Mount St Helens Intrusion from Analogue Models. Geophys Res Lett 28:639–642

Dufresne A, Salinas S, Siebe C (2010) Substrate deformation associated with the Jocotitlán edifice collapse and debris avalanche deposit, Central México. J Volcanol Geoth Res 197:133–148

Ebmeier SK, Biggs J, Muller C, Avard G (2014) Thin-skinned mass-wasting responsible for widespread deformation at Arenal volcano. Front Earth Sci 2:35

Elsworth D, Day S (1999) Flank collapse triggered by intrusion: the Canarian and Cape Verde Archipelagoes. J Volcanol Geoth Res 94:323–340

Elsworth D, Voight B (1995) Dike intrusion as a trigger for large earthquakes and the failure of volcano flanks. J Geophys Res 100:6005–6024

Famin V, Michon L (2010) Volcano destabilization by magma injections in a detachment. Geology 38:219–222

Farquharson JI, Amelung F (2020) Extreme rainfall triggered the 2018 rift eruption at Kīlauea Volcano. Nature 580:491–495

Fisher MA, Ruppert NA, White RA, Wilson FH, Comer D, Sliter RW et al (2009) A distal earthquake cluster concurrent with the 2006 explosive eruption of Augustine Volcano, Alaska. Tectonophysics 469:25–36

Froger JL, Merle O, Briole P (2001) Active spreading and regional extension at Mount Etna imaged by SAR interferometry. Earth Planet Sci Lett 187:245–258

Goto Y, Danhara T, Tomiya A (2019) Catastrophic sector collapse at Usu volcano, Hokkaido, Japan: failure of a young edifice built on soft substratum. Bull Volcanol 81:37

Green DN, Neuberg J (2006) Waveform classification of volcanic low-frequency earthquake swarms and its implication at Soufriere Hills Volcano, Montserrat. J Volcanol Geoth Res 153:51–63

Gudmundsson A (2009) Toughness and failure of volcanic edifices. Tectonophysics 471:27–35

Hill GJ, Caldwell TG, Heise W, Chertkoff DG, Bibly HM, Burgess MK et al (2009) Distribution of melt beneath Mount St Helens and Mount Adams inferred from magnetotelluric data. Nat Geosci 2:785–789

Hoek E, Marinos P, Benissi M (1998) Applicability of the geological strength index (GSI) classification for very weak and sheared rock masses. The case of the Athens Schist Formation. Bull Eng Geol Env 57:151–160

Hunt JE, Cassidy M, Talling PJ (2018) Multi-stage volcanic island flank collapses with coeval explosive caldera-forming eruptions. Sci Rep 8:1146. https://doi.org/10.1038/s41598-018-19285-2

Hurlimann M, Ledesma A, Marti J (1999) Conditions favouring catastrophic landslides on Tenerife (Canard Islands). Terra Nova 11:106–111

Iverson RM (1995) Can magma-injection and groundwater forces cause massive landslides on Hawaiian volcanoes? J Volc Geoth Res 66:295–308

Karstens J, Berndt C, Urlaub M, Watt FL, Micallef A, Ray M et al (2019) From gradual spreading to catastrophic collapse—Reconstruction of the 1888 Ritter Island volcanic sector collapse from high-resolution 3D seismic data. Earth Planet Sci Lett 517:1–13

Keating BH, McGuire WJ (2004) Instability and structural failure at volcanic oceanic islands and the climate change dimension. Adv Geophys 47:175–271

Kendrick JE, Smith R, Sammonds P, Meredith PG, Dainty M, Pallister JS (2013) The influence of thermal and cyclic stressing on the strength of rocks from Mount St. Helens, Washington. Bull Volcanol 75:728

Kiser E, Levander A, Zelt C, Schmandt B, Hansen S (2018) Focusing of melt near the top of the Mount St. Helens (USA) magma reservoir and its relationship to major volcanic eruptions. Geology 46: 775–778

Lipman PW, Sisson TW, Coombs ML, Calvert A, Kimura JI (2006) Piggyback tectonics: long-term growth of Kilauea on the south flank of Mauna Loa. J Volcanol Geoth Res 151:73–108

Lopez DL, Williams SN (1993) Catastrophic volcanic collapse: relation to hydrothermal processes. Science 260:1794–1796

Maccaferri F, Richter N, Walter TR (2017) The effect of giant lateral collapses on magma pathways and the location of volcanism. Nat Commun 8:1097. https://doi.org/10.1038/s41467-017-01256-2

Manconi A, Longprè MA, Walter TR, Troll VR, Hansteen TH (2009) The effects of flank collapses on volcano plumbing systems. Geology 37:1099–1102

Masson DG, Le Bas TP, Grevemeyer I, Weinrebe W (2008) Flank collapse and large-scale landsliding in the Cape Verde Islands, off West Africa. Geochem Geophys Geosyst 9:Q07015. https://doi.org/10.1029/2008GC001983

Masson DG, Watts AB, Gee MJR, Urgeles R, Mitchell NC, Le Bas TP et al (2002) Slope failures on the flanks of the western Canary Islands. Earth-Sci Rev 57:1–35

Mathieau L, Kervyn M, Ernst GGJ (2011) Field evidence for flank instability, basal spreading and volcano-tectonic interactions at Mt Cameroon, West Africa. Bull Volcanol 73:851–867

McGuire WJ, Pullen AD, Saunders SJ (1990) Recent dyke-induced large-scale block movement at Mount Etna and potential slope failure. Nature 343:357–359

McGuire WJ, Howarth RJ, Firth CR, Solow AR, Pullen AD, Saunders SJ et al (1997) Correlation between rate of sea-level change and frequency of explosive volcanism in the Mediterranean. Nature 389:473–476

McGuire WJ (1996) Volcano instability: a review of contemporary themes. In: McGuire WJ, Jones AP, Neuberg J (eds) Volcano instability on the earth and other planets, vol 110. Journal of the Geological Society of London Special Publication, pp 1–23

Merle O, Borgia A (1996) Scaled experiments of volcanic spreading. J Geophys Res 101:13805–13817

Miklius A, Cervelli P, Sako M, Lisowski M, Owen S, Segall P et al (2005) Global positioning system measurements on the Island of Hawaii: 1997 through 2004. USGS open-file report 2005–1425

Montanaro C, Beget J (2011) Volcano collapse along the Aleutian Ridge (western Aleutian Arc). Nat Hazards Earth Syst Sci 11:715–730

Montgomery-Brown EK, Segall P, Miklius A (2009) Kilauea slow slip events: Identification, source inversions, and relation to seismicity. J Geophys Res 114: B00A03. https://doi.org/10.1029/2008JB006074

Montgomery-Brown EK, Sinnett DK, Larson KM, Poland MP, Segall P, Miklius A (2011) Spatiotemporal evolution of dike opening and décollement slip at Kilauea Volcano, Hawaii. J Geophys Res 116: B03401. https://doi.org/10.1029/2010JB007762

Morgan JK, Moore GF, Hills DJ, Leslie S (2000) Overthrusting and sediment accretion along Kilauea's mobile south flank, Hawaii: evidence for volcanic spreading from marine seismic reflection data. Geology 28:667–670

Morgan JK, McGovern PJ (2005) Discrete element simulations of gravitational volcanic deformation: 1. Deformation structures and geometries. J Geophys Res 110:B05402. https://doi.org/10.1029/2004JB003252

Murray JB, van Wyk de Vries B, Pitty A, Sargent P, Wooller L (2018) Gravitational sliding of the Mt. Etna massif along a sloping basement. Bull Volcanol 80:40

Musumeci C, Gresta S, Malone SD (2002) Magma system recharge of Mount St. Helens from precise relative hypocenter location of microearthquakes. J Geophys Res 107:2264. https://doi.org/10.1029/2001JB000629

Neal CA, Brantley SR, Antolik L, Babb JL, Burgess M, Calles K et al (2019) The 2018 rift eruption and summit collapse of Kīlauea Volcano. Science 363:367–374

Neri M, Casu F, Acocella V, Solaro G, Pepe S, Berardino P et al (2009) Deformation and eruptions at Mt. Etna (Italy): a lesson from 15 years of observations. Geophys Res Lett 36:L02309. https://doi.org/10.1029/2008GL036151

Norini G, Acocella V (2011) Analogue modeling of flank instability at Mount Etna: understanding the driving factors. J Geophys Res 116:B07206. https://doi.org/10.1029/2011JB008216

Pallister J, Hoblitt RP, Crandell DR, Mullineaux DR (1992) Mount St. Helens a decade after the 1980 eruptions: magmatic models, chemical cycles, and a revised hazards assessment. Bull Volcanol 54:126–146

Phillips KA, Chadwell CD, Hildebrand JA (2008) Vertical deformation measurements on the submerged south flank of Kılauea volcano, Hawaii reveal seafloor motion associated with volcanic collapse. J Geophys Res 113: B05106. https://doi.org/10.1029/2007JB005124

Pinel V, Albino F (2013) Consequences of volcano sector collapse on magmatic storage zones: Insights from numerical modeling. J Volcanol Geoth Res 252:29–37

Pinel V, Jaupart C (2004) Likelihood of basaltic eruptions as a function of volatile content and volcanic edifice size. J Volcanol Geoth Res 137:201–217

Pinel V, Jaupart C (2005) Some consequences of volcanic edifice destruction for eruption conditions. J Volcanol Geoth Res 145:68–80

Ponomareva V, Melekestsev IV, Dirksen OV (2006) Sector collapses and large landslides on Late Pleistocene-Holocene volcanoes in Kamchatka, Russia. J Volcanol Geoth Res 158:117–138

Reid ME, Sisson TW, Brien DL (2001) Volcano collapse promoted by hydrothermal alteration and edifice shape, Mount Rainier, Washington. Geology 29:779–782

Reid ME (2004) Massive collapse of volcano edifices triggered by hydrothermal pressurization. Geology 32:373–376

Reid ME, Brien DL, Waythomas CF (2006) Preliminary slope-stability analysis of Augustine Volcano. In: Power JA, Coombs ML, Freymueller JT (eds) The 2006 eruption of Augustine Volcano, Alaska, vol 1769. US Geological Survey Professional Paper, pp 321–332

Reid ME, Keith TEC, Kayen RE, Iverson NR, Iverson RM, Brien DL (2010) Volcano collapse promoted by progressive strength reduction: new data from Mount St Helens. Bullet Volcanol 72:761–766

Roman DC, Neuberg J, Luckett RR (2006) Assessing the likelihood of volcanic eruption through analysis of volcanotectonic earthquake fault–plane solutions. Earth Planet Sci Lett 248:244–252

Ruch J, Acocella V, Storti F, Neri M, Pepe S, Solaro G et al (2010) Detachment depth revealed by rollover deformation: an integrated approach at Mount Etna.

Geophys Res Lett 37:L16304. https://doi.org/10.1029/2010GL044131
Rust D, Neri M (1996) The boundaries of large-scale collapse on the flanks of Mount Etna, Sicily, In: McGuire WJ, Jones AP, Neuberg J (eds) Volcano instability on the Earth and other planets, vol 110. Journal of the Geological Society of London Special Publication, pp 193–208
Schaefer LN, Lu Z, Oommen T (2015) Dramatic volcanic instability revealed by InSAR. Geology 43:743–746
Shirzaei M, Walter TR, Nankall HR, Holohan EP (2011) Gravity-driven deformation of Damavand volcano, Iran, detected through InSAR time series. Geology 39:251–254
Siebert L (1984) Large volcanic debris avalanches: characteristics of source areas, deposits, and associated eruptions. J Volcanol Geoth Res 22:163–197
Siebert L, Glicken H, Ui T (1987) Volcanic hazard from Bezymianny and Bandai-type eruptions. Bull Volcanol 49:435–459
Siebert L, Begey JE, Glicken H (1995) The 1883 and late-prehistoric eruptions of Augustine volcano, Alaska. J Volcanol Geoth Res 66:367–395
Siniscalchi A, Tripaldi S, Neri M, Balasco M, Romano G, Ruch J et al (2012) Flank instability structure of Mt. Etna inferred by a magnetotelluric survey. J Geophys Res 117: B03216, https://doi.org/10.1029/2011JB008657
Solaro G, Acocella V, Pepe S, Ruch J, Neri M, Sansosti E (2010) Anatomy of an unstable volcano from InSAR: multiple processes affecting flank instability at Mt. Etna, 1994–2008. J Geophys Res 115:B10405. https://doi.org/10.1029/2009JB000820
Sousa J, Voight B (1995) Multiple-pulsed debris avalanche emplacement at Mount St. Helens in 1980: Evidence from numerical continuum flow simulations. J Volcanol Geoth Res 66:227–250
Syracuse E, Thurber CH, Power JA (2011) The Augustine magmatic system as revealed by seismic tomography and relocated earthquake hypocenters from 1994 through 2009. J Geophys Res 116:B09306. https://doi.org/10.1029/2010JB008129
Terzaghi K (1936) Relation between soil mechanics and foundation engineering: presidential address. In: Proceedings, first international conference on soil mechanics and foundation engineering, Boston, vol 3, pp 13–18
Thomas ME, Petford N, Bromhead EN (2004) The effect of internal gas pressurization on volcanic edifice stability: evolution towards a critical state. Terra Nova 00:1–6
Tibaldi A (2001) Multiple sector collapses at Stromboli volcano, Italy: how they work. Bull Volcanol 63:112–125
Tibaldi A (2004) Major changes in volcano behaviour after a sector collapse: insights from Stromboli, Italy. Terra Nova 16:2–8
Tibaldi A, Corazzato C, Marani M, Gamberi F (2009) Subaerial-submarine evidence of structures feeding magma to Stromboli Volcano, Italy, and relations with edifice flank failure and creep. Tectonophysics 469:112–136
van Wyk deVries B, Matela R (1998) Styles of volcano-induced deformation: numerical models of substratum flexure, spreading and extrusion. J Volcanol Geothermal Res 81:1–18
van Wyk deVries B, Self S, Francis PW, Keszthelyi L (2001) A gravitational spreading origin for the Socompa debris avalanche. J Volcanol Geothermal Res 105:225–247
Verrucci L, Tommasi P, Boldini D, Graziani A, Rotonda T (2019) Modelling the instability phenomena on the NW flank of Stromboli Volcano (Italy) due to lateral dyke intrusion. J Volcanol Geoth Res 371:245–262
Vezzoli L, Tibaldi A, Renzulli A, Menna M, Flude S (2008) Faulting-assisted lateral collapses and influence on shallow magma feeding system at Ollagüe volcano (Central Volcanic Zone, Chile-Bolivia Andes). J Volcanol Geoth Res 171:137–159
Vezzoli L, Corazzato C (2016) Geological constraints of a structural model of sector collapse at Stromboli volcano, Italy. Tectonics 35:2070–2081
Villani F, Pucci S, Azzaro R, Civico R, Cinti FR, Pizzimenti L et al (2020) Surface ruptures database related to the 26 December 2018, MW 4.9 Mt. Etna earthquake, southern Italy. Scie Data 7:42 https://doi.org/10.1038/s41597-020-0383-0
Voight B, Glicken H, Janda RJ, Douglas PM (1981) Catastrophic rockslide avalanche of May 18. In: Lipman PW, Mullineaux DR (eds) The 1980 eruptions of Mount St. Helens, Washington, vol 125. US Geological Survey Professional Papers, pp 347–377
Voight B, Elsworth D (1997) Failure of volcano slopes. Geotechnique 47:1–31
Voight B (2000) Structural stability of andesite volcanoes and lava domes. Philos Trans R Soc Lond 358:1663–1703
Waite GP, Moran SC (2009) VP Structure of Mount St. Helens, Washington, USA, imaged with local earthquake tomography. J Volcanol Geoth Res 182:113–122
Walter TR, Troll VR, Cailleau B, Belousov A, Schmincke HU, Amelung F et al (2005) Rift zone reorganization through flank instability in ocean island volcanoes: an example from Tenerife, Canary Islands. Bull Volcanol 67:281–291
Walter TR, Klugel A, Munn S (2006) Gravitational spreading and formation of new rift zones on overlapping volcanoes. Terra Nova 18:26–33
Walter TR (2011) Structural architecture of the 1980 Mount St. Helens collapse: an analysis of the Rosenquist photo sequence using digital image correlation. Geology 39:767–770
Walter TR, Haghigh MH, Schneider FM, Coppola D, Motagh M, Saul J et al (2019) Complex hazard cascade culminating in the Anak Krakatau sector collapse. Nat Commun 10:4339. https://doi.org/10.1038/s41467-019-12284-5
Watt SFL (2019) The evolution of volcanic systems following sector collapse. Journ Volc Geoth Res 384:280–303

Watters RJ, Zimbelman DR, Bowman SD, Crowley JK (2000) Rock Mass Strength assessment and significance to edifice stability, Mount Rainier and Mount Hood, cascade range volcanoes. Pure Appl Geophys 157:957–976

Weaver CS, Grant WC, Shemeta JE (1987) Local crustal extension at Mount St. Helens Washington. J Geophys Res 92:10170–10178

Wright IC, Chadwick WW, de Ronde CEJ, Reymond D, Hyvernaud O, Gennerich HH, et al (2008) Collapse and reconstruction of Monowai submarine volcano, Kermadec Arc, 1998–2004. J Geophys Re 113: B08S03. https://doi.org/10.1029/2007JB005138

Zang A, Stephansson O (2010) Stress field in the Earth's crust. Springer, 322 pp

Zernack AV, Cronin SJ, Bebbington MS, Price RC, Smith IEM, Stewart RB et al (2013) Forecasting catastrophic stratovolcano collapse: a model based on Mount Taranaki, New Zealand. Geology 40:983–986

7 Shallow Magma Transfer

7.1 Introduction

This chapter focuses on the processes controlling the shallow transfer of magma below and within volcanic edifices. As such, it builds on the knowledge of the previous chapters, which considered the general mechanisms for the rise of magma in the crust (Chap. 3), its arrest and the development of magma chambers nucleating magma-filled fractures (Chap. 4), and the important topographic variations induced by calderas (Chap. 5) and sector collapse (Chap. 6). At the same time, shallow magma transfer generates signals detected by monitoring systems (Chap. 8), which can be used to understand the state of active volcanoes and ultimately improve eruption forecasting (Chap. 9).

Here "shallow" refers to crustal levels between the depth to the top of magma chambers (often ≤ 5 km) and the surface, thus also including volcanic edifices and calderas. In this zone "magma transfer" generally occurs through the propagation of magma-filled fractures, often consisting of vertically or laterally propagating dikes, although inclined sheets and sills may also be important. As these magma-filled fractures respond to the shallow stress conditions, including those imposed by topographic gradients, determining the shallow regional and local stress components allows defining magma paths, also anticipating where eruptive vents may open and providing an important contribution to hazard assessment.

The main aims of this chapter are thus to define the:

- general patterns of magma propagation at shallow crustal levels;
- specific patterns of magma propagation within central volcanic edifices and calderas;
- likely location of eruptive vents at the surface.

7.2 Overview

This chapter describes the paths of shallow magma transfer, from the shallow propagation of regional dikes of deeper origin, or from the nucleation of magma-filled fractures around magma chambers, to dikes reaching the surface and feeding eruptions. Therefore, to determine the path of magma approaching the surface, the focus in this chapter is first on the magma propagating under a regional stress field (see Sect. 7.3). With a regional stress field magma propagation mainly occurs through dikes oriented perpendicular to the minimum principal stress σ_3. These dikes may be fed by a deeper reservoir and propagate through the upper crust, or may be fed by a shallower magma chamber and propagate some distance away from the

V. Acocella, *Volcano-Tectonic Processes*, Advances in Volcanology,
https://doi.org/10.1007/978-3-030-65968-4_7

chamber, where the effect of its local stress field becomes negligible. In some cases, the local stress field induced by the magma chamber may dominate over the regional stress field, controlling the shallow propagation of magma-filled fractures. The latter may consist of: (a) sills (see Sect. 7.4), which mainly promote the lateral transfer of magma below volcanoes, although saucer-shaped sills may also enhance vertical transfer; (b) circumferential magma-filled fractures (see Sect. 7.5) which, depending on the pressure conditions within the chamber, may consist of inclined or cone sheets (with overpressurized chambers) or subvertical ring-dikes related to the formation of calderas (underpressurized chambers). All these magmatic sheets may reach the surface, feeding eruptions. While approaching the surface, the propagation path of any magma-filled fracture may be also influenced by a local and shallower stress field induced by topographic gradients, as for example the presence of calderas and other non-volcanic depressions, or volcanic edifices and non-volcanic reliefs: as a result, these magma-filled fractures may become locally deflected from their path. Among the magma-filled fractures affected by the unloading induced by a caldera are circumferential dikes feeding eruptive fissures along caldera rims (see Sect. 7.5), whereas among the magma-filled fractures affected by the load of a positive relief are radial dikes feeding fissures within volcanic edifices (see Sect. 7.6). In addition to these topographic stress components, other transient stress perturbations may affect the shallow propagation path of magma. These mainly include the pressurization of intrusions emplaced outside magma chambers and the stress perturbations induced by large earthquakes or slow slip events. Finally, when magma approaches the surface to feed an eruption some distinctive deformation features are observed, independently of the type of feeder dike, be it regional, circumferential or radial (see Sect. 7.7).

In synthesis, the stress state affecting the propagation path of magma-filled fractures at any time, t, and at any point in the crust (of coordinates x, y, z) results from the superposition of the following components (Rivalta et al. 2019):

$$\sigma_{TOT} = \sigma_T + \sigma_C + \sigma_L + \sigma_U + \sigma_I + \sigma_E \quad (7.1)$$

where σ_{TOT} is the supra-lithostatic full stress tensor, σ_T is the regional tectonic stress tensor, σ_C is the stress perturbation due to magma chamber pressurization, σ_L is the stress perturbation due to edifice load, σ_U is the stress perturbation due to topographic unloading (caused by calderas, flank collapse or icecap melting), σ_I is the stress perturbation due to previous magmatic intrusions and σ_E is the stress perturbation due to previous large earthquakes or slow slip events. Not all these components may be present at the same time or in any one location, and in the case of multiple contributions, the weight of each term may be quite different. For example, the stresses due to the gravitational loading of a volcanic edifice are generally larger than magma chamber pressurisation stresses. For realistic edifice and magma chamber pressurisation levels, edifice stresses dominate everywhere, except very close to the magma chamber walls (Roman and Jaupart 2014). In the case of multiple contributions, coupling effects may be expected, for example edifice loading influencing magma chamber shape.

Before discussing in the following sections the different patterns of shallow magma transfer resulting from several of these contributions, here a general rationale for the distinctive role of topography in inducing local stress variations is introduced. In fact, the general problem concerning the control of the stress field induced by topography on shallow magma propagation has been so far only considered in the specific case of caldera unloading in Sect. 5.9.

Topographic gradients may significantly influence the shallow propagation path of magma, producing deviations from expected patterns, with important implications for the location of eruptive vents. In general, any topographic gradient, associated with a positive or negative relief, induces a near-field stress: the larger the relief, the stronger the induced local stress, which also extends farther. The outcome of this local stress field is not only determined by its absolute value and extent, but also by its relative intensity with regard to that of other near- or far-field stresses, as well as the density, pressure and direction of propagation of magma (e.g., Albino et al. 2019).

In cross-section view, the load from a positive relief induces a local variation in the trajectories of the maximum principal stress σ_1. These pass from a subvertical direction at depth to an outward direction within the relief, converging towards its summit. This local outward inclination of σ_1 thus contrasts with the vertical (or horizontal) orientation of σ_1 resulting from uniform topography (i.e., flat) or from Andersonian stress conditions (Fig. 7.1; Dieterich 1988; Kervyn et al. 2009).

In map view, in the case of a subcircular volcanic edifice the σ_1 follows a radial pattern (with the minimum principal stress σ_3 being concentric). In the case of an elongated edifice or ridge the σ_1 becomes parallel to the direction of ridge elongation, with the σ_3 perpendicular to the axis of elongation (see Sect. 7.6). As magma-filled fractures propagate parallel to the σ_1 direction and perpendicular to the σ_3 direction, the distribution of the stress directions also allows representing the likely magma propagation trajectories. In section view, below an elongated positive relief (or ridge), magma propagating upward at some lateral distance would tend to intrude towards the ridge, developing magmatic sheets inclined towards the ridge and parallel to it. Therefore, vertically propagating magma tends to be attracted by the ridge (Fig. 7.1; Muller et al. 2001; Pinel and Jaupart 2004; Maccaferri et al. 2011). In the case of a conical edifice, this behaviour is less evident, because the concentric direction of the minimum principal stress σ_3 at the base of the edifice favours the propagation of radial dikes. This hinders the propagation of sheets parallel to the direction of the slope of the edifice (i.e., circumferential sheets, in this case), as inferred for a ridge. The three-dimensional problem of magma propagation below a conical edifice has been partially investigated (e.g., Kervyn et al. 2009; Derrien and Taisne 2019) and, for simplicity, the following discussion considers the situation only in two dimensions, that is in cross-section view. In section view, approaching the base of a positive relief, the stresses imposed by its load become progressively stronger. This behaviour may be illustrated by the distribution of the horizontal stress normalized to the load of the relief as a function of the depth normalized to the radius of the cone. This distribution indicates that

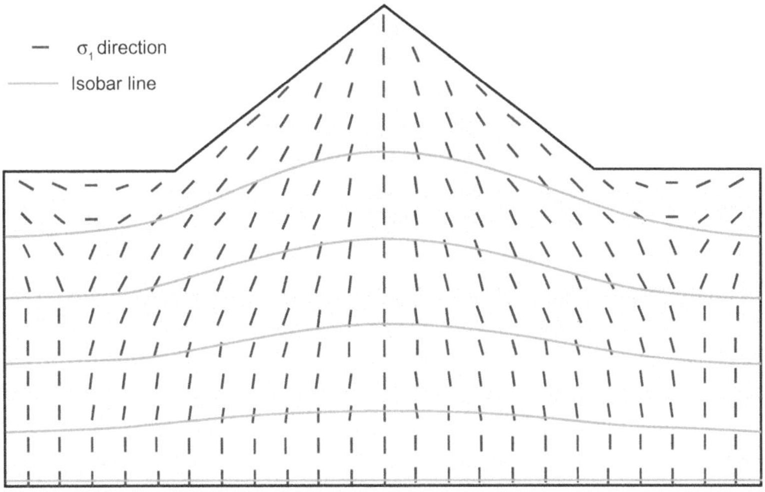

Fig. 7.1 Distribution of the maximum principal stress σ_1 and related isobar lines within a positive relief of triangular shape in section view (possibly representing a conical volcanic edifice or an elongated ridge) and its substratum (Modified after Kervyn et al. 2009)

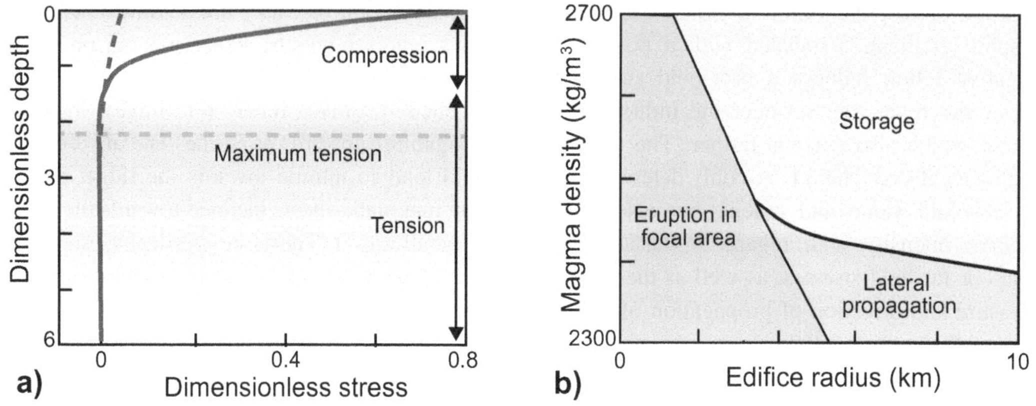

Fig. 7.2 **a** Dimensionless stress (horizontal stress normalized to the load of the relief) at the axis of an edifice (red continuous line) and at a distance two times the edifice radius (red dashed line) as a function of dimensionless depth (depth normalized to the radius of the circular relief). With increasing depth, the edifice induces compression and then tension; tensile stresses are largest at the depth of the dashed orange line. **b** Various possibilities (storage, propagation, eruption) for a dike beneath a volcanic edifice as a function of magma density and edifice size. Calculations are for a magma rising from 10 km with initial overpressure of 20 MPa and passing through lower and upper crusts of densities of 2700 and 2200 kg/m^3, respectively (Modified after Pinel and Jaupart 2004)

the maximum compression induced by the cone occurs at its base (dimensionless depth = 0), with the load of the relief becoming negligible only at a dimensionless depth of ~1.5 (Fig. 7.2a; Pinel and Jaupart 2000, 2004).

Therefore, the compression due to the load of a cone hinders the opening of shallow magma-filled fractures and thus magma propagation within the cone. For magma propagation within a conical volcanic edifice, a more buoyant magma is required. This may be attained with crystallization and differentiation, so that the residual magma tends to become lighter, favouring eruption. At the same time, the edifice may grow with eruptive and intrusive activity, counteracting these effects. Therefore, edifice growth and magmatic differentiation may be parallel although competing processes in the evolution of a volcano (Pinel and Jaupart 2000). This competition ends when large edifices, having reached a certain size, inhibit vertical dike propagation even with particularly buoyant magma. In this case, as the pressure gradient below the volcano's load can favour lateral magma propagation towards lower confining pressure (Fig. 7.1), dikes may propagate horizontally below the edifice, eventually erupting in a distal area. Therefore, depending upon the magma density and the volcanic edifice radius, three main shallow regimes for magma transport can be identified: (a) eruption within the volcanic edifice, usually with smaller edifices and buoyant magma; (b) storage beneath the edifice, usually with larger edifices and poorly buoyant magma; (c) lateral propagation feeding a distal eruptive vent, usually with larger edifices and more buoyant magma (Fig. 7.2b; Pinel and Jaupart 2004; Kervyn et al. 2009).

Any "attraction" of positive topographic features on magma propagation can manifest at very different scales, although the larger the relief, the stronger its load on the basement and the wider its attracting distance. At a crustal scale, an application to eastern Sicily (Italy) shows how dikes are deflected towards the foot wall of the ~2 km high Malta Escarpment, a major lithospheric discontinuity, explaining the location and orientation of the Hyblean and Etnean volcanism as a function of the changes in the regional stress field. At a local scale, even active normal fault scarps less than 100 m high in rift zones may attract monogenic volcanoes on their foot wall as a consequence of the shallow deflection of feeder dikes (Maccaferri et al. 2015; Neri et al. 2018).

In the case of topographic depressions, there is an unloading imposed by the mass removal. As explained in Sect. 5.9, this produces a local rotation of the minimum principal stress σ_3, which is subvertical below the depression and becomes progressively subhorizontal towards the sides. As a result, different types of magma-filled fractures are emplaced below different parts of the depression: sills predominate in the central portion, laterally passing to inclined sheets and then to dikes towards the sides. In contrast to positive reliefs, depressions tend to deflect magma propagation paths towards their sides. Magma diverges from the depression, propagating laterally and eventually reaching the surface at some distance outside the depression. Again, this behaviour is scale-independent. At a wide scale, an application to rift zones suggests that the frequently observed off-rift volcanism, as for example outside the Main Ethiopian Rift, results from the lateral propagation of dikes to the rift sides (Fig. 7.3; Maccaferri et al. 2014). At a local scale, the presence of circumferential fissure eruptions outside the rim of a caldera can be explained by the effective unloading induced by the caldera depression.

7.3 Regional Dikes

Regional dikes develop in the presence of a far-field stress and are usually aligned perpendicular to the regional minimum principal stress σ_3: in particular, the lower is the intensity of σ_3, the longer the dikes. Regional dikes may thus have variable extent depending on the tectonic setting, for example extending well beyond a volcanic edifice, as along divergent plate boundaries: here regional dikes may show a remarkably consistent strike for several tens of kilometres, with a thickness of several metres, on average larger than that of non-regional dikes (Fig. 7.4; e.g., Gudmundsson 1998).

Regional dikes may be easily recognized in eroded crustal terrain, as along river valleys, or, in absence of any erosion, by the alignment of

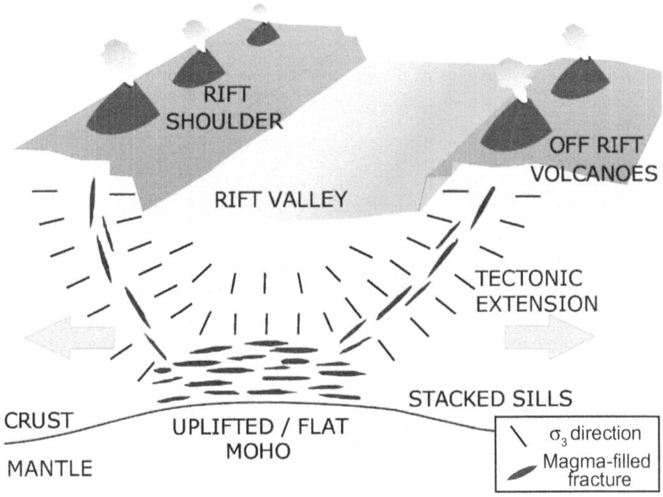

Fig. 7.3 Effect of crustal unloading on magma propagation. Magma-filled fractures become progressively steeper, passing from sills to sheets to dikes, approaching the shoulders of a rift depression as a consequence of the decrease in the plunge of the minimum principal stress σ_3 trajectories due to the unloading of the depression (Maccaferri et al. 2014)

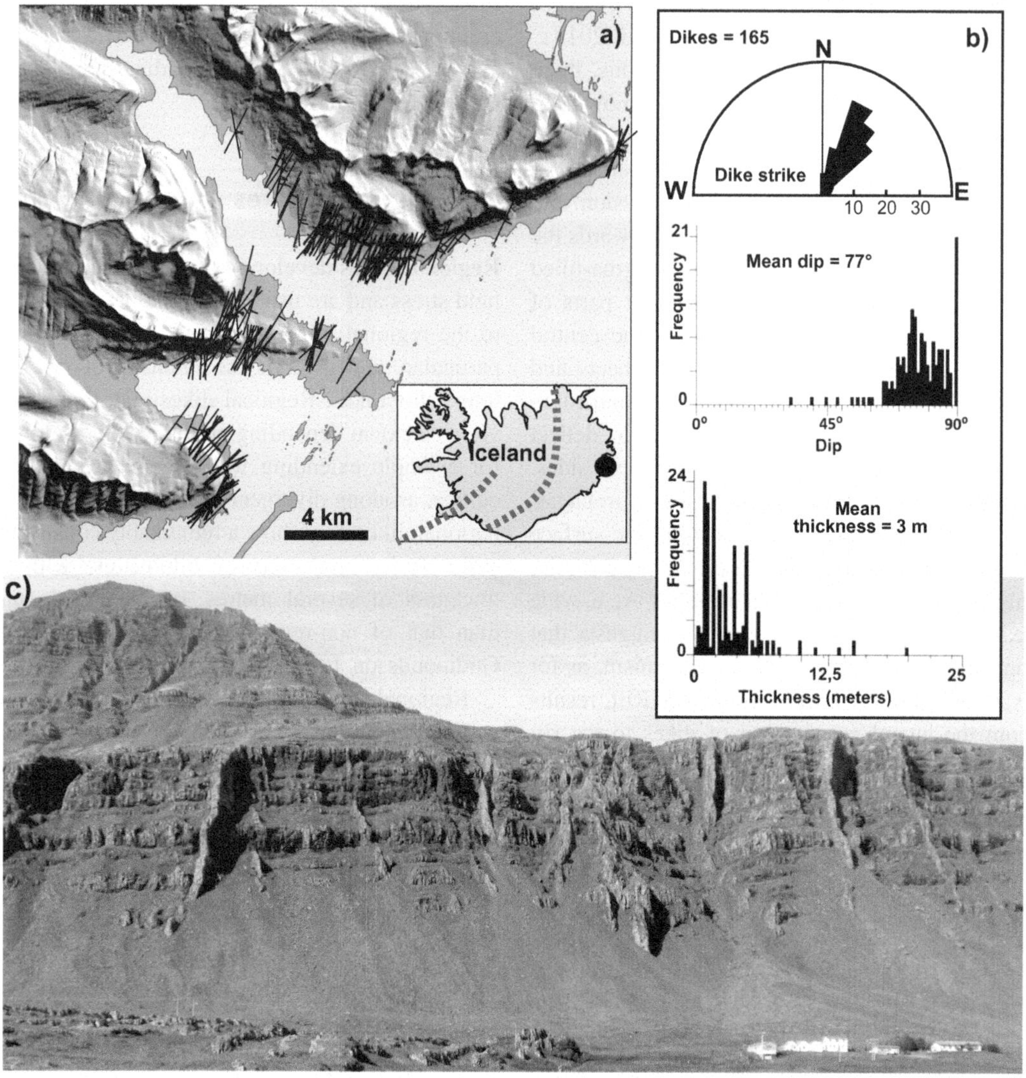

Fig. 7.4 Eroded regional dikes outcropping in the extinct Alftafjordur magmatic system (east Iceland). **a** Overview of the dike pattern, shown as black lines; black dot in inset indicates the location of the magmatic system and red dashed lines show current rift zones. **b** Dike strike, dip and thickness distribution. **c** View of a portion of the subparallel dikes; houses in the foreground provide a scale (Modified after Urbani et al. 2015)

eruptive fissures they eventually feed. Dike-fed eruptive fissures consist of aligned vents developed when the top of a propagating dike reaches the Earth's surface. Along divergent plate boundaries, the regional dikes may feed eruptive fissures for several tens of kilometres, as the ~75 km long Holocene Sveinar-Randarholar eruptive fissure (Iceland), the ~25 km long Lakagigar 1783–84 fissure (Iceland) or the ~60 km long 2005 Dabbahu intrusion (Afar) (Fig. 7.5).

The location of vents along a regional feeder dike is mainly influenced by the local topography. Vents fed by regional dikes usually develop in low relief areas, as at the base of volcanic edifices, especially in the saddle between

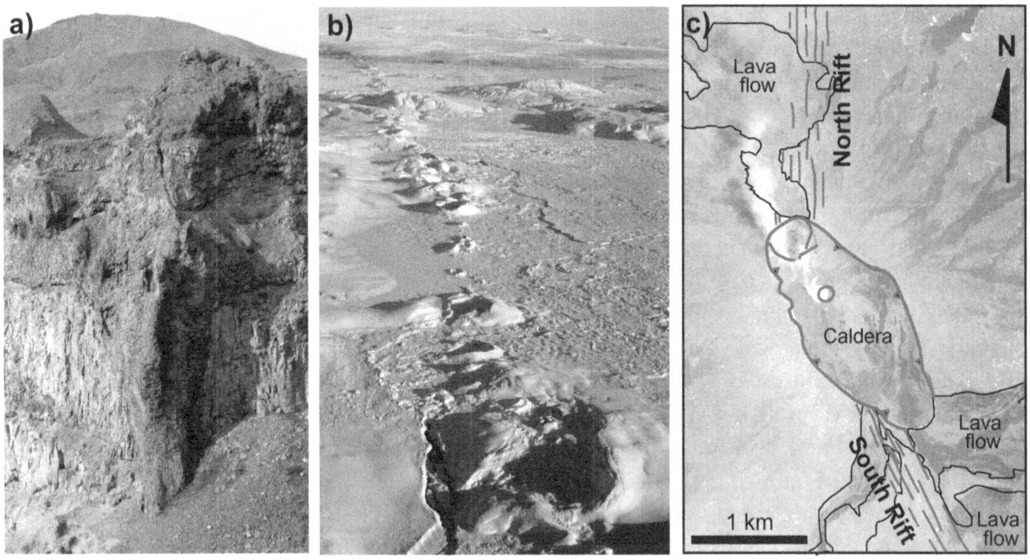

Fig. 7.5 Regional dikes, eruptive fissures and volcanic rift zones. **a** Image of a ~8 m thick regional dike, eroded in a canyon, feeding the ~75 km long Holocene Sveinar-Randarholar eruptive fissure (Iceland); above the dike are the remnants of two eruptive cones (the one in the foreground partially eroded) fed by the dike. **b** Northern portion of the ~25 km long 1783–1784 Lakagigar eruptive fissure (Iceland; photo courtesy Agust Gudmundsson). **c** Volcanic rift zones consistent with the regional tectonic setting, branching from the caldera on the summit of the Erta Ale shield volcano, Afar, Ethiopia; base image from Google Earth

consecutive edifices with prominent topography. In fact, while the load of a volcanic edifice induces a local increase in the principal stresses, including the minimum principal stress σ_3, a depression causes a local decrease in the principal stresses, including σ_3. If a regional dike propagates laterally from a volcano, the edifice load may thus hinder its opening until this reaches a flat area, where the intensity of the regional σ_3 decreases and the dike may widen and erupt. If another volcanic edifice is present to the front of the propagating dike, its load may increase the principal stresses, including σ_3, arresting the lateral propagation of the dike. Such behaviour has been postulated for the dike feeding the 2014 Bardarbunga (Iceland) eruption, which propagated laterally for tens of kilometres, erupting in the saddle between the Bardarbunga and Askja volcanoes. The dike followed a complex path, with some deviation from the regional pattern, reconstructed and explained applying the minimum energy principle, which also considers the role of topography (Heimisson et al. 2015; Urbani et al. 2017).

Geophysical evidence suggests that regional dikes along divergent plate boundaries nucleate from the base of a magma chamber replenished with basic magma and then propagate laterally, perpendicular to the regional σ_3, for tens of kilometres (see Chap. 11; Pinel and Jaupart 2004; Ebinger et al. 2010). Conversely, other petrological and volcanological models suggest predominant vertical propagation of regional dikes, fed with primitive magma derived directly from the mantle (Peccerillo et al. 2007; Hartley and Thordarson 2012). In principle, the sequence of vent opening, if established from direct observations or field data, may be used to discriminate the direction of propagation of a feeder dike in areas with relief. This derives from the intersection of the top of the propagating dike with an inclined slope on the Earth's surface, generating a set of vents. While laterally propagating dikes support a progression of dike-fed vents from high relief areas (on volcanic edifices) to low relief areas (outside volcanic edifices), vertically propagating dikes support an inverse progression of vents, from low to high relief areas.

Within large volcanic edifices, regional dikes may cluster along a narrow zone passing through the volcano summit perpendicular to the regional minimum principal stress σ_3. Here the regional dike swarm feeds a narrow but elongated zone of volcanism along the volcano slopes, that is a **volcanic rift zone**. Volcanic rift zones do not usually extend outside volcanic edifices and are thus smaller-scale analogues of the much longer rift zones observed along divergent plate boundaries. An example of a volcano with well-developed volcanic rift zones within a regional extensional setting is Erta Ale (Afar) (Fig. 7.5c). Volcanic rift zones may also be found in intraplate volcanoes, especially in oceanic volcanic islands (e.g., the Hawaiian and Canary Islands), where they are often associated with flank instability.

7.4 Magma Transfer Through Sills

After having considered the role of the regional stress field on shallow magma transfer, let us now consider the role of the local stress field. In this case, the shallow transfer of magma most commonly occurs through the propagation of dikes or, subordinately, inclined sheets, which are the focus of the following sections. Less commonly, magma is transferred through sills, although recent studies have re-evaluated the role of shallow sills in transferring laterally and even vertically magma under a local stress field, which is the focus of the present section.

Below volcanic edifices, there is frequent evidence for mafic sills transferring magma laterally and vertically over considerable distance. Sills nucleating from the side of oblate magma chambers may initiate the lateral transfer of magma, as observed during several unrest episodes at active volcanoes, ultimately explaining the location of peripheral eruptions. This is common in the western Galapagos calderas, where the transfer of magma feeding some recent eruptions initiated by the lateral propagation of sills from shallow magma chambers: here a sill may steepen, becoming a circumferential dike, or twist around a radial axis, becoming a radial dike. In some cases, narrow sills may propagate radially below volcanic edifices for considerable distances (up to 20 km), feeding distal vents, as during the 2018 eruption at Sierra Negra (Galapagos; Bagnardi et al. 2013; Bagnardi and Lundgren 2018). In several cases, the shallow plumbing system of polygenic or monogenic volcanoes may largely or predominantly consist of sills. In fact, many mafic complexes in sedimentary basins are made of interconnected sills and inclined sheets, often arranged forming saucer-shaped sills, mainly transferring magma laterally, as for example at Hopi Buttes (Arizona, USA), in the Karroo Basin (South Africa) and the Yilgan craton (Australia). These sill complexes provide an alternative perspective to feed volcanic activity (Fig. 7.6; Leat 2008; Magee et al. 2016; Muirhead et al. 2016).

The importance of sills in the shallow transfer of magma is not restricted to the local stress field of a magma chamber or a plumbing system. A striking example is represented by the sills of the Ferrar magmatic province of Antarctica, Australia and New Zealand, which show remarkably homogeneous compositions for thousands of kilometres, implying a single magma source, which coincides with the head of a mantle plume. Paleogeographic reconstructions suggest that the Ferrar sill complex extends for more than 4100 km from the proposed site of the plume head, implying that extensive lateral magma transport within mafic sill complexes may ultimately feed volcanoes even at thousands of kilometres from the melt source. Field evidence of the Ferrar magmatic province in South Victoria Land, Antarctica also shows that sill-fed dikes were the likely feeders for the voluminous (70,000 km^3) flood basalt eruptions. This sill-fed dike network contrasts with dike swarms conventionally depicted to feed flood basalt provinces, resembling a variably "cracked lid" atop a sill complex (Muirhead et al. 2014).

These examples highlight the importance of sills in propagating magma laterally and of saucer-shaped sills in propagating magma both laterally and vertically. Whether magma propagates vertically and/or laterally mainly depends on the conditions occurring at the sill edge. In the

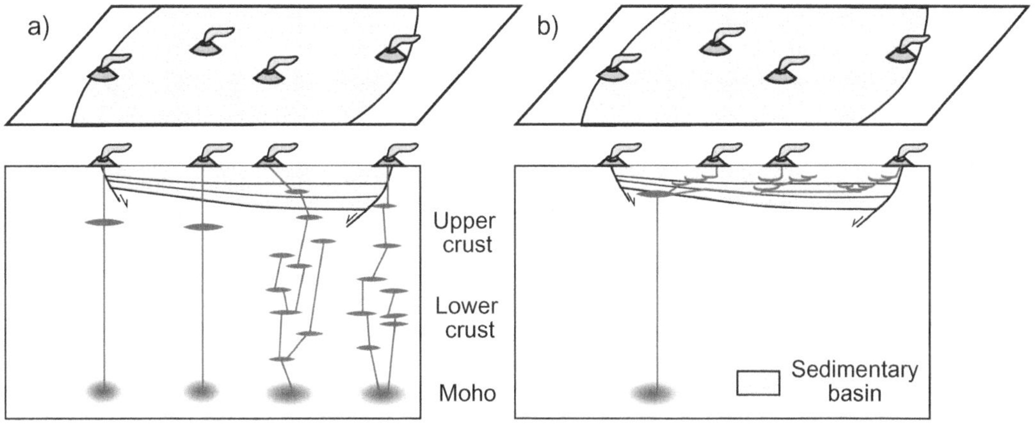

Fig. 7.6 Schematic oblique view of a volcanic province and section view representation of the related possible end-member magma plumbing systems. **a** The traditional perspective, whereby volcanoes are fed via dikes and directly over the melt source and shallow-level reservoirs. **b** Alternative model, were upon entering a sedimentary basin, a laterally extensive sill complex forms, feeding volcanoes that are also laterally offset from the melt source (modified after Magee et al. 2016)

case of sills migrating along stratigraphic bedding, a predominant lateral propagation of magma may be expected, whereas sill propagation through the bedding may develop staircase-like or saucer-shaped sills, thus allowing the magma to propagate also vertically. The propagation of a sill along or through the bedding thus ultimately depends on whether it can transition to an inclined sheet. As discussed in Sect. 4.3, this is a function of the mechanical and thermal conditions occurring at the sill tip. The mechanical conditions include the local damage, which results from the tensile and/or shear stresses in the host rock in the vicinity of the propagating sill tip. When these reach the strength of the host rock, failure occurs, guiding the upward propagation of an inclined sheet. The thermal conditions include the capability of the magma within a sill to flow along the interface or, conversely, to experience partial solidification, arresting lateral propagation, promoting thickening and eventually the breaching of magma along an inclined sheet. Despite the generally understood theoretical framework, it is difficult to determine which of these mechanisms may control the propagation of a specific sill in nature, and more field investigations are needed to document these possible processes.

In addition to the vertical propagation of magma through sills as magma-filled cracks, the vertical transfer of magma in a system of stacked sills may occur through compaction of mush and segregation of melt via porous flow. While this mechanism can operate at any crustal level, the emphasis here is on the possibility that stacked sills at the level of a magma chamber promote a percolative melt transfer able to create melt-rich regions which can move through the regions of mush typical of large magmatic systems (e.g., Cashman et al. 2017, and references therein). For example, this condition has been recently suggested for the magmatic system of Axial Seamount, Juan de Fuca Ridge. Here the shallowest melt-rich portion of the magma chamber overlies a 3–5 km wide conduit of vertically stacked melt lenses, with near-regular spacing of 300–450 m extending into an inferred mush zone of the lower crust. These near-horizontal lenses are interpreted as melt-rich layers formed via mush compaction. Melt ascent by porous flow focused within the melt lens conduit has led to the last three observed eruptions at Axial Seamount, in 1998, 2011 and 2015 (Carbotte et al. 2020).

This process is based on the **porous flow** model of melt migration, where the separation of melt from the residual solid requires that melt flows for at least some distance through a connected network of grain-size channels. In a first approximation, the flow of melt may be described by Darcy's law:

$$Q_m = \frac{K_p}{\eta}\frac{dP}{dx} \quad (7.2)$$

where Q_m is the flux of melt (i.e., the volume of melt per unit time passing a unit cross-sectional area perpendicular to the flow direction), K_p is the permeability of the network (depending on the grain size, the melt fraction and the tortuosity of melt flow), η is the viscosity of the fluid, and dP/dx is the pressure gradient that drives melt flow (Daines and Pec 2015, and references therein). The driving force for melt migration can be chemical and/or physical: chemical driving forces can be related to differences in interfacial energies, whereas physical forces can be related to the buoyancy of melt relative to residual solids or to applied deviatoric stress. Reduction of the total amount of melt, possibly due to rapid melt extraction relative to the melt production rate, implies that the solid matrix must deform to accommodate the change in fluid volume. When the melt volume entering the system is different from the volume leaving the system, Darcy's law must be modified to take this fact into account. The reduction in pore volume requires gravity-driven compaction of the matrix. Compaction may include rearrangement of matrix grains without any distortion of individual grains (granular flow) or changes in grain shape. The process can be visualized by a column of partially molten rock that compacts due to density differences between melt and solid. At its base, melt volume decreases as the matrix compacts. Above this compacting layer, the upward flux of melt from the compaction zone maintains the melt volume. Melt segregation may occur into melt channels that are larger than those present along individual grains, allowing for much more rapid melt migration. At the top of the column, melt accumulates, forming a melt-rich layer. In this process, the movement of melt due to buoyancy is resisted by the strength of the matrix and the viscosity of the melt (e.g., Daines and Pec 2015, and references therein). Therefore, stacked sills systems may also promote vertical magma transfer in the shallow crust via porous flow.

7.5 Circumferential Intrusions

The local stress field imposed by a magma chamber may be responsible for a circumferential intrusive pattern consisting of arcuate dikes and sheets concentric to the volcano. In active volcanoes, these intrusions can be easily spotted by the presence of circumferential eruptive fissures around calderas (see Sect. 7.5.1). In eroded extinct volcanoes, circumferential intrusions may be observed in the crustal volume above the magma chamber. These circumferential intrusions are distinguished, accordingly to their dip direction and mechanism of formation, in cone sheets or ring-dikes. Cone sheets are inward dipping circumferential intrusions, with variable dip, and are usually related to overpressurized magma chambers (see Sect. 7.5.2). Ring-dikes are outward dipping to subvertical circumferential dikes and usually related to underpressurized magma chambers (see Sect. 7.5.3).

In principle, circumferential eruptive fissures may be fed by both cone sheets and ring-dikes. In practice, as explained below, available evidence shows that the best-developed circumferential fissures are associated with caldera unloading, highlighting a mechanism of formation distinct from those of cone sheets and ring-dikes.

7.5.1 Circumferential Eruptive Fissures

On active volcanoes, circumferential eruptive fissures are the evidence of the emplacement of

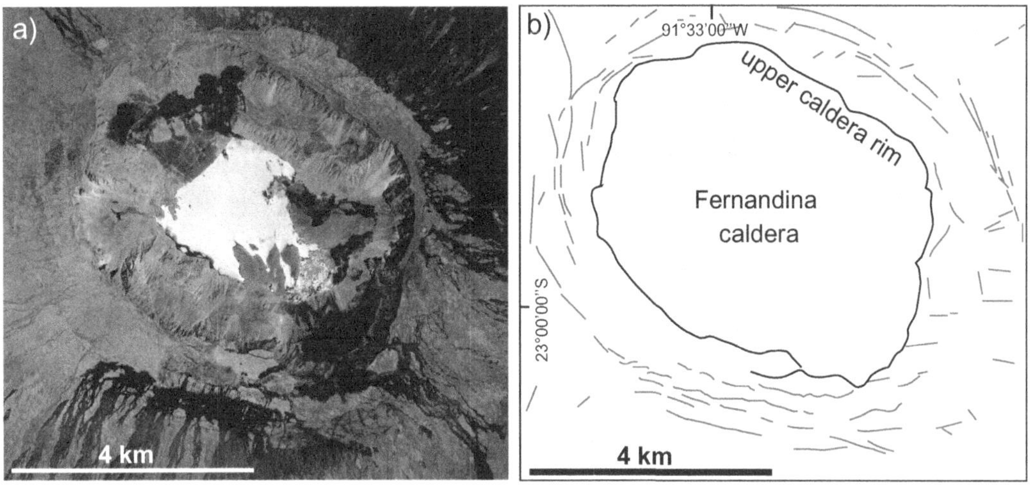

Fig. 7.7 **a** Satellite mage of Fernandina caldera (from Google Earth), showing the pattern of eruptive fissures, mostly circumferential, and related lava flows. **b** Line drawing of (**a**), highlighting the main circumferential eruptive fissures of different age (red lines)

circumferential dikes, which are arcuate magma-filled fractures developed concentrically around or within a volcano. Circumferential fissures of limited extent have been reported along the rim of a few calderas, including Medicine Lake (California, USA), Suswa (Kenya), Askja (Iceland) and Rapa Nui (Chile; Acocella and Neri 2009). However, the best-exposed circumferential systems are by far found in the active volcanoes of the western Galàpagos Archipelago, especially at Cerro Azul, Darwin, Fernandina and Wolf (Fig. 7.7; Chadwick and Howard 1991). Here dikes feed many circumferential eruptive fissures immediately outside the caldera rim. Their frequency rapidly decreases with distance from the caldera rim, transitioning to only radial fissures on the lower slopes of the volcanic edifices. One of these circumferential dikes was caught in the act of propagating immediately before the 2009 eruption of Fernandina. Geodetic observations revealed that a sill propagated laterally from the shallow magma chamber beneath the caldera and progressively steepened towards the caldera rim, to become an inward dipping sheet and then a subvertical dike feeding a circumferential eruptive fissure just outside the caldera rim (Fig. 7.8; Bagnardi et al. 2013).

A fundamental requisite to develop these circumferential dikes in the western Galapagos is the topographic depression of the caldera, which controls the orientation of the minimum principal stress σ_3. As explained in Sect. 5.9, a caldera depression creates a mass deficit responsible for a subvertical σ_3 below the caldera centre, which becomes progressively subhorizontal approaching the caldera sides. This stress field explains the progressive intrusion of a sill below the caldera, transitioning to an inclined sheet toward the caldera rim and finally becoming a subvertical dike just outside, as observed at Fernandina (Fig. 7.9; Corbi et al. 2015). The significant unloading induced by the western Galapagos calderas, which are relatively deep and large and with limited infill, explains their unique well-developed pattern of circumferential eruptive fissures.

The intensity of this unloading, or excavation stress, depends on the amount of caldera subsidence and the thickness and density of any caldera infill, including water (as lakes, sea, or glaciers). These features determine a larger or smaller mass deficit, which provides a first order control on the distal or proximal shallow transfer of magma, respectively. However, the lateral distance reached by the circumferential dikes and related

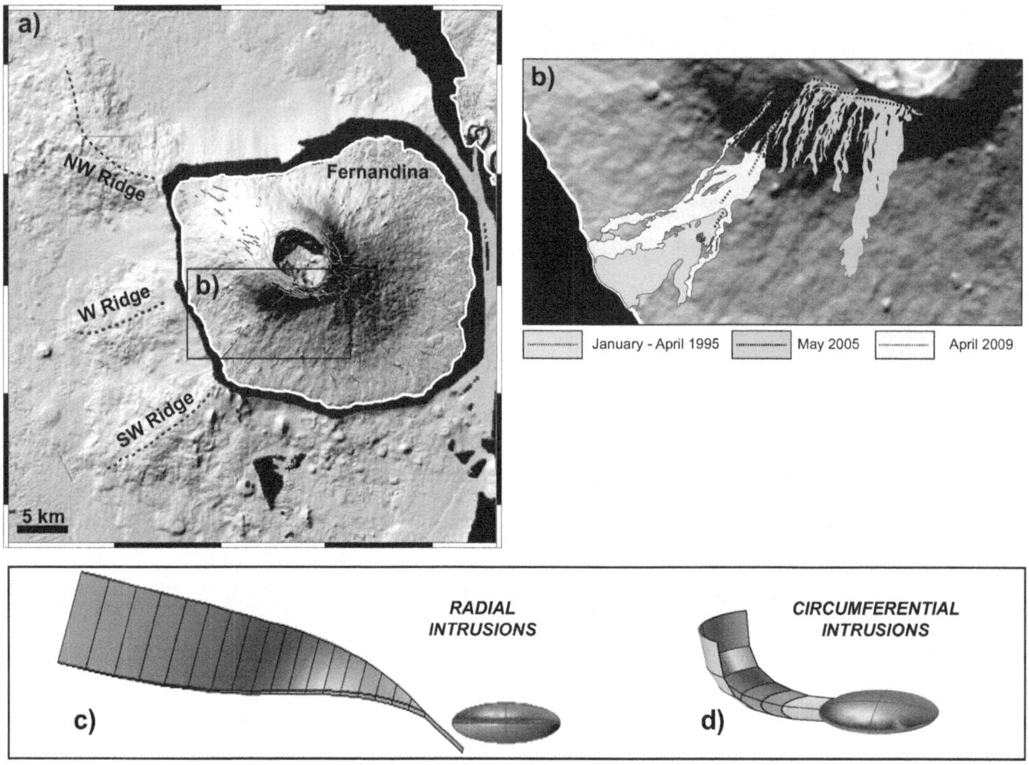

Fig. 7.8 **a** Shaded relief map of Fernandina and surrounding ocean floor. Solid lines represent circumferential (in blue) and radial (in red) eruptive fissures. Purple-dashed lines mark submarine ridges. **b** Close-up covering the location of three consecutive eruptions on Fernandina. Dashed lines mark eruptive fissures. **c, d** 3D representation of radial (**c**) and circumferential (**d**) intrusions; in purple the ~1 km deep magma reservoir, in yellow the intrusions feeding fissure eruptions (image courtesy of Marco Bagnardi; modified after Bagnardi et al. 2013)

eruptive fissures also depends on the regional stresses, the depth to the magma chamber and magma buoyancy (Fig. 7.10; Corbi et al. 2015; Acocella and Rivalta 2019). In particular, any regional extension creates steeper dikes and shifts dike trajectories inward, promoting eruptions from circumferential vents within the caldera rim, whereas any regional compression creates more gently dipping dikes and shifts dike trajectories outward, promoting eruptions from vents along or

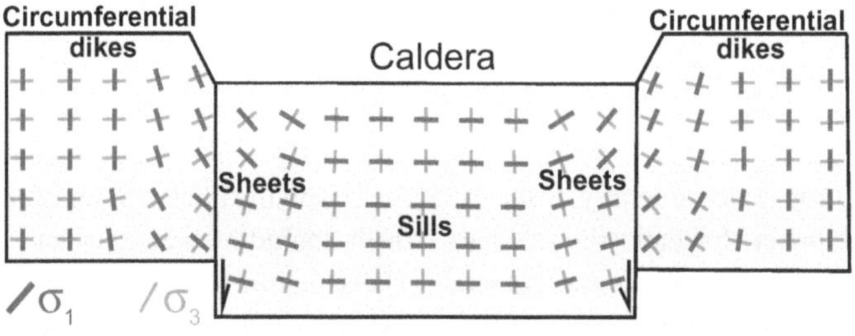

Fig. 7.9 Schematic reconstruction of the maximum and minimum principal stresses σ_1 and σ_3, respectively, induced by the unloading of a caldera depression in section view. Below the caldera, where subvertical σ_3 dominates, sills are promoted; conversely, towards the caldera rim, where subvertical σ_1 dominates, circumferential dikes are promoted; in between, sheets connect sills with dikes

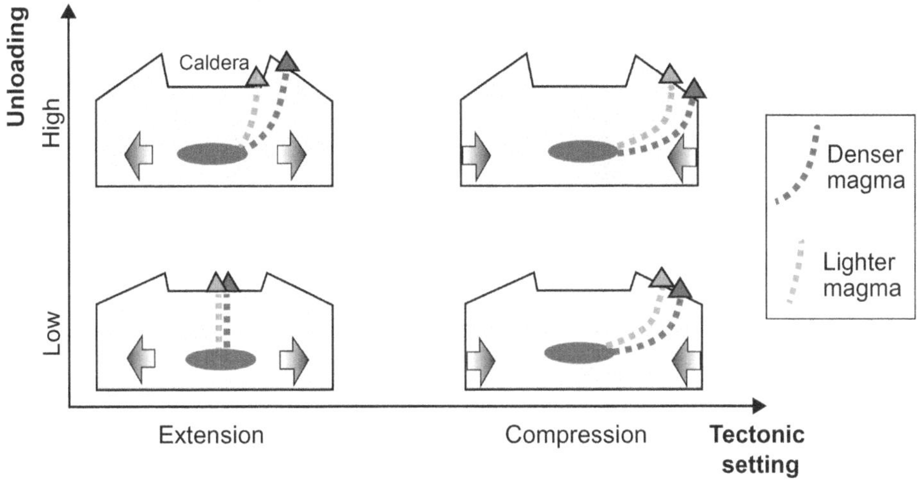

Fig. 7.10 Role of unloading, tectonic setting and magma density on selected propagation paths of magma-filled fractures feeding circumferential fissures in calderas, as shown by four end-member conditions; denser magmas may not always make it to the surface.

outside the caldera rim. Also, a deeper magma chamber requires a longer feeder dike and, in turn, a higher magmatic pressure, which inhibits the lateral propagation of sills. This feature may explain why lateral sill propagation is particularly relevant at volcanoes with shallow magma chambers, as at the western Galapagos and Piton de la Fournaise after 2007. A similar effect is also played by the buoyancy of the intruded magma, with more buoyant magma inhibiting the lateral propagation of sills, and weakly buoyant magma feeding gently dipping sheets.

While the origin of the circumferential eruptive fissures outside the caldera rim may be best explained by the steepening of inward dipping magma-filled fractures, as described above, circumferential fissures may in principle also derive from the emplacement of inward dipping cone sheets in the absence of calderas and outward dipping ring-dikes during caldera collapse, as discussed below.

7.5.2 Cone Sheets

Cone sheets are inward dipping magma-filled fractures arranged in a conical swarm above a magma chamber, and usually seen in deeply eroded crustal exposures. Within a cone sheet swarm, sheet dip may vary from that of nearby sheets and also within the same sheet, resulting in different geometries (Fig. 7.11; Ancochea et al. 2014; Tibaldi 2015): (a) concave-downward (trumpet-shaped) sheets with increasing dip closer to the magmatic source, without sheets in the central part; (b) concave-upward (bowl-shaped) sheets with decreasing dip with depth, without sheets in the centre; (c) conical sheets with increasing dip towards the centre, where intrusions show higher dip; (d) conical sheets with constant dip originated from an oblate (sill-like) magma chamber, without sheets in the centre. The steeper sheets in the centre of the swarm, dipping at high angle (up to $\sim 70°$), contribute to magma transport upward into the volcano, whereas the outer less inclined sheets, dipping as low as $\sim 30°$, tend to transfer magma laterally, eventually feeding distal eruptions. The average inclined sheet dip angle in the British Isles has been found at $\sim 45°$, at Gran Canaria (Canary Islands) at $\sim 41°$ and at some individual sheet swarms in Iceland at $\sim 34°$.

Cone sheets with the above mentioned features have been described at several eroded volcanoes. Their overall shape and distribution is important, as they can be used to reconstruct the location, depth and shape of the underlying,

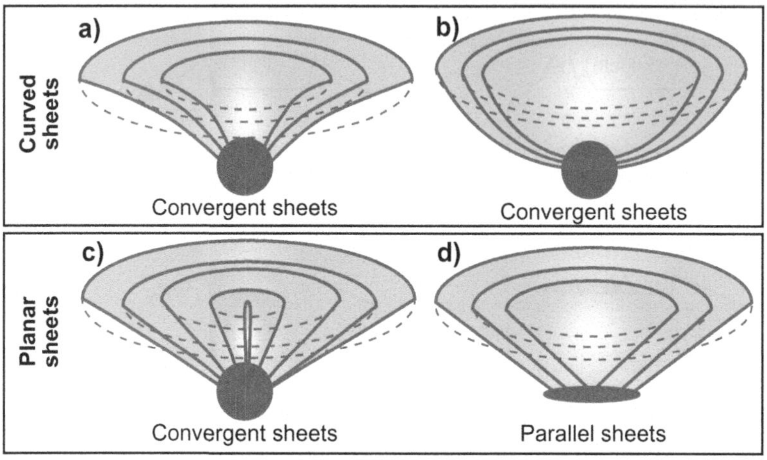

Fig. 7.11 Possible geometries of cone sheet swarms produced by an overpressurized magma chamber (red circle or ellipse): **a** concave-downward (trumpet-shaped) sheets from a spherical chamber; **b** concave-upward (bowl-shaped) sheets from a spherical chamber; **c** radial planar sheets from a spherical chamber; **d** planar parallel sheets from a sill-like chamber (Modified after Tibaldi 2015)

unexposed source magma chamber. Such a reconstruction requires careful mapping of the trace and attitudes of cone sheets. The location where the projected cone sheets meet in the subsurface is assumed to be where the source magma chamber lies. In some cases, the swarm geometry may change as the pressure source at depth changes (Burchardt 2018).

The Cuillin Complex on the island of Skye (Scotland) is an example of cone sheet system consisting of a stack of parallel concentric sheets, rather than sheets converging towards a single focus. Dip angles vary between 40° and 50° and sheet thickness ranges from a few centimetres to 5 m, with a dominant thickness of 1 m. Intrusion intensity ranges from 1 to 35%, increasing towards the central zone. Two successive magma chambers at different depths, with flat geometries and different volumes, may explain the sheet architecture and locations here (Tibaldi et al. 2011). Conversely, the Boa Vista (Cape Verde Islands) and La Gomera (Canary Islands) felsic cone sheet swarms are characterized by a decreasing dip outward, together with a constant dip of each individual sheet. At Boa Vista the average inclination of sheets is ~40° next to the centre and decreases to ~30° next to its periphery. The magmatic focus of the sheets is over 3 km deep (Ancochea et al 2003, 2014). In the Thverartindur igneous centre (Iceland) there is a ~0.1 m/km decrease in the average sheet thickness up through the swarm, with a relatively low number of less than 0.5 m thick sheets in the uppermost part of the swarm, reflecting the subsurface thermal arrest of more than a third of all sheets (Klausen 2004). In some cases, cone sheets are evidence for the shallow intrusion of a significant volume of magma, as in the Miocene Tejeda caldera on Gran Canaria (Canary Islands). Here the estimated aggregate volume of evolved cone sheet magma emplaced at less than 2000 m below sea level amounts to ~250 km^3, although the material erupted during the same period exceeds by far 500 km^3 (Fig. 7.12; Schirnick et al. 1999).

Several models have been proposed to explain the formation of cone sheets, building on the original intuition that these may be related to an increase in pressure of a magma reservoir, favouring the formation of inward dipping fractures, whose intrusion produces the cone sheets. These models largely differ in the relationships between the cone sheets and the stress field responsible for their formation. In the classic Anderson (1936) model, cone sheets propagate as tensile (Mode I) fractures within the σ_1–σ_2 plane, while in the Phillips (1974) model, cone sheets occupy shear fractures and

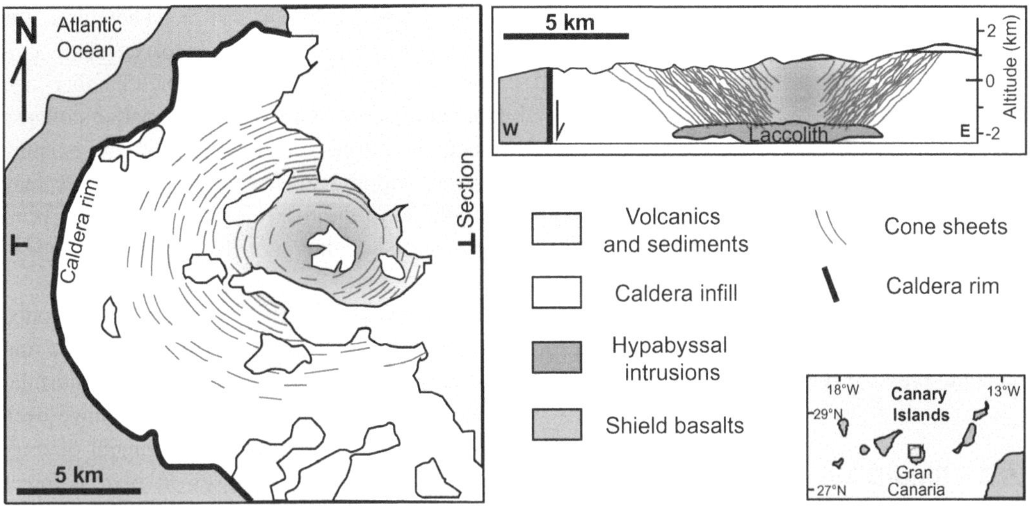

Fig. 7.12 Simplified geologic map and E-W cross section showing the inferred spatial relations and geometry of the Tejeda intrusive complex, particularly the cone sheet swarm, of Gran Canaria (Canary Islands; modified after Schirnick et al. 1999)

are oblique to the principal stresses (Fig. 7.13; Bailey et al. 1924; Anderson 1936; Phillips 1974).

Currently available data suggest that intrusive sheets mainly follow planes of fracturing that contain the maximum and intermediate principal stresses σ_1 and σ_2, thus being perpendicular to the least principal stress σ_3 and behaving as Mode I fractures (Mathieu et al. 2015; Tibaldi 2015). In some cases, as for example at the Isle of Skye (Scotland), a minor part of sheets intruded along shear planes, with lower angle planes showing normal motion and steeper planes reverse motion. This specific condition may be partly related to the intrusion of magma along pre-existing fractures misaligned with respect to the principal stress directions. Also debated is the role of the magma chamber shape in developing cone sheets, with recent studies challenging the possibility that cone sheets form

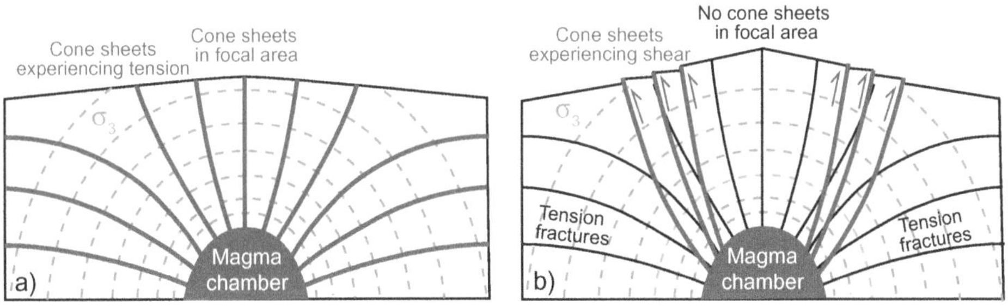

Fig. 7.13 **a**, **b** Stress field around a pressurized shallow magma chamber and resulting path of cone sheets according to **a** Anderson (1936) and **b** Phillips (1974). In **a** cone sheets propagate as tensile fractures, while in **b** occupy shear fractures (Sense of shear shown by red arrows; modified after Burchardt 2018).

from pressurized spherical or prolate magma chambers: accordingly, cone sheets would form only from inflating oblate shallow magma chambers (Bistacchi et al. 2012).

Recent studies highlight an overall similarity in the conditions promoting the development of cone sheets and circumferential dikes, both related to a pressurized magma chamber and mainly differing in the crustal unloading conditions. In particular, cone sheets may form only if the unloading pressure is much smaller than the lithostatic pressure (Gaete et al. 2019).

7.5.3 Ring-Dikes

Ring-dikes are outward dipping to subvertical annular intrusions usually found in crustal volumes between magma chambers and calderas. They are interpreted to form as a consequence of the decrease in pressure within the magma chamber, which promotes instability in the roof rock, resulting in subsidence along outward dipping annular fractures, whose intrusion by magma produces the ring-dikes (Fig. 7.14; Bailey et al. 1924).

The magma intruded within the ring-dike may eventually feed eruptions from arcuate or ring fissures at the surface, as for example inferred for the 760 ka old Bishop Tuff eruption at Long Valley (California, USA), when 700 km^3 of magma erupted associated with caldera collapse (Wilson and Hildreth 1997). Therefore, the development of ring-dikes involves some collapse of the roof of a magma chamber and magma injection within the pre-existing collapse structures. Moreover, as intruding pre-existing outward dipping reverse faults, ring-dikes may be oblique to the maximum and minimum principal stresses, thus not behaving as Mode I fractures.

Ring-dikes are best observed in the eroded portions of extinct calderas. The subvolcanic ring complexes outcropping in the western Peninsular Ranges Batholith (Mexico and USA) have been interpreted to describe the development of ring-dikes. These ring complexes involve a three-stage sequence consisting of: (1) fracturing of the roof above an overpressured magma chamber, resulting in inward dipping conical fractures partially hosting cone sheets; (2) subsequent withdrawal of magma from the chamber, which facilitates collapse of the roof along near-vertical ring faults partially hosting ring-dikes; (3) later intrusion of magma, which locally destroys evidence for the earlier history (Johnson et al. 2002). Interestingly, the ring complexes in the Peninsular Ranges Batholith were emplaced at depths of up to 18 km, suggesting either that caldera subsidence can extend to mid-crustal levels or that additional deeper processes may also produce ring complexes. Other notable ring-dike complexes include Ishizuchi (Japan), Ossipee (New Hampshire, USA) and Grizzly Peak

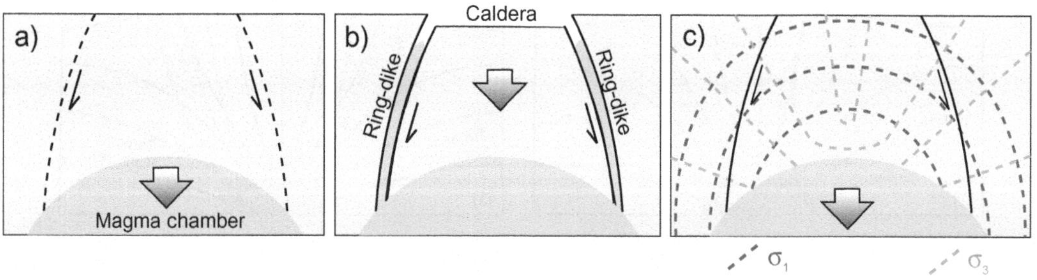

Fig. 7.14 Mechanism of formation of ring-dikes: **a** underpressure within a magma chamber generates incipient outward dipping reverse faults; **b** the subsequent collapse of the magma chamber roof dilates the previously formed faults, promoting magma intrusion to form ring-dikes eventually feeding eruptions; **c** stress conditions to develop the ring faults intruded by magma generating the ring-dikes

Caldera (Colorado, USA; Yoshida 1984; Fridrich et al. 1991; Kennedy and Styx 2007). In particular, the Ossipee ring-dike records the rejuvenation of the magma chamber and subsequent subsidence of the caldera, with the dike consisting of residual magma from a caldera-forming eruption and illustrating the interplay between the magmatic processes indicated by petrology and physical processes associated with caldera formation. In the asymmetric Grizzly Peak Caldera, welded-tuff ring-dikes are locally exposed at erosion levels below the caldera fill; these dikes are remnants of fissure vents in the outer ring fracture zone.

7.6 Radial Dikes

Radial dikes strike radially with respect to a central volcano, although radial dike swarms may be also found at a completely different scale, up to thousands of kilometres long and ~1000 m thick, when associated with the activity of mantle plumes (e.g., Ernst et al. 1995). At the local scale considered here, radial dikes do not extend significantly farther from the volcanic edifice, thus having an extent of up to a few tens of kilometres. These dikes are commonly a few metres thick, on average thinner than the regional dikes found in the same setting (Fig. 7.15). Radial dikes may develop at any level within a volcanic edifice, although they are mostly found in its shallower and deeper parts, probably reflecting two slightly different mechanisms of emplacement, both controlled by the load imposed by the topography of the volcano (Pinel and Jaupart 2004; Acocella and Neri 2009).

At the volcano base, if the load of the edifice becomes significant with regard to the regional tectonic stresses, it promotes the lateral propagation of dikes, generating radial patterns, as anticipated in Sect. 7.2. In this case, a decreasing supply rate, promoting lateral magma propagation, coupled with a decreasing magma viscosity may generate eruptive centres at increasing distance from the focal area. The 30 to 35 Ma old Summer Coon volcano (Colorado, USA) shows a remarkable eroded pattern of radial dikes departing from a central intrusive complex at the base of the eroded edifice (Poland et al. 2008). Interestingly, here the dikes thicken with distance

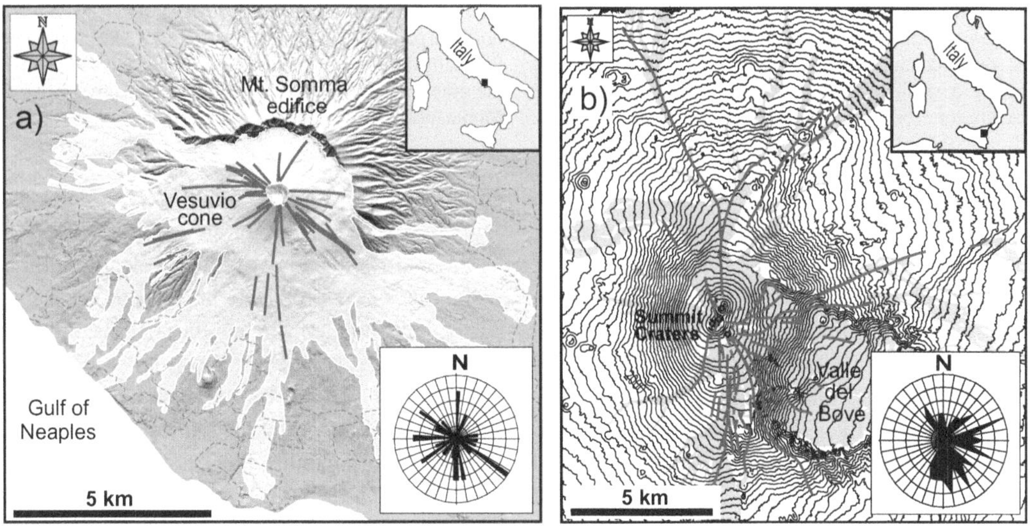

Fig. 7.15 Examples of radial dikes feeding eruptive fissures (red lines): **a** Vesuvio cone (Italy), 1631–1944 period; bottom inset shows rose diagram of fissure orientation; **b** Mount Etna (Italy), 1900–2005 period; bottom inset shows rose diagram of fissure orientation. Light grey zones in (**a**) and grey zones in (**b**) show the lava flows produced by the eruptive fissures (Modified after Acocella and Neri 2009)

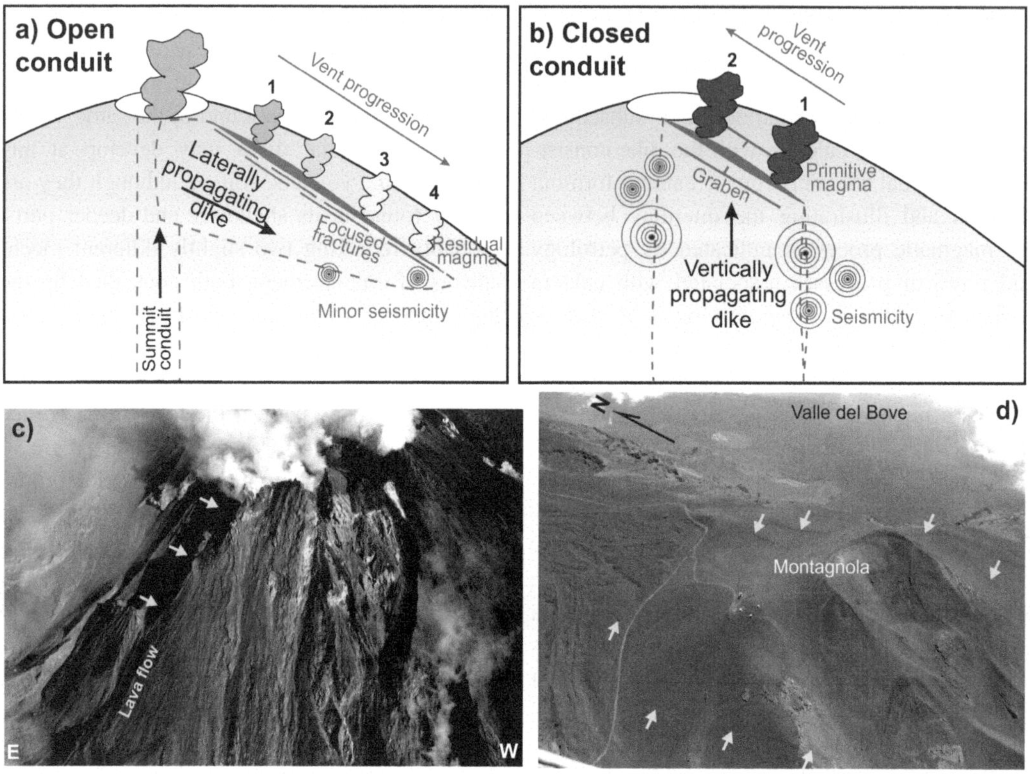

Fig. 7.16 Dike propagation as a function of the opening of the summit conduit. **a**, **b** show theoretical model. Dikes mostly propagate **a** laterally when the summit conduit is open and **b** vertically when the conduit is closed; the main observables that may allow distinguishing one mode of propagation from the other are summarized in blue. **c**, **d** Show real examples. **c** Laterally propagating dike from the open conduit summit craters of Stromboli (Italy) feeding a small lava flow along a tens of metres wide fracture zone (highlighted by yellow arrows) on August 18, 2014 (photo courtesy Marco Neri). **d** Aerial view of part of the ~500 m wide graben (border faults highlighted by yellow arrows) formed by a vertically propagating dike below Montagnola, along the South Rift of Etna (Italy) on July 17, 2001. The image was taken a few hours before the onset of the eruption, which focused in a vent within the northern part of the graben. In this case, the dike propagated vertically despite the open summit conduit and the width of the graben was locally increased by the positive relief of Montagnola (Photo courtesy Marco Neri)

from the volcano centre, a feature that has been related to an increase in driving pressure resulting from the progressively shallower depth of emplacement of the dikes below the volcano flanks. This, in line with what discussed in Sect. 7.3, confirms the importance of positive relief in hindering dike opening. On active volcanoes, radial dikes have been for example geodetically caught propagating laterally from sills at ~2 km depth below Fernandina caldera (Galapagos). Here in 1995 and 2009 a sill twisted about a radial horizontal axis below the volcanic edifice, generating a subvertical radial dike erupting on the volcano flank (Fig. 7.8; Pinel and Jaupart 2004; Bagnardi et al. 2013).

At shallower levels, within the volcanic edifice, radial dikes may propagate laterally or vertically, partly depending on whether the central conduit is open or closed (Fig. 7.16; Acocella et al. 2006a; Geshi 2008). If the central conduit is closed, magma may rise within the edifice through vertically propagating radial dikes, possibly emanating from dikes propagating laterally below the volcano. Conversely, if the central conduit is open, magma may exploit the conduit to rise, reaching its magmastatic level (often at a few hundred of metres of depth, as at Stromboli, Italy), passively degassing and in turn increasing its density. If this density increase is also accompanied by an increase in the magmatic pressure, as during deeper magma injection, the denser pressurized magma may fracture the shallower part of the conduit, where the lithostatic stress (and rock resistance) is smaller, triggering the lateral propagation of a radial dike. This dike is extremely shallow and not rooted at depth, its nucleation being limited to the topmost part of the conduit. Estimates from Mount Etna (Italy; for the last century) and Vesuvio (Italy; for the 1631–1944 period) suggest that the top of laterally propagating dikes was at a depth between a few tens to a few hundred of metres and had a mean downslope dip of 10°–15°. Laterally propagating radial dikes are very common on volcanic edifices with open conduit: for example, at Mount Etna more than 90% of the flank eruptions in the last century have been fed by laterally propagating radial dikes (Acocella and Neri 2003). These radial dikes also have the potential to feed eruptive vents on the lower flanks of volcanoes, posing a significant hazard to surrounding population and infrastructures. Note that volcanoes with open conduit may be intruded also by vertically propagating dikes, reaching the surface independently of the open conduit, as observed at Mount Etna in 2001.

Several features allow a shallow radial dike in the act of propagating laterally to be recognized from one propagating vertically below an active volcano (Fig. 7.16). Laterally-propagating shallow radial dikes are commonly associated with low levels of seismicity, as a minor amount of stress is required to fracture the shallowest portion of a volcano. Also, since a laterally propagating dike is very shallow, its induced surface deformation pattern (see Sect. 7.7) is narrow, on the order of very few tens of metres. In addition, the lateral propagation of a radial dike from a vertical conduit develops aligned vents at progressively lower elevations, as recognized for example from historical chronicles describing the 1631–1944 eruptive activity at Vesuvio. Here, except for the 1631 eruption, which opened a previously plugged conduit with a vertically propagating dike, all the successive flank eruptions were fed by dikes propagating laterally from the open summit conduit (Acocella et al. 2006a). Finally, magma erupted from laterally propagating dikes is generally residual (evolved) and degassed. In contrast, evidence for the vertical propagation of a radial dike includes upward propagating seismicity, wider (width of several hundreds of metres) fracture fields at the surface, and aligned vents opened at progressively higher elevations erupting more primitive and non-degassed magma, as observed during the 2001 eruption at Mount Etna (Acocella and Neri 2003; Geshi 2008).

Field observations on solidified and eroded dikes are commonly used to infer the direction of flow of the magma within a dike, and thus the possible direction of propagation of the dike. To this aim, striations, grooves, aligned vesicles or minerals, and the anisotropy of the magnetic susceptibility (AMS) on selected magnetic minerals are often considered. However, the direction of magma flow within a dike is not necessarily coincident with the direction of dike propagation, as the former may show a great variability, even within a dike propagating along a constant direction (Knight and Walker 1988; Porreca et al. 2006; Soriano et al. 2008).

In the absence of exposures of the eroded portions of active volcanic edifices, a radial pattern of eruptive fissures highlights the surface distribution of the feeder dikes, thus providing indirect, though incomplete, evidence of the shallow propagation of magma in a volcanic edifice. The pattern of radial eruptive fissures at some volcanoes, such as Vesuvio or Fernandina, may show a quite uniform, or nearly isotropic, distribution, with many radial directions similarly represented on the flanks of the edifice (Fig. 7.15a). In some cases, as in the Canary Islands Archipelago (Spain), radial dikes may cluster in volcanic rift zones that form triple-arms spaced $\sim 120°$ apart centred on the volcano summit. The origin of this specific configuration is debated, as it may result from an initially radial pattern of dikes partly masked by giant landslides, giving a false impression of a deep-seated three armed rift system (Carracedo et al. 1998; Acosta et al. 2003; Walter 2003; Becerril et al. 2015). In other volcanoes, including Mount Etna

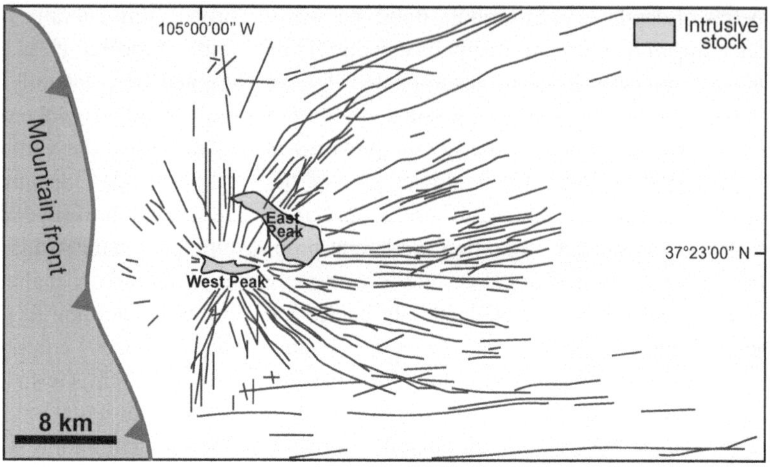

Fig. 7.17 Radial and regional dike patterns (purple lines) at the eroded Spanish Peaks volcano, Colorado (USA, modified after Odè 1957)

or Mount Fuji (Japan), the radial pattern of eruptive fissures may be anisotropic, clustering parallel to the regional maximum principal stress σ_1 and/or perpendicular to the regional minimum principal stress σ_3 (Fig. 7.15b). This configuration resembles that of regionally-controlled rift zones observed on volcanic edifices, although with a more dispersed pattern. The distribution of radial dikes clustering along a preferred orientation thus indicates the superimposition of a local stress field on a regional one. A striking example of a dike pattern produced by the superposition of a local and regional stress field is provided by the Spanish Peaks (Colorado, USA). This is an eroded central stratovolcano exposing two main intrusive stocks surrounded by a distinct pattern of $\sim$500 dikes with an overall radial distribution (Fig. 7.17; Odè 1957). The radial dikes in the distal eastward portion are on average much longer, and they progressively rotate toward an E-W trend. To the west, a N-S trending rigid mountain front created a regional E-W compression. The dike pattern has been interpreted as resulting from the superimposition of this regional stress field, responsible for the distal re-orientation of the dikes parallel to the compression direction, and a local stress field, controlling the emplacement of the proximal radial dikes.

All these cases show that the radial dike pattern inferred by the eruptive fissures at the surface can be maintained at depth within the edifice following the edifice load, although second-order factors (i.e., opening/closure of the conduit) may superimpose. Notwithstanding this, in some cases the radial pattern of feeder dikes inferred by the distribution of eruptive fissures may be extremely shallow, becoming less distinct at deeper levels within the edifice, as suggested by field data and analogue models (Fig. 7.18; Acocella et al. 2009). For example, the radial pattern of the older dikes outcropping in the eroded portions of Mount Etna and Somma-Vesuvio is much weaker than that currently observed at the surface through the distribution of the more recent eruptive fissures. This discrepancy is not related to the different periods of emplacement, as analogue models highlight a strong reorganization in the strike and dip of the dikes propagating within a conical edifice, which become radial only at very shallow levels.

It is generally possible to distinguish different types of radial dike patterns depending on the shape of the edifice. The most common shapes, (a) a conical edifice, (b) a conical edifice with sector collapse and (c) an elongated edifice, are considered below (Sects. 7.6.1–7.6.3).

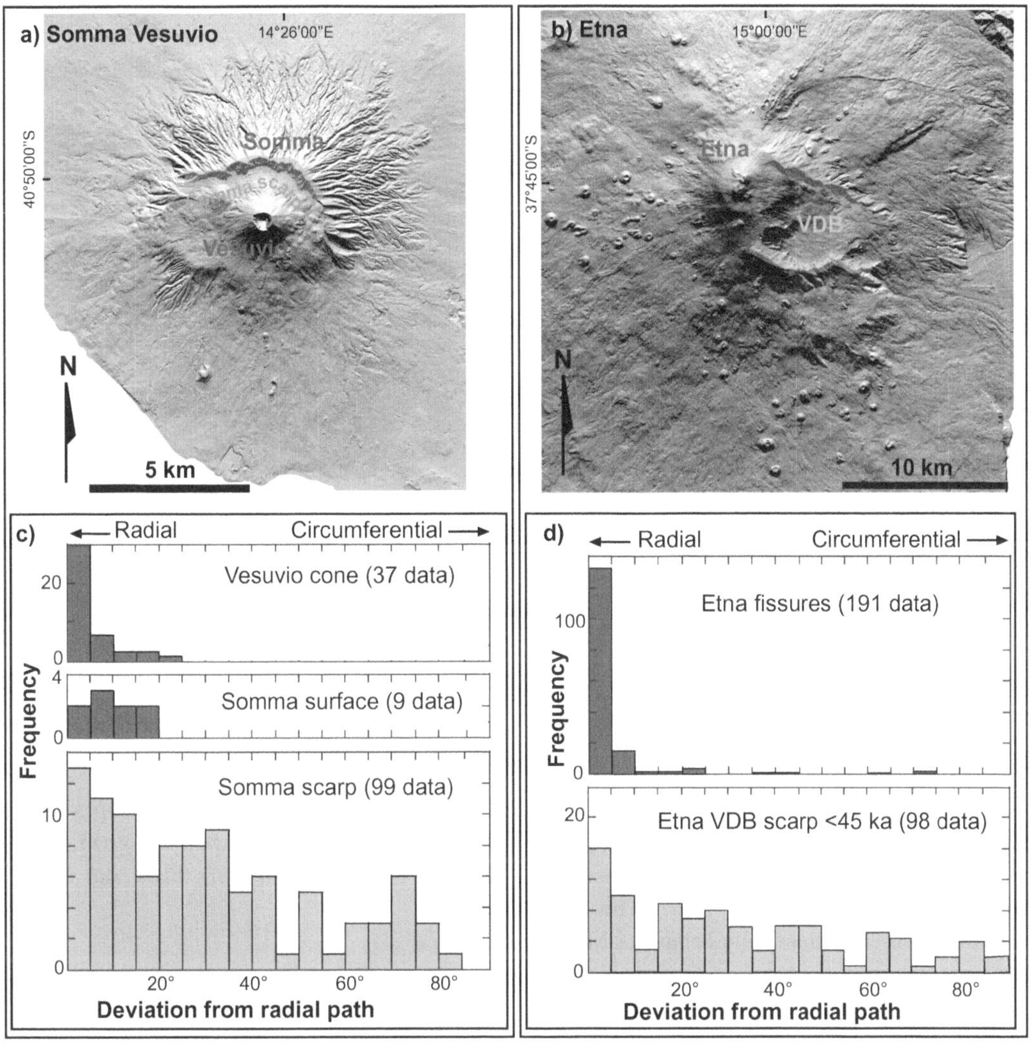

Fig. 7.18 **a**, **b** Dike fed-fissures and eroded dikes at Somma-Vesuvio (**a**) and Mount Etna (**b**) (Italy; VDB = Valle del Bove depression). The eruptive fissures, shown in Fig. 7.15, are measured at the surface; the eroded dikes are measured in the major scarps of the edifices (orange areas) at paleodepths of hundreds of meters. **c**, **d** Distribution of the fissures at the surface (red) and of the eroded dikes at depth (orange) at Somma-Vesuvio (**c**) and Mount Etna (**d**): the deviation from the radial direction is 0° for radial dikes and 90° for circumferential dikes (Modified after Acocella et al. 2009)

7.6.1 Radial Dikes in Conical Volcanic Edifices

For an idealized conical edifice, the maximum principal stress σ_1 is subvertical along the cone axis while, below the flanks, it becomes subparallel to the slope, that is radially directed and inclined outward (Fig. 7.19).

The minimum principal stress σ_3 within the edifice is circumferential, forming concentric patterns, whereas the intermediate principal stress σ_2, being perpendicular to both σ_1 and σ_3, is horizontal along the cone axis and at the sides becomes nearly perpendicular to the slopes. This stress configuration derives from the gravitational load imposed by the topography of the

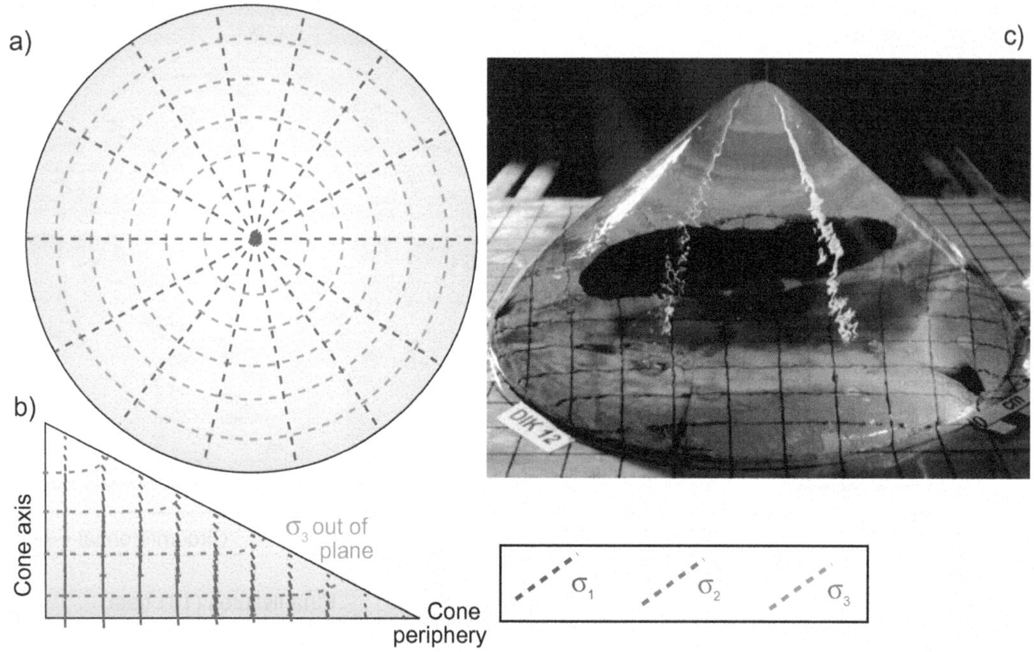

Fig. 7.19 Map (**a**) and enlarged axisymmetric section (**b**) views of the three principal stresses σ_1, σ_2 and σ_3 induced by the load of a conical volcanic edifice. **c** Analogue example of radial dike developing symmetrically at the base of a gelatine cone due to the injection of coloured water from the centre of the base of the cone.

conical edifice and therefore is a near-field stress. It promotes the formation of radial dikes perpendicular to the concentric σ_3 and containing the radial and vertical σ_1–σ_2 plane. Therefore, these radial dikes mechanically behave as Mode I fractures. Generally, the higher the volcanic edifice, the stronger its stress field. This local stress becomes particularly strong with edifices higher than ~3 km, where any far-field regional stress becomes subordinate.

In addition to the load of the edifice, a radially outward oriented maximum principal stress σ_1 may be amplified by the pressurization of a prolate (vertically elongated) magma conduit within the volcanic edifice. In this case, the pressurized conduit may push the upper part of the edifice radially outward, increasing its circumference, especially if the edifice is steep. This decreases the concentric minimum principal stress σ_3, allowing further radial dikes to accommodate the increase in circumference (Acocella and Neri 2009).

The eruptive fissures observed at Vesuvio (Fig. 7.15), at Piton de la Fournaise (La Reunion Island) and on the lower slopes of the western Galapagos volcanoes are good examples of radial patterns of dikes on conical edifices.

7.6.2 Radial Dikes in Conical Edifices with Sector Collapse

The axisymmetric (symmetrical around an axis) shape of a conical edifice may be interrupted by the presence of a scarp on a flank resulting from sector collapse. Sector collapse is common in volcanoes and may result from various external and internal processes during their growth (see Chap. 6). A sector collapse on the flank of a volcano may alter the stress distribution of conical edifices. In particular, there is a change in the stress trajectories on the side of the edifice hosting the collapse (Fig. 7.20).

This variation consists of a deflection of the direction of maximum principal stress σ_1 approaching the sector collapse: in fact, the σ_1 trajectories are mostly deflected to the sides of

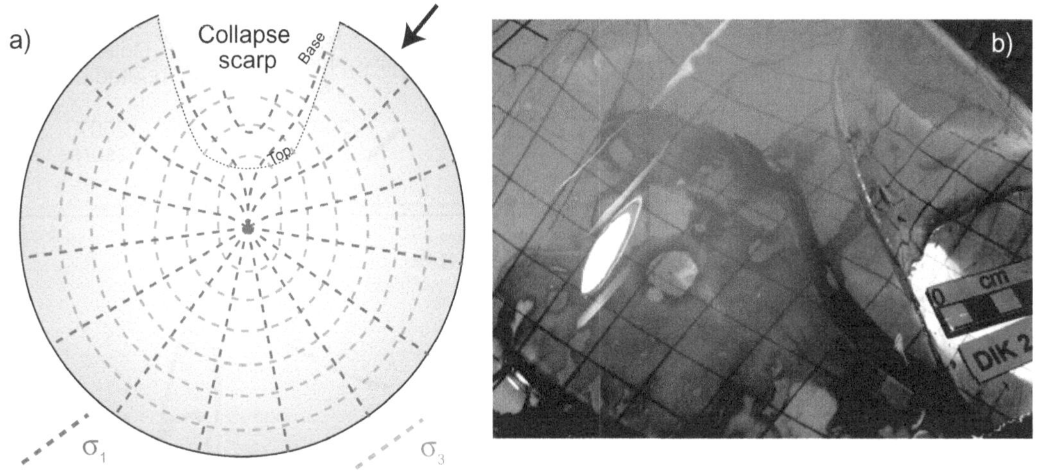

Fig. 7.20 **a** Map view of the maximum and minimum principal stresses, σ_1 and σ_3 respectively, induced by the load of a conical edifice with sector collapse (defined by top and base) on the one side. **b** Analogue example of dike developing at the base of a gelatine cone (only the central part of the cone is shown) due to the injection of coloured water from the centre of the base of the cone and deflecting towards the scarp side, visible to the right. Black arrow in (**a**) indicates the oblique point of view of the image in (**b**)

the sector collapse, where they become parallel to the lateral scarps. This deflection results from a minor deviation in the direction of the minimum principal stress σ_3 from the concentric pattern, to become locally perpendicular to the collapse sides, a consequence of the decrease in buttressing herein. Therefore, following the deflected σ_1 and σ_3 trajectories, dikes propagating laterally from outside and inside the sector collapse steer towards its lateral scarps. Only the dikes nucleating within the inner collapsed area may propagate radially, without being deflected. The effect of this stress variation is an overall focusing of the dikes along and parallel to the collapse sides.

This stress configuration explains both the long- and short-term distribution of dikes at Stromboli (Italy), characterized by a major depression on its flank, Sciara del Fuoco, resulting from repeated sector collapses in the last ~13 ka. On the longer-term, the pattern of the dikes at Stromboli is explained by a regional stress field with minimum principal stress σ_3 oriented NW–SE, promoting NE-SW trending dikes across the island, and a local stress field imposed by the Sciara del Fuoco collapse, promoting dikes along its lateral scarps (Fig. 7.21; Acocella and Tibaldi 2005). On the shorter-term, the path followed by the dikes during the 2002–2003 and 2007 eruptions is explained by the local stress distribution induced by the Sciara del Fuoco collapse. In particular, in 2002–2003 a dike propagated laterally from the summit craters along the upper scarp of the Sciara del Fuoco, whereas another dike nucleated within the collapsed area and propagated radially. In 2007 exactly the same dike patterns were observed, although the dike within the upper scarp of the Sciara propagated further downslope, deflecting and feeding vents at lower elevations (Fig. 7.21; Acocella et al. 2006b; Neri et al. 2008).

Detailed analysis of the dike pattern outcropping along the sector collapses of the NE Rift of Tenerife (Canary Islands) reveals more complex situations characterized by a dynamic interplay between gravity and magma. Here the dikes along the lateral collapse scarps dip away from the rift or embayment axes and strike oblique to the collapse sides, with an en-echelon pattern. This is consistent with the lateral spreading of the sectors before collapse, when the slump sides would create the necessary strike-slip movement to promote precursory en-echelon dike patterns (Del Camp et al. 2012). This model, in which the

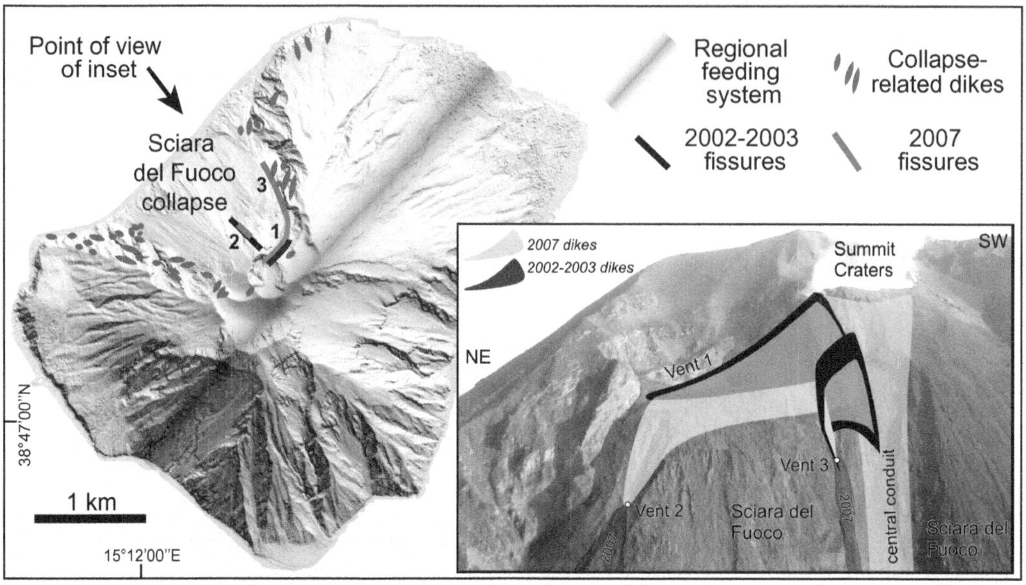

Fig. 7.21 Pattern of eroded dikes (red lenses) around the Sciara del Fuoco sector collapse at Stromboli (Italy). The distribution of the 2002–2003 and 2007 eruptive fissures and related groups of vents (numbered from 1 to 3, with 1 and 2 common to both eruptions) are also shown. Both the longer-term (eroded dikes) and shorter-term (2002–2003 and 2007 fissure eruptions) dike patterns are controlled by the sector collapse of the Sciara del Fuoco. Inset shows a reconstruction of the path of the dikes feeding the 2002–2003 and 2007 eruptions (Vents numbers are the same as in the larger map; modified after Neri et al. 2008)

dikes oblique to the scarp sides are the precursor to lateral collapse, complements the previously described one, where dikes emplace along the scarp of a previously formed sector collapse, and may find a wider applicability at other volcanoes, explaining the dike patterns also at the nearby Madeira Island and at Piton de la Fournaise.

7.6.3 Radial Dikes in Elongated Edifices

Radial dike patterns are usually associated with conical volcanoes, with a subcircular shape in map view. However, they may also be present in elongated volcanic edifices, with an elliptical shape in map view. Elongated edifices are characterized by a maximum principal stress σ_1 oriented parallel to the major axis of the ellipse approximating the map view of the edifice. However, towards the opposite distal edges of the elongated edifice the σ_1 trajectories diverge, with a radial pattern similar to what observed to the sides of conical edifices. As a result, the minimum principal stress σ_3 is mostly perpendicular to the axis of the elongated volcano, promoting dikes parallel to that direction, whereas it may bend at the distal edges, developing a local diverging pattern of dikes (Fig. 7.22a; Tibaldi et al. 2014). A similar overall pattern was highlighted by gelatine experiments, showing that intrusions in elongated ridges are consistently parallel to the ridge axis, regardless of any curvature of the ridge (Fiske and Jackson 1972). These experiments explain the coincidence between the direction of elongation of the Hawaiian volcanoes, such as Mauna Loa and Kilauea, and that of their hundreds of kilometres long volcanic rift zones, that extend also below sea level and are an expression of the clustering of dikes (Decker 1987). Other examples of well-elongated volcanoes with a similar dike distribution are Pico (Azores) and Mount Cameroon (Cameroon), both characterized by focused volcanic activity along their crest, which becomes dispersed at the distal

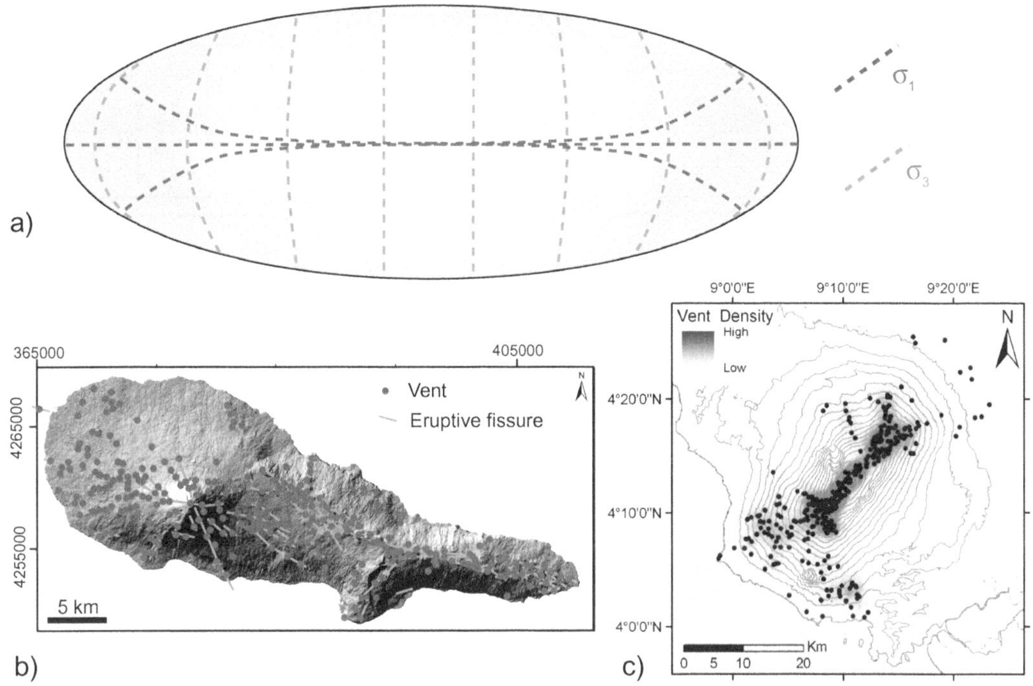

Fig. 7.22 **a** Map view of the maximum and minimum principal stresses, σ_1 and σ_3 respectively, induced by the load of an elongated edifice. **b**, **c** Examples of elongated volcanic edifices with rift zones parallel to the elongation: Pico Island (**b**; Azores; modified after Nunes et al. 2006; basal DEM courtesy of Adriano Pimentel) and Mt. Cameroon (**c**; Cameroon; Kervyn et al. 2014, the latter showing a distinctive diverging pattern of vents at its northeast and southwest sides

edges, matching the theoretical stress distribution (Fig. 7.22b, c; Nunes et al. 2006; Kervyn et al. 2014). Hekla (Iceland) also has a similar pattern of eruptive fissures aligned along the NE-SW trending ridge of the volcano, although the less dispersed fissure distribution at its northeast and southwest terminations suggests a stronger far-field stress at the volcano base.

7.7 Dikes Reaching the Surface

Dikes approaching the Earth's surface have the potential to feed eruptions. To assess eruptive hazard and mitigate the related risk, prompt recognition of these dikes is essential. This delicate task relies on the real-time analysis of monitoring parameters, especially seismicity and surface deformation. Surface deformation, in particular, exhibits a distinctive pattern during dike propagation, with the advantage of being readily recognizable. As this pattern is partly affected by the depth of the propagating dike, it may also allow tracking the location and depth of the dike. In this way, a real-time monitoring system may contribute to forecasting eruptions. The broader description and use of geodetic, geophysical and geochemical monitoring data is given in Chap. 8; here the focus is on the more specific relationships between a dike reaching the surface and the ground deformation it produces, which involves both elastic and non-elastic behaviours.

An upward propagating dike generates elastic and non-elastic surface deformation patterns which are distinct from those induced by any other magmatic source associated with the accumulation of magma. In fact, while sills, oblate or prolate ellipsoids, spheres and Mogi sources (see Sect. 8.3.2) promote a deformation pattern consistently characterized by a radial symmetry in its horizontal and vertical

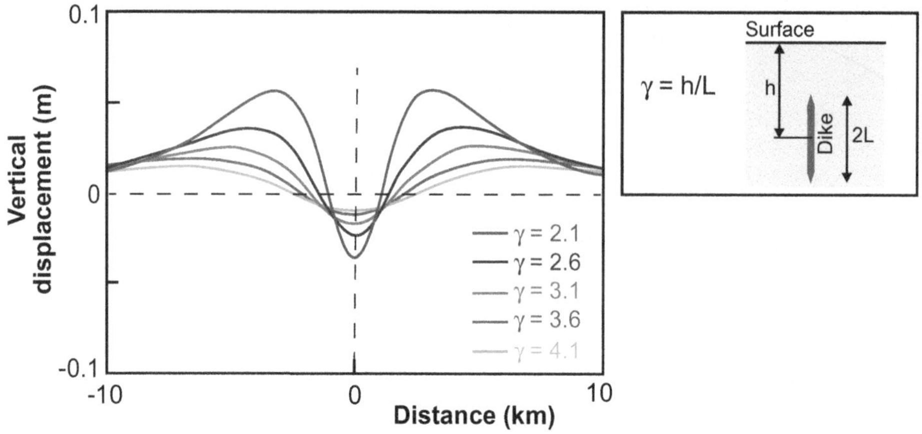

Fig. 7.23 Theoretical elastic vertical displacement at the surface due to a dike rising at different depths identified by the parameter $\gamma = h/L$ (where, as shown in the inset, h is the depth to the mid-point of the dike and L is the half-length of the dike; in these simulations L is fixed at 1 km; image courtesy of Eleonora Rivalta)

components, dikes steeper than $\sim$60° are responsible for a vertical and horizontal deformation characterized by **non-radial symmetry**: this deformation is "directional", in that the maximum vertical and horizontal deformation occurs off to the side of an opening dike, and the minimum vertical and horizontal deformation occurs along the strike of the dike.

To illustrate this, let us consider a dike propagating upward from a magma chamber a few kilometres deep and progressively approaching the Earth's surface. The resulting surface deformation pattern maintains the same non-radial symmetry, but is also dependent on the depth to the top of the dike. When the dike top is still at a few kilometres of depth, the deformation at the surface is elastic, with two uplifted lobes off to the sides of the dike axis and a weak area of subsidence in between and above the dike. The lateral extent of these uplifted and depressed areas is proportional to the along-strike length of the dike. This deformation pattern results from the fact that the opening of a dike induces lateral compression, which manifests as uplift to the sides of the dike at the surface, and upward dilation, which manifests as local subsidence above the dike (Fig. 7.23; Pollard et al. 1983; Rubin and Pollard 1988; Bonafede and Danesi 1997; Dvorak and Dzurisin 1997). As the dike rises, the amplitude of the elastic deformation increases, with larger though more focused uplift to the sides and larger subsidence above the dike.

When the dike approaches the surface it reaches a critical depth, which mainly depends on the dike thickness and host rock elasticity, at which the deformation at the surface passes the elastic threshold, and causes faulting instead. As a result, two inward-facing normal faults form in the two regions between the maximum uplifted and depressed areas, generating a graben at the surface (Fig. 7.24a, b; Pollard et al. 1983; Mastin and Pollard 1988). The two normal faults usually have similar slip, up to a few metres, accompanied by a similar dilational component. Sometimes, a normal fault may be replaced by several closely spaced normal faults with smaller displacement. Given the width b of the graben and assuming a mean dip $\delta \sim 60°$ for the normal faults (under Andersonian conditions), it is in principle possible to infer the depth h of the dike top at the moment of nucleation of the graben as:

$$h = 0.5\, b \tan \delta \qquad (7.3)$$

Following this relation, given that most dike-induced grabens from vertically propagating dikes are usually approximately 1 km wide, dikes are expected to nucleate grabens at depths on the order of $\sim$1 km (Trippanera et al. 2015). This theoretical relationship may show some

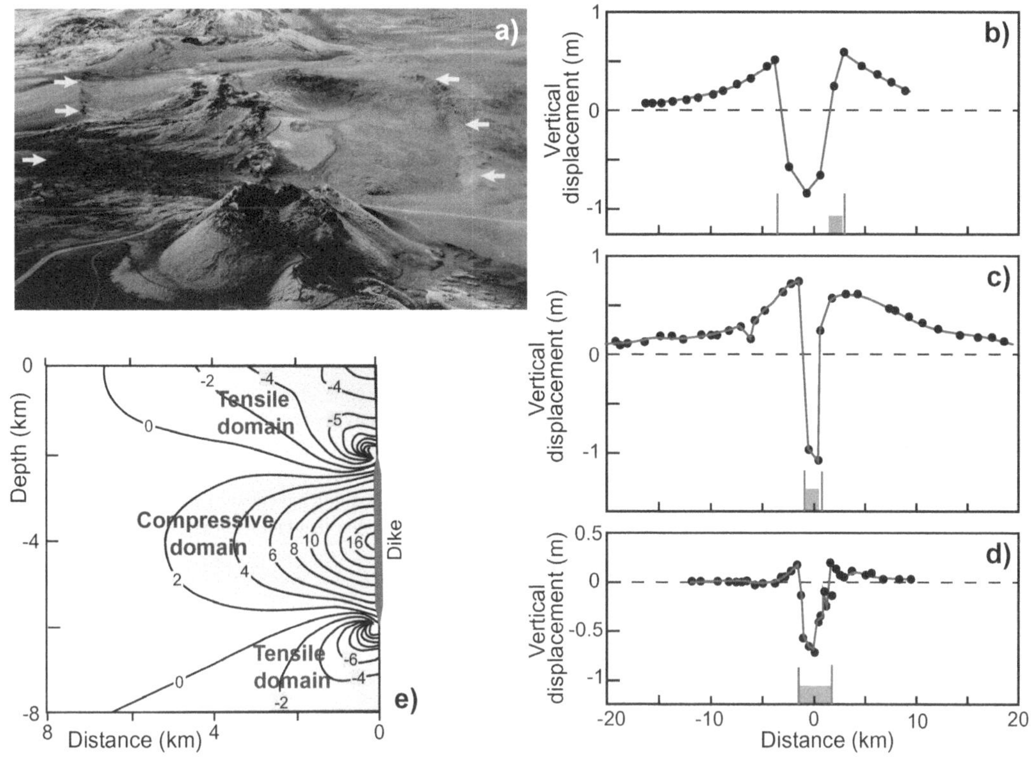

Fig. 7.24 Anelastic deformation induced by a dike approaching the surface. **a** Portion of the 1783–1784 Lakagigar eruptive fissure (Iceland), bordered by two inward dipping normal faults (highlighted by the yellow arrows) forming a graben. **b–d** Geodetic data indicating vertical displacement as a function of the distance due to dike emplacement in rift zones in Iceland (Krafla) in 1978 (**b**), in 1977 (**c**) and in Afar (Djibouti) in 1978 (**d**). Red lines at bottom of each diagram represent outermost faults activated by the dike, while reddish rectangles represent zones of intense fracturing. **e** Theoretical change in horizontal stress (in MPa) due to dike intrusion; only the stresses to the left side of the dike are shown here for simplicity. Positive stresses indicate compression (green area), and negative tension (purple areas). Inward dipping normal faults, as shown in (**a**), are expected to form within the tensile domain at the surface (Modified after Rubin and Pollard 1988)

substantial variability, as dikes may be significantly deeper or shallower than expected considering only the width of the graben at the surface and the dip of the faults. An important parameter affecting the above relationship is the mechanical layering of the host rock, when the layers above the dike show different stiffness values. Under these conditions, the boundary faults of a graben are expected to form at the tensile/shear stress peaks, and not at the location of the peaks of elastic deformation at the surface, affecting the graben width (Bazargan and Gudmundsson 2019). Therefore, any use of Eq. 7.3 in inferring the depth to the top of a dike should take into account uncertainty. In any case, a dike-induced graben marks the advanced stage of rise of a dike, and its appearance at the surface should be considered as evidence for impending eruption. To the sides of the graben, the lateral compression induced by dike opening may occasionally generate low-angle thrust faults with displacement much smaller than that of the normal faults, as observed at Kilauea in 1974 or Mount Etna in 1989. Similarly, pre-existing high-angle normal faults to the sides of a dike may also be reactivated as reverse faults, as observed in Iceland (Pollard et al. 1983; Barberi et al. 1990; Gudmundsson et al. 2008).

In some cases, the non-elastic surface deformation pattern associated with feeder dikes is

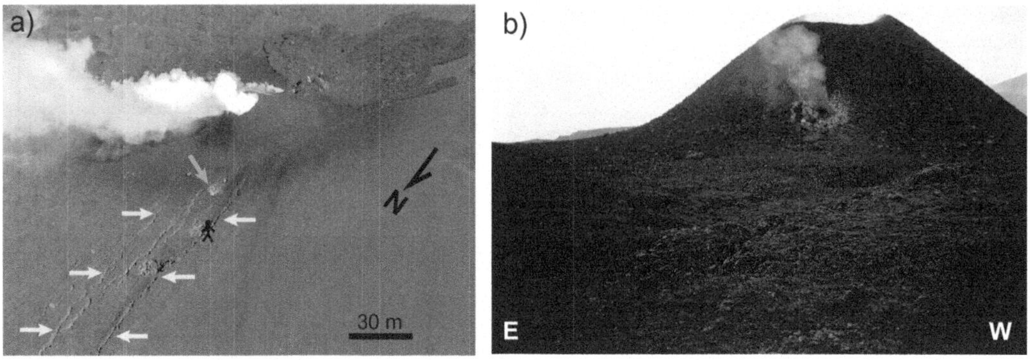

Fig. 7.25 Aerial (**a**) and ground (**b**) views of the same portion of a focused fracture zone ($\sim$30 m wide) striking towards the nearby eruptive vent and induced by a laterally propagating shallow dike during the 2001 Etna (Italy) eruption. Yellow arrows in (**a**) highlight boundaries of fracture zone, whereas orange arrow in (**a**) shows a small sag on the side of the cone, characterized by increased degassing. This same steaming vent is subsequently characterized by the incipient emission of magma, as shown in (**b**), developing a peripheral eruptive locus on the cone. The image in (**b**) is taken from the position indicated by the drawn black man in (**a**) a few hours after that taken in (**a**)

much narrower than that shown by these grabens, on the order of a few tens of metres (Fig. 7.25). For example, the submerged graben structures observed along portions of the Juan de Fuca Ridge and the East Pacific Rise (Pacific Ocean) are a few tens of metres wide (Chadwick and Embley 1998). Also, during the 2001 Etna eruption, a $\sim$30 m wide area of fracturing, mostly consisting of tension fractures a few tens of centimetres wide, connected nearby vents aligned along a fissure (Billi et al. 2003). These focused fracture zones result from the emplacement of extremely shallow (few tens of m deep) dikes. Evidence that these fracture zones are not enclosed within a wider graben suggests that either the dikes did not propagate vertically, and rather propagated laterally and shallowly from the summit open conduit, or that inelastic deformation did not occur until the dikes were very shallow, implying that higher stresses were required for failure. In both cases, the cumulative horizontal displacement of the formed extension and shear fractures at the surface is a good proxy for the thickness of the dike.

The thickness of a dike may vary along its length, that is in the vertical dimension. Observations from the inner wall of Miyakejima caldera (Japan) reveal that feeder dikes usually become thicker in the metres just before reaching the Earth's surface. The upward widening of feeder dikes is interpreted to result from free-surface effects and magmatic overpressure (driving pressure) changes during the eruption. This evidence may permit distinguishing eroded feeder dikes from non-feeder dikes, also providing the opportunity to determine the conditions which allowed some dikes to reach the surface, related to the magma viscosity, volatile content, or host rock properties (Geshi et al. 2010).

A feeder dike may also show thickness variations along strike, also associated with segmentation into en-echelon patterns (see Sect. 3.5). These thickness variations are related to irregular widening, possibly induced by heterogeneities in the mechanical resistance of the host rock during the opening of the dike. The result is that larger apertures favour flow channelling, which in turn can develop a long-lived focused vent, thus concentrating the rise of magma from a linear to a point-source feature (see Eq. 3.11). The birth of a vent above a dike is preceded by focused non-elastic deformation at the surface, soon followed by increased degassing and, subsequently, by the progressive appearance of magma in one or more spots along the fracture zone; the extrusion of magma finally produces a growing cone (Fig. 7.25). Analogue experiments suggest that, after eruption initiates, the feeder dike rapidly channelizes into a conduit geometry. If the time interval between successive events of magma migration is short, subsequent magma may follow the pre-existing conduit, with the result that monogenic vents cluster in the same area (Pansino et al. 2019). On a longer time-span, the repeated and relatively frequent emplacement of feeder dikes in the same crustal volume may develop a stable magmatic **conduit**, which becomes the pathway used by magma to ascend to the surface and erupt, developing a polygenic volcano. Even though magmatic conduits are commonly inferred to be cylindrical in shape, there is evidence that they can derive from clusters of nearby dikes, such as shown by drilling below Unzen volcano (Japan), where a dike swarm was discovered in the inferred location of the volcano's conduit (Sakuma et al. 2008). While it is appropriate to think of a conduit as originating from one or more dikes, simple thermal considerations indicate that a dike may soon solidify at its extremities, and the flow of magma may become progressively restricted to its central portion. At the same time, mechanical erosion of the dike walls, mainly due to the repeated magma and volatile flow and overpressures and associated fracturing, may further widen the most open portion of the dike. These processes can result in a conduit within a polygenic volcano progressively approaching a subcylindrical shape, usually at depth less than 1 km, whose surface expression is one or more overlapping summit craters (Fig. 7.26; see also Sects. 3.4.2 and 3.4.3; Costa et al. 2007).

In some cases, a dike approaching the Earth's surface may stall at very shallow depth. Under a continued magma supply, the upper portion of the dike, experiencing a lower confining pressure, may inflate, developing a thicker subvolcanic intrusion. This stalk-like magmatic feeder, connecting a volcano to a deeper supply of magma and with a diameter up to hundreds of metres, is commonly referred to as **neck**. Necks have a massive, often subcylindrical, shape with steep walls and homogeneous structure and fabric, as well as columnar jointing (Marsh 2015; Fittipaldi et al. 2019). Their considerable thickness poses a space problem, requiring the host rock to be pushed aside of up to hundreds of metres over a short time (weeks to months) to accommodate the intrusion. The shallow location of necks, often within volcanic edifices, suggests that topography plays an important role in allowing magma accumulation. For example, in the upper portion of volcanic edifices with reduced and/or differential buttressing conditions, the downslope shear stresses due to the

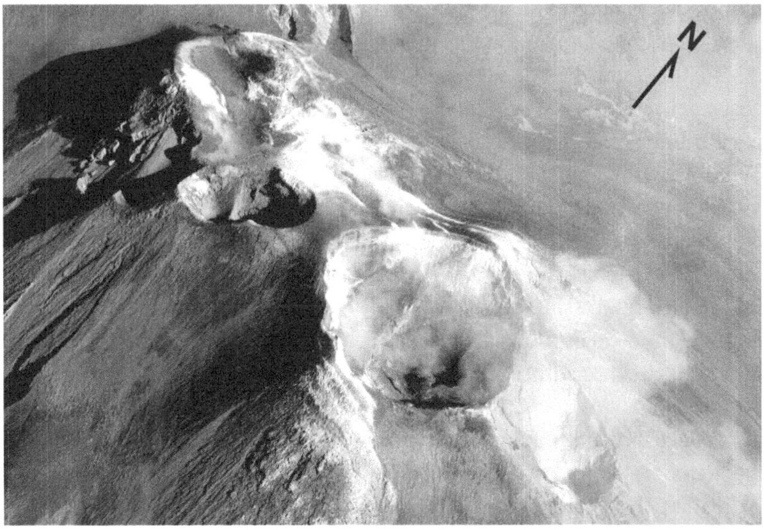

Fig. 7.26 Air image, taken on July 14, 2016, of the Southeast Crater (in the background, formed in 1971) and the New Southeast Crater (in the foreground; formed in 2011) on the summit of Mount Etna, Italy. Both vents are the surface expression of the most active portion of the locally WNW-ESE elongated summit conduit at Etna (Photo courtesy Marco Neri)

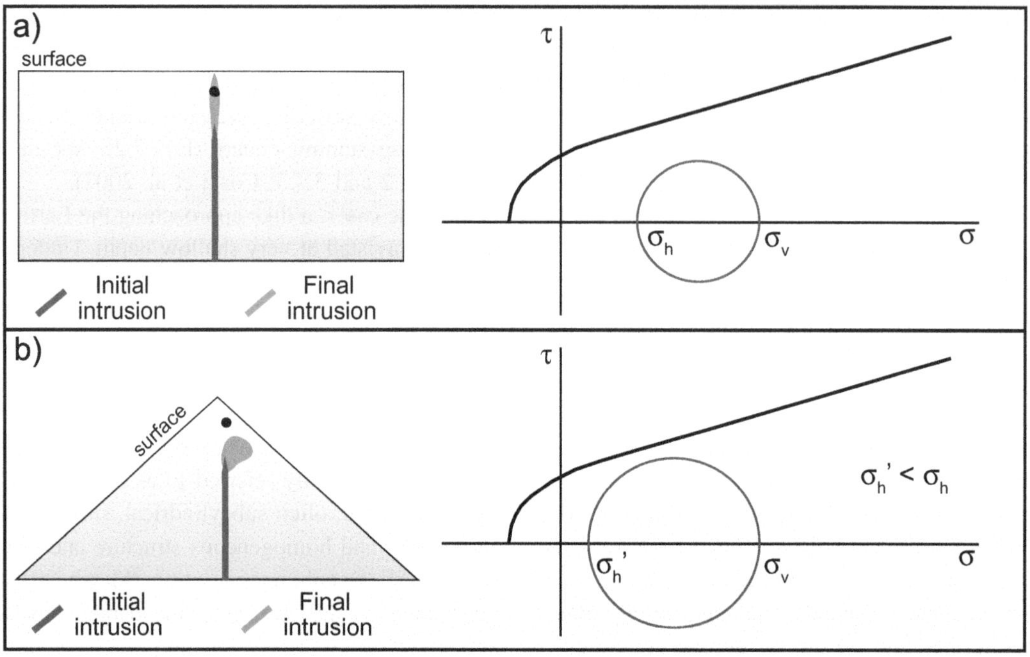

Fig. 7.27 Section views of shallow magma propagation below a flat topography (**a**) and within a cone (**b**). The Mohr diagrams to the right are indicative of the shallow stress conditions in the zones indicated by the black dots to the left, sufficiently far from the propagating intrusions (in red). **a** With a flat topography, the area above the dike indicated by the black dot is stable. If the dike propagates upward, it may reach the surface thickening because of the local effect of the free surface (in orange). **b** Within a cone, the area above the dike indicated by the black dot is in less stable conditions, following the lower confinement to the side of the cone ($\sigma_h' < \sigma_h$); the higher differential stress here develops a Mohr circle nearer to the failure envelope. Therefore, the upward propagating magma may more easily induce failure to the side nearest to the slope. Failure may provide the space to accommodate the emplacement of a thicker intrusion (in orange) (modified after Fittipaldi et al. 2019). This highlights the role of unbuttressing in accommodating the space for the growth of necks

edifice load enhance slope deformation, promoting thickening of shallow dikes (Fig. 7.27).

Similar boundary conditions may also explain the shallow and directional growth of high viscosity felsic intrusions within stratovolcanoes, or **cryptodomes**, such as those responsible for the 1956 Bezymianny (Kamchatka, Russia) and 1980 Mount St. Helens (Washington, USA) eruptions. These are interpreted to have grown towards the nearest slope of the edifice, promoted by its least buttressing conditions and then benefiting from the downslope shear stresses imposed by the edifice load (Donnadieu and Merle 2001). At times, the shear stresses may also affect the cryptodome itself, as observed in the eroded Miocene Chachahuén volcano, Argentina. Here magmatic shear zones accommodated intrusion growth, locally overprinting the earlier formed concentric fabric domes, creating weakness zones that also increase the risk of collapse (Burchardt et al. 2019).

The shallow emplacement of a cryptodome may result in an eruption, which may be followed by the growth of an extrusive **lava dome** within a cylindrical conduit, as for example at Bezymianny in 1956 and Mount St. Helens in 1980 (e.g., Belousov et al. 2007). The growth of an extrusive lava dome within the central craters of stratovolcanoes induces an overall symmetric deformation pattern, as observed at Mount St. Helens from 1980 to 1986. Here accelerating deformation of the crater floor around a growing extrusive lava dome was the earliest indicator of impending eruptions. The deformation consisted of a system of radial cracks and tangential thrust faults. Radial cracks formed first and, as some

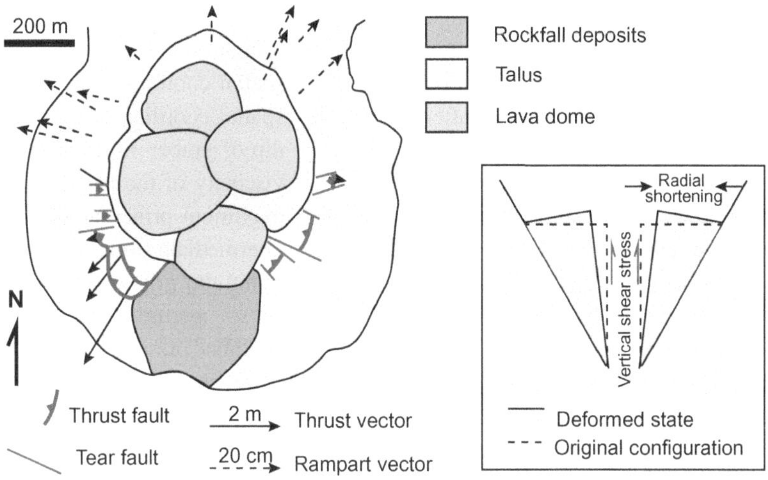

Fig. 7.28 Map view radial displacements due to the growth of an extrusive lava dome on the crater floor of Mount St. Helens (Washington, USA) as taken from aerial photographs in September 1981. Inset at bottom right shows east–west schematic cross section illustrating the mechanical model for deformation of the crater fill. Fill occupies conical vent bounded by crater walls with vertical cylindrical conduit at centre (lava dome not shown for clarity). Shear stress directed vertically along conduit walls deforms fill upward and outward and produces the radial shortening at ground surface associated with the thrust faults (Modified after Chadwick and Swanson 1989)

evolved into tear faults, influenced the subsequent geometry of thrusting. The magnitude of displacements increased before extrusions, with the hanging walls of each thrust increasingly disrupted as cumulative fault slip increased. These cracks and thrusts formed because the crater fill was displaced upward and radially outward from the feeder conduit, owing to uniform shear-stress along the conduit wall, related to the flow of viscous magma (Fig. 7.28; Chadwick et al 1988; Chadwick and Swanson 1989). As the rates of displacement accelerated before eruptions, monitoring of this deformation pattern allowed proposing a successful predictive approach with an error of a few days (Chadwick et al. 1983).

7.8 Summary

Shallow magma transfer may be characterized by different types of intrusions related to a regional and/or local stress field, with the latter commonly associated with the activity of magma chambers and topographic gradients. Regional dikes may develop when near-field stresses are negligible compared to far-field stresses, commonly away from magma chambers and topographic gradients. Conversely, the activity of a magma chamber may develop shallow sills, inclined sheets and dikes related to a near-field stress. Sills promote a predominant lateral transfer of magma, although saucer shaped sills may also allow vertical transfer, at times feeding eruptions. Inclined sheets are most commonly in the form of circumferential inward dipping cone sheets, resulting from overpressure conditions within the magma chamber. If their propagation is associated with an unloading pressure not negligible with regard to the lithostatic pressure, as for example below calderas, sheets may become circumferential dikes feeding fissure eruptions also outside caldera rims, as at the Galapagos. Dikes affected by the underpressure of a magma chamber commonly take the form of circumferential outward dipping to subvertical ring-dikes, whose formation occurs during caldera collapse. At shallower levels, the load of a volcanic edifice controls the formation of radial dikes, both at the base and within the edifice: within the edifice, this mechanism is enhanced by the opening of the magmatic conduit and by poor magma buoyancy after degassing. Sector collapse may produce deviations from the radial pattern, focusing dikes to the sides of the collapse. Also, the more elongated the volcanic edifice, the more anisotropic is the radial dike pattern, with dikes clustering along the edifice elongation. Dikes approaching the surface may evolve into point-source, long-lived magmatic conduits, necks or cryptodomes, depending upon different boundary conditions (i.e., magma temperature and viscosity, host rock strength, topography).

Unlike any other magma source, dikes create a distinctive elastic and non-elastic surface deformation pattern with non-radial symmetry, which, together with the distribution of any migrating seismicity, may allow detecting the location and depth of a dike about to feed an eruption. Understanding the signals related to shallow dike propagation may allow not only to forecast an impending eruption, but may also provide important clues on its possible style. For example, volatile-rich vertically propagating dikes are more prone to feed voluminous explosive activity, whereas dikes propagating laterally from the central conduit are usually fed by degassed residual magma, feeding minor non-explosive activity: this is illustrated in Fig. 1.28, which shows a space image of the eruptive plumes at Mount Etna in 2002 and their structural interpretation in terms of vertical and lateral dike propagation. This figure underlines the importance of capturing the patterns of shallow magma transfer, to be prepared in forecasting the possible location of any vent and its expected eruptive activity, a feature which is the topic of the next two chapters.

7.9 Main Symbols Used

b	width of the graben
dP/dx	pressure gradient
h	depth of the dike
K_p	permeability
Q_m	flux of melt
t	time
x	spatial coordinate
y	spatial coordinate
z	spatial coordinate
δ	dip of graben normal faults
η	viscosity of the fluid
σ_1	maximum principal stress
σ_2	intermediate principal stress
σ_3	minimum principal stress
σ_C	stress perturbation due to magma chamber pressurization
σ_E	stress perturbation due to previous large earthquakes or slow slip events
σ_I	stress perturbation due to previous magmatic intrusions
σ_L	stress perturbation due to edifice load
σ_T	regional tectonic stress tensor
σ_{TOT}	supra-lithostatic full stress tensor
σ_U	stress perturbation due to topographic unloading.

References

Acocella V, Neri M (2003) What makes flank eruptions? The 2001 Etna eruption and its possible triggering mechanisms. Bull Volcanol 65:517–529

Acocella V, Tibaldi A (2005) Dike propagation driven by volcano collapse: a general model tested at Stromboli, Italy. Geophys Res Lett 32:L08308. https://doi.org/10.1029/2004GL022248

Acocella V, Porreca M, Neri M, Mattei M, Funiciello R (2006a) Fissure eruptions at Mount Vesuvius (Italy): insights on the shallow propagation of dikes at volcanoes. Geology 34:673–676

Acocella V, Neri M, Scarlato P (2006b) Understanding shallow magma emplacement at volcanoes: orthogonal feeder dikes during the 2002–2003 Stromboli (Italy) eruption. Geophys Res Lett 33:L17310. https://doi.org/10.1029/2006GL026862

Acocella V, Neri M (2009) Dike propagation in volcanic edifices: overview and possible developments. Tectonophysics 471:67–77

Acocella V, Neri M, Sulpizio R (2009) Dike propagation within active central volcanic edifices: constraints from Somma-Vesuvius, Etna and analogue models. Bull Volcanol 71:219–223

Acocella V, Rivalta E (2019) Calderas: structure, unrest, magma transfer and eruptions. In: Reference module in earth systems and environmental sciences. Elsevier, 15 pp

Acosta J, Uchupi E, Smith D, Munoz A, Herranz P, Palomo C et al (2003) Comparison of volcanic rifts on La Palma and El Hierro, Canary Islands and the Island of Hawaii. Mar Geophys Res 24:59–90

Albino F, Biggs J, Syahbana DK (2019) Dyke intrusion between neighbouring arc volcanoes responsible for 2017 pre-eruptive seismic swarm at Agung. Nat Commun 10:748. https://doi.org/10.1038/s41467-019-08564-9

Ancochea E, Branlde JL, Huertas MJ, Cubas CR, Hernan F (2003) The felsic dikes of La Gomera (Canary Islands): identification of cone sheet and radial dike swarms. J Volcanol Geoth Res 120:197–206

Ancochea E, Huertas MJ, Hernan F, Brandle JL (2014) A new felsic cone-sheet swarm in the Central Atlantic Islands: The cone-sheet swarm of Boa Vista (Cape Verde). J Volcanol Geoth Res 274:1–15

Anderson EM (1936) Dynamics of formation of cone-sheets, ring-dykes and cauldron-subsidences. Proc Roy Soc Edinb 61:128–157

Bagnardi M, Lundgren P (2018) Chasing a "Magma Gopher": Insights on intricate subsurface magma pathways at Sierra Negra Volcano, Galápagos. In: American geophysical union fall meeting abstract #G14A-08

Bagnardi M, Amelung F, Poland MP (2013) A new model for the growth of basaltic shields based on deformation of Fernandina volcano, Galápagos Islands. Earth Planet Sci Lett 377–378:358–436

Bailey EB, Clough CT, Wright WB, Richey JE, Wilson GV (1924) Tertiary and post-Tertiary geology of Mull, Loch Aline, and Oban. Geological Survey of Scotland Memoir Edinburgh

Barberi F, Bertagnini A, Landi P (1990) Mt. Etna: the 1989 eruption. CNR Gruppo Nazionale per la Vulcanologia Report Giardini Editore, 77 pp

Bazargan M, Gudmundsson A (2019) Dike-induced stresses and displacements in layered volcanic zones. J Volcanol Geoth Res 384:189–205

Becerril L, Galindo I, Martì J, Gudmundsson A (2015) Three-armed rifts or masked radial pattern of eruptive fissures? The intriguing case of El Hierro volcano (Canary Islands). Tectonophysics 647–648:33–47

Belousov A, Voight B, Belousova M (2007) Directed blasts and blast-generated pyroclastic density currents: a comparison of the Bezymianny 1956, Mount St Helens 1980, and Soufrière Hills, Montserrat 1997 eruptions and deposits. Bull Volcanol 69:701–740

Billi A, Acocella V, Funiciello R, Giordano G, Lanzafame G, Neri M (2003) Mechanism for ground surface fracturing and incipient slope failure associated with the 2001 eruption of Mt. Etna, Italy: analysis of ephemeral field data. J Volcanol Geoth Res 122:281–294

Bistacchi A, Tibaldi A, Pasquarè F, Rust D (2012) The association of cone–sheets and radial dykes: data from the Isle of Skye (UK), numerical modelling, and implications for shallow magma chambers. Earth Planet Sci Lett 339–340:46–56

Bonafede M, Danesi S (1997) Near-field modifications of stress induced by dyke injection at shallow depth. Geophys J Int 130:435–448

Burchardt S (2018) Volcanic and igneous plumbing systems. In: Understanding Magma transport, storage and evolution in the earth's crust. Elsevier, 347 pp

Burchardt S, Mattsson T, Palma JO, Galland O, Almqvist B, Mair K et al (2019) Progressive growth of the Cerro Bayo cryptodome, Chachahuén volcano, Argentina—Implications for viscous magma emplacement. J Geophys Res 124:7934–7961

Carbotte SM, Arnulf A, Spiengelman M, Lee M, Harding A, Kent G et al (2020) Stacked sills forming a deep melt-mush feeder conduit beneath Axial Seamount. Geology 48:693–697

Carracedo JC, Day S, Guillou H, Rodriguez Badiola E, Canas JA, Perez Torrado FJ (1998) Hotspot volcanism close to a passive continental margin: the Canary Islands. Geol Mag 135:591–604

Cashman KV, Sparks RSJ, Blundy JD (2017) Vertically extensive and unstable magmatic systems: a unified view of igneous processes. Science 355:1280

Chadwick WW, Swanson DA, Iwatsubo IY, Heliker CC, Leighley TA (1983) Deformation monitoring at Mount St. Helens in 1981 and 1982. Science 221:1378–1380

Chadwick WW, Archuleta RJ, Swanson DA (1988) The mechanics of ground deformation precursory to dome building extrusions at Mount St. Helens: 1981–1982. J Geophys Res 93:4351–4366

Chadwick WW, Swanson DW (1989) Thrust faults and related structures in the crater floor of Mount St. Helens volcano, Washington. Geol Soc Am Bull 101:1507–1519

Chadwick WW, Howard KA (1991) The pattern of circumferential and radial eruptive fissures on the volcanoes of Fernandina and Isabela islands, Galapagos. Bull Volcanol 53:259–275

Chadwick WW, Embley RW (1998) Graben formation associated with recent dike intrusions and volcanic eruptions on the mid-ocean ridge. J Geophys Res 103:9807–9825

Corbi F, Rivalta E, Pinel V, Maccaferri F, Bagnardi M, Acocella V (2015) How caldera collapse shapes the shallow emplacement and transfer of magma in active volcanoes. Earth Planet Sci Lett 431:287–293

Costa A, Melnik O, Sparks RSJ, Voight B (2007) Control of magma flow in dykes on cyclic lava dome extrusion. Geophys Res Lett 34:L02303. https://doi.org/10.1029/2006GL027466

Daines MJ, Pec M (2015) Migration of melt. In: Sigurdsson H, Houghton B, McNutt S, Rymer H, Stix J (eds) The encyclopedia of volcanoes, 2nd edn. Elsevier Academic Press, pp 49–64

Derrien A, Taisne B (2019) 360 Intrusions in a miniature volcano: birth, growth, and evolution of an analog edifice. Front Earth Sci 7:19. https://doi.org/10.3389/feart.2019.00019

Decker RW (1987) Dynamics of Hawaiian volcanoes: an overview In: Decker RW, Wright L, Stauffer PH (eds) Volcanism in Hawaii, vol 1350. US Geological Survey Professional Paper, pp 997–1018.

Del Camp A, Troll VR, van Wyk deVries B, Carracedo JC, Petronis MS, Perez-Torrado FJ et al (2012) Dykes and structures of the NE rift of Tenerife, Canary Islands: a record of stabilisation and destabilisation of ocean island rift zones. Bull Volcanol 74:963–980

Dieterich JH (1988) Growth and persistence of Hawaiian volcanic rift zones. J Geophys Res 93:4258–4270

Donnadieu F, Merle O (2001) Geometrical constraints of the 1980 Mount St Helens intrusion from analogue models. Geophys Res Lett 28:639–642

Dvorak JJ, Dzurisin DD (1997) Volcano geodesy: the search for Magma reservoirs and the formation of eruptive vents. Rev Geophys 35:343–384

Ebinger C, Ayele A, Keir D, Rowland J, Yirgu G, Wright T et al (2010) Length and timescales of rift faulting and magma intrusion: the Afar rifting cycle from 2005 to present. Annu Rev Earth Planet Sci 38:439–466

Ernst RE, Head JW, Parfitt E, Grosfils E, Wilson L (1995) Giant radiating dyke swarms on Earth and Venus. Earth Sci Rev 39:1–58

Fiske RS, Jackson ED (1972) Orientation and growth of Hawaiian volcanic rifts. Proc R Soc Lond 329:299–326

Fittipaldi M, Urbani S, Neri M, Trippanera D, Acocella V (2019) Understanding the origin of magmatic necks: insights from Etna and analogue models. Bull Volcanol 81:11. https://doi.org/10.1007/s00445-019-1273-1

Fridrich CJ, Smith RP, De Witt ED, McKee EH (1991) Structural, eruptive and intrusive evolution of the Grizzly Peak caldera, Sawatch Range, Colorado. Geol Soc Am Bull 103:1160–1177

Gaete A, Kavanagh JL, Rivalta E, Hazimb SH, Walter TR, Dennis DJC (2019) The impact of unloading stresses on post-caldera Magma intrusions. Earth Planet Sci Lett 508:109–121

Geshi N (2008) Vertical and lateral propagation of radial dikes inferred from the flow direction analysis of the radial dike swarm in Komochi Volcano, Central Japan. J Volcanol Geoth Res 173:122–134

Geshi N, Kusumoto S, Gudmundsson A (2010) Geometric difference between non-feeder and feeder dikes. Geology 38:195–198

Gudmundsson A (1998) Magma chambers modeled as cavities explain the formation of rift zone central volcanoes and their eruption and intrusion statistics. J Geophys Res 103:7401–7412

Gudmundsson A, Friese N, Galindo I, Phillip SL (2008) Dike-induced reverse faulting in a graben. Geology 36:123–126

Hartley ME, Thordarson T (2012) Formation of Öskjuvatn caldera at Askja, North Iceland: Mechanism of caldera collapse and implications for the lateral flow hypothesis. J Volcanol Geoth Res 227–228:85–101

Heimisson ER, Hooper A, Sigmundsson F (2015) Forecasting the path of a laterally propagating dike. J Geophys Res 120:8774–8792

Johnson SE, Schmidt KL, Tate MC (2002) Ring complexes in the Peninsular Ranges Batholith, Mexico and the USA: magma plumbing systems in the middle and upper crust. Lithos 61:187–208

Kennedy B, Styx J (2007) Magmatic processes associated with caldera collapse at Ossipee ring dyke, New Hampshire. Geol Soc Am Bull 119. https://doi.org/10.1130/B25980.1

Kervyn M, Ernst GGJ, van Wyk deVries B, Mathieu L, Jacobs P (2009) Volcano load control on dyke propagation and vent distribution: insights from analogue modeling. J Geophys Res 114:B03401. https://doi.org/10.1029/2008JB005653

Kervyn M, van Wyk deVries B, Walter TR, Njome MS, Suh CE, Ernst GGJ (2014) Directional flank spreading at Mount Cameroon volcano: evidence from analogue modelling. J Geophys Res 119:7542–7563

Klausen MB (2004) Geometry and mode of emplacement of the Thverartindur cone sheet swarm, SE Iceland. J Volcanol Geoth Res 138:185–204

Knight MD, Walker GPL (1988) Magma flow directions in dikes of the Koolau complex, Ohau, determined from magnetic fabric studies. J Geophys Res 93:4308–4319

Leat PT (2008) On the long-distance transport of Ferrar magmas. In: Thomson K, Petford N (eds) Structure and emplacement of high-level magmatic systems, vol 302. Geological Society of London Special Publication, pp 45–61

Maccaferri F, Bonafede M, Rivalta E (2011) A quantitative study of the mechanisms governing dike propagation, dike arrest and sill formation. J Volcanol Geoth Res 208:39–50

Maccaferri F, Rivalta E, Keir D, Acocella A (2014) Off-rift volcanism in rift zones determined by crustal unloading. Nat Geosci 7:297–300. https://doi.org/10.1038/NGEO2110

Maccaferri F, Acocella V, Rivalta E (2015) How the differential load induced by normal fault scarps controls the distribution of monogenic volcanism. Geophys Res Lett 42. https://doi.org/10.1002/2015GL065638

Magee C, Muirhead JD, Karvelas A, Holford SP, Jackson CAL, Bastow ID et al (2016) Lateral magma flow in mafic sill complexes. Geosphere 12:809–841

Marsh BD (2015) Magma chambers. In: Sigurdsson H, Houghton B, McNutt S, Rymer H, Stix J (eds) The encyclopedia of volcanoes, 2nd edn. Elsevier Academic Press, pp 185–201

Mastin LG, Pollard DD (1988) Surface deformation and shallow dike intrusion processes at Inyo Craters, Long Valley, California. J Geophys Res 93:13221–13235

Mathieu L, Burchardt S, Troll VR, Krumbholz M, Delcamp A (2015) Geological constraints on the dynamic emplacement of cone-sheets—The Ardnamurchan cone-sheet swarm, NW Scotland. J Struct Geol 80:133–141

Muirhead JD, Airoldi G, White JDL, Rowland JV (2014) Cracking the lid: Sill-fed dikes are the likely feeders of flood basalt eruptions. Earth Planet Sci Lett 406:187–197

Muirhead JD, Van Eaton AR, Re G, White JDL, Ort MH (2016) Monogenetic volcanoes fed by interconnected dikes and sills in the Hopi Buttes volcanic field, Navajo Nation, USA. Bull Volcanol 78:11. https://doi.org/10.1007/s00445-016-1005-8

Muller JR, Ito G, Martel SJ (2001) Effects of volcano loading on dike propagation in an elastic half-space. J Geophys Res 106:11101–11113

Neri M, Lanzafame G, Acocella V (2008) Dike emplacement and related hazard in volcanoes with sector collapse: the 2007 Stromboli (Italy) eruption. J Geol Soc Lond 165:883–886

Neri M, Rivalta E, Maccaferri F, Acocella V, Cirrincione R (2018) Etnean and Hyblean volcanism shifted away from Malta Escarpment by crustal stresses. Earth Planet Sci Lett 486:15–22

Nunes JC, Camacho A, Franca Z, Montesinos FG, Alves M, Vieira R et al (2006) Gravity anomalies and crustal signature of volcano-tectonic structures of Pico Island (Azores). J Volcanol Geoth Res 156:55–70

Ode H (1957) Mechanical analysis of the dike pattern of the Spanish Peaks area, Colorado. Geol Soc Am Bull 68:567–575

Pansino S, Emadzadeh A, Taisne B (2019) Dike channelization and solidification: time scale controls on the geometry and placement of magma migration pathways. J Geophys Res 124:9580–9599

Peccerillo A, Donati C, Santo AP, Orlando A, Yirgu G, Ayalew D (2007) Petrogenesis of silicic peralkaline rocks in the Ethiopian rift: geochemical evidence and volcanological implications. J Afr Earth Sci 48:161–173

Phillips WJ (1974) The dynamic emplacement of cone sheets. Tectonophysics 24:69–84

Pinel V, Jaupart C (2000) The effect of edifice load on magma ascent beneath a volcano. Philos Trans R Soc Lond 358:1515–1532

Pinel V, Jaupart C (2004) Magma storage and horizontal dyke injection beneath a volcanic edifice. Earth Planet Sci Lett 221:245–262

Poland MP, Moats WP, Fink JH (2008) A model for radial dike emplacement in composite cones based on observations from Summer Coon volcano, Colorado, USA. Bull Volcanol. https://doi.org/10.1007/s00445-007-0175-9

Pollard DD, Delaney PT, Duffield WA, Endo ET, Okamura AT (1983) Surface deformation in volcanic rift zones. Tectonophysics 94:541–584

Porreca M, Acocella V, Massimi E, Mattei M, Funiciello R, De Benedetti AA (2006) Geometric and kinematic features of the dike complex at Mt. Somma, Vesuvio (Italy). Earth Planet Sci Lett 245:389–407

Rivalta E, Corbi F, Passarelli L, Acocella V, Davis T, Di Vito MA (2019) Stress inversions to forecast magma pathways and eruptive vent location. Science Adv 5: eaau9784

Roman A, Jaupart C (2014) The impact of a volcanic edifice on intrusive and eruptive activity. Earth Planet Sci Lett 408:1–8

Rubin AM, Pollard DD (1988) Dike-induced faulting in rift zones of Iceland and Afar. Geology 16:413–417

Sakuma S, Kajiwara T, Nakada S, Uto K, Shimizu H (2008) Drilling and logging results of USDP-4—Penetration into the volcanic conduit of Unzen Volcano, Japan. J Volcanol Geoth Res 175:1–12

Schirnick C, van den Boogard P, Schmincke H-U (1999) Cone sheet formation and intrusive growth of an oceanic island—The Miocene Tejeda complex on Gran Canaria (Canary Islands). Geology 27:207–210

Soriano C, Beamud E, Garcés M (2008) Magma flow in dikes from rift zones of the basaltic shield of Tenerife, Canary Islands: implications for the emplacement of buoyant magma. J Volcanol Geoth Res 173:55–68

Tibaldi A, Pasquare AF, Rust D (2011) New insights into the cone sheet structure of the Cuillin Complex, Isle of Skye, Scotland. J Geolo Soc Lond 168:689–704

Tibaldi A, Bonali FI, Corazzato C (2014) The diverging volcanic rift system. Tectonophysics 611:94–113

Tibaldi A (2015) Structure of volcano plumbing systems: a review of multi-parametric effects. J Volcanol Geoth Res 298:85–135

Trippanera D, Acocella V, Ruch J, Abebe B (2015) Fault and graben growth along active magmatic divergent plate boundaries in Iceland and Ethiopia. Tectonics 34. https://doi.org/10.1002/2015TC003991

Urbani S, Trippanera D, Porreca M, Kissel C, Acocella V (2015) Anatomy of an extinct magmatic system along a divergent plate boundary: Alftafjordur, Iceland. Geophys Res Lett 42. https://doi.org/10.1002/2015GL065087

Urbani S, Acocella V, Rivalta E, Corbi F (2017) Propagation and arrest of dikes under topography: models applied to the 2014 Bardarbunga (Iceland) rifting event. Geophys Res Lett 44. https://doi.org/10.1002/2017GL073130

Walter TR (2003) Buttressing and fractional spreading of Tenerife, an experimental approach on the formation of rift zones. Geophys Res Lett 30:1296. https://doi.org/10.1029/2002GL016610

Wilson CJN, Hildreth W (1997) The Bishop Tuff: new insights from eruptive stratigraphy. J Geol 105:407–439

Yoshida T (1984) Tertiary Ishizuchi Cauldron, southwestern Japan Arc: formation by ring fracture subsidence. J Geophys Res 89:8502–8510

8 Volcano Monitoring

8.1 Introduction

This chapter focuses on the monitoring techniques and signals associated with the volcano-tectonic processes presented in Chaps. 3–7 occurring at the volcano scale, with the aim to understand unrest and, ultimately, forecast eruptions, which are topics of Chap. 9. Volcano monitoring data are indeed essential to understand the real-time structure, dynamics and state of active volcanoes, which is a pre-requisite for hazard assessment, eruption forecasting and, ultimately, risk mitigation.

Volcano monitoring techniques have significantly benefited from the dramatic technological improvement of the last decades, and today rely on an extremely diversified pool of instruments, which operate in situ or remotely, allowing real-time detection of the physical and chemical changes occurring at volcanoes. This chapter provides an overview on the most common monitoring techniques, illustrating their importance in understanding the dynamics of active volcanoes, with particular attention to volcano-tectonic processes.

The main aims of this chapter are thus to:

- discuss the importance of monitoring for volcanology;
- provide an overview on the main monitoring techniques at active volcanoes, focusing on those allowing better detection of the volcano-tectonic processes described in the previous chapters, in particular deformation and geophysical monitoring;
- introduce how these monitoring techniques may allow determining the state of a volcano and help in forecasting eruptions.

8.2 General Features

Active volcanoes, with the exception of restless caldera systems, usually spend most of their lives in a state of quiescence, that can be defined as a background level characterized by a stationary state, during which measurable monitoring parameters, including deformation, seismicity and degassing, typically show little deviation from a baseline, with limited or no activity. When the measurable parameters show clear deviation from this baseline, the volcano experiences unrest. **Unrest** commonly results from the shallow emplacement or transfer of magma within a volcano, although it may be also triggered by physical-chemical variations in a hydrothermal system, if the latter is present. As discussed in Chap. 9, the shallow emplacement

V. Acocella, *Volcano-Tectonic Processes*, Advances in Volcanology,
https://doi.org/10.1007/978-3-030-65968-4_8

or transfer of magma triggering unrest is a prerequisite for eruptions, as eruptions are commonly preceded by unrest, although in some cases unrest may be difficult to be detected. While unrest may be symptomatic of impending eruption, it does not necessarily culminate into eruption, and in many cases a volcano may return to quiescence following unrest.

The states of quiescence, unrest or even impending eruption of an active volcano may be defined through a **monitoring system**, which consists of an array of different datasets devoted at detecting any variation in the physical or chemical state of a volcano, at different temporal and spatial scales and with different sampling frequency (Fig. 8.1). In particular, a monitoring system should ideally allow detecting the surface deformation, and the geophysical (seismicity and micro-gravity) and geochemical (composition and flux of released gases) variations within an active volcano. A monitoring system is indispensable for volcanologists to define the state of a volcano and understand its dynamics, during quiescence, unrest and eruption. In addition, volcano monitoring is crucial for hazard assessment and forecast, as well as risk mitigation. Therefore, volcano monitoring at its most effective is a synergy between basic science and mitigation of risk, providing crucial information for volcanologists and society. To be effective, the available monitoring data should be synthesized and merged in a coherent way, providing interpretations consistent with all observations and any other available information (e.g., Pallister and McNutt 2015; Wauthier et al. 2016).

To make useful measurements, monitoring techniques must have (a) a broad range to ensure that an entire event is recorded, (b) sufficient sensitivity to record slight and slow changes, and (c) adequate spatial coverage to capture an event. If a single technique may not satisfy all three requirements there is the need to use multiple techniques, also to measure a same parameter, as for example deformation (Dvorak and Dzurisin 1997). Indeed, it is better not to rely on only one instrument or one type of data, and rather consider measurements from several different types of instruments simultaneously, which may provide a better understanding of magmatic processes. For example, decreased gas emissions coupled with increased seismicity suggest increasing pressure and greater probability of

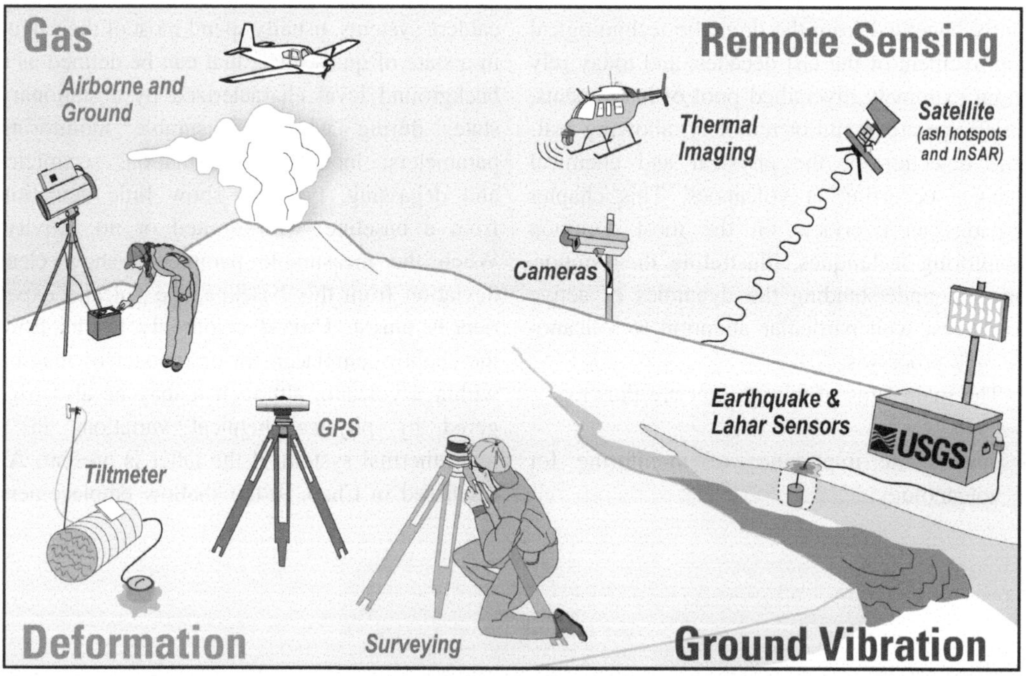

Fig. 8.1 Cartoon illustrating some frequently-used volcano monitoring techniques (from US Geological Survey, USGS; https://volcanoes.usgs.gov/vhp/monitoring.html)

explosive eruption, as experienced at Pinatubo (Philippines) in 1991. Conversely, uplift without seismicity, as detected at Three Sisters volcano (Oregon, USA), suggests deep magmatic intrusion (below the brittle-ductile transition) and consequently low probability of imminent eruption. Monitoring is particularly useful during the unrest state of a volcano, allowing detecting possible precursors of an impending eruption, ultimately improving its forecast (see Sect. 9.5). Also, monitoring data essential to understand volcanoes are often collected during the eruption itself, when important variations in the magmatic system provide valuable understanding of volcanic processes, as well as relevant information about the next eruptive stage (Pallister and McNutt 2015).

The monitoring techniques vary widely in measured parameters, instruments type, sophistication, cost, and effectiveness. Some instruments are permanent and telemetered; once installed, they collect and relay data in real-time to observatories for long periods. Other instruments may be deployed temporarily, as during unrest. Monitoring is complicated by the fact that data are acquired at different times and in different formats. Some of these data can be viewed "raw", but other data must be processed first, causing delays. Fortunately, automated data processing is decreasing these delays, and more and more monitoring information is being provided in real-time. Each monitoring technique has advantages and drawbacks of some kind, but each provides unique and complementary information, often on the same process. The merging of different techniques through a **multi-parametric monitoring** approach allows properly defining the state of a volcano and, in case, providing a warning. In fact, such integrated approach may permit to achieve a broad and objective understanding of the ongoing magmatic processes, as for example recognizing pressurization and depressurization phases within the shallow plumbing system of Mount Etna (Italy) and relate these to the observed variations in volcanic activity (Fig. 8.2; Cannata et al. 2015).

General recommendations for monitoring equipment levels at volcanoes, based on their threat, have been recently provided by the United States Geological Survey (USGS). In particular, the USGS did a threat analysis for a National Volcano Early Warning System in 2005, updated in 2018, in which volcanoes were ranked by threat, and monitoring levels were established to test if monitoring was commensurate with threat (Ewert et al. 2005; 2018). This allowed producing a report recommending instrumentation levels (Moran et al. 2008), with an approach followed also by other countries. However, except for these general recommendations, there is no established standard or common rule to deploy instruments to monitor a volcano (Fernandez 1999; Dzurisin 2007). Generally, the monitoring instrumentation to be deployed is mainly a function of the type, scale, activity, risk and scope of the monitoring tailored to a specific volcano. The scope of monitoring is to detect magma movements precisely and promptly, also to define if, and in case, where, an eruption may occur. To this aim, quiescent volcanoes should be routinely monitored to detect any unrest (background mode) and all volcanoes experiencing unrest should be more accurately monitored, with a multi-parametric network, to forecast any impending eruption (eruption mode; Wadge 2003). In practice, time, costs and resources take their share, so that the development and efficiency of monitoring systems vary widely from country to country and, even within the same country, from volcano to volcano, with priority usually given to active volcanoes in densely inhabited areas. While several volcanoes are adequately monitored in real-time, most volcanoes, especially in remote areas and in developing countries, are still monitored in a basic way, lacking multi-parametric networks, or not monitored at all. A study of 441 active volcanoes in 16 developing countries reveals that 384 have rudimentary or no monitoring, including 65 volcanoes identified as posing a high risk to large populations (Sparks et al. 2012).

Monitoring volcanic activity is usually in charge of **volcano observatories**, or organisations (including national institutions, universities or dedicated stations) whose role is to monitor active volcanoes and provide early warnings of

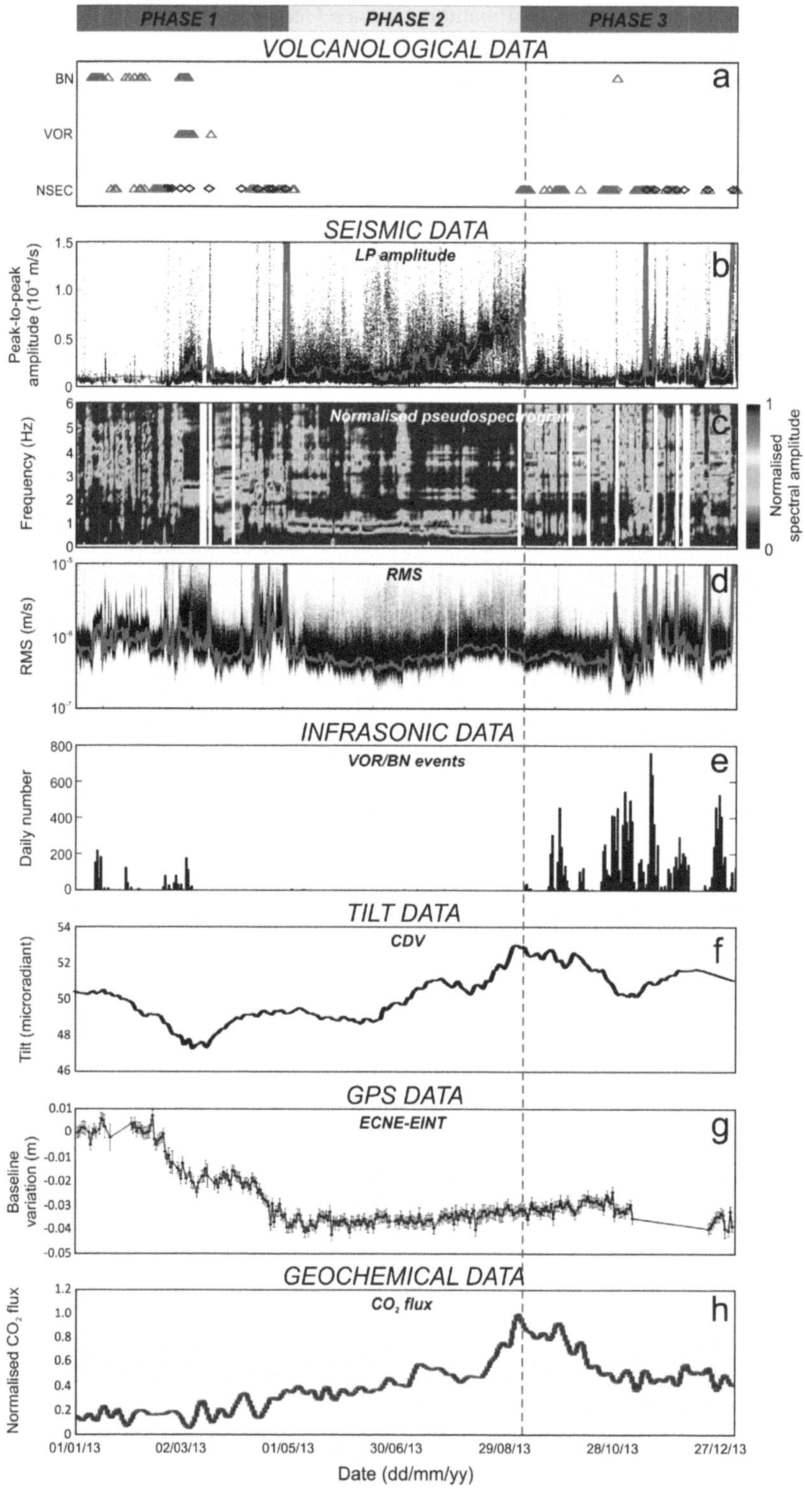
PHASE 1
PHASE 2
PHASE 3
VOLCANOLOGICAL DATA
BN
VOR
NSEC
a
SEISMIC DATA
LP amplitude
Peak-to-peak amplitude (10⁻⁴ m/s)
b
Normalised pseudospectrogram
Frequency (Hz)
Normalised spectral amplitude
c
RMS
RMS (m/s)
d
INFRASONIC DATA
VOR/BN events
Daily number
e
TILT DATA
CDV
Tilt (microradiant)
f
GPS DATA
ECNE-EINT
Baseline variation (m)
g
GEOCHEMICAL DATA
CO2 flux
Normalised CO2 flux
h
01/01/13
02/03/13
01/05/13
30/06/13
29/08/13
28/10/13
27/12/13
Date (dd/mm/yy)

◀ **Fig. 8.2** Example of multi-parametric monitoring data acquired in 2013 at Mount Etna (Italy). The vertical dashed red line indicates a major explosion at BN summit crater (5 September 2013). The top red and grey rectangles indicate main volcanic phases. **a** Eruptive activity at BN, VOR and NSEC summit craters (red triangles and black diamonds indicate Strombolian activity and lava fountaining, respectively). **b** Peak-to-peak amplitude of the detected long-period (LP) events (black dots) and relative 75° percentile (red line). **c** Pseudo-spectrogram of the LP events. **d** RMS (root mean square) amplitude of the vertical component of the seismic signals recorded by seismic station and filtered in the band 0.5–2.5 Hz (black dots) and relative 25° percentile (red line). **e** Daily number of infrasonic events located in the summit craters area. **f** Radial tilt component recorded at tiltmeter station. **g** Variation in time of the length of the baseline between two GPS stations with the 1σ associated uncertainties. **h** Total soil CO_2 flux amplitude variations (after Cannata et al. 2015; image courtesy Andrea Cannata)

anticipated volcanic activity to the authorities and the public (Fig. 8.3). A volcano observatory should thus be the source of authoritative short-term forecasts of volcanic activity and information on volcanic hazards. However, the exact constitution and responsibilities of an observatory, also in terms of decision-making, may differ country by country. Most observatories are grouped in the World Organisation of Volcano Observatories (WOVO; http://www.wovo.org/), with more than 100 members.

Current monitoring techniques vary widely, also boosted by the dramatic technological improvement of the last decades: volcano monitoring techniques include various types of measurements of surface deformation, seismicity, infrasound and acoustic emissions, microgravity, magnetic and electro-magnetic signals, thermal and infrared signals, photogrammetry, and plumes, soil, fumaroles, thermal waters and groundwater composition and fluxes. These techniques are adequately and extensively described elsewhere (e.g., Dzurisin 2007; Harris 2013; Pallister and McNutt 2015; Coppola et al. 2020). This chapter focuses on the most common monitoring techniques useful to understand volcano-tectonic processes, first considering deformation monitoring, then geophysical monitoring (in particular seismicity and gravity) and finally geochemical monitoring.

8.3 Deformation Monitoring

Volcano deformation investigates the variations in the shape and size of the Earth's surface resulting from magma accumulation and/or movement. Monitoring surface deformation in volcanic areas is essential for studying volcano-tectonic processes. In fact, understanding if and, in case, how, a volcano deforms provides important insights on its shallow structure and on magma accumulation and transfer processes, in terms of location, geometry, kinematics, volume and rate. For example, deformation time-series may allow distinguishing general but distinct

Fig. 8.3 Image of the surveillance room of Osservatorio Vesuviano (OV-INGV), in charge of monitoring Vesuvio, Campi Flegrei, Ischia and Stromboli volcanoes (photo courtesy Francesca Bianco, Head of OV). Osservatorio Vesuviano, originally established in 1841 by the King of Naples, is the oldest volcano observatory

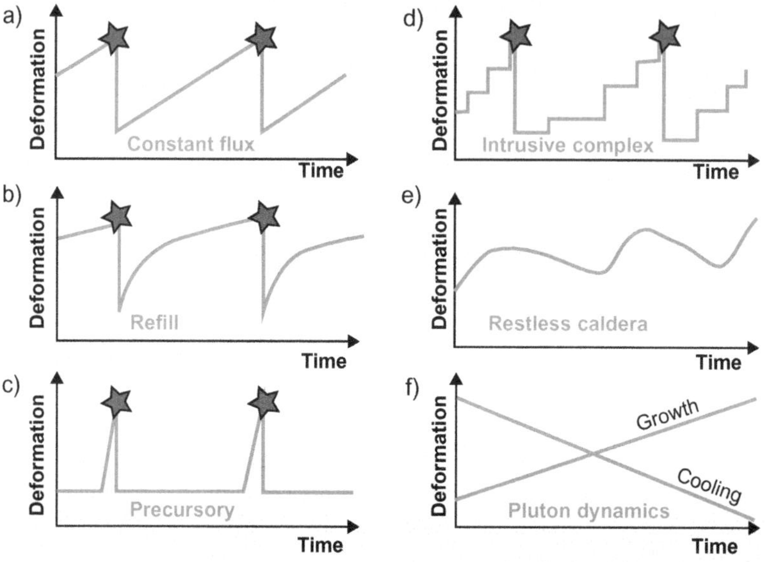

Fig. 8.4 Ideal time patterns of volcano deformation; stars represent eruptions. **a** Classic model of the volcano deformation cycle, with each step characterized by constant inflation rate, followed by sudden co-eruptive subsidence. **b** Similar to **a**, but with the refilling rate between eruptions decaying exponentially following visco-elastic effects. **c** Rapid inflation occurring only immediately before eruption. **d** Incremental uplift, due to pulsed magma supply, until a threshold is reached and eruption occurs. **e** Continuous non-eruptive unrest, with inflation and deflation, common at felsic calderas, **f** Pluton formation: constant uplift or subsidence resulting from the growth or cooling of the intrusion, respectively (modified after Biggs and Pritchard 2017)

behaviours of magmatic reservoirs, as summarized in Fig. 8.4. Deformation data may also provide fundamental clues to forecast specific hazards, as those deriving from the emplacement of a dike about to feed an eruption, or from the instability of the flank of a volcano, possibly culminating in sector collapse.

Most of the surface deformation in volcanic areas is within the threshold of elasticity of the rocks. This deformation often consists of **inflation** (i.e., expansion) of the volcanic edifice, highlighted by an overall radial distribution in the displacements, with each measurement site indicating both uplift and horizontal displacement directed radially outward. Volcano inflation commonly results from the increase in pressure of magma emplaced below the surface, occurring at any depth in the upper crust, although a shallow hydrothermal system may also contribute (Fig. 8.5a; e.g., Parks et al. 2012; Singer et al. 2014). Volcanoes may also experience **deflation** (i.e., contraction), highlighted by a radial distribution in the displacements, with each measurement site indicating both subsidence and horizontal displacement directed radially inward. Deflation may have various explanations, largely depending upon the recent activity of the volcano. It may be related to the withdrawal of magma from the chamber (as during lateral dike propagation following inflation), to the compaction of poorly consolidated deposits or porous material (as after caldera collapse or widespread effusive activity), to the crystallization and cooling of a magma chamber (as after magma intrusion), to the degassing of a magma chamber (as after an increase in crustal permeability promoted by seismicity), to the depressurization of an hydrothermal system (as following permeability variations) and/or to tectonic stretching (as following a mega-earthquake; Fig. 8.5b; e.g., Michon et al. 2009; de Zeeuw-van et al. 2013; Todesco et al. 2014; Nobile et al. 2017; Trasatti et al. 2019). For example, at Campi Flegrei (Italy) deflation between 1985 and 2005 followed the inflation of the 1969–1972 and 1982–1984 unrest episodes (Fig. 5.14a). While local contraction by cooling and regional extension may have contributed to deflation, the dominant factor was probably the slow removal of magmatic fluids as

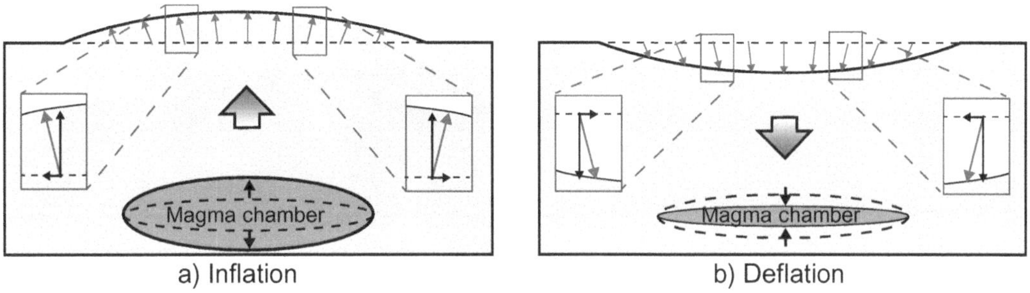

Fig. 8.5 Inflation and deflation result from magma chamber pressurization and depressurization, respectively. In the schematic section views (not to scale), the red arrows highlight surface displacement, whereas dashed black lines represent pre-inflation and pre-deflation conditions for topography and magma chambers. **a** Inflation induces uplift along the vertical component of displacement and horizontal displacement directed radially outward (black vectors in insets); **b** deflation induces subsidence along the vertical component of displacement and horizontal displacement directed radially inward (black vectors in insets)

the previously intruded magma cooled and released fluids to the overlying hydrothermal system. In fact, the 1969–1972 and 1982–1984 (Fig. 5.14a) uplift episodes were followed by subsidence at an initial rate that was larger than the average rate during the previous 140 years, and only after a few years did it return to the long-term minor subsidence rate (Dvorak and Dzurisin 1997).

Hydrothermal fluids may contribute to the inflation or deflation pattern, either amplifying or suppressing movement caused by magmatic processes. These fluids cluster within hydrothermal systems, as deep as 4–5 km: below 5 km hydrothermal circulation is usually choked off by reduced permeability and the higher temperatures, which approach magmatic conditions.

Both inflation and deflation are the expression of a radially symmetric displacement field at the surface. However, in some cases, surface deformation may consist of directional horizontal and vertical displacements, with non-radial symmetry: this characteristic pattern indicates magma transfer due to the propagation of a dike (see Sect. 7.7). While inflation and deflation are commonly centred at volcanoes, displacements with non-radial symmetry may occur at any location, within and outside volcanoes. These displacements must be detected in real-time, as they herald the shallow transfer of magma, potentially feeding an impending eruption.

Only with shallow sources and for significant pressure variations may it be possible to induce anelastic deformation at the surface: in particular, anelastic deformation usually results from intrusions (most notably dikes) approaching the surface, caldera collapse and flank instability, within and outside volcanoes. Under these conditions, extension fractures and faults are the most commonly observed deformation features. (Fig. 8.6; e.g., Pollard et al. 1983; Neri et al. 2004; Neal et al. 2019). This type of deformation may be detected through various monitoring techniques, although it can also be directly recognized and measured in the field without sophisticated instrumentation.

Surface deformation is an ubiquitous monitoring signal which characterizes unrest and often precedes eruptions. However, no deformation was detected in some unrest episodes leading to eruptions. There are several possible explanations for the lack of precursory deformation. This may depend upon a non-continuous acquisition of monitoring data. In this case, unnoticed deformation may occur only hours before the eruption onset or, alternatively, any pre-eruptive inflation may be compensated by co-eruptive subsidence, so that the net displacement approaches zero (as for example inferred for the April 2009 Llaima eruption, Chile). However, in many cases the lack of precursory deformation is not related to any limited sampling frequency, but to the absence of the deformation itself. This usually occurs with **open conduit** volcanoes, where the conduit is filled with molten magma and open (connected) to the surface. Open conduits may

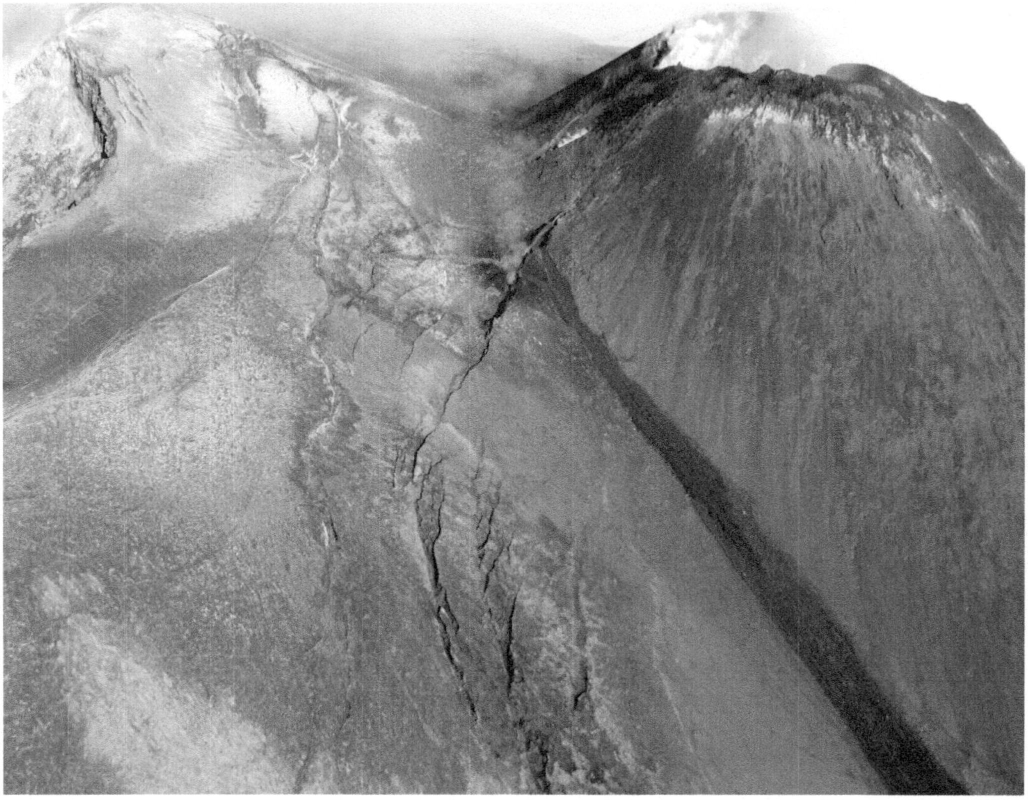

Fig. 8.6 Aerial view of the flank of the Bocca Nuova Cone (left; west) facing the Southeast Cone (right; east) of Mount Etna in 2006, showing the anelastic deformation pattern consisting of faults partly permeated by magma and gas (photo courtesy Marco Neri)

not be significantly pressurized to the extent of promoting appreciable surface deformation, as observed in March 2015 at Villarrica (Chile) and between 2007 and 2011 at Merapi (Indonesia), Colima and Popocatépetl (Mexico). Despite any lack of pre-eruptive deformation, in some open conduit volcanoes a minor but clear co-eruptive deflation may be still noticed, as at Stromboli in 2007 (Bonaccorso et al. 2008; Chaussard et al. 2013; Delgado et al. 2017).

8.3.1 Deformation Monitoring Techniques

The first deformation measurements at active volcanoes, in the form of levelling surveys, were conducted at Aira (Japan), Yellowstone (Wyoming, USA) and Campi Flegrei calderas, approximately one century ago. Today volcano deformation techniques include traditional approaches, as triangulation/trilateration, levelling, tiltmeters and strainmeters, and modern methods exploiting satellites, as Global Navigation Satellite System (GNSS) and Interferometric Synthetic Aperture Radar (InSAR), which promoted an explosion in the quantity and quality of studies on deforming volcanoes (Dzurisin 2007; Biggs and Pritchard 2017; Battaglia et al. 2019).

Triangulation/trilateration and levelling are classic monitoring techniques requiring the repeated deployment of personnel and sensors in the field to detect any deformation over multiple campaigns of measurements, not in real-time (e.g., Dzurisin 2007, and references therein).

Triangulation and trilateration are the oldest methods used to locate points and measure distance. The points whose positions are to be determined are connected in a network of triangles allowing line of sight visibility and measurement. In triangulation angles between points are measured from a reference station. In the trilateration distances between points are electronically measured from a reference station. In both cases, any deformation of the network can be measured by repeating the survey at a later date. Measurements errors are in the order of a few centimetres per kilometre of distance (or cm/km). **Electronic Distance Measurement**, or EDM, are a trilateration method consisting of an electro-optical (or electronic) distance metre, as photoelectric or laser devices. These direct a laser beam onto a distant reflector and measure the distance by comparing the phases (that is a fraction of period of the laser wave, measured through an angle) of the outgoing and reflected beams. While being very popular for decades, by the 1990s the role of EDM has been partly taken over by GNSS. However, modern laser EDMs are portable, inexpensive, and still helpful where deployment of GNSS receivers is not viable or prudent.

Levelling has been another popular and successful deformation technique for decades. It measures the height differences within a network of benchmarks and, by repeated surveys, any height change through time. The height differences can be determined by summing incremental vertical displacements of a graduated rod (differential levelling) or by measuring vertical angles (trigonometric levelling). Despite its simplicity, this technique has high accuracy, in the order of 1 mm/km.

Tiltmeters and strainmeters allow continuous monitoring with in situ sensors, also providing real-time measurements (Dzurisin 2007).

Tiltmeters measure any change in the local inclination of the Earth's surface or, in the case of borehole tiltmeters, in orientation of the borehole. Tiltmeters consist of short- and long-base instruments. Short-base tiltmeters use a pendulum or bubble as vertical reference and are more portable and less expensive. Long-base tiltmeters use the free surface of a liquid as horizontal reference measured by sensors at the extremities of a tube up to hundreds of metres long. They are more precise than the short-base instruments, reaching a sensitivity of one part per billion, or 10^{-3} μrad (one μrad, or microradian, is a unit of tilt equal to 10^{-6} radians, which corresponds to a vertical rise or fall of 1 mm over a horizontal distance of 1 km; in the SI system, 1 rad = 57.3°). Tiltmeters can be particularly useful to detect any inflation or deflation of a volcanic edifice, although topography may amplify or reduce the tilt by more than an order of magnitude (e.g., Marsden et al. 2019). **Strainmeters** commonly measure the linear or volumetric strain (that is the change in volume or shape). Linear strainmeters (or extensionmeters) measure displacement between two points, at the surface or in a borehole. Volumetric strainmeters (or dilatometers) measure the change of a certain volume in boreholes, where they can detect magma movement, inflation/deflation, pressure changes in magma body or hydrothermal system. Laser strainmeters may reach a precision of 100–0.1 parts per trillion, although their complexity and cost makes them unsuitable for most applications in volcanic areas.

GNSS and InSAR are modern satellite-based techniques allowing remote monitoring in real-time (permanent GNSS) or not (campaign-style GNSS and InSAR; e.g., Dzurisin 2007; Lu and Dzurisin 2014; Battaglia et al. 2019).

Global Navigation Satellite System (GNSS) is widely used in volcano monitoring, providing millimetric three-dimensional measurements of deformation. GNSS uses a ground-based antenna to track the signals transmitted from a constellation of satellites to provide an absolute location at a given time. Several constellations of satellites exist: **Global Positioning System (GPS)** from the United States of America is the most commonly used constellation, although other systems, as Glonass, Galileo and Beidou are in use. When observations are acquired only from GPS, the term "GPS measurements" is used, while when other constellations are tracked, "GNSS measurements" is preferred (e.g., Dzurisin 2007, and references therein). In ideal conditions, volcano observatories would install one or more reference GNSS sites 15–20 km away from the volcano or take advantage of the regional GNSS networks built for tectonic or geodetic purposes, removing any far-field deformation and obtaining only the signal from volcanic areas (Battaglia et al. 2019). The GNSS is the most versatile navigation, surveying, and geodetic tool ever devised. Unlike most other geodetic techniques, GNSS measures not just a single parameter pertaining to the relative positions of two points (e.g., height difference, line-length, or bearing), but also the full 3-D position of each point in an absolute reference frame, allowing detecting displacements remotely and continuously with millimetre accuracy. GNSS consists of a group of satellites that orbit the Earth at a certain altitude and continuously transmit information to receivers on the ground. GNSS receivers determine the travel times of satellite signals and keep track of the number of received carrier-signal cycles as a function of time. A travel time multiplied by the speed of light equals an apparent distance, which is related to the true satellite-to-receiver range by a mathematical model that includes errors and unknowns. This allows calculating the exact position of the receiver on Earth at a specific point in time. Assuming that the satellite positions are known, it is possible to determine the unknown position of a receiver by measuring at least four distances simultaneously, thus specifying three spatial coordinates for the phase centre of the receiver's antenna. These coordinates are measured in an absolute reference frame, with a few millimetres (N-S and E-W horizontal components) and ~ 10 mm (vertical component) accuracy. However, for practical purposes, the commonly used GNSS positions usually refer to a point origin of a local coordinate system, determining the distance from this point.

GNSS measurements can be obtained through permanent or continuous stations, which operate autonomously for long periods, or campaign-style surveys, whose sampling frequency is determined by the frequency of the surveys. In the continuous stations a permanent receiver is operated at a specific location and the phase centre of its antenna serves as the mark; measurements are provided with daily or subdaily resolution (Fig. 8.7; e.g., Dzurisin 2007, and references therein). Campaign measurements install antennas at specific benchmarks every few months or years. The antenna is set up directly over a physical benchmark and the height difference between the antenna and mark is measured to establish the position of the mark by way of the antenna's phase centre. The permanent stations and campaign-style surveys may both offer advantages. Continuous stations networks are ideal for real–time detection of transient deformations, especially at hazardous active volcanoes, whereas logistic costs of campaign-style surveys are usually lower than the cost of equipment, with the same receiver and antenna used at different sites.

A major shortcoming of GNSS, as well as all other point measurement techniques described so far, is the difficulty of obtaining data that are spatially dense enough to avoid under-sampling and thus emphasize a local deformation signal. This limitation may be properly addressed with areal deformation data acquired with **Interferometric Synthetic-Aperture Radar (InSAR)** techniques. Under ideal conditions, radar images taken from satellites can provide more information about a deforming volcano than even the most intensive ground-based geodetic surveys (Fig. 8.8a). InSAR is based on the remote observation of portions of the Earth's surface from satellites in descending (southward) and ascending (northward) polar orbits. A satellite sends radar waves (wavelength between 1 m to 1 mm) from its side reaching the imaged surface obliquely. These waves may be reflected by the Earth's surface and return to and be tracked by the satellite, where the SAR antenna switches between transmit and receive modes. The precise timing between the initial and return signals allows determination of the distance from the radar on the satellite to the object on the ground, through estimation of the phase (fraction of period of the radar wave, measured through an angle from 0 to 2π) of the backscattered signal. This produces a Synthetic Aperture Radar (SAR) image of an area (master image). Successive passes over the same area acquire later images (slave images). The co-registration of two radar images subtracting the phase information from corresponding pixels

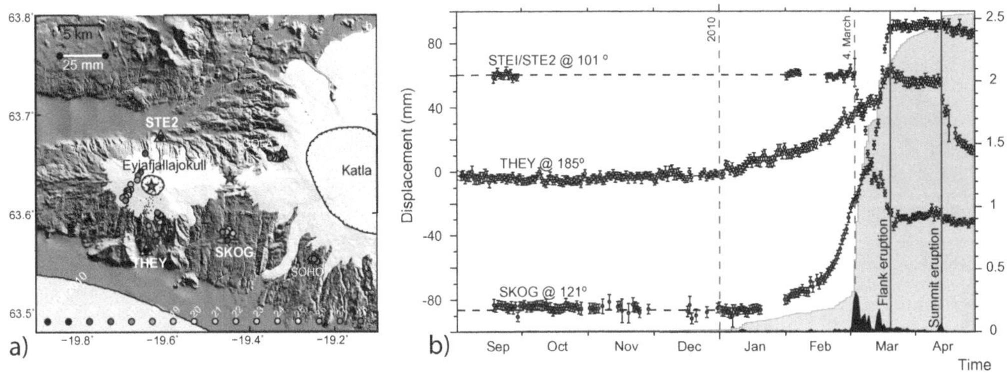

Fig. 8.7 GPS data obtained before and during the 2010 Eyjafjallajokull (Iceland) eruption (courtesy Freysteinn Sigmundsson; after Sigmundsson et al. 2010). **a** Locations of GPS stations (triangles); daily station horizontal displacements 14–28 April are shown in colour (dots at bottom show colour-coding of dates). Stations closest to the volcano contract towards it. Earthquake epicentres 14–29 April are indicated by black dots. Red stars show eruptive sites. **b** Selected displacement components from 1 September 2009 to 28 April 2010 at GPS stations THEY, SKOG and STEI/STE2. Error bars, 1σ confidence intervals. Displacement in direction N185° is shown for station THEY and direction N121° for SKOG, both away from the volcano summit. Displacement in direction N101° is shown for STEI/STE2, perpendicular to the direction to the summit. Grey shading shows the cumulative number of earthquakes and black shading the corresponding daily rate. The motions suggest overall pre-eruptive inflation and syn-eruptive deflation

allows creating their interference, or interferogram. The subtracted phase of the interferogram includes contributions from the different viewing geometries of the parent images, topography, path delays owing to different atmospheric conditions, noise, and any change caused by surface deformation between the image acquisitions. Once the other sources of phase differences in the interferogram are identified and removed, only the deformation phase remains. This is represented in terms of fringes, portrayed as a spectrum of colours from red to violet in the interferogram, each related to the half-wavelength of the emitted radar wave (Fig. 8.8 b; Dzurisin 2007; Lu and Dzurisin 2014, and references therein). The result is a concentric, coloured fringe pattern, where the number of fringes is proportional to the displacement. The sign of the displacement (towards or away from the line of sight, or LOS, of the satellite) is expressed by the inward to outward sequence of the spectrum of colours in each fringe. The frequency of the emitted radar waves encompasses the X-band (wavelength ~3 cm), C-band (wavelength ~5 cm) and L-band (wavelength ~20 cm), the latter penetrating more into vegetated areas. In general, radar signals penetrate water clouds, ash clouds and non-intense vegetation better than visible light. In addition, radar waves are equally effective in darkness and daylight, and during any weather.

There are a few limitations to the InSAR technique and its application. A general limitation is due to decorrelation, which occurs when the characteristics of the ground to reflect radar waves change rapidly, as in heavily vegetated or agricultural regions. Another limitation is due to the atmospheric delays caused by water vapour in the troposphere. Time-series methods that can combine hundreds or thousands of images are increasingly important in overcoming these limitations (Biggs and Pritchard 2017). Also, InSAR displacements are usually displayed in oblique line of sight, and the vertical and horizontal components can be evaluated only merging images from both descending and ascending orbits related to the same period. However, owing to the ~N-S orbit of the satellites, the N-S component of horizontal displacement cannot be detected. The accuracy in measuring displacements, in the case of the C-band, is usually in the order of 1 cm, that is on the upper bound

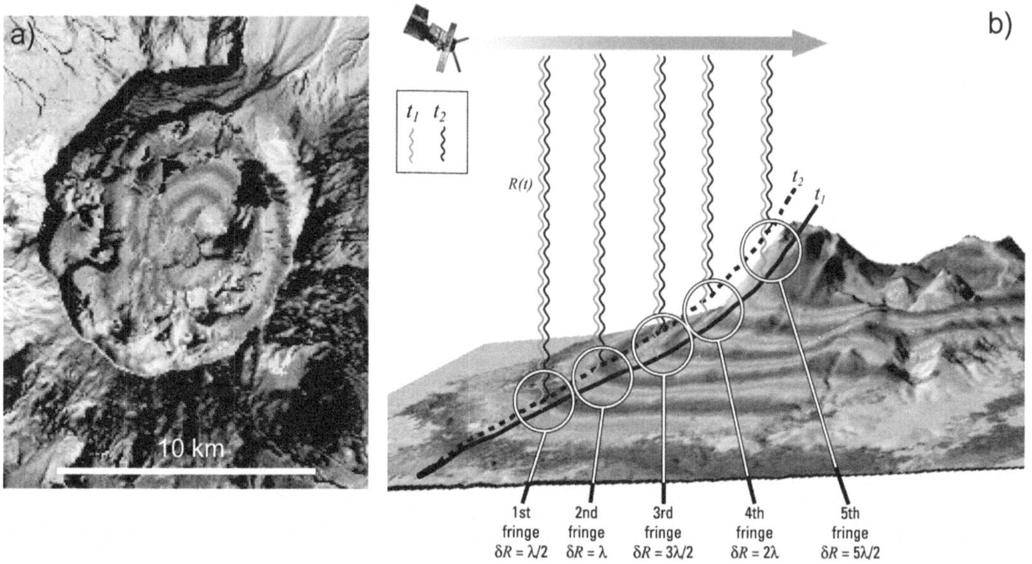

Fig. 8.8 **a** Example of interferogram illustrating the potential of InSAR data to capture areal deformation at active volcanoes. The image refers to the deformation of Okmok caldera (Aleutian Arc), as shown by Radarsat-1 descending tracks acquired between 2000 and 2006; each fringe (full colour cycle) represents 28 mm of range change between the ground and the satellite (Lu and Dzurisin 2014; image courtesy Zhong Lu). **b** Understanding interferometry: an inflating volcano produces a pattern of concentric fringes in a radar interferogram. If the volcano's height profile changes from the solid line to the dashed line between the acquisition times t_1 and t_2 of two radar images, the range $R_{(t)}$ from the SAR to the surface will decrease by half the radar wavelength $\delta R = \lambda/2$ (with λ being the wavelength) in some areas, by $\delta R = \lambda$ in other areas, $\delta R = 3\lambda/2$ in others, etc. Each half-wavelength change produces one complete interference fringe, which is portrayed as a spectrum of colours from red to violet in the interferogram detecting uplift. The result is a concentric pattern of fringes around the volcano, akin to contours of range change with a contour interval of $\lambda/2$. Subsidence produces a similar pattern of fringes, but with the opposite colour sequence (from violet to red; after Dzurisin 2007)

of that obtained from GNSS data. Finally, satellites pass over the same area in a few days, hindering real-time continuous information on deforming volcanoes and, ultimately, prompt detection of the signals of impending eruptions. Therefore, volcano SAR studies have been mainly used to characterize processes after their occurrence, which is valuable for advancing understanding of volcanic behaviour, but less useful for hazard assessment. The launching of additional satellites and their shorter passage times may provide better temporal resolution and near real-time monitoring of impending eruptions. Therefore, future application of SAR data should expand from a research tool for analysing volcanic activity after its occurrence to a real-time monitoring and research tool also capable to improve forecast during a volcanic crisis. Despite these limitations, volcano deformation studies have exploded in the last two decades thanks to InSAR. InSAR results have been reported from 620 volcanoes, with 220 observed to deform up to 2017 (Biggs and Pritchard 2017); this amount of monitoring data has given volcanology outstanding benefits.

GNSS and InSAR techniques have dramatically improved our capability to detect deformation at volcanoes in the last two decades and are now being widely used, often replacing traditional techniques. GNSS and InSAR techniques provide distinct and complementary geodetic information, best exploited through their merging (e.g., Fig. 6.18b). In fact, GNSS data are more accurate (especially on the horizontal components) and often real-time; also, the estimation of the atmospheric refraction using dual-frequency antennas provides a net gain in accuracy compared to InSAR. On the other hand,

InSAR data provide a much wider areal coverage (i.e., one GNSS site is one InSAR pixel), although neglecting the N-S horizontal component and not yet providing real-time information. Finally, the installation and maintenance of local receivers makes GNSS data expensive, whereas SAR data are commonly available to the wider scientific community. This implies that, while GNSS data are usually produced by volcano observatories with dedicated funds, SAR data may be also processed by individuals in academia or any research institution with very limited budget.

8.3.2 Deformation Source Models

Deformation data are often used in mathematical models to infer the characteristics (as the location, depth, shape, pressure change) of the source(s) responsible for the observed deformation. This procedure attempts to minimize the misfit between an observed deformation data set and synthetic idealized models of deformation sources under different boundary conditions. Therefore, once a starting model is selected, the model parameters can be adjusted iteratively to achieve the best possible fit between model results and observations; the procedure is called numerical **inversion** of deformation data. This is opposed to the **forward** modelling approach, adopted when a numerical source model with a priori fixed parameters and assumptions is used to make predictions, as for example on the resulting surface deformation field.

Most models inverting volcano deformation are kinematic, focusing on displacements. In these models, the physical processes that lead to changes in pressures acting on the source walls are ignored, and sources are limited to idealized geometries (spheres, ellipsoids, sills, dikes), although newly developed inversions are capable of resolving the pressure distribution within a source. The mathematical source models commonly represent the Earth's crust as an ideal semi-infinite elastic body, known as **elastic half-space**. This has one planar surface bounding a homogeneous and mechanically isotropic continuum that extends infinitely in all other directions. The assumption of elastic body is justified by the fact that in short-term laboratory tests at low temperatures rocks behave like linear elastic solids for strains less than 1% (e.g., Lisowski 2007). Elastic half-space analytical models often neglect many characteristics of the real Earth, but provide acceptable approximations of deformation. More complex kinematic models, often requiring numerical techniques, have been developed to account for topography (Williams and Wadge 2000), joint inversion of gravity and deformation (Battaglia and Hill 2009), layering (Crescentini and Amoruso 2007; Bazargan and Gudmundsson 2019), topography and nonlinear rheology (Trasatti et al. 2003), or visco-elastic layers (Del Negro et al. 2009; Segall 2019). For example, under some circumstances (as with long time periods or high temperatures), a visco-elastic medium may be more realistic than the elastic one. This introduces time dependence in the deformation, providing useful information if longer-term time-series displacement data are available (e.g., Lu and Dzurisin 2014; Novoa et al. 2019). In other cases, anelastic deformation should be taken into account. Using discontinuum mechanics models one can track how host rock fracturing progressively changes the shape and depth of an elastic deformation source during drainage of a magma body, simplifying the interpretation of the inversion model (e.g., Holohan et al. 2017). Also, the planar free surface of elastic half-space may constitute an important limitation with high relief volcanoes. Studies demonstrate that topographic corrections for typical composite volcanoes are of the same order as differences generated by different source geometries; effects can be significant even for shield volcanoes with low slopes (Lisowski 2007, and references therein). However, unlike other simplifying model assumptions (as material homogeneity and isotropy), topography is well known and, with a suitable method, it could be used to improve a model. Continued advances in computing power and availability in the quantity and quality of data lead to widespread use of these more realistic models.

A potential drawback in the inversion procedure is that, given enough model parameters to adjust, or the flexibility to include multiple deformation sources, it may be possible to fit any data set arbitrarily well. The modelling of surface deformation in fact does not provide a unique description of the responsible source. For example, the width of the deformed area at the surface is proportional to both the size and depth of the source, so that a deeper and smaller source may produce a deformation as wide as a shallower and larger source. Also, the amplitude of the surface deformation (maximum uplift or subsidence) increases with both the shallowing and the pressurization of the source, so that deeper sources experiencing higher pressure variations may produce an uplift similar to shallower sources experiencing lower pressure variations. As a result of these non-univocal behaviours, different types of sources may require different depths to reproduce the deformation (Fig. 8.9; e.g., Ryan et al. 1983; Pritchard and Simons 2004). To reduce the degrees of freedom in defining the shape (and depth) of the source, in addition to using statistical tests, it is important to invert both the vertical and horizontal deformation: while the former is similar for most sources, the latter provides more distinctive variations with distance as a function of the modelled source (Dieterich and Decker 1975; Lisowski 2007; Segall 2010). Also, while models may provide information about the location, size and shape of a source, they are not informative of its magmatic or hydrothermal nature. As a rule of thumb, in absence of additional information, very shallow (< 2–3 km depth) and transient sources may be usually related to hydrothermal systems, whereas deeper (>5 km) and more stable sources may be interpreted as magmatic. Realistic uncertainties should be then assigned to parameters estimated from the modelling of deformation data. Because of the nonlinearity and complexity of the models, it is necessary to estimate the errors using a statistical approach.

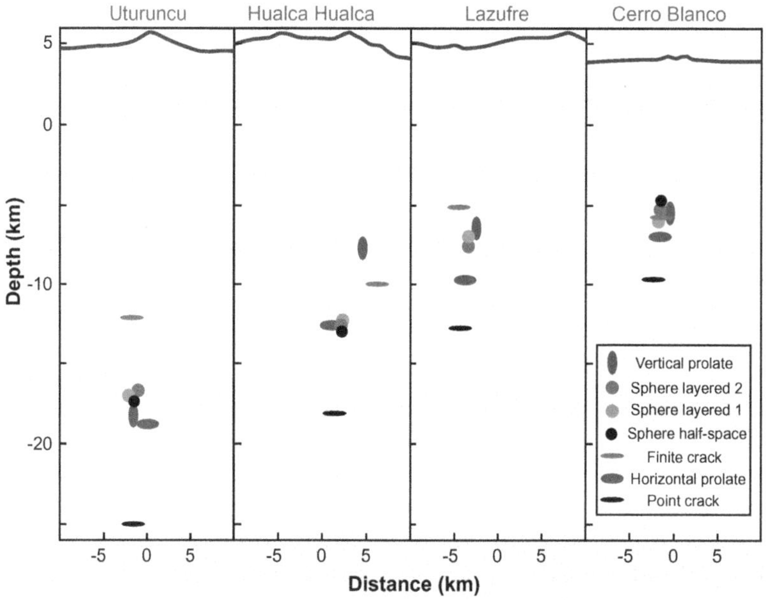

Fig. 8.9 North-South profiles at four deformation centres in the Central Andes (names in red) showing the topography and the inferred location and depths for different spherical, ellipsoidal and crack sources of deformation. The ellipsoidal and crack sources were calculated in a half-space, while the spherical sources were calculated using both layered- and half-space crustal models. The models shown are intended to be end-members and do not represent all possible sources that can explain the deformation (modified after Pritchard and Simons 2004)

A-posteriori errors for individual parameters can be determined using either a Monte-Carlo simulation technique or bootstrap (e.g., Wright et al. 1999; Battaglia and Hill 2009). Model assumptions, simplifications, and data uncertainty further complicate interpretation. To minimize or avoid any misinterpretation, one should look for the simplest deformation source model that: (a) fits the data adequately, with the smallest misfit, given the uncertainty inherent in the data and the over-simplifications inherent in the model (both of which should be adequately represented); (b) is consistent with other observations; (c) makes sense from a volcanological perspective.

This modelling approach may be used for real-time detection of magmatic processes during volcanic crises and for semi-operative purposes, as for example to forecast any impending eruption and the location of opening of a vent. However, the interpretation of the modelling results in terms of defining the potential of an eruption is subject to at least two main shortcomings (Lu and Dzurisin 2014). First, magma is a compressible fluid and its compressibility can range over several orders of magnitude, depending on different factors (including gas content, bubble density and crystal fraction). Similarly, the host rock can vary in stiffness (shear modulus). Therefore, highly compressible magma injected in a strong reservoir might produce little or no surface deformation, whereas the same amount of nearly incompressible magma added to a weak reservoir would cause much more deformation. The result is a wide variation in the relations between the amount of magma added to or withdrawn from a reservoir and the related surface deformation. A way to reduce this ambiguity is to use micro-gravity measurements (see Sect. 8.4.2), which are sensitive to changes in subsurface mass instead of volume. Second, any estimate of the volume of magma added to or withdrawn from a reservoir may not be representative of the volume or composition of the potentially eruptible magma. In fact, deformation measurements provide information only about incremental changes in the source, not about its overall size or nature. Both shortcomings limit our capability to estimate the amount of the potentially eruptible magma, and thus the size of any eruption.

Volcanic deformation sources consist of pressurizing and depressurizing bodies of various shapes and sizes, which produce changes in volume and are collectively known as **volumetric sources** (note that volumetric sources are distinct from tectonic sources, where the two sides of a fault slide by one another). To a first approximation, magma intrusions responsible for surface deformation can be simplified as idealized fluid-pressurized ellipsoidal cavities in elastic half-space, with variable axes and symmetric about the vertical axis. The deformation depends on the shape and size of the source, the increment of pressure, and the elastic properties of the medium. For any particular source depth and geometry, the surface deformation mainly scales with the ratio of the cavity pressure change to the half-space elastic shear modulus $\Delta P/G$, whose typical values range between 10^{-3} and 10^{-1} (Mogi 1958; Davis et al. 1974; Davis 1986; Yang et al. 1988; Fialko et al. 2001).

A simple magmatic source is the point pressure, or **point dilatation source**, often called the "**Mogi model**", characterized by radius r much smaller than the depth d; the limiting case of a finite-sized pressurized cavity as r/d tends to zero is indeed the point pressure source (Fig. 8.10; Mogi 1958; Lisowski 2007, and references therein). This formulation remains the best-known and most widely used to model surface deformation from deflating or inflating sources. The displacement in the three components u, v and w at the point on the surface $(x, y, 0)$, produced by pressure change within a spherical cavity embedded in an elastic half-space with radius r much smaller than its depth d ($r \ll d$, even though in practice approximation is good for $r < 0.5d$), is given by:

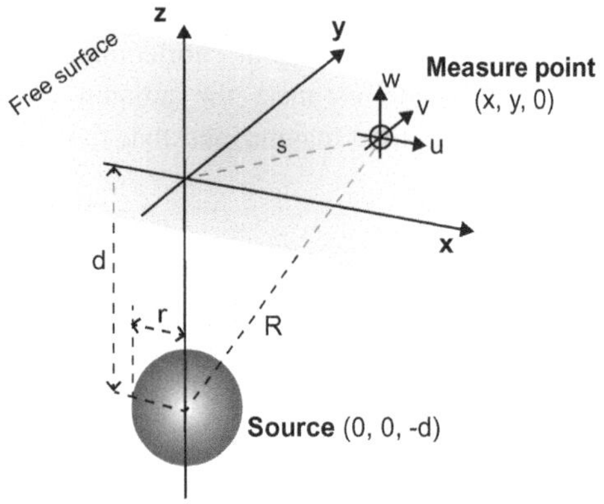

Fig. 8.10 Coordinate system and geometric relationships used to derive surface deformation from a point pressure source (red sphere)

$$\begin{pmatrix} u \\ v \\ w \end{pmatrix} = \left(r^3 \Delta P \frac{(1-v)}{G}\right) \begin{pmatrix} x/R^3 \\ y/R^3 \\ d/R^3 \end{pmatrix} \tag{8.1}$$

where ΔP is the cavity pressure change, G the elastic shear modulus, v is the Poisson's ratio and $R = \sqrt{x^2 + y^2 + d^2}$ the radial distance from the centre of the cavity to a point on the free surface (Lisowski 2007, and references therein). The individual contributions of the pressure change and radius of the source and the shear modulus and Poisson's ratio of the half-space are grouped together in the scaling coefficient, and cannot be estimated independently. This implies that, in the same elastic half-space, a small pressure change in a large cavity produces the same surface deformation as a large pressure change in a small cavity. However, assuming that the source behaves like a sphere with radius small compared to the depth of burial, it is possible to approximate the volume change of the cavity ΔV associated with the deformation, given by the integral of the radial displacement over the surface of the cavity, as (e.g., Segall 2010):

$$\Delta V \approx \frac{\Delta P}{G} \pi r^3 \tag{8.2}$$

This volume change considers only the mechanical properties of the surrounding half-space and is equivalent to the injection volume only in the case of incompressible magma. Using Eq. (8.2), Eq. (8.1) becomes:

$$\begin{pmatrix} u \\ v \\ w \end{pmatrix} = \left(\Delta V \frac{(1-v)}{\pi}\right) \begin{pmatrix} x/R^3 \\ y/R^3 \\ d/R^3 \end{pmatrix} \tag{8.3}$$

which allows relating the surface deformation due to a Mogi source to the volume changes in the source.

The surface displacements associated with the Mogi source are axisymmetric, with the vertical displacement peaking directly above the source, while the horizontal displacement reaches its maximum value at $0.7d$ (Fig. 8.11; e.g., Lisowski 2007). The maximum horizontal displacement is $\sim 38.5\%$ of the maximum vertical displacement, and it decays more slowly from its peak value, becoming larger at $d > 1$. Knowing the ratio of vertical to horizontal displacement at a point, it is possible to estimate the depth of the source by taking the product of this ratio and the approximate horizontal distance to the source. The location of the source (in terms of x, y and z coordinates) can be defined measuring two well-chosen surface displacement vectors: a surface displacement vector U is radial to the source and its magnitude

$$U = \sqrt{u^2 + v^2 + w^2} = r^3 \Delta P \frac{(1-v)}{G} \frac{1}{R^2} \tag{8.4}$$

varies with the inverse square of the distance from the centre of the source R.

The Mogi model is simple and widely used, particularly for deep intrusions. In fact, the deeper the deformation source, the more difficult it is to ascertain its shape from surface displacement data. As a consequence, all sources below ~ 10 km

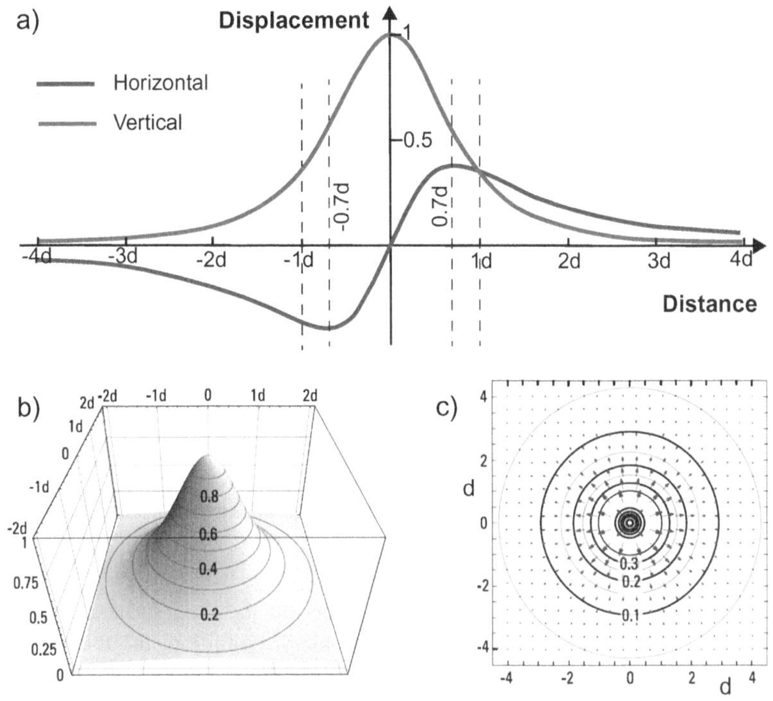

Fig. 8.11 **a** Profiles of axisymmetric surface displacements for a point pressure source (Eq. 8.1) as a function of horizontal distance in source depths d. The displacements are normalized by $r^3\Delta P((1-\nu)/Gd^2)$, the power of the source multiplied by the inverse square of its depth. The dashed vertical lines mark the maximum horizontal displacement $\pm d/\sqrt{2}$ and the distance $\pm 1d$ where horizontal displacement becomes larger than the vertical displacement. **b** Oblique three-dimensional view of vertical displacement. **c** Map view of horizontal displacement (modified after Lisowski 2007)

depth look similar (like a Mogi source) from the surface: hence the popularity of the Mogi model. At shallower depths, however, distinctive features of the deformation fields produced by sources of different shapes begin to emerge (Lu and Dzurisin 2014).

The limitation that the source radius be much smaller than its depth and the ambiguity between pressure and volume change of the Mogi source can be both resolved by using a finite **spherical source**. In the point pressure source approximation (Eq. 8.1), the radius of the source cannot be separated from the pressure change. Ideally, we would like to know the size of the cavity being pressurized and the increment of pressure. McTigue (1987) derived expressions for the surface deformation induced by a spherical source as a function of a series expansion in the ratio r/d:

$$\begin{pmatrix} u \\ v \\ w \end{pmatrix} = \left(r^3 \Delta P \frac{(1-\nu)}{G}\right)\left(1+\left(\frac{r}{d}\right)^3 \times \left(\frac{1+\nu}{2(-7+5\nu)} + \frac{15d^2(-2+\nu)}{4R^2(-7+5\nu)}\right)\right)\begin{pmatrix} x/R^3 \\ y/R^3 \\ d/R^3 \end{pmatrix} \tag{8.5}$$

The first term in the scaling coefficient is that of a point pressure source, while the second term scales this coefficient by the corrections for a finite-sized cavity. The terms of the corrections have a common factor of the cube of the ratio of cavity radius to source depth $(r/d)^3$. These terms are very small, except when the radius of the cavity is similar to its depth ($0.2 < r/d < 1$). Therefore, to a first approximation, the maximum

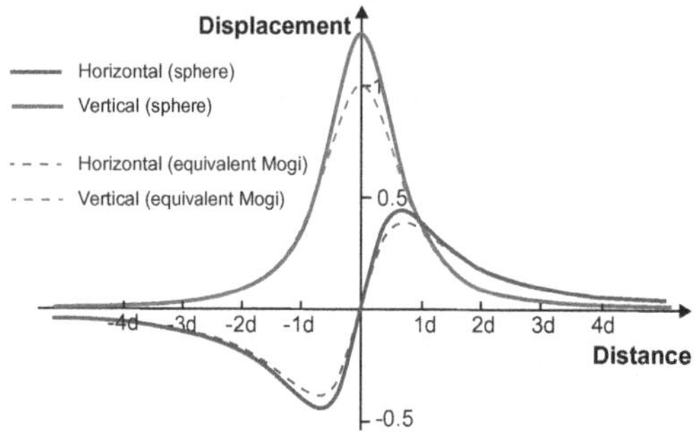

Fig. 8.12 Profiles of axisymmetric surface displacements (solid lines) for a finite spherical pressure source (Eq. 8.5; with $r/d = 0.6$) as a function of horizontal distance in source depths d. The displacements are normalized by the product of the power of the equivalent point pressure source and the inverse square of its depth. The surface displacements for an equivalent point pressure source ($r/d \rightarrow 0$) are shown for reference (dashed lines; modified after Lisowski 2007)

displacements for a finite sphere are about $1 + (r/d)^3$ times those for a point sphere. The pattern of deformation, however, is very similar and it would be difficult to distinguish a spherical source from a slightly shallower point source. Again, the source depth can be estimated from the ratio of vertical to horizontal displacement at a point (Fig. 8.12; Lisowski 2007, and references therein).

Unless the crust surrounding the spherical source is incompressible ($\nu = 0.5$), the volume of the surface uplift or subsidence V_u is not the same as the volume change in the spherical magma chamber ΔV (Delaney and McTigue 1994). In fact, it can be demonstrated that:

$$V_u = 2(1 - \nu)\Delta V \tag{8.6}$$

Therefore, quite counterintuitively, for $\nu < 0.5$ the volume of uplift is greater than the volume change of the magma chamber and this discrepancy increases with the decrease in the Poisson's ratio, that is with the crust becoming more compressible. However, this result holds only for a spherical magma chamber, as for a horizontal circular sill $V_u = \Delta V$ (see below; Fialko et al. 2001). To evaluate the relationships between the volume of intruded magma and the volume of the magma chamber one has to take into account also for the compressibility of magma (Segall 2010, and references therein).

Actual magma chambers are unlikely to be spherical. An obvious and more realistic extension of the spherical cavity model is the three-dimensional **ellipsoidal cavity** subjected to uniform internal pressure. The analytical treatment of the ellipsoid is much more challenging than that of a spherical cavity (Segall 2010, and references therein). Hence, efforts have focused on various approximate solutions, which build on the Eshelby's (1957) solution for an ellipsoidal inclusion in a full-space elastic medium. Starting from the Eshelby approach, an approximate solution for a pressurized ellipsoidal cavity in a half-space was first proposed (Davis 1986), followed by a more accurate solution for a prolate ellipsoid fitting the boundary conditions with a magma body far from the free surface (Yang et al. 1988). Both models fit the boundary conditions on the free surface, but neither fits the boundary condition on the cavity walls. The displacements due to vertical prolate ellipsoidal magma chambers depend on the aspect ratio of the ellipsoid. As the aspect ratio of the prolate ellipsoid increases, the horizontal displacement becomes more significant than the vertical displacement. Also, as the ellipsoid becomes more prolate, the vertical displacement is no longer maximal over the centre of the chamber. In fact, Poisson's effect induces a slight subsidence associated with extension on the top of the cavity. A limit condition for a prolate ellipsoid is the vertically **elongated pipe**, where the vertical axis is much longer than the two (equal) horizontal axes. This source is commonly used to simulate magma transported through a conduit closed at the top, forming a viscous intrusion. The surface deformation resulting from the inflation of a closed pipe has been formulated by Bonaccorso and Davis (1999) and, with minor modifications, Segall (2010). For $\nu = 0.25$, the maximum vertical displacement is at a distance equal to the depth of the top of the pipe (d_t) and the maximum horizontal displacement, at $2d_t$, is up to 77% of the maximum vertical displacement, similarly to very prolate ellipsoids (Fig. 8.13; Lisowski 2007). In general, the elongated closed pipe generates proportionally more horizontal displacement than a spherical source, the vertical

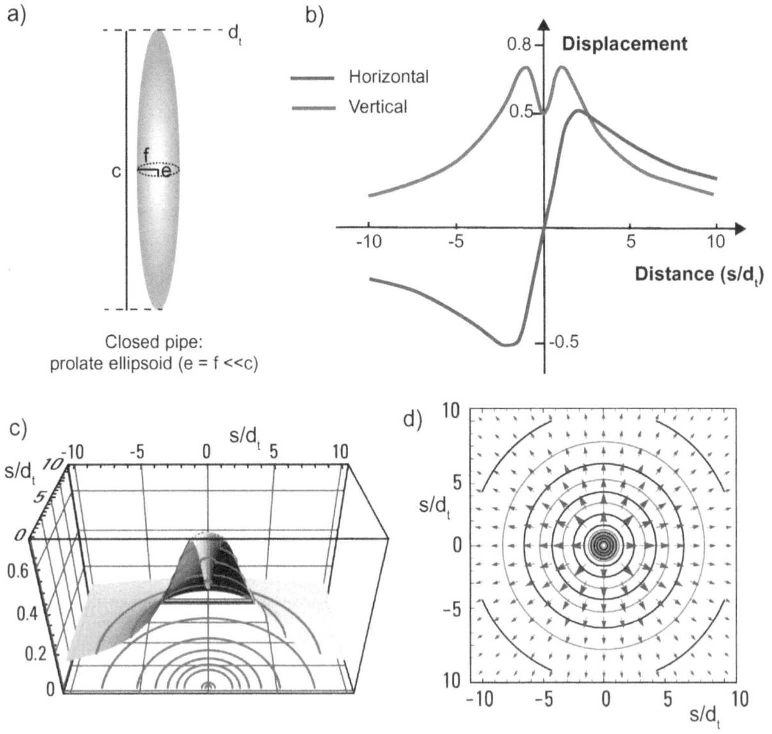

Fig. 8.13 Surface displacements for a prolate ellipsoid with high aspect ratio, as a closed pipe. **a** Geometry of the prolate ellipsoid. **b** Profiles of axisymmetric surface displacements. The displacements are normalized by the power of the source multiplied by the inverse of the top depth $r^2\Delta P/4Gd_t$, and the distance is in top depths s/d_t (s is the horizontal distance from the axis of the ellipsoid and d_t is the depth of the top of the pipe; see also Fig. 8.10). **c** Cross section of contoured vertical displacements showing central dimple. **d** Contoured horizontal vector displacements (modified after Lisowski 2007)

and horizontal displacements decay more slowly, and a profile of vertical displacement has a characteristic dimple over the pipe. Solutions may be obtained also for open pipes, simulating magma transport and eruption through an open conduit. The open pipe generates little uplift and subsidence over the source due to Poisson contraction (Segall 2010).

Sill-like magma intrusions or chambers may be represented by pressurized oblate spheroids (Davis 1986), finite rectangular tensile dislocations (Davis 1983; Yang and Davis 1986), or finite pressurized horizontal circular cracks (Fialko et al. 2001). These solutions may carry some significant error when the crack radius is similar to or exceeds its depth (Davis 1983), or if the crack radius is larger than 5 times its depth (Fialko et al. 2001). The solution for a pressurized degenerate oblate spheroid with the lengths of the major axes much less than the length of the minor axis is shown below (Davis 1986). This model is adequate when the radius of the sill is much less than its depth. The surface displacements, simplified from the equations of Okada (1992) (see below), are given by:

$$\begin{pmatrix} u \\ v \\ w \end{pmatrix} = \left(\frac{3M_0}{2G\pi}\right) \begin{pmatrix} xd^2/R^5 \\ yd^2/R^5 \\ d^3/R^5 \end{pmatrix} \tag{8.7}$$

where M_0 is the moment, equivalent to the amount of opening multiplied by the area and the half-space shear modulus G. The profiles of surface displacements from sill-like magma bodies are similar to those from a corresponding spherical source (Fig. 8.14; Lisowski 2007), although the maximum vertical displacement is ~20% greater and the maximum horizontal displacement ~10% smaller. The deformation area is smaller, with the maximum horizontal displacement occurring at distance $d/2$. Sills are difficult to distinguish from other more equidimensional bodies given only vertical deformation data, and for this a combination of vertical and horizontal deformation data is needed. As anticipated, the volume of surface uplift or subsidence V_u (integral of vertical displacements) for a pressurized sill is equal to the change in the volume of the cavity ΔV (Delaney and McTigue 1994; Fialko et al. 2001; Lisowski 2007).

The solution in Eq. (8.7) is a special case of a more general class of deformation source models used to represent planar intrusions, including dikes, sheets and sills. A dipping rectangular tensile dislocation within an elastic half-space may allow modelling the deformation associated with a planar tension fracture, including a dike,

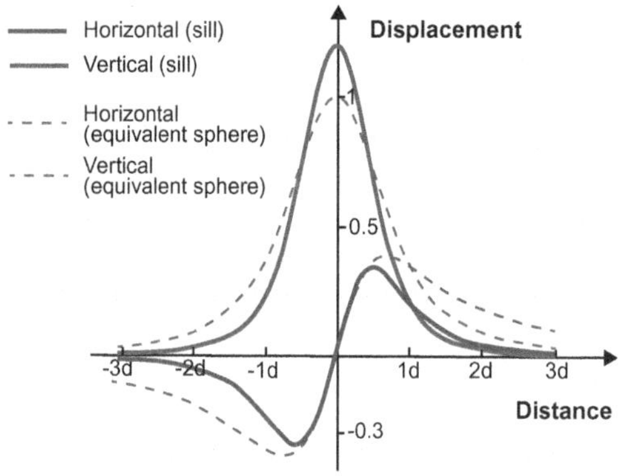

Fig. 8.14 Profiles of axisymmetric surface displacements (solid lines) from a pressurized sill-like magma body, compared to those from an equivalent spherical pressure source (dashed lines). The displacements are normalized by the power of the equivalent spherical source multiplied by the inverse square of the depth $r^3\Delta P/4Gd^2$, and the distance is in source depths (modified after Lisowski 2007)

similarly to solutions considering elliptical cracks (Pollard and Holzhausen 1979; Davis 1983). The "**Okada model**" (Okada 1985, 1992) is commonly used to represent the deformation field at the surface produced by a finite displacement applied on a rectangular dislocation. This model can be discretized into a number of smaller rectangular elements to obtain a distribution of displacement over the plane. Here the solutions of Okada (1992) to calculate surface deformation from dipping rectangular tensile dislocations are used to illustrate how the deformation across the centre of the crack varies with the crack's dip angle δ (Fig. 8.15; Lisowski 2007). For $\delta = 0°$ (subhorizontal crack, or sill) the deformation is axisymmetric, as already shown in Fig. 8.14. As the dip increases, the maximum uplift decreases and its centre moves away from the origin in the down-dip direction, and for $\delta > 60°$ a depression develops along the up-dip projection of the crack at the surface (Fig. 8.15). The vertical deformation becomes again symmetric across the crack at $\delta = 90°$, that is the case of a vertical crack, or **dike**, with a depression centred over the crack and flanked by two lobes of uplift. For vertical cracks or dikes that are very long (in map view) compared to their height (in section view), the vertical displacement immediately above the intrusion tends to zero and the depression disappears. As explained in Sect. 7.7, the surface uplift to the sides of the dike results from the lateral compression induced by the dike opening, whereas the subsidence above results from the dilation induced by the same opening. As for the horizontal displacements, with increasing dip of the crack these become concentrated in the down-dip direction, becoming equal in magnitude again when $\delta = 90°$, that is in the case of a vertical crack, or dike (Fig. 8.15). As the dike grows, the amplitude of the uplift increases: this is due to the fact that the dike gets shallower and, at the same time, its opening increases. As the dike propagates upward, the boundary between the zones of extension and compression shifts inward. Once the dike breaches the surface, the maximum vertical and horizontal displacements occur at the dike, and the strain is everywhere compressive

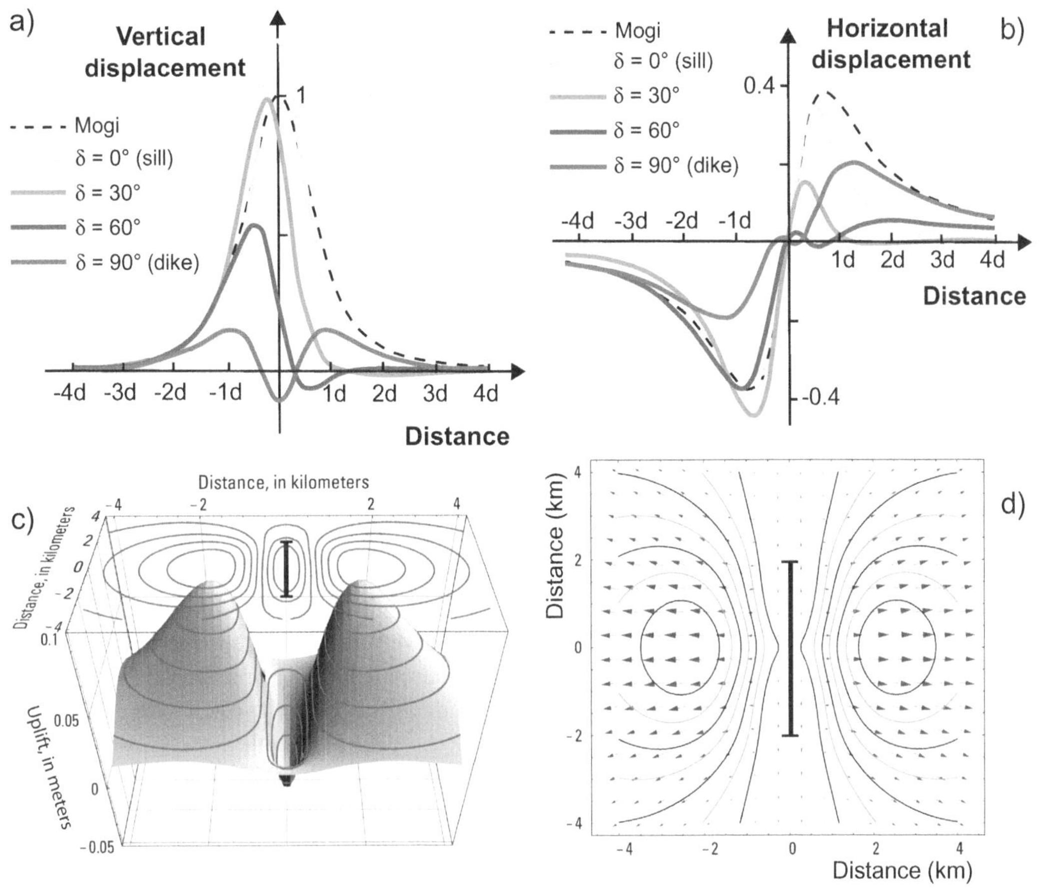

Fig. 8.15 Profiles of normalized vertical (**a**) and horizontal (**b**) displacements across a point tension crack dipping leftward with dip $\delta = 0°$ (sill), 30°, 60° and 90° (dike). The distance is in source depths *d*. A profile for a Mogi point source with equivalent moment is shown with the dashed black line for comparison. **c** Contoured 3-D vertical displacement surface and **d** contoured horizontal vector displacements (map view) for a vertical rectangular opening dislocation (dike) whose surface projection is shown with a solid black line (modified after Lisowski 2007)

perpendicular to the dike (Segall 2010). The characteristic non-axisymmetric (i.e., non-radial) surface displacement field for a dike can be easily distinguished from that of more equidimensional magma bodies, which are responsible for axisymmetric (i.e., radial) deformations; this distinctive feature has important implications in the possibility of forecasting impending eruptions and their location (see Sect. 9.2).

8.3.3 Deformation Forecasting Potential

The last decades have been characterized by a dramatic increase in the quantity and quality of volcano deformation monitoring data. In particular, as traditional ground-based survey methods have been complemented with satellite technology, the number of known deforming volcanoes has increased from 44 in 1997 to over 220 in 2016. This increase is not the result of any rise in volcanic activity, but simply a consequence of improved observation and reporting, particularly in developing countries and remote areas (Biggs and Pritchard 2017). Wide portions of the Earth with limited or difficult access have been well monitored through InSAR. For example, in the Aleutian Arc deformation is more prevalent than previously thought: here only 12 of the 52 volcanoes (23%) showed no evidence of surface deformation from 1992 to 2010, making this arc one of the most active (Dzurisin et al 2019).

These measurements, combined with deformation inversion models, allow estimating the location, size and shape of the sources responsible for the deformation. In addition, distinctive time-deformation patterns may be interpreted as diagnostic of specific magmatic process (Fig. 8.4; Biggs and Pritchard 2017). This association between time patterns and processes may enhance understanding volcanoes and, ultimately, forecasting eruptions. This latter feature is particularly relevant, as the kinematic models of Sect. 8.3.2 alone provide no predictive capability (e.g., Segall 2013). The huge amount of deformation data acquired in the last decades can be exploited for forecasting purposes with appropriate computational approaches. For example, sequential data assimilation methods are being created for merging multiple deformation observations to try to forecast impending activity. These methods are based on artificial intelligence, capable to glean subtle patterns in vast quantities of data and automatically create an alert when a volcano's unrest may raise concern (Bato et al. 2017; Zhan and Gregg 2017; Gaddes et al. 2019).

The occurrence of deformation and of specific deformation patterns has been used to try to forecast eruptions in different ways. From a general point of view, of the 198 volcanoes systematically observed by InSAR between 1995 and 2013, 54 deformed, of which 25 also erupted. For assessing eruption potential, this high proportion of deforming volcanoes that also erupted (46%), together with the proportion of non-deforming volcanoes that did not erupt (94%), represent indicators with significant evidential worth, although deformation alone should not be considered as a strong diagnostic of imminent eruption (Biggs et al. 2014). The magnitude of the deformation of volcanoes shows an overall inverse correlation with the duration of the deformation, with the strongest events being the fastest. Also, events that had large magnitudes typically led to an eruption, even though there does not seem to be any common threshold at which deformation duration and/or magnitude should cause concern (Fig. 8.16; Biggs and Pritchard 2017). These studies are useful to understand the general behaviour and outcome of deforming volcanoes. However, for more helpful forecast, deformation results should be tailored to specific volcanoes, or even groups of volcanoes with similar features and behaviour (**analogue volcanoes**). In fact, a first and very general distinction to forecast the outcome of deforming volcanoes is to consider the composition of the involved magma, as mafic and felsic deforming volcanoes show quite distinct behaviours (Newhall and Dzurisin 1988; Phillipson et al. 2013; Biggs et al. 2014; Acocella et al. 2015). Mafic systems are usually more

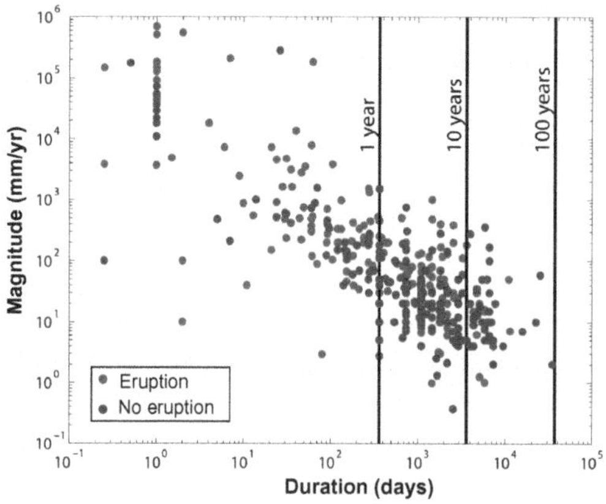

Fig. 8.16 Rates of volcano deformation as a function of the duration of the deformation from a global compilation of 485 events from 221 volcanoes. Deformation magnitudes indicate maximum displacements (horizontal or vertical for GPS data; along the line of sight for InSAR data). Eruption (red) or no eruption (blue) points indicate where as deformation occurs before, during or immediately after an eruption, or that is not obviously related to an eruption (after Biggs and Pritchard 2017)

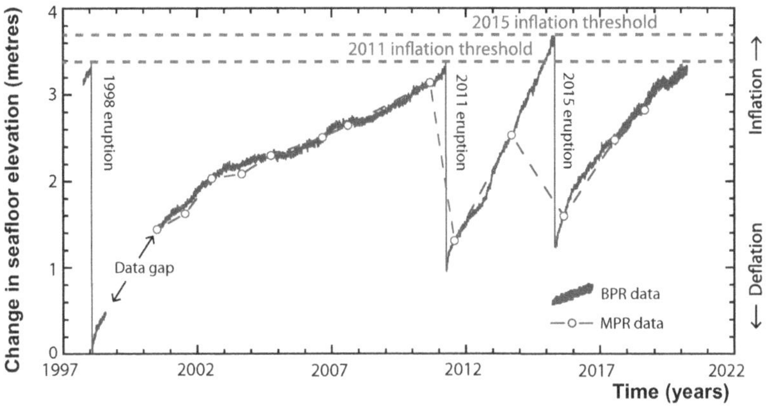

Fig. 8.17 Deformation time series (from 1997 to 2020) at the centre of the submarine Axial Seamount caldera (Juan de Fuca Ridge), including the 1998, 2011 and 2015 eruptions. Purple dots represent Mobile Pressure Recorders (MPR) measurements (error bars indicate 1 standard deviation); blue curves show Bottom Pressure Recorders (BPR) data (diagram courtesy William Chadwick)

prone to eruptive unrest, that is to experience unrest culminating into eruption. This behaviour may be summarized with the model of the **volcano deformation cycle**, based on the common evidence that a volcano progressively inflates, often exponentially, until it erupts and suddenly deflates, to start inflating again after the eruption (Biggs and Pritchard 2017). This behaviour leads to the "saw tooth" deformation style often observed at many mafic and repeatedly erupting calderas, as Fernandina (Galapagos) or Axial Seamount (Juan de Fuca Ridge). Axial Seamount, in particular, shows a distinctive behaviour, with a remarkably constant deformation threshold that has systematically anticipated the last three eruptions, in 1998, 2011 and 2015 (Fig. 8.17; Nooner and Chadwick 2016). This "**inflation predictable**" behaviour is evident from multiple eruptions, each occurring when a certain threshold of inflation recovers the deflation of the preceding eruption. In this behaviour, the stress relative to the threshold value is proportional to the displacement accumulated since the most recent eruption (see Sect. 9.7.1). Such a threshold of predictability requires (Segall 2013): (a) elastic deformation; (b) that the geometry of the magma system and location of failure remain invariant with time; (c) that the tectonic stress and failure stress remain constant. This predictable behaviour represents the easiest condition to forecast impending eruptions based on surface deformation, but it is difficult to detect (the monitored volcano must have erupted several times) and, mostly, rare to occur, even at mafic volcanoes. Commonly the deformation at an active volcano leads to eruption without any clear threshold.

More generally, the recent explosion in volcano monitoring data has demonstrated that the classic model of the volcano deformation cycle is an oversimplification. Departures from the model have been observed at mafic volcanoes, where inflation was not followed by any eruption, as at Alcedo (Galapagos) between 2007 and 2011 (Galetto et al. 2019) and, mostly, at felsic volcanoes. In particular, felsic calderas are characterized by repeated and prolonged deformation, not necessarily culminating in eruption. Emblematic is the comparison between deformation behaviour of Campi Flegrei and Rabaul (Papua New Guinea) calderas (Acocella et al. 2015; Kilburn et al. 2017). Campi Flegrei experienced three main uplift episodes between 1950 and 1952, 1969 and 1972, and 1982 and 1984, with maximum uplift in the caldera centre of ~0.7, ~1.7, ~1.8 m, respectively. After two decades of deflation, the caldera floor uplifted again and from 2005 to 2020 it experienced ~0.7 m of uplift (Fig. 5.14). The current uplift, as of 2020, is thus occurring at a much slower rate and also with minor seismicity compared to the previous episodes, particularly the 1982–1984 one. In principle, it may be expected that the ongoing minor uplift, with rates of cm/yr, is not worrisome, as the previous stronger uplifts, with rates of m/yr, did not produce any eruption. However, the recent history of Rabaul brings less reassuring insights. Rabaul experienced uplift of ~1 m between 1971 and 1983. Between 1983 and 1985 both the uplift rate and seismicity dramatically increased, with uplift of nearly 1 m accompanied by several tens of thousands of earthquakes. From 1985 to 1992, the uplift rate and seismicity significantly decreased, with only ~35 cm of uplift in seven years. From 1992 to 1994 the uplift rate increased again, with net uplift of ~20 centimetres, and seismicity moderately increased relative to the 1985–1992 period (Fig. 5.24b). In September 1994, after only 27 h of sustained seismicity, a VEI 4 eruption began, generating an ash cloud reaching ~20 km in altitude. Therefore, Rabaul showed higher rates of deformation (and seismicity) not leading to any eruption, and lower rates preceding the moderate eruption, with very different outcome from Campi Flegrei, where no eruption occurred so far. This apparently non-linear behaviour of Rabaul may be understood only by considering the second unrest phase and eruption as a continuation and dependent on the first; the first primed the system, the second set off the eruption. This suggests that the longer-term, cumulative effect of successive uplift episodes should be also considered in forecasting any eruption, although this is still not a guarantee for eruption. Taking into account for the cumulative role played by subsequent uplift episodes allows interpreting the eruption of Rabaul as resulting from the **progressive failure** of the system, a condition also supported by the trend of the pre-eruptive seismicity at Mount St. Helens (Washington, USA) in 1980, Pinatubo in 1991, and at Sierra Negra (Galapagos) in 2005 (see Sects. 8.4.1 and 9.6; Voight 1988; Voight and Cornelius 1991; Chiodini et al. 2016; Kilburn et al. 2017).

In other cases characterized by lateral magma transport through dikes, as at Krafla (Iceland), Kilauea and Mauna Loa (Hawaii, USA), the surface uplift increases with time towards a limiting value followed by a decaying exponential with characteristic timescale t, approximately equal to the duration of inflation. This allows forecasting the probability that a deflation associated with lateral intrusion of magma and, possibly, an eruption will start within a certain period (Blake and Cortes 2018). In addition to the vertical displacement, the tilt of the volcano flanks may also reveal a predictive capacity, as around the crater rim of Soufriere Hills (Montserrat) in 1996–1997, where inflation cycles caused by shallow magma pressurization, correlated with seismicity, explosions, and pyroclastic flow activity, were used to forecast times of increased volcanic hazard (Voight et al. 1998).

The possible location of a vent may be also forecast before an impending eruption, based on the progression of the deformation. For example, the progressive inversion of the deformation data may allow tracking magma migration in real-time, as in 2008 at Mount Etna (Cannavò et al. 2015). However, deformation sources and associated eruptive vents are not necessarily coincident, and offsets can be up to several kilometres, as observed in 2005 at Sierra Negra (see Sect. 5.10.1; Chadwick et al. 2006).

As anticipated, eruptions may not be always preceded by evident deformation of the volcanic edifice, as at volcanoes that erupt frequently and are characterized by an hot and ductile open conduit from the shallow magma chamber to the surface. Open conduits allow magma to rise to the surface without significantly pressurizing and deforming the shallow plumbing system, also with limited or no seismicity deriving from the fracturing of the host rock. Non-deforming erupting volcanoes with open conduit are mainly observed in mafic systems, including Stromboli (Italy) and some Aleutian Arc volcanoes. However, these may be found also in felsic systems, as for example Sakurajima (Japan), and include calderas, as Aso (Japan; Acocella et al. 2015; Dzurisin et al. 2019).

In summary, despite some apparently non-linear behaviour, deformation data remain one of the most valuable indicators for the state of an active volcano and, although with limited forecasting potential (as any other monitoring indicator), they should be carefully considered during unrest.

8.4 Geophysical Monitoring

Geophysical monitoring is an essential component to understand active volcanoes. It relies on different approaches and techniques, including seismicity, infrasound waves, acoustic emissions, microgravity, magnetic and electro-magnetic signals, thermal and infrared signals. In this section, the focus is on two widely used geophysical techniques, as seismicity and gravity measurements, which are both popular in volcano-tectonic studies.

8.4.1 Monitoring Volcano Seismicity

An **earthquake** is a sudden release of energy in the Earth's lithosphere that creates seismic waves. These consist of P (primary) and S (secondary) **waves**, which are both body waves, thus propagating through the Earth. P-waves are compressional elastic waves with primarily longitudinal particle motion parallel to the propagation direction. These travel faster than any other type of wave, although they are significantly slower in completely molten bodies. S-waves are shear elastic waves with particle motion perpendicular to the direction of propagation. They are 50–60% slower than the P-waves and cannot pass through liquids, including totally molten bodies. Because of these distinctive behaviours, tomographic studies (aiming at defining the structure of the interior of the Earth) are based on the variations in the velocities of the P- and S-waves, defined as Vp and Vs respectively, and on their ratio Vp/Vs, which is a key-parameter to evaluate the content of fluids and melts. For example, the presence of magma or molten material, including magma chambers, is typically characterised by high Vp/Vs. Once the body waves reach the Earth's surface, surface waves are generated at the interface between the Earth and atmosphere, propagating along the surface and rapidly decaying with depth. Surface waves mainly consist of the faster "Love" and slower "Rayleigh" waves and are usually responsible for most of the damage during earthquakes.

The crustal volume where the earthquake originates is called **hypocentre**, and its projection at the surface is the **epicentre**. Most earthquakes are due to the shear failure of rocks and have a regional tectonic origin. The attitude and slip motion of the failure plane are commonly expressed through fault-plane solutions, or **focal mechanism**, which can be derived from observing the pattern of "first motions," that is, whether the first-arriving P-waves break up or down at a given seismic station (Fig. 8.18a). The graphical representation of a focal mechanism is a "beachball" plot, where two orthogonal great circles, or nodal planes, in a lower hemisphere stereographic projection divide the compressive (first motion up) from the tensional (first motion down) signals. By convention, the compressional quadrants are colour-filled (usually in black) and the tensional are in white. The compressional P- and tensional T-axes responsible for the earthquake lie in the centre of the white portion and of the colour-filled portion, respectively (Fig. 8.18b). The fault generating the earthquake is parallel to one of the two nodal planes, and may be identified using additional geological or geophysical data.

There are different measures for the size of an earthquake. The most popular is the **magnitude** M, which is the instrumentally recorded amplitude of the elastic waves corrected for distance and is the basis of the Richter scale, where a change of one unit in magnitude corresponds to a change in radiated energy by a factor of approximately 30. Another measure is the **seismic moment** M_s, which is the product of fault area A, fault slip Δs, and the shear modulus of the rock G, as:

$$M_s = A\ \Delta s\ G \tag{8.8}$$

The distribution of the frequency of the earthquakes versus their magnitude follows a

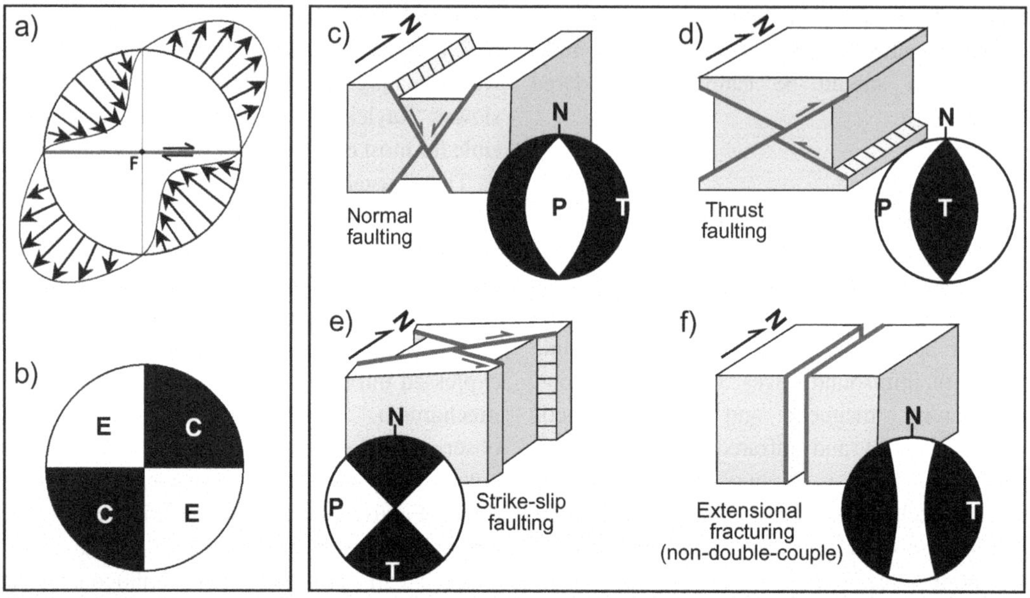

Fig. 8.18 Focal mechanism of earthquakes. **a** Diagram showing the direction of initial movement in a crustal volume (map view) surrounding an E-W trending dextral fault generating an earthquake with focus F and **b** equivalent zones of compressional (C) and tensional (E) sense first motion in the seismic waves radiating outward. **c–e** Focal mechanisms associated with the main types of conjugate faults (all double-couple mechanisms), schematically shown in the block diagrams; **f** focal mechanisms associated with extension fracturing (non-double-couple mechanism) shown in the block diagram. The corresponding compressional P and tensional T axes generating the earthquakes are also shown

power law, known as the frequency-magnitude distribution (or **Gutenberg-Richter distribution**), which indicates that there are many more small earthquakes than large ones. The relation is:

$$\log_{10} N = a - b\,M \tag{8.9}$$

where M is the magnitude, N is the cumulative number of earthquakes greater than or equal to M, and a and b are constants. The slope b is known as the **b-value**, which is a quantification of the proportion between smaller and larger earthquakes: higher b-values are associated with more frequent smaller earthquakes with regard to larger ones. The b-value is commonly of 0.9–1.0 for tectonic earthquake sequences. The constant a, or **a-value**, quantifies the lateral shift of the Gutenberg-Richter distribution.

Volcanic earthquakes cluster within a few tens of kilometres below volcanoes and are usually related to magmatic processes, prior to (unrest phase) and during eruptions (McNutt and Roman 2015). As nearly every eruption from closed conduit volcanoes has been preceded by an increase in earthquake activity and accompanied by varying levels of seismicity, seismology has become one of the most useful tools for eruption forecasting. At present, approximately 200 of the world's volcanoes are seismically monitored, although in very different ways. Volcanic earthquakes often occur in **swarms**, or groups of several events with similar location and magnitude, without a main shock. Typical background rates are a few to a few tens of events per day, also depending on the location and sensitivity of the seismic stations. Conversely, the rates of seismicity before and during eruptions are typically several tens to several hundreds or more events per day and include larger magnitude events. Also, very high b-values (usually between 1 and 2, at times reaching 3) have been observed in volcanic areas (McNutt 2005). The steeper slope of the power law Eq. (8.9) associated with these high b-values implies a larger proportion of smaller earthquakes, consistently with their occurrence in swarms and the lack of main shocks.

There are different types of volcanic earthquakes, mainly distinguished by their **frequency**, which consists of cycles of rise and fall of a wave per second, expressed in Hertz (Hz). These types include high-frequency earthquakes (also known as volcano-tectonic, or A-type earthquakes), low frequency (also known as long-period, or B-type events) and volcanic tremor (Fig. 8.19; McNutt and Roman 2015).

Volcano-tectonic (VT) or **high-frequency (HF)** earthquakes have clear P- and S-wave onsets, and dominant frequencies are 5–15 Hz. These earthquakes are caused by shear failure along faults, commonly due to the pressure of intruding magma (as during dike propagation) or associated hydrothermal fluids, and differ from their tectonic counterparts in: (a) their maximum magnitudes, which are usually smaller in VT earthquakes, where magnitudes are typically $M \leq 3.0$; (b) patterns of occurrence, as volcanic earthquakes cluster in swarms, rather than main shock-aftershock sequences. Volcano-tectonic earthquake swarms may have variable duration and frequently last from weeks to months.

Long-Period (LP) or **low-frequency (LF)** events often have emergent P-waves, lack distinct S-waves, and have dominant frequencies between 1 and 5 Hz. The nature of these events is still not well understood, although LP events are thought to be mainly caused by fluid pressurization processes, such as bubble formation and collapse,

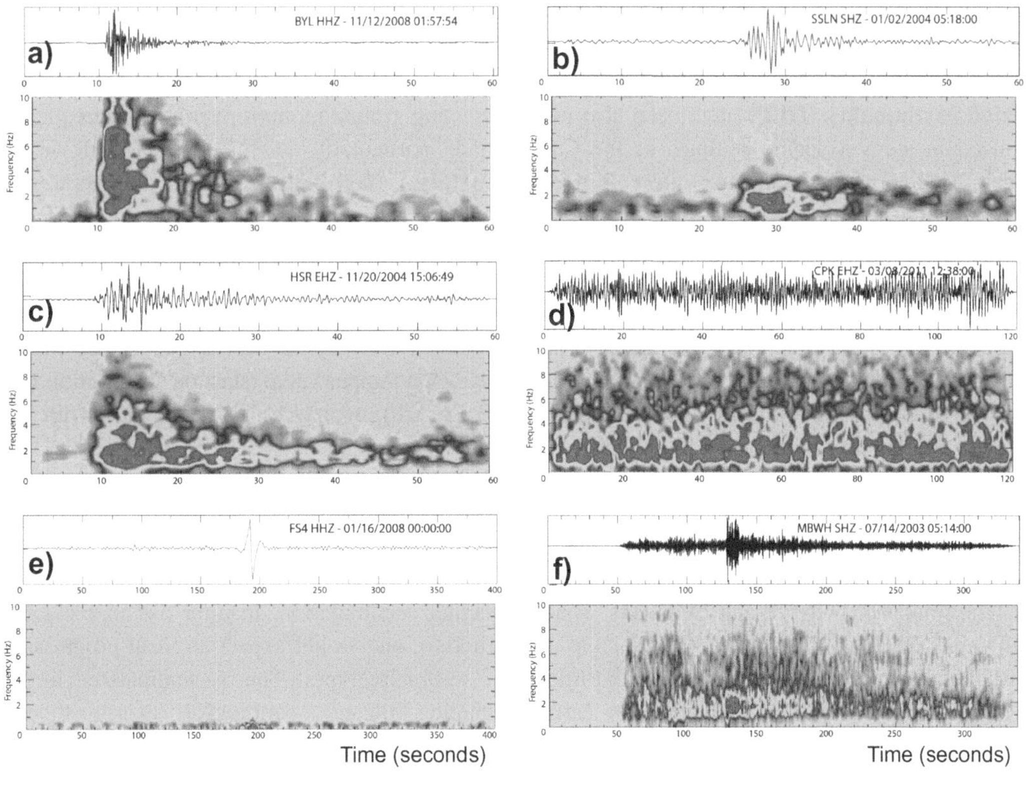

Fig. 8.19 Example waveforms (top) and spectrograms (bottom) of volcanic seismicity. **a** Volcano-tectonic earthquake at Kilauea volcano (Hawaii, USA). **b** Long-period event at Shishaldin volcano (Alaska, USA). **c** Hybrid event at Mount St. Helens (Washington, USA). **d** Volcanic tremor at Kilauea. **e** Very long-period earthquake at Fuego volcano, (Guatemala). **f** Explosion earthquake at Soufriere Hills volcano (Montserrat; this event begins with a dome collapse and the explosion begins at ~130 s). Station name, channel code, and event date and time are given on each waveform (modified after McNutt and Roman 2015)

and also by nonlinear flow processes that occur at very shallow depths, for which attenuation and path effects also play an important role. Shallow LP seismicity may also signal the pressure-induced disruption of the steam-dominated region of the volcano, and can accordingly be an indicator of impending eruption (Chouet 1996; Chouet and Dawson 2016; Shapiro et al 2017). For these events there is some ambiguity in defining the source, in particular on whether this is the source of mechanical energy alone or the ensemble of mechanical source and fluid-filled conduit that radiates energy (McNutt 2005). Recent studies challenge the established interpretation of shallow, long-period events being generated by resonance in fluid-filled cracks or conduits. Rather, the short-duration long-period events may mark the deformation in the upper volcanic edifice, forming part of the spectrum between slow slip earthquakes and fast dynamic rupture, similar to what observed in non-volcanic environments (Bean et al. 2013). **Deep long-period earthquakes (DLP)** have been also recognized in many volcanic settings, as the Cascades, Aleutian (USA) and Japan arc volcanoes, and Hawaii. These events show a low-frequency energy and location below the crustal seismogenic zone, down to ~40 km of depth. Several hypotheses have been proposed to interpret their source process, although magmatic and/or fluid activity seems the most likely. Understanding the physical mechanism of DLP could provide insights on the magma/fluid migration process, forearc mantle wedge dynamics and eruption forecasting (McNutt 2005; Han et al. 2018, and references therein).

Some earthquakes, called **hybrid** events, share attributes of both HF and LF events. These may represent a mixture of processes, such as an earthquake occurring adjacent to a fluid-filled cavity and setting it into oscillation, or earthquakes shallower than LF events, preserving most of the HF energy that is attenuated for deeper events (Benson et al. 2010).

Volcanic **tremor** is a continuous signal with duration of minutes to days, or longer. The dominant frequencies of tremor are 1–5 Hz, similar to LF events, suggesting that tremor is a series of LF events occurring at intervals of a few seconds. However, volcanic tremor obeys a different type of relation from the frequency-magnitude relation of Eq. 8.9, showing an exponential law distribution, rather than a power law (Benoit et al. 2003; McNutt 2005). This implies that, unlike earthquakes, there is a scale bound for tremor, such as a fixed conduit length or a constant pressure. Volcanic tremor is often interpreted as an indicator of impending eruption, as it is inferred to result from the magma propagating within a shallow dike. However, this association is not straightforward, as in many cases tremor does not culminate into eruption. Tremor is also useful for estimating the size of eruptions while they are in progress: this is achieved by measuring tremor on a normalized amplitude scale, known as reduced displacement, which is proportional to the VEI of an eruption (McNutt 2005).

In the last decades new types of seismometers, known as broadband seismometers, capable of detecting ground motions over a wider frequency band, particularly at the LF end (down to 0.016 Hz), have been deployed at several volcanoes. These have revealed a new class of earthquakes, called **very-long-period (VLP)** events, associated with either eruptions or vigorous fumarolic activity. These events may be produced as magma moves in distinct pulses through a flapper valve (such as a restriction in a dike or sill), or may be also associated with gas release.

Finally, **explosion earthquakes** accompany explosive eruptions, commonly characterized by an air-shock phase on the seismogram.

In a sequence characterized by shallow magma accumulation, magma rise and magma eruption, one would expect an ideal progression of seismicity types. This is summarized in the generic volcanic earthquake swarm model, which, despite the unicity in the seismic behaviour of a volcano, may account for the main differences highlighted by the variations in some parameters (Fig. 8.20; McNutt 1996; McNutt et al. 2015). Seismicity would be first characterized by HF events related to magma emplacement, occurring at a certain depth in the

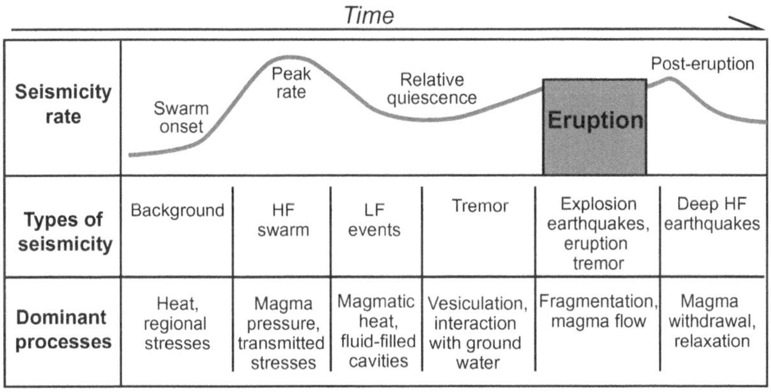

Fig. 8.20 Schematic diagram of the time history of a generic volcanic earthquake swarm model. Seismicity rates are shown along with the main types of events observed at each stage and the related possible dominant processes (modified after McNutt 1996)

upper crustal volume and to the sides of the intrusion, as for example a laterally propagating sill. This seismicity may eventually evolve into an upward migrating pattern of HF events, as produced by the vertical propagation of a dike. When the dike reaches shallower levels (1–3 km), gas exsolution increases the viscosity of magma, lowering the acoustic velocity and, with seismic events generated by resonance of magma-filled cavities, promoting lower frequencies. Most low-frequency events and tremor occur at shallow depths, and several models for their origin involve fluids, as magma, exsolved volatiles and water. A pre-eruptive phase of relative seismic quiescence may follow the seismicity peak: this may be related to strain hardening, higher temperatures, or the shallower depth of magma.

In addition to the frequency and duration, seismic events may be distinguished based on the mechanisms acting at the seismic source. In particular, four main types of seismic sources can be identified: shear failure ("double-couple" mechanisms), tensile failure ("non-double-couple" mechanisms), single forces (resulting from explosive injection of material into the atmosphere), and passive and active fluid involvement (producing LF events and volcanic tremor). Non-double-couple earthquakes are common in volcanic and geothermal areas, where their tensile component may be associated with high-pressure, high-temperature fluids. In some cases, non-double-couple components may be also explained by ring-faulting mechanisms (Fig. 8.18f; Shuler et al. 2013; McNutt et al. 2015, and references therein).

Finally, among the important seismic signals to be detected in volcanic areas is **seismic anisotropy**. This is a more general and intrinsic property of the crust highlighted by the passage of seismic waves and is caused by mechanical anisotropies, as faults or aligned melt pockets, including fluid-filled micro-cracks, where the polarization direction is parallel to the trend of structural features. When a stress is applied to the crust, micro-cracks selectively open to align with the direction of the maximum principal stress, inducing variations in the seismic anisotropy. As a result, seismic anisotropy may monitor deformation in the magma plumbing systems of active volcanoes (e.g., Illsley-Kemp et al. 2018, and references therein).

Eruptions have been forecast and evacuations have been called often relying on seismology, as for example for the effusive dome-building eruptions at Mount St. Helens from mid-1980 to 1986 and at Pinatubo in 1991. Indeed, seismicity remains an essential indicator to understand the state of a volcano and to forecast impending eruptions (Fig. 8.21; McNutt and Roman 2015; White and McCausland 2019).

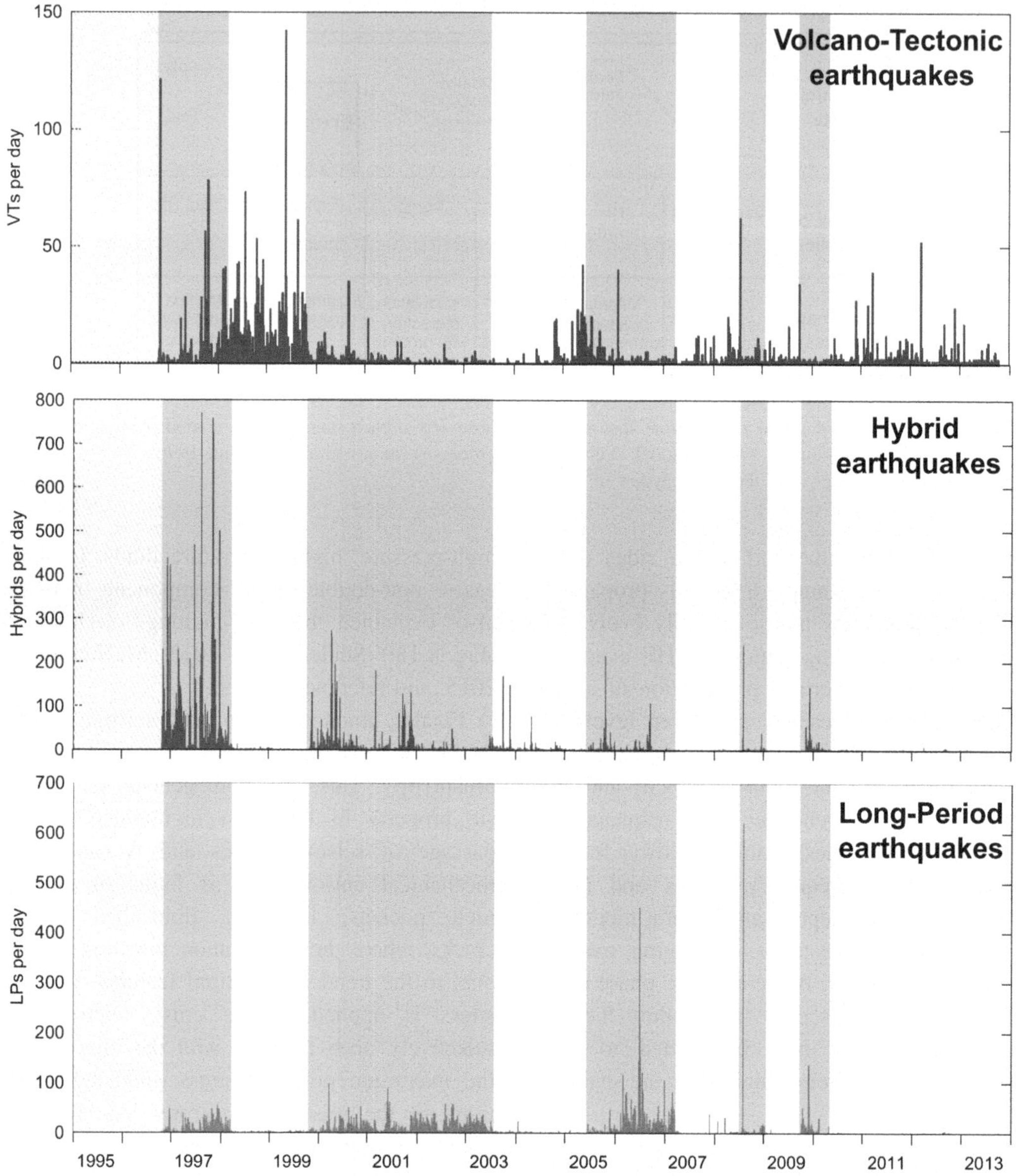

Fig. 8.21 Histograms of the number of volcano-tectonic (VT) (top), hybrid (middle), and long-period (LP) (bottom) events per day at the Soufriere Hills volcano (Montserrat) from 1996 to 2013. Grey stippled area indicates pre-monitoring period. Solid grey areas indicate eruption and white areas inter-eruptive periods. LP and, to a lesser extent, hybrid activity occurs during eruptions, while VT activity declines during eruptions and is strongest during inter-eruptive periods and immediately before the onset of each eruption (after McNutt and Roman 2015)

The forecast mainly relies on the increase in seismicity above previously recorded background levels, based on the assumption that the level of seismicity reflects the level of magmatic activity. In principle, large eruptions may be forecast more easily than smaller ones, as the amount of magma involved is larger, and precursory earthquakes tend to be more frequent and distributed over a large volume. In practice, if the unrest is followed by eruption, the seismic energy released during unrest, parameterized by the duration of the swarm and the maximum magnitude recorded, is not always indicative of the magnitude of the following eruption (Sandri et al. 2004).

Theoretical and seismicity studies (as at Mount St. Helens between 1980 and 1986, Pinatubo in 1991 and Soufriere Hills in 1995) focused on forecasting eruptions considering the progressive failure of a volcanic system as deduced from the evolution of seismicity (Kilburn 2003; Kilburn and Sammonds 2005; Smith et al. 2007; Smith and Kilburn 2010; Kilburn et al. 2017). This approach is based on the assumption that a volcanic system undergoes progressive mechanical failure until reaching rupture, that is when eruption occurs. Although this method may be used to warn an impending eruption a few days in advance, specific forecasts may be subject to an uncertainty of weeks or more. The forecasting potential of this approach is discussed in detail in Sect. 9.6.

The forecast of an impending eruption also relies on the presence of a migrating pattern in the seismicity, which may allow detecting the path of a propagating dike. However, there are surprisingly few well-documented cases of such a trend associated with eruptions: among these, are the vertically propagating dike at Etna in 2001 and the laterally propagating dike at Bardarbunga (Iceland) in 2014 (Patanè et al. 2002; Green et al. 2015; McNutt and Roman 2015; Eibl et al. 2017). Also, as mafic caldera collapse is often related to lateral dike propagation, at Piton de la Fournaise (La Reunion Island) the location and timing of pre-eruptive seismic swarms on the ring-shaped collapse structure have repeatedly brought information on the distance of the future dike-fed eruptive site. In addition, here very-long period signals prior to and during the caldera formation highlight the possibility to detect the beginning of caldera collapse at depth and anticipate its surface formation (Duputel et al. 2019; Fontaine et al. 2019).

An interesting line of research on the forecasting potential of volcano seismicity has focused on the signals emitted by viscous magma emplacing within volcanic edifices, as lava domes. Here micro-earthquake swarms have coincided with the emplacement of lava spines, leading to models of seismogenic stick-slip along shallow shear zones in the magma. Laboratory experiments support this model, revealing that high temperature magma fracture is itself seismogenic at eruptive temperatures, regardless of the crystal content (Lavalllee et al. 2008; Tuffen et al. 2008). Application of these experiments to emplacing lava domes may allow using magma seismicity patterns to forecast dome-building eruptions.

Seismicity does not necessarily correlate to surface deformation during unrest, as it may be expected. Some volcanoes, as for example Campi Flegrei, reveal an almost predictable correlation between deformation and seismicity rates, with the stronger uplift episodes accompanied by more intense seismicity (Chiodini et al. 2012; Kilburn et al. 2017). However, other volcanoes show a quite surprising inverse correlation between inflation and seismicity, where significant deformation is nearly aseismic, as Westdah, Peulik, Okmok and Three Sisters (USA), plus additional cases in South America and Kamchatka (Russia; McNutt 2005). These discrepant behaviours appear indicative of different thermal and mechanical states of the deforming crust during magma emplacement. Despite any nearly aseismic inflation, the rise of magma towards the surface at volcanoes with closed conduit is commonly accompanied by the fracturing of the host rock. Therefore, seismicity is a constant pre-eruptive manifestation at closed conduit volcanoes, where it may guide eruption forecast. However, volcanoes with open conduit show much less, if any, precursory seismicity, as fracturing of the host rock is not required. As a result, eruptions and also phreatic explosions at open-system volcanoes occasionally take place with little or no apparent seismic warning, as occurred in 2014 at Ontake (Japan) or in 2019 at Stromboli (Kato et al. 2015; Pallister and McNutt 2015; Giudicepietro et al. 2019).

8.4.2 Monitoring Volcano Gravity

In recent decades volcano gravimetry has shown potential for imaging processes at active volcanoes. Gravity monitoring supplies information on changes in the distribution of mass at depth over time, a tool that under some circumstances represents the only available insight to constrain magmatic processes (Carbone et al. 2017). For instance, magma accumulation in a network of fractures may not produce the changes in pressure that would result in measureable ground deformation, but its mass signature may be recognized through gravity changes (e.g., Rymer et al. 1993; Bagnardi et al. 2014). Microgravity monitoring can also provide essential information on the cause of unrest, allowing discrimination between the nature of the source responsible for volcano inflation, as on whether this results from the pressurization of (lighter) hydrothermal fluids or (denser) magma. Discriminating between the magmatic or hydrothermal nature of the source of unrest allows properly defining the expected hazards (Battaglia et al. 1999; Carbone et al. 2017; Gottsmann et al. 2007; Amoruso et al. 2008). In addition to understanding processes that induce mass/density changes (or time-varying gravimetry, which is the focus of this section), gravity is also used to image the subsurface density structure of a volcano resulting from longer-term processes (static gravimetry). In the case of time-varying gravimetry, measurements must be repeated over time, in either campaign or continuous mode, to detect any gravity change. With static gravimetry, a dense array of measurements over a broad area allows mapping subsurface density distributions.

In time-varying gravimetry, measurements may remain ambiguous to interpret without simultaneous elevation (ground deformation) observations (Pallister and McNutt 2015). In fact, the measured gravity at any point at the surface can vary as a result of changes in height and/or subsurface mass distribution, in the case the subsurface mass has a different density with regard to the host rock. As a magmatic process responsible for mass variations at depth may also produce simultaneous surface displacements, to avoid any ambiguity in the interpretation gravity and deformation data must be considered jointly. Fluctuations in the depth of the groundwater table and the Earth's tides may affect the gravity measurements, and should be evaluated. When the variations in the height of the Earth's surface, the fluctuations of the groundwater table and the effect of the Earth's tides are considered and removed, the remaining gravity signal can be interpreted in terms of subsurface density changes caused by magmatic processes. Therefore, the measured gravity changes Δg reflect the superposition of several contributions (Battaglia et al. 2019):

$$\Delta g = \Delta g_{FA} + \Delta g_E + \Delta g_D + \Delta g_R \qquad (8.10)$$

where Δg_{FA} is the free-air effect (proportional to the vertical displacement h and the **normal free-air gradient** FAG = −0.3086 mGal/m, which is the rate of change of gravity by change of elevation), Δg_E is the environmental factors correction (which includes any contribution from changes in the water table level, in the sea level as due to tides and in the mass/topography of any volcano's glacier) and Δg_D is the deformation effect (which takes into account for coupling effects between gravity and elastic deformation, and depends upon the shape of the source). Once the estimates of the different effects/corrections are evaluated, the residual gravity Δg_R resulting from the subsurface mass flow (as magma and/or water flow) can be determined. The magnitude of time-varying gravity changes associated with active volcanic systems is generally between 0.01 and 1 mGal (1 mGal = 10^{-5} m/s^2), that is about 10^{-8} to 10^{-6} times the magnitude of the acceleration due to gravity at the surface ($g = 9.81 \times 10^5$ mGal). Therefore, stable and high-precision instruments, with accuracy of 0.001 mGal, are needed to detect such small signals (Battaglia et al. 2019). Inverse methods are commonly used to investigate spatio-temporal gravity variations and constrain the source responsible for these changes. The joint inversion of deformation and residual gravity measurements allows for inferring the density of the intrusion and better constraining the cause of the observed variations. Two major features affecting the density estimate are mechanical and density discontinuities in the subsurface and the geometry employed to approximate the source of unrest (Carbone et al. 2017; Battaglia et al. 2019).

The wide nature of the magmatic processes can result in different relations between gravity change and ground deformation. When gravity and deformation data are analysed jointly, it is possible to recognize, for example, densification of stored magma, filling of void space by magma, compression of bubbles, the efficiency of long-term eruptions and magma drainage (Carbone et al. 2017, and references therein). In general, uplift causes a decrease in gravity, and subsidence an increase, due to the changing distance between the surface and the centre of the Earth, following the above-mentioned FAG (Fig. 8.22). Subsurface mass changes overprint the FAG, such that magma accumulation in a reservoir may cause both surface uplift (gravity decrease following the FAG) and subsurface mass increase (gravity increase), resulting in a ratio between gravity and elevation changes $\Delta g/\Delta h$ different from the FAG. In addition, the FAG assumes that the only material between the station of measure at a certain height and the reference surface is air (free-air). In reality, in volcanological applications a rock column is always present between the station of measure and the reference surface: hence the need to include a further approximation to the free-air correction, that is the **Bouguer corrected free-air gradient** (BCFAG), which takes into account the rock column with certain thickness and density. The $\Delta g/\Delta h$ ratio provides a framework to assess the likelihood and type of the magmatic processes associated with these changes (Williams-Jones and Rymer 2002; Carbone et al. 2017). If the increases (or decreases) in gravity that share a common source with surface deformation are driven only by magma accumulation (or withdrawal), the $\Delta g/\Delta h$ will approximate the BCFAG. In this case, there will be gravity changes due to both elevation changes and mass-related effects, without significant change in subsurface density. In other cases, explaining the observed gravity and elevation change would require a density

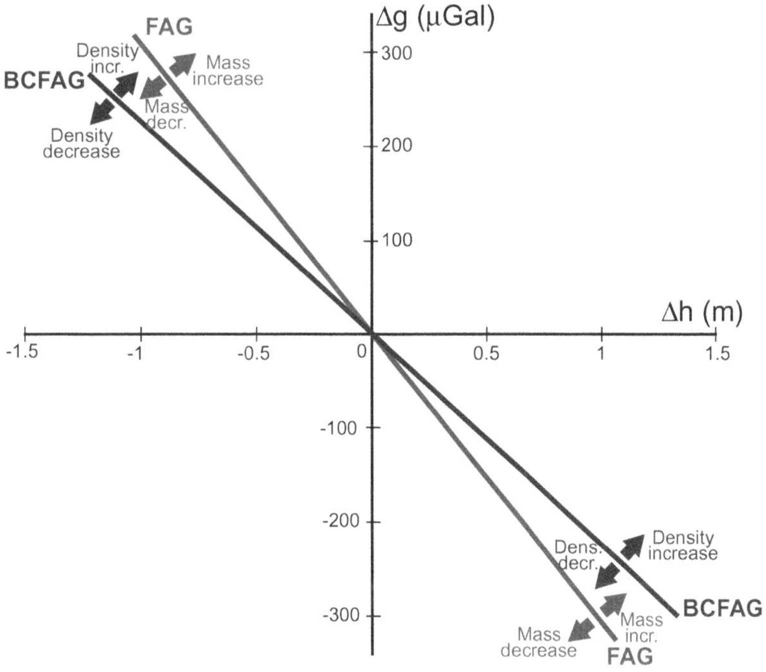

Fig. 8.22 Changes in gravity (Δg) and elevation (Δh) can be plotted in terms of Free-air (FAG) and Bouguer-corrected free-air (BCFAG) gradients. Points falling along the FAG line indicate gravity changes due solely to changes in elevation, while points falling along the BCFAG indicate gravity changes due to both elevation changes and mass-related effects (e.g., intrusion or withdrawal of magma to/from the reservoir; modified after Carbone et al. 2017)

higher or lower than that expected for subsurface magma ($\Delta g/\Delta h$ values above or below, respectively, the BCFAG curve in Fig. 8.22). In particular, in some cases, the mass gain or loss is less than expected if the mechanism of gravity change and deformation was driven by magma accumulation or withdrawal alone, and the $\Delta g/\Delta h$ values plot between the FAG and the BCFAG curves in Fig. 8.22. This indicates that not enough mass is gained or lost. This may be the case of a strong gravity decrease during inflation, pointing to a density decrease in the magma reservoir, perhaps due to pre-eruptive magma vesiculation, as postulated for Sierra Negra in 2005. In other cases, the amount of mass required to explain a gravity signal may be far more than that expected from the deformation signal, with the $\Delta g/\Delta h$ values falling outside the FAG-BCFAG curves of Fig. 8.22, indicating that too much mass is gained or lost. This large gravity change associated with minor deformation may be explained by several mechanisms, including replacement of gas-rich magma with denser, degassed magma, filling of subsurface void space by magma and compression of bubbles in the resident magma, as postulated for Etna in 1981, Piton de la Fournaise in 1998 and Kilauea in 2011–2012 (Carbone et al. 2017, and references therein).

These cases briefly illustrate how the $\Delta g/\Delta h$ framework summarized in Fig. 8.22 may help to provide a window into magmatic processes, also increasing the forecasting capability during a volcanic crisis. Despite its potential, time-variable volcano gravimetry is still an underexploited method, not widely adopted by volcano researchers or observatories. The cost of instrumentation and the difficulty in using it under harsh or noisy environmental conditions are an impediment to a common exploitation.

8.5 Geochemical Monitoring

Geochemical monitoring aims to measure the volatile species degassing from volcanoes, as well as detecting their variation in time. Volcanoes degas significantly during eruptions; however, eruptions are relatively rare in the life cycle of most volcanoes, with only some tens of volcanoes erupting each year. At any given time, most gases are released through persistent **passive degassing** (not associated with eruption), either as plumes from summit vents at open conduit volcanoes or as diffuse emissions from fumaroles, springs and volcano flanks. As the number of passively degassing volcanoes is five to seven times higher than that of erupting volcanoes, passive degassing likely emits 90–99% of the total volatiles discharged from volcanoes. Indeed, data on passive degassing compositions dominate the literature (Carn 2015; Fischer and Chiodini 2015).

When considering the major components of volcanic gases, one should distinguish between magmatic and hydrothermal compositions (Fig. 8.23; Fischer and Chiodini 2015). **Magmatic gases** are directly released from the magma, whereas **hydrothermal gases** are related to volcano-hosted hydrothermal systems and their composition results from the interaction of magmatic gases with an external (meteoric) contribution. The most common magmatic gases in order of abundance are water (H_2O, 30–90 mol%), carbon dioxide (CO_2, 5–40 mol%), sulphur dioxide (SO_2, 5–50 mol%), hydrogen (H_2, <2 mol%), hydrogen sulfide (H_2S, <2 mol%),

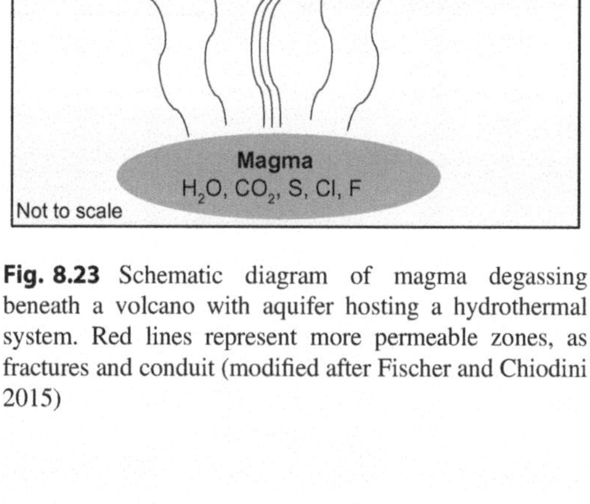

Fig. 8.23 Schematic diagram of magma degassing beneath a volcano with aquifer hosting a hydrothermal system. Red lines represent more permeable zones, as fractures and conduit (modified after Fischer and Chiodini 2015)

and carbon monoxide (CO, <0.5 mol%; Williams-Jones and Rymer 2015, and references therein). Hydrothermal gases are generally more water-rich and contain CH_4 (methane) and H_2S. Magmatic and hydrothermal gases are emitted from several sources, most commonly from craters, vents and lava lakes on the volcano summit, fumaroles on the volcano flanks, as well as more diffuse degassing through soils. Gases may also discharge from bubbling springs, mud pots and along fractures. Generally, magmatic gases mainly discharge from the summit of volcanoes, whereas hydrothermal gases are mostly emitted from the flanks, after interaction with shallow aquifers; however, each gas component may be a mixture of various sources. The location of the gas discharges affects their chemical composition, where high-temperature (>400 °C) fumaroles contain acid gases, such as SO_2, HCl, and HF that are generally less abundant in medium-temperature (100–400 °C) and absent in low-temperature (<100 °C) fumaroles (Fischer and Chiodini 2015, and references therein).

Gas **solubility** within magma is a fundamental feature to explain the degassed species and their concentrations at the surface. At the great pressures of magma generation, all the volatile species are completely dissolved in the melt. As magma ascends to lower pressures, volatiles begin to exsolve from the melt and form bubbles, initiating degassing. Volatile exsolution is directly related to volatile solubility in melts, which depends on pressure, temperature, and magma composition: overall, during the rise of a magma, pressure exerts the dominant control upon volatile solubility, because of the large volume expansion that gases experience when exsolving. Of the major volatile species, CO_2 has the lowest solubility (roughly one order of magnitude lower than water), SO_2 has intermediate solubility, and water and the halogen species (HCl, HF) are the most soluble. This implies that a rising magma will initially degas CO_2 and, subsequently, SO_2, while keeping water contents relatively constant. Only at low pressures and near-surface conditions does H_2O degases substantially, decreasing in the melt. Finally, HCl and HF are exsolved approaching the surface (Carn 2015; Fischer and Chiodini 2015, and references therein).

Two other features may affect the appearance of magmatic gases at the surface (Carn 2015, and references therein). One is the reactivity of volatile species: the species that form strong acids (e.g., SO_2, HCl, and HF) are more water-soluble and susceptible to dissolution in hydrothermal fluids or groundwater. Their depletion via dissolution is referred to as "**scrubbing**": SO_2 is particularly prone to scrubbing and hence SO_2 degassed from magma will only reach the surface in presence of a dry pathway. Other volatile species (e.g., CO_2 and H_2S) are more inert, and their measurements are particularly important at volcanoes with hydrothermal systems, groundwater or surface water. The other major factor that determines whether degassed volatiles reach the surface or not is the **permeability** of the crustal volume between the magma and the surface. In general, this volume may be characterized by open, semi-plugged, or fully-plugged conditions. Open conditions occur with open conduits, where magma and gas can rise and erupt freely. Semi-plugged conditions are met when magma does not reach the surface, as the conduit is closed, but it still allows gases to escape through fractures. Fully-plugged conditions occur when neither magma nor gas escape, and result in the lack of degassing at the surface. Semi-plugged conditions are most common at volcanoes, determining a dominant fracture-controlled permeability.

Several sampling techniques, both ground and space based, allow detecting the degassing species and relative concentrations at volcanoes, depending on the gas temperature, expected composition and the subsequent analytical protocol. These techniques mostly fall into the following categories (Sutton et al. 1992; Carn 2015; Fischer and Chiodini 2015, and references therein). Direct, in situ sampling of volcanic gases at the point of emission provides the complete chemical and isotopic composition of the gases. However, requiring laboratory analysis of the collected samples, this approach does not provide real-time information. This limitation can be overcome with continuous on-site monitoring of the concentrations of selected volcanic gases. This allows real-time detection of concentrations of gas species at summit craters, in fumaroles and in air near active fumaroles. In addition to the conspicuous, focused volcanic emissions from summit craters and fumaroles, volcanoes also release significant amounts of gas through their flanks, as diffuse or soil gas emissions. Since soil gases must traverse considerable crustal thickness, they are dominated by inert gases that are not scrubbed by groundwater, like CO_2. The amount of volcanic hydrothermal CO_2 discharged by diffuse soil degassing can be significant compared to the CO_2 released from fumaroles and can in some cases also be comparable to crater plume emissions. Soil degassing of deeply derived CO_2 commonly focuses along fractures, where it can be generally detected using both the survey and the continuous monitoring modes. Finally, the emission rate of magmatic gases in volcanic plumes can be also measured. Plume composition is often measured remotely using spectroscopy. Land-based or airborne remote sensing of the composition of volcanic plumes has been performed for decades using the correlation spectrometer (COSPEC), developed for measuring SO_2. The instrument measures the absorption of ultraviolet light by SO_2 in the volcanic plume. The emission rate of SO_2 has been used to infer the volume of degassing magma and magma supply rates during both eruptive and non-eruptive periods (Casadevall et al. 1983, 1987; Allard et al. 2006). For comprehensive measurements of plume components, such as CO_2, H_2O, SO_2, HCl, and HF, an Open-Path Fourier transform infrared spectrometer (FTIR) is used nowadays. Satellite measurements of SO_2 emissions are also used, particularly when gas plumes reach high into the atmosphere.

Since gases are more mobile than magma, geochemical monitoring aims to capture precursor signals of magma ascent by tracking changes in the chemical composition and flux of emitted gases. In fact, for volcano monitoring and hazards assessment purposes, more important than absolute gas fluxes are any changes in the emission rates, as well as their ratios (Dzurisin 2007, and references therein). In general, the interpretation of the geochemical signals is case-dependent and may require a volcano-specific conceptual model, especially at semi-plugged felsic volcanoes. At open conduit volcanoes, the detected gas variations from the summit conduit allow faster recognition of rising magma, in turn enhancing the understanding of the state of the volcano. Among the various volatile species, CO_2 and SO_2 variations, as well as their CO_2/SO_2 ratio, have often acted as indicators of impending volcanic activity, mainly at open conduit volcanoes.

At Stromboli, continuous CO_2 flux monitoring may allow large explosions to be forecast, like the paroxysm that occurred during the 2007 effusive eruption, which possibly resulted from degassing following deeper magma depressurization during the first and main part of the eruption (Aiuppa et al. 2010). At Turrialba (Costa Rica), also characterized by an open magmatic system, pulses of deeply derived CO_2-rich gas preceded explosive activity in 2014 and 2015, providing a precursor to eruptive periods of up to 2 weeks before eruptions (De Moor et al. 2016). On a longer time span, on the order of several years, increased CO_2 emissions at Kilauea couple with increased magma supply from the mantle source (Poland et al. 2012). Interestingly, precursory CO_2 flux increase may be also detected in soils at closed-conduit volcanoes prior to eruption, as occurred in the 2011 submarine eruption at El Hierro (Canary Islands; Fig. 8.24a; Perez et al. 2012).

Volcanic SO_2 is relatively easy to quantify, because it is spectroscopically active and present in air at a much lower concentration than in a volcanic gas/plume. There is evidence of precursor variations in SO_2 fluxes being detectable prior to the eruption of felsic volcanoes, as at Pinatubo in 1991. Here the cataclysmal Plinian eruption was preceded by the SO_2 flux increasing by an order of magnitude over $\sim$2 weeks, indicating the shallow intrusion of magma. Precursory SO_2 variations prior to eruptions have been more difficult to identify at basaltic volcanoes, probably because open vent basaltic volcanoes are persistent SO_2 emitters during non-eruptive

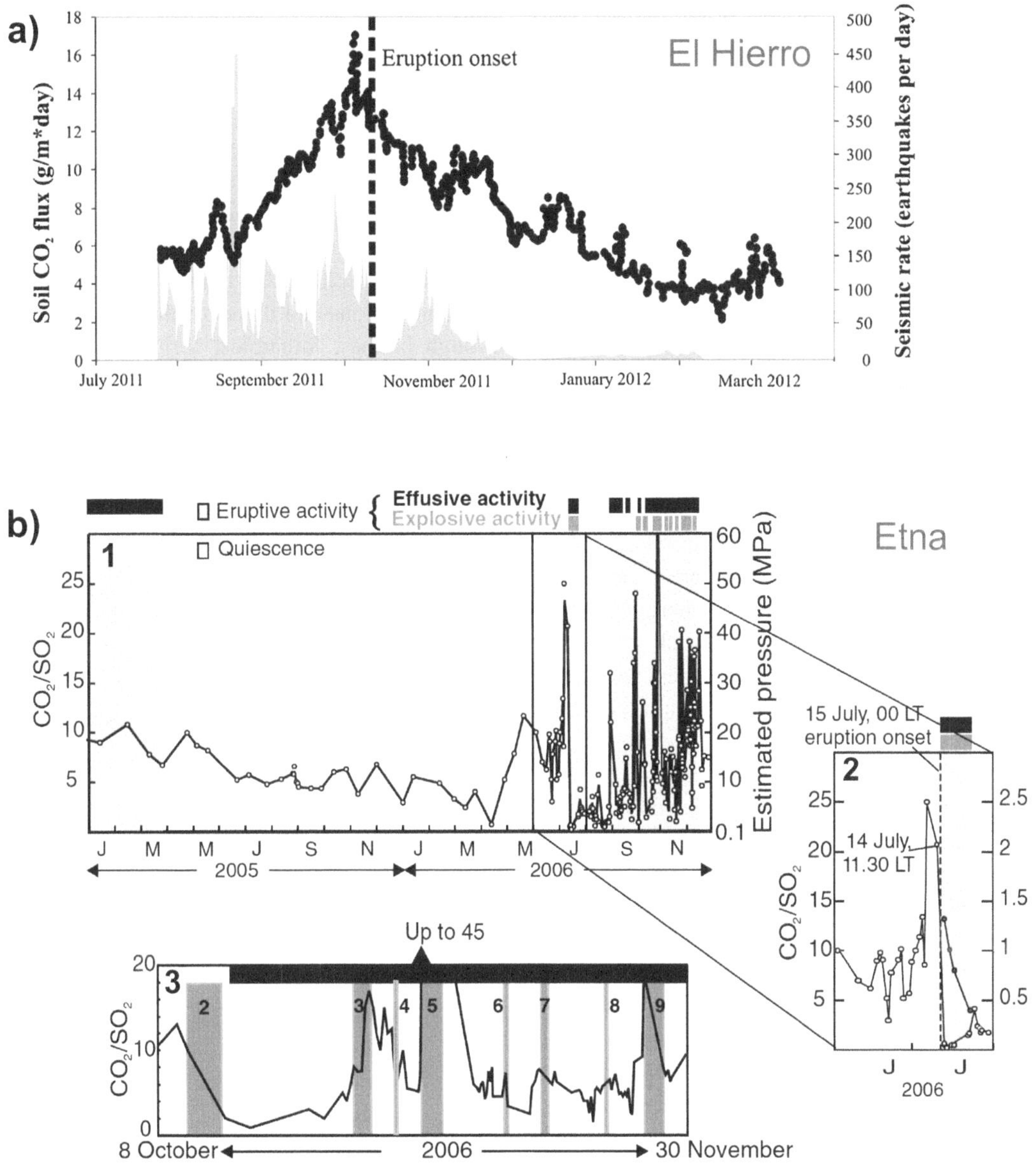

Fig. 8.24 **a** Precursory diffuse soil CO_2 flux increases prior to the 2011–2012 submarine eruption of El Hierro (Canary Islands; after Saccorotti et al. 2015). **b** Degassing at Mount Etna in 2005–2006. 1: Time evolution of CO_2/SO_2 molar ratios; right axis shows estimated pressures (in MPa). 2: Detail of June–July 2006 period: grey dots refer to composition of plume released at eruptive vent (right scale); LT = local time. 3: Detail of the October–November 2006 period; dark grey bars indicate timing of Strombolian events numbered from 2 to 9 (after Aiuppa et al. 2007)

phases, masking any signal related to pre-eruptive decompressing magma. In some cases, SO_2 scrubbing has been invoked to explain the pre-eruptive decrease in the SO_2 flux, as in the 2001 eruption at Masaya volcano (Nicaragua; Duffell et al. 2003; Saccorotti et al. 2015; Aiuppa 2015).

Finally, variations in the CO_2/SO_2 ratios have allowed for detecting the pre-eruptive degassing of rising magmas at open conduit volcanoes, as at Mount Etna in 2006 and at Villarrica (Chile) before the 2015 eruption (Fig. 8.24b; Aiuppa et al. 2007, 2017). In particular, the observed CO_2/SO_2 variations highlight a marked increase, followed by a sudden decrease immediately before the eruption. The increase, mostly due to the increase in the CO_2 component, is due to the replenishment of the reservoir with new magma, whereas the decrease, due to the increase in the SO_2 component, is related to the ascent and depressurization of the magma.

8.6 Summary

Monitoring, especially if multi-parametric, allows understanding active volcanoes and, at the same time, is a prerequisite to forecasting eruptions: as such, it provides a synergy between basic science and mitigation of risk. There are many methods to monitor volcanoes, which can be broadly classified into deformation, geophysical and geochemical techniques.

Deformation monitoring, of most interest to volcano-tectonics, relies on the detection of surface deformation through various techniques, of which GNSS and InSAR are currently the most popular. Deformation data are commonly inverted through mathematical models to infer the location, size variation and shape of the possible hydrothermal or, most commonly, magmatic source. These models usually consider point, spherical, ellipsoid and crack (mostly sill- and dike-like) sources. Of these, only dike-like sources show a non-radial vertical and horizontal displacement pattern at the surface, a distinctive feature that may allow for real-time recognition of the rise of magma. Although mathematical models have no predictive value, the deformation data, in addition to being used to understand magmatic processes, may under favourable conditions allow forecasting of impending eruptions and the possible location of vent opening.

Of the many geophysical techniques, those detecting volcano seismicity and volcano gravity variations are of most interest for volcano-tectonics. Seismicity monitoring considers the frequency, mechanisms, distribution and migration of the volcanic earthquakes, allowing for understanding magmatic processes and with promising potential in forecasting the occurrence and location of eruptions. Gravity monitoring allows for detecting subsurface mass variations, as well as capturing the accompanying processes.

Finally, geochemical monitoring focuses on the compositions and fluxes of the gases emitted through plumes, fumaroles, springs and diffuse soil degassing. Geochemical monitoring may have predictive capability mainly at open conduit volcanoes.

8.7 Main Symbols Used

a	a-value
A	Area
b	b-value
d	Depth to the source
d_t	Depth to the top of the elongated pipe
G	Elastic shear modulus
h	Uplift
M	Magnitude
M_0	Moment
M_s	Seismic moment
N	Cumulative number of earthquakes
R	Radial distance from the centre of the cavity
r	Radius
s	Horizontal distance
u	Component of displacement
U	Surface displacement vector
v	Component of displacement
V	Intruded magma volume
Vp	Velocity of P-waves
Vs	Velocity of S-waves
V_u	Volume of the uplift
w	Component of displacement
x	Spatial coordinate
y	Spatial coordinate
z	Spatial coordinate
δ	Crack's dip angle
Δg	Measured gravity change
Δg_D	Deformation effect

Δg_E	Environmental factors correction
Δg_{FA}	Free-air effect
Δg_R	Residual gravity
Δh	Measured elevation change
ΔP	Cavity pressure change
Δs	Fault slip
ΔV	Volume change of the cavity
ν	Poisson's ratio

References

Acocella V, Di Lorenzo R, Newhall C, Scandone R (2015) An overview of recent (1988 to 2014) caldera unrest: knowledge and perspectives. Rev Geophys 53:896–955. https://doi.org/10.1002/2015RG000492

Aiuppa A (2015) Volcano gas monitoring. In: Schmidt A, Fristad KE, Elkins-Tanton LT (eds) Volcanism and global environmental change. Cambridge University Press, Cambridge, pp 81–96

Aiuppa A, Moretti R, Federico C, Giudice G, Gurrieri S, Liuzzo M et al (2007) Forecasting Etna eruptions by real-time observation of volcanic gas composition. Geology 35:1115–1118

Aiuppa A, Bertagnini A, Metrich N, Moretti R, Di Muro A, Liuzzo M (2010) A model of degassing for Stromboli volcano. Earth Planet Sci Lett 295:195–204

Aiuppa A, Bitetto M, Francofonte V, Velasquez G, Bucarey Parra C, Giudice G et al (2017) A CO_2-gas precursor to the March 2015 Villarrica volcano eruption. Geochem Geophys Geosyst. https://doi.org/10.1002/2017GC006892

Allard P, Behncke B, D'Amico S, Neri M, Gambino S (2006) Mount Etna 1993–2005: anatomy of an evolving eruptive cycle. Earth-Sci Rev 78:85–114

Amoruso A, Crescentini L, Berrino G (2008) Simultaneous inversion of deformation and gravity changes in a horizontally layered half-space: evidences for magma intrusion during the 1982–1984 unrest at Campi Flegrei caldera (Italy). Earth Planet Sci Lett 272:181–188

Bagnardi M, Poland MP, Carbone D, Baker S, Battaglia M, Amelung F (2014) Gravity changes and deformation at Kīlauea Volcano, Hawaii, associated with summit eruptive activity, 2009–2012. J Geophys Res 119:7288–7305

Bato MG, Pinel V, Yan Y (2017) Assimilation of deformation data for eruption forecasting: potentiality assessment based on synthetic cases. Front Earth Sci 5:48. https://doi.org/10.3389/feart.2017.00048

Battaglia M, Hill DP (2009) Analytical modeling of gravity changes and crustal deformation at volcanoes: the Long Valley Caldera, California, case study. Tectonophysics 471:45–57

Battaglia M, Roberts CW, Segall P (1999) Magma intrusion beneath Long Valley caldera confirmed by temporal changes in gravity. Science 285:2119–2122

Battaglia M, Alpata J, Alpata R, Angarita M, Arcos D, Euillades L et al (2019) Monitoring volcanic deformation. Reference module in earth systems and environmental sciences. Elsevier, 42 pp

Bazargan M, Gudmundsson A (2019) Dike-induced stresses and displacements in layered volcanic zones. J Volcanol Geoth Res 384:189–205

Bean CJ, De Barros L, Lokmer I, Metaxian J-P, O'Brien G, Murphy S (2013) Long-period seismicity in the shallow volcanic edifice formed from slow-rupture earthquakes. Nat Geosci 7:71–75

Benoit JP, McNutt SR, Barboza V (2003) The duration-amplitude distribution of volcanic tremor. J Geophys Res. https://doi.org/10.1029/2001JB001520

Benson PM, Vinciguerra S, Meredith PG, Young RP (2010) Spatio-temporal evolution of volcano seismicity: a laboratory study. Earth Planet Sci Lett 297:315–323

Biggs J, Pritchard ME (2017) Global volcano monitoring: what does it mean when volcanoes deform? Elements 13:17–22

Biggs J, Ebmeier SK, Aspinall WP, Lu Z, Pritchard ME, Sparks RSJ (2014) Global link between deformation and volcanic eruption quantified by satellite imagery. Nat Commun 5:3471. https://doi.org/10.1038/ncomms4471

Blake S, Cortes JA (2018) Forecasting deflation, intrusion and eruption at inflating volcanoes. Earth Planet Sci Lett 481:246–254

Bonaccorso A, Davis PM (1999) Models of ground deformation from vertical volcanic conduits with application to eruptions of Mount St. Helens and Mount Etna. J Geophys Res 104:10531–10542

Bonaccorso A, Gambino S, Guglielmino F, Mattia M, Puglisi G, Boschi E (2008) Stromboli 2007 eruption: deflation modeling to infer shallow-intermediate plumbing system. Geophys Res Lett 35:L06311. https://doi.org/10.1029/2007GL032921

Cannata A, Spedalieri G, Behncke B, Cannavò F, Di Grazia G, Gambino S et al (2015) Pressurization and depressurization phases inside the plumbing system of Mount Etna volcano: evidence from a multiparametric approach. J Geophys Res 120:5965–5982

Cannavò F, Camacho AG, Gonzales PJ, Mattia M, Puglisi G, Fernandez J (2015) Real time tracking of magmatic intrusions by means of ground deformation modeling during volcanic crises. Sci Rep 5:10970. https://doi.org/10.1038/srep10970

Carbone D, Poland MP, Diament M, Greco F (2017) The added value of time-variable microgravimetry to the understanding of how volcanoes work. Earth Sci Rev 169:146–179

Carn SA (2015) Gas, plume, and thermal monitoring. In: Sigurdsson H, Houghton B, McNutt S, Rymer H,

Stix J (eds) The encyclopedia of volcanoes, 2nd edn. Elsevier Academic Press, pp 1125–1150
Casadevall TJ, Rose WI, Gerlach TM, Greenland LP, Ewert J, Wunderman R et al (1983) Gas emissions and the eruptions of Mount St. Helens through 1982. Science 221:1383–1385
Casadevall TJ, Stokes JB, Greenland LP, Malinconico LL, Casadevall JR, Furukawa BT (1987) SO_2 and CO_2 emission rates at Kilauea Volcano, 1979–1984. In: Decker RW, Wright TL, Stauffer PH (eds) Volcanism in Hawaii. USGS Professional Paper 1350 US Geological Survey Reston VA, pp 771–780
Chadwick WW, Geist DJ, Jonsson S, Poland M, Johnson DJ, Meertens CM (2006) A volcano bursting at the seams: inflation, faulting, and eruption at Sierra Negra volcano, Galapagos. Geology 34:1025–1028
Chaussard E, Amelung F, Aoki Y (2013) Characterization of open and closed volcanic systems in Indonesia and Mexico using InSAR time series. J Geophys Res 118:3957–3969
Chiodini G, Caliro S, De Martino P, Avino R, Gherardi F (2012) Early signals of new volcanic unrest at Campi Flegrei caldera? Insights from geochemical data and physical simulations. Geology 40:943–946
Chiodini G, Paonita A, Aiuppa A, Costa A, Caliro S, De Martino P et al (2016) Hotter volcanic unrest for magmas near the critical degassing pressure. Nat Commun 7:13712. https://doi.org/10.1038/ncomms13712
Chouet B (1996) Long-period volcano seismicity: its source and use in eruption forecasting. Nature 380:309–316
Chouet BA, Dawson PB (2016) Origin of the pulse-like signature of shallow long-period volcano seismicity. J Geophys Res 121:5931–5941
Coppola D, Laiolo M, Cigolini C, Massimetti F, Delle Donne D, Ripepe M et al (2020) Thermal remote sensing for global volcano monitoring: experiences from the MIROVA system. Front Earth Sci 7:362. https://doi.org/10.3389/feart.2019.00362
Crescentini L, Amoruso A (2007) Effects of crustal layering on the inversion of deformation and gravity data in volcanic areas: an application to the Campi Flegrei caldera, Italy. Geophys Res Lett 34:L09303. https://doi.org/10.1029/2007GL029919
Davis PM (1983) Surface deformation associated with a dipping hydrofracture. J Geophys Res 88:5826–5834
Davis PM (1986) Surface deformation due to inflation of an arbitrarily oriented triaxial ellipsoidal cavity in an elastic half-space, with reference to Kilauea Volcano, Hawaii. J Geophys Res 91:7429–7438
Davis PM, Hastie LM, Stacey FD (1974) Stresses within an active volcano: with particular reference to Kilauea. Tectonophysics 22:363–375
de Moor JM, Aiuppa A, Avard G, Wehrmann H, Dunbar N, Muller C et al (2016) Turmoil at Turrialba Volcano (Costa Rica): degassing and eruptive processes inferred from high-frequency gas monitoring. J Geophys Res 121:5761–5775
de Zeeuw-van Dalfsen E, Rymer H, Sturkell E, Pedersen R, Hooper A, Sigmundsson F et al (2013) Geodetic data shed light on ongoing caldera subsidence at Askja, Iceland. Bull Volcanol 75:709
Del Negro C, Currenti G, Scandurra D (2009) Temperature-dependent viscoelastic modeling of ground deformation: application to Etna volcano during the 1993–1997 inflation period. Phys Earth Planet Inter 172:299–309
Delaney PT, McTigue DF (1994) Volume of magma accumulation or withdrawal estimated from surface uplift or subsidence, with application to the 1960 collapse of Kilauea Volcano. Bull Volc 56:417–424
Delgado F, Pritchard ME, Ebmeier S, Gonzalez P, Lara L (2017) Recent unrest (2002–2015) imaged by space geodesy at the highest risk Chilean volcanoes: Villarrica, Llaima, and Calbuco (Southern Andes). J Volcanol Geoth Res 344:270–288
Dieterich JH, Decker RW (1975) Finite element modeling of surface deformation associated with volcanism. J Geophys Res 80:4095–4102
Duffell HJ, Oppenheimer C, Pyle DM, Galle B, McGonigle AJS, Burton MR (2003) Changes in gas composition prior to a minor explosive eruption at Masaya volcano, Nicaragua. J Volcanol Geoth Res 126:327–339
Duputel Z, Lengline LO, Ferrazzini V (2019) Constraining spatiotemporal characteristics of magma migration at Piton de la Fournaise volcano from pre-eruptive seismicity. Geophys Res Lett 46:119–127
Dvorak JJ, Dzurisin DD (1997) Volcano geodesy: the search for magma reservoirs and the formation of eruptive vents. Rev Geophys 35:343–384
Dzurisin DD (2007) Volcano deformation geodetic monitoring techniques. Springer–Praxis Books in Geophysical Sciences, 475 p
Dzurisin DD, Lu Z, Poland MP, Wicks CW (2019) Space-based imaging radar studies of U.S. volcanoes. Front Earth Sci 6:249. https://doi.org/10.3389/feart.2018.00249
Eibl EPS, Bean CJ, Vogfjord KS, Ying Y, Lokmer I, Mollhoff M et al (2017) Tremor-rich shallow dyke formation followed by silent magma flow at Bardarbunga in Iceland. Nat Geosci 10:299–304
Eshelby JD (1957) The determination of the elastic field of an ellipsoidal inclusion and related problems. Proc R Soc Lond A 241:376–396
Ewert JW, Guffanti M, Murray TL (2005) An assessment of volcanic threat and monitoring capabilities in the United States-Framework for a National Volcano Early Warning System. US Geological Survey Open-File Report 1164:62 p
Ewert JW, Diefenbach AK, Ramsey DW (2018) Update to the U.S. geological survey national volcanic threat assessment. US Geol Surv Sci Investig Rep 5140:40. https://doi.org/10.3133/sir20185140
Fernandez J (1999) Geodetic methods for detecting volcanic unrest: a theoretical approach. Bull Volc 60:534–544

Fialko Y, Khazan Y, Simons M (2001) Deformation due to a pressurized horizontal circular crack in an elastic half-space, with applications to volcano geodesy. Geophys J Int 146:181–190

Fischer TP, Chiodini G (2015) Volcanic, magmatic and hydrothermal gases. In: Sigurdsson H, Houghton B, McNutt S, Rymer H, Stix J (eds) The encyclopedia of volcanoes, 2nd edn. Elsevier Academic Press, Cambridge, pp 779–798

Fontaine FR, Roult G, Hejrani B, Michon L, Ferrazzini V, Barruol G (2019) Very- and ultra-long-period seismic signals prior to and during caldera formation on La Réunion Island. Sci Rep 9:8068. https://doi.org/10.1038/s41598-019-44439-1

Gaddes ME, Hooper A, Bagnardi M (2019) Using machine learning to automatically detect volcanic unrest in a time series of interferograms. J Geophys Res 124:12304–12322. https://doi.org/10.1029/2019JB017519

Galetto F, Bagnardi M, Acocella V, Hooper A (2019) Noneruptive unrest at the caldera of Alcedo Volcano (Galápagos Islands) revealed by InSAR data and geodetic modelling. J Geophys Res 124:3365–3381. https://doi.org/10.1029/2018JB017103

Giudicepietro F, Calvari S, Alparone S, Bianco F, Bonaccorso A, Bruno V et al (2019) Integration of ground-based remote-sensing and in situ multidisciplinary monitoring data to analyze the eruptive activity of stromboli volcano in 2017–2018. Remote Sens 11:1813. https://doi.org/10.3390/rs11151813

Gottsmann J, Carniel R, Coppo N, Wooller L, Hautmann S, Rymer H (2007) Oscillations in hydrothermal systems as a source of periodic unrest at caldera volcanoes: Multiparameter insights from Nisyros, Greece. Geophys Res Lett 34:L07307. https://doi.org/10.1029/2007GL029594

Green RG, Greenfield T, White RS (2015) Triggered earthquakes suppressed by an evolving stress shadow from a propagating dyke. Nat Geosci 8:629–632

Han J, Vidale JE, Houston H, Schmidt DA, Creager KC (2018) Deep long-period earthquakes beneath Mount St. Helens: their relationship to tidal stress, episodic tremor and slip, and regular earthquakes. Geophys Res Lett 45:2241–2247

Harris A (2013) Thermal remote sensing of active volcanoes: a user's manual. Cambridge University Press, Cambridge, 736 pp

Holohan EP, Sudhaus H, Walter TR, Schopfer MPJ, Walsh JJ (2017) Effects of host-rock fracturing on elastic-deformation source models of volcano deflation. Sci Rep 7:10970. https://doi.org/10.1038/s41598-017-10009-6

Kato A, Terakawa T, Yamanaka Y, Maeda Y, Horikawa S, Matsuhiro K et al (2015) Preparatory and precursory processes leading up to the 2014 phreatic eruption of Mount Ontake, Japan. Earth Planets Space 67:111. https://doi.org/10.1186/s40623-015-0288-x

Kilburn CRJ (2003) Multiscale fracturing as a key to forecasting volcanic eruptions. J Volcanol Geoth Res 125:271–289

Kilburn CRJ, Sammonds PR (2005) Maximum warning times for imminent volcanic eruptions. Geophys Res Lett 32:L24313. https://doi.org/10.1029/2005GL024184

Kilburn RJ, de Natale G, Carlino S (2017) Progressive approach to eruption at Campi Flegrei caldera in southern Italy. Nat Commun 8:15312. https://doi.org/10.1038/ncomms15312

Lavalllee Y, Meredith PG, Dingwell DB, Hess K-U, Wassemann J, Cordonnier B (2008) Seismogenic lavas and explosive eruption forecasting. Nature 453:507–510

Lisowski M (2007) Analytical volcano deformation source models. In: Dzurisin DD (ed) Volcano deformation geodetic monitoring techniques. Springer–Praxis Books in Geophysical Sciences, pp 279–304

Lu Z, Dzurisin DD (2014) InSAR imaging of aleutian volcanoes, Monitoring a volcanic arc from space. Springer, Berlin, p 390

Marsden LH, Neuberg JW, Thomas ME (2019) Topography and tilt at volcanoes. Front Earth Sci 7:317. https://doi.org/10.3389/feart.2019.00317

McNutt SR (1996) Seismic monitoring and eruption forecasting of volcanoes: a review of the state-of-the-art and case histories. In: Scarpa R, Tilling R (eds) Monitoring and mitigation of volcano hazards. Springer, Berlin/Heidelberg, pp 99–146

McNutt SR (2005) Volcanic seismology. Ann Rev Earth Planet Sci 32:461–491

McNutt SR, Roman DC (2015) Volcanic seismicity. In: Sigurdsson H, Houghton B, McNutt S, Rymer H, Stix J (eds) The encyclopedia of volcanoes, 2nd edn. Elsevier Academic Press, Cambridge, pp 1011–1034

McNutt SR, Johnson J, De Angelis S, Fee D (2015) Seismic and infrasonic monitoring. In: Sigurdsson H, Houghton B, McNutt S, Rymer H, Stix J (eds) The encyclopedia of volcanoes, 2nd edn. Elsevier Academic Press, Cambridge, pp 1071–1100

McTigue DF (1987) Elastic stress and deformation near a finite spherical magma body: resolution of the point source paradox. J Geophys Res 92:12931–12940

Michon L, Villenueve N, Catry T, Merle O (2009) How summit calderas collapse on basaltic volcanoes: new insights from the April 2007 caldera collapse of Piton de la Fournaise volcano. J Volcanol Geoth Res 184:138–151

Mogi K (1958) Relations between the eruptions of various volcanoes and the deformation of the ground surfaces around them. Bull Earthq Res Inst U Tokyo 36:99–134

Moran SC, Freymueller JT, LaHusen RG, McGee KA, Poland MP, Power JA et al (2008) Instrumentation recommendations for volcano monitoring at U.S. volcanoes under the National Volcano Early Warning System. US Geol Surv Sci Investig Rep 5114:47

Neal CA, Brantley SR, Antolik L, Babb JL, Burgess M, Calles K et al (2019) The 2018 rift eruption and summit collapse of Kīlauea Volcano. Science 363:367–374

Neri M, Acocella V, Behncke B (2004) The role of the Pernicana Fault System in the spreading of Mt. Etna (Italy) during the 2002–2003 eruption. Bull Volc 66:417–430. https://doi.org/10.1007/s00445-003-0322-x

Newhall CG, Dzurisin DD (1988) Historical unrest at large calderas of the world. US Geological Survey Professional Paper, 1109 pp

Nobile A, Acocella V, Ruch J, Aoki Y, Borgstrom S, Siniscalchi V et al (2017) Steady subsidence of a repeatedly erupting caldera through InSAR observations: Aso, Japan. Bull Volcanol 79:32. https://doi.org/10.1007/s00445-017-1112-1

Nooner SL, Chadwick WW (2016) Inflation-predictable behavior and co-eruption deformation at Axial Seamount. Science 354:1399–1403

Novoa C, Remy D, Gerbault M, Baez JC, Tassara A, Cordova I (2019) Viscoelastic relaxation: a mechanism to explain the decennial large surface displacements at the Laguna del Maule silicic volcanic complex. Earth Planet Sci Lett 521:46–59

Okada Y (1985) Surface deformation due to shear and tensile faults in a half-space. Bull Seismol Soc Am 75:1135–1154

Okada Y (1992) Internal deformation due to shear and tensile faults in a half-space. Bull Seismol Soc Am 82:1018–1040

Pallister J, McNutt SR (2015) Synthesis of volcano monitoring. In: Sigurdsson H, Houghton B, McNutt S, Rymer H, Stix J (eds) The encyclopedia of volcanoes, 2nd edn. Elsevier Academic Press, Cambridge, pp 1151–1172

Parks MM, Biggs J, England P, Mather TA, Nomikou P, Palamartchouk K (2012) Evolution of Santorini Volcano dominated by episodic and rapid fluxes of melt from depth. Nat Geosci 5:749–754

Patanè D, Chiarabba C, Cocina O, De Gori P, Moretti M, Boschi E (2002) Tomographic images and 3D earthquake locations of the seismic swarm preceding the 2001 Mt. Etna eruption: evidence for a dyke intrusion. Geophys Res Lett 29. https://doi.org/10.1029/2001gl014391

Perez NM, Padilla GD, Padron E, Hernandez PA, Melian GV, Barrancos J (2012) Precursory diffuse CO_2 and H_2S emission signatures of the 2011–2012 El Hierro submarine eruption, Canary Islands. Geophys Res Lett 39:L16311. https://doi.org/10.1029/2012GL052410

Phillipson G, Sobradelo R, Gottsmann J (2013) Global volcanic unrest in the 21st century: an analysis of the first decade. J Volcanol Geoth Res 264:183–196

Poland M, Miklius A, Sutton J, Thornber CR (2012) A mantle-driven surge in magma supply to Kılauea Volcano during 2003–2007. Nat Geosci 5:295–300

Pollard DD, Holzhausen G (1979) On the mechanical interaction between fluid-filled fracture and the Earth's surface. Tectonophysics 53:27–57

Pollard DD, Delaney PT, Duffield WA, Endo ET, Okamura AT (1983) Surface deformation in volcanic rift zones. Tectonophysics 94:541–584

Pritchard ME, Simons M (2004) An InSAR-based survey of volcanic deformation in the central Andes. Geochem Geophys Geosyst 5:Q02002. https://doi.org/10.1029/2003GC000610

Ryan MP, Blevins JYK, Okamura AT, Koyanagi RY (1983) Magma reservoir mechanics: theoretical summary and application to Kilauea Volcano. J Geophys Res 88:4147–4182

Rymer H, Murray JB, Brown GC, Ferrucci F, McGuire WJ (1993) Mechanisms of magma eruption and emplacement at Mt. Etna between 1989 and 1992. Nature 361:439–441

Saccorotti G, Iguchi M, Aiuppa A (2015) In situ volcano monitoring: present and future. In: Papale P, Shroder JF (eds) Volcanic hazards, risks and disasters. Elsevier, Amsterdam, pp 169–202

Sandri L, Marzocchi W, Zaccarelli L (2004) A new perspective in identifying the precursory patterns of eruptions. Bull Volcanol 66:263–275

Segall P (2010) Earthquake and volcano deformation. Princeton University Press, Princeton, 457 pp

Segall P (2013) Volcano deformation and eruption forecasting. In: Pyle DM, Mather TA, Biggs J (eds) Remote sensing of volcanoes and volcanic processes: integrating observation and modelling. Geological Society London, Special Publications, pp 380. https://doi.org/10.1144/sp380.4

Segall P (2019) Magma chambers: what we can, and cannot, learn from volcano geodesy. Philos Trans R Soc A 377:20180158. https://doi.org/10.1098/rsta.2018.0158

Shapiro NM, Droznin DV, Droznina SA, Senyukov SL, Gusev AA, Gordeev II (2017) Deep and shallow long-period volcanic seismicity linked by fluid-pressure transfer. Nat Geosci 10:442–445

Shuler A, Ekstrom G, Nettles M (2013) Physical mechanisms for vertical-CLVD earthquakes at active volcanoes. J Geophys Res 118:1569–1586

Sigmundsson F, Hreinsdottir S, Hooper A, Arnadottir T, Pedersen R, Roberts MJ et al (2010) Intrusion triggering of the 2010 Eyjafjallajokull explosive eruption. Nature 468:426–430

Singer BD, Andersen NL, Le Mevel H, Feigl KL, DeMets C, Tikoff B et al (2014) Dynamics of a large, restless, rhyolitic magma system at Laguna del Maule, southern Andes, Chile. GSA Today. https://doi.org/10.1130/GSATG216A

Smith R, Kilburn CRJ (2010) recasting eruptions after long repose intervals from accelerating rates of rock fracture: the June 1991 eruption of Mount Pinatubo, Philippines. J Volcanol Geoth Res 191:129–136

Smith R, Kilburn CRJ, Sammonds PR (2007) Rock fracture as a precursor to lava dome eruptions at Mount St Helens from June 1980 to October 1986. Bull Volc 69:681–693

Sparks RSJ, Biggs J, Neuberg JW (2012) Monitoring volcanoes. Science 335:1310. https://doi.org/10.1126/science.1219485

Sutton AJ, McGee KA, Casadevall TJ, Stokes JB (1992) Fundamental volcano-gas-study techniques: an integrated approach to monitoring. In: Ewert J, Swanson DA (eds) Monitoring volcanoes: techniques and strategies used by the staff of the cascades volcano observatory, 1980–1990. US Geological Survey Bull, 1966, Reston VA, pp 181–188

Todesco M, Costa A, Comastri A, Colleoni F, Spada G, Quareni F (2014) Vertical ground displacement at Campi Flegrei (Italy) in the fifth century: rapid subsidence driven by pore pressure drop. Geophys Res Lett 41:1471–1478. https://doi.org/10.1002/2013GL059083

Trasatti E, Giunchi C, Bonafede M (2003) Effects of topography and rheological layering on ground deformation in volcanic regions. J Volcanol Geoth Res 122:89–110

Trasatti E, Acocella V, Di Vito MA, Del Gaudio C, Weber G, Aquino I et al (2019) Magma degassing as a source of long-term seismicity at volcanoes: the Ischia island (Italy) case. Geophys Res Lett. https://doi.org/10.1029/2019GL085371

Tuffen H, Smith R, Sammonds PR (2008) Evidence for seismogenic fracture of silicic magma. Nature 453:511–514

Voight B (1988) A method for prediction of volcanic eruptions. Nature 332:125–130

Voight B, Cornelius RR (1991) Prospects for eruption prediction in near real-time. Nature 350:695–698

Voight B, Hobblitt RP, Clarke AB, Lockart AB, Miller AD, Lynch L (1998) Remarkable cyclic ground deformation monitored in real-time on Montserrat, and its use in eruption forecasting. Geophys Res Lett 25:3405–3408

Wadge G (2003) A strategy for the observation of volcanism on Earth from space. Philos Trans R Soc Lond A 361:145–156

Wauthier C, Roman DC, Poland MP (2016) Joint analysis of geodetic and earthquake fault-plane solution data to constrain magmatic sources: a case study from Kılauea Volcano. Earth Planet Sci Lett 455:38–48

White RA, McCausland WA (2019) A process-based model of pre-eruption seismicity patterns and its use for eruption forecasting at dormant stratovolcanoes. J Volcanol Geoth Res 382:267–297

Williams CA, Wadge G (2000) An accurate and efficient method for including the effects of topography in three-dimensional elastic models of ground deformation with applications to radar interferometry. J Geophys Res 105:8103–8120

Williams-Jones G, Rymer H (2002) Detecting volcanic eruption precursors: a new method using gravity and deformation measurements. J Volcanol Geoth Res 113:379–389

Williams-Jones G, Rymer H (2015) Hazards of volcanic gases. In: Sigurdsson H, Houghton B, McNutt S, Rymer H, Stix J (eds) The encyclopedia of volcanoes, 2nd edn. Elsevier Academic Press, Cambridge, pp 985–992

Wright TJ, Parsons BE, Jackson JA, Haynes M, Fielding EJ, England PC (1999) Source parameters of the 1 October 1995 Dinar (Turkey) earthquake from SAR interferometry and seismic bodywave modelling. Earth Planet Sci Lett 172:23–37

Yang X, Davis PM (1986) Deformation due to a rectangular tension crack in an elastic half-space. Bull Seismol Soc Am 76:865–881

Yang X, Davis PM, Dieterich JH (1988) Deformation from inflation of a dipping finite prolate spheroid in an elastic half-space as a model for volcanic stressing. J Geophys Res 93:4289–4257

Zhan Y, Gregg PM (2017) Data assimilation strategies for volcano geodesy. J Volcanol Geoth Res 344:13–25

9 Unrest and Eruption Forecasting

9.1 Introduction

The shallow accumulation and/or transfer of magma commonly result in anomalous geodetic, geophysical or geochemical activity of the volcano, named unrest, which is detectable through a monitoring system. At closed conduit volcanoes, every eruption is preceded by unrest. However, not all unrest culminates into eruption, as unrest may also vanish back into quiescence. Therefore, deciphering the monitoring signals captured during unrest is probably the main challenge for volcanology. This in fact allows not only understanding magmatic processes, but also forecasting impending eruptions and the possible location for the opening of vents. The recent boost in the quantity and quality of monitoring data, based on key multi-parameter sets, significantly contributes to face this challenge towards enhanced understanding and more accurate and reliable eruptive forecast: the future for this exciting field looks definitely promising.

This chapter merges knowledge from all the previous chapters, highlighting the contribution of volcano-tectonics in defining unrest, its triggering mechanisms, processes and outcomes. Then it discusses how this knowledge merges into hazard assessment for risk mitigation and, in particular, in the art of forecasting eruptions, with an overview of current forecasting approaches.

The main aims of this chapter are to:

- define unrest behaviours and related processes;
- discuss the conditions triggering unrest;
- merge this knowledge into hazard assessment, also introducing key principles;
- summarize current long- and short-term forecasting approaches.

9.2 General Features

Volcanoes spend most of their lives in a state of **quiescence**, that can be defined as a background or baseline level characterized by a stationary state, during which measurable parameters including deformation, seismicity, or degassing activity show little or no variation. Quiescence can last for years, decades, centuries or more, depending on the type of volcano, and may be even accompanied by different manifestations in different volcanoes. For example, at a given stratovolcano with closed conduit the quiescence state is usually characterized by little or no seismicity, deformation and degassing: this may be the case of Vesuvio (Italy), from 1944 to the present. At calderas, the baseline may be defined less easily. In fact, many calderas, especially those defined as restless, do not really experience

V. Acocella, *Volcano-Tectonic Processes*, Advances in Volcanology,
https://doi.org/10.1007/978-3-030-65968-4_9

a truly quiescent state, being characterized by minor vertical deformation (often subsidence), sporadic seismicity and minor degassing activity. These features have been for example observed at Askja (Iceland) in most of the last century or at Campi Flegrei (Italy) between 1985 and 2005. These minor monitoring variations may be interpreted as the baseline for quiescence of restless calderas, provided that they are not indicative of shallow magma accumulation or transfer, or of any hydrothermal change. In fact, the fundamental point is that, regardless of the type of volcano, the baseline for quiescence should represent a state characterized by the absence of magma accumulation and/or transfer within the shallow plumbing system, as well as by the absence of important variations in any hydrothermal system.

Unrest is a state of the volcano, lasting from weeks to decades, characterized by an evident deviation of the monitoring parameters from the baseline (Newhall and Dzurisin 1988). This deviation may be the result of several processes, of magmatic, hydrothermal or tectonic nature, although it is most commonly related to the shallow emplacement or accumulation of magma or of magmatic and hydrothermal fluids, or to shallow magma transfer. The term "shallow" here refers to crustal depths located between the magma chamber and the surface, thus including the plumbing system in between. The shallow accumulation or transfer of magma during unrest induces stress perturbations which may promote several phenomena within the host rock. In the crustal volume surrounding the intrusion, fractures may be generated and/or reactivated, and their formation detected at the surface as swarms of volcano-tectonic earthquakes. Fracturing also changes the permeability of the host rock, affecting the shallow migration of magmatic fluids and their interaction with fluids of meteoric origin. This may induce pressure changes in any hydrothermal system, ultimately resulting in fluctuations of the composition and fluxes of the gases detected at the surface. Shallow magma and fluid emplacement or transfer also generates surface displacement. The more voluminous the intrusion, the larger the displacement, while deeper intrusions induce a weaker, albeit wider, displacement. Therefore, shallow magma accumulation or transfer can be detected at the surface through the variations in the geodetic, geophysical and geochemical monitoring signals, identifying the state of unrest (Fig. 9.1; Hill et al. 2002). These phenomena represent a simple and common, although not exclusive, sequence of magmatic and hydrothermal processes possibly occurring during unrest. In turn, unrest processes may be characterized by a wide variability of manifestations, occurring in isolated or multiple discrete events, being weaker, more intense, shorter, longer or delayed by the response of the system. Given this variability, and given the difficulty in detecting and interpreting the produced signals, the definition and duration of unrest encompass a broad spectrum of conditions which are usually defined on subjective basis. Nevertheless, efforts have been also made to codify these conditions, for communication among volcanologists as well as to authorities. A result is the definition of the Volcanic Unrest Index (VUI), whose purpose is to provide a semi-quantitative rating of unrest intensity relative to each volcano's past level of unrest and to that of analogous volcanoes (Potter et al. 2015). The VUI, calculated using a worksheet of observed phenomena, can be also determined retrospectively for historical unrest, based on qualitative information, as well as for recent episodes, based on monitoring data.

As related to shallow magma and fluid accumulation or transfer, unrest is a necessary pre-eruptive step. Indeed, any volcano usually experiences some sort of unrest before erupting, although in a few cases (especially at open conduit volcanoes, see below) this may not be easily detected. Despite the common occurrence of unrest before eruptions, unrest culminates into eruption (**eruptive unrest**) only in some cases, while in others the accumulated or transferred magma does not feed any eruption and the system returns to quiescence (**non-eruptive unrest**).

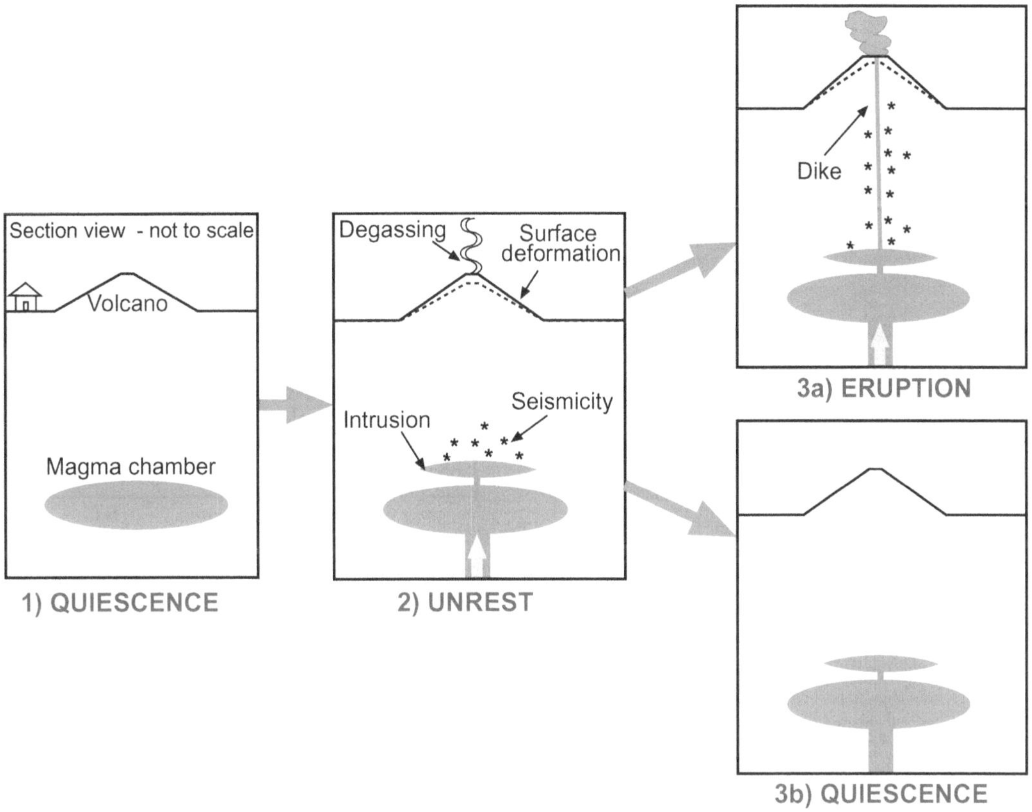

Fig. 9.1 Unrest commonly results from shallow magma accumulation (intrusion) and is associated with surface deformation, seismicity and degassing. Unrest may culminate in eruption, if the accumulated magma reaches the surface through a dike, or turn back into quiescence, if the accumulated magma solidifies

For example, the island of El Hierro (Canary Islands) experienced more than 3 months of eruptive unrest in 2011 (Fig. 9.2; Lopez et al. 2012). In mid-July, tens of seismic events per day were detected. These then increased to hundreds of seismic events per day (maximum magnitude was 2.7) below the northern part of the island, at 10–15 km of depth. Anomalous CO_2 degassing and slight inflation were detected at the surface. Throughout September seismicity increased considerably (maximum magnitude M was 3.3) and migrated southward, accompanied by increased inflation rate and ^{222}Rn emission. In late September-early October seismicity accelerated in frequency and magnitude (maximum magnitude was 3.7), focusing at 12–14 km of depth. Deformation also accelerated, with sudden inflation, deflation and re-inflation, accompanied by an increase in the ^{222}Rn emission. In the two days before the eruption deformation and seismicity continued to increase. The maximum magnitude reached 4.7 at 12 km depth, shallowing, a few hours later, between 6 and 1 km to the south of the island, suggesting dike ascent. Seismicity then turned into tremor, suggesting the onset of a submarine eruption a few kilometres off the south coast, along the feeding system of the southern ridge of the island. The submarine eruption lasted for 5 months and was then followed, for nearly 2 years after the eruption, by several sill-like intrusions, each emplaced at 13–16 km of depth between 3 and 20 days (Benito-Saz et al. 2019).

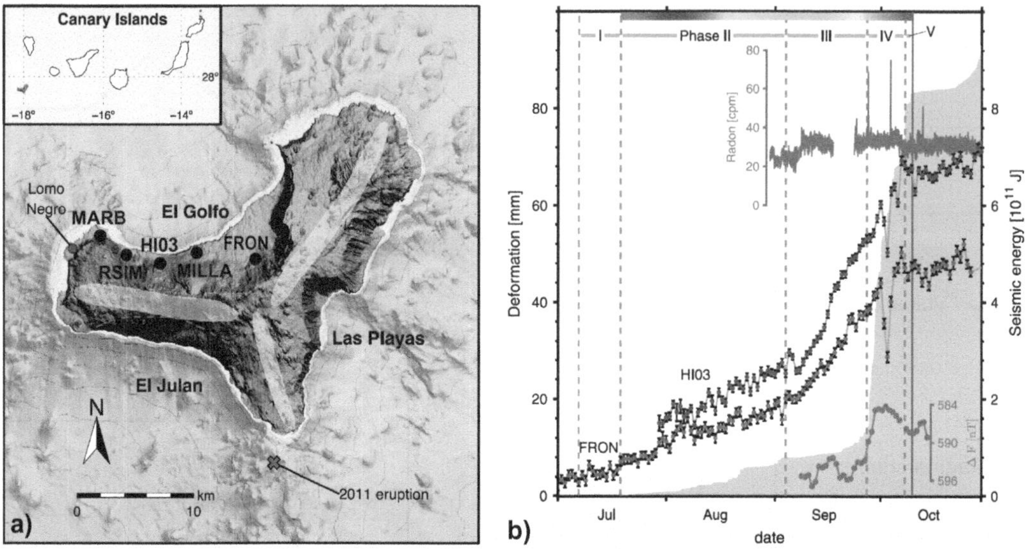

Fig. 9.2 The 2011 El Hierro (Canary Islands) unrest. a) DEM of El Hierro and its surrounding bathymetry. The three main rift zones (yellow) and collapse scarps (red lines), and the location of the last inland eruption (blue dot) and the 2011 eruption (red cross) are shown. b) Evolution of monitored parameters from July to October 2011. Grey shading shows accumulated seismic energy. Red and blue points show he northing displacement of GPS stations with error bars (1σ). The difference between MLLA and MARB magnetic stations is plotted as green dots. The ^{222}Rn temporal variation in RSIM is shown in magenta. Red vertical line marks eruption onset (modified after Lopez et al., 2012)

Conversely, Alcedo caldera (Galapagos) experienced non-eruptive unrest between 2007 and 2011 (Fig. 9.3; Galetto et al. 2019). An initial asymmetric uplift of $\sim$30 cm of the southern caldera floor from 2007 to 2009 was due to the inflation of a sill and the activation of an inner ring fault. This was followed by subsidence and contemporary uplift of the northwestern caldera rim in early 2010, compatible with the withdrawal of magma from the previously inflated sill through its lateral migration to the northwest. Then, from June 2010 through March 2011, caldera uplift resumed, consistent with the repressurization of the previously inflated sill, although without promoting further shallow magma transfer. This evolution suggests episodic magma accumulation in a shallow reservoir beneath the caldera, with aborted lateral magma migration due to discontinuous or weak magma supply.

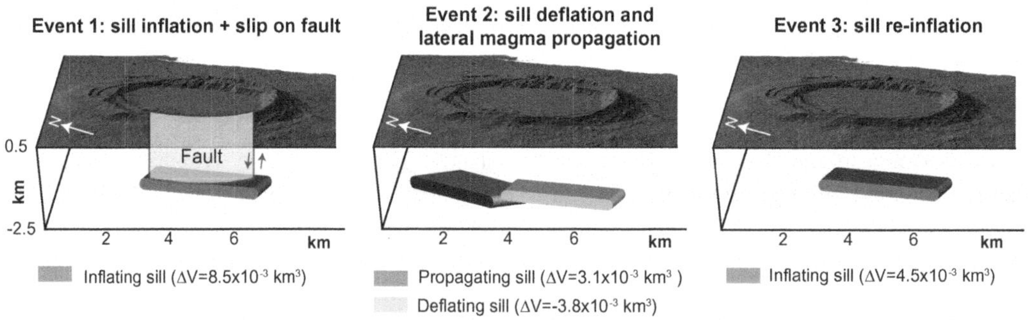

Fig. 9.3 The three deformation events due to magma emplacement and transfer, with the related volume variations ΔV, during the 2007–2011 non-eruptive unrest at Alcedo caldera (Galapagos; modified after Galetto et al. 2019)

More in general, on average 36% of unrest episodes at 228 volcanoes between 2000 and 2011 have not culminated into eruptions, a percentage lowering to 14% for shield volcanoes and rising to 48% for calderas (Phillipson et al. 2013). Therefore, as eruptions are usually preceded by unrest, but unrest does not necessary culminate into eruption, unrest represents a necessary but not sufficient condition for eruption. In fact, to erupt, the magma responsible for unrest must be able to reach the surface, a condition difficult to define a priori, causing significant uncertainty in forecasting any impending eruption. For a first-order understanding and also forecast, a fundamental distinction is to try to define whether unrest mainly results from (a) shallow storage (emplacement or accumulation) of magma or, conversely, also from (b) shallow transfer (including the rise) of magma towards the surface, typically through a dike. Based on the identification of the dominant process, it may be in principle possible to forecast an impending eruption. Diagnostic criteria to determine whether magma is accumulating or transferring during unrest may be given by the monitoring data. As anticipated in Sect. 8.3.2, the surface deformation pattern resulting from shallow magma accumulation (induced by spherical or elliptical reservoirs and sills) shows radial symmetry, where the similar vertical or horizontal component of the radial displacements at the surface identifies an axisymmetric pattern. Conversely, the surface deformation pattern induced by shallow magma transfer (i.e., a propagating dike) shows non-radial symmetry (non-axisymmetric displacement), being the vertical and horizontal displacements directional. Seismicity may also provide important clues, in terms of spatial patterns and seismic signals. The seismicity due to the shallow rise of magma may be highlighted by an upward migrating pattern of volcano-tectonic earthquakes becoming progressively dominated by long-period and very long-period earthquakes, and tremor.

Therefore, it is possible to distinguish a quiescent volcano from a volcano experiencing unrest, and within unrest, it may be possible to distinguish shallow magma accumulation from shallow magma transfer, the latter heralding an impending eruption. The conditions related to **magma accumulation** may show larger variability in intensity than those related to shallow magma transfer (Fig. 9.4). For example, unrest may be characterized by small amplitude and low rate deformation, with overall radial symmetry, low-level seismicity, and degassing anomalies related, in composition and rate, to the less soluble fluids (as CO_2). This condition has been occurring in many, probably most, unrest episodes, especially at felsic calderas, as for example at Santorini (Greece) in 2011 or Campi Flegrei between 2012 and 2013 (Parks et al. 2012; D'Auria et al. 2015; Chiodini et al. 2015). In other cases, unrest may be characterized by a larger deformation (in terms of both amplitude and rate) with radial symmetry at the surface, more intense seismicity (both in number and magnitude of earthquakes) and an increase in the degassed fluxes, possibly also characterized by more soluble species (as SO_2). These more intense conditions may be recognized a few weeks before an eruption, when a significant amount of magma accumulates at shallow levels fracturing the host rock, inducing seismicity and gas release, as for example observed at Mount St. Helens (Washington, USA) in 1980 or at Pinatubo (Philippines) in 1991 (Lipman and Mullineaux 1981; Newhall and Punongbayan 1996). Nevertheless, in some cases these intense conditions may even last for a few years without necessarily culminating into eruption, as for example observed between 1982 and 1984 at Campi Flegrei and 1983 and 1985 at Rabaul (Papua New Guinea; Acocella et al. 2015, and references therein). Despite the variability in the intensity of the monitoring parameters, or in their accelerating or decelerating trends, the radial symmetry of the deformation and the persistence of the seismicity in a specific crustal volume remain the features which may allow interpreting unrest as resulting from shallow magma accumulation. Available data suggest that such a magma accumulation commonly occurs in sill-like bodies at the top of the magma chamber.

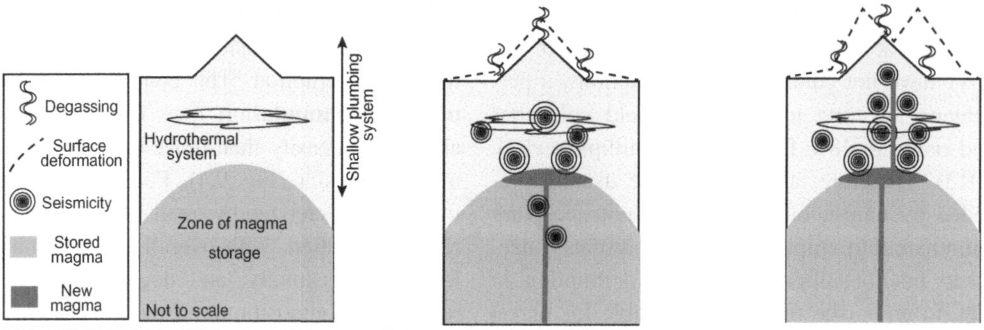

Facts	DEFORMATION	None, or subsidence	Radial symmetry Uplift mm/yr ⟷ Uplift m/yr	Non-radial symmetry Uplift ≥ m/yr
Facts	SEISMICITY	None	Above intrusion, stationary Few events/month ⟷ 10^3 events/month VT, deep LP → VT, deep LP, shallow LP? VLP? ULP? Tremor?	Above intrusion, widening / migrating VT, shallow LP, VLP, ULP, tremor
Facts	DEGASSING	CO_2	CO_2 (magmatic component), SO_2?	CO_2, SO_2, HCL, HF
Facts	DURATION	Years to thousand years	Months to years	Days
Interpretation	PROCESS	No shallow magma accumulation nor transfer	Shallow magma accumulation	Shallow magma transfer
Interpretation	STATE	Quiescence	Unrest: magma accumulation	Unrest: impending eruption

Fig. 9.4 The states of a volcano while not erupting: quiescence, magma accumulation and magma transfer. Indicative monitoring parameters to define these states are usually expressed as orders of magnitude. Seismicity: VT (volcano-tectonic), LP (long-period), VLP (very long-period), ULP (ultra long-period); the orange area in the diagrams indicates a magma reservoir with mush and melt zones. See text for details

Conversely, the appearance of deformation with non-radial symmetry and any variation in the distribution of seismicity, possibly evolving in a migrating pattern, are indicative of a shallow **magma transfer** during unrest. As this state likely anticipates eruption, it may be defined of impending eruption (Fig. 9.4). This is commonly characterized by dike propagation: dikes may propagate vertically, feeding eruptions within a volcanic edifice, or horizontally, erupting at a distal location, as observed on the slopes of (mainly mafic) volcanoes or along rift zones. Dikes usually nucleate from magma chambers, or other zones of shallow accumulation, and reach the surface within a few days. However, their nucleation and propagation do not necessarily guarantee that magma will reach the surface and erupt, as dikes may stall at depth. Therefore, the state of impending eruption represents more precisely a state of probable eruption. As it is difficult to determine if a nucleated dike will reach the surface and feed an eruption, the fundamental point here is that dike propagation marks a change in the state of a volcano, which passes from shallow magma accumulation to magma transfer towards the surface. From the risk mitigation perspective, these two states significantly differ, as magma transfer should require the activation of all the operational procedures needed to mitigate risk, including evacuation.

The transitions between these quiescence, accumulation and transfer states, though grounded on variations of observables and/or monitoring parameters, cannot be simply defined detecting pre-defined values or thresholds. In fact, there is usually no default and unique value which defines a priori a specific state of a volcano, and even less for groups of volcanoes. Rather, a range of monitoring values, also varying over an order of magnitude, may better characterize the transitions between the volcano states (Fig. 9.4).

As anticipated, in some cases it is difficult to detect unrest. While unrest is more intense and evident at monitored volcanoes with closed conduit, at some closed conduit volcanoes unrest may consist of extremely short (hours or days) and subtle variations. This condition is sometimes explained by retrograde or **second boiling** of the previously accumulated magma, which cools and crystallizes, enriching the residual liquid in gas and promoting gas exsolution. The result may be a sudden explosive eruption, as proposed for Kalud (Indonesia) in 2014 and Calbuco (Chile) in 2015, both explained by an internal trigger, suggesting that closed-conduit systems can evolve into an explosive eruption with very little precursory warning (i.e., unrest; Arzilli et al. 2019; Cassidy et al. 2019). Conversely, at open conduit volcanoes, the magma usually emplaces and rises to the surface without significant pressure build up and thus with limited deformation of the host rock. Therefore, here both the surface deformation and seismicity may be limited or absent, providing subtle information to detect any unrest and even impending eruption. In case the vertical open conduit resembles a pipe and becomes pressurized, any deformation may also have radial symmetry, similar to that of accumulating sources and thus potentially masking the rise of magma. However, as open conduit volcanoes degas significantly, the composition and flux of the degassed species may be indicative of the depth of the magma, as gas solubility within magma mainly depends on pressure: therefore, degassing variations may reveal the rise of magma at open conduit volcanoes (see Sect. 8.5).

9.3 Unrest Triggers

Unrest may result from magmatic, hydrothermal and tectonic processes (e.g., Newhall and Dzurisin 1988). Magmatic processes are those involving magma emplacement or transfer and in which magma, thermal energy and magmatic volatiles enter the shallow plumbing system (see Sect. 9.3.1). Hydrothermal processes are those involving the dynamics of the hydrothermal system and subsurface aquifers, in which the physical or chemical state of fluids is changed by interaction with magmatic fluids or through heating driven by magma-related thermal anomalies (see Sect. 9.3.2). In principle, variations in the microgravity field during unrest may allow discriminating the density of the responsible source, and thus its magmatic or hydrothermal nature (see Sect. 8.4.2; Battaglia et al. 1999; Tizzani et al. 2009). Tectonic processes occur in country rocks and dominantly involve changes in mechanical energy, with little or no movement of mass into or out of the subvolcano environment: these commonly derive from external forcing. A tectonically-driven unrest typically results from the shorter-term (transient passage of seismic waves) and/or longer-term (stress accumulation) variations associated with nearby regional tectonic earthquakes. These variations may affect the hydrothermal and/or the magmatic system of the volcano, eventually leading to unrest and eruption (see Sect. 9.3.3; Hill et al. 2002; Manga and Brodsky 2006).

Probably, the anomalies signalling unrest are never purely magmatic, purely hydrothermal or purely tectonic, as for example a magmatic unrest may affect also the hydrothermal system, whereas a tectonically-driven unrest may affect both the magmatic and hydrothermal systems. However, when clearly established, the cause of unrest usually involves the accumulation and/or transfer of magma (or magmatic fluids) within the shallow plumbing system of the volcano (Newhall and Dzurisin 1988; Phillipson et al. 2013; Acocella et al. 2015). The possible triggering factors of unrest, as well as the related processes, are described separately below.

9.3.1 Magmatic Trigger

Shallow magma accumulation is probably the most common cause of unrest. In fact, available evidence suggests that the direct rise of magma via dikes from deeper reservoirs without any shallow storage is not frequent, and usually the deep magma stops ascending and emplaces at shallow levels. In particular, following vertical transfer from the deeper plumbing system, magma often accumulates at a few kilometres of depth, often at the top of a magma chamber. The deeper transfer may pass unnoticed, because occurring at greater depth (limiting surface deformation and degassing), at higher temperature (limiting seismicity) and masked by shallower magma emplacement (limiting again surface deformation). Magma commonly emplaces through sills and eventually accumulates developing laccoliths. These intrusion shapes are justified by available field evidence of shallow plumbing systems and eroded magma chambers, as well as deformation and geophysical data (see Chap. 4). Sill emplacement is commonly accompanied by fracturing of the nearby host rock, focused above the sill and around its lateral terminations, detected as a swarm of VT earthquakes (Fig. 9.5). As fracturing increases the permeability of the host rock to magmatic fluids, it may also increase the degassed fluxes at the surface. In addition to seismicity and degassing, magma accumulation usually induces surface deformation with radial symmetry and microgravity changes suggesting magma addition. A hydrothermal system may amplify or buffer these monitoring signals. In principle, as previously anticipated, it is expected that the increased permeability following the fracturing generated by the accumulation of magma releases magmatic gases. These gases may reach and pressurize the hydrothermal system, in turn triggering shallower seismicity, increased degassing (where the magmatic component becomes diluted) and surface deformation, amplifying the effects due to magma accumulation alone. However, a hydrothermal system may also buffer the signals deriving from magma accumulation, for example, acting as a high permeability medium, releasing the fluids and relaxing the surface deformation. Also, some magmatic gases indicating the rise of magma, as SO_2, may be scrubbed by hydrothermal systems and be undetected at the surface. These mixed possibilities warn against assuming an a priori expected role of the hydrothermal system during shallow magma accumulation.

A crucial feature controlling the fate of a magmatically-triggered unrest is the eruptibility of the accumulated magma. Under which

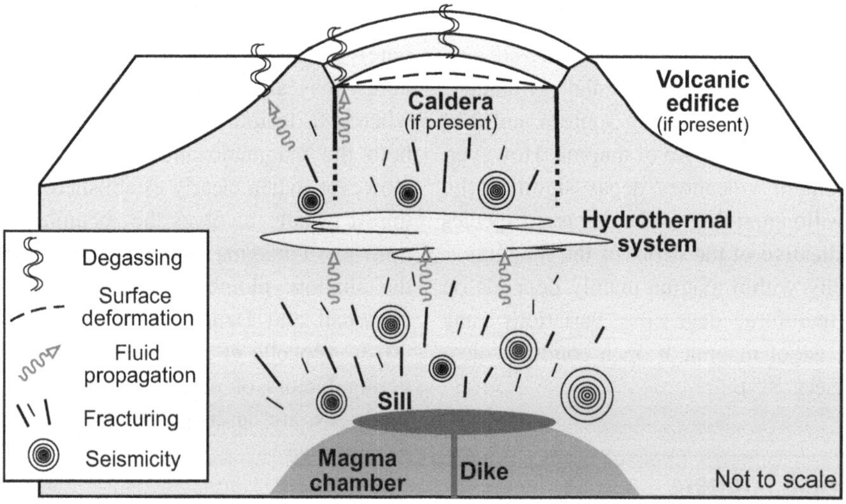

Fig. 9.5 Possible processes involved in unrest triggered by magma accumulation through an inflated sill. Magma accumulation may induce fracturing and thus variations in the permeability of the host rock, also resulting in fluid migration and, in turn, pressure variations within a hydrothermal system. At the surface, these processes manifest through seismicity, deformation, and degassing (modified after Acocella et al. 2015)

conditions does the magma rising from depth arrest, emplace, accumulate, and then eventually rise again towards the surface? This behaviour, which ultimately distinguishes eruptive from non-eruptive unrest, may depend upon several conditions, which can be largely reconciled with those promoting dike propagation, sill formation and dike nucleation (see Sects. 3.4, 3.5, 4.3 and 4.6.3). As explained in the previous chapters, unless a dike propagating from depth has significant overpressure, it will likely stall at some shallow crustal level and there possibly emplace feeding a sill, promoting unrest. Magma emplacement may preferably occur in correspondence of mechanical and thermal barriers, as for example along the major rigidity and viscosity contrasts at the top of magma chambers. The stronger the barrier with regard to the magma overpressure, the longer the magma accumulates. However, if magma supply is low or discontinuous, magma accumulation may cease, with the volcano returning to quiescence. Conversely, the stronger the magma overpressure with regard to a barrier, the easier is for the emplaced magma to propagate towards the surface, feeding an eruption. These balances should be also considered through time, with evidence suggesting that magma eruptibility decreases with the duration of unrest (see Sect. 9.4). Determining and quantifying the conditions controlling magma eruptibility and establishing their relative importance in general and specific situations should be a major topic for future research.

9.3.2 Hydrothermal Trigger

In some instances, magma does not appear directly involved as a cause of unrest, either because it does not directly participate in the process or because it is not possible to prove its involvement. In this case, in absence of remote tectonic triggers, a leading role of the hydrothermal component in unrest may be postulated. Hydrothermally-triggered unrest does not necessarily require external input of fluids and heat, as even an internal decrease in the permeability may perturb the system. The decrease in permeability may result from the precipitation of hydrothermal minerals within the fracture network, which becomes partly obstructed (**self-sealing**), determining a transient build up in the fluids pressure. Self-sealing of hydrothermal systems may thus generate seismicity, surface deformation, degassing and, in extreme cases, also phreatic eruptions. However, as hydrothermal systems in volcanoes derive most of their energy from the underlying magma, the involvement of the hydrothermal system alone in unrest is not common. More often, a supply of heat from depth, probably related to some magmatic input or possibly only related to increased deeper crustal permeability, may pressurize the hydrothermal system, determining unrest. These conditions may for example explain the phreatic eruption occurred without evident warning at Ontake (Japan) in 2014, causing 67 victims (Fig. 9.6; Kato et al. 2015; Oikawa et al. 2016). Retrospective re-examination of the minor seismicity before the phreatic eruption allowed detecting volcano-tectonic and long-period earthquakes for at least one month before the event. This seismicity aligned along a subvertical path beneath the craters, suggesting the rise of fluids along the summit conduit. A few minutes before the phreatic eruption the seismicity migrated upward and laterally, accompanied by an accelerated increase of tremor amplitude and anomalous tiltmeter signals indicating summit upheaval. This behaviour is explained through the filling of the conduit with pressurized fluids, which propagated to the surface immediately before the eruption. While it may be argued that the unrest signals have been so subtle to be detected only retrospectively, and that the fluids derived at least in part from the magma reservoir (Miyagi et al. 2020), the Ontake event still marks a dramatic episode of rapid unrest with major involvement of the hydrothermal system. In a similar fashion, the 2004–2006 unrest at Mt. Spurr (Alaska, USA) and the 2016 unrest at Tenerife (Canary Islands) have been characterized by the rise of magmatic fluids into the hydrothermal system (Koulakov et al. 2018; D'Auria et al. 2019).

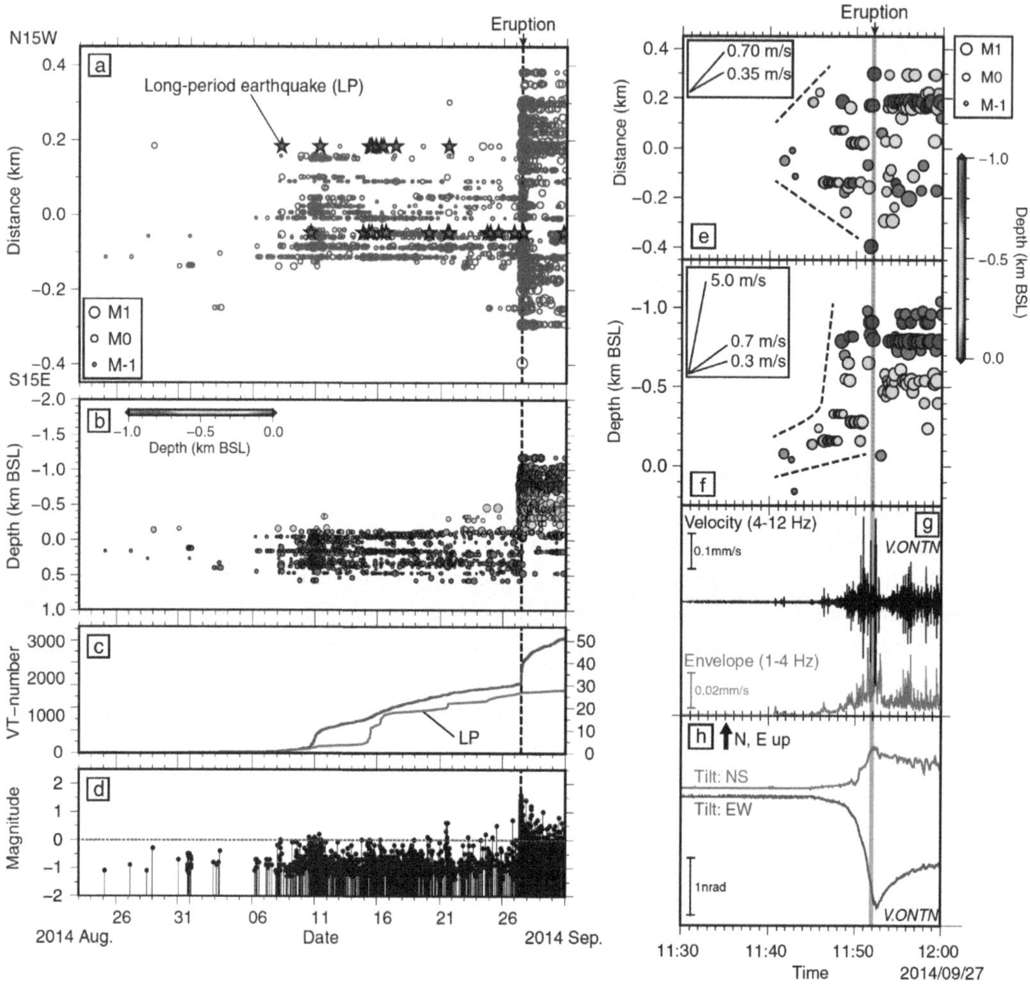

Fig. 9.6 The 2014 Ontake (Japan) phreatic eruption. **a** to **d** Longer-term seismicity. **a** Space–time diagram of seismic events before and after the eruption (23 August-30 September), with earthquake times and locations projected onto the strike of the hypocentre alignments; blue circles = VT earthquakes; red stars = LP events. **b** Depth variations in VT earthquakes over time. Circles are scaled to magnitude and color-coded to depth, with positive values below sea level (BSL). **c** Cumulative number of VT earthquakes (blue curve) and LP events (red curve) over time. **d** Magnitude versus time. **e–h** Shorter-term (minutes) precursors. **e** Horizontal distance of VT earthquakes from 11:30 to 12:00 JST on 27 September, projected onto the strike of the hypocentre alignments (N15°W-S15°E). Circles are scaled to magnitude and color-coded to depth. Black dashed lines approximate the locations of the fronts of earthquake migrations. **f** Depth variations in VT earthquakes, coloured to depth. **g** Band-pass filtered waveform between 4 and 12 Hz (black curve) and envelope between 1 and 4 Hz (red curve) recorded at V.ONTN station. **h** Time series of tiltmeter records observed at V.ONTN station. NS and EW components denote the northward and eastward ground-up components, respectively (Kato et al. 2015)

9.3.3 Seismic Trigger

Hydrologic responses to earthquakes, including liquefaction, changes in stream and spring discharge, in the properties of groundwater, in the water level in wells, and the eruption of mud volcanoes have been all documented for thousands of years. This response has been observed also at volcanoes, where the above-mentioned processes have been commonly associated with unrest affecting the hydrothermal system and even the magma chamber, suggesting that earthquakes influence the state of volcanoes (Wang and Manga 2010, and references therein). Indeed, studies in the last two decades have shown that far-field (i.e., regional) seismicity may trigger unrest and magmatic eruptions at nearby volcanoes. For example, a perturbation in the deformation field and increase in seismicity were detected at Long Valley caldera (California, USA) during and after the 1992 Landers earthquake, which occurred ~420 km away. This variation has been interpreted as resulting from pressure increase owing to gas bubbles, which may have been shaken loose during the passage of the seismic waves (Fig. 9.7; Linde et al. 1994).

Similarly, in September 2002 an earthquake occurring along a fault of the unstable flank of Mount Etna (Italy) was followed, one month later, by the eruption along the nearby Northeast Rift, which had been inactive for more than 50 years. The eruption in turn seismically reactivated the same fault initially responsible for the earthquake, in a context of kinematic coupling between seismicity and magmatic activity (Acocella et al. 2003). Evidence for regional earthquakes promoting volcanic or hydrothermal activity has been provided in several other cases, including the eruptions of Vesuvio in the last ~1000 years (Nostro et al. 1998), the largest historical 1707 eruption at Mount Fuji (Japan; Chesley et al. 2012) and the 2015 unrest at Deception Island (Antarctica; Almendros et al. 2018). Evidence has been also collected for a single earthquake triggering eruptions at different volcanoes, as immediately after mega-earthquakes (which are characterized by

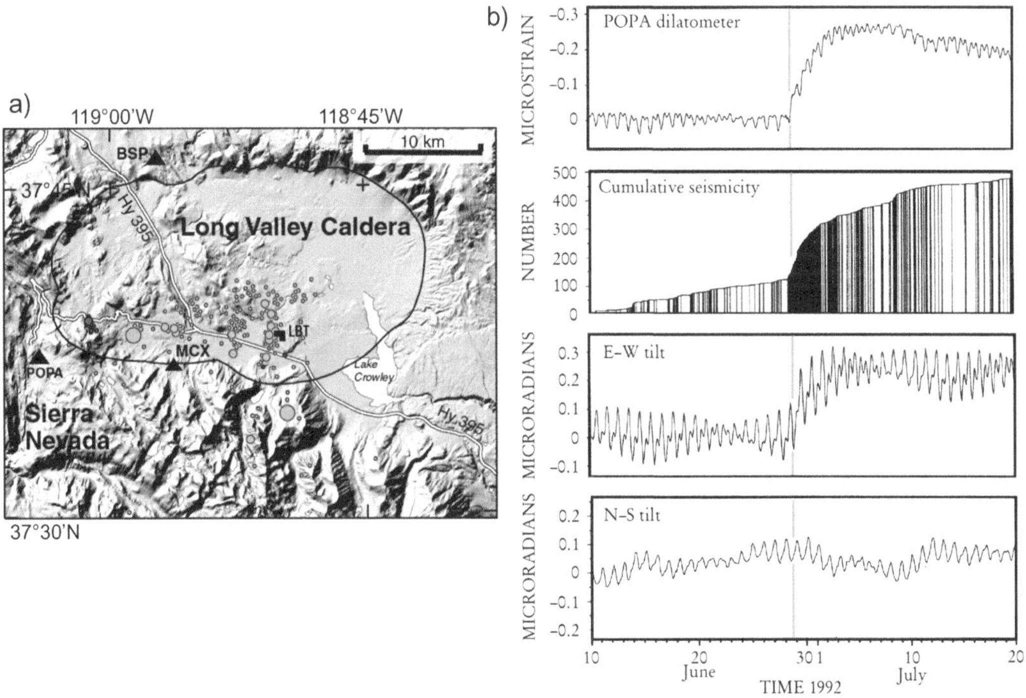

Fig. 9.7 **a** Seismicity induced at Long Valley caldera (California, USA) by the *M7.3* Landers earthquake (28 June 1992; orange circles) and the *M7.1* Hector Mine earthquake (16 October 1999; red circles); borehole dilatometers = solid triangles; long-base tiltmeter (LBT) = inverted L. **b** Volumetric strain recorded by POPA dilatometer, cumulative number of earthquakes, and E-W and N-S components of LBT from 10 June to 20 July 1992. Negative strain is compressional; positive tilt is down to the east and north; high-frequency wiggles on the dilatometer and tiltmeter records are solid-earth tides. Orange line marks the Landers earthquake (modified after Hill et al. 2002; Linde et al. 1994)

magnitude $M > 8.5$) in Kamchatka (Russia) in 1952, in Chile in 1960, in Alaska (USA) in 1964 and in Sumatra (Indonesia) in 2004. In all these cases, the post-seismic eruptive frequency along the portion of volcanic arc closer to the hypocentre increased sensitively with regard to the pre-seismic frequency (Walter and Amelung 2007). However, not all mega-earthquakes trigger eruptions, as observed after the 2010 Maule (Chile; *M8.8*) and 2011 Tohoku (Japan; *M9.0*) earthquakes, where only minor seismicity and deflation were observed at nearby volcanoes in the months following the earthquakes. The deflation may be related to the enhanced degassing from the hydrothermal systems after the co-seismic increase in crustal permeability and/or to crustal stretching (Pritchard et al. 2013; Takada and Fukushima 2013). Recent experiments suggest that the possibility to have post-seismic eruption or instead deflation results from the different patterns (upward or downward-lateral, respectively) of the migrating fluids during an earthquake, as constrained by the oscillation frequency of the volcanic edifice with regard to its resonance frequency (Namiki et al. 2019).

In general, it has been estimated that ~0.4% of explosive volcanic eruptions occur within a few days of large, distant earthquakes. This many “triggered” eruptions is much greater than expected by chance. This behaviour is not restricted to magma, as a widespread hydrologic response to seismicity is also observed (Fig. 9.8; Linde and Sacks 1998; Montgomery and Manga 2003; Manga and Brodsky 2006; Wang and Manga 2010). In particular, the stronger the earthquake, the farther a certain magmatic or hydrologic effect can manifest, with the largest magnitude earthquakes potentially affecting magmatic activity at volcanoes nearly 1000 km distant. The highest correlation coefficients between earthquakes and eruptions are found for regions along the Pacific subduction zones, where the largest earthquakes occur (Eggert and Walter 2009). The state of the affected magma may provide another important influencing factor. Evidence from the 2015 Ambrym

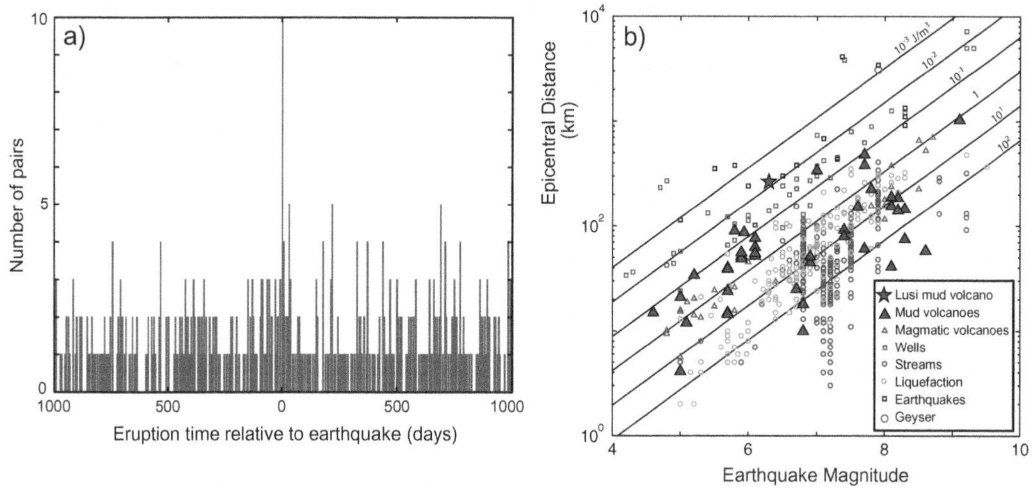

Fig. 9.8 Time and space relationships between earthquakes and eruptions. **a** Histogram showing the number of eruptions as a function of time relative to $M > 8$ earthquakes. Negative times correspond to eruption prior to the earthquake. Only eruptions located within 800 km of the earthquake epicentre are included; bins are 5 days wide (modified after Manga and Brodsky 2006). **b** Distribution of earthquake-triggered hydrologic changes and earthquake-induced magmatic volcanic eruptions as a function of earthquake magnitude and distance. Also plotted are the log distance versus magnitude contours of constant seismic energy density, which is the seismic energy in a unit volume in the seismic wave train: this represents the maximum seismic energy available to do work at a given location during the earthquake (modified after Wang and Manga 2010; image courtesy Michael Manga)

(Vanuatu) eruption, triggered by a *M6.4* earthquake which occurred 30 h earlier, suggests that partially cooled and crystallized (i.e., H_2O-saturated) basaltic magmas are more susceptible to triggering from earthquakes. In fact, if the magma is too hot the stress drop required to grow the bubbles is too large, whereas if it is too cold the magma can no longer flow (Hamling and Kilgour 2020).

Several mechanisms may explain the remote triggering of unrest and volcanic eruptions, as the failure of rocks surrounding stored magma, a decrease in the deviatoric stress promoting dike propagation, changes in magma overpressure, including volatile exsolution, the growth of bubbles, the advection of large pressures by rising bubbles and overturn of magma chambers. In the case of mud volcanoes and geysers, liquefaction caused by shaking and changes in permeability by opening or creating new fractures may explain the observed hydrologic changes. All these mechanisms require a process that enhances small static stress changes caused by earthquakes or that can convert (the larger) transient, dynamic strains into permanent changes in pressure (Belardinelli et al. 2003; Manga and Brodsky 2006; Walter and Amelung 2007).

Static stress changes develop as a consequence of the slip of a fault during an earthquake and result in permanent perturbation of the nearby stress field. As the two sides of the fault move in opposite directions, stress is exerted parallel and perpendicular to the fault plane. When the shear stress exceeds the frictional resistance on the fault, or when the perpendicular stress is eased, the rocks on either side slip past each other suddenly, generating an earthquake. Both components of stress, which when added together are called **Coulomb stress**, diminish along the segment of the fault that slips and become redistributed to inactive zones along the same fault or to nearby faults (Fig. 9.9; Stein 2003). The local increase in Coulomb stress at these inactive zones could be sufficient to trigger successive earthquakes. To estimate if slip on a new fault is encouraged or discouraged after an earthquake, the change in Coulomb failure stress ΔC_{FS} is calculated as:

$$\Delta C_{FS} = \Delta\sigma_S + \mu(\Delta\sigma_N + \Delta P) \quad (9.1)$$

where $\Delta\sigma_S$ is the shear stress change on a fault, $\Delta\sigma_N$ the normal stress change, ΔP the pore pressure change in the fault zone, and μ the coefficient of friction; in this approach a constant friction model is assumed (e.g. King et al. 1994; Harris 1998; Walter et al. 2005). Fault slip is encouraged if the change in Coulomb failure

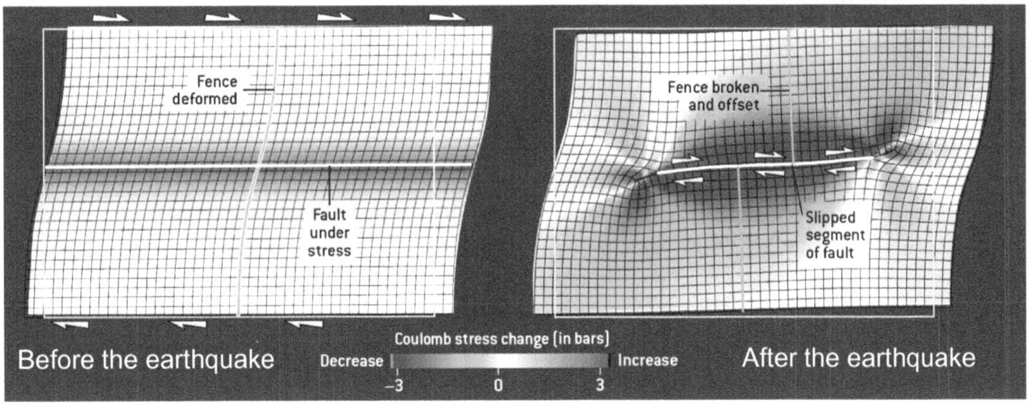

Fig. 9.9 Conceptual model of variation of the stress accumulated along a dextral fault before and after an earthquake occurring in the central part of the fault (map view). Before the earthquake (left) the long-term accumulated stress focuses along the entire fault length. After the earthquake (right) the portion of fault that has slipped has released the stress (in blue), whereas at its lateral terminations the stresses have increased (in red) because of stress transfer (Stein 2003)

stress induced by the earthquake is positive, and discouraged if the change is negative. A change of the Coulomb failure stress by 0.1 MPa is considered significant, although stress triggering may occur even below 0.01 MPa (Stein 2004). Therefore, one can estimate if the stress perturbations induced by an earthquake may trigger slip (and thus further seismicity) on nearby faults. Indeed, the same process may also affect magmatic activity. In fact, Coulomb stresses may likewise induce pressure changes in a magma body below a volcano. Increased compressional stress in the crust surrounding a magma chamber that is close to its critical state may squeeze magma upward, whereas a decrease in compressional stress can promote magma depressurization, additional melting, the formation of bubbles as volatiles exsolve, the "unclamping" of conduits above the magma chamber, dike nucleation and propagation (Hill et al. 2002; Walter and Amelung 2007). Static stresses are, by definition, time independent and their effects may manifest immediately after the stress perturbation or even after years. The observed delays of months to years between the earthquake and the eruption are poorly understood, although diffusion of interstitial crustal fluid, mostly water, in response to static stress changes may be important.

Dynamic stress changes develop only during the passage of seismic waves induced by an earthquake and are thus transient, lasting from seconds to minutes. Dynamic stresses propagating through seismic waves from large earthquakes are capable of triggering other earthquakes or, in the case of volcanoes, unrest and eruption (Linde et al. 1994; Hill and Prejan 2007). Models for dynamic triggering fall under two broad groups, one appealing to Coulomb failure with various friction laws and the other appealing to the activation of hydrous or magmatic fluids. For dynamic triggering under the frictional models, the stress state in the crust must differ from the Coulomb failure stress by less than the peak amplitudes of the dynamic stresses (typically <0.1 MPa for remote triggering). These models are generally consistent with the onset of triggered seismicity during the dynamic stresses, followed by decay. Fluid excitation models involve fluid transport or a phase change. These fluid-based models, whether hydrous or magmatic, involve some crustal deformation through intrusion, pressure changes (in the case of bubble excitation), convection (in a magma body) or poroelasticity (in the case of fluid diffusion). These models admit the possibility of delayed onsets of triggered seismicity and increasing or sustained activity rates for extended periods following passage of the dynamic stresses. Extensional stress regimes hosting geothermal and volcanic activity seem more susceptible to remote dynamic triggering than compressional stress regimes, although remote triggering is not limited to extension (Hill and Prejan 2006). Dynamic models also explain most of the observed hydrologic responses, both within and beyond the near-field (Wang and Manga 2010).

Despite the common triggering capacity, static and dynamic stress changes may induce quite different effects. While a static stress change is able to advance or delay an instability depending on its sign, a dynamic stress pulse promotes nearly instantaneous failure, in case its amplitude is positive and large enough (Belardinelli et al. 2003). Also, static and dynamic stresses have different value and decay with distance. For a same earthquake, static stress changes are one order of magnitude smaller than dynamic stresses within 10^2 km from the hypocentre (Table 9.1; Manga and Brodsky 2006). This difference increases with distance s, so that static stress changes become four orders of magnitude smaller than dynamic stresses at 10^4 km from the hypocentre. This follows from the fact that while static stresses fall off as $1/s^3$, dynamic stresses fall off as $1/s^{1.66}$. Therefore, dynamic stresses may propagate to longer distances and may be still important where static stress become negligible. For example, the simultaneous occurrence of two eruptions (Stromboli and Etna) and one unrest (Panarea Island) in late 2002 in the

Table 9.1 Static and dynamic stress changes induced by a *M8* earthquake as a function of the distance from the hypocentre (modified after Manga and Brodsky 2006)

Forcing	Stress (MPa)			
	Distance	10^2 km	10^3 km	10^4 km
Static stress changes, M8		10^{-1}	10^{-4}	10^{-7}
Dynamic stress changes, M8		3	0.06	10^{-3}

Southern Tyrrhenian Sea (Italy) could be explained by the dynamic stresses imposed by a *M5.9* earthquake at ~130 km of distance, and not by the weaker static stress changes (Walter et al. 2009). Recent studies relax the importance of Coulomb stresses at long distance from the earthquake. In particular, for volcanoes within 200 km of earthquakes of $M \geq 7.5$, the eruption occurrence probability increases by ~50% for 5 years after the earthquake, likely triggered by static stress changes and/or strong ground motions. However, no significant increase in the occurrence probability of new eruptions was observed at more distant volcanoes or for smaller earthquakes (Nishimura 2017).

The fraction of eruptions that are triggered by earthquakes X_t is (Manga and Brodsky 2006):

$$X_t = \frac{\Delta P_{EQ} T_V}{\Delta P_m T_{EQ}} \quad (9.2)$$

where ΔP_{EQ} is the extra pressure generated by the earthquake, T_v is the ordinary recurrence time of volcanic eruptions, T_{EQ} is the time between large earthquakes affecting the volcano and ΔP_m is the overpressure required to generate tensile deviatoric stresses to nucleate and propagate a dike to the surface. The latter is estimated as 10–100 MPa for silicic magmas and less than 1 MPa for basaltic magmas (Tait et al. 1989; McLeod and Tait 1999; Jellinek and DePaolo 2003). As the static and dynamic stresses ΔP_{EQ} caused by earthquakes are typically 10^{-2}–10^{-1} MPa (Table 9.1), the overpressure ΔP_m required to generate tensile deviatoric stresses sufficient to allow a dike to form must be within 99–99.9% of the maximum overpressure for the earthquake to initiate an eruption. Also, typical recurrence intervals T_v for VEI 2 and 3 eruptions are 1–10^2 years, whereas the recurrence time T_{EQ} for $M > 8$ earthquakes near a given volcano is 10^2–10^3 years. Thus, Eq. (9.2) indicates that only a very small fraction ($\ll$1%) of eruptions will be seismically triggered at a given volcano (Manga and Brodsky 2006). This fraction likely concerns those volcanic systems already approaching the unstable state, where a stress change may anticipate a forthcoming unrest or eruption.

Viscoelastic relaxation of earthquake-induced stresses may explain not only delayed (time lag of years to decades) earthquake-earthquake triggering, but also delayed eruptions to distances of 10^3 km (Marzocchi 2002; Marzocchi et al. 2004a). However, as the stress diffusion caused by viscoelastic relaxation results in a non-linear spatial and temporal evolution, quantifying these relationship between earthquakes and eruptions from observations remains challenging (Manga and Brodsky 2006).

Finally, in addition to earthquake-earthquake and earthquake-eruption interactions, eruption-eruption interactions between different volcanoes have been recognized. Such a volcanic coupling is largely controlled by the distance between nearby magma reservoirs, as magmatic sources that are spaced less than about 10 km apart interact, whereas those spaced more than about 25 km apart usually do not (Biggs et al. 2016). In particular, the interactions between the most closely spaced magmatic systems are controlled by the extent of shallow crystal mush layers, whereas stress changes can couple large eruptions over distances of 20–40 km, and only large dike intrusions or subduction earthquakes could generate coupled eruptions over distances of 50–100 km.

9.4 Understanding Unrest

Unrest may be triggered by multiple factors, acting independently or simultaneously, eventually culminating into eruption. Given the variability of the conditions leading to unrest and controlling the evolution of magmatic and hydrothermal systems, understanding what happens during unrest at a given volcano, in terms of processes and outcome, remains challenging. Nevertheless, this represents a crucial question in volcanology. In fact, understanding unrest may allow defining the conditions determining the shallow emplacement and accumulation of magma, as well as the nucleation of any feeder dike, finally providing the opportunity to better forecasting eruptions, which is the ultimate challenge for volcanology (e.g., Acocella 2014).

To determine the fate of unrest, and thus the probability of eruption, it is equally important to investigate why feeder dikes propagate (leading to eruptive unrest) or magma becomes stalled at depth (leading to non-eruptive unrest). These problems have been discussed, mainly from a theoretical perspective, in Chaps. 3 and 4. Nevertheless, studies have been also focusing on real cases concerning the propagation of feeder dikes and the stalling of magma ascending towards the surface, the latter condition also termed "failed eruption" (e.g., Crider et al. 2011; Gardine et al. 2011; Moran et al. 2011; Nishimura and Ueki 2011; Roman and Power 2011; Werner et al. 2011). In these studies on real cases, a first and common difficulty in understanding unrest and determining its fate is the general scarcity of monitoring, geophysical and geological constraints on the specific unrest episode, also to allow comparison to similar episodes. Also, even when monitoring data have been collected, a further difficulty derives from their accessibility and exploitation, as these data are often dispersed, fragmented and inhomogeneous. These limitations highlight a basic need to collect and organize data, creating large accessible databases, to understand unrest (e.g., Moran et al. 2011).

A few studies have created and investigated databases of unrest episodes (e.g., Newhall and Dzurisin 1988; Sandri et al. 2004, 2017a; Phillipson et al. 2013; Acocella et al. 2015). The description of the conditions and manifestations accompanying unrest, their statistical analysis and the identification of recurrent features provide in fact an approach to identify general behaviours, patterns, thresholds and relationships, in turn allowing to infer the possible related processes, to be understood also with the support of models (e.g., Sandri et al. 2004). In general, these studies have shown that volcanoes do not respond in the same way to unrest. The duration and outcome of unrest largely depend on the composition and size of the magmatic reservoir, which also affect the type of volcanic edifice. In particular, in the decade between 2000 and 2011 the duration of pre-eruptive unrest differed across various volcano types (Fig. 9.10; Phillipson et al. 2013). Approximately 50% of stratovolcanoes erupted after about one month of reported unrest. At large calderas this average duration of pre-eruptive unrest was about twice as long. At almost five months, shield volcanoes had a significantly longer unrest period before the onset of eruption. Also, while unrest at shield volcanoes is largely eruptive, at calderas there is a large uncertainty ($\sim$50%) on whether it culminates into eruption. In addition, calderas very frequently experience unrest, as all the calderas monitored over approximately 2 decades have shown at least one unrest episode. Therefore, despite their complexity, calderas host frequent unrest, commonly non-eruptive, and are thus ideal to observe, investigate and understand both eruptive and non-eruptive unrest processes (Acocella et al. 2015).

A few studies focused on caldera unrest. The first systematic study was a comprehensive work which included at least 1299 historical and relatively recent monitored episodes of unrest that have occurred at 138 calderas (Newhall and Dzurisin 1988). This study showed that calderas are dynamic and delicately balanced systems, which can be disturbed by even small stimuli, most notably minor tectonic strain or small-volume basaltic underplating. Different processes with different implications for hazards may

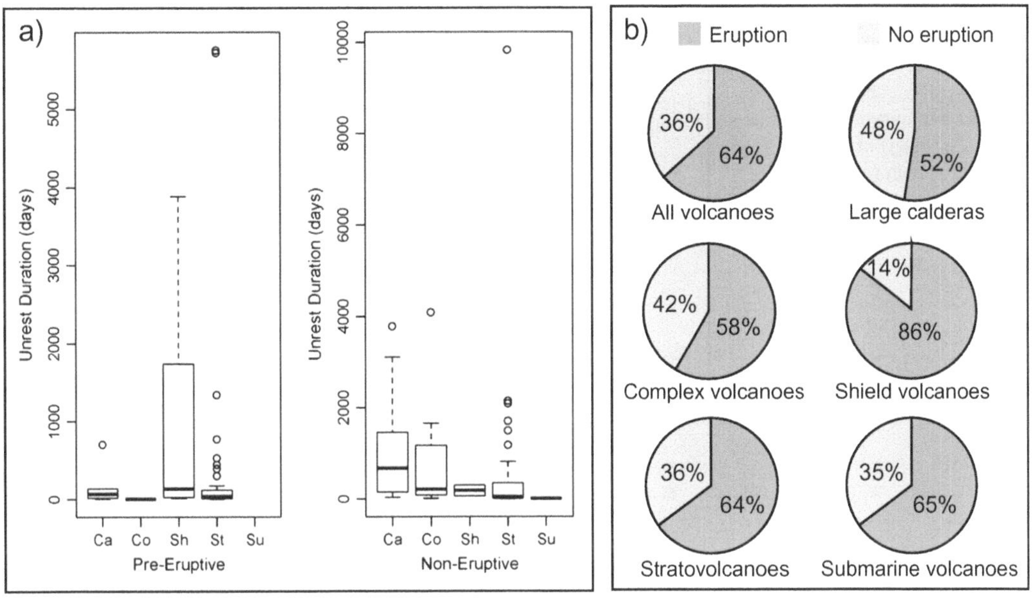

Fig. 9.10 Unrest at volcanoes. **a** Boxplots of pre-eruptive and non-eruptive unrest duration (days), segmented by volcano type (Ca = Caldera, Co = Complex, Sh = Shield, St = Strato, Su = Submarine); note the different scales in the y axes. **b** Pie charts of the proportions of volcanoes with unrest leading and not leading to eruption (modified after Phillipson et al. 2013)

produce similar symptoms of unrest, so that even a well-understood hybrid of processes can lead to many different outcomes. This results in unrest which is likely to persist for months to years, sometimes even for decades to centuries, and to be intermittent rather than showing a systematic increase until culmination. Because of this variability, unrest alone is not a reliable indicator of impending eruption. Changes that are more diagnostic of impending eruptions are: harmonic tremor; an exponentially escalating rate of seismic energy release or a sudden, pronounced drop in seismic energy release; a sudden, pronounced increase in the uplift rate or sudden deflation after an extended period of inflation; opening of, and intense fumarolic emissions from, new fissures; or a sudden, sharp increase in hot spring discharge. Importantly, the outcome of a specific episode of unrest cannot be forecast solely on the basis of patterns of unrest at other calderas, given the uncertainties in generalizing from one caldera (or a group of) to another. When caldera unrest culminates into eruption, the latter is usually small, despite the large potential for catastrophic eruptions. Overall, this wealth of information suggests no simple solution to the interpretation and forecasting of complex events at calderas (Newhall and Dzurisin 1988).

This study has been later expanded with available monitoring data for the 1988–2014 period, providing an updated database and revealing different types of first-order behaviours depending upon the composition of the volcano (mafic or felsic) and the degree of opening of the magmatic system. The latter varies from open, through semi-plugged, to fully-plugged. In open conduits magma and gas can rise and erupt freely, whereas in semi-plugged conduits only gas can escape, and in fully-plugged (or plugged) conduits neither magma nor gas can escape (Fig. 9.11; Acocella et al. 2015). In general, unrest in mafic calderas is subtler, less pronounced and repeated, especially with open systems, which ensure a continuous, aseismic and moderate release of magma. Plugged felsic calderas are quite rare and erupt infrequently, anticipated by isolated, short and seismically active unrest. Semi-plugged felsic calderas are

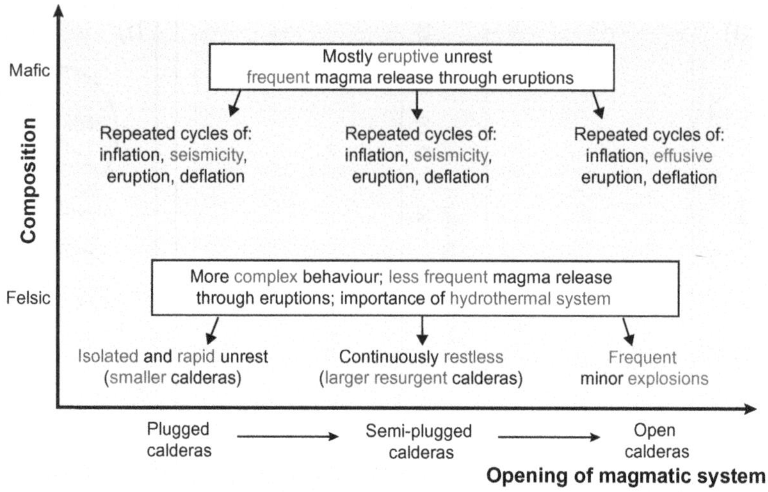

Fig. 9.11 Types of caldera unrest as a function of the composition of the caldera (mafic vs. felsic) and the state of opening (plugged, semi-plugged, or open) of the magmatic system (modified after Acocella et al. 2015). See text for details

more common and also erupt infrequently, being often restless over decades or centuries, with uplift, seismicity and degassing and, on the longer-term, resurgence, suggesting repeated stalled intrusions. The statistical analysis of these data has highlighted some difference between eruptive and non-eruptive unrest. Eruptive unrest is shorter than non-eruptive unrest, with 72% of eruptive unrest, mainly at mafic calderas, lasting less than 10 months and showing high seismicity and degassing. The remaining 28% lasts between 10 and 18 months, with seismicity and degassing constituting a longer-duration tail (11% of the cases), or is essentially aseismic in calderas with open conduit (17%).

The limited duration of eruptive unrest (up to several months), conversely to non-eruptive unrest (up to decades) suggests that magma withstands only a limited period of "eruptibility", before becoming stored in the upper crust (Fig. 9.12; Sandri et al. 2017a). This may provide an important clue to focus research on the parameters controlling the eruptibility through time of the magma accumulated in a volcano. Preliminary studies suggest that this inferred critical period may be related to: (a) the amount of degassing of the intrusion, where higher and longer degassing reduces the volatile content of the intrusion and, at the same time, increases its cooling and viscosity (Sandri et al. 2017a); (b) the decay of the tensile stresses induced by the intrusion within the host rock, inhibiting further fracturing and magma propagation (Giudicepietro et al. 2017).

If forecasting the outcome of an isolated unrest episode is difficult, the occurrence of additional unrest episodes in the previous decades at a same volcano makes a forecast even more challenging. In fact, the intrusions emplaced in the previous unrest episodes may be activated by the last intrusion with a cumulative response, and even under a minor final

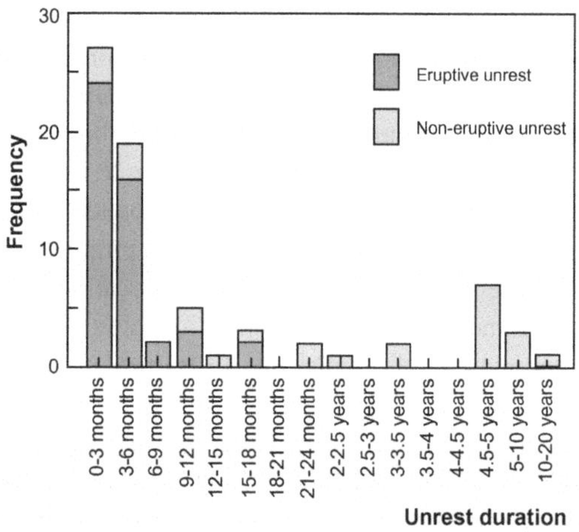

Fig. 9.12 Histogram of the duration of unrest episodes at calderas between 1988 and 2014, considering their eruptive or non-eruptive outcome (modified after Sandri et al. 2017a)

perturbation, which alone would not have allowed the system reaching criticality. This has likely been the case of Rabaul erupting in 1994 during a minor unrest episode following a major one (Fig. 5.24b). The possibility that only a minor final perturbation leads to critical conditions with a cumulative response is particularly worrisome at restless calderas, as for example Campi Flegrei. Here the ongoing minor unrest episode (2005 to the time of writing in 2020) follows three major unrest episodes in the previous decades that may have promoted thermal weakening of the caldera, raising concern for hazard assessment and risk mitigation. A similar multiple disturbance over decades to months before eruptions has been recognized at Santorini before the ~1600 BC major eruption (Fig. 5.14a; Druitt et al. 2012; Chiodini et al. 2016).

While the unrest phase may help in understanding the state of the volcano and forecasting any impeding eruption, any estimate of the size of a possible eruption currently remains purely speculative, possibly even in the end-member case of super-eruptions. In fact, what is known about super-eruptions, in terms of physical processes, represents a logical extension of the activity observed and inferred from smaller eruptions of the same composition, the only difference consisting of the volume of released magma. This implies that there are no grounds for supposing that super-eruptions involve novel processes, and thus unrest behaviours, that do not also apply to other eruptions (Wilson 2008).

In addition to isolated studies providing and analysing databases, important international initiatives are managing more comprehensive databases. These include WOVODAT (https://www.wovodat.org/), which aims at collecting homogeneous monitoring data on volcanoes from a constellation of volcano observatories worldwide (with data from > 900 unrest episodes covering > 75 volcanoes already accessible; Newhall et al. 2017), and Geohazards Supersites (https://geo-gsnl.org/supersites/permanent-supersites/), developed to encourage scientific exploration of earthquake and volcano hazards to improve disaster risk management at specific sites around the globe.

9.5 Assessing Volcanic Hazard and Forecasting Eruptions

As introduced in Sect. 1.10, volcanic risk is defined as the product of volcanic hazard, exposure and vulnerability. Volcanic hazard is the competence of volcanologists, who are in charge of studying and monitoring volcanoes, to define their state, define and forecast any hazard and communicate their warnings to civil defence, authorities and public. The definition and reduction of exposure and vulnerability is the competence of experts with complementary knowledge (engineers, architects, urban planners, modellers, doctors) and, most importantly, authorities (including civil defence), whose duty is to manage a crisis with appropriate mitigation procedures (e.g., Tilling 1989). Therefore, volcanologists are required to provide appropriate hazard assessment, largely in collaboration with civil defence and authorities, to mitigate risk. In practical terms, this includes the definition of hazardous scenarios (including the reference eruption), the establishment of early warning systems, the definition of volcano alert levels and the capability to forecast eruptions.

Hazardous scenarios are expected hazardous events that may occur at a certain volcano, based on its current state. Defining hazardous scenarios is helpful for civil defence and authorities to take actions and mitigate the related risks. A volcano may be the source of very different types of hazards, directly triggered by volcanic activity (as pyroclastic flows, ash fall, lava flows, debris avalanches, surface deformation, degassing, seismicity) or indirectly triggered (as lahars, floods, tsunami, acid rain, famine, diseases, climate change), manifesting at different times from the onset of an eruption and impacting areas at various distances. These hazards are often related to each other, so that the manifestation of a specific hazard may trigger another hazard, amplifying the impact, in a **multi-hazard** combination. Multi-hazards should be adequately investigated establishing the ranking of the different types of hazard and taking into account for their possible interactions (Marzocchi et al.

2012). Multi-hazard assessment may be supported by a **conceptual model**, which consists of a theoretical frame based on any type of first-order evidence that describes the general evolution of a volcano in a reference period. In order to be useful, the conceptual model has to be kept as simple as possible, with its main constituents being shared by the wider scientific community. An example of volcano involving a potential chain of multi-hazards is Ischia island (Italy), characterized by notable resurgence ($\sim$1000 m of uplift in the last $\sim$35 ka) associated with widespread gravitational instability, minor eruptions outside the resurgent block and recurrent seismicity along a border of the block (De Vita et al. 2010; Trasatti et al. 2019). Here a multi-hazard assessment has relied on a conceptual model, valid for the last $\sim$10 ka, based on the general understanding of the behaviour of a magmatic system experiencing resurgence. This conceptual model has allowed zonation of the origin of the different hazards, as well as the related hazardous scenarios (for eruptions, seismicity, landslides, tsunamis), also integrated in a multi-hazard perspective (Fig. 9.13; Galetto et al. 2017; Selva et al. 2019).

In high-risk volcanoes a reference eruptive scenario should be considered. The **reference eruption** allows civil defence and authorities to prepare appropriate mitigation procedures and emergency plans, so that established and effective protocols may be used in case of impending eruption. The reference eruption should meet a balance between likelihood and severity and should correspond to a reasonable evaluation of the "acceptable risk", as based on probabilistic studies. For example, the reference eruption for Vesuvio (Italy) has been evaluated considering the past eruptive history of the volcano (mainly consisting of violent Strombolian VEI 3 eruptions, sub-Plinian VEI 4 eruptions and Plinian VEI 5 eruptions), as well as their conditional (conditioned to the occurrence of an eruption) probabilities in a time frame of 200 years. Accordingly, the probability of a VEI 3 eruption is the highest (72%), whereas the

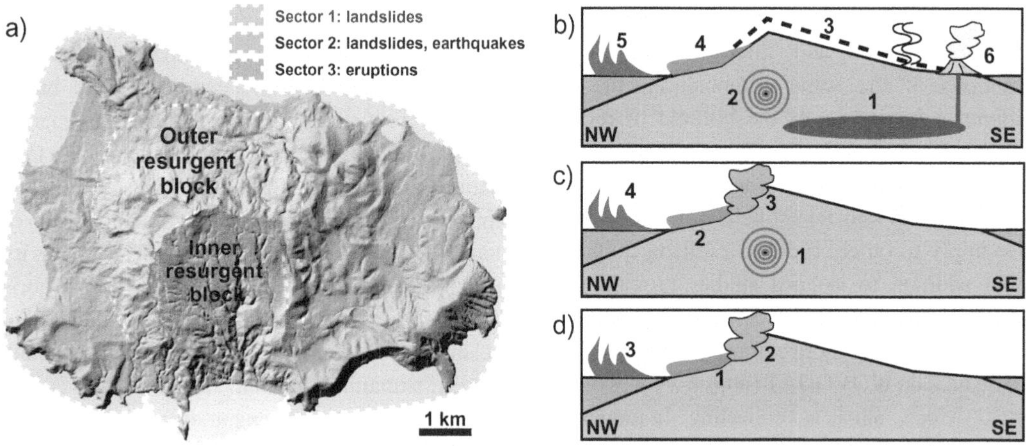

Fig. 9.13 Example of multi-hazard approach at Ischia (Italy; modified after Selva et al. 2019). **a** Schematic view of the areas source of the main hazards; dashed boundaries are meant to represent broad transitions zones. **b–d** NW–SE section views through the resurgent block illustrating the possible multi-hazard scenarios; not to scale. **b** Magma-induced multi-hazard: the emplacement of a shallow intrusion (1) induces seismicity (2) and deformation (3), possibly destabilizing the resurgence border (4) and, in case the landslide products enter the sea, triggering a tsunami (5); eventually, the intruded magma may erupt (6). **c** Seismicity-induced multi-hazard: an earthquake (1) may trigger the collapse along the resurgent block (2), eventually promoting phreatic eruptions through decompression (3) and, if the collapsed deposits enter the sea, triggering a tsunami (4). **d** Landslide induced multi-hazard: collapse along the border of the resurgent block (1) may promote phreatic eruptions through decompression (2) and, if the collapsed deposits enter the sea, may trigger a tsunami (3)

Table 9.2 Conditional probabilities for expected eruptions with different size at Vesuvio (Italy) for time frames between the next 60 and 200 years and from 60 years onward (modified after Marzocchi et al. 2004b)

Eruptive activity	VEI	Conditional probability (%)	
		(>60 to <200 years)	(>60 years)
Violent Strombolian	3	72	65
Sub-Plinian	4	27	24
Plinian	5	1	11

probability of a VEI 4 eruption is lower, but not negligible (27%) and the probability of a VEI 5 eruption is much smaller (1%) (Table 9.2; Marzocchi et al. 2004b). Choosing a VEI 3 as the reference eruption considers the most likely eruptive event, but neglects a still relatively probable VEI 4 eruption. Choosing a VEI 4 as the reference eruption allows to include a VEI 3 event and to be on a safer side including a relatively probable VEI 4 eruption. Choosing a VEI 5 eruption brings to an even safer approach, but requires much more demanding mitigation measures for an unlikely event. Based on these considerations, the reference eruption for Vesuvio has been taken as a VEI 4 sub-Plinian eruption. This leaves out an unlikely VEI 5 Plinian eruption as the **acceptable risk** that a community is willing to take considering the benefits and costs (including feasibility) of risk mitigation. Therefore, a reference scenario is neither necessarily the most likely nor the most hazardous event, its choice resulting from a trade-off between size, probability of occurrence and acceptable risk. Note that the acceptable risk is usually not defined by the volcanological community, as requiring knowledge on the expected impact of a hazardous event and the capability of a community to face it. Recent hazard studies go beyond the concept of eruptive scenario, to take into account for full variability of events in addition to a specific size class (Sandri et al. 2016).

An **Early Warning System** (EWS) is used by volcanologists to inform authorities about the occurrence of any impending hazard. It consists of capacities to generate timely warning information to enable those threatened by a hazard to prepare and act to reduce harm or loss. An EWS is released by volcano observatories and concerns information that often describes the state of a volcano, the expected hazards, a time frame for specific activity and generic recommendations (Gregg et al. 2015). This information is disseminated to agencies linked to emergency management/civil defence, aviation, media and the public. EWSs are a key risk reduction tool as part of a programme of volcanic hazard assessment that brings together the physical and social sciences to enable effective decision-making.

A **Volcano Alert Level** system (VAL) is a specific key tool or subsystem of a volcano EWS that simplifies the communication of volcanologists' interpretation of data (Newhall 2000). A VAL consists of a series of levels that correspond to increasing stages of volcanic unrest. Ideally, the VAL should increase progressively before eruption or increase and then decrease where unrest does not proceed to eruption (Table 9.3; Winson et al. 2014). Most VALs are country-specific and, although based on different schemes, include a quick description of the activity occurring at a volcano, with indication of the potential time frame before an event, and a colour scale to communicate an escalating (or decreasing) volcanic hazard. Colours are in the "traffic light" scheme, where green is equated with normal conditions and red is the most dangerous level (usually an impending eruption). These features respond to the original meaning and issuance of the VALs, which date back to the late 1980s and were aimed at defining the state of a volcano, although implying forecast windows used by officials to decide on mitigation actions. In the last decades, some VALs have included recommended actions, as information on exclusion zones and **evacuation**, that is the temporary

Table 9.3 Example of alert levels used for the 1991 Pinatubo (Philippines) crisis; colours allow linking alert levels to mitigation measures (modified after Newhall 2000)

Alert level	*Criteria*	*Interpretation*
No alert	*Background*; quiet	No eruption in foreseeable future
Level 1	*Low level* seismicity; other unrest	Magmatic, tetonic, or hydrothermal disturbance; no eruption imminent
Level 2 (May 13)	*Moderate level* of seismicity, other unrest, with positive evidence for involvement of magma	Probable magmatic intrusion; could eventually lead to an eruption
Level 3 (June 5)	*Relatively high and increasing unrest* including numerous b-type earthquakes, accelerating ground deformation, increased vigor of fumaroles, gas emissions	If trend of increasing unrest continues, eruption possible within 2 weeks
Level 4 (June 7)	*Intense unrest*, including harmonic tremor and (or) many "long-period" (low-frequency) earthquakes	Eruption possible within 24 hours
Level 5 (June 9)	*Eruption in progress*	Eruption in progress
	(Note that large-scale explosive eruptions began June 12; climactic eruption occurred on June 15)	

Mitigation measures as a function of alert level

10,000 evacuated on June 5	40,000 evacuated on June 10
25,000 evacuated on June 9 (20 km zone)	85,000 evacuated on June 14 (40 km zone)

transfer of the potentially affected population. This extension of the original concept of VAL has brought to some uncertainty on whether volcanologists should be the de facto decision makers or not, with an assumption of responsibility that may go beyond the expertise of the scientist and which, in most countries, is not granted by a corresponding societal mandate (Papale 2017). Despite this, VALs remain a useful communication tool to codify the state of a volcano during unrest, and their changes at volcanoes experiencing unrest have been also used to track the response of volcanologists during volcanic crises (Fig. 9.14; Winson et al. 2014). In particular, only 19% of the VALs issued between 1990 and 2013 for 194 events that ended with eruption accurately reflect the hazard before eruption. This increases to $\sim$30% considering eruptions with VEI $\geq$ 3, with VALs of eruptions from closed conduit volcanoes more appropriately issued than those from open conduits. Considering also the alerts for unrest without eruption (in the 30 days following the change in the VAL) the number of accurate VALs increases from 19 to 55%. These results show that the forecasting capability of the volcanological community is still far from being optimal, with a substantial number of "missed" (onset of eruption not preceded by increase in the VAL) and "too late" (VAL increased for the first time after the eruption began) change of alert levels, while the "premature" changes (VAL increased but subsequently decreased to lower levels prior to the onset of eruption changes) remain low. Both the "missed" and the "too late" cases imply a **missed alarm** for evacuation before eruption: missed alarms can easily lead to tragedy. "Premature" changes in VAL may be also problematic, as potentially leading to a **false alarm** for evacuation, which undermines trust in scientists and authorities. Quite surprisingly, the success rate for all alerts (with or without eruption) is only moderately improving over time, suggesting that the implementation of monitoring systems and our enhanced understanding of volcanoes have carried limited benefits so far. In facing a volcanic crisis, an additional useful tool is the cost–benefit analysis. This, based on the forecast of a certain event (made by volcanologists), allows decision makers to assess the costs and possible benefits of specific mitigating actions (e.g., Marzocchi and Woo 2007).

Eruption forecast is the capability to anticipate the occurrence of an eruption in order to provide the possibility to take appropriate mitigation measures, most commonly to evacuate the area potentially affected by the eruption. Eruption forecast is probably the ultimate challenge

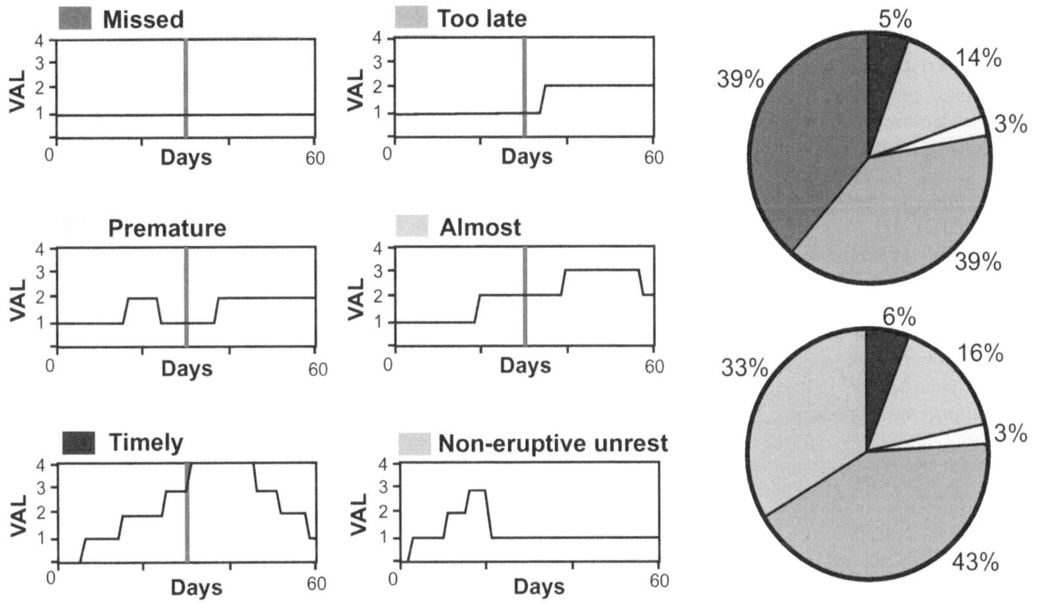

Fig. 9.14 Left diagrams: classification of the issued alert levels relative to eruptions for 194 events that have occurred between 1990 and 2013. The VALs are plotted against time and the red line denotes the onset of the eruption. Right diagrams: pie charts showing the proportion of the issued alert levels relative to eruptions in eruptive unrest episodes (top) and including non-eruptive unrest as a percentage of all cases except "missed" (bottom; modified after Winson et al. 2014)

for volcanologists, in which all the knowledge acquired on volcanic and magmatic processes is directed towards the benefit of society. Similarly important is also forecasting what happens after the eruption onset, for example if an eruption may consist of multiple phases defined by different styles of activity (e.g., effusive and/or explosive) and/or quiescent periods between them (Bebbington and Jenkins 2019). In the last decades there has been a variable preparedness and capability of the volcanological community to forecast impending eruptions. Some eruptions have been successfully forecast, while others not. This has led to successful evacuations of the potentially affected population, failed evacuations (no eruption occurred) or missed evacuations (eruptions have occurred before). As a result, volcanology has experienced successes and confidence, as well as disasters and frustrations, in forecasting eruptions.

Successful forecasts have been made at Mount St. Helens in 1980 and Pinatubo in 1991. Both stratovolcanoes passed from a centuries-long quiescence to unrest, characterized by the progressive increase in the intensity of the monitoring signals and occurrence of phreatic eruptions, leading to eruption within just 2–3 months. At Mount St. Helens the emplacement of a cryptodome within the edifice produced the bulging of ~ 100 m of the northern flank, which evolved in sector collapse triggering a lateral eruptive blast in May 1980. Several eruptions between June 1980 and December 1982 were also successfully forecast, mainly using seismicity and deformation data (see Sect. 6.8.2; Lipman and Mullineaux 1981; Swanson et al. 1983). At Pinatubo the magma intruded below the edifice, probably following a nearby *M7.8* regional earthquake. The gradual increase in activity during unrest (increased and shallowing seismicity, increased SO_2 emissions and phreatic eruptions) called for the evacuation of progressively wider areas until the day before the eruption climax (Fig. 9.15; Newhall and Punongbayan 1996). Despite the large size of the eruptions (VEI 5 for Mount St. Helens and 6 for Pinatubo), the successful forecasts and

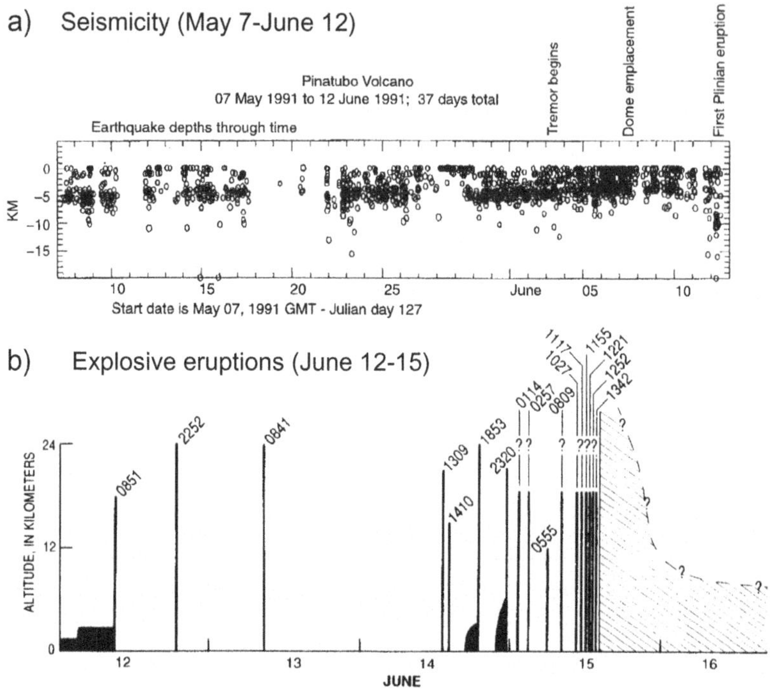

Fig. 9.15 Pre-eruptive and eruptive events associated with the 1991 Pinatubo (Philippines) eruption, the largest in decades. **a** Earthquake depths from May 7 to June 12, 1991; data gaps occurred on May 10–11, 17–19, and 21–22. **b** Chronology of explosive eruptions and related maximum altitude in June 12–15; four-digit number is local time of eruption onset; emission indicated by diagonal pattern is schematic (McNutt 2000)

evacuations limited the death toll to 57 victims at Mount St. Helens and 847 (mostly for measles in the evacuation camps) at the much more densely inhabited Pinatubo.

Volcanology has also witnessed disasters in recent times, as in 1985 at Nevado del Ruiz (Colombia), where unrest was characterized by the progressive increase in the seismicity, degassing and phreatic activity for nearly one year, culminating in a magmatic eruption in November. This eruption melted the ice on the volcano summit, producing lahars travelling at 60 km/hr within the valleys around the volcano, one of which hit the ~60 km distant village of Armero, causing ~23,000 victims. While the potential hazard deriving from lahars was foreseen by volcanologists months in advance, failure to take actions before and during the eruption in a wider context of improper management resulted in tragedy, despite the minor size of the VEI 3 eruption, emitting only 5×10^6 m^3 of magma (Barberi et al. 1990; Voight 1990).

In other situations, the response of scientists and authorities to a volcanic crisis may be considered premature, at least in retrospective. This is the case of La Soufriere volcano (Guadeloupe, Antilles) in 1976, where an explosion in July marked the onset of nine months of emission of 10^6 m^3 of non-juvenile tephra. An increase in the VT seismicity in August called for a contentious evacuation, not agreed on by all the scientists, of 74,000 people in less than 24 h. The evacuation ended in December, following the decrease in the intensity of the monitored signals. The 4-months long evacuation of 74,000 people made this crisis one of the most costly of twentieth century, although without loss of life, causing severe socio-economical difficulties on the island. Later understanding allowed defining the crisis as resulting from an aborted magmatic episode, with the unrest signals produced by the migration of magmatic gases into the hydrothermal system. These gases dissipated heat, inhibited magmatic eruption and promoted phreatic eruptions (Hincks et al. 2014).

Even an early alarm may cause difficulties to volcanologists, authorities and population; this is the case of the 2017 Agung (Bali, Indonesia) eruption, which was preceded by 5 months of unrest. An early increase in seismicity prompted the evacuation of 140,000 people (including 70,000 who spontaneously evacuated) for one month at 2 months before the onset of the eruption. The delay between intense unrest and eruption caused considerable challenges to emergency responders, local and national governmental agencies, and population near the volcano (Fig. 9.16; Syahbana et al. 2019).

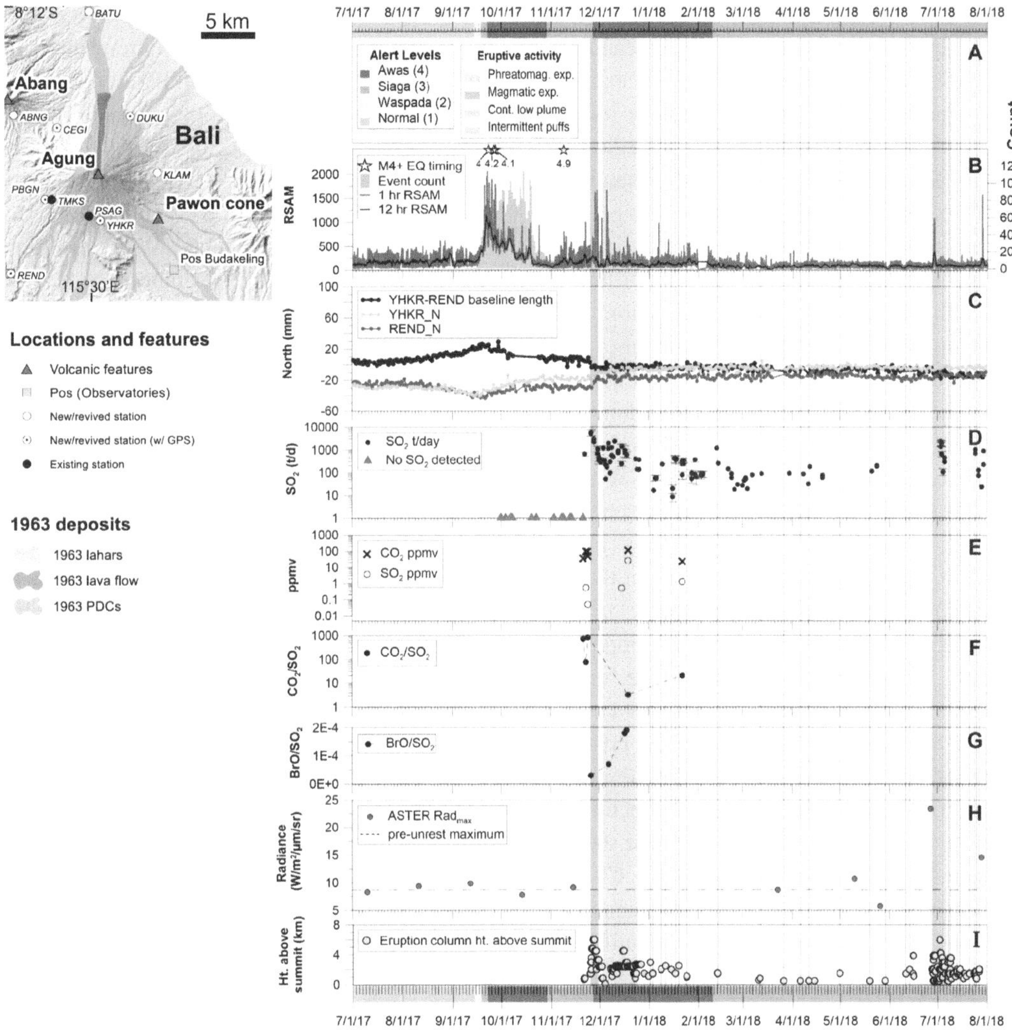

Fig. 9.16 Timeline of the 2017–2018 unrest and eruption at Mount Agung (Bali, Indonesia, shown in left inset). **a** Alert level changes. **b** Real-time Seismic-Amplitude Measurement (RSAM) from TMKS, and daily seismic event counts. $M \geq 4$ earthquakes displayed as stars. **c** GNSS displacements and baseline length between YHKR and REND. **d** SO_2 emission rates from ground-based mobile DOAS. **e** CO_2 and SO_2 mixing ratios from drone-transported Multi-GAS. **f** CO_2/SO_2 ratios (molar) from Multi-GAS. **g** BrO/SO_2 ratio from mobile DOAS. **h** Advanced spaceborne thermal emission and reflection Radiometer (ASTER) maximum radiance values from the crater, with pre-unrest maximum radiance plotted as dashed line. **l** Eruption column heights. Running across the entire graph are phreatomagmatic (blue) and magmatic (pink) explosions, periods of continuous ash venting (grey) and intermittent ash puffing (purple; modified after Syahbana et al. 2019)

These representative cases show how forecasting eruptions in the recent past has brought to very different outcomes, which may depend on several features. Among these are the different degree of knowledge and consciousness of the state of a volcano and its processes, as derived from geology and monitoring data, and the experience of the scientific community involved. The pre-eruptive behaviour of the volcano may also show very different features. For example, unrest characterized by the progressive increase of the monitoring signals is more "predictable" (Mount St. Helens in 1980 and Pinatubo in 1991). Conversely, unrest showing a significant decrease in the monitoring signals anticipating eruption leads to non-linear behaviours that are more difficult to interpret (Rabaul before 1994 and Agung in 2018). Also, in some cases, possible **precursors,** or specific anomalies that may anticipate an eruption within a certain time frame, may be recognized and used for forecast, while in other cases this is not possible. With this regard, it should be made clear that it is currently extremely difficult to identify eruptive precursors. This difficulty mainly results from the limited knowledge and measurement of the parameters potentially controlling an impending eruption, which may depend on the magma properties (as the viscosity, density, gas content, crystallization and excess pressure), the host rock properties (as the stress state around the magma chamber), and external processes affecting the response of a system in a nearly critical state (as regional seismicity and rainfall; e.g., Chiodini et al. 2016; Albright et al. 2019; Farquharson and Amelung 2020; Manga 2020). The identification of promising eruptive precursors has been limited to specific (largely open conduit) volcanoes with frequent and regular eruptive behaviour, as Mount St. Helens between 1980 and 1982 (Swanson et al. 1983) or Mount Etna. At Etna, an infrasound array has been detecting explosive eruptions in the last decade, generating an automated pre-alert with a 97% of success rate (Ripepe et al. 2018). While providing the first case of automatic operational early warning system, and finding potential application also to other volcanoes, this precursors based approach remains limited to open conduit volcanoes characterized by lava fountaining. Several recent approaches rely on the use of artificial intelligence to more accurately monitor and eventually forecast eruptions, from using algorithms to initially sort through the huge amount of available space monitoring data and then focus on the volcanoes of most interest, to developing systems able to detect potential signs of unrest automatically. While the current aim of these efforts is to process all available monitoring data of volcanoes, a longer-term perspective may allow issuing warnings of volcanic activity on societally relevant time frames, although eruption forecasts are not expected to be reliable as weather forecasts (Gaddes et al. 2018; Albino et al. 2019; Witze 2019; Palmer 2020; Poland and Anderson 2020).

Despite any rare case limited to favourable conditions, currently there is no unique and reliable eruptive precursor in volcanology, and the forecasting capability still largely results from the non-unique interpretation of multi-parametric data. As anticipated in Sect. 9.2, the most reliable indicators of an impending eruption in closed conduit volcanoes remain the multiparametric data allowing detection of a dike propagating from a zone of shallow magma accumulation towards the surface. Notwithstanding this possibility, as mentioned, there is still no reliable precursor to infer the size of an impending eruption. Phreatic eruptions are possibly even harder to forecast, even if the volcano is thoroughly monitored, as not being anticipated by dike propagation and with any unrest phase showing extremely variable duration, from minutes to years, with subtle variations in the monitoring parameters (e.g., Kato et al. 2015; Caudron et al. 2019).

In forecasting eruptions it is essential to refer to the **time window** of the forecast. This is mainly determined by the time span and resolution of the available data, as a forecast can look no farther forward than the time span of data on which it is based, nor can it have any greater resolution than that of the data on which it is based. A geologic record of millennia and resolution of centuries applies to coming centuries

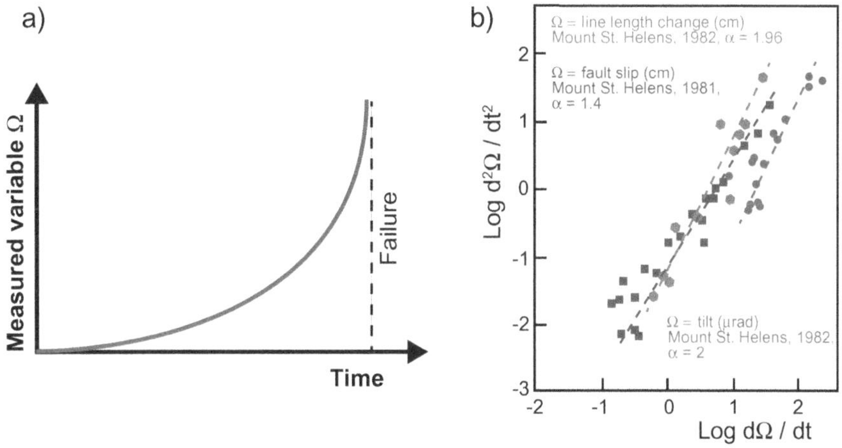

Fig. 9.17 **a** Principle of the Failure Forecast Method, where a measured variable Ω increases approaching failure: in this case an exponential increase is shown. **b** Application to erupting volcanoes: the diagram shows the log–log relationship between the acceleration of the monitored variable Ω and its rate for three eruptions at Mount St. Helens, in which Ω consists of different deformation parameters (modified after Voight 1988)

and millennia, and a record of last week's monitoring applies to next week. Therefore, forecasting eruptions at quiescent volcanoes relies on the knowledge of the past eruptive frequency and size of the volcano, requiring a longer-term time window, from years to centuries. Conversely, forecasting eruptions during ongoing unrest relies on the knowledge of the recent variation of the monitoring parameters, requiring a shorter time window, from days to months. Eruption forecast has moved from a deterministic (focused on the short-term) towards a probabilistic (concerning both the long- and short-term) approach, both illustrated below.

9.6 Deterministic Forecasting

Deterministic eruption forecasting is based on the notion that an eruption can be satisfactorily predicted in time and space. The deterministic approach was popular at the end of the last century, following the successful management of the 1980 Mount St. Helens and 1991 Pinatubo eruptions. The progression of the activity observed in both pre-eruptive periods highlighted a linear behaviour, giving volcanologists the confidence that eruptions could be predicted ahead of a few days-weeks. The deterministic approach considers that, in analogy to what observed during the deformation of a specimen in a laboratory test, the volcanic system undergoes progressive mechanical failure until reaching rupture, that is when eruption occurs (Fig. 9.17a). This progressive failure may be recognized through monitoring data mainly for short-term predictions, although decade-long time-series of monitoring data may be also used.

The theoretical rationale of this approach, known as the **Failure Forecast Method** (FFM), is summarized by the equation which links the acceleration of any physical (or even chemical) measured variable Ω to its rate, as:

$$\frac{d^2\Omega}{dt^2} = K\left(\frac{d\Omega}{dt}\right)^{\alpha} \tag{9.3}$$

where, K and α are constants and t is the time (Voight 1988; Voight and Cornelius 1991). In particular, α (with $1 \leq \alpha \leq 2$) controls the type of acceleration involved: $\alpha = 1$ corresponds to an exponential increase, whereas $\alpha = 2$ to a hyperbolic increase. As the rate of the variable increases, the acceleration becomes larger and increases the rate more quickly than before. The deterministic forecasting potential of this relation depends on estimating the time at which the rate tends to infinity, that is the failure, or eruption. In the general case that $\alpha \neq 1$, solutions to Eq. (9.3) take the form of a power law increase in the rate of the monitored variable with time:

$$\frac{d\Omega}{dt} = j(t_f - t)^{-p} \quad (9.4)$$

where t_f is the time of failure, j is a multiplicative amplitude term and $p = 1/(\alpha - 1)$ is a power law exponent. Values of j and α may be empirically estimated from a plot, such as that in Fig. 9.17b. The accuracy of this approach is determined by the precision and frequency of observations and the regularity and continuity of the observed process. Equation (9.4) can be solved by linearizing the problem in the form:

$$\left(\frac{d\Omega}{dt}\right)^{\frac{-1}{p}} = j^{\frac{-1}{p}}(t_f - t) \quad (9.5)$$

and using standard least squares regression to determine the failure time (Voight 1988). Commonly $p \sim 1$ (Kilburn 2003), in which case the solution is a straightforward regression of inverse rate against time. The time of failure can be best ascertained graphically by extrapolation of the concave curve $(d\Omega/dt)^{-1}$ versus time to a predetermined intercept. In fact, forecast accuracy is aided by inverse representation, as large differences and trends in such sequences may be recognized earlier (Voight 1988).

While having raised hopes of making deterministic and objective forecasts of eruptions and having provided clues to developing a formal physical model, this simple approach presents several limitations. First, the method is not quantitatively linked to a physical process, as it reveals preferred types of acceleration in whatever is the controlling process (Eq. 9.3). Second, the best-fit trends for individual time series of FFM are not unique and statistical fits can yield ambiguous results. Also, since a week or more is required to identify an accelerating trend, any forecast is unlikely to be reliable more than a few days at most in advance (Kilburn and Sammonds 2005). In addition, closed conduit volcanoes and rock physics experiments show that rates of fracturing can accelerate while the deformation rate remains constant. Under such conditions, Eq. (9.3) may accommodate a seismicity rate, but not the contemporaneous rate of deformation, so describing only part of the precursory processes that lead to eruption (Kilburn 2018).

These limitations have brought to the modification of the FFM or the formulation of other deterministic approaches (Bell et al. 2011; Robertson and Kilburn 2016; Kilburn 2018). Among these, a modified physical model (known as the **parent model**) shows that precursory time series are governed by a parent relation between seismicity and deformation. This relation postulates that, in the failure of a material, the applied differential stress increases from zero and the deformation evolves from the elastic, through quasi-elastic to inelastic regimes (Fig. 9.18; Kilburn et al. 2017; Kilburn 2018). The start of the inelastic regime coincides with deformation under a constant maintained stress. Here the deviation from elastic behaviour is caused by faulting, and faulting may be expressed by the cumulative number of volcano-tectonic (VT) seismic event. The number of VT events increases exponentially with deformation in the quasi-elastic regime, but linearly with deformation in the inelastic regime. This theoretical model has been applied to pre-eruptive seismicity-deformation sequences from different volcanoes, where the monitoring data confirmed the expected evolution from quasi-elastic to inelastic behaviour, with eruption appearing within the inelastic domain. This occurs independent of the type of volcano, the duration of precursor, magma composition and style of eruption. Therefore, the parent model may provide a tool to predict eruption based only on the seismicity and deformation patterns. This model also underlines the importance of considering the cumulative, even if non-linear, unrest history at a certain volcano, rather than the isolated unrest events, as highlighted in Sect. 8.3.3 comparing the recent unrest episodes at Rabaul and Campi Flegrei. In particular, the cumulative seismicity-deformation from consecutive unrest episodes indicates that the inelastic threshold was reached at Rabaul just before the eruption. Conversely, at Campi Flegrei, experiencing a partial recovery of the deformation after the major uplift of 1982–1984,

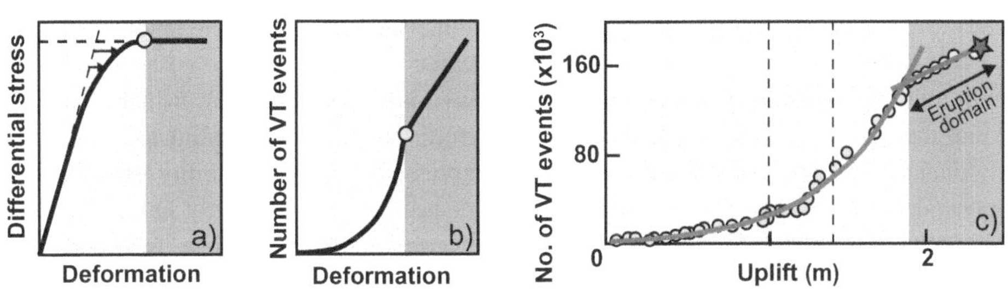

Fig. 9.18 Theory of the parent model illustrating the relation between elastic deformation and faulting (modified after Kilburn et al. 2017). **a** In the failure of a material, the applied differential stress increases from zero and the deformation deviates from the elastic (white), through quasi-elastic (yellow) to inelastic (purple) regimes; blue dot shows onset of inelastic regime under a constant maintained stress. **b** The deviation from elastic behaviour is caused by faulting, here expressed through the cumulative number of volcano-tectonic (VT) earthquakes. The number of VT events increases exponentially with deformation in the quasi-elastic regime, but linearly with deformation in the inelastic regime, where eruption is expected. **c** The evolution from quasi-elastic to inelastic deformation was observed (blue curve and line fitting data from grey dots) during the 23-year unrest before the 1994 eruption at Rabaul caldera (Papua New Guinea); red square denotes eruption, black dashed lines refer to the 1983–1985 episode of accelerated uplift

the inelastic threshold has not been achieved yet (Kilburn et al. 2017). These results also suggest that the FFM is a particular form of the parent model when rates of stress supply are constant. The outcome of the parent method yields deterministic criteria that can be incorporated into existing operational procedures for evaluating the probability of crustal failure at closed volcanoes. This model, however, is still an incomplete description of pre-eruptive conditions at volcanoes. It identifies conditions for bulk failure in the crust, which, although necessary to open a new pathway for magma ascent, do not guarantee that magma reaches the surface. Also, magma may reach the surface soon after the inelastic behaviour is reached (as at Kilauea, Hawaii, in 1971–1972) or with some further delay (as at El Hierro in 2011), thus adding uncertainty in the prediction. The parent model thus provides a promising starting platform for identifying additional precursory trends and pre-eruptive criteria through a deterministic approach (Kilburn 2018).

9.7 Probabilistic Forecasting

While some unrest episodes show a progressive increase in the intensity of the monitoring signals, culminating in eruption, as at Pinatubo in 1991, many other unrest episodes show more complex and **non-linear** behaviour, in which the variation of the output is not proportional to the variation of the input (Tilling 1988). In general, this behaviour results from the fact that material failure is typically non-linear and volcanoes are highly complex systems hosting a wide range of kinetic and dynamic magmatic processes, governed by many degrees of freedom at the same time, and leading to a wide range of potential behaviours intrinsically unpredictable. This condition is complicated by the limited access to and capability of understanding and quantifying these processes, which are also accompanied by intrinsic uncertainties that may be quantified and reduced, but cannot be completely removed. In fact, natural processes usually own a scientific uncertainty (**epistemic uncertainty**), due to the availability of limited data and limited understanding, which may be in principle reduced, and an irreducible unpredictability intrinsic in the system (**aleatory uncertainty**), which cannot be eliminated (Marzocchi and Bebbington 2012). Due to the uncertainties and the complexity of non-linear systems, a deterministic prediction of eruptions is, in most cases, not achievable. Rather, in the last two decades there has been a shift towards probabilistic forecast of eruptions that takes also into account for uncertainties (e.g., Sparks 2003; Sparks and Aspinall 2004). In forecasting volcanic hazards and assessing risks, one needs to estimate the probability that a hazardous event will happen, the probability that the event will affect a particular place and the probability that the effects will include fatalities and property damage.

In general, there are two main classes of probability: the frequentist (objective probability) and the degree of belief (subjective probability). The **frequentist** approach views probability as the long-run expected frequency of occurrence, with the probability P of occurrence of an event A being:

$$P(A) = n/N \tag{9.6}$$

where n is the number of times the event A occurs in N opportunities. An estimate of objective probability can be obtained through a stochastic model or empirical analysis of frequencies in datasets. **Stochastic** models possess some inherent randomness (the same set of parameter values and initial conditions will lead to an ensemble of different outputs) and are widely used in long-term eruption forecast, where the primary information comes from historical and geological catalogues of eruption onsets and, in some cases, sizes. Stochastic models can also take into account for a **multivariate analysis**, which involves observation and analysis of more than one statistical variable at a time (Jaquet et al. 2006; Bebbington 2009; Marzocchi and Bebbington 2012). In this context, eruptions are considered as the outcome of stochastic point processes. This approach simply calculates the frequency of past events, assuming that this frequency is an estimator of the true, and unknown, probability for the future events. This analysis implies the absence of a reference model, and faithfully reproduces the random variation from the catalogue in forecasts. Unless the catalogue is a large one, which is unlikely in volcanology, the resulting forecasts tend not to be smooth and can be biased (Marzocchi and Bebbington 2012).

The **degree of belief** is a level of confidence or credence that expresses in statistical terms how much a person, or a group, believes that a proposition is true: its use is commonly required in the case of a limited amount of data. The degree of belief is at the base of subjectivist, or **Bayesian,** probability in which, rather than a "true" probability to estimate, any quantity is expressed by a probability distribution. This perspective can be directly linked to the aleatory uncertainty (described by a "best-guess" value of the distribution, such as the mean or the median) and epistemic uncertainty (described by the dispersion around the best-guess in the probability distribution). The probabilities are combined through Bayes' theorem, which is a formula that describes how to update the probabilities of hypotheses when given evidence, following from the axioms of conditional probability, that is the probability of one thing being true given that another thing is true. Given a hypothesis H and evidence E, Bayes' theorem states that the relationship between the probability of the hypothesis before getting the evidence $P(H)$ and the probability of the hypothesis after getting the evidence $P(H|E)$ is:

$$P(H|E) = \frac{P(E|H)}{P(E)} P(H) \tag{9.7}$$

where the quotient $P(E|H)/P(E)$ represents the support E provides for H. Note that the Bayes' theorem holds also if the probabilities are represented by probability density functions, as postulated in the Bayesian approach. Here the probability is no longer an expected frequency and rather represents the degree of belief, or measure of plausibility, about the occurrence of the next event given incomplete knowledge. This may well be the case of an unrest episode at a volcano. Subjective probability is estimated in a different way from the frequentist probability. The best procedure is through the formalization of the degree of belief of a group of scientists (intersubjective probability), which tends to be much more coherent than that of a single researcher and also evaluates the epistemic uncertainty from multiple perspectives, increasing the likelihood of considering a fuller range of information. The most accepted procedure of eliciting a degree of belief is the Delphi method, which relies on a structured panel of experts where information is fed back in summary form, allowing the panel to discuss and revise assessments several times, with opinions usually kept anonymous. Experts' opinions may be weighted according to the experience of the participants (Aspinall 2006; Marzocchi and Bebbington 2012).

Therefore, objective and subjective probabilities correspond to two different approaches to estimate uncertainty, which are expected to converge towards the same results when many data are available. The latter condition is often not met, hindering the use of the frequentist approach and hence requiring the use of Bayesian probability.

Probabilistic forecast considers the possibility of occurrence of an eruption both on the long- and short-term. The shift from long- to short-term forecast is determined by whether or not there are anomalous monitoring observations. Without anomalous monitoring data, probabilistic forecasts are made on the basis of the past eruption history of the volcano. Both time windows are discussed below.

9.7.1 Long-Term Forecasting

Long-term eruption forecast commonly relies on a frequentist approach that is concerned with the quality and quantity of the available data on the past eruptions at a volcano. More data on the past eruptions do not necessarily lead to a better forecast, as inhomogeneity in the catalogue complicates the approach, and different eruption catalogues can lead to different eruption forecasts. The used data are historical and geological. Historical records are usually short, lasting no longer than a few centuries and often incomplete, particularly in the earlier part of the record, for small-scale eruptions or for unrest episodes (which are unrecorded in geological data), in case they failed to erupt. Geological data can go back to tens of thousands of years, although the geologic record also incompletely preserves evidence of smaller eruptions and burial of older deposits is common.

A first-order requisite in determining the type of long-term forecast model to be adopted is dictated by any evidence that the history of the volcano has been characterized by activity at statistically different levels (or rates) in different intervals. If this condition is not met, a **stationary** model (whose parameters do not change with time) may be used. Otherwise, a **non-stationary** model, which involves trend(s) or level changes in parameters and hence in activity, may be preferred (Marzocchi and Bebbington 2012). Stationary models constitute a sort of maximum ignorance alternative, as lacking any information about the temporal evolution of the activity. Moreover, as incorporating an incorrect non-stationary model leads to greater bias, stationary models are also more robust. The **Poisson process** is an example of stationary model that assumes that the distribution of the remainder of the repose length is independent of the elapsed repose length. This is a memoryless property and implies that nothing is known about the temporal structure of the process, except the mean return period. The Poisson process is a good model for describing phenomena where the probability of occurrence is small and constant. In the Poisson process with n events in τ unit time intervals, the parameter $\lambda = n/\tau$ is the expected average rate of events, i.e., the reciprocal of the average return period. For a Poisson distribution, the probability of observing k events (for example eruptions) in a time interval t is given by:

$$P(k, \lambda) = \frac{\lambda^k e^{-\lambda}}{k!} \tag{9.8}$$

The model also indicates that the probability of an event occurring at time T within a time interval δt is:

$$P(t < T < t + \delta t) = 1 - e^{-\lambda \delta t} \tag{9.9}$$

providing a useful approach to define the probability of an eruption in a given time.

A basic principle of the Poisson process is that events occur independently. However, in several cases a dependency between events can be postulated: for example, one can assume that the previous eruption influences the timing of the next eruption onset. In this case, taking into account for the size of the previous event leads to a **time-predictable model**, which assumes a constant rate of magma input, so that eruption occurs when a critical level of magma is reached (Burt et al. 1994). Thus, the repose length depends on the volume of only the most recent eruption. An example of time-predictable behaviour is shown by the recent eruptive history of the Axial Seamount caldera (Juan de Fuca Ridge; Fig. 8.17; Nooner and Chadwick 2016). Here, given a critical threshold (the level of inflation) for eruptions, one can evaluate the time needed to reach again the threshold. With a size-dependent model the possible size of future events is conditioned on the prior eruptive history. In the case of a **size-predictable behaviour**, given a general volumetric threshold, one can consider the elapsed time from the last eruption and estimate the expected size of the next eruption as a function of time on a cumulative curve, as for example illustrated for the recent eruptive history of Akan volcano, Japan (Fig. 9.19; Hasegawa and Nakagawa 2016).

Long-term eruption forecast usually focuses on the eruptive record of a single volcano. However, in some cases, regional or global estimates are made, including the probability to have eruption as a function of its VEI for different exposure times (Pyle 1998; Martin et al. 2004; Mason et al. 2004; Sandri et al. 2004; Papale and Marzocchi 2019). The global probabilities are based on the frequency of the eruptions as a function of the VEI, which follows an overall power law behaviour, with many smaller-size events and fewer larger-size events (Fig. 1.14b). In this frame, the largest explosive eruptions have a longer repose time (or time elapsed from the previous eruption) than moderate eruptions. This implies that large explosive eruptions require a longer time to recharge the magmatic system and accumulate a sufficiently large amount of gas. Interestingly, the global probability to have a VEI 8 eruption, which can be considered the most destructive natural event of our planet, in our lifetime may be orders of magnitude larger than other events considered relatively remote in our everyday life (Fig. 9.20).

These global long-term statistical approaches also reveal different behaviours for volcanoes with open conduit compared to those with closed conduit. Open conduit systems seem to follow a time-predictable model, with a marked time clustering of events, whereas closed conduit systems seem to have no significant tendency

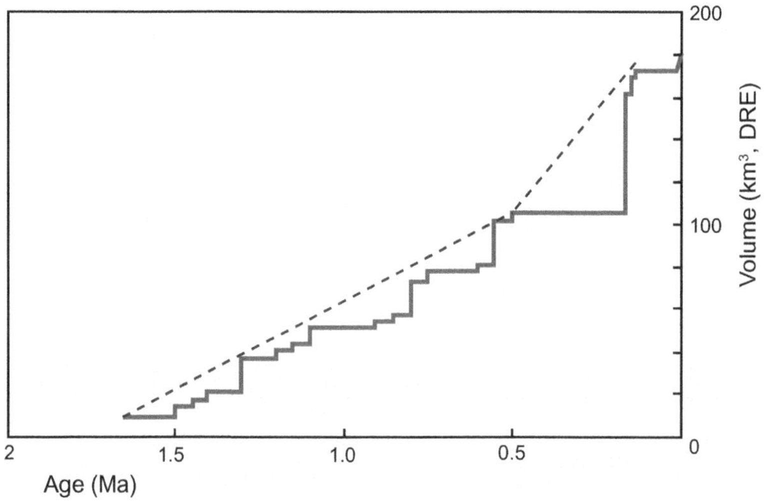

Fig. 9.19 Age versus cumulative volume of eruptive groups and post-caldera explosive eruptions of Akan volcano (Japan). Dashed line is the volume-predictable line (modified after Hasegawa and Nakagawa 2016). In this case, the line shows a variation through time

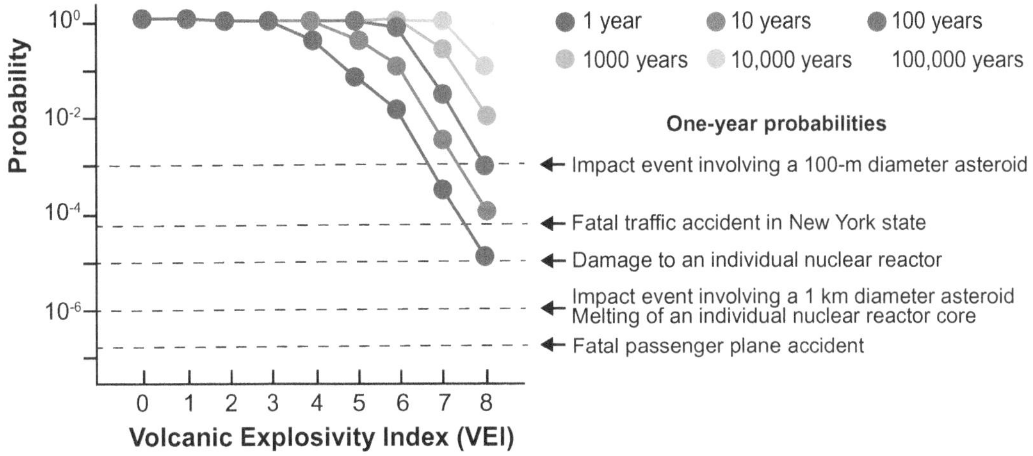

Fig. 9.20 The probability of at least one eruption of a given VEI size on Earth is shown over different time intervals from 1 to 100,000 years. The 1-year probabilities of these events are compared to those for other threats (modified after Papale and Marzocchi 2019)

toward a size- or a time-predictable model, and the eruptions mostly follow a Poisson distribution (Marzocchi and Zaccarelli 2006). However, these global-scale results may be only partly consistent with evidence at regional scale. For example, while Indonesian open conduit volcanoes show no evidence of size-predictable behaviour, closed conduit volcanoes show a significant probability (>0.999) that the VEI of the next eruption increases with increasing repose length, supporting size-predictability (Bebbington 2014). Despite any regional variation, these results are useful in building general probabilistic models for volcanic hazard assessment of open and closed conduit systems. Magma composition is a further feature statistically affecting repose and unrest times. In particular, high silica, and thus high viscosity, felsic systems have longer repose and duration of precursory activity and tend to erupt larger quantities of material (Passarelli and Brodsky 2005).

A limitation of long-term eruption forecast is that this is commonly achieved only considering the reported (description-based) eruptive history of a volcano, rather than understanding the possible processes (physics-based) responsible for the eruptive record. This limitation, while not introducing any bias, may significantly affect forecast estimates, not only in time, but also in space, concerning the forecast of the location of the possible eruptive vent (Connor et al. 2003). This latter possibility holds especially true for volcanoes without central conduit, as calderas, where the location of a future vent is more uncertain. Nevertheless, new approaches are considering the mechanical conditions controlling the shallow transfer of magma below calderas, allowing moving from empirical pattern recognition to models based on the physics behind the patterns to define the location of future vents (see Sect. 5.9; Rivalta et al. 2019).

9.7.2 Short-Term Forecasting

For obvious reasons, most of the attention in forecasting eruptions has been given to the short-term forecast, which is based on the monitoring data collected during unrest. In fact, during unrest any long-term probability based on the past frequency of eruptions becomes less important than what is being observed from the monitoring signals.

In the case of frequently erupting volcanoes (that is, with open conduit) and available monitoring record, a frequentist approach may be used (Marzocchi and Bebbington 2012). For example, the probability $P(T)$ of occurrence of eruption based on the appearance of a certain monitoring precursor (or hit rate) is:

$$P(T) = n_p/N \quad (9.10)$$

where n_p is the number of eruptions anticipated by the precursor in an arbitrary time window t and N is the total number of eruptions. The probability $P(F)$ to have a false alarm using the same precursor (false alarm rate) is given by:

$$P(F) = (m - n_p)/m \quad (9.11)$$

where m is the number of times in which the precursor has been observed. Retrospective testing for this approach has shown a large variability in the results, with $P(T)$ from 60 to 100%, and $P(F)$ from 20 to 60%.

However, most volcanoes, including the high-risk ones, have closed conduit, and, not having erupted in the last decades, have hindered the creation of the eruptive unrest database required by the frequentist approach. Vesuvio, which last erupted in 1944 and has been in a continuous state of quiescence ever since, is an example of volcano with this behaviour. As a result of the lack of eruptive unrest databases, short-term eruption forecast at closed conduit volcanoes has to make use of unavoidable expert opinion and relies on a Bayesian approach based on current monitoring data, or on data from analogue volcanoes (i.e., volcanoes with similar behaviour; e.g., Marzocchi et al. 2004b). In this context, the monitoring anomalies acquire a status of eruptive precursors. While precursors in science are often interpreted in a deterministic (certain) sense, in volcanology, because of the general lack of a one-to-one correlation between precursors and eruptions, the possible observation of precursors should be more realistically translated into an increased probability of eruption (Marzocchi and Bebbington 2012). As anticipated, in volcanoes with closed conduit the most reliable indicators of impending eruption remain the multi-parametric data indicating shallow dike propagation.

Converting the detection of possible precursors into the probability of eruption is challenging (e.g., Anderson and Poland 2016). The scarce amount of past data and the limited knowledge of the pre-eruptive physical processes hinder solving the hazard/risk problem with a rigorous and testable scientific model. Nevertheless, the structured solving approach required by the risk associated with eruption may be supported by treating scientific uncertainty in a fully systematized manner, using Bayesian statistics. This is achieved driving the opinion of experts in a formal and transparent probabilistic procedure with quantitative decision-making protocols, which are prepared before a crisis and consist of quantitative rules that can justify each step of the decision-making process (Marzocchi et al. 2004b; Marzocchi and Bebbington 2012). This procedure is based on the evidence that, when scientists asked to make a forecast limit their estimates to those that they think are really likely, the final results are usually within a single order of magnitude.

Examples of these structured approaches translating the observation of one or more possible precursors into a probabilistic assessment using expert opinion are elicitations, Bayesian belief networks and Bayesian event trees.

Elicitations are techniques to gather information from scientists who give their opinion on a specific topic. In volcanology this usually regards the state of a volcano during unrest, the probability to have eruption and the identification of thresholds, usually as orders of magnitude, in the monitoring parameters to define specific behaviours, to be also included in Bayesian event trees (Cooke et al. 1991; Aspinall 2006; Selva et al. 2012). Elicitations may be run in meetings or online, and are usually preceded by an informative meeting and followed by a discussion meeting.

Bayesian belief networks and event trees have a similar structure, representing a general quantitative framework where all relevant monitoring observations are embedded into a probabilistic scheme through expert opinion, conceptual models and any monitoring data. In particular, a **Bayesian Belief Network** is a graphical representation of relevant observations (nodes) and their causal links (Fig. 9.21; Aspinall et al. 2003, 2006; Hincks et al. 2014). Associated with each node is a set of conditional probabilities that describe the relationship between the states of the variable at the node with the states of the other variables at connected nodes. These conditional probabilities are then combined through the Bayes' theorem to get the probability of any specific event. Bayesian Belief Networks are also used for retrospective analysis, as for the explosion at Galeras (Colombia) volcano in 1993 and for the 1975–1977 unrest at La Soufriere, in the latter case demonstrating that a formal evidential frame could have supported the authorities' concerns about public safety and decision to evacuate (Hincks et al. 2014).

A **Bayesian Event Trees** (BET) is a framework for discussing probabilities of possible outcomes of unrest, with each branch of the tree leading from a necessary prior event to a more specific outcome (e.g., Newhall and Hoblitt 2002). In particular, a BET is a graphical, tree-like representation of events in which branches are logical steps from a general prior event through increasingly specific subsequent events (intermediate outcomes) to final outcomes. These events, or nodes, include the inferred nature and outcome of unrest, and the inferred size and type of the possible eruption. Due to their Bayesian nature, at each node the probability density functions for the successive nodes are quantified, and aleatoric and epistemic uncertainties acknowledged and, if possible, quantified (Fig. 9.22). Therefore, a BET does not rule out any possibility, but it shapes the probability distribution of the event considered around the most likely outcome, accounting for all available data. Indeed, a BET is structured to merge any kind of information, including theoretical models of the eruptive process, past (historical and geological) and monitoring data (Marzocchi et al., 2004b,2008). Bayesian Event Trees have been used for several decades at many volcanoes, including Mount St. Helens, Pinatubo, Soufrière Hills (Montserrat), Popocatépetl (Mexico), Sinabung (Indonesia), Guagua Pichincha and Tungurahua (Ecuador; Newhall and Hoblitt 2002; Wright et al. 2019).

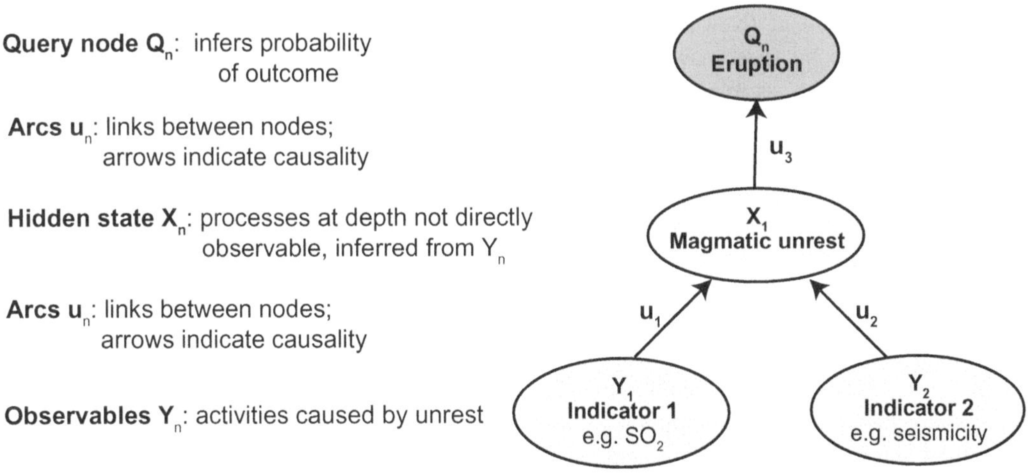

Fig. 9.21 Elementary Bayesian Belief Network (BNN) representing a simple volcanological model to infer the probability of eruption resulting from unrest. Unrest must be inferred from observables Y (e.g., seismicity and SO_2 emission). The query node (eruption) is the outcome of interest. Y = set of observables, X = set of unobserved states, u = set of directed links between nodes. Arrows indicate direction of causality or influence. For the simple case where nodes are assigned discrete states, node relationships are described by conditional probability tables (modified after Hincks et al. 2014)

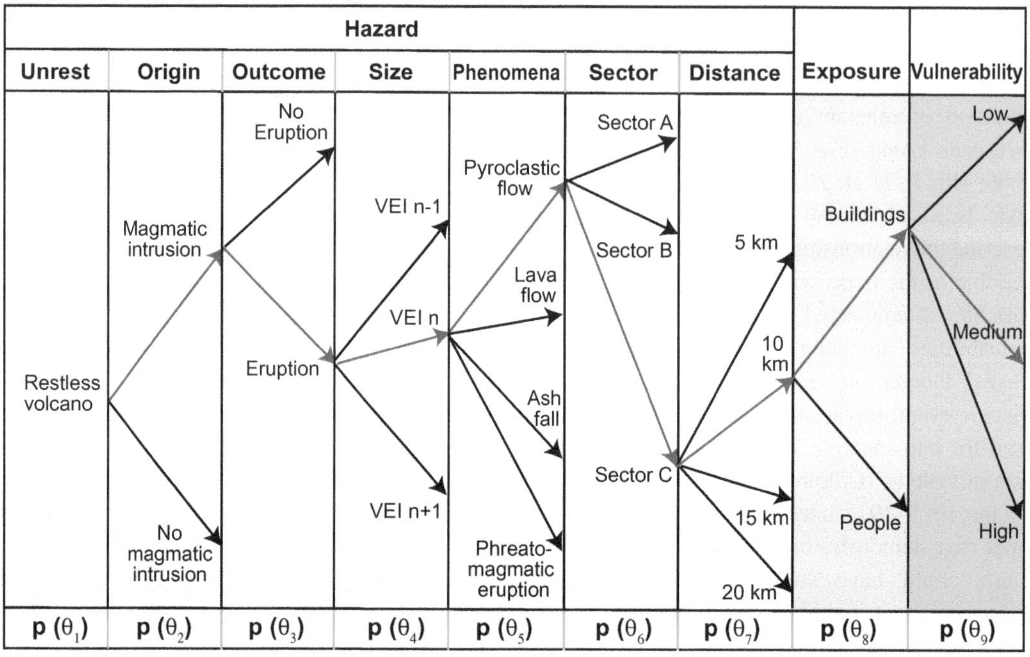

Fig. 9.22 Sketch of a Bayesian Event Tree (BET). The nine steps progress from general to more specific events, involving the hazard, exposure and vulnerability components. The selected path (red arrows) is characterized by a probability $p(\theta_n)$ at each step (modified after Marzocchi et al. 2004b)

Additionally, BETs have been used for real-time eruption forecast of quiescent volcanic areas without historical activity, for which otherwise limited prior information is available, as the Auckland Volcanic Field (New Zealand; Lindsay et al. 2010). Event trees may be also used for long-term forecast, as for example at the Teide-Pico Vejo volcanoes of Tenerife (Canary Islands): here the BET was based on the geological, historical, geophysical and volcanological knowledge, with the possibility of including short-term monitoring data (Martì et al. 2008). Finally, BETs have been applied in real-time during simulation exercises (as in New Zealand and Dominica; Lindsay et al. 2010; Sandri et al. 2017b) or have been even applied retrospectively, as for the 1631 Vesuvio eruption and the 1982–1984 unrest at Campi Flegrei (Sandri et al. 2009; Selva et al. 2012). Application to Vesuvio suggests that the pre-eruptive signals that significantly increased the probability of the 1631 eruption could have been likely detected more than 2 months in advance, providing satisfactory forecast. At Campi Flegrei the probabilities associated with the first three nodes of the BET have been estimated for the 1982–1984 unrest, when the appearance of surface fractures in 1983 increased the eruptive probability to ~40% as best-guess value, although affected by a large uncertainty (the 80% confidence interval was 0.3–91%).

9.8 Summary

Unrest is defined by evident geodetic, geophysical and geochemical variations in the monitoring signals, which may anticipate a possible eruption. Unrest commonly results from shallow magma accumulation or transfer, although a remote seismic trigger, responsible for dynamic and static stress variations, or the pressurization of a hydrothermal system may also promote unrest. Defining the processes occurring during unrest is fundamental to understand volcanoes and may also allow effective eruption forecast.

However, the limited knowledge of unrest processes and the lack of real precursors confine any forecast to the identification of the states of an active volcano and the statistical analysis of the monitoring parameters. In the first case, the monitoring parameters may indicate a state of magma accumulation or magma transfer. Magma transfer is related to shallow dike propagation and accompanied by distinctive monitoring signals, as deformation with non-radial symmetry and possible upward propagating seismicity. Because shallow dike propagation may feed eruption, its real-time detection should be the ultimate target of any monitoring system. In the second case, the statistical analysis of unrest monitoring data feeds short-term eruption forecast models. These have progressed from a deterministic to a dominantly probabilistic approach as a result of the overall lack of reliable and ubiquitous precursors. Multi-parametric monitoring data are currently incorporated, especially at closed conduit volcanoes, in several Bayesian forecasting tools, as elicitations, Bayesian Belief Networks and Bayesian Event Trees. Short-term forecast may also exploit a frequentist probabilistic approach at open conduit volcanoes, if monitoring data on previous eruptive unrest are available. The frequentist approach is otherwise widely used for long-term forecast, relying on historical or geologic data. In addition to forecasting eruptions, volcanologists take other actions to assess volcanic hazard to mitigate risk. These include the definition of hazardous scenarios (including the reference eruption), the establishment of early warning systems and the issuance of volcano alert levels.

9.9 Main Symbols Used

A	event
s	distance
E	evidence
H	hypothesis
j	amplitude term
K	constant
k	number of events
M	magnitude
m	number of times in which the precursor has been observed
N	number of opportunities or eruptions
n	number of times an event occurs
n_p	number of eruptions anticipated by the precursor
p	power law exponent
P	probability
$P(F)$	probability of false alarm rate
$P(T)$	probability of hit rate
P_{EQ}	extra pressure generated by the earthquake
t	time
t_f	time of failure
T	time of occurrence
T_{EQ}	time between large earthquakes
T_v	ordinary recurrence time of volcanic eruptions
X_t	fraction of eruptions triggered by earthquakes
α	constant
ΔC_{FS}	Coulomb failure stress
ΔP	pore pressure change
ΔP_{m}	overpressure required to nucleate a dike
$\Delta\sigma_N$	normal stress change
$\Delta\sigma_S$	shear stress change
λ	average rate of events
μ	coefficient of friction
τ	unit time intervals
Ω	monitoring variable

References

Acocella V, Behncke B, Neri M, D'Amico S (2003) Link between major flank slip and 2002–2003 eruption at Mt. Etna (Italy). Geophys Res Lett 30:2286. https://doi.org/10.1029/2003GL018642

Acocella V (2014) Great challenges in volcanology: how does the volcano factory work? Front Volcanol 2:1–10. https://doi.org/10.3389/feart.2014.00004

Acocella V, Di Lorenzo R, Newhall C, Scandone R (2015) An overview of recent (1988 to 2014) caldera unrest: knowledge and perspectives. Rev Geophys 53: https://doi.org/https://doi.org/10.1002/2015RG000492

Albino F, Biggs J, Syahbana DK (2019) Dyke intrusion between neighbouring arc volcanoes responsible for 2017 pre-eruptive seismic swarm at Agung. Nat Commun 10:748. https://doi.org/10.1038/s41467-019-08564-9

Albright JA, Gregg PM, Lu Z, Freymueller JT (2019) Hindcasting magma reservoir stability preceding the 2008 eruption of Okmok, Alaska. Geophys Res Lett 46:8801–8808

Almendros J, Carmona E, Jimenez V, Diaz-Moreno A, Lorenzo F (2018) Volcano-tectonic activity at Deception Island volcano following a seismic swarm in the Bransfield Rift (2014–2015). Geophys Res Lett 45:4788–4798

Anderson KR, Poland MP (2016) Bayesian estimation of magma supply, storage, and eruption rates using a multiphysical volcano model: Kılauea Volcano, 2000–2012. Earth Planet Sci Lett 447:161–171

Arzilli F, Morgavi D, Petrelli M, Polacci M, Burton M, Di Genova D (2019) The unexpected explosive sub-Plinian eruption of Calbuco volcano (22–23 April 2015; southern Chile): triggering mechanism implications. J Volcanol Geoth Res 378:35–50

Aspinall WP, Woo G, Voight B, Baxter PJ (2003) Evidence-based volcanology: application to eruption crises. J Volcanol Geoth Res 128:273–285

Aspinall W (2006) Structured elicitation of expert judgment for probabilistic hazard and risk assessment in volcanic eruptions. In: Mader H, Coles S, Connor C, Connor L (eds) Statistics in volcanology. Geological Society of London on behalf of IAVCEI London, pp 15–30

Aspinall WP, Carniel R, Jaquet O, Woo G, Hincks T (2006) Using hidden multi-state Markov models with multi-parameter volcanic data to provide empirical evidence for alert level decision-support. J Volcanol Geoth Res 153:112–124

Barberi F, Martini M, Rosi M (1990) Nevado del Ruiz volcano (Colombia): pre-eruption observations and the November 13, 1985 catastrophic event. J Volcanol Geoth Res 42:1–12

Battaglia M, Roberts CW, Segall P (1999) Magma intrusion beneath Long Valley caldera confirmed by temporal changes in gravity. Science 285:2119–2122

Bebbington M (2009) Volcanic eruptions: stochastic models of occurrence patterns. In: Meyers B (ed) Encyclopedia of complexity and system science, vol 9. Springer New York, pp 9831–9861

Bebbington MS (2014) Long-term forecasting of volcanic explosivity. Geophys J Int 197:1500–1515

Bebbington MS, Jenkins FS (2019) Intra-eruption forecasting. Bull Volcanol 81:34

Belardinelli ME, Bozzarri A, Cocco M (2003) Earthquake triggering by static and dynamic stress changes. J Geophys Res 108:2135. https://doi.org/10.1029/2002JB001779

Bell AF, Greenhough J, Heap MJ, Main IG (2011) Challenges for forecasting based on accelerating rates of earthquakes at volcanoes and laboratory analogues. Geophys J Int 185:718–723

Benito-Saz MA, Sigmundsson F, Charco M, Hooper A, Parks M (2019) Magma flow rates and temporal evolution of the 2012–2014 post eruptive intrusions at El Hierro, Canary Islands. J Geophys Res 124 https://doi.org/https://doi.org/10.1029/2019JB018219

Biggs J, Robertson E, Cashman K (2016) The lateral extent of volcanic interactions during unrest and eruption. Nat Geosci 9:308–311

Burt ML, Wadge G, Scott WA (1994) Simple stochastic modelling of eruption history of basaltic volcano: Nyamuragira, Zaire. Bull Volcanol 56:87–97

Cassidy M, Ebmeier SK, Helo C, Watt SFL, Caudron C, Odell A et al (2019) Explosive eruptions with little warning: experimental petrology and volcano monitoring observations from the 2014 eruption of Kelud, Indonesia. Geochem Geophys Geosyst 20:4218–4247

Caudron C, Girona T, Taisne B, Suparjan, Gunawan H, Kristianto et al (2019) Change in seismic attenuation as a long-term precursor of gas-driven eruptions. Geology 47:632–636

Chesley C, LaFemina PC, Puskas C, Kobayashi D (2012) The 1707 Mw8.7 Hoei earthquake triggered the largest historical eruption of Mt. Fuji. Geophys Res Lett 39:L24309. https://doi.org/https://doi.org/10.1029/2012GL053868

Chiodini G, Vandemeulebrouck J, Caliro S, D'Auria L, De Martino P, Mangiacapra A et al (2015) Evidence of thermal-driven processes triggering the 2005–2014 unrest at Campi Flegrei caldera. Earth Planet Sci Lett 414:58–67

Chiodini G, Paonita A, Aiuppa A, Costa A, Caliro S, De Martino P et al (2016) Hotter volcanic unrest for magmas near the critical degassing pressure. Nat Commun 7:13712. https://doi.org/10.1038/ncomms13712

Connor CB, Sparks RSJ, Mason RM, Bonadonna C, Young SR (2003) Exploring links between physical and probabilistic models of volcanic eruptions: the Soufriere Hills Volcano. Montserrat. Geophys Res Lett 30:1701. https://doi.org/10.1029/2003GL017384

Cooke RM (1991) Experts in uncertainty: opinion and subjective probability in science. Oxford Univ Press, New York, p 336

Crider JG, Frank D, Malone SD, Poland MP, Werner C, Caplan J (2011) Magma at depth: a retrospective analysis of the 1975 unrest at Mount Baker, Washington, USA. Bull Volcanol 73:175–189

D'Auria L, Pepe S, Castaldo R, Giudicepietro F, Macedonio G, Ricciolino P et al (2015) Magma injection beneath the urban area of Naples: a new mechanism for the 2012–2013 volcanic unrest at Campi Flegrei caldera. Sci Rep 5:13100. https://doi.org/10.1038/srep13100

D'Auria L, Barrancos J, Padilla GD, Pérez NM, Hernández PA, Melián G et al (2019) The 2016 Tenerife (Canary Islands) long-period seismic swarm. J Geophys Res 124:8739–8752

de Vita S, Sansivero F, Orsi G, Marotta E, Piochi M (2010) Volcanological and structural evolution of the Ischia resurgent caldera (Italy) over the past 10 ka. In: Groppelli G, Viereck L (eds) Stratigraphy and geology in volcanic areas. GSA Book Series Special Paper, vol 464, pp 193–239

Druitt TH, Costa F, Deloule E, Dungan M, Scaillet (2012) Decadal to monthly timescales of magma transfer and reservoir growth at a caldera volcano. Nature 482:77–80

Eggert S, Walter TR (2009) Volcanic activity before and after large tectonic earthquakes: observations and statistical significance. Tectonophysics 471:14–26

Farquharson JI, Amelung F (2020) Extreme rainfall triggered the 2018 rift eruption at Kīlauea Volcano. Nature 580:491–495

Gaddes M, Hooper A, Bagnardi M, Inman H, Albino F (2018) Blind signal separation methods for InSAR: the potential to automatically detect and monitor signals of volcanic deformation. J Geophys Res 123:10226–10251

Galetto F, Acocella V, Caricchi L (2017) Caldera resurgence driven by magma viscosity contrasts. Nat Commun 8:1750. https://doi.org/10.1038/s41467-017-01632-y

Galetto F, Bagnardi M, Acocella V, Hooper A (2019) Noneruptive unrest at the caldera of Alcedo Volcano (Galápagos Islands) revealed by InSAR data and geodetic modelling. J Geophys Res 124. https://doi.org/https://doi.org/10.1029/2018JB017103

Gardine M, West M, Werner C, Doukas M (2011) Evidence of magma intrusion at Fourpeaked volcano, Alaska in 2006–2007 from a rapid-response seismic network and volcanic gases. J Volcanol Geoth Res 200:192–200

Giudicepietro F, Macedonio G, Martini M (2017) A physical model of sill expansion to explain the dynamics of unrest at Calderas with application to Campi Flegrei. Front Earth Sci 5:54. https://doi.org/10.3389/feart.2017.00054

Gregg CE, Houghton B, Ewert JW (2015) Volcano warning systems. In: Sigurdsson H, Houghton B, Rymer H, Stix J (eds) The encyclopaedia of volcanoes, 2nd edn. Academic Press, pp 1173–1186

Hamling IJ, Kilgour G (2020) Goldilocks conditions required for earthquakes to trigger basaltic eruptions: evidence from the 2015 Ambrym eruption. Sci Adv 6: eaaz5261

Harris RA (1998) Introduction to a special section: stress triggers, stress shadows, and implications for seismic hazards. J Geophys Res 103:24347–24358

Hasegawa T, Nakagawa M (2016) Large scale explosive eruptions of Akan volcano, eastern Hokkaido, Japan: a geological and petrological case study for establishing tephro-stratigraphy and -chronology around a caldera cluster. Quatern Int 397:39–51

Hincks TK, Komorowski J-C, Sparks RSJ, Aspinall WP (2014) Retrospective analysis of uncertain eruption precursors at La Soufrière volcano, Guadeloupe, 1975–77: volcanic hazard assessment using a Bayesian belief network approach. J Appl Volcanol 3:3

Hill DP, Pollitz F, Newhall C (2002) Earthquake-volcano interactions. Phys Today 55:41–47

Hill DP, Prejean SG (2007) Dynamic triggering Treatise of Geophysics, vol 4. Elsevier Amsterdam, pp 293–320

Jaquet O, Carniel R, Sparks S, Thompson G, Namar R, Di Cecca M (2006) DEVIN: A forecasting approach using stochastic methods applied to the Soufriere Hills Volcano. J Volcanol Geoth Res 153:97–111

Jellinek AM, DePaolo DJ (2003) A model for the origin of large silicic magma chambers: precursors of caldera-forming eruptions. Bull Volcanol 65:363–381

Kato A, Terkawa T, Yamanaka Y, Maeda Y, Horikawa S, Matsuhiro K (2015) Preparatory and precursory processes leading up to the 2014 phreatic eruption of Mount Ontake. Japan. Earth Plan Space 67:111

Kilburn CRJ (2003) Multiscale fracturing as a key to forecasting volcanic eruptions. J Volcanol Geoth Res 125:271–289

Kilburn CRJ, Sammonds PR (2005) Maximum warning times for imminent volcanic eruptions. Geophys Res Lett 32:L24313. https://doi.org/10.1029/2005GL024184

Kilburn RJ, de Natale G, Carlino S (2017) Progressive approach to eruption at Campi Flegrei caldera in southern Italy. Nat Commun 8:15312. https://doi.org/10.1038/ncomms15312

Kilburn CRJ (2018) Forecasting volcanic eruptions: beyond the failure forecast method. Front Earth Sci 6:133. https://doi.org/10.3389/feart.2018.00133

King C-Y, Basler D, Presser TS, Evans CW, White LD, Minissale AD (1994) In search of earthquake-related hydrologic and chemical changes along the Hayward fault. Appl Geochem 9:83–91

Koulakov I, Smirnov SZ, Gladkov V, Kasaktina E, West M, El Khrepy S et al (2018) Causes of volcanic unrest at Mt. Spurr in 2004–2005 inferred from repeated tomography. Sci Rep 8:17482. https://doi.org/https://doi.org/10.1038/s41598-018-35453-w

Linde AT, Sacks IS, Johnston MJS, Hill DP, Bilham RG (1994) Increased pressure from rising bubbles as a mechanism for remotely triggered seismicity. Nature 371:408–410

Linde AT, Sacks IS (1998) Triggering of volcanic eruptions. Nature 395:888–890

Lindsay J, Marzocchi W, Jolly G, Constantinescu R, Selva J, Sandri L (2010) Towards real-time eruption forecasting in the Auckland Volcanic Field: application of BET_EF during the New Zealand National Disaster Exercise 'Ruaumoko.' Bull Volcanol 72:185–204

Lipman PW, Mullineaux DR (eds) (1981) The 1980 eruptions of Mount St. Helens, Washington. US Geological Survey Professional Paper 1250, 844 p

Lopez C, Blanco MJ, Abella R, Brenes B, Cabrera Rodriguez VM, Casas B (2012) Monitoring the volcanic unrest of El Hierro (Canary Islands) before the onset of the 2011–2012 submarine eruption. Geophys Res Lett 39:L13303. https://doi.org/10.1029/2012GL051846

Manga M, Brodsky E (2006) Seismic triggering of eruptions in the far field: volcanoes and geysers. Ann Rev Earth Planet Sci 34:263–291

Manga M (2020) When it rains, lava pours. Nature 580:457–458

Martì J, Aspinall WP, Sobradelo R, Felpeto A, Geyer A, Ortiz R et al (2008) A long-term volcanic hazard event

tree for Teide-Pico Viejo stratovolcanoes (Tenerife, Canary Islands). J Volcanol Geoth Res 178:543–552
Martin AJ, Umeda K, Connor CB, Weller JN, Zhao D, Takahashi M (2004) Modeling long-term volcanic hazards through Bayesian inference: an example from the Tohoku volcanic arc. Japan. J Geophys Res 109: B10208. https://doi.org/10.1029/2004JB003201
Marzocchi W (2002) Remote seismic influence on large explosive eruptions. J Geophys Res 107:2018
Marzocchi W, Zaccarelli L, Boschi E (2004a) Phenomenological evidence in favour of a remote seismic coupling for large volcanic eruptions. Geophys Res Lett 31:L04601
Marzocchi W, Sandri L, Gasparini P, Newhall C, Boschi E (2004b) Quantifying probabilities of volcanic events: the example of volcanic hazard at Mt. Vesuvius. J Geophys Res 109:B11201. https://doi.org/10.1029/2004JB003155
Marzocchi W, Zaccarelli L (2006) A quantitative model for the time-size distribution of eruptions. J Geophys Res 111:B04204. https://doi.org/10.1029/2005JB003709
Marzocchi W, Woo G (2007) Probabilistic eruption forecasting and the call for an evacuation. Geophys Res Lett 34:L22310
Marzocchi W, Sandri L, Selva J (2008) BET_EF: a probabilistic tool for long- and short-term eruption forecasting. Bull Volcanol 70:623–632
Marzocchi W, Bebbington MS (2012) Probabilistic eruption forecasting at short and long time scales. Bull Volcanol 74:1777–1805
Marzocchi W, Garcia-Aristizabal A, Gasparini P, Mastellone ML, di Ruocco A (2012) Basic principles of multi-risk assessment: a case study in Italy. Nat Hazards 62:551–573
Mason BG, Pyle DM, Oppenheimer C (2004) The size and frequency of the largest explosive eruptions on Earth. Bull Volcanol 66:735–748
McNutt SR (2000) Volcanic seismicity. In: Sigurdsson H, Houghton B, McNutt S, Rymer H, Stix J (eds) The encyclopedia of volcanoes, 1st edn. Elsevier Academic Press, pp 1015–1034
McLeod P, Tait S (1999) The growth of dykes from magma chambers. J Volcanol Geoth Res 92:231–245
Montgomery DR, Manga M (2003) Streamflow and water well responses to earthquakes. Science 300:2047. https://doi.org/10.1126/science.1082980
Moran SC, Newhall C, Roman DC (2011) Failed magmatic eruptions: late-stage cessation of magma Ascent. Bull Volcanol 73:115–122
Namiki A, Rivalta E, Woith H, Willey T, Parolai S, Walter TR (2019) Volcanic activities triggered or inhibited by resonance of volcanic edifices to large earthquakes. Geology 47:67–70
Newhall CG, Dzurisin DD (1988) Historical unrest at large calderas of the world. US Geological Survey Professional Paper, 1109 p
Newhall CG, Punongbayan RS (1996) Fire and mud: eruptions and lahars of Mount Pinatubo, Philippines. Philippine Institute of Volcanology and Seismology, Quezon City and University of Washington Press Seattle and London, 1126 p
Newhall CG (2000) Volcano warnings In: Sigurdsson H, Houghton B, McNutt S, Rymer H, Stix J (eds) The encyclopaedia of volcanoes, 1st edn. Elsevier Academic Press, pp 1185–1198
Newhall CG, Hoblitt RP (2002) Constructing event trees for volcanic crises. Bull Volcanol 64:3–20
Newhall CG, Costa F, Ratdomopurbo A, Venezky DY, Widiwijayanti C, Thin Zar Win N (2017) WOVOdat—an online, growing library of worldwide volcanic unrest. J Volcanol Geoth Res 345:184–199
Nishimura T, Ueki S (2011) Seismicity and magma supply rate of the 1998 failed eruption at Iwate volcano, Japan. Bull Volcanol 73:133–142
Nishimura T (2017) Triggering of volcanic eruptions by large earthquakes. Geophys Res Lett 44:7750–7756
Nooner SL, Chadwick WW (2016) Inflation-predictable behavior and co-eruption deformation at Axial Seamount. Science 354:1399–1403
Nostro C, Stein RS, Cocco M, Belardinelli ME, Marzocchi W (1998) Two-way coupling between Vesuvius eruptions and southern Apennine earthquakes, Italy, by elastic stress transfer. J Geophys Res 103:24487–24504
Oikawa T, Yoshimoto M, Nakada S, Maeno F, Komori J, Shimano T et al (2016) Reconstruction of the 2014 eruption sequence of Ontake Volcano from recorded images and interviews. Earth Planets Space 68:79
Palmer J (2020) The volcanology revolution. Nature 581:256–259
Papale P (2017) Rational volcanic hazard forecasts and the use of volcanic alert levels. J Appl Volcanol 6:13
Papale P, Marzocchi W (2019) Volcanic threats to global society. Science 363:1275–1276
Parks MM, Biggs J, England P, Mather TA, Nomikou P, Palamartchouk K (2012) Evolution of Santorini Volcano dominated by episodic and rapid fluxes of melt from depth. Nat Geosci 5:749–754
Passarelli L, Brodsky EE (2005) The correlation between run-up and repose times of volcanic eruptions. Geophys J Int 188:1025–1045
Phillipson G, Sobradelo R, Gottsmann J (2013) Global volcanic unrest in the 21st century: an analysis of the first decade. J Volcanol Geoth Res 264:183–196
Poland MP, Anderson KR (2020) Partly cloudy with a chance of lava flows: forecasting volcanic eruptions in the twenty-first century. J Geophys Res 125: e2018JB016974. https://doi.org/https://doi.org/10.1029/2018JB016974
Potter SH, Scott BJ, Jolly GE, Neall VE, Johnston DM (2015) Introducing the Volcanic Unrest Index (VUI): a tool to quantify and communicate the intensity of volcanic unrest. Bull Volcanol 77:77
Pritchard ME, Jay JA, Aron F, Henderson ST, Lara LE (2013) Subsidence at southern Andes volcanoes induced by the 2010 Maule, Chile earthquake. Nat Geosci 6:632–636
Pyle DM (1998) Forecasting sizes and repose times of future extreme volcanic events. Geology 26:367–370

Ripepe M, Marchetti E, Delle Donne D, Genco R, Innocenti L, Lacanna G et al (2018) Infrasonic early warning system for explosive eruptions. J Geophys Res 123:9570–9585

Rivalta E, Corbi F, Passarelli L, Acocella V, Davis T, Di Vito MA (2019) Stress inversions to forecast magma pathways and eruptive vent location. Sci Adv 5: eaau9784

Robertson RM, Kilburn CRJ (2016) Deformation regime and longterm precursors to eruption at large calderas: Rabaul, Papua New Guinea. Earth Planet Sci Lett 438:86–94

Roman DC, Power JA (2011) Mechanism of the 1996–97 non-eruptive volcano-tectonic earthquake swarm at Iliamna Volcano, Alaska. Bull Volcanol 73:143–153

Sandri L, Marzocchi W, Zaccarelli L (2004) A new perspective in identifying the precursory patterns of eruptions. Bull Volcanol 66:263–275

Sandri L, Guidoboni E, Marzocchi W, Selva J (2009) Bayesian event tree for eruption forecasting (BET_EF) at Vesuvius, Italy: a retrospective forward application to the 1631 eruption. Bull Volcanol 71:729–745

Sandri L, Costa A, Selva J, Tonini R, Macedonio G, Folch A et al (2016) Beyond eruptive scenarios: assessing tephra fallout hazard from Neapolitan volcanoes. Sci Rep 6:24271. https://doi.org/10.1038/srep24271

Sandri L, Acocella V, Newhall C (2017a) Searching for patterns in caldera unrest. Geochem Geophys Geosyst 18. https://doi.org/https://doi.org/10.1002/2017GC006870

Sandri L, Tonini R, Rouwet D, Constantinescu R, Mendoza-Rosas AT, Andrade D et al (2017b) The need to quantify hazard related to non-magmatic unrest: from BET_EF to BET_UNREST. In: Gottsmann J, Neuberg, J, Scheu B (eds) Volcanic unrest: from science to society. IAVCEI Advances in Volcanology Springer, pp 63–82

Selva J, Orsi G, Di Vito MA, Marzocchi W, Sandri L (2012) Probability hazard map for future vent opening at the Campi Flegrei caldera, Italy. Bull Volcanol 74:497–510

Selva J, Acocella V, Bisson M, Costa A, Caliro S, De Martino P et al (2019) Volcanic and related hazards at Ischia (Italy): state of knowledge and future perspectives. J Appl Volcanol 8:5. https://doi.org/10.1186/s13617-019-0086-4

Sparks RSJ (2003) Frontiers: forecasting volcanic eruptions. Earth Planet Sci Lett 210:1–15

Sparks RSJ, Aspinall WP (2004) Volcanic activity: frontiers and challenges in forecasting, prediction and risk assessment. State Planet Front Challenges Geophys. Geophys Monogr 150. https://doi.org/https://doi.org/10.1029/150GM28

Stein RS (2003) Earthquake conversations. Sci Am 288:72–79

Stein RS (2004) Tidal triggering caught in the Act. Science 305:1248–1249

Syahbana DK, Kasbani K, Suantika G, Prambada O, Andreas AS, Saing UB (2019) The 2017–19 activity at Mount Agung in Bali (Indonesia): intense unrest, monitoring, crisis response, evacuation, and eruption. Sci Rep 9:8848. https://doi.org/10.1038/s41598-019-45295-9

Swanson DA, Casadevall TJ, Dzurisin D, Malone SD, Newhall CG, Weaver CS (1983) Predicting Eruptions at Mount St. Helens, June 1980 through December 1982. Science 221:1369–1376

Tait S, Jaupart C, Vergniolle S (1989) Pressure, gas content and eruption periodicity of a shallow crystallizing magma chamber. Earth Planet Sci Lett 92:107–123

Takada Y, Fukushima Y (2013) Volcanic subsidence triggered by the 2011 Tohoku earthquake in Japan. Nat Geosci 6:637–641

Tilling RI (1988) Lessons from materials science. Nature 332:108–109

Tilling RI (1989) Volcanic hazards and their mitigation: Progress and problems. Rev Geophys 27:237–269

Tizzani P, Battaglia M, Zeni G, Atzori S, Berardino P, Lanari R (2009) Uplift and magma intrusion at Long Valley caldera from InSAR and gravity measurements. Geology 37:63–66

Trasatti E, Acocella V, Di Vito MA, Del Gaudio C, Weber G, Aquino I et al (2019) Magma degassing as a source of long-term seismicity at volcanoes: the Ischia island (Italy) case. Geophys Res Lett 46. https://doi.org/https://doi.org/10.1029/2019GL085371

Voight B (1988) A method for prediction of volcanic eruptions. Nature 332:125–130

Voight B (1990) The 1985 Nevado del Ruiz volcano catastrophe: anatomy and retrospection. J Volcanol Geoth Res 44:349–386

Voight B, Cornelius RR (1991) Prospects for eruption prediction in near real-time. Nature 350:695–698

Walter TR, Troll VR, Cailleau B, Belousov A, Schmincke HU, Amelung F et al (2005) Rift zone reorganization through flank instability in ocean island volcanoes: an example from Tenerife, Canary Islands. Bull Volcanol 67:281–291

Walter TR, Amelung F (2007) Volcanic eruptions following M $\geq$ 9 megathrust earthquakes: implications for the Sumatra-Andaman volcanoes. Geology 35:539–542

Walter TR, Wang R, Acocella V, Neri M, Grosser H, Zschau J (2009) Simultaneous magma and gas eruptions at three volcanoes in southern Italy: an earthquake trigger? Geology 37:251–254

Wang CY, Manga M (2010) Hydrologic responses to earthquakes and a general metric. Geofluids 10:206–216

Werner CA, Doukas MP, Kelly PJ (2011) Gas emissions from failed and actual eruptions from Cook Inlet Volcanoes, Alaska, 1989–2006. Bull Volcanol 73:155–173

Wilson CJN (2008) Supereruptions and supervolcanoes: processes and products. Elements 4:29–34

Winson AEG, Costa F, Newhall CG, Woo G (2014) An analysis of the issuance of volcanic alert levels during volcanic crises. J Appl Volcanol 3:14

Witze A (2019) AI could help to predict eruptions. Nature 567:156–157
Wright HMN, Pallister JS, McCausland WA, Griswold JP, Andreastuti S, Budianto A (2019) Construction of probabilistic event trees for eruption forecasting at Sinabung volcano, Indonesia 2013–14. J Volcanol Geoth Res 382:233–252

10 Volcanoes and Plate Tectonics

10.1 Introduction

This chapter introduces the third part of the book, presenting the regional tectonic frame of volcanoes. In this third part the previously described volcano-tectonic processes are considered at a wider scale, highlighting the interaction of the volcanoes with the regional tectonic context. The focus here is on how the tectonic setting may affect the location, distribution, style, type and frequency of volcanic activity, on both the longer-term (i.e., thousands to millions of years) and, as suggested by recent studies, the shorter-term (i.e., years or less, as in the co- and post-seismic cycles of regional earthquakes). The change in scale in this part of the book permits to appreciate first-order magmatic processes related to plate-tectonics mechanisms, in an exciting journey around key regions of our planet. This journey allows explaining the major differences among the volcanic provinces, as well as showing how magma plays a leading role not only in promoting intra-plate processes but, quite unexpectedly, also in shaping plate boundaries.

In particular, this chapter introduces the main tectono-magmatic processes occurring in the frame of plate tectonics at the planetary scale, along divergent and convergent plate boundaries, and at hot spots; transform plate boundaries, as usually not generating magma, are not considered. In the following chapters, key regional examples are illustrated for an in-depth understanding. In more detail, Chap. 11 describes regional examples of divergent plate boundaries, including continental, transitional and oceanic magmatic rifts, providing a final discussion. Chapter 12 describes regional examples of subducting convergent plate boundaries, experiencing extensional, strike-slip and contractional regimes along the volcanic arc, with a final discussion. Finally, Chap. 13 describes the regional intraplate volcanism of oceanic and, subordinately, continental hot spots, with a final discussion.

The main aims of this chapter are to:

- provide an overview of plate tectonics, presenting the main tectonic settings where volcanic activity occurs;
- describe the main tectono-magmatic processes responsible for the observed differences in volcanic activity in each setting;
- discuss the general causal relationships between magmatism and plate tectonics.

10.2 The Plate Tectonics Frame

Plate tectonics is the paradigm that aims to describe most of the solid Earth processes. The theory has been formulated in the late 1960s to explain in a coherent frame several geological

V. Acocella, *Volcano-Tectonic Processes*, Advances in Volcanology,
https://doi.org/10.1007/978-3-030-65968-4_10

and geophysical data acquired in the previous decades, building on the concepts of continental drift and seafloor spreading (Wegener 1912; Vine and Matthews 1963; Vine 1966). The theory postulates that the outer portion of our planet consists of seven major rigid plates (North American, Pacific, Eurasian, African, Indo-Australian, South American and Antarctic) and several minor plates which are slowly moving with regard to each other (Fig. 10.1; e.g., McKenzie and Parker 1967; Isacks et al. 1968; Le Pichon 1968; Morgan 1968; Wilson 1968).

These mobile plates involve not only the Earth's crust, but also the underlying upper mantle, with both crust and upper mantle forming the lithosphere, which is generally defined as the outer portion of the Earth with homogeneous elastic behaviour on time scales of thousands of years at least. However, depending on the considered problem, there are different definitions of lithosphere. The **seismic lithosphere** is defined by a zone of higher seismic velocity, contrasting with the underlying low velocity zone (or asthenosphere, see below). In the ocean basins its thickness increases from less than 20 km at oceanic ridges to more than 60 km approaching the continents. The temperature at the base of the seismic lithosphere is estimated as 600–650 °C. The **elastic lithosphere** is derived from the flexural behaviour of the Earth's surface when subjected to loading and unloading (such as by huge volcanic edifices, as in Hawaii, USA). Its thickness in ocean basins is similar to that of the seismic lithosphere, although thinner (30–40 km) under continents. In the ocean basins, the base of the elastic lithosphere roughly corresponds to a temperature of approximately 500 °C for dry olivine rheology. The **thermal lithosphere** is determined by a conductive gradient with a basal temperature of about 1280 °C. Therefore, the thermal lithosphere is about twice as thick as the seismic and elastic lithosphere. In the ocean basins, its thickness increases from a few

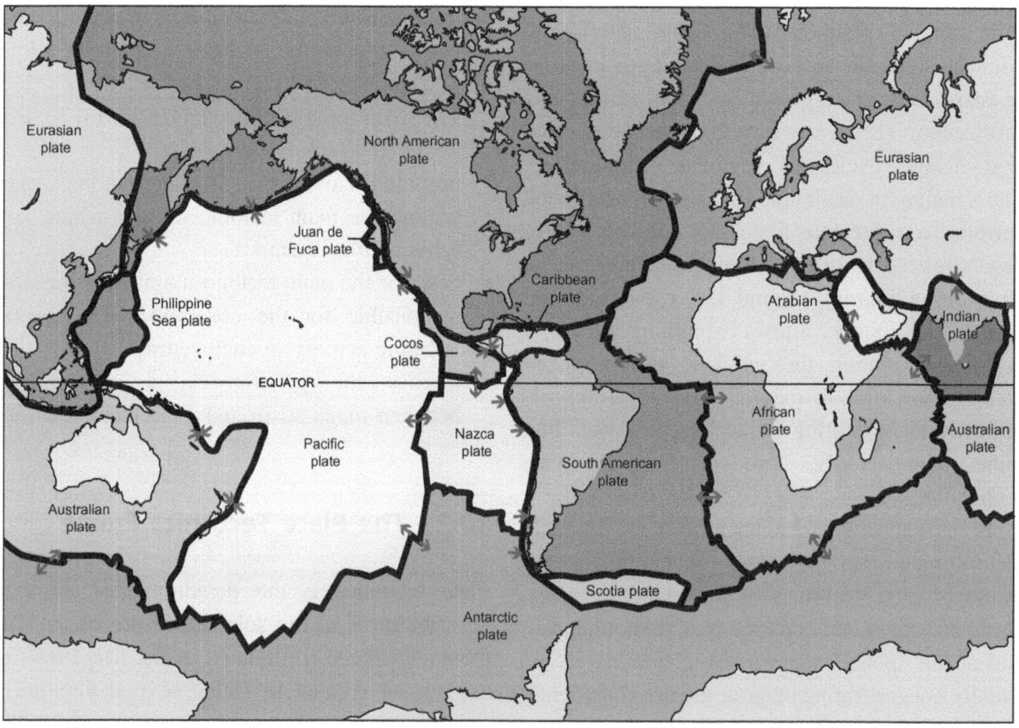

Fig. 10.1 Overview of the tectonic plates composing the Earth lithosphere and relative motion along their boundaries (red arrows; from USGS; https://commons.wikimedia.org/w/index.php?curid=535201)

kilometres at oceanic ridges to approximately 100 km on older oceanic lithosphere, thickening to approximately 280 km beneath the continents (Parsons and McKenzie 1979). Based on their composition, the lithospheric plates may be oceanic or continental. The **oceanic lithosphere** includes, in its upper part, the basaltic crust generated along oceanic ridges. The thickness of the oceanic lithosphere varies between a few kilometres at the ridge axis to 50–140 km away from the ridge, when the lithosphere becomes older, colder and denser. The **continental lithosphere** is formed through magmatism and accretion of continental terranes (which are fault-bounded regions with distinctive geological history), and its crust has a bulk granitic composition. The thickness of the continental lithosphere varies between ~40 and ~280 km. The basaltic oceanic crust is denser (average density of ~2900 kg/m^3) than the granitic continental crust (~2700 kg/m^3), thus determining the higher density of the oceanic lithosphere with regard to the continental lithosphere.

Below the lithosphere lies the **asthenosphere**, which is the highly viscous portion of upper mantle with variable portion of partial melt (up to ~10%). Lithospheric plates move relative to or with the asthenosphere, driven by body forces or traction forces, respectively. Body forces acting within the lithosphere are generated by gravity and density contrasts between the plates and the mantle. The two main body forces are ridge push and slab pull (Forsyth and Uyeda 1975; McKenzie 1977; Turcotte and Schubert 1982; and references therein; Stuwe 2007, and references therein). **Ridge push** is the result of gravitational forces acting on the young, raised oceanic lithosphere around ocean ridges, causing it to slide down the similarly raised but weaker asthenosphere and push on lithospheric material farther from the ridges. **Slab pull** is the pulling force exerted by a cold, denser oceanic plate plunging into the mantle due to its own weight: in fact, the subducting plate is also named slab. Because the oceanic plate is denser than the hotter mantle beneath it, the density contrast causes the plate to sink into the mantle. Traction forces are imparted by the convecting asthenosphere along the base of the lithosphere and along the slab at depth. The convecting asthenosphere is the shallow manifestation of a larger convection occurring within the upper mantle, driven by the density contrast between its hotter and lighter rising portions and the colder and denser sinking portions. Plate tectonics is thus driven by thermal processes acting not only in the outer portion of the Earth, but also within. The latter include internal cooling related to primordial heat associated with initial planetary formation and heat generated by radioactive decay. Magmatic and, in particular, volcanic activity are the final result of these thermal processes and their interaction with the lithospheric plates. The relative motion between the lithospheric plates is on the order of a few cm/year, with the fastest rates of ~15 cm/year observed along portions of the ridge in the Pacific Ocean (East Pacific Rise; Gordon and Jurdy 1986; Gripp and Gordon 2002; Steinberger et al. 2004; Kreemer et al. 2014).

An assumption of plate tectonics is that lithospheric plates behave in a rigid way. This implies that the motion at any portion within the same plate is consistent and that the deformation resulting from the relative motion of nearby plates mainly focuses along the edge of the plates, or plate boundaries (Fig. 10.2). **Plate boundaries** thus consist of a zone of deformation resulting from the relative motion of the plates. This zone may have an extremely variable width, in some cases reaching up a thousand of kilometres, as for example in the Himalayan–Tibetan collisional zone. Depending upon the relative motion between neighbouring plates, three main types of plate boundaries are distinguished: divergent, convergent and transform (Fig. 10.3; e.g., van der Pluijm and Marshak 2004; Fossen 2010, and references therein).

Divergent plate boundaries develop along the contact between plates that move away from each other (Fig. 10.3a). This motion determines lithospheric thinning at the plates contact; thinning in turn promotes decompression and magma generation. The rate of divergence is proportional to the flux of magma through the plate boundary, determining its structure, magmatic and volcanic

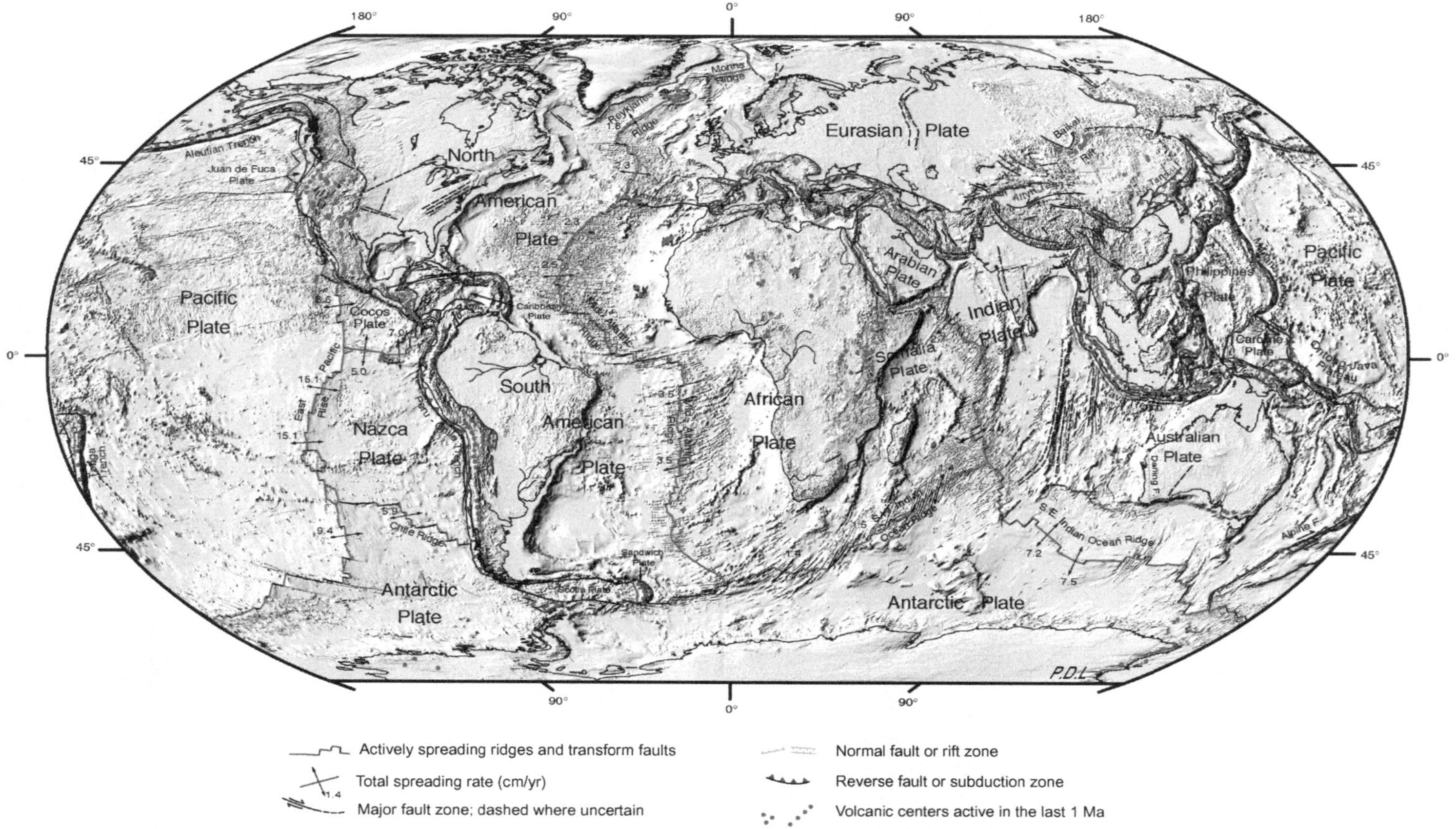

Fig. 10.2 Tectonic and volcanic activity of the Earth in the last one million years, highlighting the overall structure of the plate boundaries and the distribution of volcanoes (from NASA; https://commons.wikimedia.org/w/index.php?curid=61568)

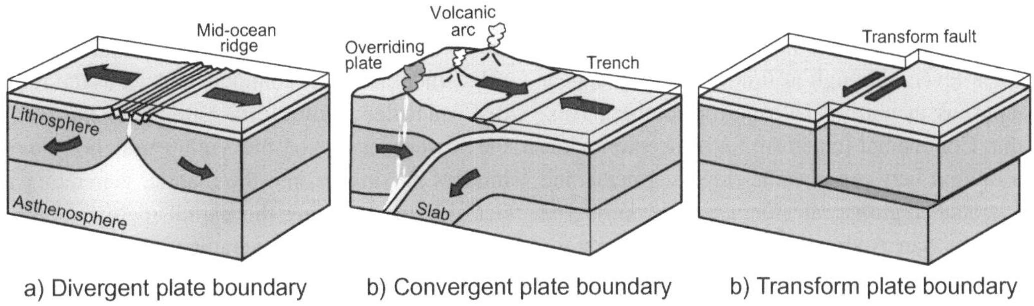

Fig. 10.3 Sketches of a **a** divergent, **b** convergent and **c** transform plate boundary

features. Divergent plate boundaries are found in continental lithosphere (as in the East African Rift System), oceanic lithosphere (as along the Mid-Atlantic Ridge or the East Pacific Rise) and any transitional lithosphere in between (as in Afar, Ethiopia). Continental divergent plate boundaries mark the onset of the divergence between two plates, and are characterized by extension rates of a few mm/yr. Oceanic divergent plate boundaries are the expression of the mature stage of divergence, with extension rates ranging from several mm/year to ~15 cm/year. These oceanic boundaries, coinciding with the oceanic ridges, are by far the most frequent type of divergent plate boundary. The general processes occurring at divergent plate boundaries are introduced in Sect. 10.3, whereas their regional features are described in Chap. 11.

Convergent plate boundaries develop when two nearby lithospheric plates move toward each other, determining overall contraction along the boundary (Fig. 10.3b). If an oceanic plate is involved in the convergence, its higher density determines its subduction into the mantle below the lighter overriding plate; an oceanic **trench** marks the submarine contact between the subducting and overriding plates. The subduction may occur with a variable dip of the slab, which depends upon its density, itself a function of its temperature and age, and the general tectonic setting. For example, younger and/or warmer subducting lithosphere creates a lighter slab gently dipping in the mantle, conversely to older and colder lithosphere, associated with a denser and steeper slab. At a depth of 80–150 km, the slab experiences the partial melting of its least stable components, which mainly consist of hydrated minerals. The fluids associated with these minerals rise in the overlying mantle wedge, decreasing the melting point of the rocks and promoting the generation of mafic magma. This magma rises and interacts with the overlying crust, becoming more evolved and felsic, eventually erupting along volcanic arcs. **Volcanic arcs** are the surface expression of magmatism along convergent plate boundaries, consisting of chains of aligned and regularly spaced volcanoes, mostly stratovolcanoes, lying on the plate that is overriding the slab, as observed along the circum-Pacific Ring of Fire. Conversely, if no oceanic plate is involved in the convergence process, as in the case of two approaching continental plates, no dense plate sinks and **continental collision** replaces subduction. Collision leads to the thickening of the continental lithosphere and to orogenesis (i.e., mountain building) along the plate boundary, as for example along the Alpine-Himalayan region. In this case, volcanism is negligible or absent. The general processes occurring at convergent plate boundaries are introduced in Sect. 10.4, whereas their regional features are described in Chap. 12.

Transform plate boundaries develop when two nearby plates move laterally past each other, without significant convergence or divergence component (Fig. 10.3c). Transform plate boundaries are thus characterized by a dominant component of horizontal motion, accommodated by the development of strike-slip fault systems. Transform boundaries may affect oceanic and continental lithosphere. Oceanic transform faults accommodate motion between active oceanic ridge segments, as along the Mid-Atlantic Ridge. These transform faults are usually found below the sea level, although at times they may appear onland, as in northern Iceland or southern Australia. Continental transform faults accommodate the motion between oceanic ridge segments and collisional regions, as for example observed along the San Andreas Fault (California) or the Alpine Fault (New Zealand), respectively. Conversely to divergent and convergent plate boundaries, the horizontal motion along transform boundaries does not usually involve any important process associated with the generation of magma. Indeed, in the few cases of volcanism along transform zones there is no evidence or model that magma generation results from the activity of the transform. Rather, volcanism appears controlled by other processes, most notably the activity of hot spots (see below), as for example in the Azores Archipelago. As magma is usually not generated along transform plate boundaries, this type of boundary is not considered further, although the effect of a transform setting on magmatism is considered in the Azores hot spot (Sect. 13.7).

At the global scale, mantle convection determines plate motion, with upwelling regions of mantle flux coinciding with divergent plate boundaries and downwelling regions with convergent boundaries, so that the extension created at divergent boundaries is compensated by the contraction at convergent boundaries. As a result, new plates may grow, or be even created, whereas old plates may consume and become ultimately destroyed along subduction or collision zones. These features merge into an imaginary evolutionary cycle, or **Wilson cycle** (Wilson 1972; e.g., Stuwe 2007). This ideally begins with the lithospheric extension of a continent (as along the East African Rift System), evolving in continental break-up and the development of an immature spreading centre producing oceanic lithosphere (as in the Red Sea). This condition may then lead to an ocean with mature ridge, as observed in the Atlantic or in the Pacific. This divergent motion is balanced by plate convergence, which begins with the onset of subduction of an oceanic plate underneath a continental plate, as observed along the west coast of South America. This condition may also involve the subduction of an oceanic ridge underneath the continental plate, as the Juan de Fuca Ridge below North America: in this case, the lower density of the young and hot ridge hinders its sinking into the mantle, generating a flat slab not reaching the partial melting conditions, and thus not generating any volcanism. Alternatively, all the oceanic lithosphere may be consumed in the subduction process, leading to the collision between two lower-density continents and the cessation of subduction, as observed between India and Asia.

Therefore, most of the magmatic activity on our planet develops along divergent and convergent plate boundaries. Nevertheless, a minor though significant portion of magma is erupted away from plate boundaries, in focused areas, or **hot spots**, whose activity is independent of plate motion (Fig. 10.4; Davies 1999). In particular, hot spot volcanism results from the rise through the mantle and the crust of hot and narrow plumes, which often originate at the mantle-core boundary. Being independent of any plate motion, mantle plumes and their surface manifestation hot spots are considered to be stationary and, for this reason, they have been commonly used as a reference frame in an otherwise ever moving Earth's surface. Hot spots are best identified in intraplate domains, that is distant from plate boundaries, as for example the Hawaiian, Reunion or Yellowstone hot spots. However, several major hot spots lie along plate boundaries, especially divergent boundaries, as for example Iceland and Afar (the latter occurring at a triple junction between three divergent boundaries). These less evident hot spots may be recognized by several features, including the locally anomalous production of magma with slightly distinct composition that generates alkali basalts, rather than the tholeiitic basalts commonly produced along divergent plate boundaries. Since in many cases the activity of a hot spot along a divergent plate boundary predates the development of that boundary (as in Afar), it is likely that the thermal weakening induced by the hot spot activity in the lithosphere "captured" the plate boundary. In these cases, the upwelling

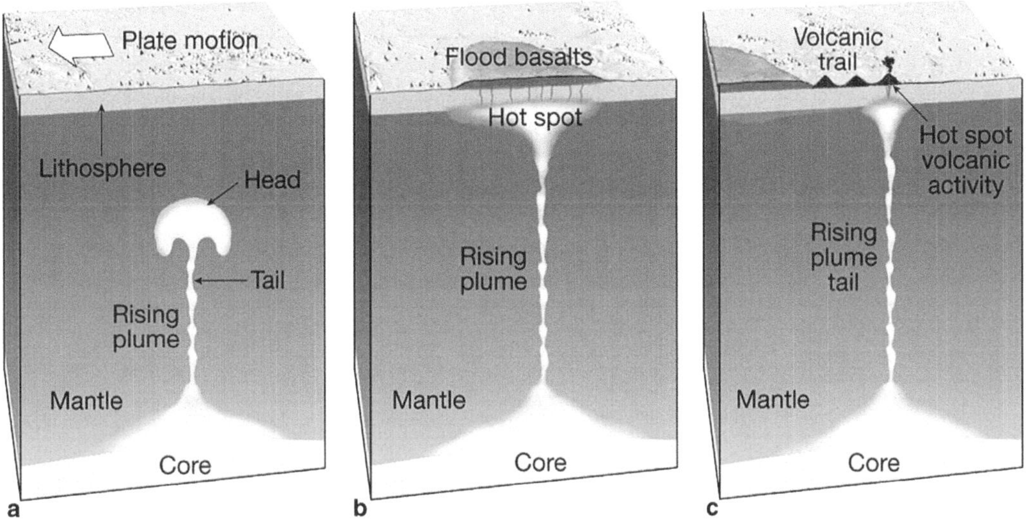

Fig. 10.4 Stages of mantle plume development: **a** a plume rises form the mantle-core boundary; **b** the head of the plume generates massive volcanism at the surface; **c** the plume forms a hot spot that generates a chain of volcanoes parallel to plate motion (Jaffe and Taylor 2018; Image from Tasa Graphics)

of the upper mantle along the boundary couples with the upwelling of the plume. The general processes occurring at hot spots are introduced in Sect. 10.5, whereas their regional features are described in Chap. 13.

In summary, divergent and convergent plate boundaries, as well as hot spots, directly or indirectly explain the distribution of most of the volcanoes on our planet through different processes, which are discussed below. This distribution closely matches that of the Earth's seismicity, as seismic activity also focuses along plate boundaries. This seismicity generally results from regional tectonic processes related to the deformation of the plate boundary (as along transform faults, slabs or in collisional zones), although in some cases shallow seismicity may be also associated with magmatic activity (as along some divergent plate boundaries).

There are three main modes of obtaining partial melting of a peridotite rock in the upper mantle: pressure decrease, temperature increase and addition of volatiles (as H_2O and CO_2). Through one or more of these conditions a solid peridotite just below the dry melting curve (solidus) may experience partial melting (Fig. 10.5; e.g., Schmincke 2004). These conditions controlling the generation of magma and the associated geothermal gradients may significantly differ at divergent and convergent plate boundaries and hot spots (Fig. 10.5). Under ordinary intraplate conditions, and away from hot spots, partial melting is not expected to occur at any lithospheric and asthenospheric level: as a result, the geotherm here is below the solidus of the rocks at any depth. Along divergent plate boundaries, as for example in an oceanic ridge, lithospheric thinning induces decompression within the asthenosphere, which becomes shallower and experiences partial melting. This creation of melt in conditions of rapid decrease of pressure and at a constant temperature is known as **decompression melting**. The generated magma rises and heats the overlying lithosphere, so that here the geotherm overcomes the solidus of the rocks. Along convergent plate boundaries, the fluids released from the partially melting slab shift the solidus of the rocks towards lower temperatures. In particular, in the lower lithosphere the rocks solidus shifts below the geotherm, promoting partial melting and the generation of magma above the subducting lithosphere. Below hot spots, the rise of hot mantle plumes increases the geothermal gradient,

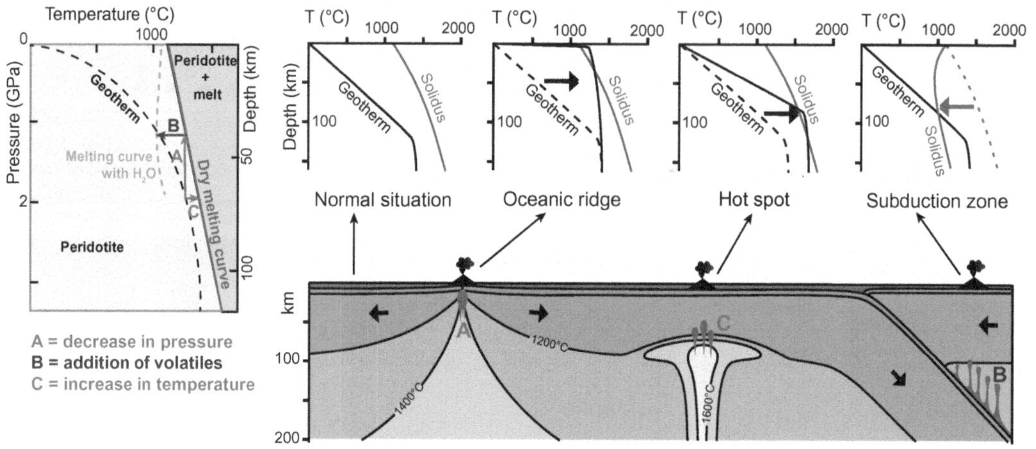

Fig. 10.5 Left: modes of partial melting of a peridotite rock through pressure decrease (A), addition of volatiles (as H_2O and CO_2; B) and temperature increase (C). Through one or more of these conditions, a solid peridotite just below the dry melting curve (solidus) may experience partial melting (modified after Schmincke 2004). Right: these three conditions predominate at divergent boundaries (as oceanic ridges), convergent boundaries (as subduction zones) and hot spots, respectively (see A, B and C in bottom panel). In each of these domains the predominant condition results in the geotherm partially overcoming the rock solidus, producing melt (see top diagrams; modified after: https://commons.wikimedia.org/w/index.php?curid=10433862)

which overcomes the solidus of the rocks in the lower lithosphere: this also results in partial melting and magma generation.

The distribution of the relative volumes erupted along divergent and convergent plate boundaries, as well as hot spots, shows that approximately two thirds of the magma are erupted along divergent plate boundaries, approximately one quarter is erupted along volcanic arcs, and more than one tenth along hot spots Fig. 10.6 (Fisher and Schmincke 1984). Clearly, most of the magma is erupted along oceanic ridges, where new crust is being continuously created over tens of thousands of kilometres. The creation of new crust does not only occur at the surface, through volcanism, but also at depth. Indeed, the amount of erupted magma is usually one tenth of the amount of magma permanently intruded in the crust. Similar ratios between erupted and intruded magma are also inferred for convergent plate boundaries and hot spots. This indicates that volcanic activity is the final and limited manifestation of much larger processes ultimately responsible for magma accumulation at depth, whose release at the surface is largely independent of the specific tectonic setting.

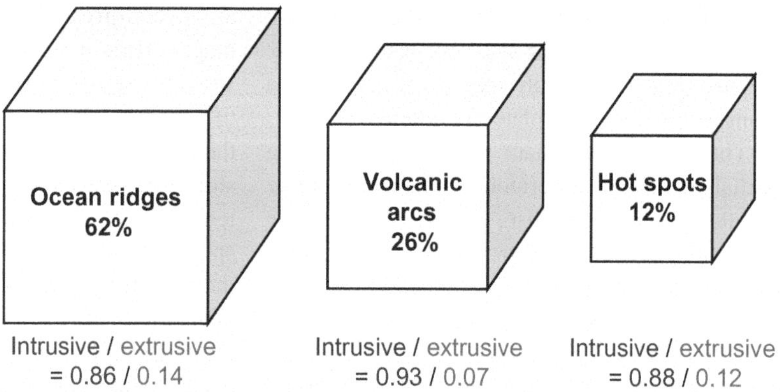

Fig. 10.6 Relative volumes of magma, and their inferred intrusive to eruptive ratios, erupted at ocean ridges, volcanic arcs and hot spots (modified after Fisher and Schmincke 1984)

The following sections introduce the general tectono-magmatic processes representative of the relationships between the regional tectonic context and volcanic activity occurring along divergent and convergent plate boundaries, as well as at hot spots.

10.3 Magmatic Processes Along Divergent Plate Boundaries

A divergent plate boundary identifies a **rift zone**, or rift, which is the lithospheric volume experiencing extension at the contact between the two diverging plates. Note that the term rift is more broadly used for any zone experiencing extension, also in intraplate settings, as for example the Rhine Graben (western Europe), the Baikal Rift (central Asia) and the Rio Grande Rift (southern USA).

Rift zones may be perpendicular or at some angle to the direction of plate divergence. **Orthogonal rifting** occurs when the direction of divergence is orthogonal to the trend of the rift axis, and thus to the trend of the major normal faults identifying the rift zone. However, this condition is rarely verified, as approximately 80% of the plate boundaries are oblique with regard to plate motion, forming an angle between 10° and 80° and highlighting a predominant **oblique rifting**. In particular, the direction perpendicular to the rift trend and the direction of plate divergence define the angle δ of obliquity of the rift, with orthogonal rifts having $\delta \sim 0°$ and oblique rifts $\delta > 0°$ (Fig. 10.7; Withjack and Jamison 1986; Tron and Brun 1991; Brune et al. 2018; Sani et al. 2019). In an oblique rift three structural directions are usefully defined: (a) the rift trend, (b) the direction of relative motion between the two plates (or direction of divergence), and (c) the trend of the greatest principal horizontal strain axis ε_1 of the finite strain ellipsoid. For instantaneous strain, the latter coincides with the minimum principal stress σ_3, or the local extension direction. In orthogonal rifts, ε_1 is perpendicular to the normal faults and parallel to the divergence direction. In oblique rifts, for low amounts of extension, ε_1 is approximately bisector of the angle between the regional direction of divergence and the normal to the rift axis; only for large stretching factor β (with $\beta \geq 5$) ε_1 tends to become parallel to the direction of divergence (Withjack and Jamison 1986; Tikoff and Teyssier 1994; Fournier and Petit 2007). Typically, fault systems generated during oblique rifting show an en-echelon arrangement, with fault segments oblique to the rift axis and to the direction of divergence. As oblique rifts are characterized by a shear component along the rift, faults striking orthogonal to the local extension direction display dominant dip-slip kinematics, whereas those oblique have a

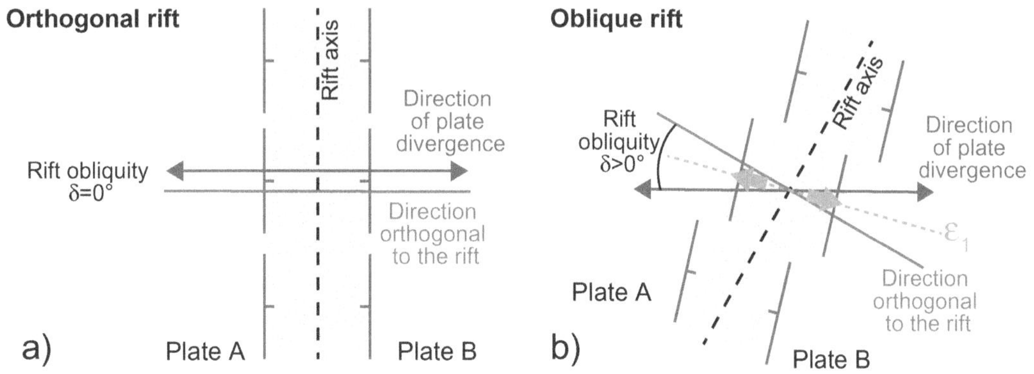

Fig. 10.7 Geometrical relationships (in map view) between the rift axis (and its perpendicular, in blue), the direction of relative motion between two plates (in purple) and the trend of the greatest horizontal principal strain ε_1 (or local extension direction, in orange) during orthogonal (**a**) and oblique rifting (**b**); yellow areas schematically indicate rift valleys (modified after Sani et al. 2019)

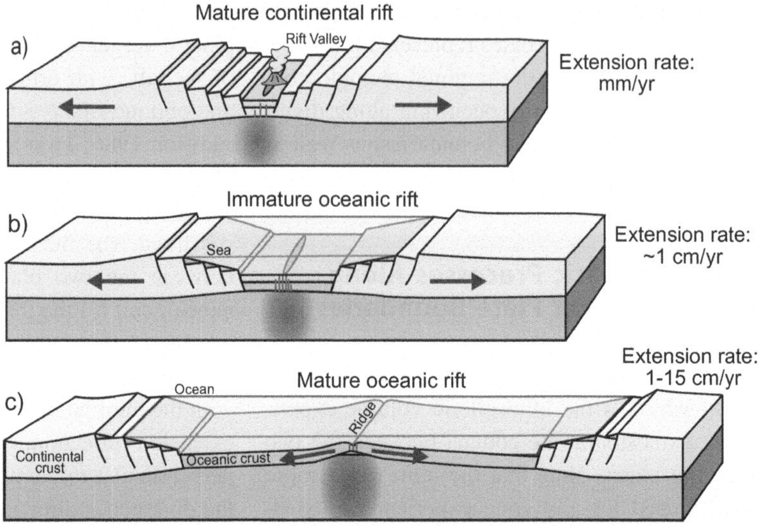

Fig. 10.8 Main stages of rifting, from mature continental rifting (**a**) to immature oceanic rifting (**b**) and oceanic spreading developing an oceanic ridge (**c**)

strike-slip component of motion that depends upon the degree of obliquity (e.g., Sani et al. 2019, and references therein).

In an ideal progression consistent with the Wilson cycle, rifting and thus plate divergence first develops on continental domains (**continental rifting**), determining the split of a larger plate into two or more smaller plates, as observed in the East African Rift System, which generates the larger Nubia (African) and Somali plates, as well as additional microplates. The persistence of rifting promotes the progressive thinning of the continental lithosphere, with augmented decompression, partial melting and magma generation (Fig. 10.8). As a consequence of crustal thinning and the progressive rise of more primitive mantle melts, the erupted magma becomes finally characterized by an increasing mafic component, evolving into transitional crust. Then mafic magmatism produces progressively newer crust and lithosphere within the rift, eventually replacing the previous lithosphere that shifts to the sides following plate divergence. This results in an **oceanic rifting** stage characterized by seafloor spreading, which may be initially immature (as observed in the Red Sea) and then mature (as in the Mid-Atlantic Ridge). This evolutionary continuum describes an ideal full spectrum of development of a divergent plate boundary, as not all continental rifts create oceanic lithosphere and evolve into seafloor spreading. The main stages of this ideal progression are described below.

The initiation of a divergent plate boundary may derive from two main processes. In fact, the extensional stresses within the lithosphere required for plate divergence may be initially imparted by either relative plate motion or doming of the lithosphere by a mantle plume. On these premises, two models have been proposed to explain the origin of rifting: passive and active rifting (Fig. 10.9; Bott 1995; Corti et al. 2003; Ziegler and Cloetingh 2004; van der Pluijm and Marshak 2004, and references therein). **Passive rifts** are initiated along plate boundaries and are tectonically driven: here regional extension generated by plate boundary forces and/or convective drag at the base of the lithosphere is responsible for thinning. Thinning reduces the lithostatic load on the asthenosphere rocks, enhancing decompression and partial melting herein. The asthenospheric melts replace the previously unstretched lithosphere, promoting magma emplacement and rise at the base and within the extended crust. Passive rifts thus undergo an earlier phase of tectonic extension and subsequent magmatism. In other cases, the initiation of continental rifting may be aided by a mantle plume, which provides the geodynamic forces to drive extension within the lithosphere

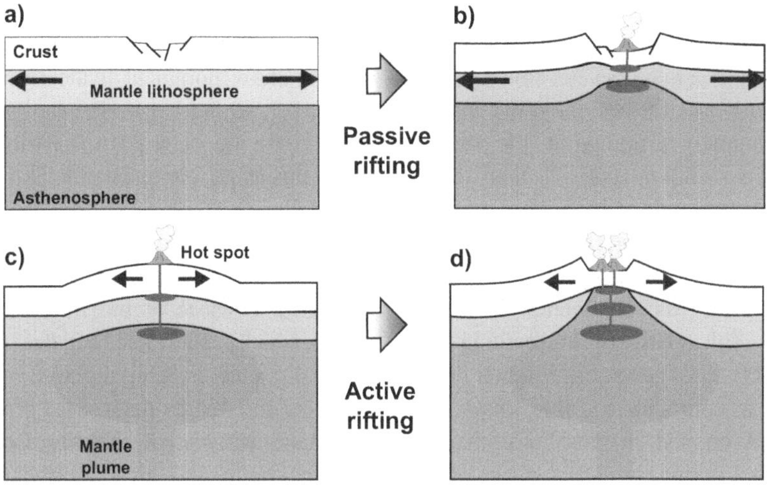

Fig. 10.9 Schematic diagrams illustrating the passive (**a**, **b**) and active (**c**, **d**) hypotheses for the initiation of rifts. In passive rifting extension is driven by a tensional regional stress originating from remote plate boundary forces: generation of a hot, low-density region in the asthenosphere (orange area) results from extension, followed by magmatic underplating (red areas) and volcanism. In active rifting the ascent and emplacement of a hot, low-density body in the sublithospheric mantle (orange area) is a primary cause for extension: this accounts for pre-rift uplift, magmatic underplating (red areas) and volcanism (not to scale; modified after Corti et al. 2003)

and the magma to accommodate the extension. These **active rifts** are thus initiated and driven in the mantle by the activity of plumes, emplacing hotter material below the lithosphere, inducing uplift (doming) and failure. Doming, much wider than the area to be affected by extension, and widespread volcanism predating extension of several million years, are distinctive features of active rifts. In particular, the volcanism fed by the mantle plume initiates the active rift erupting high volumes of basaltic lavas, or flood basalts (see below), as for example observed in Afar. This phase of doming and widespread volcanism anticipates the phase of lithospheric extension associated with more limited volcanism. Therefore, the relative timing of three main components (uplift/subsidence, extension, volcanism) determines whether rifting can be attributed to passive or active processes, although the application of this procedure to real settings is often difficult (Ruppel 1995).

Considering in detail the development of a divergent plate boundary, stretching associated with rifting thins the lithosphere, decreasing the lithostatic pressure in the asthenosphere below. The asthenosphere is so hot (>1280 °C) that, when such decompression occurs, partial melting begins, promoting magma production. The proportion between the amount of extension and percentage of created melt is not linear, with an effective melt production being already available for moderate extension, the latter quantified through the stretching factor β (White and McKenzie, 1989). The produced magma is less dense than the overlying lithosphere, and thus rises through it. A part of this rising magma may get trapped and solidify at the base of the crust, a process called **magmatic underplating**, which thickens the crust by adding mass to its base. In the case of emplacement of a hot mantle plume beneath a rifting lithosphere, as with active rifts, the higher temperature of the plume induces a higher amount of partial melting in the asthenosphere, and thus a higher production of magma. This condition may ultimately explain the generation of **flood basalts**, which are very voluminous (≥ 1000 km^3) amounts of erupted basaltic lavas. Flood basalts likely form where a mantle plume underlies a rift, for the peridotite in the plume is significantly hotter than the peridotite comprising normal asthenosphere, and thus undergoes a greater degree of partial melting during decompression (see Sect. 10.5; e.g., Rooney 2020a, and references therein).

The mafic magma that does not solidify at the base of the crust rises further, intruding into the crust and eventually erupting at the surface. Mafic magma is so hot (>1000 °C) that when it intrudes the continental crust it also conducts enough heat into the adjacent crust to cause partial melting. This melting takes place because the rock comprising the continental crust contains minerals with relatively low melting temperatures (<900 °C), generating lighter felsic magmas. As a consequence, the source and composition of erupted magmas also evolves. Magmas erupted during the initial stages of rifting are typically tholeiitic basalts formed through decompression melting of the mantle. As the rift evolves, magma differentiation towards felsic compositions may occur in the continental crust, forming the entire petrologic compositional spectrum of magmas. In some cases, the crustal emplacement of large volumes of relatively hot basaltic magma from a mantle source and the generation of crustal granitic magmas by partial melting may result in the eruption of a **bimodal volcanic suite**, characterized by the association of rhyolitic and basaltic products. Bimodal volcanism is for example observed in the mature portion of the continental Main Ethiopian Rift (see Chap. 11; e.g., Rooney 2020b, and references therein).

With the increased generation and rise of melts, magmatism plays a progressively more important role and the continental plate boundary system begins to switch from one that is dominated by tectonic processes to one where magmatism accommodates divergence. In fact, although regional tectonic structures (as normal faults) dominate in the early stage of continental rift growth, once magmatism is initiated these structures become less important in accommodating plate motion, as exemplified by the East African Rift System. Here the decrease in the activity of regional normal faults is accompanied by the increase in magmatic activity through repeated dike emplacement and dike-induced normal faults, which progressively accommodate plate motion. Following the increased crustal thinning and the encouraged injection of deeper feeder dikes, the intruded and erupted magmas start to show dominant mafic composition (see Sect. 10.7; Ebinger and Hayward 1996; Ebinger et al. 2010; Acocella 2014; LaFemina 2015).

At this stage, continental break-up may occur, especially if the progressively dominating magmatic activity within the rift is supported by the flux of an underlying mantle plume. **Continental break-up** consists of the fragmentation of the plate, which splits into smaller plates whose original contact is being replaced by the creation of a new and mainly mafic crust promoted by the magmatic activity of the rift and the plume. During continental break-up the crust evolves from continental to transitional to proto-oceanic, as exemplified in the Afar region (e.g., Ebinger and Hayward 1996).

Increased rates and amount of extension and increasing magmatic activity develop an oceanic divergent plate boundary with oceanic lithosphere, whose crust is mainly made of basalt and gabbro, and the upper mantle is made of ultramafic rock, as peridotite. The oceanic lithosphere is thinner than the continental lithosphere, although its thickness is a function of its age. As oceanic lithosphere moves away from the ridge axis, it cools and thickens to between 50 and 140 km, following the simplified relation:

$$z \approx 2\sqrt{\alpha t} \qquad (10.1)$$

where z is the thickness of the lithosphere, α is its thermal diffusivity, and t its age: the latter can be approximated by the ratio between the distance from the ridge axis and the relative rate of plate motion (e.g., LaFemina 2015, and references therein). In addition to becoming thicker with distance from the ridge axis, the lithospheric mantle becomes denser. After ~15 Ma, oceanic lithosphere becomes denser than the asthenospheric mantle and can therefore subduct into the mantle, a crucial condition for convergent plate boundaries (see Sect. 10.4). Oceanic divergent plate boundaries are defined by the relative rate of plate motion between the diverging plates and are subdivided into ultrafast (full spreading rate of >12 cm/year), fast (<12 to >8 cm/year), intermediate (<8 to >5 cm/year), slow (<5 to >2 cm/year), and ultraslow (<2 cm/year). These differences in spreading rate give rise to distinctive morphologies, structures and magmatic activities, as discussed in Chap. 11.

As anticipated in Chap. 2, the architecture of the rift zone characterizing the divergent plate boundary may be associated with a symmetric pure shear or an asymmetric simple shear mode of extension, or a combination thereof (McKenzie 1978; Wernicke 1985). Nevertheless, magmatically productive divergent plate boundaries are usually associated with a symmetric pure shear type of extension, rather than an asymmetric simple shear type. This follows from the fact that focused lithospheric extension, as determined by pure shear conditions, promotes melt generation much more than delocalized extension associated with simple shear (see Sect. 2.7.1; Latin and White 1990). Asymmetric rifting may be more common in the early continental stages, where the rift is mostly characterized by half-graben structures. Conversely in the advanced oceanic stages rifts develop a more symmetric architecture, mainly consisting of graben-like structures.

The rise of magma along divergent plate boundaries commonly occurs through the emplacement of regional dikes perpendicular to the local direction of extension, or minimum principal stress σ_3. In the case of rifts experiencing orthogonal extension, the dikes strike parallel to the rift axis, with an overall collinear configuration; in the case of rifts experiencing oblique extension, the dikes strike at some angle with the rift zone, with an overall en-echelon configuration (Fig. 10.10).

With the possible exception of portions of fast oceanic ridges (as the East Pacific Rise), dikes, and more in general magma, are not uniformly distributed along the entire rift length. Indeed, the rise and storage of magma along both continental and oceanic divergent boundaries focuses in elongated crustal portions, called **magmatic systems** or magmatic segments (Gudmundsson 1995; Ebinger and Casey 2001). These are usually located along the rift axis, and their length depends upon the extension rate of that portion of rift: the higher the extension rate, the longer the magmatic segment, with typical lengths in the order of 10^1 km. Each magmatic system consists of a dominant polygenic volcano and several monogenic vents aligned parallel (in case of orthogonal extension) or slightly oblique (in case of oblique extension) to the direction of the rift (Fig. 10.10). Volcanic activity is accompanied by diffuse fracturing, mostly consisting of extension (tensile) fractures and normal faults striking parallel to the alignment of the volcanoes. Magmatic segments indicate that magmatic activity and extension are not uniformly distributed along the plate boundary. Rather, they show that the diverging lithosphere is split along

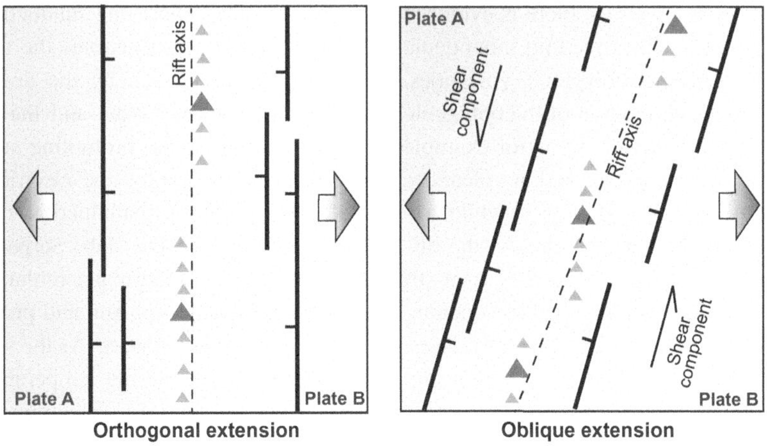

Fig. 10.10 Schematic structure (map view) of a portion of a divergent plate boundary experiencing orthogonal (left) and oblique (right) extension: magmatic systems are represented by yellow ellipses, polygenic volcanoes by red triangles and monogenic volcanoes by orange triangles. See text for details

$\sim 10^1$ km long wedges of intruded magma, which are periodically activated, determining a temporally and spatially incremental mode of opening of the plates. In the fast oceanic ridge of the East Pacific Rise the magmatic contribution is more stable in time and continuous in space: as a result, the magmatic systems are longer, on the order of 10^2 km, and more frequently active (Acocella 2014, and references therein).

In some cases, significant volcanic activity may be found also off-rift, or along the sides of the rift marking the plate boundary. This is for example the case of the extinct polygenic trachytic volcanoes, up to 1500 m high, lying on the flanks or the Main Ethiopian Rift. The unusual location of these volcanoes is interpreted to result from the deflection of the magma propagation paths during the early stages of rifting, under the unloading induced by the rift depression; before the establishment of a central (in-rift) propagation path, the unloading determines the lateral propagation of dikes, feeding the off-rift volcanoes (Fig. 7.3; Maccaferri et al. 2014).

In synthesis, in a divergent plate boundary tectonics and volcanism focus in magmatic systems along the rift axis (Fig. 10.10): tectonic structures consists of extension (tension) fractures and normal faults, whereas volcanism is represented by polygenic and monogenic volcanoes. While monogenic volcanoes are fed by dikes perpendicular to the opening direction, the feeding system of polygenic edifices is less straightforward. In many cases, there is evidence that the same regional dikes feeding monogenic volcanoes also feed the polygenic volcanoes. This may explain the elongation of the polygenic volcanoes parallel to the rift axis, as for example Boseti, Erta Ale (Ethiopia) and Hekla volcanoes (Iceland; Nakamura 1977). However, while frequent, this condition is not the rule and the different orientation of the volcanic edifices in rift zones, for example parallel to the opening direction, may be explained by other processes, as discussed in Sect. 5.5. The composition and type of volcanism associated with divergent plate boundaries is largely dependent on the amount of extension. Lower extension rates (and amounts of extension) are usually found on continental divergent boundaries. These are associated with a thicker continental granitoid crust (thicker than 20–30 km), which affects the composition of the erupted magma through storage, assimilation and differentiation towards felsic compositions: volcanism may be explosive, and ignimbritic eruptions are not uncommon. The associated volcanic edifices include polygenic, dominantly felsic stratovolcanoes and calderas. Conversely, higher extension rates and amounts of extension are typical of oceanic divergent boundaries, where the thinner and mafic crust does not affect the original composition of the basaltic magmas, determining a primitive and predominantly effusive volcanism. In this case, volcanic activity is more frequent and commonly associated with fissure eruptions, although mafic calderas and central volcanoes may be present.

10.4 Magmatic Processes Along Convergent Plate Boundaries

Magmatic activity along convergent plate boundaries is related to the development of subduction zones, with volcanic arcs located above slabs. Subduction results from the density difference between the two converging plates, where the denser oceanic lithosphere underthrusts the lighter lithosphere and sinks into the mantle: this may occur between an oceanic and a continental plate, as well as between two oceanic plates, with the denser one subducting. Once the oceanic lithosphere penetrates the upper mantle, the body forces driven by the density contrast between the oceanic plate and the mantle (slab pull forces) take over, promoting subduction.

Seismic tomography and electrical resistivity studies across several subduction zones indicate serpentinization of the slab. Serpentinization is the result of hydrothermal circulation and low-temperature metamorphism, and produces a suite of hydrated mineral phases. As the slab sinks into the mantle, the increasing temperature and pressure drive metamorphic reactions that release these fluids (mainly H_2O and CO_2) from the crust into the overlying mantle wedge. This addition of fluids to the asthenosphere (hydration) has the

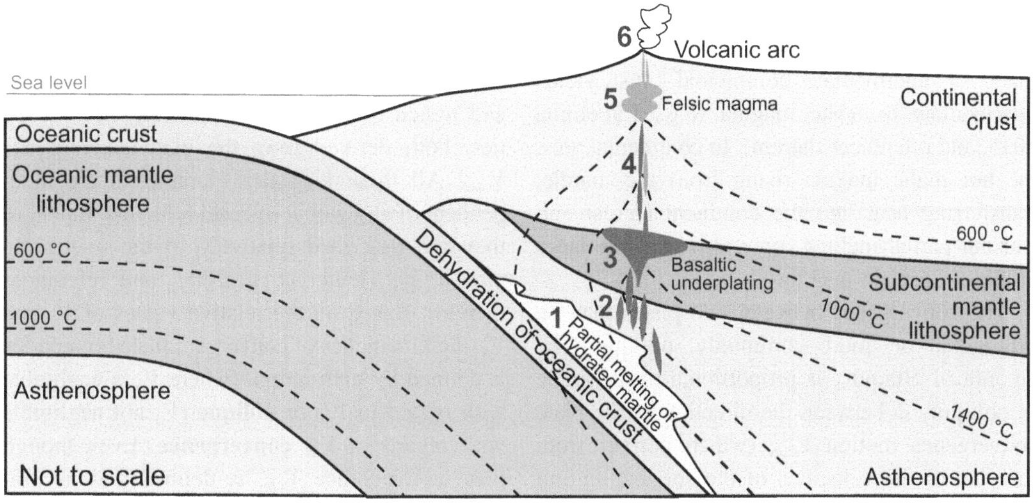

Fig. 10.11 Processes accompanying the generation and rise of magma in subducting convergent plate boundaries (section view), from volatile flux melting above the slab (1) to magma generation (2), magma underplating (3), magma rise through the crust (4), magma differentiation (5) and magma eruption (6)

effect of lowering its solidus temperature, developing a melting front above the slab and generating mafic magma. This process, termed **volatile flux melting**, typically occurs between depths of 80 and 150 km (Fig. 10.11; Turner et al. 2000; Cagnioncle et al. 2007; LaFemina 2015, and references therein). Once these melts are formed in the mantle wedge, they rise for buoyancy through the wedge and to the overriding plate. Here the magmas may accumulate at the Moho (underplating), heating the base of the crust and promoting partial melting and the production of more evolved felsic and lighter magmas, which will tend to rise. At the same time, the mafic magma may also continue to rise through dikes in the lower crust. While rising and emplacing in the crust, this magma may undergo a number of processes that lead to its differentiation towards more evolved felsic (andesitic to rhyolitic) compositions. These processes include assimilation and melting of crust, fractional crystallization, and magma mixing with more evolved magmas. In particular, the mixing of basaltic and rhyolitic magmas is the main process responsible for the formation of andesitic magma, with andesite being the most common rock type along volcanic arcs. Eventually, the magma rising in the crust reaches the surface and erupts along the arc. The location of the volcanic arc results from the location of the thermal anomaly within the wedge above the slab, which in turn is constrained by the depth of release of the volatiles from the slab, itself a function of the slab dip. Therefore, steeper slabs develop volcanic arcs nearer to the trench and shallower slabs develop volcanic arcs farther from the trench. Very shallow dipping (i.e., flat) slabs may not reach the temperature and pressure conditions to release volatiles, thus inhibiting magmatism: this is currently the case of the Peruvian portion of the Andes and the western portion of Honshu, Japan.

Volcanic arcs may consist of island or continental arcs. **Island arcs** form where the denser oceanic plate subducts beneath the lighter oceanic one (as for example the Marianas), or where the volcanic arc grows on a sliver of continental crust that has separated from a continent (as for example Japan). **Continental arcs** grow where an oceanic plate subducts beneath continental lithosphere, as the Andes or the Cascades Arc. Volcanism at island arcs formed on oceanic plates tends to produce mostly mafic and intermediate igneous rocks, whereas volcanism at continental arcs may also produce intermediate and felsic igneous rocks, including granitic batholiths. In fact, while partial melting of mantle peridotite yields mafic magma, partial melting of mafic or intermediate continental crust yields intermediate to felsic magma (e.g., LaFemina 2015, and references therein). In continental arcs, the hot mafic magma rising from the mantle, transferring heat into the continental crust and causing partial melting, may also generate large volumes of felsic magmas.

The convergence between two plates may be orthogonal or, more commonly, oblique. The amount of obliquity is proportional to the angle of obliquity ϕ between the direction of the plate convergence motion V_{con} (which derives from the sum of the velocities of the subducting and overriding plates) and the direction perpendicular to the contact between the two plates, or to the trench (Fig. 10.12).

In addition to the angle ϕ, the degree of convergence between the two plates can be expressed through the ratio between the trench parallel (V_p) and trench orthogonal (V_n) convergence velocities, both derived from the plate convergence V_{con}. All these kinematic parameters are independent of any global selected reference frame, as they are described relatively to the overriding plate (e.g., Heuret et al. 2007, and references therein). Based on the relative values of V_p and V_n, the kinematics of convergent plate boundaries is defined by **orthogonal** (where V_p is negligible with regard to V_n) or **oblique** (V_p not negligible with regard to V_n) **convergence**. Even though plate convergence V_{con} is defined and accommodated at the contact between the two plates, a minor part of this motion is commonly transferred to the front of the overriding plate, including the region of the volcanic arc. Indeed, the overall

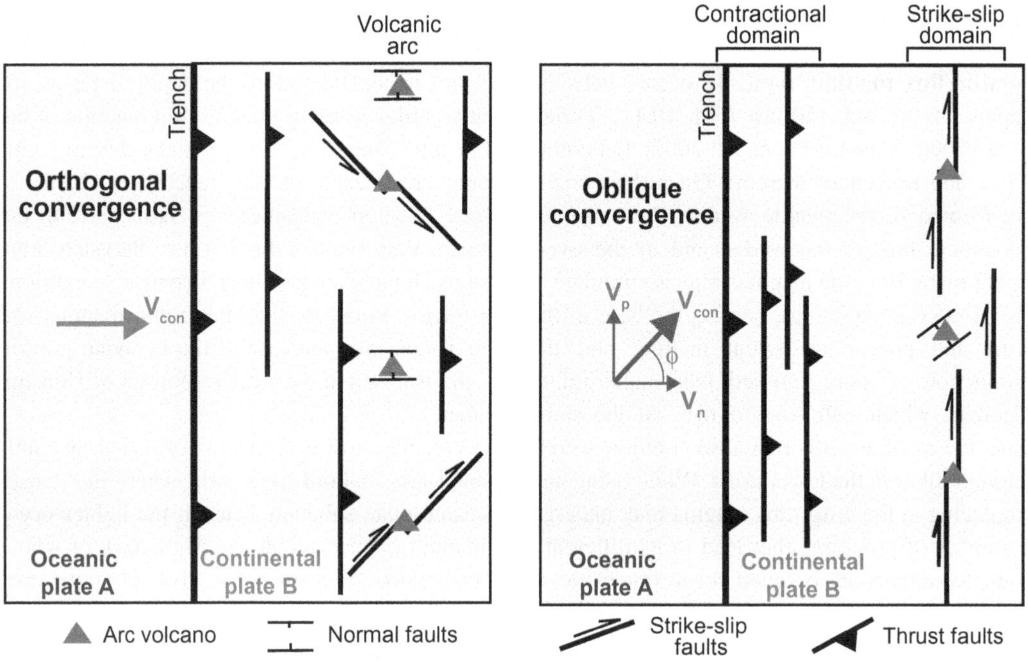

Fig. 10.12 Orthogonal and oblique convergence along portions of convergent plate boundaries experiencing subduction (map view). In orthogonal convergence, the motion resulting from both plates defines the convergence V_{con}, which is orthogonal to the contact between the plates, or trench. In oblique convergence the motion resulting from both plates V_{con} makes an angle ϕ (angle of obliquity) with regard to the direction orthogonal to the trench, defining a trench parallel V_p and trench orthogonal V_n component of convergence. Orthogonal and oblique convergence result in distinct structural styles on the overriding plate: in particular, the oblique motion is partitioned between predominant contraction on the accretionary prism and predominant strike-slip motion along the magmatic arc (see text for details)

motion of the volcanic arc usually reflects that deriving from the convergence between the two plates. If the convergence is accommodated through a predominant orthogonal component, this is present also in the arc region, as for example observed in Northeast Honshu (Japan). This arc-perpendicular motion may be characterized by extension or contraction, as a function of the variation (decrease or increase, respectively) in the trench-normal convergence velocity (e.g., Acocella and Funiciello 2010). Conversely, in presence of an important oblique convergence component, the volcanic arc may experience a predominant trench parallel motion, as for example observed at Sumatra (Indonesia). This behaviour is interpreted to result from a process of strain partitioning (Fig. 10.12). **Strain partitioning** here refers to the separation of the two components of the overall oblique motion of convergence across the portion of the overriding plate between the trench and volcanic arc. As a result, the trench portion of the overriding plate accommodates predominant arc-perpendicular motion, mainly developing thrust faults, whereas the arc portion mainly accommodates the arc-parallel motion, developing strike-slip systems consistent with the left- or right-lateral component of convergence between the plates. Depending on the obliquity angle and the convergence velocity V_{con}, the strike-slip faults along the arc may also display a contractional or extensional component, creating transpression or transtension. All these thrust and strike-slip fault systems are parallel to the trench (McCaffrey 1992, 1996). The spatial coincidence in the location of the volcanic arc and the zone of strike-slip faulting likely results from magma-induced thermal weakening, which focuses the deformation in the upper plate. A possible consequence of oblique convergence and strain partitioning is the creation of a microplate, or **sliver** in the forearc, with arc-parallel motion. The forearc side of the approaching sliver experiences contraction, whereas the forearc side of the departing sliver undergoes extension (Fig. 10.13; DeMets 1992; e.g., Acocella et al. 2018). More in general, available structural data show that, depending on the convergence values and their arc-normal and arc-parallel components, any type of motion may be found along volcanic arcs, including extension, contraction, oblique (transtension or transpression) and strike-slip motions. Therefore, in addition to the overall plate motion, a crucial parameter to consider for understanding volcanic processes along arcs is the overall arc kinematics (Acocella and Funiciello 2010).

The arc-perpendicular amount of convergence may be used to define two ideal and end-member behaviours of subducting plate boundaries (e.g., van der Pluijm and Marshak 2004). In a **coupled**

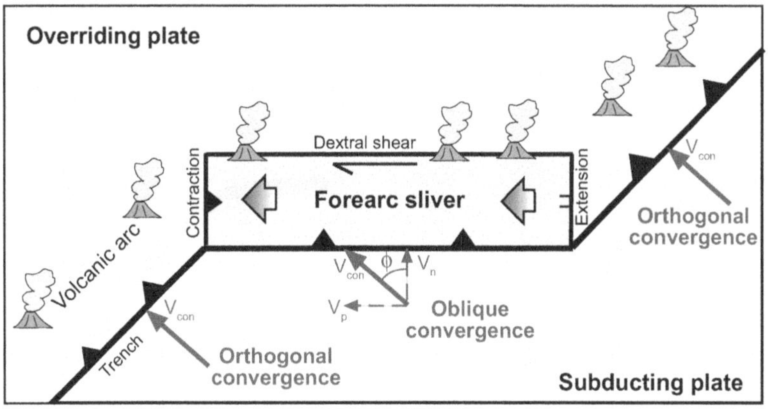

Fig. 10.13 Tectonic features associated with a portion of convergent plate boundary with strong obliquity (map view): in this case the forearc may be decoupled with regard to the subducting plate and a forearc sliver (in orange), characterized by overall arc-parallel motion, may develop (modified after DeMets 1992). A similar situation may apply to the southern Kuril Arc

convergent margin, the downgoing plate pushes tightly against the overriding plate, so the plate boundary overall is under compression (Fig. 10.14a). As a consequence, significant stresses develop across the contact, triggering large magnitude earthquakes; even the back-arc region may experience contraction. Because of the compression, the rise of magma is slower and its storage is not shallow. Therefore, magma has time to fractionate and/or cause partial melting of adjacent continental crust before intruding at shallower depth or erupting. Since partial melting of continental crust produces intermediate to felsic magma, and since fractionation removes mafic minerals from a melt, intermediate to felsic igneous rocks predominate at coupled convergent margins. The difficulty in the rise and shallow emplacement of magma may also result in lower eruptive rates. In an **uncoupled convergent margin**, the slab does not push tightly against the overriding plate, so compression across the margin is limited (Fig. 10.14a). As a consequence, the stress coupling across the plate boundary is smaller and earthquakes have smaller magnitude; the back-arc region may even experience extension. Because of the lower compression within the volcanic arc, mantle-derived magmas may rise more easily to the surface and intrude at shallower levels before any significant fractionation or crustal contamination; thus, mafic rocks are more common. Also, the enhanced conditions for the rise and shallow storage of magma may result in higher eruptive rates (Acocella and Funiciello 2010; Chaussard and Amelung 2012). While Northeast Japan and the Taupo Volcanic Zone of New Zealand may resemble coupled and uncoupled convergent margins, respectively, these two types of margins should be better seen as simplified end-members, as there is no textbook example of arc showing a purely coupled or uncoupled behaviour.

A further source of complexity in the structure of the arc, not necessarily depending from plate coupling, derives from the presence of extension

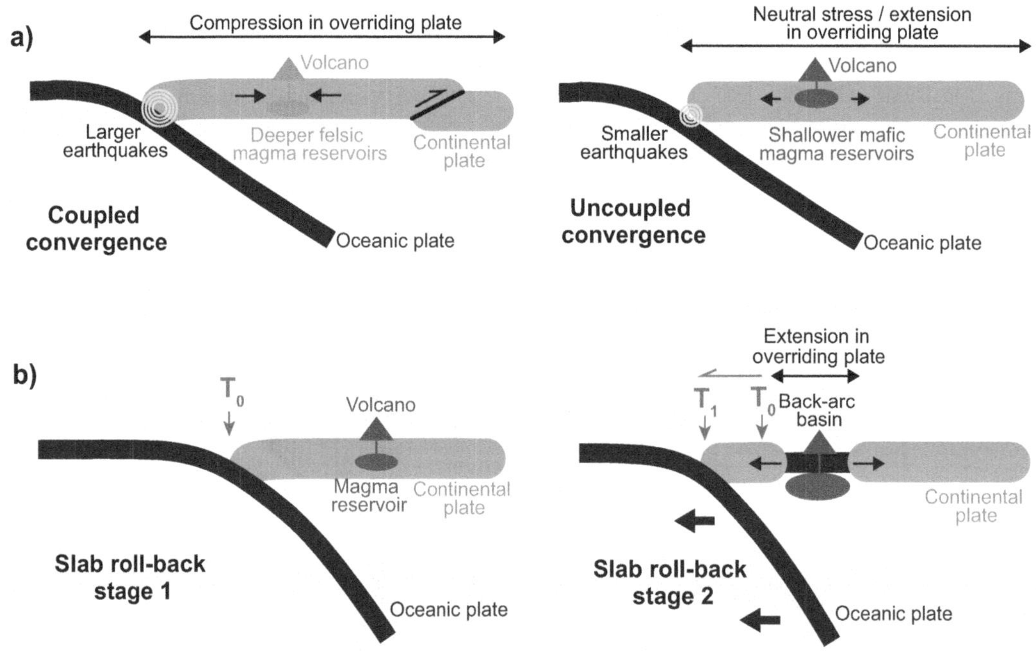

Fig. 10.14 **a** Main features of a coupled (left) and uncoupled (right) convergent plate boundary, illustrating the seismicity along the plate boundary, the state of stress within the overriding plate and depth and composition of magma reservoirs (section view). **b** Two progressive stages of slab roll-back, with the trench moving backward from position T_0 to T_1, determining the opening of a back-arc basin with oceanic crust and widespread mafic volcanism

in the arc associated with a wider area of back-arc opening, which develops a back-arc basin. **Back-arc opening** results from the differential kinematics of the trench, the subducting plate and the overriding plate. This condition is often related to the **roll-back** of the negatively buoyant subducting lithosphere, which occurs when the slab migrates oceanward, inducing the oceanward migration of the trench (Fig. 10.14b; e.g., Uyeda and Kanamori 1979; Sdrolias and Muller 2006). During the development of a back-arc basin, extension of the overriding plate enhances the decompression melting of the asthenosphere and, in turn, magma generation. Slab roll-back is thus responsible for an extensional back-arc basin with significantly thinned crust, which may also experience oceanization, as observed for example in the Tyrrhenian Sea (Italy), the Japan Sea and the Havre Trough-Lau Basin in the southwest Pacific.

In general, extension along a volcanic arc is the ultimate requisite for significant volcanic activity, so that the higher the amount of extension, the higher is the arc output rate. The extension along the arc may be regional, as along extensional arcs, or local, as in arcs where more limited extension may be related to the activity of strike-slip structures in releasing bends, pull-apart structures, dilational jogs or horsetail structures along principal displacement zones. Contractional conditions may also accompany volcanism, as testified by volcanic arcs undergoing predominant shortening. Different mechanisms may be invoked to explain the emplacement and rise of magma in arcs experiencing shortening. For example, contraction promotes the emplacement and accumulation of magma, as through sills feeding stacked systems or tabular intrusions (as laccoliths, lopoliths) at different crustal levels. Also, stacked magma may locally increase the vertical component of the principal stress, which may pass from the least component σ_3 to the intermediate component σ_2 and finally to the maximum component σ_1, promoting strike-slip and extensional structures, respectively. The latter may also enhance local arc-parallel extension. In addition, transient variations in the stress field induced by major regional earthquakes (or mega-earthquakes, with magnitude $M > 8.5$) may induce temporary extension in arcs otherwise experiencing contraction or strike-slip motions, also encouraging volcanic activity (see Chap. 12). Moreover, the activity of strike-slip or transtensive fault systems related to variations in the amount and direction of shortening in contractional arcs may locally decrease the minimum principal stress σ_3, increasing the deviatoric stress and further encouraging the rise of magma through dikes (Fig. 10.12; Tibaldi 2005; Galland et al. 2007; Acocella et al. 2008; Shabanian et al. 2012).

In some cases, the location and/or relevant activity of large volcanoes may be not directly related to the presence of a slab in a convergent setting, as inferred for example for the volcanism occurring along the northern Tonga Arc (southwest Pacific Ocean), Mount Etna and the cluster of large volcanoes within the Kamchatka depression (Russia), which includes the imposing Klyuchevskoy and Sheveluch volcanoes. In fact, the location of these volcanoes and, for the Etna and Kamchatka cases, also the high erupted volumes have been related to their peripheral position at the slab edge, a condition that promotes the suction of asthenospheric material from under the neighbouring plate to the side. Such a lateral flow is expected when descending slabs migrate backwards in the mantle (roll-back), leaving low-pressure regions to the front. In this case, the asthenospheric return flow from the back to the front position at the edge of the slab induces mantle upwelling, eventually driving decompression melting and promoting volcanism. In particular, the generation of magma at Mount Etna, in a clear off-arc location, is largely explained by this process (Fig. 10.15; Wendt et al. 1997; Gvirtzman and Nur 1999; Peyton et al. 2001; Faccenna et al. 2011).

Magma may be also produced through different processes in non-subducting collisional orogens. When subduction ceases and continental collision takes over, slab breakoff may promote magma generation. **Slab breakoff** consists of the tearing of the subducting plate due to the buoyancy-driven detachment of the denser subducted oceanic lithosphere from the lighter continental

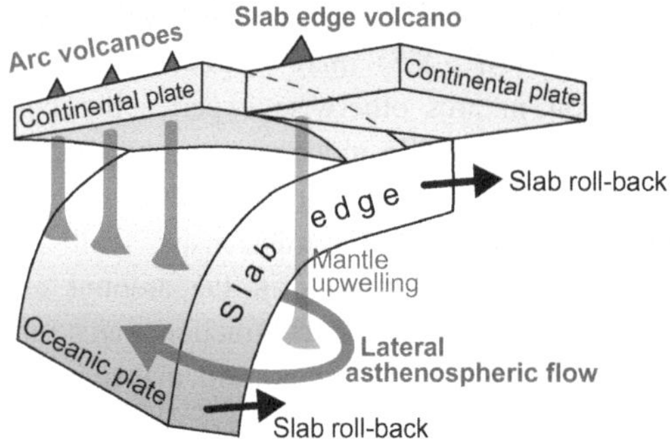

Fig. 10.15 Scheme of asthenospheric window (in purple) formed at the lateral edge of a slab experiencing roll-back, oblique view. The lateral asthenospheric flow induces local mantle upwelling at the slab edge; the asthenospheric material penetrates the plate boundary and reaches the crust in the forearc region, generating volcanism above the slab edge

lithosphere that follows it during continental collision (Fig. 10.16; e.g., Davies and von Blanckenburg 1995). Breakoff leads to heating of the overriding lithospheric mantle by the upwelling asthenosphere intruded within the tear and also by decompression due to the removal of the slab. This thermal perturbation induces the melting of the metasomatised overriding mantle lithosphere, producing basaltic magmatism that leads to granitic magmatism in the crust (bimodal magmatism). Slab breakoff has been invoked to explain the existence of anomalous volcanism within collisional and subduction domains, as in Eastern Anatolia, Central Italy and at Toba (Sumatra; Keskin 2003; Rosenbaum et al. 2008; Hall and Spakman 2015; Koulakov et al. 2016).

Melting and thus volcanism may also be caused by **lithosphere delamination**, when the deep keel of lithosphere that develops during thickening drops off (Fig. 10.17; e.g., van der Pluijm and Marshak 2004; Fossen 2010). Warm asthenosphere rises to take its place and heats the remaining lithosphere, possibly partially melting a broad portion of the mid crust. Lithospheric delamination has been invoked to explain the presence of the isolated but imposing Damavand volcano, 5600 m high, within the Alborz collisional orogen of Iran. At a different scale, piecemeal lithospheric delamination has been also suggested to explain part of the impressive

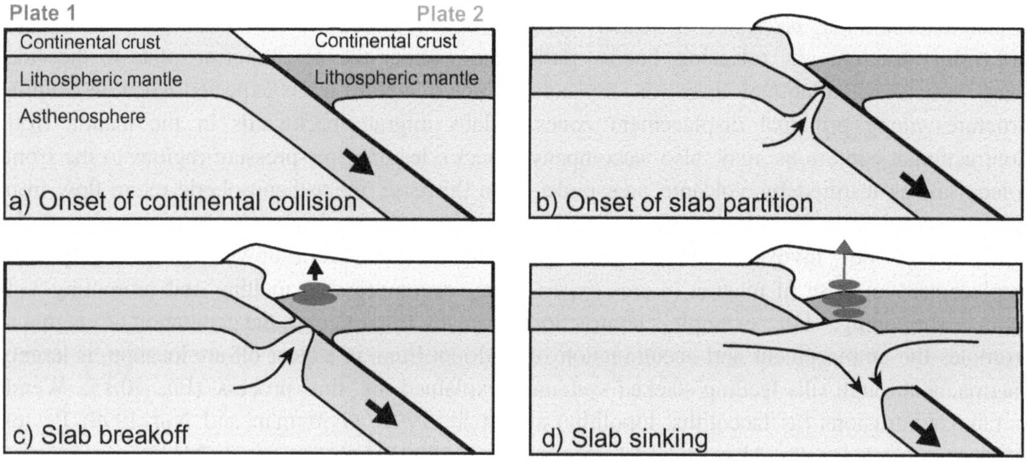

Fig. 10.16 Schematic slab breakoff process, section view. **a** Continental collision begins when all the oceanic lithosphere of plate 1 has already sunk and the continental lithosphere of both plates comes into contact. **b** The density difference between the heavier oceanic slab at depth and the lighter continental lithosphere at the surface determines the partitioning of the slab. **c** The slab undergoes breakoff; the impingement of hot asthenosphere at the base of metasomatised continental lithosphere leads to magmatism (in red) and the release of the load leads to uplift. **d** As the slab sinks away, a deep return flow develops above, maintaining a high temperature at the base of the lithosphere, promoting magmatism (modified after Davies and von Blanckenburg 1995)

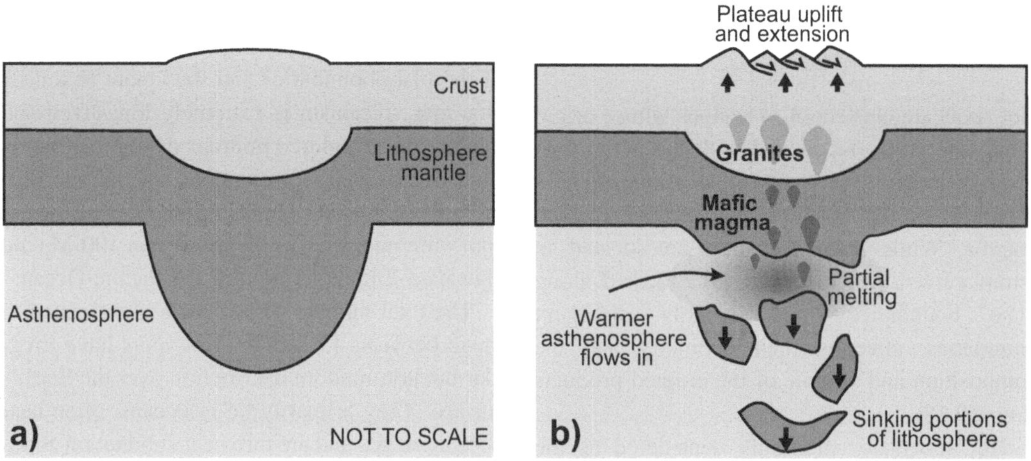

Fig. 10.17 Lithospheric delamination, section view. **a** Thickening of lithosphere forms a keel-shaped mass of cool lithosphere to protrude down into the asthenosphere. **b** The keel drops off and is replaced by warm asthenosphere, causing partial melting and formation of post-orogenic plutons. As a consequence, the crust may uplift (modified after van der Pluijm and Marshak 2004)

volcanism of the Central Andes, in a subduction setting (Kay and Kay 1993; Beck and Zandt 2002; Schurr et al. 2006; Liotard et al. 2008; Shabanian et al. 2012). Lithosphere delamination may also explain the anomalous uplift of portions of both subducting and collisional convergent boundaries, as the Central Andean and the Tibetan plateaus, respectively. In fact, the hot asthenosphere at the base of the remaining lithosphere would cause the lithosphere to heat up. Thus, to maintain isostatic equilibrium, the surface of the lithosphere would have to rise. In other cases, very thick orogens collapse under their own weight, spreading laterally (**orogenic collapse**). Because the extension associated with the collapse thins the upper crust, it causes decompression, which may promote partial melting of the deep crust and the underlying asthenosphere, producing magmas. Magma may also form and intrude the upper crust after the collision has ceased: in this case, the magma is **post-orogenic**.

Volcanic arcs mainly consist of andesitic stratovolcanoes, with subordinate but large rhyolitic caldera systems. The latter constitute the largest volcanic systems on Earth and are often associated with specific tectonic conditions promoting deep partial melting and inducing the pervasive rise of asthenospheric material, as explained in the following chapters. Arc volcanoes do not show frequent unrest, volcanic activity and open conduits, as magma is usually stored at higher depth with regard to divergent plate boundaries and occasionally reaches the surface to erupt (e.g., Chaussard and Amelung 2012). As a result of a moderately to highly viscous magma experiencing infrequent eruptions, explosive volcanism predominates along volcanic arcs. The volcanic edifices often show evidence of the regional convergence conditions. In fact, many volcanic edifices are elongated parallel to the direction of the maximum principal stress σ_1. In addition, several arc volcanoes show a local alignment parallel to the σ_1 direction. These features, as for example observed along the arcs of the Andes and Japan, result from the preferred development of feeder dikes parallel to the regional σ_1 direction, determining the elongation and alignment of the volcanic edifices (Nakamura 1977; Nakamura et al. 1977).

10.5 Magmatic Processes at Hot Spots

Hot spots are clusters of volcanoes whose origin cannot be easily reconciled with the activity of plate boundaries, in terms of location and distribution, or composition and flux of erupted magma. While most hot spots are located in intraplate settings, many are also located along plate boundaries, most notably divergent boundaries, affecting their geometry and the composition and volume of the erupted products (Fig. 10.18).

Hot spots are commonly considered as the surface expression of mantle plumes. **Mantle plumes** may originate at different depths within the mantle and consist of the localized rise of material hotter and less dense than the surrounding mantle rocks. Mantle plumes play a key role in transferring heat from the mantle to the lithosphere, also locally influencing the geometry of the plate boundaries and thus plate tectonics. Hot spot volcanism is extremely long-lived and may last for a hundred millions of years, or more. This is indicated, for example, by the age progression of volcanic chains formed during the hot spot's lifetime, such as the more than 100 Ma old Hawaiian-Emperor Chain in the Pacific Ocean.

The total number of hot spots is estimated to range between 45 and 70. Hot spots have irregular but non-random distribution over the Earth's surface. They primarily lie in oceans, often near oceanic ridges and are rare near subduction zones (Courtillot et al. 2003; Zhao 2004). This irregular distribution may result from several factors. First, as oceans cover most of the Earth's surface, most hot spots are expected to lie in oceans. Second, mantle plumes can weaken the continental

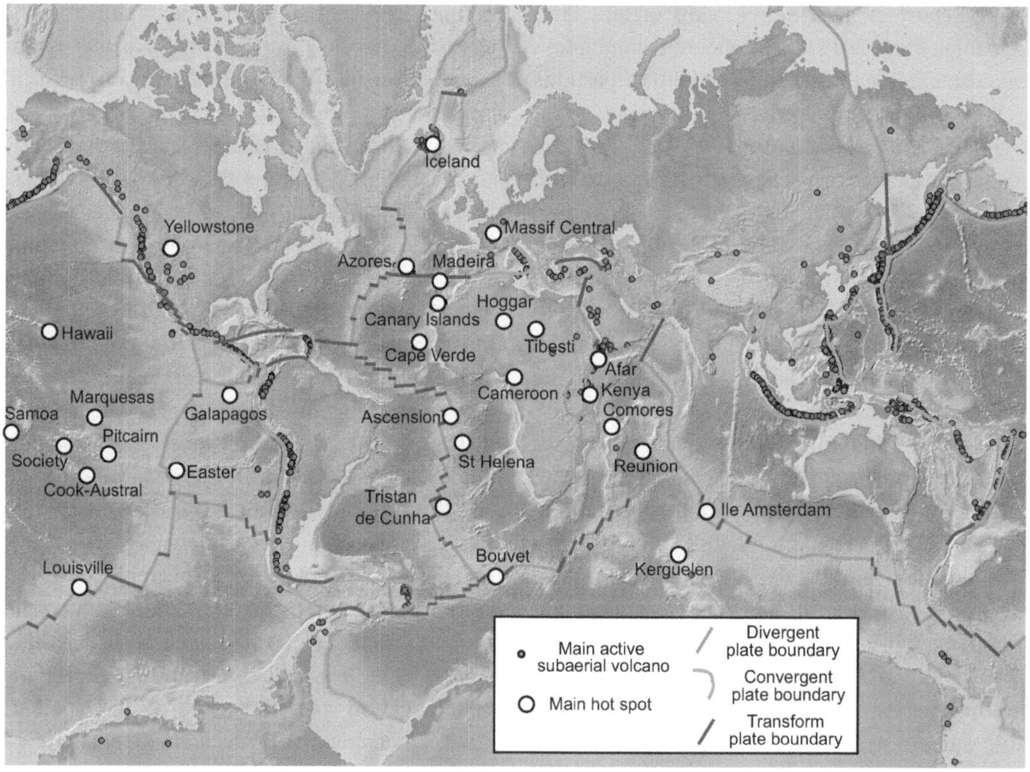

Fig. 10.18 Schematic distribution of the major hot spots in the frame of plate tectonics, with divergent (orange lines), convergent (purple lines) and transform (red lines) plate boundaries and major Holocene volcanoes. Base DEM provided by https://www.geomapapp.org

lithosphere and enhance its break-up, promoting the activity of divergent plate boundaries ultimately evolving in oceanic domains. Therefore, although many hot spots may have originated on continents, they currently appear on oceans. Third, the difference in thickness (100–150 km) between the oceanic and continental lithosphere may control partial melting. While weak plumes may not reach the conditions for melt generation beneath the thicker continental lithosphere, they may do so beneath a thinner oceanic lithosphere, at lower pressure.

The best-defined hot spots appear relatively stationary over time, as their respective motions (≤ 1 cm/year) are lower than the plate velocities. This feature, together with the long-lived hot spot volcanism, has been used to exploit hot spots as a near-fixed reference frame to determine absolute plate motions (Courtillot et al. 2003). Typically, the direction of plate motion is given by the direction of migration of the aligned hot spot volcanoes; in some cases, as in the Hawaiian-Emperor Chain, the alignment of volcanoes shows major changes in direction, indicating important variations in plate motion. The average rate of plate motion is given by the distance of a certain volcano from the active site of the hot spot with regard to the oldest age of activity of the volcano.

Plume-like slow seismic velocity anomalies are tomographically detected under major hot spot regions, as Hawaii, Iceland, South Pacific, and Africa. The slow anomalies under South Pacific and East Africa extend more than 1000 km, representing two **superplumes** in opposite locations of the Earth's mantle. The slow anomalies under hot spots normally do not show a straight pillar shape, but exhibit winding images, suggesting that even plumes can be deflected by mantle flow (Fig. 10.19; McNutt and Judge 1990; Romanowicz and Gung 2002; Zhao 2004; O'Connor et al. 2019).

Data acquired in the last decades have improved the understanding of the depth of

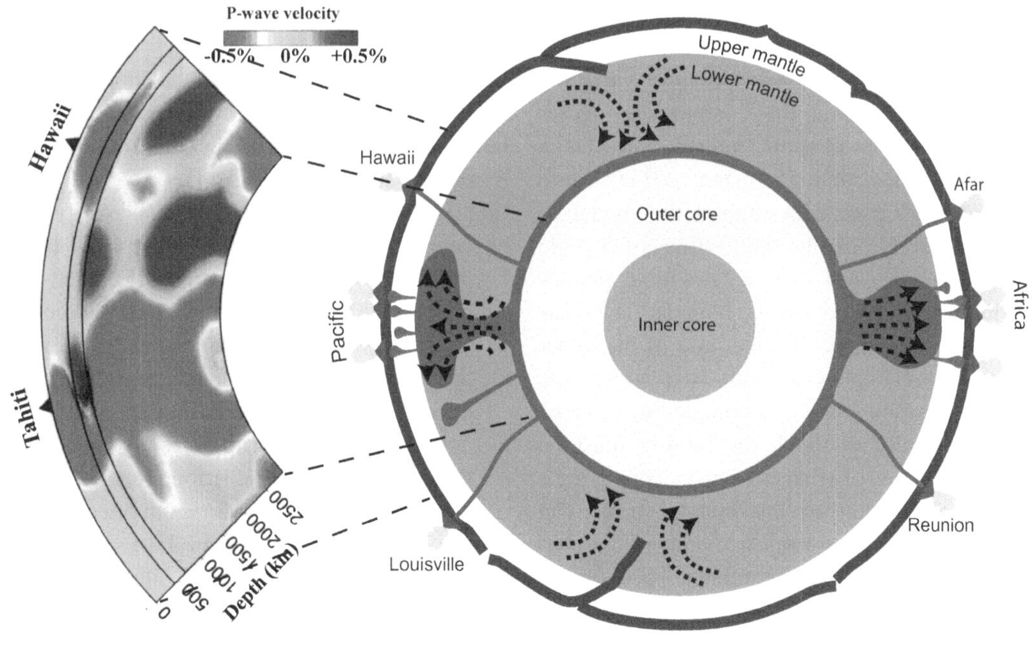

Fig. 10.19 Mantle plumes feeding hot spots. Right: schematic cross section of the Earth, outlining the sources of the different types of plumes/hot spots and the two antipodal superplumes upwelling below the South Pacific and East Africa (Courtillot et al. 2003). Left: north–south cross-section of P-wave velocity image along a profile passing through Hawaii and the South Pacific superswell; red and blue colours denote slow and fast velocities, respectively (Courtesy of Dapeng Zhao; Zhao 2004)

generation and shape of mantle plumes. The mantle plumes are thought to be generated by Rayleigh–Taylor instabilities from **thermal boundary layers**, or zones with higher temperature and viscosity gradient within the mantle. The most important of these layers, named *D″*, lies at the core-mantle boundary. Seven primary hot spots, namely Hawaii, Easter, Louisville, Iceland, Afar, Reunion, and Tristan, were initially inferred to have such a deep origin: all these hot spots, except Iceland, are close to or at the margins of the two superplumes beneath East Africa and South Pacific. The remaining, non-primary hot spots may form in the transition zone at the bottom of the upper mantle or at shallower levels, and then be linked to lithospheric break-up (Courtillot et al. 2003). Also, fluid dynamics studies have suggested narrower cylindrical conduits of radius of 50–150 km feeding large spherical heads, requiring important viscosity variations in the thermal boundary layer, with temperature differences between plumes and the surrounding mantle of a few hundred degrees. These large heads would reflect the dynamical requirements for a plume to ascend into a higher-viscosity environment, whereas the narrow trailing conduits would form because the low-viscosity fluid within the conduit rises faster than the plume head itself (Jellinek and Manga 2004; Farnetani and Hofmann 2011). More recent tomographic data support the model of multiple levels for the generation of mantle plumes, and suggest that at least nine plumes originate near the core-mantle boundary, whereas eight or more originate at the base of the upper mantle. This implies at least two convecting regions, possibly linked in a complex way. To be tomographically resolved, the plumes must be several hundreds of kilometres wide, conveying a substantial fraction of the internal heat. The imaged plumes show thick and subvertical conduits beneath many prominent hot spots, extending from the core–mantle boundary to about 1000 km below Earth's surface. These conduits are thus much broader than classical thermal plume tails, suggesting that they can carry more heat away from the core than it was previously thought (Montelli et al. 2004; Zhao 2004; French and Romanowicz 2015; Hand 2015).

The longevity and structure of the mantle plumes result from interactions between core cooling, dense low-viscosity material at the thermal boundary layer and plate tectonics. As anticipated, plumes may be affected by the flow regimes in the upper mantle. Weak plumes may be sufficiently deflected by large-scale mantle flow driven by subduction and plate tectonics, so that they may break up into diapirs (Jellinek and Manga 2004; Kerr and Meriaux 2004). Also, a mantle plume head can separate from its trailing conduit upon passing the interface between the upper and the lower mantle. The detached plume head may eventually trigger a first volcanic (usually flood basalt, see below) event at the surface. The remaining conduit forms a new plume head, which may cause a second flood basalt event at least 10 Ma after the first one. This interaction between the plume and the mantle discontinuities may explain distinct episodes of major eruptions, commonly separated by between 20 and 90 Ma within several of the world's flood basalt provinces (Bercovici and Mahoney 1994).

At shallower levels, the interaction between ascending plumes and the lithosphere includes, but it is not restricted to, the following aspects (Farnetani and Hofmann 2011).

(a) The control of the thickness, elasticity and age of the lithosphere on the spatial distribution of plume magmatism (Ebinger and Sleep 1998; Hieronymus and Bercovici 2001). As regards the lithosphere thickness, lateral flow from a plume impinged beneath a relatively thin lithosphere may be channelled along pre-existing rift-zones, explaining a wider and selective distribution of magmatism at the surface, as for example observed above the Afar plume in East Africa. As regards the elasticity of the lithosphere, the interaction of magma transport with lithospheric flexure may affect the spacing of volcanoes in hot spot chains through the generation of flexure-induced tensile stresses that determine where magma preferentially extrudes. In turn, plumes may also affect the structure and mechanical properties of the lithosphere, especially through magmatic underplating. As regards the age of the lithosphere, this is a main feature controlling the density of oceanic plates, in turn affecting the topography of hot spot tracks. In fact, the prolonged hot spot activity over a relatively fast moving plate develops a volcanic chain, or hot spot track, parallel to the plate motion direction. Topographically smooth and broad hot spot tracks, as the Galapagos, lie on young and lighter lithosphere, which favours magma intrusion. Conversely, rough and discontinuous hot spot tracks (as Hawaii) lie on older, thicker and denser lithosphere, which encourages volcanism (Orellana-Rovirosa and Richards 2017).

(b) A dynamic topography induced by the arrival of a plume head, commonly testified by a ~ 1000 km wide and ~ 1 km anomalously high topographic swell. Usually, the continental swells are much smaller than the oceanic ones. Hot spot swells have been used to estimate the flux of material of the mantle plumes (i.e., buoyancy flux) and this flux of material has been in turn used to estimate the amount of heat transported to the surface. The **buoyancy flux** of hot spots is a measure of the strength of the plume, which depends upon the buoyancy driven volume flux. The buoyancy flux B is defined as:

$$B = LH_e(\rho_{ma} - \rho_w)\, V_h \qquad (10.2)$$

where L is the width of the swell, H_e is the excess elevation averaged across the swell, ρ_{ma} is the mantle density, ρ_w is the water density (in the case of oceanic hot spots) and V_h is the plate velocity in the hot spot frame. The buoyancy flux, measured in Mg/s, ranges over a factor of 20, Hawaii being the largest; generally, the largest buoyancy fluxes are found in the Pacific region (e.g., Sleep 1990; King and Adam 2014). The topographic swell induced by the hot spot implies a rapid surface uplift, which may be often followed by flood volcanism by 1–2 Ma. Subsequently, there is a slower subsidence and, eventually (if the plume is below a continent), continental break-up, as in Afar (Farnetani and Richards 1994). The buoyancy flux allows quantifying hot spot activity, although its use becomes irrelevant over long periods, due to the own subsidence of the hot spot swell. In this case, to quantify hot spot activity it may be more relevant to determine the production rate of volcanic material, although this refers only to the surface manifestation of the hot spot (Vidal and Bonneville 2004).

(c) Hot spot-ridge interaction(s): there are strong feedbacks between the dynamics of slowly migrating ridges and deeply sourced plumes, also able to explain the repeated formation of large igneous provinces (see point d below). At least 21 hot spots are situated near a spreading ridge, with evident topographic, thermal and crustal thickness variations. The Oceanic Island Basalts (OIBs) produced by the hot spot are geochemically distinct from Mid-Ocean Ridge Basalts (MORBs). This suggests that the lower mantle, the inferred source of plumes, is compositionally different from the upper mantle, source of MORBs (Ito et al. 2003; Farnetani and Hofmann 2011; Whittaker et al. 2015).

(d) The onset of massive hot spot magmatism, often associated with a large igneous province, which results from the melting of the plume head, while the subsequent hot spot activity is associated with the long-lasting and narrower plume tail, as for example observed with the Yellowstone hot spot (see Sect. 13.8).

In particular, **Large Igneous Provinces**, or LIPs, are exceptional volcanic events responsible for a large total volume of dominantly mafic magma (up to millions of km^3) erupted over a brief period (usually < 1 Ma). Furthermore, the volume of magma erupted during each of the individual eruptions that make up a LIP (frequently 10^3–10^4 km^3) is also exceptional. Without LIP-forming igneous events, basaltic super-eruptions would not have occurred, nor indeed many of Earth's largest volcanic deposits: for example, the largest continental flood basalts province, the 252 Ma old Siberian Traps, might have totalled 4–5 million km^3 of lava. These large erupted volumes and the relatively frequent eruptions lead to the geologically rapid construction of extensive lava plateaus, up to several kilometres thick. Large igneous provinces are found on continental crust, in ocean basins and in transitional domains, and their rocks are distinguished from the products of other types of magmatism on the basis of petrologic, geochemical, geochronological, geophysical and volcanological data. LIPs come in two broad compositional varieties: basaltic LIPs, which form flood basalt provinces and are the most frequent, and silicic LIPs, which are few in number and form some of the major ignimbrite provinces (Fig. 10.20; Coffin and Eldholm 1994; Bryan et al. 2010; Self et al. 2015).

Owing to their surface exposure, continental flood basalts are the most intensively studied large igneous provinces. These are dominated by tholeiitic mafic lavas most likely originated in less than 100,000 years from prolonged eruptions, each lasting years to decades, with the largest flow fields fed by long fissures. A short-lived and isolated recent example of such fissure activity may be the 1783–84 Lakagigar eruption (Iceland). This, erupting 14 km^3 of basaltic magma in 8 months, provides the largest historical lava flow eruption. The voluminous lava flow-fields associated with large igneous provinces do not require rapid extrusion of mafic magma at rates much higher than the 1783–84 Lakagigar type of eruptions, nor do long fissures imply high eruption rates, as only single segments of the LIPs fissures were likely active at any one time. Taking the 1783–84 Lakagigar fissure eruption as a reference, the time required to emplace a large 1000–2000 km^3 lava flow at the range of its peak output rates would be approximately 10–20 years.

Large igneous provinces are not uniformly distributed over the Earth's history. Statistical studies suggest periodicities in LIP production at approximately 30–35 and 60 million years (Self et al. 2015). Such episodicity probably reflects variations in rates of mantle circulation. Large igneous provinces repeatedly formed at locations where oceanic ridges and plumes interact. In particular, slowly migrating ridge systems that have been stabilized by upwelling plumes have extracted large volumes of material from the same part of the upper mantle over periods up to hundreds of million years. This indicates a feedback between the dynamics of slowly migrating ridges and deeply sourced plumes (Whittaker et al. 2015).

The most common hypothesis to explain large igneous provinces is that these are generated when a rising mantle plume impacts the lithosphere, although alternative hypotheses also exist (Campbell 2005). As in the case of hot spot volcanoes, large igneous provinces are commonly attributed to decompression melting of hot and low-density mantle material ascending in mantle plumes. These plumes would initially transfer huge volumes (10^5–10^7 km^3) of mafic rock into localized regions of the crust over short intervals (10^5–10^6 years). This large-volume magmatism during LIP formation is commonly attributed to mantle plume heads reaching the crust, whereas the persistent (hot spot) magmatism is considered to result from mantle plume tails penetrating the lithosphere, which may be moving relative to the plume. Modelling of the impingement of a plume head below the lithosphere shows that the large head spreads out to form a thermal dome with the potential to form basaltic eruptions by decompression melting over an area that may be as much as 2000 km in diameter (Campbell 2005). The unusually large melt fractions formed in the mantle by this process cause the high extrusion rates that typify flood basalt provinces. As lithospheric plates move across a plume, the plume head disperses, leaving only the plume tail erupting basaltic magma at a much lower rate. These tails create chains of volcanic islands that terminate in active volcanoes marking the current position of the mantle plume, as at Hawaii or La Reunion.

The fact that not all large igneous provinces have obvious connections with mantle plumes, or even hot spot tracks, suggests that more than one

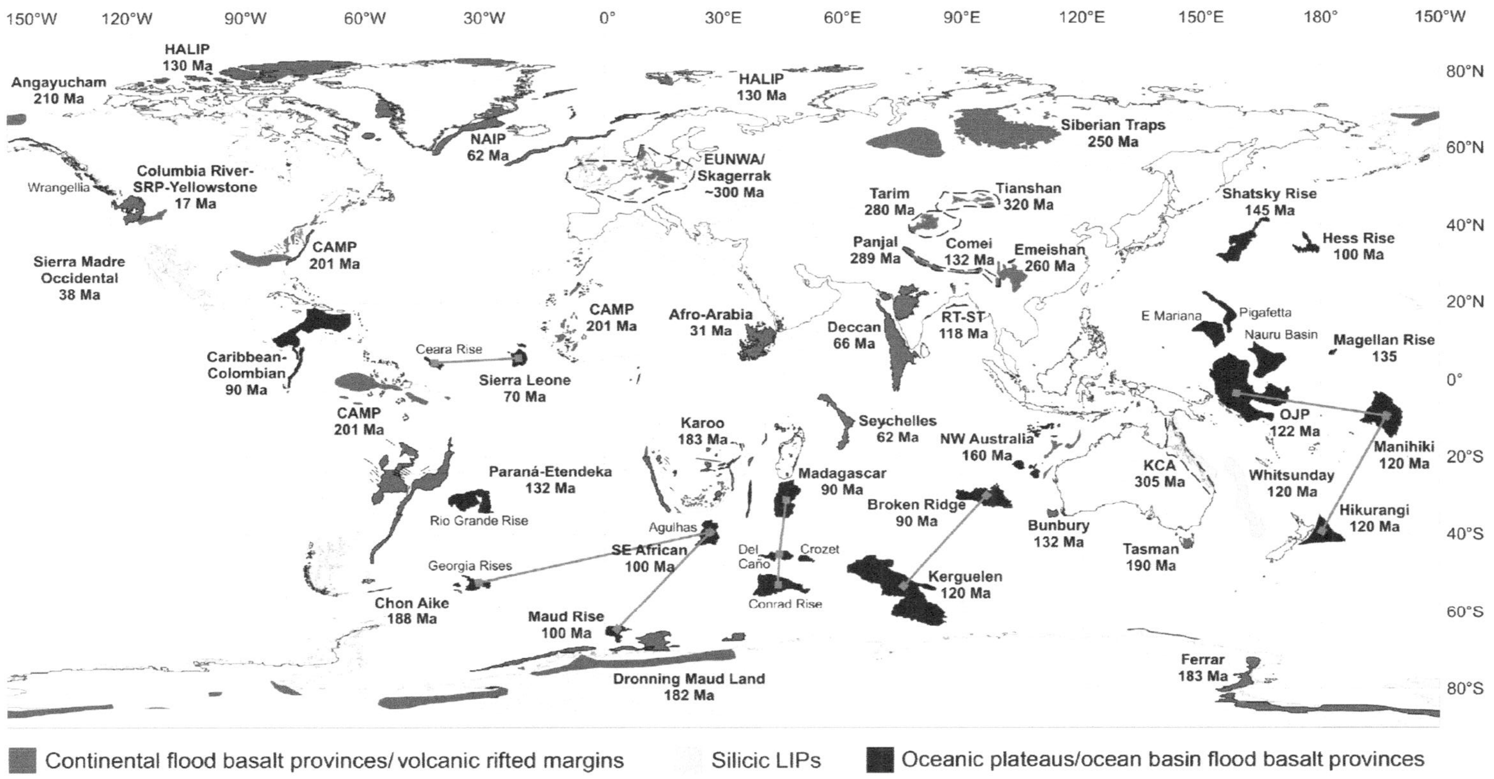

Fig. 10.20 Large igneous provinces (LIPs) of the world, including continental flood basalt provinces, oceanic plateaus, and silicic LIPs. Ages are those of main phase or pulse of volcanism. Inferred extent of some provinces shown by dashed lines; tie-lines join parts of same original province (Self et al. 2015)

source model is required for their formation (e.g., Self et al. 2015, and references therein).

10.6 Polygenic and Monogenic Volcanism

In any of the above-mentioned tectonic settings, magmatism may manifest at the Earth's surface through polygenic and/or monogenic volcanoes. Polygenic volcanoes consist of larger and long-lived volcanic edifices growing from repeated eruptions, whereas monogenic volcanoes consist of smaller volcanic edifices resulting from a single eruptive event (Fig. 10.21a, b).

The local relationships between polygenic and monogenic volcanoes, including the conditions controlling the development of satellite monogenic volcanoes on the flanks of polygenic volcanoes, have been broadly discussed throughout Chap. 7. Here the focus is mainly on the regional context controlling the development of polygenic and/or monogenic volcanoes, including the monogenic volcanoes formed outside and independently of polygenic volcanoes. These volcanoes produce widespread **monogenic fields**, consisting of a set of vents, each of which has erupted once or only a few times, and whose number varies from tens to several hundreds. The composition of these monogenic fields is usually mafic (as for example the Auckland Volcanic Field, New Zealand, the Chaine des Puy, France, and Al-Madinah, Saudi Arabia), although felsic monogenic vents, often related to lava domes, are also found (as in the Central Andes; e.g., Le Corvec et al. 2013; de Silva and Lindsay 2015). While it is usually believed that monogenic volcanoes in a field erupt independently from each other, with only the nearest volcanoes being possibly related to a same eruptive fissure or event, there is also repeated evidence, as in the Auckland Volcanic Field, of contemporaneous but separated and structurally unrelated volcanoes. This suggests that spatial and temporal

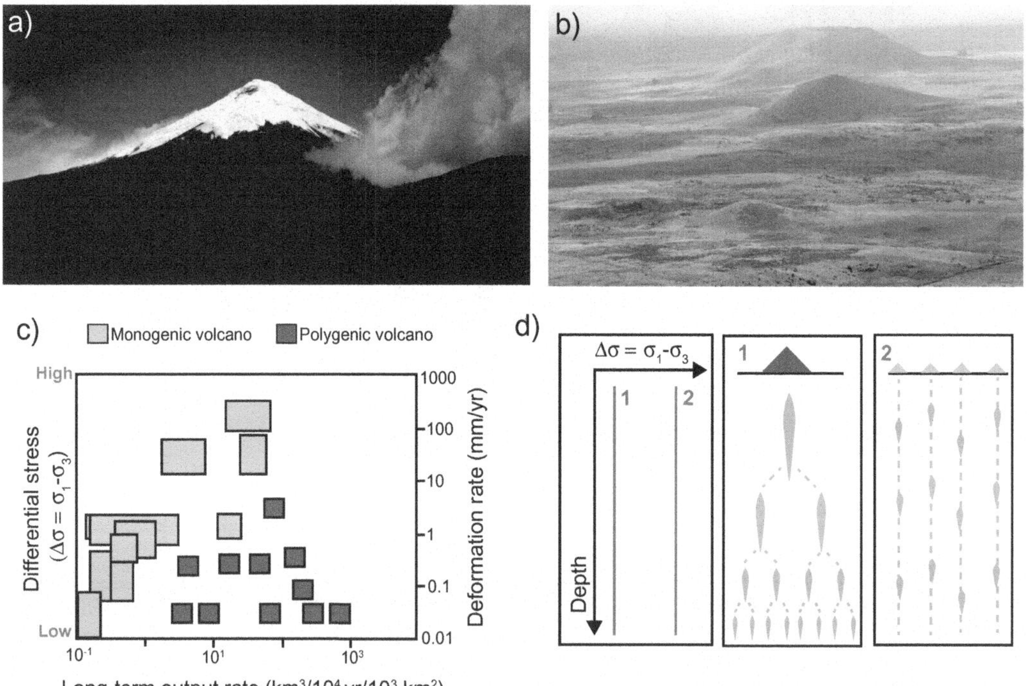

Fig. 10.21 Polygenic and monogenic volcanism. **a** View of the polygenic Cotopaxi volcano, Ecuador. **b** Aligned monogenic volcanoes at Rapa Nui (Easter Island, Chile). **c** Stress and magma flux conditions promoting polygenic or monogenic volcanism and **d** related theoretical model based on the crack interaction mechanism (Takada 1994)

heterogeneities can occur in monogenic fields, with important implications for statistically based hazard assessments (Cassidy and Locke 2010). The development of monogenic volcanic fields is mainly controlled by the regional tectonic setting: although found in any tectonic environment, most monogenic fields are in fact related to extensional regimes.

More in general, the regional tectonic setting controls the occurrence and the distribution, in terms of spacing and alignment, of polygenic or monogenic volcanoes. In fact, the possibility to develop polygenic or monogenic volcanoes has been inferred to depend upon the differential stress $\Delta\sigma$, that is the difference between the maximum and minimum principal stresses σ_1 and σ_3, respectively (with $\Delta\sigma = \sigma_1 - \sigma_3$). In particular, monogenic volcanoes focus in settings with higher differential stress, whereas polygenic volcanoes focus in settings with lower differential stress (Fig. 10.21c, d; Takada 1994). This follows from the fact that a higher differential stress is associated with a minimum principal stress σ_3 that is much smaller than the maximum principal stress σ_1, a condition that promotes the propagation of independent and non-coalescing dike paths, preferably feeding a monogenic volcanic field. Conversely, a lower differential stress results from a σ_3 approaching σ_1 in magnitude, a condition that promotes the coalescence of multiple dikes and magmatic sheets that tap magma over a wider area below a polygenic volcano. The possibility to develop polygenic or monogenic volcanoes depends, in addition to the tectonic setting, on magma availability (Takada 1994). In fact, a higher availability of magma, or magma flux, is more likely to result in the build up of a large and long-lived magma chamber that is usually required to feed a polygenic volcano. These two conditions suggest that monogenic volcanoes result from lower magma fluxes mainly in extensional settings and polygenic volcanoes result from higher magma fluxes mainly in neutral (neither compressive nor extensional) tectonic settings. This explains why monogenic volcanic fields are more frequent along divergent plate boundaries, although a same tectonic setting may show both polygenic and monogenic volcanoes, as observed on divergent (as the East African Rift System and the oceanic ridge of Iceland) and convergent plate boundaries (as the Trans Mexican Volcanic Belt and the Central Andes). Note that, while monogenic volcanism is commonly fed by dikes, there is also evidence, as for example at the Hopi Buttes Volcanic Field (Arizona, USA), of saucer-shaped sills feeding a significant amount of monogenic volcanoes. This indicates that shallow feeders in monogenic fields can form geometrically complex networks, particularly those intruding poorly consolidated sedimentary rocks (Muirhead et al. 2016).

The distribution of polygenic and monogenic volcanoes, as expressed through their spacing and alignment, also depends upon the tectonic context. The distance or **spacing** between nearby polygenic volcanoes may be related to several first-order factors. In general, the thinner the crust or the lithosphere, the more closely spaced are the volcanoes, as observed in the Main Ethiopian Rift—Afar area (Fig. 10.22a; Mohr and Wood 1976). This behaviour, supported by analogue models, is described by the development of Rayleigh–Taylor instabilities, where the wavelength associated with a dominant disturbance is proportional to the thickness of the upper layer, indicating that the thicker is the upper layer the more distant are the instabilities (see Sect. 3.2.1; Fig. 10.22b; e.g., Ramberg 1981; e.g., Turcotte and Schubert 1982). In a similar fashion to polygenic volcanoes, the distribution of the spacing of monogenic vents has been related to the thickness of the crust below various volcanic fields in different extensional tectonic settings. This highlights a relationship between the distribution of monogenic vents and the depth of origin of the magma (e.g., Mazzarini 2007; Mazzarini et al. 2010). These behaviours provide a convenient framework to explain volcano spacing as a consequence of variations in lithospheric or crustal thickness due to regional extension. As regards convergent plate boundaries, the spacing of the polygenic arc volcanoes does not show characteristic values and is more difficult to interpret, although probably related to random distributions

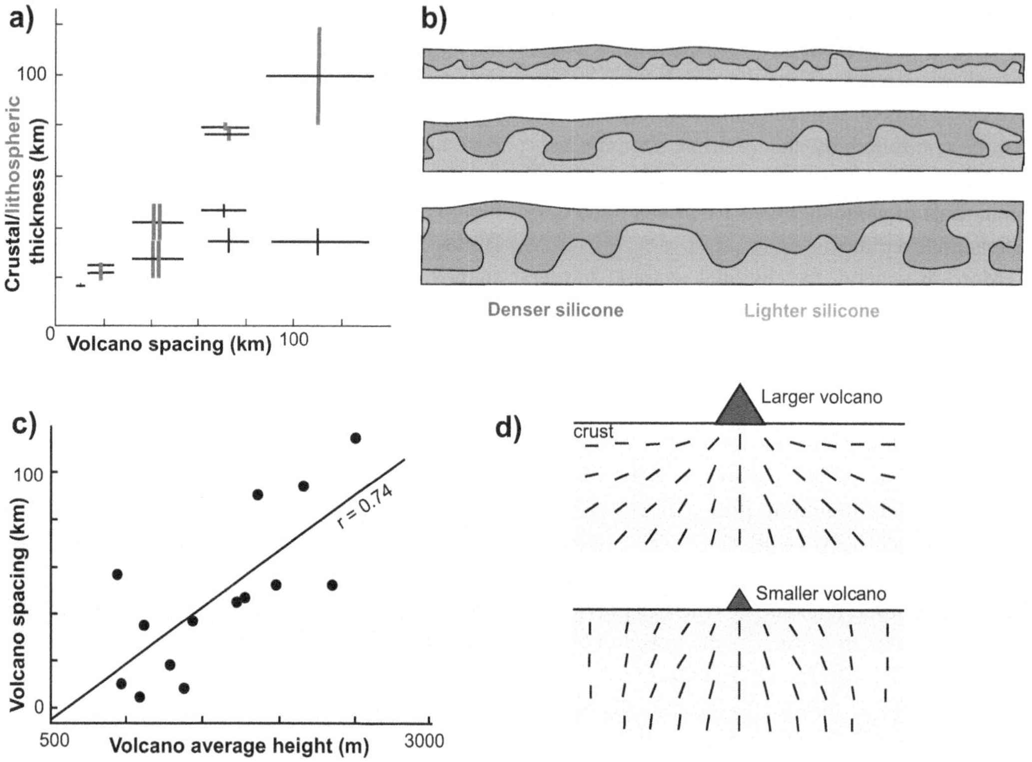

Fig. 10.22 Two models **a**, **b** and **c**, **d** used to explain polygenic volcano spacing. **a** The spacing of volcanoes along the Afar-Main Ethiopian Rift is proportional to the crustal (black lines) or lithospheric (blue lines) thickness (Mohr and Wood 1976). This behaviour is consistent with the three sections of analogue models in (**b**), characterized by two layers of silicone (viscosity of 10^4–10^5 Pa s), with the upper denser layer increasing its thickness from top to bottom (Ramberg 1981): the spacing between nearby intrusions of the lighter silicone increases with the thickness of the upper layer. Both behaviours are explained by the Rayleigh–Taylor theory, where the disturbance wavelength is proportional to the upper layer thickness. An alternative mechanism **c** relates volcano height to volcano spacing, considering the volcanoes of the Cascades Arc. This behaviour is explained in **d** by the fact that higher (or larger) volcanoes impose a higher load (load trajectories shown as small black lines in section view) on the crust, capable of attracting magma over a wider region (Muller et al. 2001)

resulting from repeated instability events associated with the rise of magma in different locations (de Bremond d'Ars et al. 1995). Other studies relate polygenic volcano spacing to the load of a volcanic edifice: larger volcanoes impose a higher load on the crust, inducing a stronger local stress field capable to attract the magma below in the form of dikes curving from progressively wider regions and focusing towards the volcano. This capability to attract magma over wider areas may explain why larger volcanoes are also more distantly spaced, as in the Cascades Volcanic Arc (Fig. 10.22c; Muller et al. 2001). The flexural response of the lithosphere due to the load of a volcanic edifice may also control the distance between nearby polygenic volcanoes forming in progression, as on plates moving above mantle plumes. In particular, compressive stresses within a crust flexured by the load of a volcano prevent new upwelling nearby, forcing a new volcano to develop at a minimum distance that is equal to the distance in which the radial stresses change from compressional to tensile, that is the inflection point of a flexured lithosphere. The distance to this inflection point is proportional to the thickness of the plate (e.g., Tenbrink 1991; Bianco et al. 2005, and references therein).

Alignments of active polygenic volcanoes are related to the tectonically-assisted availability of magma. For example, aligned polygenic volcanoes in divergent plate boundaries form as a consequence of localized decompression along the axis of rift zones (see Chap. 11). Similarly, aligned polygenic volcanoes in convergent plate boundaries result from the availability of magma above the zone of partial melting of the slab, with the location of the volcanic arc ultimately controlled by the slab dip (see Chap. 12). At a more detailed scale, the alignment of monogenic volcanoes along a given direction, as often witnessed by eruptive fissures up to several tens of kilometres long, commonly results from the shallow emplacement of a dike, whose direction is controlled by the tectonic and gravitational stresses and, partly, by pre-existing tectonic structures. In the case of tectonic stresses, dikes tend to orient perpendicularly to the direction of the least principal stress σ_3. In the case of gravitational stresses, the stresses due to topography at very shallow crustal levels may cause a dike to deflect from its path, in order to experience the local least buttressing conditions (see Sect. 7.2). Numerical modelling indicates that such near-surface effects are important to a depth of approximately 500 times the dike width. In the case of pre-existing tectonic structures, a dike may in theory deviate from its orientation in favour of a more energy-efficient path along a pre-existing discontinuity, as a fault plane. As discussed in Sect. 3.5, the orientation of the pre-existing structure relative to the regional principal stresses (σ_3 in particular) mainly determines whether a fracture is more or less likely to dilate in response to dike injection and redirect the ascending dike. However, even in the most favourable conditions, as with fractures perpendicular to σ_3, dikes may not need to reactivate pre-existing discontinuities. This is supported by the slight but systematic offset in the location of several tens of eruptive vents from important normal faults along the East African Rift System, suggesting that other processes have a stronger control on dike propagation (Maccaferri et al. 2015; Valentine and Connor 2015). Therefore, most monogenic fields display a statistically clustered distribution of their vents, with the alignment of volcanoes depending on the tectonic environment and other factors (i.e., topography, pre-existing structures, local stress changes due to older intrusions; Le Corvec et al. 2013). However, volcanoes in monogenic fields may also not show any evident major alignment, displaying an apparently random orientation, or minor and subtler orientations, as for example in the Auckland Volcanic Field. Here, although minor alignments may be still identified, most vents do not seem to follow evident preferred trends. In these cases, statistics can be applied to estimate eruptive recurrence rates, while geochemistry can characterize any variability in the magma source characteristics and magmatic processes. Vent clusters within volcanic fields may be explained by spatial magma supply variations, reflecting the scale of temperature, pressure, and compositional variations within the underlying mantle, as well as the development of shallow magmatic plumbing systems. In synthesis, the overall distribution of vent clusters within volcanic fields suggests that the propagation of the magma is the product of the interplay between deeper level influences (i.e., nature of the magma source) and shallower level influences of the crust (i.e., stress field, pre-existing crustal structures; Le Corvec et al. 2013; Valentine and Connor 2015, and references therein).

10.7 Magma Versus Regional Tectonics

Regional studies from the last decades, which are presented in Chaps. 11 and 12, have allowed gaining important insights on the role of magmatic processes with regard to regional tectonics in the development of plate boundaries, in some cases also changing significantly established perspectives. The general conclusions arising from these studies are anticipated and summarized below.

As regards divergent plate boundaries, as specified in Chap. 11, the established perspective that regional tectonics associated with plate motion controls the morphology, structure and development of rift zones has been repeatedly questioned by recent **rifting episodes** and **rifting events** (where a rifting episode consists of several rifting events clustering in time in the same rift portion) associated with the emplacement of tens of kilometres long dikes in continental, transitional and oceanic rifts. These rifting episodes and events include Krafla (Iceland, 1975–1984), Asal-Ghoubbet (Afar, Djibouti, 1978–1979), Dallol (Afar, Ethiopia, 2004), East Pacific Rise (2005–2009), Dabbahu (Afar, Ethiopia, 2005–2010), Lake Natron (Tanzania, 2007), Harrat Lunayyir (Saudi Arabia, 2009) and Bardarbunga (Iceland 2014; Jacques et al. 1996; Ruegg et al. 1979; Tarantola et al. 1979; Buck et al. 2006; Tolstoy et al. 2006; Wright et al. 2006; Rowland et al. 2007; Calais et al. 2008; Biggs et al. 2009; Pallister et al. 2010; Nobile et al. 2012; Sigmundsson et al. 2015). Each of these rifting episodes or events produced the rapid (lasting a few weeks) and significant (a few metres) opening of a portion of a magmatic system along the plate boundary and, with the possible exception of Lake Natron, was characterized by the lack of faulting due to regional tectonics. In fact, the observed generation or reactivation of normal faults at the surface, usually forming graben-like structures parallel to the rift axis, resulted from dike emplacement, and was thus magma-induced. The Lake Natron event, on continental crust, was the only observed "hybrid" rifting event possibly accompanied by significant regional seismicity promoting normal faulting in addition to dike emplacement. However, even in this case, the opening associated with the intruded dike was larger than the extension induced by regional faulting. Therefore, all these cases showed that dike-induced extension is the dominant mechanism responsible for splitting the plates along divergent plate boundaries. With the exception of the Harrat Lunayyir event, dike emplacement occurred along magmatic systems, emphasizing their central role in the activity of these boundaries. In addition, in several episodes (as at Krafla, Asal-Ghoubbet and Dabbahu) multiple rifting events insisted in the same area within a few years, highlighting repeated dike emplacement. As the emplaced dikes produced graben-like structures at the surface, repeated diking episodes may explain on the long-term most of the surface deformation and the rift morphology of divergent plate boundaries. This possibility is supported by the evidence that the relatively high frequency of diking and the associated deformation along divergent boundaries often hinders extension to be accommodated through regional tectonics, be it through creep, seismic or aseismic faulting (Fig. 10.23; Sigmundsson 2006; Ebinger et al. 2010; Acocella and Trippanera 2016). Possible exceptions to this behaviour are found in immature continental rifts (as in the above-

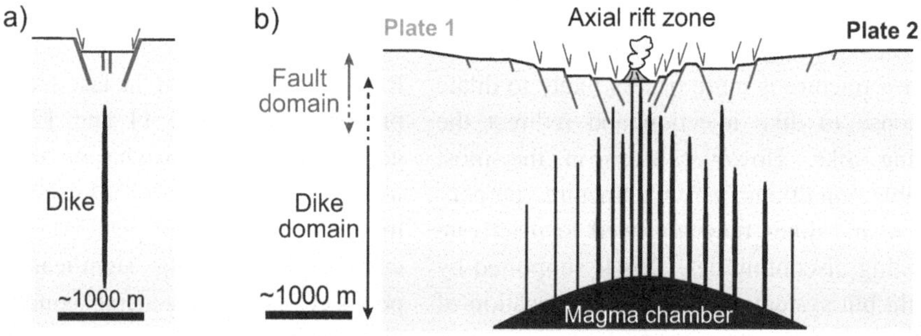

Fig. 10.23 Magmatic shaping of rift morphology and structure along divergent plate boundaries. **a** Example of surface deformation induced by a dike. **b** Upper crustal structure of the axial portion of a divergent plate boundary, as derived from field data in Iceland. Extension due to normal faults (propagating downward from the surface) becomes progressively less important at ~1 km depth, whereas at greater depths (>1 km) extension occurs almost exclusively through dike emplacement. Thus, the overall graben-like and depressed morphology of the plate boundary at the surface results from repeated dike emplacement at depth (modified after Acocella and Trippanera 2016)

mentioned Lake Natron event), where the regional tectonic component may be still relevant, and in the non-magmatic portions of oceanic rifts, as those experiencing detachment faulting (see Sect. 11.4). Therefore, it appears that while regional tectonic processes control the far-field plate motion and divergence, magmatic processes control the near-field evolution of mature plate boundaries and shape their structure and morphology.

There is evidence that magma plays an important role also along convergent plate boundaries, as specified in Chap. 12. Convergent plate boundaries characterized by extension along the arc, as the Taupo Volcanic Zone of New Zealand, have tectono-magmatic features similar to those of continental rifts, including proto-magmatic systems. This suggests an important, although probably not predominant, role of magmatic activity in controlling the evolution of the extensional arc and in shaping its structure (Rowland et al. 2010; Allan et al. 2012). These features are not observed along magmatic arcs experiencing weak extension, strike-slip motions, or contraction, where volcanic activity is restricted to the area of the volcanic edifice and magmatic systems are lacking. Despite this lack, at least two features highlight the non-negligible role of magmatism in strike-slip and contracting arcs (Fig. 10.24). The first is that, despite the lack of magmatic systems in strike-slip arcs, the magma intruded in the crust remains able to thermally weaken it, focusing along the volcanic arc the strike-slip deformation induced by strain partitioning. This shows how magma may control the structural architecture of the overriding plate. The second feature, as mentioned in Sect. 10.4, is related to the evidence that magma in contractional arcs finds its way to intrude from sills to stacked tabular intrusions, and even to rise to the surface, by locally increasing the principal vertical stress, which may switch from the minimum principal stress σ_3 to the maximum principal stress σ_1. This supports the idea that magmatic rates may locally overcome tectonic rates also in convergent plate boundaries, so that magma may intrude, accumulate and erupt largely independently of any regional tectonic context.

Nevertheless, despite the overall independence of magmatic processes from regional tectonics, magma may still benefit from favourable shorter-term tectonic variations occurring along volcanic arcs. This is exemplified by the volcanic activity (in terms of both unrest and eruptions) promoted by transient stress changes due to mega-earthquakes in various convergent settings. These changes are in fact able to temporarily affect the kinematics of a volcanic arc, even including shifts from overall inter-seismic contraction to co- and post-seismic extension (see Sect. 12.5; Walter and Amelung 2007; Pritchard et al. 2013; Takada and Fukushima 2013).

Independently of any specific tectonic setting, a further and more general question regards the role of pre-existing regional structures on the rise of magma, where a dike may intrude discontinuities deviating from its orientation perpendicular to the minimum principal stress in favour of a more energy-efficient path along a fracture or fault plane. The approach to this topic has shifted from an earlier perspective of fracture-controlled magmatism, where rising magma is captured by a pre-existing structure, to a modern vision of relative independence of dike ascent from pre-existing discontinuities. In fact, as discussed in Sect. 3.5 and mentioned in Sect. 10.6, only fractures oriented perpendicular to the least principal stress σ_3 seem likely to provide a low energy path to the surface and thus be intruded by magma, although this may still not be a sufficient condition (e.g., Rubin 1995; Maccaferri et al. 2015; Valentine and Connor 2015).

All these features indicate that magmatic activity, mostly in the form of dike propagation and subordinately as sill emplacement, is largely independent although not insensitive to the regional tectonic context and, under favourable conditions, it may even overcome non-magmatic regional tectonic processes. The details on this important role of magma are presented in the next two chapters.

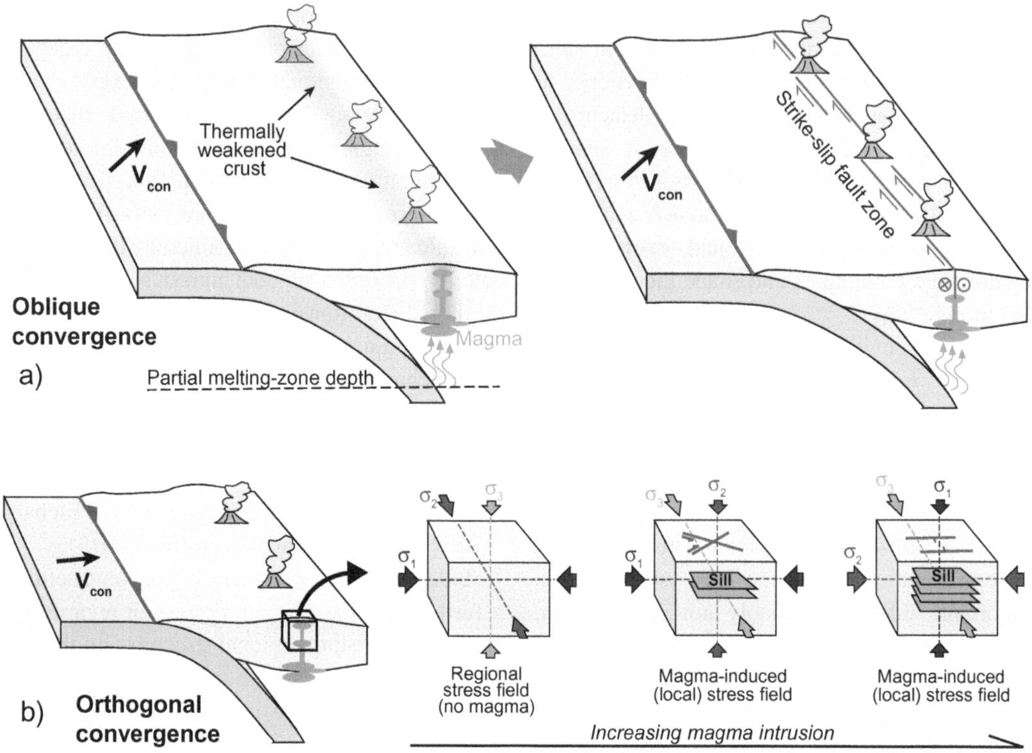

Fig. 10.24 Control of magma on the evolution of magmatic arcs in convergent plate boundaries. **a** Under oblique convergence, a crust thermally weakened by magma (dark yellow portion, left diagram) focuses the strain in the overriding plate experiencing strain partitioning, developing a strike-slip zone along the volcanic arc (right). **b** Under orthogonal convergence, the repeated emplacement of magma as stacked sills below the arc (in the zone highlighted by the cube, left diagram) increases the vertical component of the stress field, which passes from σ_3 to σ_1, promoting local arc-parallel extension and enhancing volcanism (right)

10.8 Summary

Plate tectonics theory provides the framework to understand the global distribution of volcanism, including the location and shape of volcanoes, the size and frequency of the eruptions and the composition of the erupted products. Most of the magma is being erupted along divergent plate boundaries, in continental, transitional and, especially, oceanic domains. In particular, the tectonically and magmatically active portions of divergent plate boundaries are the magmatic systems, where dikes actively split the plates. Convergent plate boundaries account for approximately a quarter of the erupted volumes. The structure of the volcanic arc overlying the slab may be characterized by various kinematics, mainly depending on the convergence motion between the two plates. In some cases, processes not requiring active subduction, as slab break-off and lithospheric delamination, may explain collisional magmatism. Hot spot volcanism explains a relatively lower amount of erupted volumes through the activity of mantle plumes. This volcanism is largely found in intraplate settings and along divergent plate boundaries, where the plumes interact with the rift segments. Despite the relatively low erupted volumes, volcanism from mantle plumes can be at times volumetrically impressive, as in the case of LIPs. The regional tectonic setting also controls, through different mechanisms, the occurrence and the distribution, in terms of spacing and alignment, of polygenic or monogenic volcanoes. Nevertheless, there is ample evidence of the predominant role of magmatism on regional tectonics in shaping the structure, morphology and evolution of divergent plate boundaries, whereas this role becomes less evident, although still affecting different processes at various scales, along convergent plate boundaries.

10.9 Main Symbols Used

B	Buoyancy flux
D''	Thermal boundary layer
H_e	Excess elevation averaged across the swell
L	Width of the swell
M	Magnitude
t	Age of lithosphere
V_{con}	Plate convergence motion
V_h	Plate velocity
V_n	Trench orthogonal convergence velocity
V_p	Trench parallel convergence velocity
z	Thickness of the lithosphere
α	Thermal diffusivity
β	Stretching factor
δ	Angle of obliquity of the rift
ε_1	Principal horizontal strain axis
ρ_{ma}	Mantle density
ρ_w	Water density
σ_1	Maximum principal stress
σ_2	Intermediate principal stress
σ_3	Minimum principal stress
ϕ	Angle of obliquity of convergence

References

Acocella V (2014) Structural control on magmatism along divergent and convergent plate boundaries: overview, model, problems. Earth-Sci Rev 136:226–288

Acocella V, Funiciello F (2010) Kinematic setting and structural control of arc volcanism. Earth Planet Sci Lett 289:43–53

Acocella V, Trippanera D (2016) How diking affects the tectono-magmatic evolution of slow spreading plate boundaries: overview and model. Geosphere 12:1–17

Acocella V, Yoshida T, Yamada R, Funiciello F (2008) Structural control on Late Miocene to quaternary volcanism in the NE Honshu arc, Japan. Tectonics 27: TC5008. https://doi.org/10.1029/2008TC002296

Acocella V, Bellier O, Sandri L, Sébrier M, Pramumijoyo S (2018) Tectono-magmatic relationships along an obliquely convergent plate boundary: Sumatra, Indonesia. Front Ear Sci 6:3. https://doi.org/10.3389/feart.2018.00003

Allan ASR, Wilson CJN, Millet MA, Wysoczansk RJ (2012) The invisible hand: tectonic triggering and modulation of a rhyolitic supereruption. Geology 40:563–566

Beck SL, Zandt G (2002) The nature of orogenic crust in the central Andes. J Geophys Res 107:2230. https://doi.org/10.1029/2000JB000124

Bercovici D, Mahoney J (1994) Double flood basalts and plume head separation at the 660-kilometer discontinuity. Science 266:1367–1369

Bianco TA, Ito G, Beker JM, Garcia MO (2005) Secondary Hawaiian volcanism formed by flexural arch decompression. Geochem Geophys Geosyst 6: Q08009. https://doi.org/10.1029/2005GC000945

Biggs J, Anthony EY, Ebinger CJ (2009) Multiple inflation and deflation events at Kenyan volcanoes, East African Rift. Geology 37:979–982

Bott MHP (1995) Mechanisms of rifting: geodynamic modelling of continental rift systems. In: Olsen KH (ed) Continental rifts: evolution, structure, tectonics. developments in geotectonics, vol 25, pp 27–43

Brune S, Williams SE, Dietmar Mulle R (2018) Oblique rifting: the rule, not the exception. Solid Earth 9:1187–1206

Bryan SE, Peate IU, Peate DW, Self S, Jerram DA, Mawby MR et al (2010) The largest volcanic eruptions on Earth. Earth Sci Rev 102:207–229

Buck RW, Einarsson P, Brandsdottir B (2006) Tectonic stress and magma chamber size as controls on dike propagation: constraints from the 1975–1984 Krafla rifting episode. J Geophys Res 111:B12404. https://doi.org/10.1029/2005JB003879

Cagnioncle AM, Parmentier EM, Elkins-Tanton LT (2007) Effect of solid flow above a subducting slab on water distribution and melting at convergent plate boundaries. J Geophys Res 112:9402. https://doi.org/10.1029/2007JB004934

Calais E, d'Oreye N, Albaric J, Deschamps A, Delvaux D, Deverchere J (2008) Strain accommodation by dyking in a youthful continental rift, East Africa. Nature 456:783–787

Campbell IH (2005) Large igneous provinces and the mantle plume hypothesis. Elements 1:265–268

Cassidy J, Locke CA (2010) The Auckland volcanic field, New Zealand: geophysical evidence for structural and

spatio-temporal relationships. J Volcanol Geoth Res 195:127–137
Chaussard E, Amelung F (2012) Precursory inflation of shallow magma reservoirs at west Sunda volcanoes detected by InSAR. Geophys Res Lett 39:L21311. https://doi.org/10.1029/2012GL053817
Coffin MF, Eldholm O (1994) Large igneous provinces: crustal structure, dimensions, and external consequences. Rev Geophys 32:1–36
Corti G, Bonini M, Conticelli S, Innocenti F, Manetti P, Sokoutis D (2003) Analogue modelling of continental extension: a review focused on the relations between the patterns of deformation and the presence of magma. Earth-Sci Rev 63:169–247
Courtillot V, Davaille A, Besse J, Stock J (2003) Three distinct types of hotspots in the Earth's mantle. Earth Planet Sci Lett 205:295–308
Davies GF (1999) Dynamic earth plates, plumes and mantle convection. Cambridge University Press, 470 pp
Davies JH, von Blanckenburg F (1995) Slab breakoff: A model of lithosphere detachment and its test in the magmatism and deformation of collisional orogens. Earth Planet Sci Lett 129:85–102
De Bremond d'Ars, J, Jaupart C, Sparks RSJ (1995) Distribution of volcanoes in active margins. J Geophys Res 100:20421–20432
DeMets C (1992) Oblique convergence and deformation along the Kuril and Japanese trenches. J Geophys Res 97:17615–17625
de Silva S, Lindsay JM (2015) Primary volcanic landforms In: Sigurdsson H, Houghton B, McNutt S, Rymer H, Stix J (eds) The encyclopedia of volcanoes, 2nd edn. Elsevier Academic Press, pp 273–297
Ebinger CJ, Casey M (2001) Continental breakup in magmatic provinces: an Ethiopian example. Geology 29:527–530
Ebinger CJ, Hayward NJ (1996) Soft plates and hot spots: views from Afar. J Geophys Res 101:21859–21876
Ebinger CJ, Sleep NH (1998) Cenozoic magmatism throughout east Africa resulting from impact of a single plume. Nature 395:788–791
Ebinger CJ, Ayele A, Keir D, Rowland J, Yirgu G, Wright T et al (2010) Length and timescales of rift faulting and magma intrusion: the Afar rifting cycle from 2005 to present. Ann Rev Earth Planet Sci 38:439–466
Faccenna C, Molin P, Orecchio B, Olivetti V, Bellier O, Funiciello F et al (2011) Topography of the Calabria subduction zone (southern Italy): clues for the origin of Mt. Etna. Tectonics 30:TC1003. https://doi.org/10.1029/2010TC002694
Farnetani CG, Hofmann AW (2011) Mantle plumes. In: Encyclopedia of solid earth geophysics. Springer. https://doi.org/10.1007/978-90-481-8702-7
Farnetani CG, Richards MA (1994) Numerical investigation of the mantle plume initiation model for flood basalt events. J Geophys Res 99:13813–13883
Fisher RV, Schmincke H-U (1984) Pyroclastic rocks. Springer, Berlin, p 472
Forsyth D, Uyeda S (1975) On the relative importance of the driving forces of plate motion. Geophys J Int 43:163–200
Fossen H (2010) Structural geology. Cambridge University Press, 481 pp
Fournier M, Petit C (2007) Oblique rifting at oceanic ridges: relationship between spreading and stretching directions from earthquake focal mechanisms. J Struct Geol 29:201–208
French SW, Romanowicz B (2015) Broad plumes rooted at the base of the Earth's mantle beneath major hotspots. Nature 525:95–99
Galland O, Cobbold PR, de Bremond DJ, Hallot E (2007) Rise and emplacement of magma during horizontal shortening of the brittle crust: insights from experimental modelling. J Geophys Res 112:B06402. https://doi.org/10.1029/2006JB004604
Gordon RG, Jurdy DM (1986) Cenozoic global plate motions. J Geophys Res 91:12389–12406
Gripp AE, Gordon RG (2002) Young tracks of hot spot and current plate velocities. Geophys J Int 150:321–361
Gudmundsson A (1995) Infrastructure and mechanics of volcanic systems in Iceland. J Volcanol Geoth Res 64:1–22
Gvirtzman Z, Nur A (1999) The formation of Mount Etna as the consequence of slab rollback. Nature 401:782–785
Hall R, Spakman W (2015) Mantle structure and tectonic history of SE Asia. Tectonophysics 658:14–45
Hand E (2015) Mantle plumes seen rising from Earth's core. Science 349:1032–1033
Heuret A, Funiciello F, Faccenna C, Lallemand S (2007) Plate kinematics, slab shape and back-arc stress: a comparison between laboratory models and current subduction zones. Ear Planet Sci Lett 256:473–483
Hieronymus CF, Bercovici D (2001) A theoretical model of hotspot volcanism: control on volcanic spacing and patterns via magma dynamics and lithospheric stresses. J Geophys Res 106:683–702
Isacks B, Oliver J, Sykes LR (1968) Seismology and the new global tectonics. J Geophys Res 73:5855–5899
Ito G, Lin J, Graham D (2003) Observational and theoretical studies of the dynamics of mantle plume-Mid-Ocean Ridge interaction. Rev Geophys 41:1017. https://doi.org/10.1029/2002RG000117
Jacques E, King GCP, Tapponnier P, Ruegg JC, Manighetti I (1996) Seismic activity triggered by stress changes after the 1978 events in the Asal Rift, Djibouti. Geophys Res Lett 23:2481–2484
Jaffe RL, Taylor W (2018) The physics of energy. Cambridge University Press, 874 pp
Jellinek MA, Manga M (2004) Links between long-lived hot spots, mantle plumes, "D", and plate tectonics. Rev Geophys 42:RG3002. https://doi.org/10.1029/2003RG000144
Kay RW, Kay SM (1993) Delamination and delamination magmatism. Tectonophysics 219:177–189

Kerr RC, Meriaux C (2004) Structure and dynamics of sheared mantle plumes. Geochem Geophys Geosyst 5: Q12009. https://doi.org/10.1029/2004GC000749
Keskin M (2003) Magma generation by slab steepening and breakoff beneath a subduction-accretion complex: an alternative model for collision-related volcanism in Eastern Anatolia, Turkey. Geophys Res Lett 30:8046. https://doi.org/10.1029/2003GL018019
King SD, Adam C (2014) Hotspot swells revisited. Phys Earth Planet Inter 235:66–83
Koulakov I, Kasatkina E, Shapiro NM, Jaupart C, Vasilevsky A, El Khrepy S et al (2016) The feeder system of the Toba supervolcano from the slab to the shallow reservoir. Nat Commun 7:12228. https://doi.org/10.1038/ncomms12228
Kreemer C, Blewitt G, Klein EC (2014) A geodetic plate motion and global strain rate model. Geochem Geophys Geosyst 15:3849–3889
LaFemina PC (2015) Plate tectonics and volcanism. In: Sigurdsson H, Houghton B, McNutt S, Rymer H, Stix J (eds) The encyclopedia of volcanoes, 2nd edn. Elsevier Academic Press, pp 65–91
Latin D, White N (1990) Generating melt during lithospheric extension: pure shear vs. simple shear. Geology 18:327–331
Le Corvec N, Menand T, Lindsay J (2013) Interaction of ascending magma with pre-existing crustal fractures in monogenetic basaltic volcanism: an experimental approach. J Geophys Res 118:968–984
Le Pichon X (1968) Sea-floor spreading and continental drift. J Geophys Res 73:3661–3697
Liotard JM, Dautria JM, Bosch D, Condomines M, Mehdizadeh M, Ritz J-F (2008) Origin of the absarokite–banakite association of the Damavand volcano (Iran): trace elements and Sr, Nd, Pb isotope constraints. Int J Ear Sci 97:89–102
Maccaferri F, Rivalta E, Keir D, Acocella A (2014) Off-rift volcanism in rift zones determined by crustal unloading. Nat Geosci 7:297–300
Maccaferri F, Acocella V, Rivalta E (2015) How the differential load induced by normal fault scarps controls the distribution of monogenic volcanism. Geophys Res Lett 42. https://doi.org/10.1002/2015GL065638
Mazzarini F (2007) Vent distribution and crustal thickness in stretched continental crust: the case of the Afar Depression (Ethiopia). Geosphere 3:152–162
Mazzarini F, Ferrari L, Isola I (2010) Self-similar clustering of cinder cones and crust thickness in the Michoacan-Guanajuato and Sierra de Chichinautzin volcanic fields, Trans-Mexican Volcanic Belt. Tectonophysics 486:55–64
McCaffrey R (1992) Oblique plate convergence, slip vectors and forearc deformation. J Geophys Res 97:8905–8915
McCaffrey R (1996) Estimates of modern arc-parallel strain rates in fore arcs. Geology 24:27–30
McKenzie DP (1977) The initiation of trenches: a finite amplitude instability. In: Talwani M, Pitman WC (eds) Island arcs deep sea trenches and back-arc basins, vol 1. Maurice Ewing Ser AGU, Washington, pp 57–61
McKenzie D (1978) Some remarks on the development of sedimentary basins. Earth Planet Sci Lett 40:25–32
McKenzie DP, Parker RL (1967) The North Pacific: an example of tectonics on a sphere. Nature 216:1276–1280
McNutt MK, Judge AV (1990) The superswell and mantle dynamics beneath the South Pacific. Science 248:969–975
Mohr PA, Wood CA (1976) Volcano spacing and lithospheric attenuation in the Eastern Rift of Africa. Earth Planet Sci Lett 33:126–144
Montelli R, Nolet G, Dahlen FA, Masters G, Engdahl ER, Hung SH (2004) Finite-frequency tomography reveals a variety of plumes in the mantle. Science 303:338–343
Morgan WJ (1968) Rises, trenches, great faults and crustal blocks. J Geophys Res 73:1959–1982
Muirhead J, Van Eaton AR, Re G, White JDL, Ort MH (2016) Monogenetic volcanoes fed by interconnected dikes and sills in the Hopi Buttes volcanic field, Navajo Nation, USA. Bull Volcanol 78:11. https://doi.org/10.1007/s00445-016-1005-8
Muller JR, Ito G, Martel SJ (2001) Effects of volcano loading on dike propagation in an elastic half-space. J Geophys Res 106:11101–11113
Nakamura K (1977) Volcanoes as possible indicators of tectonic stress orientation: principle and proposal. J Volcanol Geoth Res 2:1–16
Nakamura K, Jacob KH, Davies JN (1977) Volcanoes as possible indicators of tectonic stress orientation: aleutians and Alaska. In: Pageoph, vol 115. Birkhauser Verlag, Basel, pp 87–112
Nobile A, Pagli C, Keir D, Wright TJ, Ayele A, Ruch J et al (2012) Dyke-fault interaction during the 2004 Dallol intrusion at the northern edge of the Erta Ale Ridge (Afar, Ethiopia). Geophys Res Lett 39:L19305. https://doi.org/10.1029/2012GL053152
O'Connor JM, Jokat W, Regelous M, Kuiper KF, Miggins DP, Koppers AAP (2019) Superplume mantle tracked isotopically the length of Africa from the Indian ocean to the Red Sea. Nat Commun 10:5493. https://doi.org/10.1038/s41467-019-13181-7
Orellana-Rovirosa F, Richards M (2017) Rough versus smooth topography along oceanic hotspot tracks: observations and scaling analysis. Geophys Res Lett 44:4074–4081
Pallister JS, McCausland WA, Jonsson S, Lu Z, Zahran HM, El Hadidy S et al (2010) Broad accommodation of rift-related extension recorder by dyke intrusion in Saudi Arabia. Nat Geosci 3:708–712
Parsons B, McKenzie D (1979) Mantle convection and the thermal structure of the plates. J Geophys Res 83:4485–4496
Peyton V, Levin V, Park J, Brandon M, Lees J, Gordeev E et al (2001) Mantle flow at a slab edge: seismic anisotropy in the Kamchatka region. Geophys Res Lett 28:379–382

Pritchard ME, Jay JA, Aron F, Henderson ST, Lara LE (2013) Subsidence at southern Andes volcanoes induced by the 2010 Maule, Chile earthquake. Nat Geosci 6:632–636
Ramberg H (1981) Gravity, deformation and the Earth's crust in theory, experiments and geological applications. Academic Press, London, 452 pp
Romanowicz B, Gung Y (2002) Superplumes from the core-mantle boundary to the lithosphere: implications for heat flux. Science 296:513–516
Rooney TO (2020a) The cenozoic magmatism of East-Africa: part i—flood basalts and pulsed magmatism. Lithos 286–287:264–301
Rooney TO (2020b) The Cenozoic magmatism of East Africa: part ii—rifting of the mobile belt. Lithos 360–361:105291
Rosenbaum G, Gasparon M, Lucente FP, Peccerillo A, Miller MS (2008) Kinematics of slab tear faults during subduction segmentation and implications for Italian magmatism. Tectonics 27:TC2008. https://doi.org/10.1029/2007TC002143
Rowland JV, Baker E, Ebinger CJ, Keir D, Kidane T, Biggs J et al (2007) Fault growth at a nascent slow-spreading ridge: the 2005 Dabbahu rifting episode, Afar. Geophys J Int 171:1226–1246
Rowland JV, Wilson CJN, Gravley DM (2010) Spatial and temporal variations in magma-assisted rifting, Taupo Volcanic Zone, New Zealand. J Volcanol Geoth Res 190:89–108
Rubin AM (1995) Propagation of magma-filled cracks. Annu Rev Earth Planet Sci 23:287–336
Ruegg JC, Kasser M, Lépine JC, Tarantola A (1979) Geodetic measurements of rifting associated with a seismo–volcanic crisis in Afar. Geophys Res Lett 6:817–820
Ruppel C (1995) Extensional processes in continental lithosphere. J Geophys Res 100:24187–24215
Sani F, Bonini M, Corti G, Moratti G (2019) Extension direction re-orientation in the oceanic rift of Iceland, and comparison with continental rifts. Tectonophysics 756:25–42
Schmincke H-U (2004) Volcanism. Springer, Berlin, p 290
Schurr B, Rietbrock A, Asch G, Kind R, Oncken O (2006) Evidence for lithospheric detachment in the central Andes from local earthquake tomography. Tectonophysics 415:203–223
Sdrolias M, Muller RD (2006) Controls on back-arc basin formation. Geochem Geophys Geosyst 7:Q04016. https://doi.org/10.1029/2005GC001090
Self S, Coffin MF, Rampino MR, Wolff JA (2015) Large Igneous Provinces and Flood Basalt Volcanism In: Sigurdsson H, Houghton B, McNutt S, Rymer H, Stix J (eds) The encyclopedia of volcanoes, 2nd edn. Elsevier Academic Press, pp 441–458
Shabanian E, Acocella V, Gioncada A, Ghasemi H, Bellier O (2012) Structural control on magmatism in intraplate collisional settings: extinct example from NE Iran and current analogues. Tectonics 31:TC3013. https://doi.org/10.1029/2011TC003042
Sigmundsson F (2006) Magma does the splits. Nature 442:251–252
Sigmundsson F, Hooper A, Hreinsdottir S, Vogfjord KS, Ofeigsson DG, Heimisson et al (2015) Segmented lateral dyke growth in a rifting event at Barðarbunga volcanic system, Iceland. Nature 517:191–195
Sleep NH (1990) Hotspots and mantle plumes some phenomenology. J Geophys Res 95:6715–6736
Steinberger B, Sutherland R, O'Connell RJ (2004) Prediction of Hawaiian-Emperor seamount locations from a revised model of global plate motion and mantle flow. Nature 430:167–173
Stuwe K (2007) Geodynamics of the lithosphere. Springer, 493 pp.
Takada A (1994) The influence of regional stress and magmatic input on styles of monogenetic and polygenetic volcanism. J Geophys Res 99:13563–13573
Takada Y, Fukushima Y (2013) Volcanic subsidence triggered by the 2011 Tohoku earthquake in Japan. Nat Geosci 6:637–641
Tarantola A, Ruegg JC, Lepine JC (1979) Geodetic evidence for rifting in Afar. A brittle-elastic model of the behavior of the lithosphere. Earth Planet Sci Lett 45:435–444
Tenbrink UT (1991) Volcano spacing and plate rigidity. Geology 19:397–400
Tibaldi A (2005) Volcanism in compressional tectonic settings. Is it possible? Geophys Res Lett 32. https://doi.org/10.1029/2004GL021798
Tikoff B, Teyssier C (1994) Strain modeling of displacement-field partitioning in transpression orogens. J Struct Geol 16:1575–1588
Tolstoy M, Cowen JP, Baker ET, Fornari DJ, Rubin KH, Shank TM et al (2006) A sea-floor spreading event captured by seismometers. Science 314:1920–1922
Tron V, Brun JP (1991) Experiments on oblique rifting in brittle–ductile systems. Tectonophysics 188:71–84
Turcotte DL, Schubert G (1982) Geodynamics: application of continuum physics to geological problems. Wiley, NY, 450 pp
Turner SP, George RM, Evans PJ, Hawkesworthy CJ, Zellmer GF (2000) Time-scales of magma formation, ascent and storage beneath subduction-zone volcanoes. Philos Trans R Soc Lond A 358:1443–1464
Uyeda S, Kanamori H (1979) Back-arc opening and the mode of subduction. J Geophys Res 84:1049–1061
Valentine GA, Connor CB (2015) Basaltic volcanic fields In: Sigurdsson H, Houghton B, McNutt S, Rymer H, Stix J (eds) The encyclopedia of volcanoes, 2nd edn. Elsevier Academic Press, pp 423–440
van der Pluijm BA, Marshak S (2004) Earth structure: an introduction to structural geology and tectonics. WW Norton & Company, 673 pp
Vidal V, Bonneville A (2004) Variations of the Hawaiian hot spot activity revealed by variations in the magma production rate. J Geophys Res 109:B03104. https://doi.org/10.1029/2003JB002559
Vine FJ (1966) Spreading of the ocean floor: new evidence. Science 154:1405–1415

Vine FJ, Matthews DH (1963) Magnetic anomalies over oceanic ridges. Nature 199:947–949

Walter TR, Amelung F (2007) Volcanic eruptions following M $\geq$ 9 megathrust earthquakes: implications for the Sumatra-Andaman volcanoes. Geology 35:539–542

Wegener A (1912) Die Entstehung der Kontinente. Geol Rundsch 3:276–292

Wendt JI, Regelous M, Collerson KD, Ewart A (1997) Evidence for a contribution from two mantle plumes to island-arc lavas from northern Tonga. Geology 25:611–614

Wernicke B (1985) Uniform-sense normal simple shear of the continental lithosphere. Can J Earth Sci 22:108–125

White R, McKenzie D (1989) Magmatism at rift zones. The generation of volcanic continental margins and flood basalts. J Geophys Res 94:7685–7729

Whittaker JM, Afonso JC, Masterton S, Müller RD, Wessel P, Williams SE et al (2015) Long-term interaction between mid-ocean ridges and mantle plumes. Nat Geosci 8:479–483

Wilson JT (1968) Static or mobile Earth: the current scientific revolution. Proc Am Philos Soc 112:309–320

Wilson JT (1972) Continents adrift. Readings from Scientific American Freeman WH and Company San Francisco, 172 pp.

Withjack MO, Jamison WR (1986) Deformation produced by oblique rifting. Tectonophysics 126:99–124

Wright TJ, Ebinger C, Biggs J, Ayele A, Yirgu G, Keir D et al (2006) Magma maintained rift segmentation at continental rupture in the 2005 Afar dyking episode. Nature 442:291–294

Zhao D (2004) Global tomographic images of mantle plumes and subducting slabs: insight into deep Earth dynamics. Phys Earth Planet Inter 146:3–34

Ziegler P, Cloetingh S (2004) Dynamic processes controlling evolution of rifted basins. Earth-Sci Rev 64:1–50

11 Volcanoes at Divergent Plate Boundaries

11.1 Introduction

This chapter focuses on divergent plate boundaries, that is where two lithospheric plates drift away from each other. Plate divergence may occur on continental, transitional and oceanic lithosphere, with spreading rates varying over nearly 2 orders of magnitude, from a few mm/year to ~15 cm/year. Except for the continental and oceanic boundaries characterized by the lowest bound of spreading rates, divergent plate boundaries are commonly associated with widespread magmatic activity. For decades it has been assumed that such a magmatic activity is a mere product of the regional tectonic processes responsible for plate divergence. Recent geological, geophysical and geodetic data have highlighted a much more active role of magma in separating the plates, independently of the nature of the rifted crust, allowing redefining the tectono-magmatic relationships of divergent plate boundaries.

This chapter follows the steps of the Wilson cycle for divergent plate boundaries describing the progression from slowly spreading immature continental rifts to fast spreading oceanic ridges. The main aims of this chapter are to:

- describe the tectono-magmatic features of representative cases of divergent plate boundaries;
- highlight similarities and differences in the magmatic activity as a function of the spreading rate on continental, transitional and oceanic crust;
- propose a general model summarizing the role of magmatic activity in the evolution of divergent plate boundaries.

11.2 Continental Rifts: The East African Rift System

Several rift zones can be found on the continents. As described in Chap. 2, those associated with magma are usually narrow rifts, such as the Rio Grande Rift (southern USA), the Rhine Graben (western Europe), the Baikal Rift (central Asia) and the East African Rift System. Of these, only the latter is a magmatic continental rift along a divergent plate boundary: therefore, this section focuses on its tectonic and magmatic features.

The East African Rift System (EARS) separates the Nubian plate from the Somali plate, stretching over nearly 5000 km onland, from Mozambique to the triple junction of Afar (Ethiopia), where the EARS meets the Gulf of Aden and Red Sea oceanic rifts (Fig. 11.1). Despite its overall structural continuity, the EARS is not homogeneous, consisting of several

V. Acocella, *Volcano-Tectonic Processes*, Advances in Volcanology,
https://doi.org/10.1007/978-3-030-65968-4_11

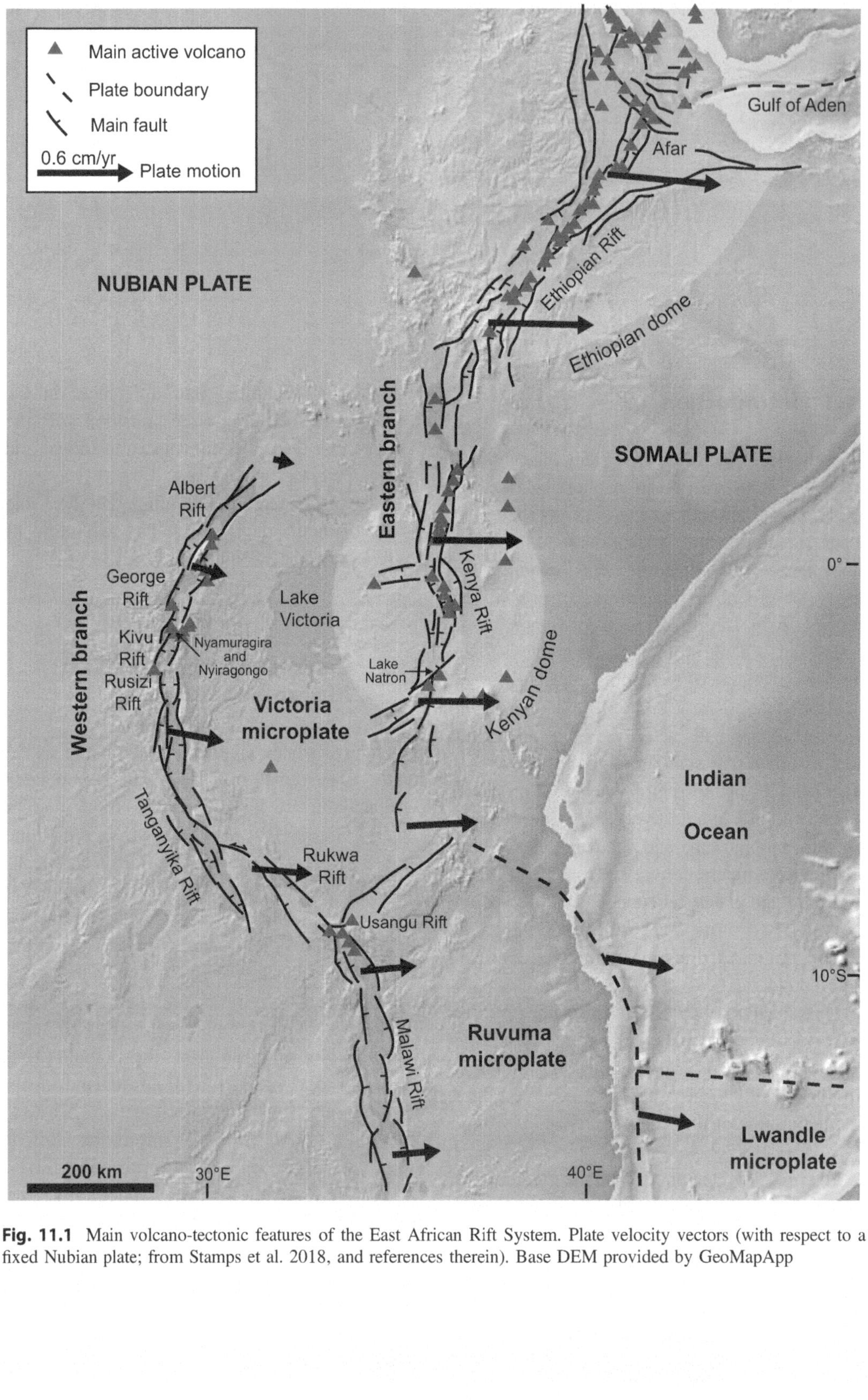

Fig. 11.1 Main volcano-tectonic features of the East African Rift System. Plate velocity vectors (with respect to a fixed Nubian plate; from Stamps et al. 2018, and references therein). Base DEM provided by GeoMapApp

branches forming three microplates: Victoria, Ruvuma and Lwandle. There is an overall northward increase in the ~E–W trending amount of extension, passing from 0.2 to 0.3 cm/year to the south to ~0.7 cm/year to the north, before reaching Afar. This corresponds to an overall northward decrease in the crustal thickness, an increase in the magmatism and volcanic activity and the shallowing of earthquakes. In particular, the southern predominantly amagmatic portion shows seismicity down to 30–40 km of depth, suggesting rift border faults are active at lower crustal levels, whereas the northern predominantly magmatic portion of the EARS shows seismicity mostly shallower than 15 km (Keir et al. 2006; Craig et al. 2011; Perez-Gussinye et al. 2009; Ebinger et al. 2017; Stamps et al. 2018). These features highlight an interdependence among the amount of extension, the resulting crustal thinning and decompression melting, and magma generation and volcanic activity, ultimately also thinning the seismogenic crust. In addition, the ages of the erupted products indicate an overall southward migration of volcanic activity, from Ethiopia to Tanzania, suggesting that the most extended northern rift portion is also older. This tectonic and volcanic gradient along the EARS may result from several processes, including plumes and superplumes initiating rifting, as below the Ethiopian and Kenyan portions, and the reactivation of pre-existing mobile belts (Ebinger and Sleep 1998; Corti et al. 2007; O'Connor et al. 2019; Rooney 2020a).

The southernmost portion of the EARS consists of the onshore western system (the Malawi Rift) and the offshore eastern system (the Mozambique Basin) and lacks volcanism (Fig. 11.1). The Malawi Rift has an overall asymmetric structure, with interacting half-grabens with alternating polarity, each forming a ~100 km long and ~50 km wide basin. Pronounced heat flow focuses in regions of asthenospheric upwelling and in ancient magmatic rift segments. The offshore eastern system enters the oceanic lithosphere, with an overall amount of extension between 5 (to the south) and 12 (to the north) kilometres and sporadic magmatic intrusions (Ebinger et al. 1984; Franke et al. 2015; Deville et al. 2018; Njinju et al. 2019).

North of the Malawi Rift, the EARS branches into an eastern and western portion, both initiated at ~25 Ma (Fig. 11.1). The connection with the offset western branch occurs through the highly oblique Rukwa Rift, with localized volcanism. Along the northwest continuation of the Rukwa Rift, the western branch of the EARS bends to a N–S direction, reflecting the perturbation of the tensional stress field along the west side of the stronger Tanzanian craton (Ebinger et al. 1989; Corti et al. 2007). The western branch of EARS is 40–70 km wide and has experienced crustal extension of less than 15%, with widespread seismicity and asymmetric structure, consisting of interacting half-grabens. Volcanism has been mainly occurring in the last 11 Ma between the half-grabens, as in the Nyiragongo dome, hosting the Nyamuragira and Nyiragongo volcanoes (Ebinger 1989a, b; Upcott et al. 1996; Wadge et al. 2016).

The eastern branch of the EARS is magmatically more active and consists of two main portions, centred on the Kenyan and Ethiopian domes (Fig. 11.1; e.g., Rooney 2020a, b and references therein). Both domes are uplifted crust resulting from the thermal buoyancy of mantle plumes below. The Kenyan dome, which uplifted to more than 1400 m by at least 13.5 Ma, lies on a long-lived low seismic velocity zone, extending to ~150 km depth, possibly feeding mafic underplating in the lowermost crust. Above, a ~N–S trending asymmetric rift consisting of interacting half-grabens reactivates pre-existing structures. The rift thins the ~40 km thick crust to ~35 (to the south) and ~20 km (to the north), with extension varying from 5–10 to 35–40 km, respectively (Dugda et al. 2005; Park and Nyblade 2006). Many Kenyan volcanoes have ~NW–SE elongated summit calderas parallel to the direction of the minimum principal stress σ_3, suggesting a far-field control on the elongation of the magma chambers below (Bosworth et al. 2003). The Kenya Rift shows a more advanced stage of tectonic and magmatic activity with regard to the western branch, at the same latitude. This activity also manifests through unrest and eruptions at several volcanoes and through rifting events involving seismicity along regional normal faults and dike injections, as observed at Lake Natron in 2007. During this event, the opening of the rift was achieved through two contemporaneous processes. One was regional normal faulting, responsible for a nearly 3 months long seismic swarm with the seismic moment much higher than the geodetic moment, and culminating in its early phase with a magnitude *M5.9* earthquake. This seismicity may have been induced by a pressurized deep magma chamber. The other was the emplacement of a ~1.5 m thick dike between 2 and 6 km of depth, which followed the peak in seismicity and was responsible for most of the measured rift opening (Calais et al. 2008; Reiss et al. 2021 and references therein).

To the north of the Kenyan dome is the Ethiopian dome, which includes the continental Main Ethiopian Rift, described in Sect. 11.2.1, and the transitional Afar triple junction, described in Sect. 11.3.

11.2.1 The Main Ethiopian Rift

The continental Main Ethiopian Rift (MER) and the adjacent transitional crust of Afar are the ideal sites to study the tectono-magmatic features of a nascent divergent plate boundary. Indeed, within a thousand of kilometres, incipient continental rifting (southern MER) evolves into mature continental rifting (northern MER) and continental break-up (central Afar) to proto-oceanic rifting (northern Afar; Fig. 11.2). These variations allow investigating and comparing contiguous rift portions at progressively more advanced evolutionary stages, providing important insights on the early phases of plate divergence.

The MER-Afar system lies on the Ethiopian dome, resulting from a broad uplift due to the activity of the Afar mantle plume, presently identified as a more than 400 km deep and more than 500 km wide low-wave speed anomaly beneath Afar and the MER (Mohr and Wood 1976; Ebinger and Sleep 1998; Benoit et al. 2006). Shallow tomographic data allow capturing the current structure of the MER-Afar area at a depth of several tens of kilometres, suggesting pervasive partial melt, with focused upwelling and melt storage beneath the MER, where the slowest velocities are observed. Average crustal shear velocity is faster beneath Afar than the MER, albeit Afar has localized slow velocities beneath active volcanic centres. These slow-velocity regions in the MER-Afar systems are interpreted as due to magmatic intrusions and crustal heating (Chambers et al. 2019).

As observed along the EARS, in the MER-Afar system there is also a progressive northward increase in extension rate and a parallel decrease in lithospheric thickness: as a result, the frequency of volcanoes also increases northward. The NE-SW trending MER developed in two main stages under predominant oblique opening, with an overall ~E–W direction of divergence. Mio-Pliocene continental rifting first activated the NE–SW striking border faults, creating up to 5 km of subsidence, and was accompanied by diffuse magmatic activity, also off-rift. During the Quaternary, tectonic and magmatic activity focused in NNE-SSW trending narrow zones within the rift, deactivating the border faults (Fig. 11.2; Mohr 1967; Corti 2009). The two evolutionary stages can be also recognized along different portions of the current MER. The less-extended southern MER has a ~40 km thick crust and is in incipient continental rifting, with most of the deformation accommodated along border faults and minor internal deformation and volcanic activity; these features are reconciled with the first evolutionary stage. The northern MER has thinner (~30 km thick) crust and is in mature incipient continental rupture, with the deformation focused within the rift and minor activity along border faults, therefore representing the second evolutionary stage (Agostini et al. 2011).

The portions of the rift focusing Quaternary-Recent volcanic and tectonic activity are identified by NNE-SSW striking right stepping en-echelon zones, ~20 km wide and ~60 km long, constituting **magmatic systems**, Most magmatic systems lie within the rift and along its axis, although striking at a slightly oblique angle to the rift axis. Active surface deformation in magmatic systems

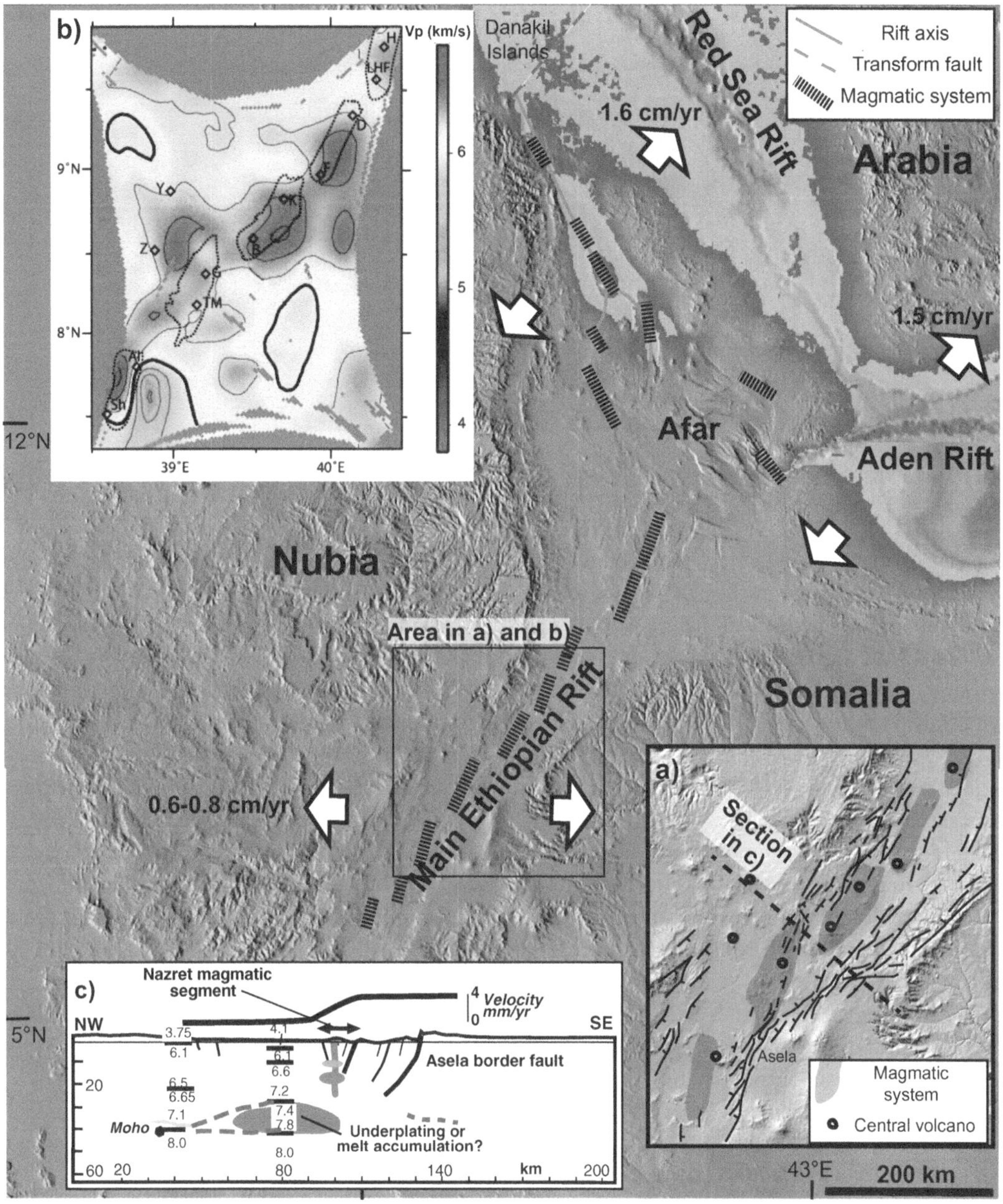

Fig. 11.2 Overview of the Main Ethiopian Rift and Afar triple junction, highlighting the rift axis, magmatic systems and possible transform faults (red dashed line). Insets **a**, **b** enlarge the same area of the MER. Inset a) shows the main magmatic systems, faults and polygenic volcanoes (Ebinger and Casey, 2001). Inset **b** shows velocity anomalies in a horizontal slice at 10 km depth below rift floor; high-velocity (Vp) bodies below rift axis are interpreted as solidified magmatic intrusions; magmatic systems are shown by dotted lines (Keranen et al. 2004). Inset **c** shows a NW–SE trending crustal section of the central portion of the MER (location in **a**); (Ebinger and Casey 2001): numbers represent seismic velocities (in km/s); dot is crustal thickness; bold line above is geodetically determined extensional velocity along line of section (Bilham et al. 1999). Base DEM provided by GeoMapApp

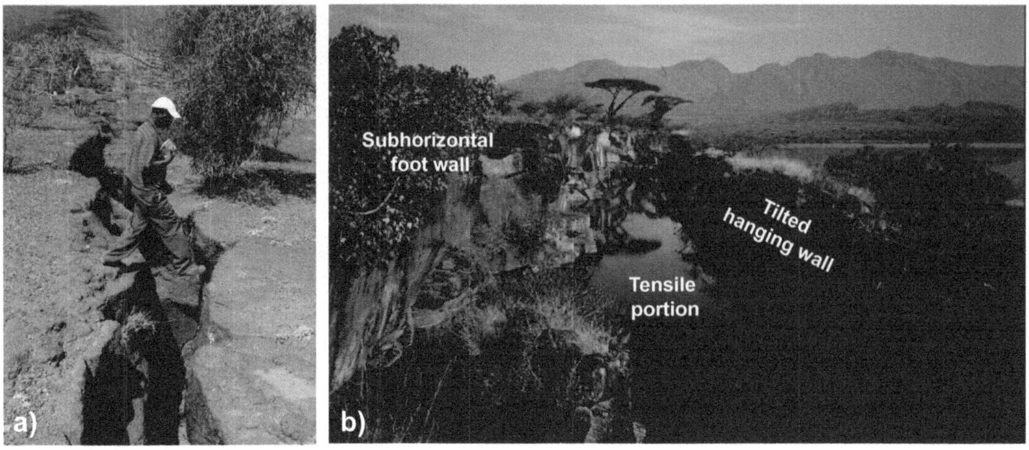

Fig. 11.3 Structural features along the axis of the Main Ethiopian Rift. **a** Extension fracture (photo courtesy Joel Ruch); **b** open normal fault with tilted hanging wall: Fantale volcano is visible in the background, to the north

consists of metres-wide extension fractures and "open normal faults", showing a distinctive tensile portion between their tilted hanging wall and horizontal foot wall (Fig. 11.3). Volcanic activity in a magmatic system includes a dominant polygenic felsic volcano, often with summit caldera, and several mafic monogenic volcanoes, broadly aligned parallel to the system. The magma within the reservoir of the polygenic volcano may intrude laterally (along the system) through dikes, eventually feeding the monogenic fissures, or intrude vertically and erupt within the volcano. In the first case, the mafic primitive composition suggests a provenance of the dikes from the deep, less evolved portion of the magma chamber, or of the reservoir. In the second case, the deeper magma may interact with the felsic magma chamber, differentiating and producing more evolved compositions. The magmatic systems, focusing magmatic and tectonic activity, are responsible for rift opening through dikes mainly at depth and through normal faults and extension fractures mainly at the surface, accommodating more than 80% of the strain across the MER at depths shallower than 10 km (Fig. 11.4; Ebinger and Casey 2001).

Geodetic measurements between 1992 and 2010 suggest that the MER has an overall divergent motion of ~0.7 cm/year along an east–west direction, with a combination of localized high strain rate and adjacent distributed low strain rate zones over a wide range of length and time scales in different settings. In addition, the oblique opening of the Nubian and Somali plates contrasts with the orthogonal opening (NW–SE oriented) of the extension fractures observed along the rift axis, suggesting kinematic partitioning across rift. The geodetic estimates for opening rate are at least one order of magnitude larger than the geological ones over the last ~7 ka, suggesting that, along the rift axis, most extension is accommodated at depth or that the spreading rate is not uniform (Williams et al. 2004; Acocella et al. 2011; Kogan et al. 2012; Birhanu et al. 2016). Geodetic data also highlight deformation of polygenic volcanoes, mainly in the southern MER, as at Corbetti, Aluto and Tullu Moye (Biggs et al. 2011; Greenfield et al. 2019).

Volcanism along the MER is mainly bimodal, with mafic and felsic compositions and scarce intermediate rocks. Fractional crystallization generated zoned magma chambers with felsic melts accumulating at the top and feeding the polygenic volcanoes, often with explosive rhyolitic eruptions generating calderas. Mafic magmas were likely erupted from the bottom of these reservoirs. Considerable volumes of crystal cumulate (>100 km^3) are inferred to be stored beneath these felsic magmatic systems, filling at least 16–30% of the volume generated by crustal extension (Peccerillo et al. 2007; Hutchison et al. 2018). The calderas accompanying the

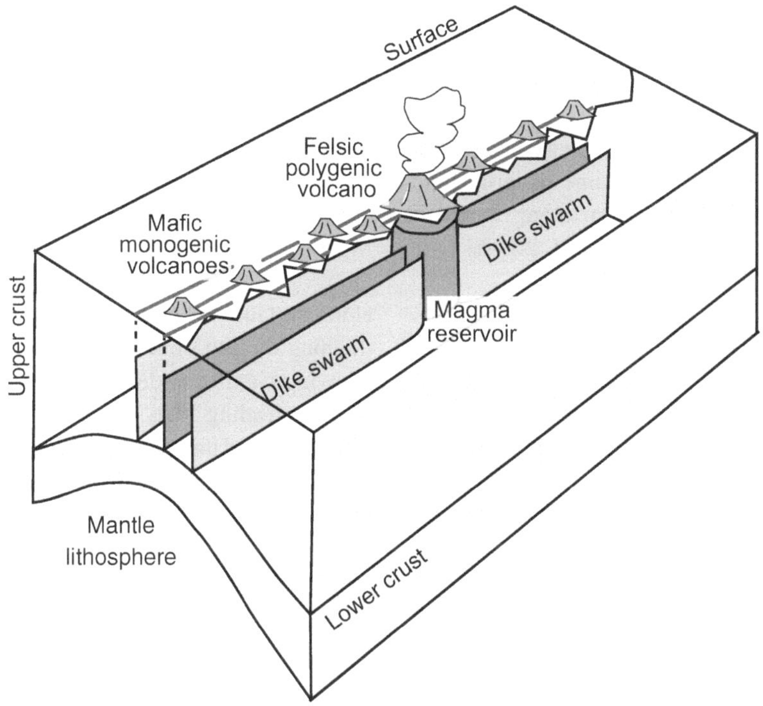

Fig. 11.4 Scheme of a magmatic system along a continental divergent boundary. Magma is fed from a central magma reservoir (cylinder) by dikes propagating mainly laterally in the upper crust (modified after Ebinger and Casey 2001; original image courtesy Cynthia Ebinger)

development of these felsic reservoirs are usually elongated parallel to the extension direction, or to pre-existing regional structures which may be intruded by magma, suggesting in both cases some control of the regional far-field stress. Trachythic volcanism focuses off-rift, likely related to a deeper source, which prevents melts from reaching rhyolitic compositions. The widespread off-rift volcanism, building elongated volcanic edifices up to 1.5 km high, is structurally controlled by the unloading of the rift depression, which alters the local stress field below the rift, steering the magma from beneath the rift axis towards the rift flanks. At a much smaller scale, the activity of normal fault scarps within the rift, responsible for local unloading on their hanging wall, may also explain the clustering of the monogenic vents on their foot wall, at a few hundred of metres from the scarp (Maccaferri et al. 2014, 2015).

At depth, geophysical data show that the crustal thickness along the MER varies between 38 (to the south) and 27 km (to the north) and that of the Ethiopian Plateau to the sides of the rift varies between 44 and 33 km (Mackenzie et al. 2005). The MER at depths between 7 and 25 km hosts pervasive cooled magmatic intrusions, largely dikes, accommodating extension. These intrusions form segmented, ~20 km wide and ~50 km long bodies in a right stepping en-echelon pattern, approximately mimicking the surface segmentation of the magmatic systems. Therefore, rifting obliquity localizes intrusions into the crust within the en-echelon magmatic segments. Seismicity also focuses along the magmatic systems, providing deeper evidence for faulting and diking (Fig. 11.2; Keranen et al. 2004; Dugda et al. 2005; Keir et al. 2006, 2015). At further depth, a segmented low velocity zone suggesting partial melt reaches the upper mantle and extends to the rift sides, connecting northward with low velocity structures under Afar. At even higher depth, this low velocity zone merges with the upper mantle continuation of the plume

(Hammond et al. 2013). Shear-wave splitting in the upper mantle indicates a strong melt-induced anisotropy, supporting a magma-assisted rifted lithosphere. In synthesis, the timing of rift sector development, the three-dimensional focusing of melt, and the ponding of plume material influence intrusions and volcanism along the MER (Bastow et al. 2008; Kendall et al. 2005).

11.3 Transitional Rifts: Afar

The transition between continental and oceanic lithosphere along a divergent plate boundary can be observed, although partly complicated by the activity of the mantle plume, in the Afar area. This area connects the above-mentioned continental MER with the Gulf of Aden and Red Sea oceanic rifts, forming a triple junction, that is where the Nubian, Somali and Arabian plates meet (for an overview see Varet 2018).

The evolution of the Afar area is mainly related to the emplacement of a mantle plume below East Africa. The arrival of the plume head induced a broad uplift of the lithosphere from the late Eocene to the early Oligocene, accompanied by the emplacement and eruption of more than 0.5×10^6 km^3 of flood basalts (Ethiopian Traps), mainly within 1 Ma, at ~30 Ma (Mohr 1983; Hofmann et al. 1997; Furman et al. 2016). In the last ~23 Ma the Arabian-Nubian shield fragmented, separating the Nubian, Arabian and Somali plates along the Gulf of Aden, the Red Sea and the MER, thinning the lithosphere from 100 to 50 km and allowing significant decompression melting. The mean spreading rates of the Gulf of Aden (or Aden) Rift and the southern portion of the Red Sea Rift are ~1.1 to ~1.6 cm/year and ~1.6 cm/year respectively, higher than the 0.7–0.8 cm/year of the northern MER, implying that most of the deformation in Afar results from the activity and interaction of the oceanic Aden and Red Sea rifts (Fig. 11.5; Tapponnier et al. 1990; Vigny et al. 2007; McClusky et al. 2010; Kogan et al. 2012). In the easternmost Aden Rift the onset of seafloor spreading is dated at ~20 Ma, with a

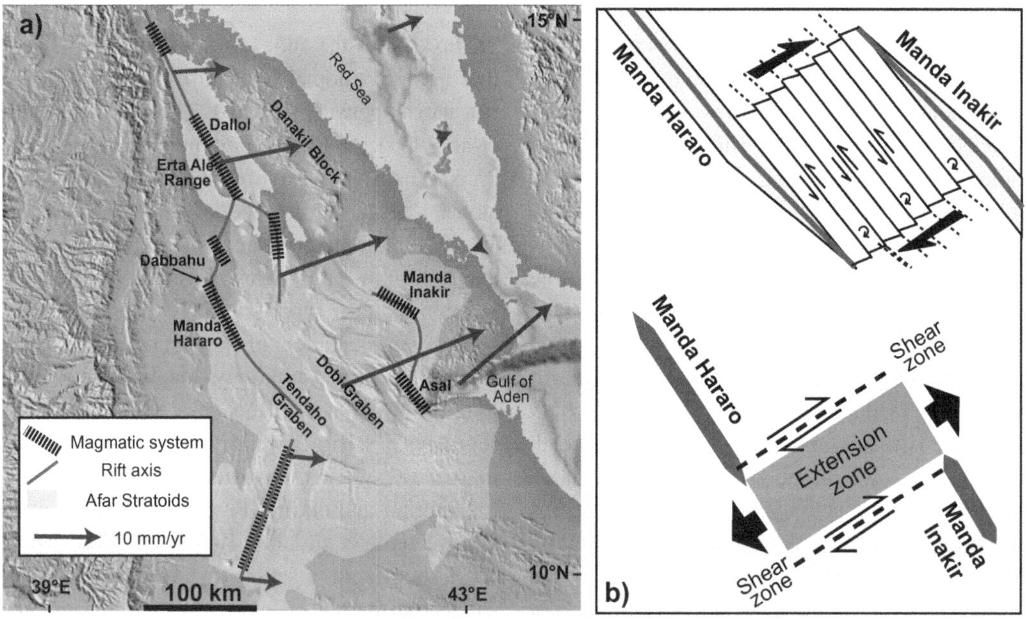

Fig. 11.5 **a** Overview of Afar, highlighting the main magmatic systems, the extent of the Afar Stratoids and the plate motion from GPS measurements: purple arrows show predicted motion on block boundaries (McClusky et al. 2010; base DEM provided by GeoMapApp). **b** End-members kinematic models proposed for central Afar. Top: schematic model of clockwise rotation in central Afar due to "bookshelf" left-lateral faulting between the left-stepping overlapping Manda Hararo and Manda Inakir rifts (modified after Tapponnier et al. 1990). Bottom: schematic model of broadly extending area, delimited by strike-slip fault zones, between the left-stepping overlapping rifts (modified after Pagli et al. 2019)

westward progression. Conversely, in the Red Sea the first pulse of seafloor spreading occurred in its southern part (~17°N), then propagating southwards, separating the Danakil microplate from Arabia, and northwards (Schettino et al. 2016). Below Afar the northward and eastward shallowing of the Moho accompanies increased crustal thinning and segmented melt supply responsible for continental break-up. The thinning shows a northward decrease in the plate strength, in the length of basin-bounding faults, in the length and width of the basins, in the separation of the magmatic centres, and an increase in magma supply, in the volume of Quaternary basalts and the appearance of ~50 to ~80 km long volcanic ridges (Hayward and Ebinger 1996; Bastow and Keir 2011; Gallacher et al. 2016).

Magmatic activity fed by a heterogeneous mantle source has accompanied the recent development of the Afar depression. This recent volcanism can be summarized through 3 main stages (Barberi and Varet 1977; Lahitte et al. 2003; Rooney 2020b). (a) The emplacement of the widespread and more than 1500 m thick "Afar Stratoids" sequence, made up of flood basalts and ignimbrites, underlain by large amounts of silicic lava, marking a major magmatic phase during continental break-up, from 4 to 1 Ma. This trap-like sequence covers approximately two thirds of central Afar, indicating widespread volcanism (Fig. 11.5). (b) The development of polygenic silicic volcanoes, as precursors to rift propagation in the last 2 Ma, prior to the main extensional phase associated with basaltic fissure eruptions. These evolved volcanoes and associated magma chambers form zones of localized lithospheric weakness, concentrating stress and guiding the development of fissure eruptions (Fig. 11.6). (c) The late Quaternary oceanic-type basaltic volcanism, mainly occurring through fissure eruptions along the magmatic systems of the active onland portions of the Red Sea and Aden rifts. The repeated and predominantly intrusive events of the last decades, whose overview is summarized in Fig. 11.7, highlight the current activity of these same axial portions.

The structure of the Afar region is presented in more detail below.

The ~WSW–ENE trending Aden Rift is highly oblique, showing an overall NE-SW

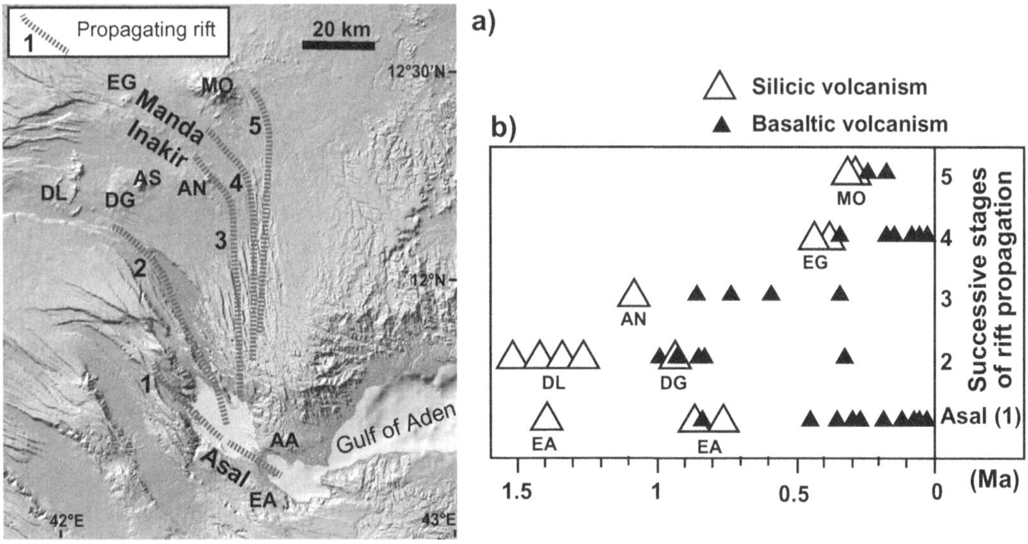

Fig. 11.6 **a** Simplified structural evolution of the onland portion of the Aden Rift in eastern Afar, showing the five successive stages (numbered) of propagation of the Asal Rift to the Manda Inakir Rift. **b** Distribution of silicic (large yellow triangles) and basaltic (small black triangles) volcanic activity as a function of time for the main rift segments associated with the propagation of the Aden ridges; location of volcanoes identified by acronyms is shown in a) (modified after Lahitte et al. 2003). Base DEM provided by GeoMapApp

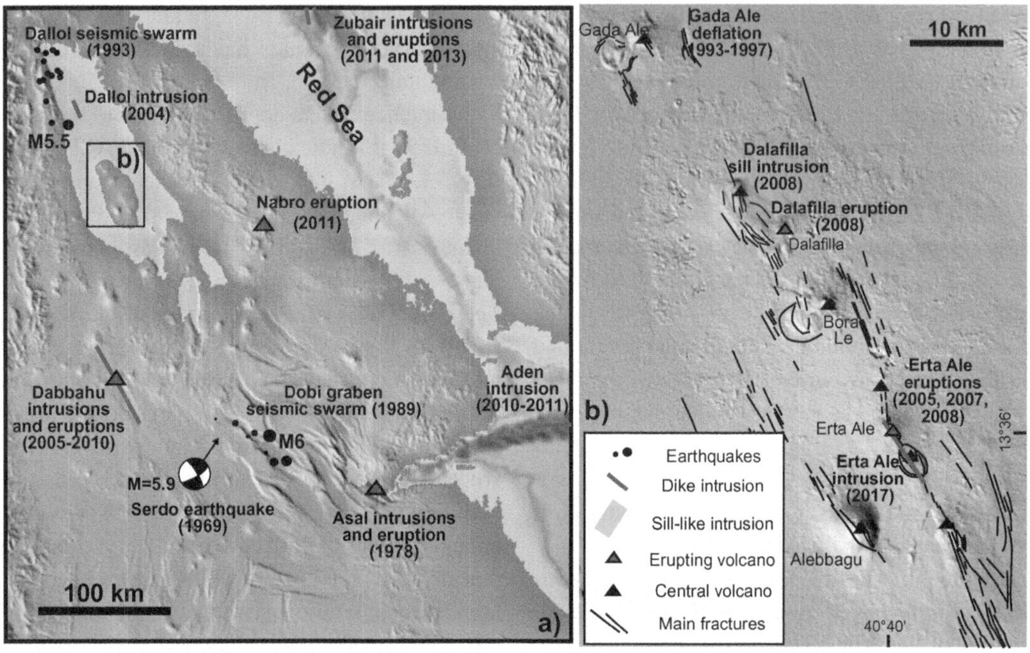

Fig. 11.7 **a** Overview of the magmatic and seismic (approximate location, maximum magnitude reported; from Jacques et al. 1999; Nobile et al. 2012) activity of Afar in the last decades; **b** detail of the recent activity along the Erta Ale Range, Red Sea Rift; see text for details. Additional diking events (not indicated here) occurred in 2000 at the junction between Afar and the northern MER, a few tens of kilometres to the south of the map (Keir et al. 2011) and in 2007 at Jebel at Tair Island, 50 km to the northwest of the Zubair Islands in the southern Red Sea (Jonsson and Xu 2015). Base DEM provided by GeoMapApp

extension direction. This oceanic rift is segmented, with ~E–W trending spreading centres interrupted by NE-SW trending transform faults. One of the easternmost segments experienced an event of seafloor spreading related to the lateral propagation of a dike from November 2010 to March 2011. This dike, responsible for widespread seismicity with magnitude clustering between 2.1 and 5.6, produced an estimated horizontal opening between 0.6 and 2.9 m (Ahmed et al. 2016). The onland continuation of the Aden Rift consists of the NW–SE trending Asal-Ghoubbet Rift and, to the north, the parallel but offset Manda Inakir Rift (Fig. 11.5). These two magmatic systems are connected by a transfer zone forming an extensional faulted monocline due to crustal downwarping undergoing rotations of small rigid blocks about a vertical axis. Each magmatic system consists of subvertical to riftward-dipping active normal faults and an axial portion of active tensile fracturing and volcanic activity, with fast fault propagation and block rotation rates, implying transient strain transfer (Tapponnier et al. 1990; Manighetti et al. 2001). The Asal magmatic system underwent a rifting episode in 1978, due to the along emplacement of a 4.5 km long and 2.2 m thick dike accompanied by widespread seismicity and eruption at Ardoukoba, in the northwest part. At the same time, a ~8 km long and ~4 m thick dike emplaced below the offshore Ghoubbet Rift, ~15 km to the southeast. The emplacement of these dikes was associated with the deflation of the mid-segment Fieale caldera, suggesting that its magmatic reservoir laterally fed the dikes and the eruption (Tarantola et al. 1979; Smittarello et al. 2016; Fig. 11.7). Between 1978 and 1986 the Asal Rift opened at a fast rate, mainly magmatically and aseismically, whereas after 1986 the opening rate and seismicity decreased. Over more than 23 years following the stretching episode, the Asal Rift has sustained mass input and post-rifting unsteady opening, higher than the large scale Arabia-Somalia motion. The InSAR data from 1997 to 2005 show slip along the Asal Rift faults, suggesting that these are in a critical failure state and respond instantly to small fluid pressure changes. More generally, the whole fault activity and distribution along the Asal Rift appears dike-controlled (Doubre et al. 2007; Vigny et al. 2007; Pinzuti et al. 2010).

Northern Afar is built on the onland appearance of a segment of the Red Sea Rift propagating southwards, probably promoting an incipient transform fault passing through the Danakil Islands (Fig. 11.2). The onland portion of the Red Sea Rift consists of two main NW–SE trending en-echelon segments, nearly 500 km long (Fig. 11.5). The northern segment separates the Ethiopian Plateau, on the Nubian plate, from the Danakil Block, or microplate, which experienced an overall eastward shift and counter clockwise rotation. The crust in between these plates, lying below sea level, is thinner (~ 15 km thick) than the average crustal thickness of Afar (~ 25 km), suggesting an advanced, proto-oceanic rifting stage (Eagles et al. 2002). To generate such a large amount of subsidence, upper-crustal extension should be dominated by faulting, coupled with ductile extension at depth, and only locally by magmatic intrusions. Here seismicity focuses along the rift axis and the western marginal graben. The former occurs in conjunction with magmatic intrusion, while the latter is either caused by upper crustal faulting accommodating crustal thinning or by flexural faulting between the rift and plateau (Bastow et al. 2018). The most dramatic manifestation of volcanic activity along the northern portion of the onland Red Sea Rift is the Erta Ale Range, consisting of largely basaltic $\sim$NW–SE aligned central volcanoes, fissure eruptions and hydrothermal spots, associated with fracturing and diking. Here InSAR and seismicity data have spotted repeated activity: this includes episodes of magma withdrawal at Gada Ale volcano between 1993 and 1996, along-rift dike emplacement below the Dallol hydrothermal spot in 2004, sill deflation below Dalafilla volcano in 2008 and along-rift dike propagation in 2017 below Erta Ale caldera, which hosts a long-lasting lava lake (Fig. 11.7; Barberi and Varet 1970; Amelung et al. 2000; Nobile et al. 2012; Pagli et al. 2012; Xu et al. 2017).

To the south, the onland portion of the Red Sea Rift branches into the Tat'Ali Rift and, to the west, the more developed NW–SE trending Manda Hararo Rift, which reaches the NW–SE trending Tendaho Graben into central Afar. The upper crust below the Manda Hararo Rift has the most significant low velocity dispersion anomalies and highest conductivities of Afar. The conductivity anomaly has been interpreted to indicate that more than 500 km^3 of magma is present, mainly near the crust-mantle boundary. The volume of magma is potentially sufficient to feed crustal intrusions tens of thousands of years (Buck 2013; Desissa et al. 2013). A rifting episode occurred at Dabbahu, in the northern Manda Hararo Rift, from 2005 to 2010. The major intrusive event, in 2005, emplaced at least two laterally propagating dikes, up to 8 m thick, over a length of 60 km from a source below the Ado'Ale volcanic complex, in the centre of a magmatic system with the dikes at a depth between 2.5 and 10 km. This event was accompanied by a minor eruption, magnitude $M > 5$ earthquakes and surface deformation with normal fault displacement of up to 3 m and extension fractures (Fig. 11.8; Wright et al. 2006; Rowland et al. 2007; Ayele et al. 2009). Normal fault reactivation during this event suggests that the faults along the rift axis result from the upward propagation of the stress during diking and thus magma may be responsible for the relief of the magmatic system. Thirteen more dike intrusions, for a total volume of magma larger than 3 km^3, occurred between 2005 and 2010. Their cumulative opening becomes largest in the central part of the intrusion, where smoothing deficits from the first intrusion were observed. Most dikes were again sourced from the centre of the magmatic system. During these diking events, the seismicity was largely released at the dike tip, during dike

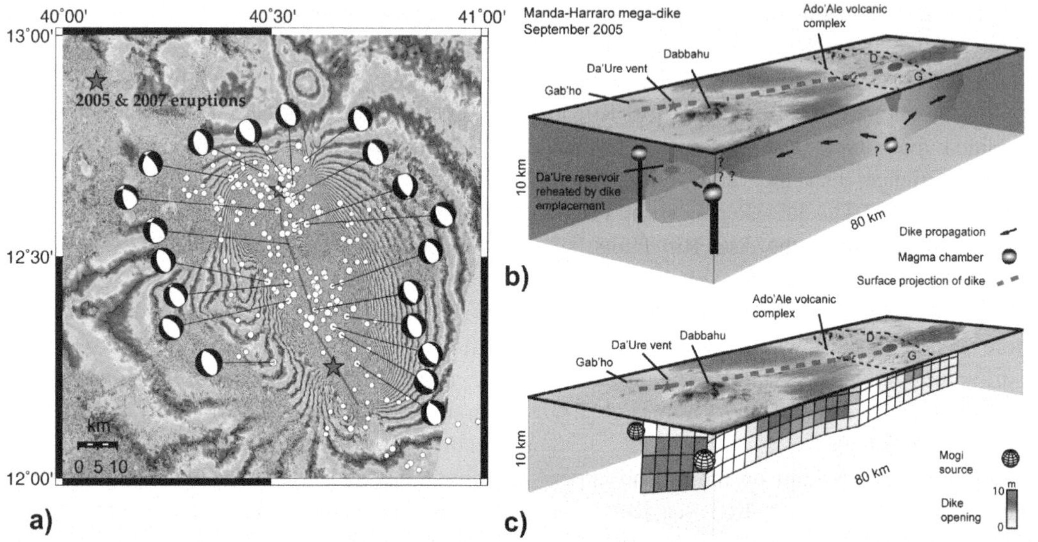

Fig. 11.8 The 2005–2010 Dabbahu rifting episode, Afar (Ayele et al. 2009). **a** Fault plane solutions of earthquakes during the onset of the rifting cycle in the Dabbahu-Manda Hararo segment between September 4–October 4, 2005. Solutions are superposed on an interferogram formed by satellite radar images acquired on May 6 and October 28, 2005. Red stars indicate sites of the 2005 silicic eruption and the 2007–2009 fissural basalt eruptions. **b** Scaled model for the September 2005 intrusive event. Red dashed line indicates the surface trace of the dikes. Arrows indicate magma movement directions. **c** Topography and model of dike opening that reproduces vertical and horizontal crustal movements determined from InSAR data, constrained by seismic observations. Spheres are Mogi sources that simulate magma pressure sources added to the dike opening to produce the observed surface deformation patterns

propagation. These diking events are followed by a decadal-scale period with extension rates faster than the secular divergent plate motion, as observed in the Aden Rift (Fig. 11.9; Ebinger et al. 2010; Belachew et al. 2011; Pagli et al. 2014). Integrated studies reveal that, on the longer-term, an intrusive episode like that observed at Dabbahu between 2005 and 2010 has occurred roughly every 10 ka. More generally, the 2005 intrusion at Dabbahu has confirmed the importance of

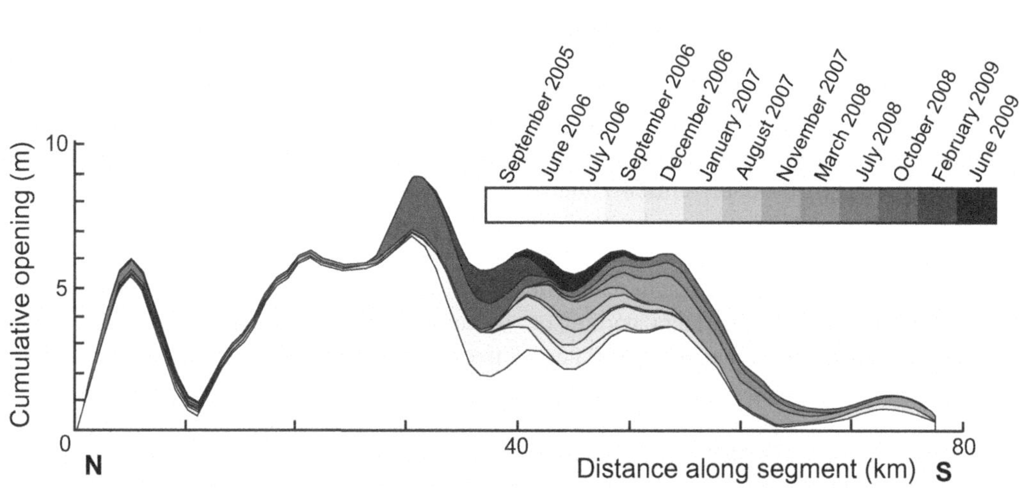

Fig. 11.9 Depth-averaged dike opening from 2005 to 2009 for the top 10 km of the crust beneath the Dabbahu-Manda Hararo segment plotted against distance along the segment from north to south (modified after Ebinger et al. 2010)

episodic and rapid magma emplacement in the opening of a rift and thus in plate boundary deformation (Sigmundsson 2006a; Medynski et al. 2016). This event has been strengthened by the successive dike intrusions, highlighting how focused accretion at magmatic systems occurs through lateral magma propagation. Therefore, the Dabbahu episode has shown that, while tectonic forces may prepare or encourage rifting over longer-time scales (thousands of years), magmatic injection promotes major rifting episodes lasting shorter time spans (years).

The Manda Hararo Rift terminates into the Tendaho Graben, which formed in the last ~1.8 Ma and abruptly interrupts the MER structures. The Tendaho Graben is several tens of kilometres wide and a few hundred kilometres long, and hosts eruptive fissures and central volcanoes. The graben is the southernmost expression of the onland portion of the Red Sea Rift and experienced a decrease in tectonic and magmatic activity once the Aden Rift appeared onland, suggesting a mutual interdependence in the activity of these two rifts. Its distinctive structure with inward tilted margins may result from two main magma-related processes: (a) the collapse induced by magma withdrawal during the eruption of the voluminous Afar Stratoids from the graben axis, showing how large and repeated fissure eruptions form collapsed rifts; (b) the load of the axial mafic intrusions in a thinned and heated plate that decreases its strength (Acocella 2010; Corti et al. 2015). The Tendaho Graben is the largest of a series of NW–SE trending basins in central Afar, between the Manda Hararo and the Asal-Ghoubbet-Manda Inakir rifts. This area is characterized by pervasive faulting, developing tens of rigid blocks separated by ~NW–SE striking faults whose kinematics may be consistent with two main models (Fig. 11.5). (a) A bookshelf faulting model, based on the reactivation of the NW–SE trending normal faults as left-lateral structures, due to the interaction between the Aden and Red Sea overlapping rifts: this is consistent with the measured clockwise rotations of crustal blocks about a horizontal axis in the last 1.8 Ma in central Afar. A modification of this bookshelf model, based on seismicity data, emphasizes the current extension of the NW–SE trending faults (Tapponnier et al. 1990; Sigmundsson 1992). (b) An overall distributed extension on rift parallel normal faults, driving rift perpendicular shearing at the tips of spreading rifts and finally achieving plate boundary linkage during incipient seafloor spreading. The geodetic behaviour of this extended area highlights a central microblock evolving separately from the three surrounding plates, although depending on the availability of magma supply within the nearby rifts (Doubre et al. 2017; Pagli et al. 2019). These two models are not mutually exclusive, as the former relies mainly on paleomagnetic data in the last ~2 Ma and the latter on present-day seismicity and geodetic data.

11.4 Oceanic Rifts

Prolonged extension leads to further lithospheric thinning and the progressive formation of new oceanic crust, which is produced at an oceanic ridge along the plate boundary. Here extensional stress accumulates at a steady rate, to be episodically released in discrete events of magma intrusion, volcanism and associated faulting, whose magnitude is proportional to the time elapsed since the last event. The prolonged activity of an oceanic ridge is responsible for **seafloor spreading**, that is the formation of new oceanic crust through magmatic activity.

Oceanic ridges are segmented features: their continuity is most notably interrupted by transform faults or overlapping spreading centres (Fig. 11.10). **Transform faults** are hard-linkage transverse structures, striking parallel to the spreading direction of the two plates and connecting offset ridge segments. Because of the opposite sense of motion of the plates between the two offset ridges, transform faults display predominant strike-slip kinematics. This motion may extend also outside the zone between the ridge segments, along fracture zones, provided that the ridge segments have different spreading velocity. **Overlapping Spreading Centres** are soft-linkage interaction zones between the extremities of oceanic ridges, showing an overlapping zone with

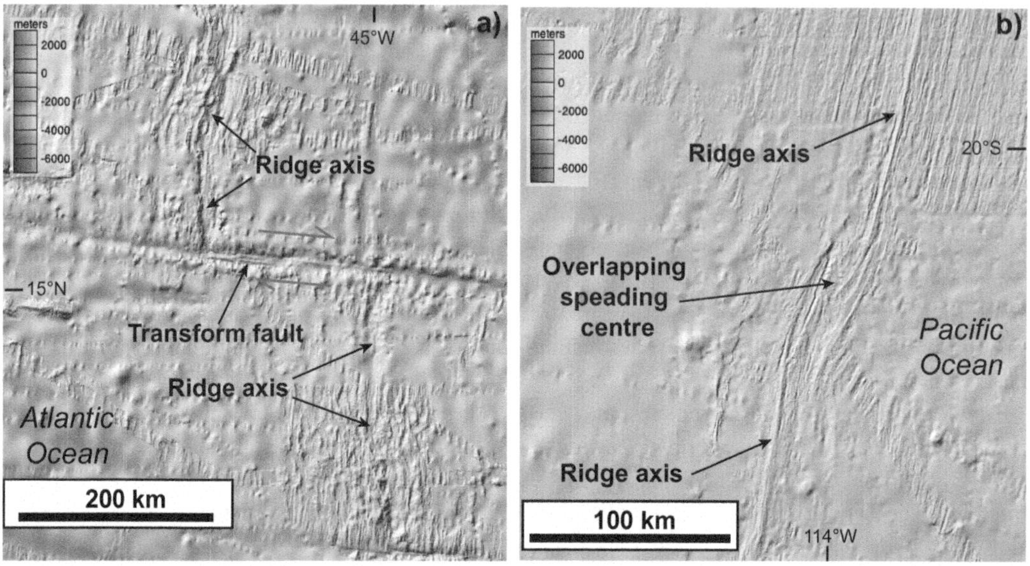

Fig. 11.10 Types of linkage between segmented oceanic ridges. **a** Hard-linkage, through a transform fault; red arrows show the relative motion along the fault. **b** Soft-linkage, through an overlapping spreading centre. Base DEM provided by GeoMapApp

characteristic hook-shaped geometry explained by the variation in the ridge propagation force as two ridges approach and pass each other (e.g., Pollard and Aydin 1984). The segments of the oceanic ridges present a four-order hierarchy. At higher orders, segments are shorter (less than 10 km for fourth-order and up to 1000 km for first-order) and more short-lived (from 10^2 to 10^4 years for fourth-order to 10^7 years for first-order; Macdonald et al. 1991). First-order ridge segments are limited by transform faults, whereas second-order segments terminate at Overlapping Spreading Centres. Several hypotheses may explain the segmentation and along-strike propagation of the ridges, as summarized below (Macdonald et al. 1991; Vanderbeek et al. 2016). Oceanic ridges may be divided into stable spreading cells, each several tens of kilometres long, linked to uniformly spaced Rayleigh–Taylor gravitational instabilities in the upper mantle; in this model, ridge segmentation is driven by mantle buoyancy. Alternatively, oblique mantle flow beneath the ridges may reorient the spreading segments and promote ridge-axis discontinuities; these would decrease the efficiency of upward melt transport, thus defining segment scale variations in magmatic processes. Gravitational spreading forces due to excess ridge axis topography, where the elevation of the axial region of the ridge above its flanks represents gravitational potential energy, may also promote along-strike growth; this model works well where hot spot magmatism has produced significant ridge axis elevations and variations in near-field stress. Another mechanism is related to the crack propagation force caused by far-field plate stresses, similar to what observed in fracture mechanics, where the propagation force at the crack tip increases with crack length, with the longer segment lengthening at the expense of the shorter; this mechanism may explain why long ridge segments tend to lengthen and prevail over shorter neighbouring segments. Finally, changes in spreading direction may also provide a mechanism for propagation and segment lengthening, although the force to drive ridge propagation is limited.

Depending upon their spreading velocity, ridges are classified as **ultrafast** (full spreading rate of >12 cm/year; as the fast portions of the East Pacific Rise), **fast** (<12 to >8 cm/year; as the slow portions of the East Pacific Rise), **intermediate** (<8 to >5 cm/year, as the Juan de Fuca Ridge, in the north Pacific), **slow** (<5 to >2 cm/year; as the Mid-Atlantic Ridge), and **ultraslow** (<2 cm/year, as the Southwest Indian

Ridge; Fig. 11.11). The spreading rate strongly influences the extent of mantle melting beneath the ridge, magma supply, hydrothermal heath loss and crustal deformation, although the extent of mantle melting may also depend on along-axis variations in the mantle temperatures and/or mantle chemistry (Michael et al. 2003). In general, faster ridges are more linear, with axial rise and narrow (less than 1 km) and elongated summit depression. Their higher magmatic productivity is supported by a continuous and shallower (approximately 1 km of depth) axial melt reservoir, resulting in frequent, relatively homogeneous and small volume eruptions. Faster ridges thus tend to approach spatially and temporally uniform along-axis mantle flow, with fluctuations of the along-axis characteristics considerably more subdued than at slower ridges. Conversely, slower ridges are more tortuous, with a deeper and wider (several kilometres) axial depression or graben, bordered by faults with higher displacement-length ratios associated with uplifted shoulders. Their magmatic productivity results from discontinuous, unstable and deeper (approximately 3 km of depth) melt pockets, feeding infrequent, heterogeneous and larger volume eruptions, also off-rift. At the end of the spectrum, ultraslow ridges are

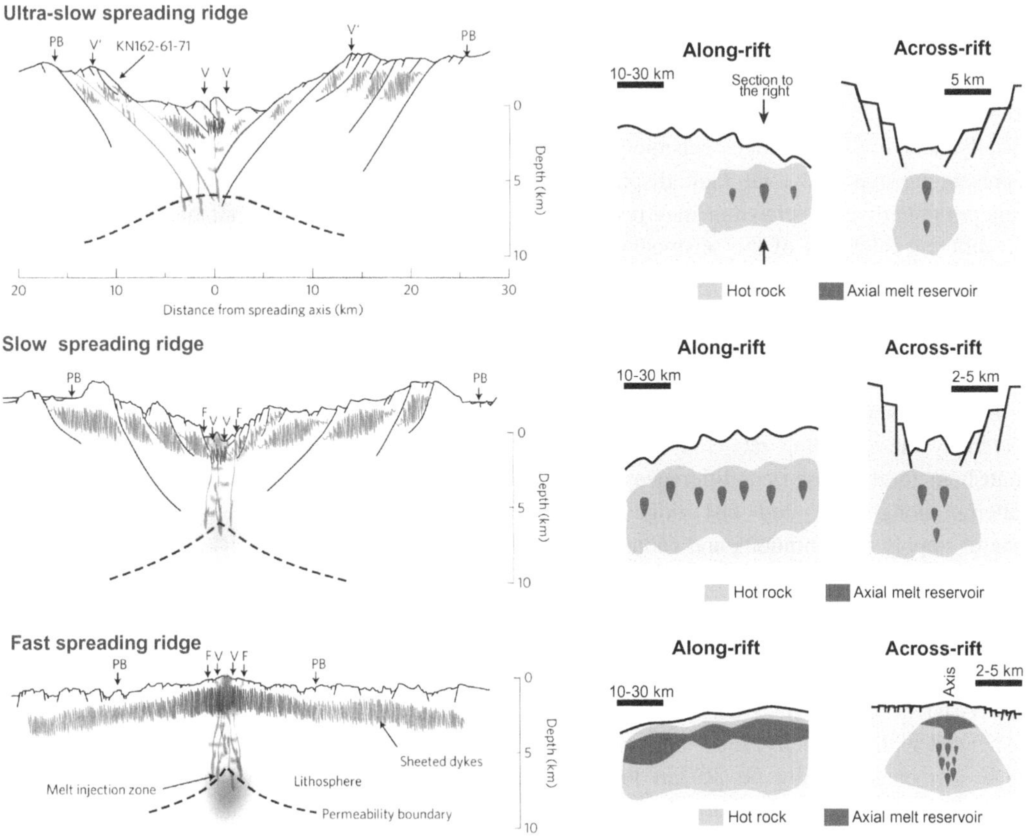

Fig. 11.11 Left: cross-axis bathymetric profiles for ultraslow, slow and fast spreading ridges, including the neovolcanic zone (V), the zone of fissuring (F) and the plate boundary zone (PB, active faulting). The intensity and width of the melt injection zone (red; that is, melt volume) and the thickness of the sheeted dikes (grey lines; dike focusing) indicate the robustness of melt generation and the relative roles of magmatic versus tectonic accretion during extension (modified after Standish and Sims, 2010). Right: schematic enlargements to show fine-scale segmentation of along- and across-rift sections for the ultraslow, slow and fast spreading examples (modified after Macdonald et al. 1991)

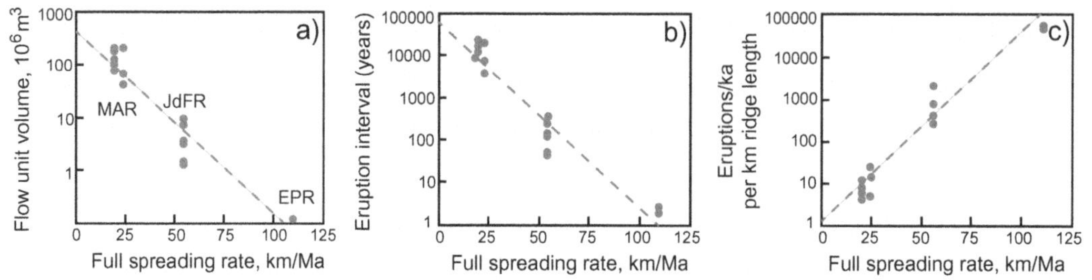

Fig. 11.12 Temporal and spatial intervals and volumes of eruptions against spreading rate (MAR: Mid-Atlantic Ridge; JdFR: Juan de Fuca Ridge; EPR: East Pacific Rise; Perfit and Chadwick 1998)

characterized by non-magmatic portions, where regional tectonic activity dominates (Fig. 11.12; Macdonald et al. 1991; Purdy et al. 1991; Calvert 1995; Niu and Hekinian 1997; Macdonald 1998; Perfit and Chadwick 1998; Sibrant et al. 2018). These features may also affect the seismic behaviour, as in faster ridges most fault slip seems to occur with minor seismicity or aseismically, and in slower ridges seismic slip may represent most of the total fault displacement. This possible diverse behaviour may result from the different thickness of the seismogenic zone, which is proportional to the seismicity during fault slip: this zone, being smaller in faster oceanic ridges, may produce a lower seismic slip (Mark et al. 2018).

Differences in the magmatic input to dikes, driven by the spreading rate, may significantly contribute to the observed differences in fault patterns along slow and fast ridges. Where magma supply is continuous and robust (steady state), volcanic output dominates over regional tectonic processes, which are largely aseismic, along ridge segments typically 100–1000 km long. Where the supply is intermittent (non-steady-state), regional tectonic activity, largely associated with seismicity, may dominate along ridge segments typically 50–80 km long, with deep earthquakes, major normal faults and exposures of lower crustal rocks. The critical spreading rate between steady-state and non-steady-state supply is inferred to be 5–6 cm/year (Mutter and Karson 1992; Perfit and Chadwick 1998; Sandwell and Smith 2009). However, investigations on the intermediate spreading Costa Rica Rift reveal that its evolution is also sensitive to relatively small changes in the full spreading rate. These fluctuations correspond to different processes occurring at the ridge axis, so that even at intermediate spreading ridges there appear to be two end-member modes of crustal formation: one magmatic, occurring at the faster end of the spectrum, and the other magma-dominated, but accompanied by enhanced tectonic extension that occurs at the slower end. These magma-dominated and faulting-enhanced modes of spreading are not mutually exclusive, as the oceanic crust can form via their combination, with varying predominance and/or durations of each, resulting in a fine balance. These modes of spreading may be influenced by tectonic events associated with the plate boundary, changes in the spreading rate, or patterns of spreading of adjacent plates (Wilson et al. 2019).

The spreading rate is thus a crucial parameter in controlling the morphology, structure, magmatic supply, productivity and eruptive frequency of oceanic ridges. This parameter in turn may also depend on the general far-field tectonics at the other ends of the oceanic lithospheric plates, that is along subduction zones. Not only this condition modulates the spreading rates, but it also alters the convection regime by obstructing the circulation of plates, which in turn modifies the surface kinematic conditions for the convecting mantle. Therefore, the spreading rate of a ridge may mirror its status in the global plate tectonics framework (Husson et al. 2015).

The main features of ultraslow (Sect. 11.4.1), slow (Sect. 11.4.2) and fast and ultrafast (Sect. 11.4.3) oceanic ridges are discussed below, considering representative examples.

11.4.1 Ultraslow Ridges: The Red Sea and the Southwest Indian Ridge

Ultraslow spreading ridges, characterized by a full spreading velocity lower than 2 cm/year, have been recently recognized as an important class of oceanic ridges, also sharing similarities in the processes that govern the break-up of continents along non-volcanic rifted margins (Snow and Edmonds 2007). Ultraslow ridges consist of linked magmatic and amagmatic accretionary segments, the latter being a previously unrecognized class of accretionary plate boundary structure. As a result, these ridges can produce magmatic crust or exhume (i.e., bring to the surface) highly serpentinized peridotite from the upper mantle to the seafloor (Dick et al. 2003; Schlindwein and Schmid 2016; Grevemeyer et al. 2018; Reston 2018). Exhumation occurs through oceanic low-angle, or detachment, faulting, which may be a major mode of seafloor accretion also at some slow ridges. This is associated with dramatic changes in seafloor morphology, as detachments form expansive dome structures with corrugated surfaces, known as oceanic core complexes. These areas lack shallow seismicity in the upper 15 km of the lithosphere, but unusually contain earthquakes down to depths of 35 km, implying a cold, thick lithosphere. The detachments often transition laterally to segments producing magmatic crust, consisting of multiple regularly-spaced normal faults that form volcanic abyssal hills parallel to the spreading axis. These magmatic regions thin dramatically under volcanic centres, allowing along-axis melt flow. The switch between magma-rich and magma-poor oceanic crustal accretion is attributed to along-axis gradients in lithospheric strength or, more likely, magma supply. Indeed, highly punctuated volcanism active for tens of millions of years suggests that first-order ridge segmentation is controlled by stable mantle processes of melting and melt segregation, with mantle temperatures varying significantly along-axis suggesting that the extent of mantle melting is not a simple function of the spreading rate. The transition from slow to ultraslow ridges may also be independent of melt productivity, and rather result in the efficiency of vertical melt extraction (Jokat et al. 2003; Michael et al. 2003; Montesi and Behn 2007).

Examples of ultraslow spreading ridges include the Mid-Cayman spreading centre (Caribbean), the Gakkel Ridge (Arctic Ocean), the Red Sea and the Southwest Indian Ridge; the last two cases are described below.

Considering the Red Sea allows completing the regional continuum from continental (MER) to transitional (Afar) and oceanic plate divergence. Moreover, the Red Sea displays lithosphere that is in transition from rifting to seafloor spreading, providing the opportunity to describe incipient oceanic floor formation. The Red Sea is in fact a newly-formed oceanic rift with only parts having entered the seafloor spreading stage. Magnetic data suggest that the Red Sea has been widening from 0.7–0.8 to 1.6 cm/year at the northern and southern portions, respectively, consistent with geodetic constraints of current opening at 0.7 and 1.5 cm/year (Fig. 11.13; Bosworth et al. 2006; Almalki et al. 2015). After a first rifting pulse at $\sim$34 Ma, the southern portion of the Red Sea started to open at $\sim$26 Ma, soon after the eruptive climax of the Ethiopian Traps. A first phase of axial spreading created oceanic lithosphere between $\sim$26 and $\sim$24 Ma, propagating segments of oceanic crust northward above areas of localized rise of mantle melts, fading into steady passive crustal accretion with slower spreading: rifting, rift flank uplift and volcanism occurred nearly simultaneously (Bonatti 1985; Omar and Steckler 1995; Ligi et al. 2011; Almalki et al. 2015, and references therein). Then, between 22 and 6 Ma, localized extension and mafic diking at the edge of the Arabian platform heralded a pause in oceanic basin formation and a shift in the locus of extension from the central spreading ridge to the eastern continental–transitional crust zone. Re-initiation of ridge activity and spreading occurred at $\sim$5 Ma, resulting from a plate reorganization attributed to the switch from subduction to collision and oblique motion between Eurasia and Arabia, driven by the extrusion tectonics in Anatolia, which allowed

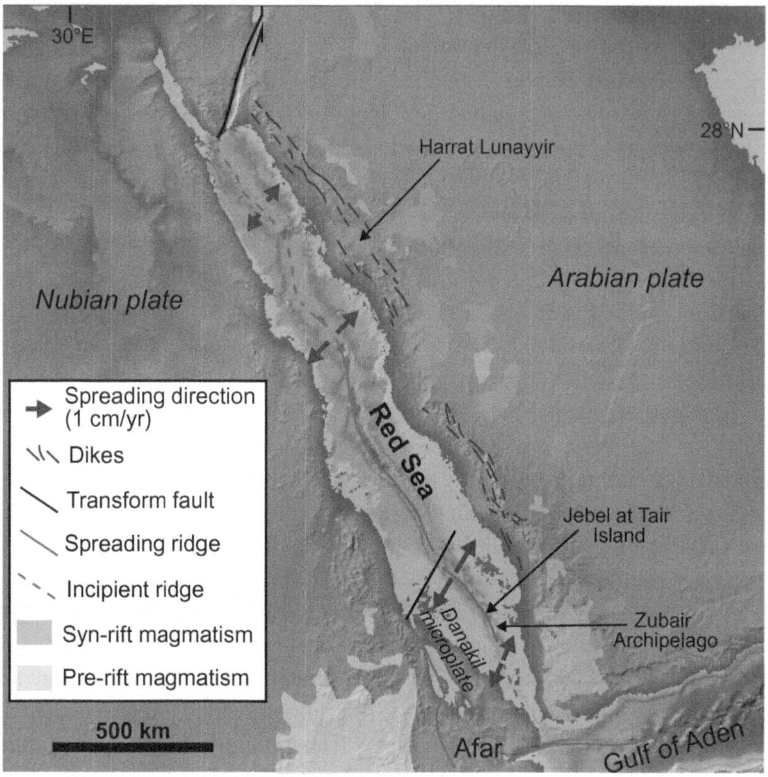

Fig. 11.13 Main tectono-magmatic features of the Red Sea area, illustrating the opening vectors, the spreading ridges in the southern part and the incipient ridges in the northern part, as well as the main volcanic products erupted before (mainly Oligocene) and during rifting (modified after Almalki et al. 2015). DEM provided by GeoMapApp

renewed and faster motion of Arabia relative to Africa (e.g., Cochran 1983; Almalki et al. 2015). Nevertheless, such a discontinuous and episodic ridge spreading activity in the southern Red Sea indicates a limited role in accommodating crustal extension.

This two-stage spreading model is applicable to the southern and perhaps the central Red Sea, whereas in the less extended northern Red Sea diffuse extensional models are better applicable, because there is debated evidence for the existence of magnetic stripes and any oceanic crust. The current Red Sea can be thus separated into three distinct portions. The southern Red Sea, experiencing seafloor spreading, or drifting. The northern Red Sea, still largely experiencing continental rifting, with thinning continental crust, intruded mafic melts and limited oceanic crust and seismicity. The central Red Sea, a transition zone where isolated bathymetric deeps mark the oceanic onset of divergence (Bonatti 1985; Ligi et al. 2011, 2019; Mitchell and Park 2014; Schettino et al. 2016; Mitchell and Stewart 2018). These variations may be ascribed to the rheologically weaker behaviour of the southern Red Sea under the influence of the nearby Afar plume, which started the Red Sea rifting (Almalki et al. 2015).

Red Sea tectonic models consider both asymmetric and symmetric extension. Asymmetric extension has been proposed based on simple shear along a regional scale low angle normal fault that breaks away at the surface on the African continent and projects to sublithospheric levels beneath the Arabian plate. This model was used to explain the asymmetric distribution of volcanic rocks along the edge of the Arabian plate and the higher topography on the Arabian side of the Red Sea (Fig. 11.13). However, such a structural asymmetry has not been confirmed by geological or geophysical data (Wernicke 1985; Almalki et al. 2015, and references therein). Indeed, there is an overall difficulty in applying tectonic models, as none of the models proposed so far is applicable to the entire Red Sea basin.

The lithospheric thickness along the eastern margin of the Red Sea varies considerably, suggesting an important role of asthenospheric flow related to the significant off-rift magmatism on the Arabian side, the last event being the non-eruptive Harrat Lunayyir (Saudi Arabia) dike intrusion in 2009, associated with seismicity and surface fracturing. The dike rose from 10–11 km of depth and almost reached the surface, forming two diverging fracture zones, whose distance increase is due to the progressive deepening of the dike top (Pallister et al. 2010; Chang and Van der Lee 2011; Trippanera et al. 2019).

Volcanic activity has been recently focusing in the southernmost part of the Red Sea. The first volcanic eruption known to occur in the southern Red Sea in over a century started on Jebel at Tair Island in September 2007. The Jebel at Tair activity was followed by two more eruptions within the Zubair Archipelago, about 50 km to the southeast, in 2011–2012 and 2013, both of which started on the seafloor and resulted in the formation of new islands. In particular, the 2011 eruptive event was fed by a 12 km long dike parallel to the Red Sea axis, associated with seismicity and normal faulting, which fed a submarine and then a subaerial eruption. These eruptions suggest that this portion of the Red Sea, where the evidence for a linear spreading centre disappears, is magmatically more active than expected (Jonsson and Xu 2015; Xu et al. 2015; Eyles et al. 2018).

The ultraslow Southwest Indian Ridge allows considering the transition from quasi-melt-free detachment-dominated spreading to magmatic spreading near prominent axial volcanoes. Detachments extend along-axis for several tens of kilometres, with lifetime of ~0.6 to ~1.5 Ma. Once a detachment becomes inactive, the successor cuts into its predecessor's foot wall with opposite polarity, causing part of the foot wall lithosphere to experience further deformation, hydrothermal alteration and, possibly, sparse magmatism (Fig. 11.14). The accretion of the oceanic lithosphere therefore occurs over the lifetime of successive detachment faults with flipping polarity. The transition from this nearly amagmatic detachment-dominated mode to the more common magmatic mode of spreading occurs over along-axis distances of a very few

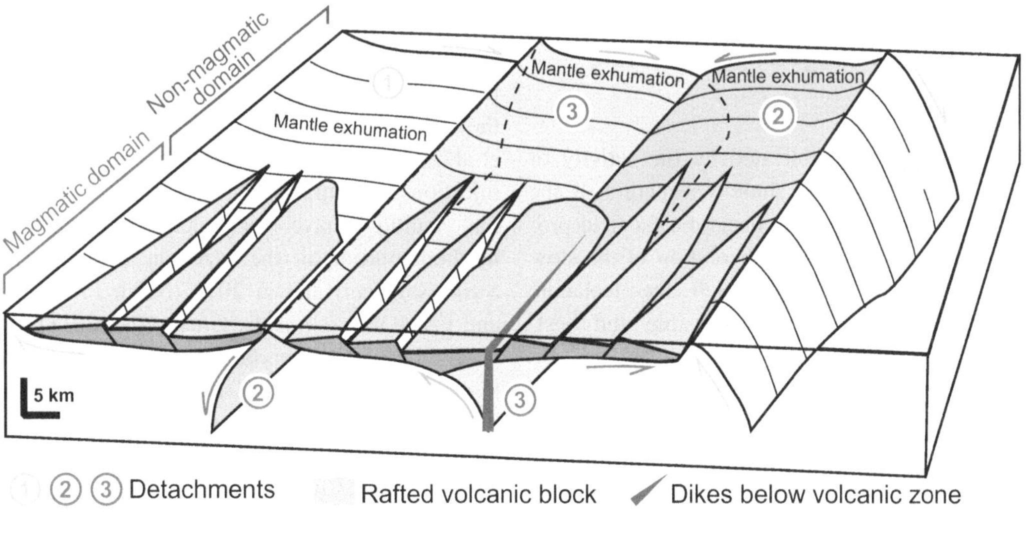

Fig. 11.14 Tectonic model of ultraslow seafloor spreading applied to the Southwest Indian Ridge. Mantle (shown transparent) is in general exhumed to form smooth seafloor by slip on successive detachments with alternating polarity: light green (1) to dark green (2) and then blue (3), the only currently active in the diagram. Block arrows show movement direction. The detachments continue laterally also beneath rafted volcanic blocks (grey; foreground), where increasing magmatic contribution to divergence occurs (modified after Reston 2018)

tens of kilometres. As the magmatic contribution to divergence increases, this transition involves a significant thinning of the axial lithosphere and a gradual decrease of the amount of displacement on faults. Axial magma chambers below the Southwest Indian Ridge have been imaged between 4 and 9 km of depth, suggesting that the ~10 km thick crust is mainly formed by magmatic process. Here magma fills the gap between the diverging plates, thus reducing fault activity and modifying the thickness, thermal state and rheology of the plate boundary. Increased magmatic divergence requires the detachments to root at shallower depths, consistent with the seismicity-defined shallowing of the base of the brittle lithosphere moving along the ridge axis towards the volcanic centres (Jian et al. 2017; Reston 2018; Cannat et al. 2019). The activation of the magmatic or of the detachment-dominated modes of spreading depends on the volume of magma supplied per increment of plate separation and the rheology of the axial lithosphere. Magmatic accretion is not confined to a narrow central spreading axis, as indicated by the young volcanic eruptions broadly dispersed throughout the rift valley (Buck et al. 2006; Standish and Sims 2010).

11.4.2 Slow Ridges: The Icelandic Ridge

Although partly complicated by the activity of the mantle plume, the shallow structure of the Icelandic Ridge largely reflects the far-field processes associated with the evolution of the slow Mid-Atlantic Ridge, of which the Icelandic Ridge remains the most accessible and best-studied portion.

The Mid-Atlantic Ridge (MAR) is the most notable example of slow spreading ridge, stretching from pole to pole along the Atlantic Ocean. The MAR is frequently dissected by transform faults, especially in the equatorial region. It shares some features with ultraslow ridges, including important portions with detachment faulting exhuming lower crust to upper mantle, as between 22 and 24 °N, where ~40% of the seafloor is made of extended terrain and detachment faults. Overall, the input of magma at the spreading axis is complementary to the development of detachment faulting, but the relationship is not strong, as for example between 14 and 16 °N where detachments continued to slip despite hosting magmatic intrusions, suggesting partially concomitant activity (Cann et al. 2015; Parnell-Turner et al. 2018, and references teherein). Magmatic activity along the MAR may be also manifest in a subtler way. For example, large scale corrugations, or megamullions, of the detachments in different portions of the MAR have been interpreted to result from significant magmatic accretion below, as due to the emplacement of shallow intrusions. Therefore, an important portion of the total extension in correspondence of the detachments may be accommodated by otherwise unnoticed magmatic accretion. Magmatic accretion may also occur off-axis, explaining the much lower distribution of magmatism than that predicted by the spreading along the ridge axis alone (Tucholke et al. 2008; Searle et al. 2010). Magma chambers have been seismically imaged in different portions of the MAR, as at 23 and 37 °N, consistent with a focused melt supply at the segment centre and steep across-axis thermal gradients, indicated by the proximity with nearby active faults. These chambers commonly lie at ~3 km of depth below the sea bottom, although in some cases they may reach ~1.2 km (Calvert 1995; Singh et al. 2006; Combier et al. 2015). Events of dike injection, accompanied with seismicity and normal faulting, have been detected along the northern portion of the MAR in 2001 (Lucky Strike segment) and in 2010 (North FAMOUS and FAMOUS segments; Giusti et al. 2018).

The best-known portions of the MAR are the highly oblique-spreading Reykjanes Ridge and the emerged Icelandic Ridge, the latter being a well-documented example of the influence of a mantle plume on a rifted lithosphere. The Reykjanes Ridge is an oblique spreading ridge consisting of en-echelon magmatic systems in a dextral configuration, partly fed by the Icelandic hotspot, which induces buoyancy-driven upwelling in the mantle below and increases the spreading rate (Searle et al. 1998; Gaherty 2001). Each of these magmatic systems constitutes a third-order oceanic segment, while multiple systems constitute second-order segments. Magma from the mantle is initially focused towards these second-order segments and then redistributed along individual magmatic systems, which may be abandoned, indicating a finite life span. The abandonment occurs within ~2 Ma, when the system has migrated sufficiently far to become isolated from its melt supply (Peirce and Sinha 2008).

Emersion of the Iceland portion of the Mid-Atlantic Ridge results from increased magma productivity due to the plume below, with low upper mantle velocities to at least 400 km depth. The plume may be confined to the upper mantle or extend down to the core-mantle boundary. As a result of the excessive production of magma, estimated as 0.2 km^3/year, and of the related underplating the crustal thickness in Iceland varies between 15 and 46 km (Sigmundsson 2006b, and references therein; Jenkins et al. 2018). Spreading between the Eurasian and North American plates along a ~N100° direction and up to 1.8 cm/year is taken up within a ~100 km wide plate boundary along two main neovolcanic zones (Fig. 11.15; Geirsson et al. 2006; Sigmundsson et al. 2020, and references therein). To the west, is the Reykjanes-Langjokull Volcanic Zone, including the obliquely spreading Reykjanes Peninsula rift to the southwest, and the Western Volcanic Zone (WVZ) to the northeast, comprising the Hengill-Langjokull systems. The WVZ has been the main locus for crustal spreading in South Iceland for the last 6–7 Ma. To the east, the neovolcanic zone comprises the Northern Volcanic Zone (NVZ), active for 6–7 Ma, to the north of the Vatnajokull icecap, and the Eastern Volcanic Zone (EVZ), active for 2–3 Ma, to the south of it. The EVZ is propagating southwards, and significant crustal spreading has only developed north of Torfajokull volcano. The EVZ and WVZ overlap in Southern Iceland, so that the spreading rate of the EVZ decreases southward, while along the WVZ it increases southwards, maintaining an overall constant total rate. The area enclosed by the EVZ and the WVZ forms the South Iceland Seismic Zone, producing

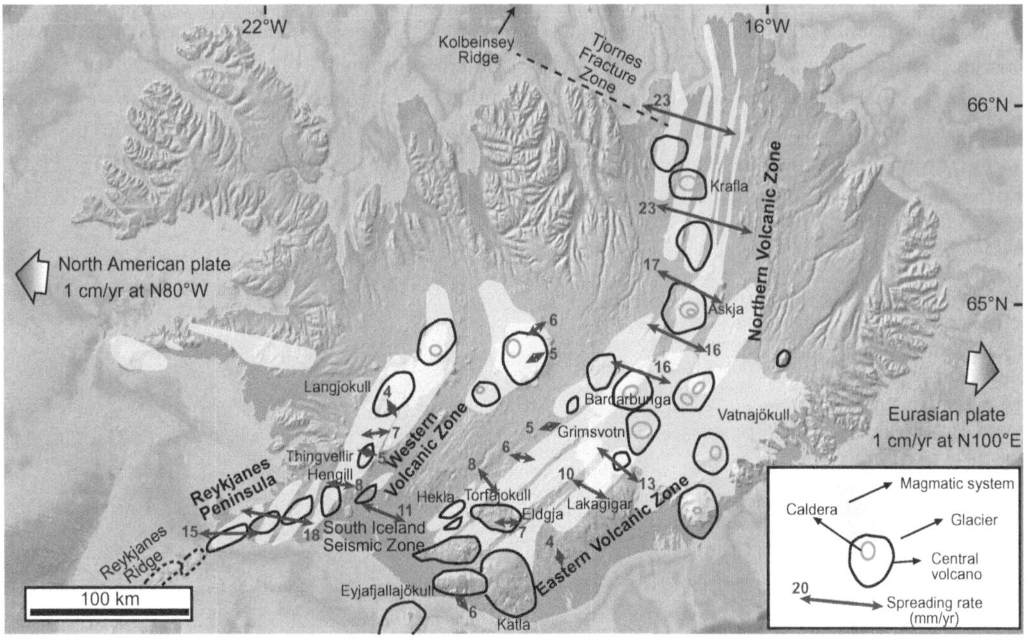

Fig. 11.15 Main tectono-magmatic features of Iceland, including the magmatic systems and the 1986–2002 GPS spreading rates (double arrows; Perlt et al. 2008). Base DEM provided by GeoMapApp

Fig. 11.16 Examples of **a** open normal fault with tilted hanging wall and **b** metre- wide extension fracture at Thingvellir, southwest Iceland

$M > 7$ earthquakes, with a tectonic frame consistent with the early stages of development of a transform-like feature (Perlt et al. 2008; Angelier et al. 2008).

Surface deformation along the active rift zone consists of extensional (tension) fractures and normal faults, commonly with a dilational component between the tilted hanging wall and the horizontal foot wall, as observed along the MER and Afar (Fig. 11.16). The normal faults may form with two main mechanisms. (a) From the coalescence of joints within the lava pile, growing into extensional fractures under a tensional σ_3 and propagating downward to become shear fractures (normal faults) at several hundreds of metres of depth: this scenario implies a regional tectonic origin for the normal faults. (b) Following the emplacement of dikes, with the normal faults propagating downward from the surface and forming graben-like structures above the dikes. Recent evidence, also from diking episodes outside Iceland, suggests that mechanism (b) is the most likely and also generally applicable to magmatic divergent plate boundaries (Gudmundsson 1992; Grant and Kattenhorn 2004; Rowland et al. 2007; Acocella and Trippanera 2016).

Volcanism focuses along the rift axis, in 40–150 km long and 5–20 km wide magmatic systems, each with a dominant volcano often with evolved composition and summit caldera. Outside the dominant volcano, eruptive fissures with monogenic cones, associated with extension fractures and normal faults, highlight the continuation of the magmatic system (Fig. 11.17; Gudmundsson 1995). The eroded Late Tertiary

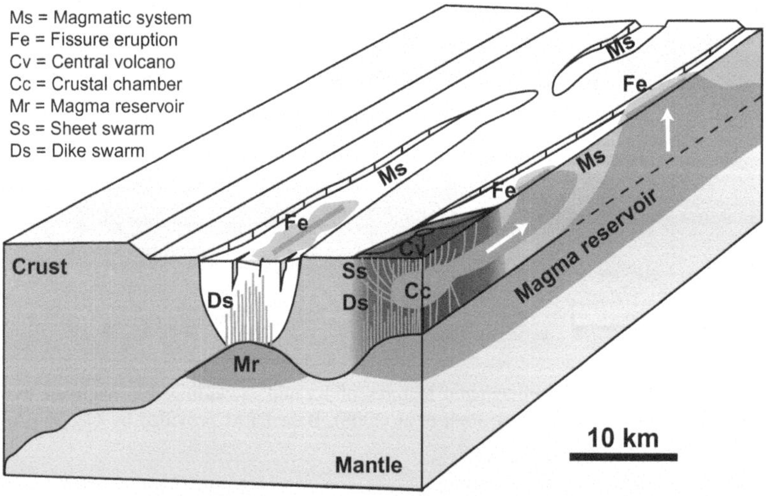

Fig. 11.17 Structure of magmatic systems along the Icelandic ridge. Fissure eruptions may be fed by the lateral propagation of dikes from a central volcano (dark area) and/or vertical propagation of dikes from the mantle below the system (white arrows; modified after Gudmundsson 1995)

and Pleistocene lava piles at the eastern and western sides of Iceland show that magmatic systems at depth consist of local sheet swarms and regional dike swarms. The sheet swarms are confined to extinct volcanoes and are usually 45°–65° inward dipping, circular to elliptical in distribution, with radius of several kilometres. Most sheet swarms are associated with plutons at 1–2 km of paleodepth. The regional dikes occur outside the eroded central volcanoes in 50 km long and 5–10 km wide swarms. These dikes are usually parallel, vertical and thicker than the sheets (see Fig. 7.4). The sheets follow the trajectories of the near-field stress around the source chambers feeding frequent, small volume proximal eruptions, whereas the regional dikes are controlled by the far-field stress, feeding larger but less frequent distal eruptions. Also, while the sheets are supplied by the shallower portion of the plumbing system, the regional dikes are supplied by the deeper portion. Magma chamber formation is facilitated by stress barriers leading to the formation of sills. The magma chamber acts like a trap for upward propagating regional dikes from mantle reservoirs and channels magma, through sheets, towards the surface (see Sect. 4.6). The eroded portions of Iceland also show that dike swarms at depths greater than ~1 km take up most of the crustal extension, with negligible contribution from normal faults. At shallower depths, the frequency of the dikes decreases, while the contribution of normal faults and, subordinately, extension fractures on extension progressively increases (Fig. 10.23; Helgason and Zentilli 1985; Gudmundsson 1986; Forslund and Gudmundsson 1991).

Within the period of geodetic measurements, most of the 35 active volcanoes in Iceland have not deformed. The deformed volcanoes inflated and deflated with rates from mm/year to cm/year (Sigmundsson et al. 2006b; Einarsson 2018). The most active and erupting Holocene volcanoes are Hekla, Katla and Grimsvotn, in the EVZ. Hekla shows the most regular behaviour, as the volume of its eruptions scales with the preceding eruptive period, producing ~1 km^3 of magma each century. Magma chambers lie at ~3 km of depth below Krafla, Katla, Askja and Grimsvotn volcanoes; other volcanoes, such as Hekla, do not have evident magma chamber shallower than 14 km depth. Although basaltic volcanism dominates, historical activity also features andesites, dacites and rhyolites. About 80% of historical eruptions occurred along EVZ. Magma productivity in the last 1100 years equals ~87 km^3 DRE, with 71 km^3 in the EVZ, with an average of 20–25 eruptions per century. Collectively, the EVZ and WVZ only erupted ~15% of total intruded volume in 1130 years (Thordarson and Larsen 2007). An overview of the tectonic and magmatic features of the main portions of the Icelandic Ridge is provided below.

The Reykjanes Peninsula consists of 5 magmatic systems in a dextral en-echelon configuration, highly oblique to the rift axis and undergoing significant left-lateral motion (1.1–1.8 cm/year). Along the margins of the magmatic systems, longer faults are subparallel to the rift axis, whereas in the centre of the system the shorter fractures strike approximately parallel to the eruptive fissures and perpendicular to the direction of minimum principal stress σ_3. The amount of geodetic extension is variable: while absent in 1993–1998, an extension of ~0.7 cm/year has been recognized between 2000 and 2006, confirming episodic rift opening. Focal mechanisms from earthquakes show a local variation of the extension direction (from N120° to N140°) along the Peninsula with regard to the regional one (N105°). In the last 4 ka, rifting episodes occurred in several portions of the Reykjanes Peninsula with average duration of ~500 years and recurrence period of ~700 years (Hreinsdottir et al. 2001; Angelier et al. 2004; Keiding et al. 2009; Saemundsson et al. 2020). To the north, the WVZ undergoes nearly orthogonal extension, alternating volcanic events and faulting on a time scale of 10^3 years, as at Thingvellir. In particular, the 40 km long and 5–7 km wide Thingvellir Graben dissects the 9000 year old pahoehoe basaltic lava flow with extension rates between 0.3 and 0.8 cm/year (Sinton et al. 2005).

In the Northern Volcanic Zone, the long-term spreading axis passes through and changes direction at the main central volcanoes, and does not follow the general direction of each fissure swarm (Drouin et al. 2017). The Krafla magmatic system is one of the most developed and active, having experienced a rifting episode between 1975 and 1984 (Fig. 11.18). In this episode, the central magma chamber fed ~20 dikes that propagated over ~80 km laterally, inducing surface fracturing and faulting with up to 2 m of vertical displacement. During the rifting sequence, magma from depths higher than 5 km accumulated in the Krafla magma chamber, at ~3 km of depth, above which deflation was observed during lateral dike injection (Bjornsson et al. 1977; Sigurdsson 1980; Arnadottir et al. 1998). Approximately 250 × 10^6 m^3 of basaltic lavas were erupted from eruptive fissures during the rifting episode, but a substantially larger volume, ~1 km^3, remained intruded, Maximum cumulated widening of ~9 m occurred 10–12 km north of Krafla (Fig. 11.19). Between 1992 to 1995, a subsidence of 2.4 cm/year above the Krafla magma chamber and of 0.7 cm/year along the magmatic system have been interpreted as due to cooling contraction and ductile flow of material away from the spreading axis. Similar processes have been invoked to explain the continuous deflation of the nearby Askja caldera between 1983 and 1998. The post-rifting spreading rate of the NVZ between 1987 and 1992 was higher than 3.0 cm/year and between 1993 and 2004 was 2.3 cm/year, slowing but still higher than the 1.8–2.0 cm/year predicted spreading rate, with a behaviour similar to that observed at Asal and

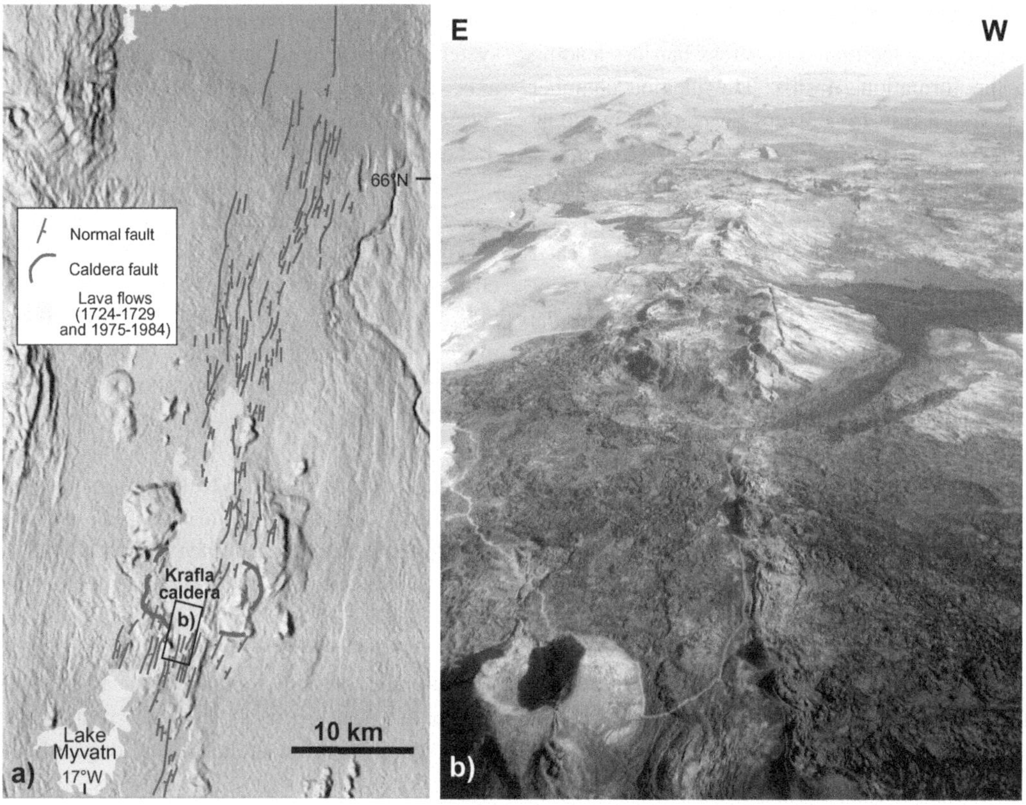

Fig. 11.18 The Krafla magmatic system, northern Iceland. **a** Overview of the magmatic system; base DEM provided by GeoMapApp. **b** View of the portion of the 1975–1984 eruptive fissure within Krafla caldera, Iceland, associated with normal faults, scoria cones and lava flows (location shown in a; photo courtesy Joel Ruch)

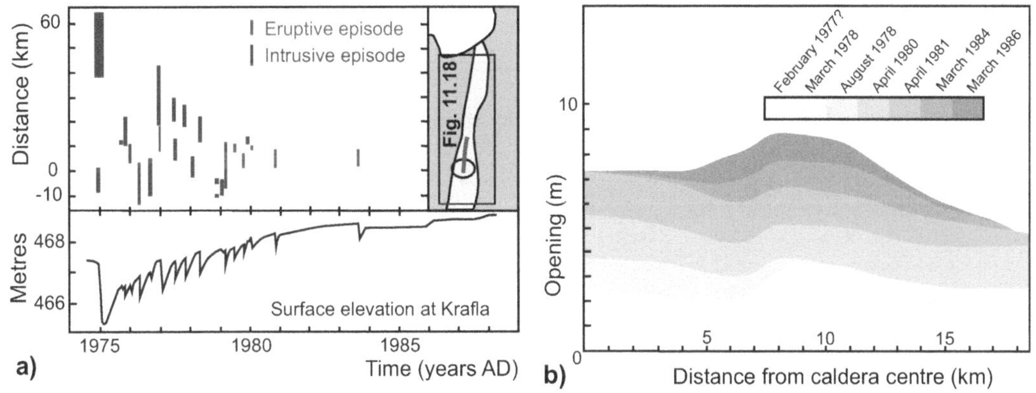

Fig. 11.19 The 1975–1984 Krafla rifting episode, Iceland. **a** Observations at Krafla between 1974 and 1989. Top: distance ranges for surface fissuring (blue) or extrusion of lava (red) measured from Krafla caldera (map, right). Thickness of the vertical lines gives duration of the activity. Bottom: changes in the elevation near the centre of inflation in the Krafla caldera; tick marks are for the beginning of each year (modified after Buck et al. 2006). **b** Opening across the Krafla fissure swarm during the rifting episode. Contributions of individual events are shown versus northward distance from Krafla caldera (modified after Sigmundsson 2006b)

Manda Hararo, in Afar (Tryggvason 1984; Sigmundsson et al. 1997; Sturkell and Sigmundsson 2000; Buck et al. 2006; Arnadottir et al. 2009). A rifting event occurred to the south, at Bardarbunga, in 2014. Here a segmented dike propagated laterally over 14 days for more than 45 km along the Bardarbunga magmatic system at a variable rate, with topography influencing the direction of propagation (Fig. 11.20; Sigmundsson et al. 2015). The strike of the dike segments varied from an initially radial direction away from the Bardarbunga caldera, towards alignment with that expected from regional stress at the distal end. Seismicity, dike opening and magma eruption focused at the distal portion, and were simultaneous with magma source deflation and slow collapse at the Bardarbunga caldera, accompanied by *M5* earthquakes. In the NVZ, rifting episodes or events appear to occur in the same magmatic system with a frequency of approximately every few hundred of years, with previous rifting sequences occurring at Krafla in 1724–1729, at Bardarbunga in 1797 and at Askja in 1874–1875. In particular, the 1875 formation of the Oskjuvatn caldera was part of a rifting episode at the Askja magmatic system that reactivated the Sveinagja Graben to the north, possibly under lateral flow of magma from the Askja reservoir, although the magma erupted along the Sveinagja Graben was different from that erupted in the caldera area and volumetric calculations do not require lateral magma propagation (Gudmundsson and Backstrom 1991; Sigmundsson 2006b; Hartley and Thordarson 2012). The EVZ has also produced several rifting episodes in historical times. Among these is the ~934 AD rifting episode at Eldgjà, feeding a nearly 70 km long eruptive fissure, and the 1783–1784 rifting episode at Lakagigar. The latter opened a 27 km long dike-fed eruptive fissure, producing one of the largest historic basaltic lava flows (nearly 15 km^3) propagating towards and activating the nearby Grimsvotn central volcano (Fig. 1.12b; Thordarson and Self 1993). Available evidence indicates that these rifting episodes are usually related to the lateral propagation of the dike(s), feeding the eruption from the magma chamber of the dominant volcano of the magmatic system. The magma feeding these fissures may rise from Moho depths within a few days (Mutch et al. 2019).

11.4.3 Fast and Ultrafast Ridges: The East Pacific Rise

Fast (spreading rate >8 to <12 cm/year) and ultrafast (>12 cm/year) oceanic ridges constitute the end-member type of divergent plate

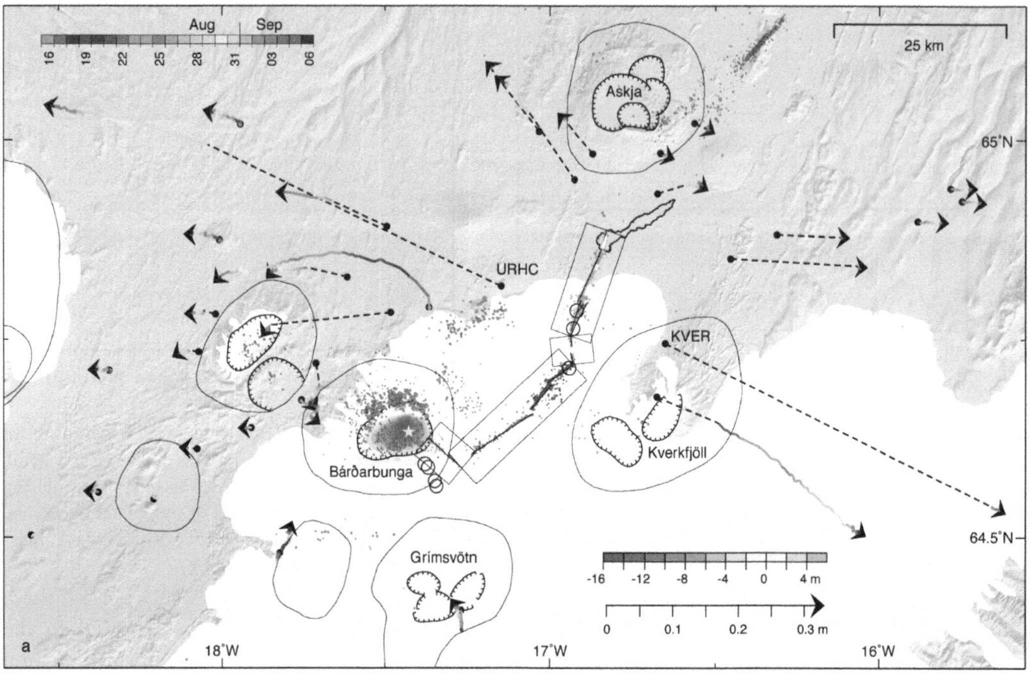

Fig. 11.20 The 2014 Bardabunga spreading event, Iceland (Sigmundsson et al. 2015). Earthquakes from 16 August to 6 September 2014 (dots) and GPS horizontal ground displacements (arrows) with central volcanoes (elliptical outlines), calderas (hatched) and northern Vatnajokull glacier (white). Epicentres and displacements are colour coded according to time of occurrence (key at top left); other single earthquake locations are in grey. Rectangles show areas characterized by dike injections and thin lines within show inferred dike segments. The red shading at Bardarbunga caldera shows subsidence up to 16 m inferred from radar profiling on 5 September. The star marks the location of the magma source inferred from modelling. Also shown are ice cauldrons formed (circles), outline of lava flow mapped from radar image on 6 September, and eruptive fissures (white). Image courtesy Freysteinn Sigmundsson

boundary. The fastest oceanic ridge, approaching a full spreading rate of 15 cm/year, is the East Pacific Rise (EPR), running from the Gulf of California to off Antarctica in the southeastern Pacific Ocean. The EPR is segmented, with the longevity of its segments increasing with their hierarchical order. While first-order volcanic segmentation remains fairly stable, the fourth-order segmentation, including the patterns of local hydrothermal activity, may undergo repeated reorganization. The EPR has prominent bathymetry, with a focused **axial rise**, or topographic high, 5–20 km wide and 200–500 m above the background slope, tapering along rift towards transform faults and overlapping spreading centres. The axial rise may be generated by the dynamic extensional stresses relieved by diffuse diking in the upper lithosphere, with dikes usually dipping away from the axis, as well as by variations in the bulk density of oceanic crust caused by magmatic differentiation, attributed to the geometry of mantle upwelling (Macdonald 1998; White et al. 2002; Toomey and Hooft 2008).

The zone of active faulting, or plate boundary zone, along a fast spreading ridge is a few tens of kilometres wide, with widespread inward and outward dipping faults. For example, along the East Pacific Rise between 8° 30′ N and 10° 00′ N, the zone of active faulting is approximately 90 km wide. Here both inward and outward dipping fault scarps increase in height away from the ridge, reaching average heights of ~100 m at ~45 km from the ridge axis; beyond this distance, there is no significant increase in scarp height (Crowder and Macdonald 2000). Although widespread, here surface faulting accounts only for a limited portion (5–10%) of the total spreading and seems to occur mainly with minor seismicity or aseismically. The development of part of these inward and outward dipping normal faults over a zone much wider than the ridge axis results from the bending of the plate at the axial rise. Removal of the topographic effect of faulting from the original seafloor bathymetry reveals a gently inward dipping surface (towards the ridge axis), possibly resulting from subsurface magma withdrawal associated with dike injection and eruption (Carbotte et al. 2003; Shah and Buck 2003).

In the central portion of the axial rise, some parts (approximately 15–20%) of the EPR show a narrow (usually up to two kilometres wide) and deep (usually up to 100 m) **axial trough** (Fig. 11.21). The trough is commonly interpreted as the surface effect of a narrow and highly active dike intrusion zone, which produces a graben-like structure, consistently with the mechanisms discussed in Sect. 7.7. In particular, the narrower and deeper structure of the axial trough with regard to grabens formed during dike intrusions along slower divergent plate boundaries implies a very shallow emplacement for the responsible dikes, which are inferred to have a relatively high internal magma pressure, thus largely representing feeder dikes (Chadwick and Embley 1998; Fornari et al. 1998). The structure of the axial trough has been related not simply to dike emplacement, but also to the intrusive to extrusive frequencies affecting a portion of the ridge. In particular, extremely narrow axial troughs (a few tens of m wide) have been interpreted to

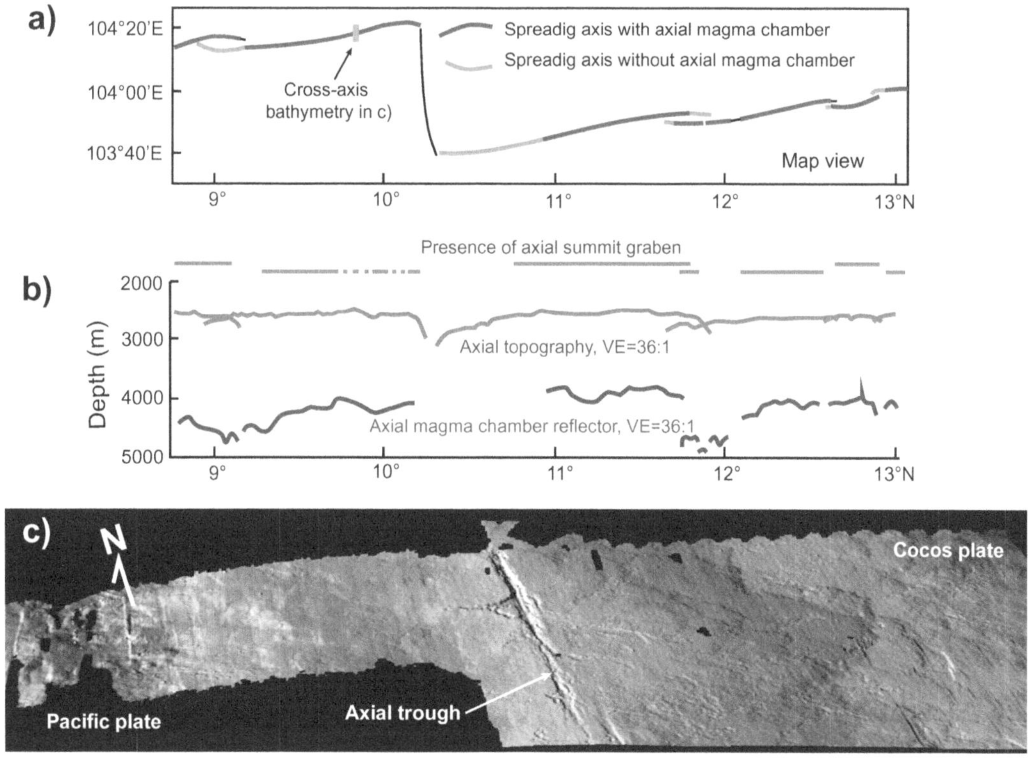

Fig. 11.21 **a** Schematic map view of portion of the fast spreading East Pacific Rise and **b** its main topographic, structural and magmatic features in section view (Macdonald and Fox 1988). **c** Oblique bathymetric view of the EPR at 9° 50′ N (location in **a**; from GeoMapApp). VE = vertical exaggeration

result from recent dike-induced volcanism, whereas the absence of axial troughs may result from a decrease in the eruptive frequency (Soule et al. 2009). Alternatively, the axial trough has been interpreted as an extremely elongated and narrow collapse caldera structure related to the waning activity of the underlying magma chamber (Lagabrielle and Cormier 1999).

Below the axial trough or, more in general, the ridge axis there may be pipe-like zones of vigorous hydrothermal activity, directly fed from underlying magma chambers and generating massive sulphide ore deposits. The seismicity associated with these zones suggests that hydrothermal circulation is strongly aligned along, and not across, the ridge axis (Tolstoy et al. 2008). Most importantly, the portion below the trough also coincides with robust and focused magma supply, through dikes, from the frequently seismically imaged shallow axial reservoirs, which commonly show a sill-like shape. In particular, along the EPR between 8° 20′ N and 10° 10′ N several sill-like magma reservoirs have been imaged at a mean depth of 1.6 km below the seafloor, with a ~6 km dominant length possibly reflecting the spacing of local sites of ascending magma from discrete melt reservoirs pooled beneath the crust. Multi-level complexes of magma lenses have been also imaged, contributing to the formation of both the upper and lower crust (as between 9° 20′ and 9° 57′ N), or resulting from the overlap of two ridge segments (as at the 9° N overlapping spreading centre; Canales et al. 2009; Arnulf et al. 2014; Marjanovic et al. 2014; Marjanovic et al. 2018). Studies based on the anisotropy of magnetic susceptibility (AMS) on samples indicate distinct compositional sources that feed melt injected into dikes with no preference for subhorizontal or subvertical magma flow, consistently with AMS data from ophiolites in Oman or Cyprus. Moreover, the well-imaged sill-like magma reservoirs between 8° 20′ N and 10° 10′ N at ~1.6 km below the seafloor form 5–15 km long segments that coincide with the fine-scale tectonic segmentation at the seafloor. Here transitions in composition, volume and morphology of erupted lavas coincide with disruptions in the lens that define magmatic systems. This suggests that eruptions at the EPR are associated with the vertical ascent of magma from lenses that are mostly physically isolated, leading to the eruption of distinct lavas at the surface that coincide with fine-scale segmentation (Fig. 11.22; Carbotte et al. 2013; Horst et al. 2014).

The mechanisms of lower-crustal accretion may vary along the EPR, as inferred between 9° 42′ and 9° 57′ N, but the volume of melt delivered to the crust appears mostly uniform. In particular, the continuous and shallow along-axis melt bodies of the EPR provide a robust magma supply, fed by narrow vertical transport down to 70 km within the mantle (Forsyth et al. 1998; Kent et al. 2000). This area is overlying a deeper vertical flow, associated with long-lived and deep-seated (down to the lower mantle) upwelling. This mantle-wide upwelling is inferred to drive horizontal components of asthenospheric flows beneath the plates that are faster than the overlying surface plates, thereby contributing to plate motions through viscous tractions in the Pacific region. Therefore, rather than being correlated to slab pull in the Circum-Pacific area, the high spreading rate of the EPR seems more related to stable mantle-scale upwelling (Gu et al. 2005; Rowley et al. 2016).

Off-axis volcanism may be present along fast spreading ridges, as from 9° 37′ to 9° 57′ N. This volcanism does not necessarily derive from a sub-axial magma body, being rather sourced directly from off-axis magma bodies with longer period of melt replenishment, lower eruption recurrence rates, and lower eruption volumes than the axial magma lens. Such off-axis magmatism may occur preferentially where pre-existing fractures inherited from offsets of the spreading axis promote melt transport from the mantle into the crust (Canales et al. 2012; Waters et al. 2013).

Seafloor spreading episodes have been also documented. In 2005–2006, a progressive build up in seismicity opened the EPR at 9° 50′ N (location in Fig. 11.22; Tolstoy et al. 2006; Dziak et al. 2009; Fundis et al. 2010; Xu et al. 2014). The resulting submarine eruptions

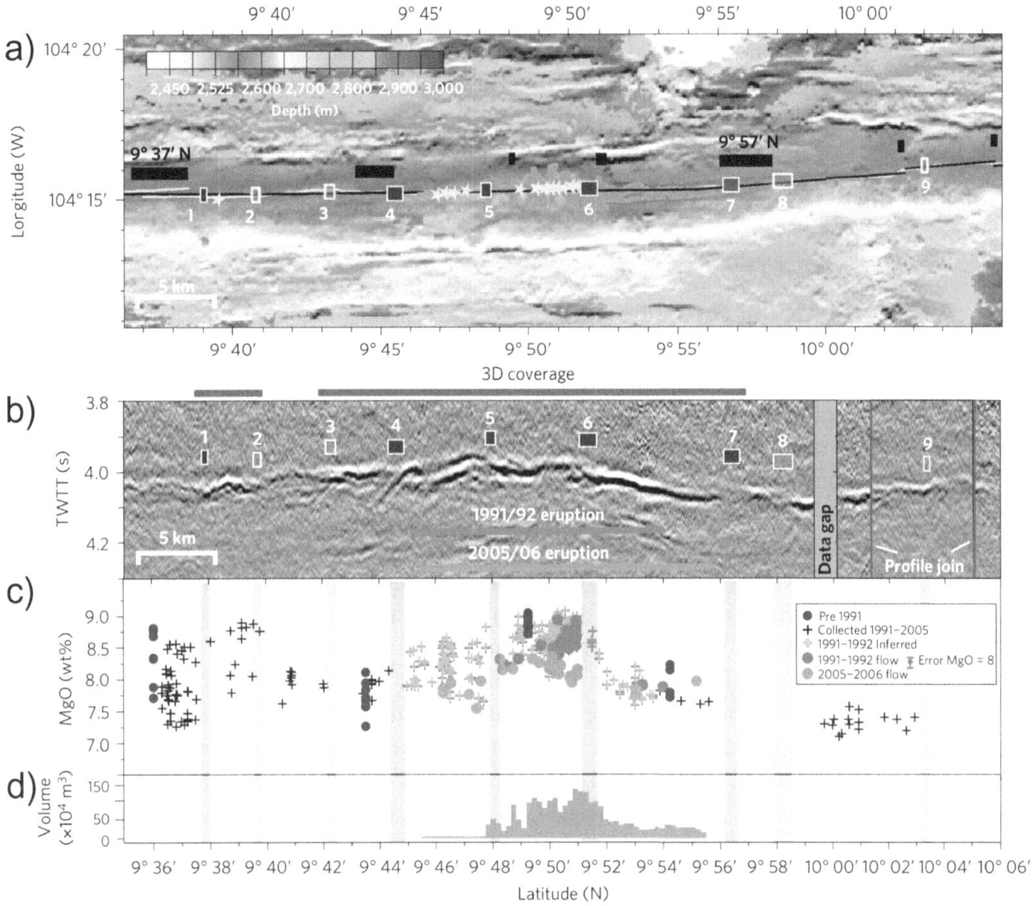

Fig. 11.22 Segmentation in seafloor structure, axial magma lens, lava geochemistry and eruption volume along the East Pacific Rise 9° 35′–10° 06′ N. **a** Bathymetry showing location of axial eruptive zone (yellow line) and composite axial seismic profile (black). Black rectangles: third- (labelled) and fourth-order tectonic discontinuities. Yellow stars: hydrothermal vents; green region: 2005–2006 lava flow. **b** Composite axial seismic reflection section showing magma lens reflection and interpreted disruptions. TWTT, two-way travel time of seismic waves. Numbered rectangles in **a**, **b** indicate magma lens disruptions identified from seismic data (purple, data from 3D seismic volume). **c** MgO composition of seafloor lavas located within 500 m of the axis colour-coded for eruption period. **d** Volume of erupted 2005–2006 lavas. Vertical bars (translucent purple and orange) mark magma lens disruptions from **a** and **b** (Carbotte et al. 2013)

activated multiple vents, with either rapid bilateral propagation of feeder dikes along the rift or near simultaneous vertical injection of melt. The stress changes due to dike intrusion along the ridge promoted normal faulting and strike-slip motion within the nearby transform zone. A total magma volume of 9–83 $\times$ 10^6 m^3 was extracted from a 5 km long section of the ridge axis, with a maximum of 71 $\times$ 10^6 m^3 left unerupted in the crust as dikes. An eruption of similar dimensions may have a frequency of years to decades to sustain the long-term average seafloor spreading rate at this location. The pre-eruptive increase in seismicity has been interpreted as mainly resulting from tectonic stress building up to a critical level along most of the ridge segment, rather than magma overpressure in the underlying magma lenses (Tan et al. 2016). This implies that, if plate pull may dominate at magmatically robust fast-spreading ridges, it could also dominate at intermediate and slow spreading ridges, with relatively lower magma supplies. From December 2009 to October 2011, the same area underwent up to 12 cm of inflation, induced by a point source along the ridge axis at a depth of 2.7 km, possibly recharged from a deeper source (Nooner et al. 2014).

11.5 A General Model for Divergent Plate Boundaries

This chapter has highlighted an intensification in the importance of magmatic processes along divergent plate boundaries with progressively higher extension rates. In immature continental rifts, as the non-volcanic portions of the EARS and the southern MER, as well as in non-magmatic portions of ultraslow oceanic ridges, regional extensional tectonics is the main mechanism responsible for divergence. Here the role of any magmatic activity on rifting appears subordinate, at least at upper crustal levels. Prolonged regional extension or, more effectively, increased extension rates may enhance decompression melting and the rise and shallow intrusion of magma (Fig. 11.23; Ebinger and Hayward 1996). This promotes a mature and magmatically active continental rift, such as the northern MER, where magma assists rifting. This condition eventually evolves into a transitional rift supported by an increased magmatic budget partially replacing the continental crust, as observed in Afar. Further extension leads to focused and magma-dominated rifting on oceanic crust, with variations in the continuity, structure, shape and stability of the active spreading zone mainly depending on the spreading rate.

In this ideal evolution, the contribution of magmatic activity on rifting manifests at progressively shallower levels, first intruding dikes mainly within the lower crust (continental rifts) and subsequently intruding the upper crust with sills generating magma chambers and propagating dikes feeding longer magmatic systems (transitional and oceanic rifts). The effectiveness of magmatic activity in separating the plates on any crustal domain is supported by the above-mentioned recent dike-induced spreading episodes and events in the Afar region, the Aden

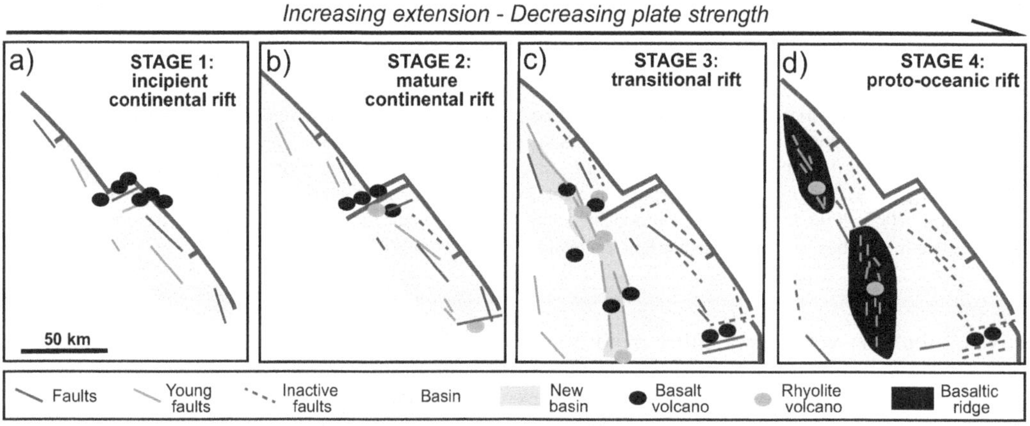

Fig. 11.23 Schematic variations in rift architecture and volcanism as the spreading rate increases in a continental to transitional rift (modified after Ebinger and Hayward 1996). **a** Immature continental rifts: faults and basins, predominantly asymmetric, within relatively strong lithosphere and basaltic volcanism along border faults. **b** Mature continental rifts: as extension increases, the rift becomes more symmetric and eruptive centres (including rhyolitic ones) move into the rift valley. **c** Rift on transitional crust: increasing the extension induces a broader rift zone, with shorter faults and an inner rift valley, with bimodal (basaltic and rhyolitic) volcanism. **d** Proto-oceanic rifts: continued extension may develop basaltic ridges along the rift axis

and Red Sea Rifts, the MAR, Iceland and the East Pacific Rise (Table 11.1). Studies on the rifting episodes in Iceland and Afar (Krafla and Manda Hararo episodes) highlight that the initial and largest dike intrusion within each episode was followed by a series of smaller intrusions. All these dike intrusions obey scaling relationships similar to earthquakes, with the dimensions of the dikes following a power law behaviour analogous to the Gutenberg-Richter relation (Passarelli et al. 2014). Each of these episodes represents the culmination of a **rifting cycle**, where the diking events (co-rifting phase) open the plate boundary by several metres, releasing the stresses built up through continuous longer-term stretching (inter-rifting phase). The co-rifting displacement decreases with across-strike distance from the rift axis, where the opening due to the intrusion superimposes over the steady state tectonic rate. For a few decades after the rifting event (post-rifting phase), the spreading rate along the rifted portion of the plate boundary

Table 11.1 Comparison among major instrumentally-detected diking episodes or events along continental and oceanic divergent plate boundaries, including the magnitude range of associated earthquakes (*M*), the seismic (SM) and geodetic (GM) moment, their ratio (SM/GM; representing the seismic component associated with the total displacement) and the volume of involved magma

Area	Period	*M*	Seism. Mom. (*SM*) (10^{17} Nm)	Geod. Mom. (*GM*) (10^{17} Nm)	*SM/GM* (%)	Volume (km^3)	Ref.
Krafla	1975–1984	6.3	58	440–900	8.6	1.6	a
Dallol	Oct–Nov 2004	2.6–5.5	2.28	22.0	10.4	0.058	b
EPR (9° 50′ N)	2005–2006	~2–3.5				~0.12	c, d
Dabbahu	Sep 2005	1.8–5.5	25–34	800–896	3.1–3.8	1.8–2.5	b, e, f
	Jun 2006	2.5–4.7	1.80	54.4	3.3	0.120	g
	Jul 2006	2–3	0.02	32.4	0.1	0.042	g
	Sep 2006	2.6–3.4		32.0	0.1	0.088	b
	Aug 2007	<3	0.01	24.1	0.0	0.048	g
	Nov 2007	2.9–4.5	1.03	60.1	1.7	0.15	g
	Mar–Apr 2008	<3.0	16	37.2	0.4	0.088	g
	Jul 2008	<3	0.08	32.1	0.2	0.066	g
	Oct 2008	1.8–4.6	1.78	78.8	2.3	0.17	g
	Feb 2009	2–3.5	0.27	30.2	0.9	0.077	g
	Jun 2009	<4	0.01	18.1	0.1	0.046	g
Lake Natron	Jul–Aug 2007	3–5.9	14.00	40.0	35.0	0.09	h
Harrat Lunayyir	Apr–Jun 2009	3–5.4	2.79	44.1	6.3	0.13	i
Bardarbunga	Aug-Dec 2014	≤5.7	51 (*); 0.2 (**)	190	26.9	~0.5	k, l, m

Ref references. [a]Hollingsworth et al. (2013); [b]Nobile et al. (2012), and references therein; [c]Dziak et al. (2009); [d]Xu et al. (2014); [e]Wright et al. (2006); [f]Grandin et al. (2009); [g]Belachew et al. (2011); [h]Calais et al. (2008); [i]Pallister et al. (2010); [k]Icelandic Met Office; [l]Kristin Vogfjord and Simone Cesca (personal communications); [m]Andy Hooper (personal communication); * at the caldera; ** along the dike

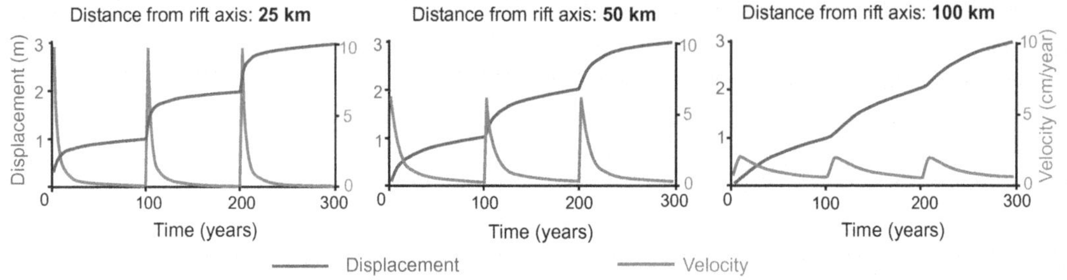

Fig. 11.24 Near- and far-field kinematic behaviours of a spreading plate boundary. Displacement (red lines) and velocity (blue lines) versus time, shown for different distances from the boundary axis: **a** 25 km; **b** 50 km; **c** 100 km (modified after Sigmundsson 2006b)

remains higher than the predicted one, as repeatedly observed in Afar and Iceland (Fig. 11.24). Oblique spreading may increase the time recurrence between major rifting episodes, as a consequence of a decrease in the amount of extension and increase in the shear component. Such behaviour is observed in the Reykjanes Peninsula (Iceland), where magma supply is limited and widening of the plate boundary may need up to a thousand years (Sigmundsson 2006b; Saemundsson et al. 2020). These diking episodes imply a subordinate role of regional tectonics in extending the crust, supported by the theoretical evidence that the stress required to tectonically separate the lithosphere by means of normal faulting (in the order of ~65 MPa) is higher than that required to magmatically rift it by means of dikes (that is less than 10 MPa) (Fig. 11.25; Buck 2006). Nevertheless, the 2005–2006 rifting episode at 9° 50′ N along the East Pacific Rise, likely triggered by the build up in the tectonic stress, implies that magma may be assisted by regional tectonics even in the most favourable conditions for its propagation, that is in an ultrafast-spreading ridge (Tan et al. 2016). This suggests that, while the increase of the spreading rate increases the importance of magmatic activity in promoting rifting, regional tectonics may play a direct role also at the upper end of the spectrum, under the fastest spreading rates.

In this context, regional tectonic extension mainly, but not exclusively, operates on the longer-term and at the wider scale, especially imposing a stable stress field that relaxes the transient lateral compression induced by the intrusion of the dike(s), thus allowing the emplacement of further dikes. In fact, the emplacement of a dike induces at its sides a temporary increase in the rift-perpendicular horizontal component, which may pass from the minimum principal stress σ_3 that promoted dike injection to the intermediate principal stress σ_2

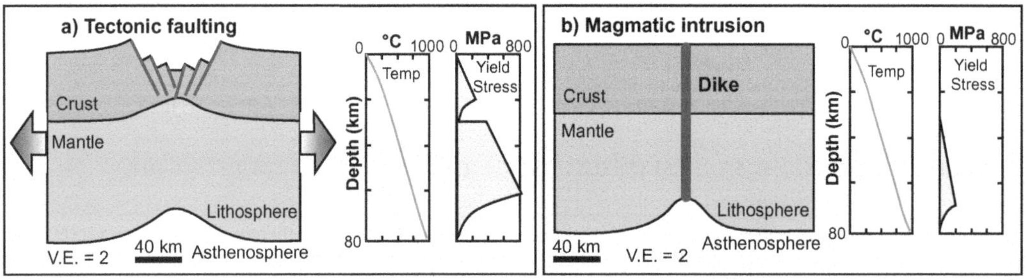

Fig. 11.25 Extension of normal thickness continental lithosphere through regional tectonic faulting (**a**) and magmatic intrusions (**b**). Note the large difference in the yield stress, or the stress difference needed to get extensional separation of two lithospheric blocks, in the two cases (modified after Buck 2006). V.E. = vertical exaggeration

and even to the maximum principal stress σ_1. Although transient, a horizontal σ_1 perpendicular to the rift may favour sill intrusion, encouraging shallow magma accumulation, until the regional stress relieves the dike-induced compression and the regional σ_3 replaces the σ_1, re-promoting dike emplacement. The repetition of this cycle, where diking episodes (controlled by the horizontal minimum principal stress σ_3) alternate to reservoir growth (controlled by the horizontal maximum principal stress σ_1), results in a feedback encouraging episodic shallow magma rise and emplacement.

Dike-fed **magmatic systems** are the fundamental unit responsible for rifting, independently of the nature of the crust (Gudmundsson 1995; Ebinger and Casey 2001). These systems may be sporadically present in immature continental rifts characterized by slow spreading rate and incipient volcanic activity, as the western branch of the EARS, or in ultraslow spreading ridges. Here magmatic systems appear as a local alternative, with limited lateral extent, complementary to the upper crustal extension resulting from regional tectonic faulting, which may manifest through normal faults bordering half-grabens (in immature continental rifts) or low-angle detachments (in ultraslow ridges). Conversely, magmatic systems are ubiquitous in divergent plate boundaries characterized by predominant magmatic activity, where they display distinctive geometric features. Along the MER, Afar, Iceland and the EPR, the larger the spreading rate, the more elongated and narrower, with lower aspect ratio A_r (where A_r = width w/length L) is the magmatic system (Fig. 11.26; Acocella 2014). This indicates that higher spreading rates are related to longer magmatic systems, where dike propagation is promoted by a weaker σ_3 and thus a higher deviatoric stress. These dikes typically feed monogenic mafic volcanism, whereas the dominant polygenic volcano of the magmatic system may erupt also felsic magma. In particular, felsic magma predominates on thicker continental crust, whereas mafic magma predominates on thinner oceanic crust. The ratio of polygenic to monogenic volcanism in a magmatic system appears also controlled by the

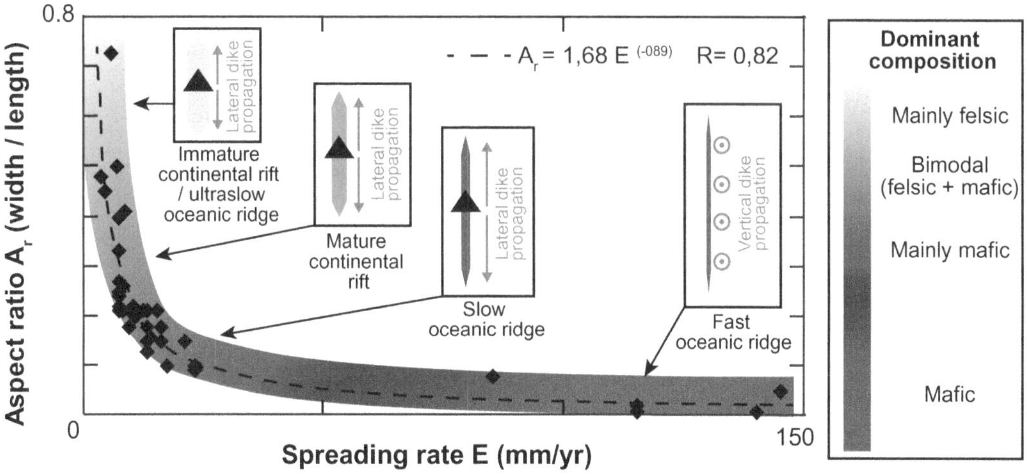

Fig. 11.26 Summary of the main features of magmatic systems along divergent plate boundaries. There is an inverse and non-linear proportion between the spreading rate of the plate boundary and the aspect ratio A_r (width/length) of the magmatic systems in immature continental rifts (and ultraslow oceanic ridges), mature continental rifts, slow oceanic ridges and fast and ultrafast oceanic ridges. Increasing the spreading rate develops progressively longer and narrower magmatic systems, also decreasing the felsic component. Insets show schematic map views of magmatic systems with different aspect ratio, also including deeper magma distribution (in blue): black triangles indicate dominant polygenic volcanoes, whereas monogenic volcanoes are not shown for simplicity. In magmatic systems with dominant polygenic volcano, the volcano magma chamber favours lateral dike propagation (blue arrows); in magmatic systems without dominant polygenic volcano (in fast oceanic ridges), the more continuous magma distribution encourages vertical dike propagation (blue circles with dots; modified after Acocella 2014)

spreading rate: polygenic volcanoes are more common in slower spreading settings and monogenic volcanoes in faster spreading settings, with higher differential stress $\Delta\sigma = \sigma_1 - \sigma_3$ and higher magmatic output (see Sect. 10.6; Takada 1994). This different distribution of volcanoes has also implications for the mode of delivery of magma along the magmatic system. In fact, in a magmatic system with dominant polygenic volcano the latter mainly feeds the monogenic fissures through laterally propagating dikes from the base of its magma chamber, as proposed for Krafla (in 1975–1984), Dabbahu (in 2005–2010) and Bardarbunga (in 2014). Where the dominant polygenic volcano is absent, as at fast spreading oceanic ridges, available evidence suggests a mainly vertical propagation of the dikes promoting rifting (Wright et al. 2012; Carbotte et al. 2013). This different mode of magma propagation derives from the different along-rift extent of the magma chamber feeding the dikes. This is more continuous in fast spreading ridges, enhancing vertical dike propagation, and of more limited extent below the polygenic volcano of slow spreading systems, where lateral dike propagation favours rifting.

The architecture of a magmatic system, in terms of aspect ratio, magma distribution, composition and delivery, is thus diagnostic to evaluate the role of magmatism on rifting along divergent plate boundaries. Magmatic systems are indeed the primary product of a divergent plate boundary defined by a given spreading rate, providing a convenient framework to evaluate its tectonic and magmatic maturity.

11.6 Summary

Divergent plate boundaries are found on continental, transitional and, mostly, oceanic crust. Regional examples show that, where magmatism is limited or absent, plate divergence occurs mainly by normal faulting, forming half-grabens or detachments, as at immature continental rifts (southern portion and western branch of the EARS) and ultraslow spreading ridges, respectively. Increasing the extension rate leads to a progressively more important and shallower magmatic contribution in rifting. This is observed in mature continental rifts (northern MER) and, especially, in transitional (Afar) and oceanic (both slow and fast) ridges, where magmatic activity via diking becomes most effective in spreading the plates. In these rifts activity focuses in magmatic systems, whose architecture, summarized by their aspect ratio, provides a framework to relate the tectonic and magmatic maturity of a divergent plate boundary, in terms of magma distribution, composition and delivery, to its spreading rate. In this frame, the laterally propagating dikes, mainly associated with the limited extent of magma chambers in slow spreading boundaries, complement the vertically propagating dikes associated with along-rift more continuous magma chambers in fast spreading boundaries.

11.7 Main Symbols Used

A_r	Aspect ratio of magmatic system
L	Length of magmatic system
M	Magnitude
w	Width of magmatic system
$\Delta\sigma$	Differential stress
σ_1	Maximum principal stress
σ_2	Intermediate principal stress
σ_3	Minimum principal stress

References

Acocella V, Trippanera D (2016) How diking affects the tectono-magmatic evolution of slow spreading plate boundaries: overview and model. Geosphere 12:1–17

Acocella V (2010) Coupling volcanism and tectonics along divergent boundaries: collapsed rifts from Central Afar, Ethiopia. Geol Soc Am Bull 122:1717–1728

Acocella V, Abebe B, Korme T (2011) Holocene opening directions along the axes of Red Sea (Afar) and Main Ethiopian rifts: an overview. In: Beccaluva L, Bianchini G, Wilson M (eds) Volcanism and evolution of the African lithosphere. Geological Society of America Special Paper, vol 478, pp 25–35

Acocella V (2014) Structural control on magmatism along divergent and convergent plate boundaries: overview, model, problems. Earth-Sci Rev 136:226–288

Agostini A, Bonini M, Corti G, Sani F, Mazzarini F (2011) Fault architecture in the main Ethiopian Rift and comparison with experimental models: implications for rift evolution and Nubia-Somalia kinematics. Earth Planet Sci Lett 301:479–492

Ahmed A, Doubrè C, Leroy S, Kassim M, Keir D, Abayazid A et al (2016) Seafloor spreading event in western Gulf of Aden during the November 2010–March 2011 period captured by regional seismic networks: evidence for diking events and interactions with a nascent transform zone. Geophys J Int 205:1244–1266

Almalki KA, Betts PG, Ailleres L (2015) The red sea—50 years of geological and geophysical research. Earth Sci Rev 147:109–140

Amelung F, Jonsson S, Zebker H, Segall P (2000) Widespread uplift and 'trapdoor' faulting on Galapagos volcanoes observed with radar interferometry. Nature 407:993–996

Angelier J, Slunga R, Bergerat F, Stefansson R, Homberg C (2004) Perturbations of stress and oceanic rift extension across transform faults shown by earthquake focal mechanisms in Iceland. Earth Planet Sci Lett 219:271–284

Angelier J, Bergerat F, Stefansson R, Bellou M (2008) Seismotectonics of a newly formed transform zone near a hotspot: earthquake mechanisms and regional stress in the South Iceland Seismic Zone. Tectonophysics 447:95–116

Arnadottir T, Sigmundsson F, Delaney PT (1998) Sources of crustal deformation associated with the Krafla, Iceland, eruption of September 1984. Geophys Res Lett 25:1043–1046

Arnadottir T, Lund B, Jiang W, Geirsson H, Bjornsson H, Einarsson P et al (2009) Glacial rebound and plate spreading: results from the first countrywide GPS observations in Iceland. Geophys J Int 177:691–716

Arnulf AF, Singh SC, Pye JW (2014) Seismic evidence of a complex multi-lens melt reservoir beneath the 9° N overlapping spreading center at the East Pacific Rise. Geophys Res Lett 41:6109–6115

Ayele A, Keir D, Ebinger CJ, Wright TJ, Stuart GW, Buck WR et al (2009) The September 2005 mega-dike emplacement in the Manda-Hararo (Afar) nascent oceanic rift. Geophys Res Lett 36:L20306. https://doi.org/10.1029/2009GL039605

Barberi F, Varet J (1970) The Erta Ale Volcanic Range (Danakil depression, northern Afar, Ethiopia). Bull Volcanologique 34:848–917

Barberi F, Varet J (1977) Volcanism of Afar: small-scale plate tectonics implications. Geol Soc Am Bull 88:1251–1266

Bastow ID, Nyblade AA, Stuart GW, Rooney TO, Benoit MH (2008) Upper mantle seismic structure beneath the Ethiopian hot spot: rifting at the edge of the African low-velocity anomaly. Geochem Geophys Geosyst 9:Q12022. https://doi.org/10.1029/2008GC002107

Bastow ID, Keir D (2011) The protracted development of the continent–ocean transition in Afar. Nat Geosci 4:248–250

Bastow ID, Booth AD, Corti G, Keir D, Magee C, Jackson CAL et al (2018) The development of late-stage continental breakup: seismic reflection and borehole evidence from the Danakil Depression, Ethiopia. Tectonics 37:2848–2862

Belachew M, Ebinger C, Coté D, Keir D, Rowland JV, Hammond JOS et al (2011) Comparison of dike intrusions in an incipient seafloor-spreading segment in Afar, Ethiopia: Seismicity perspectives. J Geophys Res 116:B06405. https://doi.org/10.1029/2010JB007908

Benoit MH, Nyblade AA, VanDecar JC (2006) Upper mantle P-wave speed variations beneath Ethiopia and the origin of the Afar hotspot. Geology 34:329–332

Biggs J, Bastow ID, Keir D, Lewi E (2011) Pulses of deformation reveal frequently recurring shallow magmatic activity beneath the Main Ethiopian Rift. Geochem Geophys Geosyst 12:Q0AB10. https://doi.org/10.1029/2011GC003662

Bilham R, Bendick R, Larson K, Mohr P, Braun J, Tesfaye S et al (1999) Secular and tidal strain across the main Ethiopian Rift. Geophys Res Lett 26:2789–2792

Birhanu Y, Bendick R, Fisseha S, Lewi E, Floyd M, King R et al (2016) GPS constraints on broad scale extension in the Ethiopian Highlands and Main Ethiopian Rift. Geophys Res Lett 43:6844–6851

Bjornsson A, Saemnundsson K, Einarsson P, Tryggvason E, Gronvald K (1977) Current rifting episode in North Iceland. Nature 266:318–323

Bonatti E (1985) Punctiform initiation of seafloor spreading in the Red Sea during transition from a continental to an oceanic rift. Nature 316:33–37

Bosworth W, Burke K, Strecker M (2003) Effect of stress fields on magma chamber stability and the formation of collapse calderas. Tectonics 22:1042. https://doi.org/10.1029/2002TC001369

Bosworth W, Huchon P, McClay K (2006) The Red Sea and Gulf of Aden Basins. J Afr Earth Sc 43:334–378

Buck WR (2006) The role of magma in the development of the Afro-Arabian Rift System. In Yirgu G, Ebinger CJ, Maguire PKH eds The Afar Volcanic Province within the East African Rift System. Journ Geol Soc London Sp Pub 259:43–54

Buck RW, Einarsson P, Brandsdottir B (2006) Tectonic stress and magma chamber size as controls on dike propagation: constraints from the 1975–1984 Krafla rifting episode. J Geophys Res 111:B12404. https://doi.org/10.1029/2005JB003879

Buck RW (2013) Magma for 50,000 years. Nat Geosci 6:811–812

Calais E, d'Oreye N, Albaric J, Deschamps A, Delvaux D, Deverchere J et al (2008) Strain accommodation by dyking in a youthful continental rift, East Africa. Nature 456:783–787

Calvert AJ (1995) Seismic evidence for a magma chamber beneath the slow-spreading Mid-Atlantic ridge. Nature 377:410–413

Canales JP, Nedimovic MR, Kent GM, Carbotte SM, Detrick RS (2009) Seismic reflection images of a near-axis melt sill within the lower crust at the Juan de Fuca ridge. Nature 460:89–94

Canales P, Carton H, Carbotte SM, Mutter JC, Nedomovic NR, Xu M et al (2012) Network of off-axis melt bodies at the East Pacific Rise. Nat Geosci 5:279–282

Cann JR, Smith DK, Escartin J, Schouten H (2015) Tectonic evolution of 200 km of Mid-Atlantic Ridge over 10 million years: interplay of volcanism and faulting. Geochem Geophys Geosyst 16:2303–2321

Cannat M, Sauter D, Lavier L, Bickert M, Momoh E, Leroy S (2019) On spreading modes and magma supply at slow and ultraslow mid-ocean ridges. Earth Planet Sci Lett 519:223–233

Carbotte SM, Tuyan WBF, Jin W, Cormier MH, Bergmanis E, Sinton J et al (2003) Magmatic subsidence of the East Pacific Rise (EPR) at 18°14'S revealed through fault restoration of ridge crest bathymetry. Geochem Geophys Geosyst 4:1008. https://doi.org/10.1029/2002GC000337

Carbotte SM, Marjanovic M, Carton H, Mutter JC, Canales JP, Nedimovic MR et al (2013) Fine-scale segmentation of the crustal magma reservoir beneath the East Pacific Rise. Nat Geosci 6:866–870

Chadwick WW, Embley RW (1998) Graben formation associated with recent dike intrusions and volcanic eruptions on the mid-ocean ridge. J Geophys Res 103:9807–9825

Chambers EL, Harmon N, Keir D, Rychert CA (2019) Using ambient noise to image the northern East African Rift. Geochem Geophys Geosyst 20:2091–2109

Chang SJ, van der Lee S (2011) Mantle plumes and associated flow beneath Arabia and East Africa. Earth Planet Sci Lett 302:448–545

Cochran JR (1983) Model for development of the Red Sea. Am Assoc Pet Geol Bull 67:41–69

Combier V, Seher T, Singh SC, Crawford WC, Cannat M, Escartín J et al (2015) Three-dimensional geometry of axial magma chamber roof and faults at Lucky Strike volcano on the Mid-Atlantic Ridge. J Geophys Res 120:5379–5400

Corti G, van Wijk J, Cloething S, Morley CK (2007) Tectonic inheritance and continental rift architecture: numerical and analogue models of the East African Rift system. Tectonics 26:TC6006. https://doi.org/10.1029/2006TC002086

Corti G (2009) Continental rift evolution: from rift initiation to incipient break-up in the Main Ethiopian Rift, East Africa. Earth Sci Rev 96:1–53

Corti G, Agostini A, Keir D, van Wijk V, Bastow ID, Ranalli G (2015) Magma-induced axial subsidence during final-stage rifting: implications for the development of seaward-dipping reflectors. Geosphere 11:563–571

Craig T, Jackson J, Priestley K, McKenzie D (2011) Earthquake distribution patterns in Africa: their relationship to variations in lithospheric and geological structure, and their rheological implications. Geophys J Int 185:403–434

Crowder LK, Macdonald KC (2000) New constraints on the width of the zone of active faulting on the East Pacific Rise 8°30'N–10°00'N from Sea Beam Bathymetry and SeaMARC II Side-scan Sonar. Mar Geophys Res 21:513–527

Desissa M, Johnson NE, Whaler KA, Hautot S, Fisseha S, Dawes GJK (2013) A mantle magma reservoir beneath an incipient mid-ocean ridge in Afar, Ethiopia. Nat Geosci 6:861–865

Deville E, Marsset T, Courgeon S, Jatiault R, Ponte J-P, Thereau E et al (2018) Active fault system across the oceanic lithosphere of the Mozambique Channel: implications for the Nubia-Somalia southern plate boundary. Earth Planet Sci Lett 502:210–220

Dick HJB, Lin J, Schouten H (2003) An ultraslow-spreading class of ocean ridge. Nature 426:405–412

Doubre C, Manighetti I, Dorbath L, Dorbath C, Bertil D, Delmond JC (2007) Crustal structure and magmatotectonic processes in an active rift (Asal-Ghoubbet, Afar, East Africa): 2. Insights from the 23-year recording of seismicity since the last rifting event. J Geophys Res 112:B05406. https://doi.org/10.1029/2006JB004333

Doubre C, Deprez A, Masson F, Socquet A, Lewi E, Grandin R (2017) Current deformation in Central Afar and triple junction kinematics deduced from GPS and InSAR measurements. Geophys J Int 208:936–953

Drouin V, Sigmundsson F, Ofeigsson BG, Hreinsdottir S, Sturkell E, Einarsson P (2017) Deformation in the Northern Volcanic Zone of Iceland 2008–2014: An interplay of tectonic, magmatic, and glacial isostatic deformation. J Geophys Res 122:3158–3178

Dugda M, Nyblade AA, Julia J, Langston CA, Ammon CJ, Simiyu S (2005) Crustal structure in Ethiopia and Kenya from receiver function analysis: implications for rift development in eastern Africa. J Geophys Res 110:B01303. https://doi.org/10.1029/2004JB003065

Dziak RP, Bohnenstiehl DR, Matsumoto H, Fowler MJ, Haxel JH, Tolstoy M et al (2009) January 2006 seafloor spreading event at 9°50'N, East Pacific Rise: ridge dike intrusion and transform fault interactions from regional hydroacoustic data. Geochem Geophys Geosyst 10. https://doi.org/10.1029/2009GC002388

Eagles G, Gloaguen R, Ebinger C (2002) Kinematics of the Danakil microplate. Earth Planet Sci Lett 203:607–620

Ebinger CJ, Crow MJ, Rosendahl BR, Livingstone DA, LeFournier J (1984) Structural evolution of Lake Malawi, Africa. Nature 308:627–629

Ebinger CJ (1989a) Tectonic development of the western branch of the East African Rift System. Geol Soc Am Bull 101:885–903

Ebinger CJ (1989b) Geometric and kinematic development of border faults and accommodation zones, Kivu-Rusizi Rift, Africa. Tectonics 8:117–133

Ebinger CJ, Deino AL, Drake RE, Tesha AL (1989) Chronology of volcanism and rift basin propagation: Rungwe Volcanic Province, East Africa. J Geophys Res 94:15785–15803

Ebinger CJ, Hayward NJ (1996) Soft plates and hot spots: views from Afar. J Geophys Res 101:21859–21876
Ebinger CJ, Sleep NH (1998) Cenozoic magmatism throughout east Africa resulting from impact of a single plume. Nature 395:788–791
Ebinger CJ, Casey M (2001) Continental breakup in magmatic provinces: an Ethiopian example. Geology 29:527–530
Ebinger C, Ayele A, Keir D, Rowland J, Yirgu G, Wright T et al (2010) Length and timescales of rift faulting and magma intrusion: the Afar rifting cycle from 2005 to present. Ann Rev Earth Planet Sci 38:439–466
Ebinger CJ, Keir D, Bastow ID, Whaler K, Hammond JOS, Ayele A et al (2017) Crustal structure of active deformation zones in Africa: implications for global crustal processes. Tectonics 36:3298–3332
Einarsson P (2018) Short-term seismic precursors to icelandic Eruptions 1973–2014. Front Earth Sci 6:45. https://doi.org/10.3389/feart.2018.00045
Eyles JHW, Illsley-Kemp F, Keir D, Ruch J, Jónsson S (2018) Seismicity associated with the formation of a New Island in the Southern Red Sea. Front Earth Sci 6:141. https://doi.org/10.3389/feart.2018.00141
Fornari DJ, Haymon RM, Pefit MR, Gregg TKP, Edwards MH (1998) Axial summit trough of the East Pacific Rise 9°–10°N: geological characteristics and evolution of the axial zone on fast spreading mid-ocean ridges. J Geophys Res 103:9827–9855
Forslund T, Gudmundsson A (1991) Crustal spreading due to dike and faults in southwest Iceland. J Struct Geol 13:443–457
Forsyth DW, Webb SC, Dorman LM, Shen Y (1998) Phase velocities of Rayleigh waves on the MELT experiment on the East Pacific Rise. Science 280:1235–1238
Franke D, Jokat W, Ladage S, Stollhofen H, Klimke J, Lutz R et al (2015) The offshore East African Rift System: structural framework at the toe of a juvenile rift. Tectonics 34:2086–2104
Fundis AT, Soule SA, Fornari DJ, Perfit MR (2010) Paving the seafloor: volcanic emplacement processes during the 2005-2006 eruptions at the fast spreading East Pacific Rise, 9°50′N. Geochem Geophys Geosyst 11:Q08024. https://doi.org/10.1029/2010GC003058
Furman T, Nelson WR, Elkins-Tanton LT (2016) Evolution of the East African rift: drip magmatism, lithospheric thinning and mafic volcanism. Geochim Cosmochim Acta 185:418–434
Gaherty JB (2001) Seismic evidence for hotspot-induced buoyant flow beneath the Reykjanes Ridge. Science 293:1645–1647
Gallacher R, Keir D, Harmon N, Stuart G, Leroy S, Hammond JOS et al (2016) The initiation of segmented buoyancy-driven melting during continental breakup. Nat Commun 7:13110. https://doi.org/10.1038/ncomms13110
Geirsson H, Arnadottir T, Volksen C, Jiang W, Sturkell E, Villemin T et al (2006) Current plate movements across the Mid-Atlantic Ridge determined from 5 years of continuous GPS measurements in Iceland. J Geophys Res 111:B09407. https://doi.org/10.1029/2005JB003717
Giusti M, Perrot J, Dziak RP, Sukhovic A, Maia M (2018) The August 2010 earthquake swarm at North FAMOUS–FAMOUS segments, Mid-Atlantic Ridge: geophysical evidence of dike intrusion. Geophys J Int 215:181–195
Grandin R, Socquet A, Binet R, Klinger R, Jacques E, de Chabalier JB et al (2009) September 2005 Manda Hararo-Dabbahu rifting event, Afar (Ethiopia): Constraints provided by geodetic data. J Geophys Res 114: B08404, https://doi.org/10.1029/2008JB005843
Grant JV, Kattenhorn SA (2004) Evolution of vertical faults at an extensional plate boundary, Southwest Iceland. J Struct Geol 26:537–557
Greenfield T, Keir D, Kendall J-M, Ayele A (2019) Low-frequency earthquakes beneath Tullu Moye volcano, Ethiopia, reveal fluid pulses from shallow magma chamber. Earth Planet Sci Lett 526:115782
Grevemeyer I, Hayman NW, Peirce C, Schwardt M, Van Avendonk HJA, Dannowski A et al (2018) Episodic magmatism and serpentinized mantle exhumation at an ultraslow-spreading centre. Nat Geosci 11:444–448
Gu YJ, Lerner-Lam AL, Dziewonski AM, Ekstrom G (2005) Deep structure and seismic anisotropy beneath the East Pacific Rise. Earth Planet Sci Lett 232:259–272
Gudmundsson A (1986) Formation of crustal magma chambers in Iceland. Geology 14:164–166
Gudmundsson A, Backstrom K (1991) Structure and development of the Sveinagja Graben, northeast Iceland. Tectonophysics 200:111–125
Gudmundsson A (1992) Formation and growth of normal faults at the divergent plate boundary in Iceland. Terra Nova 4:464–471
Gudmundsson A (1995) Infrastructure and mechanics of volcanic systems in Iceland. J Volcanol Geoth Res 64:1–22
Hammond JOS, Kendall J-M, Stuart GW, Ebinger CJ, Bastow ID, Keir D et al (2013) Mantle upwelling and initiation of rift segmentation beneath the Afar Depression. Geology 41:635–638
Hartley ME, Thordarson T (2012) Formation of Öskjuvatn caldera at Askja, North Iceland: mechanism of caldera collapse and implications for the lateral flow hypothesis. J Volcanol Geoth Res 227–228:85–101
Hayward NJ, Ebinger CJ (1996) Variation in the along-axis segmentation of the Afar Rift system. Tectonics 15:244–257
Helgason J, Zentilli M (1985) Field characteristics of laterally emplaced dikes: anatomy of an exhumed Miocene dike swarm in Reydarfjordur, Eastern Iceland. Tectonophysics 115:247–274
Hofmann C, Courtillot V, Feraud G, Rochette P, Yirgu G, Ketefo E et al (1997) Timing of the Ethiopian flood basalt event and implications for plume birth and global change. Nature 389:838–841
Hollingsworth J, Leprince S, Ayoub F, Avouac J-P (2013) New constraints on dike injection and fault slip during the 1975–1984 Krafla rift crisis, NE Iceland. J Geophys Res 118:3707–3727

Horst AJ, Varga RJ, Gee JS, Karson JA (2014) Diverse magma flow directions during construction of sheeted dike complexes at fast-to superfast-spreading centers. Earth Planet Sci Lett 408:119–131

Hreinsdottir S, Einarsson P, Sigmundsson F (2001) Crustal deformation at the oblique spreading Reykjanes Peninsula, SW Iceland: GPS measurements from 1993 to 1998. J Geophys Res 106:13803–13816

Husson L, Yamato P, Bezos A (2015) Ultraslow, slow, or fast spreading ridges: arm wrestling between mantle convection and far-field tectonics. Earth Planet Sci Lett 429:205–215

Hutchison W, Mather TA, Pyle DM, Boyce AJ, Gleeson MLM, Yirgu G et al (2018) The evolution of magma during continental rifting: New constraints from the isotopic and trace element signatures of silicic magmas from Ethiopian volcanoes. Earth Planet Sci Lett 489:203–218

Jacques E, Ruegg JC, Lepine JC, Tapponnier P, King GCP, Omar A (1999) Relocation of M $\geq$ 2 events of the 1989 Dobi seismic sequence in Afar: evidence for earthquake migration. Geophys J Int 138:447–469

Jenkins J, Maclennan J, Green RG, Cottaar S, Deuss AF, White RS (2018) Crustal formation on a spreading ridge above a mantle plume: receiver function imaging of the Icelandic crust. J Geophys Res 123:5190–5208

Jian H, Singh SC, Chen YJ, Li J (2017) Evidence of an axial magma chamber beneath the ultraslow spreading Southwest Indian Ridge. Geology 45:143–146

Jokat W, Ritzmann O, Schmidt-Aursch MC, Drachev S, Gauger S, Snow J (2003) Geophysical evidence for reduced melt production on the Arctic ultraslow Gakkel mid-ocean ridge. Nature 423:962–965

Jónsson S, Xu W (2015) Volcanic eruptions in the southern Red Sea during 2007–2013. In: Rasul NMA, Stewart ICF (eds) The Red Sea: the formation, morphology, oceanography and environment of a young ocean basin. Springer Earth System Sciences Heidelberg, pp 175–186

Keiding M, Lund B, Arnadottir T (2009) Earthquakes, stress and strain along an obliquely divergent plate boundary: Reykjanes Peninsula, southwest Iceland. J Geophys Res 114:B09306. https://doi.org/10.1029/2008JB006253

Keir D, Ebinger CJ, Stuart GW, Daly E, Ayele A (2006) Strain accommodation by magmatism and faulting as rifting proceeds to breakup: seismicity of the northern Ethiopian Rift. J Geophys Res 111:B05314. https://doi.org/10.1029/2005JB003748

Keir D, Belachew M, Ebinger CJ, Kendall JM, Hammond JOS, Stuart GW et al (2011) Mapping the evolving strain field during continental breakup from crustal anisotropy in the Afar depression. Nat Commun 2:1–7

Keir D, Bastow ID, Corti G, Mazzarini F, Rooney TO (2015) The origin of along-rift variations in faulting and magmatism in the Ethiopian Rift. Tectonics 34:464–477

Kendall JM, Stuart GW, Ebinger CJ, Bastow ID, Keir D (2005) Magma-assisted rifting in Ethiopia. Nature 433:146–214

Kent GM, Singh SC, Harding AJ, Sinha MC, Orcutt JA, Barton PJ et al (2000) Evidence from three-dimensional seismic reflectivity images for enhanced melt supply beneath mid-ocean ridge discontinuities. Nature 406:614–618

Keranen K, Klemperer SL, Gloaguen R, Eagle Working Group (2004) Three-dimensional seismic imaging of a protoridge axis in the Main Ethiopian Rift. Geology 32:949–952

Kogan L, Fisseha S, Bendick R, Reilinger R, McClusky S, King R (2012) Lithospheric strength and strain localization in continental extension from observations of the East African Rift. J Geophys Res 117:B03402. https://doi.org/10.1029/2011JB008516

Lagabrielle I, Cormier MH (1999) Formation of large summit troughs along the East Pacific Rise as collapse calderas: an evolutionary model. J Geophys Res 104:12971–12988

Lahitte P, Gillot PY, Courtillot V (2003) Silicic central volcanoes as precursors to rift propagation: the Afar case. Earth Planet Sci Lett 207:103–116

Ligi M, Bonatti E, Caratori M, Tontini F, Cipriani A, Cocchi L et al (2011) Initial burst of oceanic crust accretion in the Red Sea due to edge-driven mantle convection. Geology 39:1019–1022

Ligi M, Bonatti E, Bosworth W, Ronca S (2019) Oceanization starts at depth during continental rupturing in the Northern Red Sea. In: Rasul NMA, Stewart ICF (eds) Geological setting, palaeoenvironment and archaeology of the Red Sea. Springer Nature, Switzerland, pp 131–157

Maccaferri F, Rivalta E, Keir D, Acocella A (2014) Off-rift volcanism in rift zones determined by crustal unloading. Nat Geosci 7:297–300

Maccaferri F, Acocella V, Rivalta E (2015) How the differential load induced by normal fault scarps controls the distribution of monogenic volcanism. Geophys Res Lett 42. https://doi.org/10.1002/2015GL065638

Macdonald KC, Fox PJ (1988) The axial summit graben and cross-sectional shape of the East Pacific Rise as indicators of axial magma chambers and recent volcanic eruptions. Earth Planet Sci Lett 88:119–131

Macdonald K, Scheirer D, Carbotte S (1991) Mid-ocean ridges: discontinuities, segments and giant cracks. Science 253:986–994

Macdonald K (1998) Linkages between faulting, volcanism, hydrothermal activity and segmentation on fast spreading centres. In: Buck WR, Delaney PT, Karson JA, Lagabrielle YU (eds) Faulting and magmatism at mid-ocean ridges. AGU Monograph Washington, pp 27–58

Mackenzie GD, Thybo H, Maguire PKH (2005) Crustal velocity structure across the Main Ethiopian Rift: results from two-dimensional wide-angle seismic modelling. Geophys J Int 162:994–1006

Manighetti I, Tapponnier P, Courtillot V, Gallet Y, Jacques E, Gillot PY (2001) Strain transfer between disconnected, propagating rifts in Afar. J Geophys Res 106:13613–13665

Marjanovic M, Carbotte SM, Carton H, Nedimovic NR, Mutter JC, Canales JP (2014) A multi-sill magma plumbing system beneath the axis of the East Pacific Rise. Nat Geosci 7:825–829

Marjanović M, Carbotte SM, Carton HD, Nedimović MR, Canales JP, Mutter JC (2018) Crustal magmatic system beneath the East Pacific Rise (8°20' to 10° 10'N): implications for tectonomagmatic segmentation and crustal melt transport at fast-spreading ridges. Geochem Geophys Geosyst 19:4584–4611

Mark HF, Behn MD, Olive JA, Liu Y (2018) Controls on mid-ocean ridge normal fault seismicity across spreading rates from rate-and-state friction models. J Geophys Res 123:6719–6733

McClusky S, Reilinger R, Ogubazghi G, Amleson A, Healeb B, Vernant P et al (2010) Kinematics of the southern Red Sea-Afar triple junction and implications for plate dynamics. Geophys Res Lett 37:L05301. https://doi.org/10.1029/2009GL041127

Medynski S, Pik R, Burnard P, Dumont S, Grandin R, Williams A et al (2016) Magmatic cycles pace tectonic and morphological expression of rifting (Afar depression, Ethiopia). Earth Planet Sci Lett 446:77–88

Michael PJ, Langmuir CH, Dick HJB, Snow JE, Goldstein SL, Graham DW et al (2003) Magmatic and amagmatic seafloor generation at the ultraslow-spreading Gakkel ridge, Arctic Ocean. Nature 425:956–961

Mitchell NC, Park Y (2014) Nature of crust in the central Red Sea. Tectonophysics 628:123–139

Mitchell NC, Stewart ICF (2018) The modest seismicity of the northern Red Sea rift. Geophys J Int 214:1507–1523

Mohr P (1967) Major volcano-tectonic lineament in the Ethiopian rift system. Nature 213:664–665

Mohr PA, Wood CA (1976) Volcano spacing and lithospheric attenuation in the Eastern Rift of Africa. Earth Planet Sci Lett 33:126–144

Mohr P (1983) Ethiopian flood basalt province. Nature 303:577–584

Montesi LGJ, Behn MD (2007) Mantle flow and melting underneath oblique and ultraslow mid-ocean Ridges. Geophys Res Lett 34:L24307. https://doi.org/10.1029/2007GL031067

Mutch JF, Maclennan J, Shorttle O, Edmonds M, Rudge JF (2019) Rapid transcrustal magma movement under Iceland. Nat Geosci 12(569):573

Mutter JC, Karson JA (1992) Structural processes at slow-spreading ridges. Science 257:627–634

Njinju EA, Kolawole F, Atekwana EA, Stamps DS, Atekwana EA, Abdelsalam MG et al (2019) Terrestrial heat flow in the Malawi Rifted Zone, East Africa: implications for tectono-thermal inheritance in continental rift basins. J Volcanol Geoth Res 387:106656

Niu Y, Hekinian R (1997) Spreading rate dependence of the extent of mantle melting beneath ocean ridges. Nature 385:326–329

Nobile A, Pagli C, Keir D, Wright TJ, Ayele A, Ruch J et al (2012) Dyke-fault interaction during the 2004 Dallol intrusion at the northern edge of the Erta Ale Ridge (Afar, Ethiopia). Geophys Res Lett 39:L19305. https://doi.org/10.1029/2012GL053152

Nooner SL, Webb SC, Buck WR, Cormier M-H (2014) Post eruption inflation of the East Pacific Rise at 9°50' N. Geochem Geophys Geosyst 15:2676–2688

O'Connor JM, Jokat W, Regelous M, Kuiper KF, Miggins DP, Koppers AAP (2019) Superplume mantle tracked isotopically the length of Africa from the Indian Ocean to the Red Sea. Nat Commun 10:5493. https://doi.org/10.1038/s41467-019-13181-7

Omar GI, Steckler MS (1995) Fission track evidence on the initial rifting of the Red Sea: two pulses, no propagation. Science 270:1341–1344

Pagli C, Wright TJ, Ebinger CJ, Yun SH, Cann JR, Barnie T et al (2012) Shallow axial magma chamber at the slow spreading Erta Ale Ridge. Nat Geosci. https://doi.org/10.1038/NGEO1414

Pagli C, Wang H, Wright TJ, Calais E, Lewi E (2014) Current plate boundary deformation of the Afar rift from a 3-D velocity field inversion of InSAR and GPS. J Geophys Res 119:8562–8575

Pagli C, Yun S-H, Ebinger C, Keir D, Wang H (2019) Strike-slip tectonics during rift linkage. Geology 47:31–34

Pallister JS, McCausland WA, Jonsson S, Lu Z, Zahran HM, El Hadidy S et al (2010) Broad accommodation of rift-related extension recorder by dyke intrusion in Saudi Arabia. Nat Geosci 3:708–712

Park Y, Nyblade AA (2006) P-wave tomography reveals a westward dipping low velocity zone beneath the Kenya Rift. Geophys Res Lett 33:L07311. https://doi.org/10.1029/2005GL025605

Parnell-Turner RE, Mittelstaedt E, Kurz MD, Jones MR, Soule SA, Klein F et al (2018) The final stages of slip and volcanism on an oceanic detachment fault at 13° 48'N, Mid-Atlantic Ridge. Geochem Geophys Geosyst 19:3115–3127

Passarelli L, Rivalta E, Shuler A (2014) Dike intrusions during rifting episodes obey scaling relationships similar to earthquakes. Sci Rep 4:3886. https://doi.org/10.1038/srep03886

Peccerillo A, Donati C, Santo AP, Orlando A, Yirgu G, Ayalew D (2007) Petrogenesis of silicic peralkaline rocks in the Ethiopian rift: geochemical evidence and volcanological implications. J Afr Earth Sci 48:161–173

Peirce C, Sinha MC (2008) Life and death of axial volcanic ridges: segmentation and crustal accretion at the Reykjanes Ridge. Earth Planet Sci Lett 274:112–120

Perez-Gussinyé M, Metois M, Fernandez M, Verges J, Fullea J, Lowry A (2009) Effective elastic thickness of Africa and its relationship to other proxies for lithospheric structure and surface tectonics. Earth Planet Sci Lett 287:152–167

Perfit MR, Chadwick WW (1998) Magmatism at mid-ocean ridges: constraints from volcanological and geochemical investigations. In: Buck WR, Delaney PT, Karson JA, Lagabrielle Y (eds) Faulting and magmatism at mid-ocean ridges. AGU Geophysical Monograph Series Washington, vol 106, pp 59–115

Perlt J, Heinert M, Niemeier W (2008) The continental margin in Iceland—a snapshot derived from combined GPS networks. Tectonophysics 447:155–166

Pinzuti P, Mignan A, King GCP (2010) Surface morphology of active normal faults in hard rock: implications for the mechanics of the Asal Rift, Djibouti. Earth Planet Sci Lett 299:169–179

Pollard DD, Aydin A (1984) Propagation and linkage of oceanic ridge segments. J Geophys Res 89:10017–10028

Purdy GM, Kong LSL, Christenson GL, Solomon SC (1991) Relationship between spreading rate and the seismic structure of mid-ocean ridges. Nature 355:815–817

Reiss MC, Muirhead JD, Laizer AS, Link F, Kazimoto EO, Ebinger CJ et al (2021) The impact of complex volcanic plumbing on the nature of seismicity in the developing magmatic natron rift, Tanzania. Front Earth Sci 8:609805, https://doi.org/10.3389/feart.2020.609805

Reston T (2018) Flipping detachments: the kinematics of ultraslow spreading ridges. Earth Planet Sci Lett 503:144–157

Rooney TO (2020a) The Cenozoic magmatism of East-Africa: Part I—flood basalts and pulsed magmatism. Lithos 286–287:264–301

Rooney TO (2020b) The cenozoic magmatism of East Africa: Part II—rifting of the mobile belt. Lithos 360–361:105291

Rowland JV, Baker E, Ebinger CJ, Keir D, Kidane T, Biggs J et al (2007) Fault growth at a nascent slow-spreading ridge: the 2005 Dabbahu rifting episode, Afar. Geophys J Int 171:1226–1246

Rowley DB, Forte AM, Rowan CJ, Glišović P, Moucha R, Grand SP (2016) Kinematics and dynamics of the East Pacific Rise linked to a stable, deep mantle upwelling. Sci Adv 2:e1601107

Saemundsson K, Sigugeirsson MA, Fridleifsson GO (2020) Geology and structure of the Reykjanes volcanic system, Iceland. J Volcanol Geoth Res 391:106501

Sandwell DT, Smith WHF (2009) Global marine gravity from retracked Geosat and ERS-1 altimetry: ridge segmentation versus spreading rate. J Geophys Res 114:B01411. https://doi.org/10.1029/2008JB006008

Schettino A, Macchiavelli C, Pierantoni PP, Zanoni D, Rasul N (2016) Recent kinematics of the tectonic plates surrounding the Red Sea and Gulf of Aden. Geophys J Int 207:457–480

Schlindwein V, Schmid F (2016) Mid-ocean-ridge seismicity reveals extreme types of ocean lithosphere. Nature 535:276–279

Searle RC, Keeton JA, Owens RB, White RS, Mecklenburgh R, Parsons B et al (1998) The Reykjanes ridge: structure and tectonics of a hot-spot influenced slow spreading ridge, from multibeam bathymetry, gravity and magnetic investigations. Earth Planet Sci Lett 160:463–478

Searle RC, Murton BJ, Achenbach K, LeBas T, Tivey M, Yeo I et al (2010) Structure and development of an axial volcanic ridge: Mid-Atlantic Ridge, 45°N. Earth Planet Sci Lett 299:228–241

Shah AK, Buck WR (2003) Plate bending stresses at axial highs, and implications for faulting behaviour. Earth Planet Sci Lett 211:343–356

Sibrant ALR, Mittelstaedt E, Davaille A, Pauchard L, Aubertin A, Auffray L et al (2018) Accretion mode of oceanic ridges governed by axial mechanical strength. Nat Geosci 11:274–279

Sigmundsson F (1992) Tectonic implications of the 1989 Afar earthquake sequence. Geophys Res Lett 19:877–880

Sigmundsson F, Vadon H, Massonnet D (1997) Readjustment of the Krafla spreading segment to crustal rifting measured by satellite radar interferometry. Geophys Res Lett 24:1843–1846

Sigmundsson F (2006a) Magma does the splits. Nature 442:251–252

Sigmundsson F (2006b) Iceland geodynamics: crustal deformation and divergent plate tectonics. Springer, Berlin, 209 pp

Sigmundsson F, Hooper A, Hreinsdottir S, Vogfjord KS, Ofeigsson DG, Heimisson et al (2015) Segmented lateral dyke growth in a rifting event at Barðarbunga volcanic system, Iceland. Nature 517:191–195

Sigmundsson F, Einarsson P, Hjartardottir AR, Drouin V, Jonsdottir K, Arnadottir T et al (2020) Geodynamics of Iceland and the signatures of plate spreading. J Volcanol Geoth Res 391:106436

Sigurdsson O (1980) Surface deformation of the Krafla fissure swarm in two rifting events. J Geophys Res 47:154–159

Singh SC, Crawford WC, Carton H, Seher T, Combier V, Cannat M (2006) Discovery of a magma chamber and faults beneath a Mid-Atlantic Ridge hydrothermal field. Nature 442:1029–1032

Sinton J, Gronvole K, Saemundsson K (2005) Postglacial eruptive history of the Western Volcanic Zone, Iceland. Geochem Geophys Geosyst 6:Q12009. https://doi.org/10.1029/2005GC001021

Smittarello D, Grandin R, De Chabalier J-B, Doubre C, Deprez A, Masson F et al (2016) Transient deformation in the Asal-Ghoubbet Rift (Djibouti) since the 1978 diking event: is deformation controlled by magma supply rates? J Geophys Res 121:6030–6052

Snow JE, Edmonds HN (2007) Ultraslow-spreading ridges rapid paradigm changes. Oceanography 20:90–101

Soule SA, Escartin J, Fornari DJ (2009) A record of eruption and intrusion at a fast spreading ridge axis: axial summit trough of the East pacific Rise at 9–10°N. Geochem Geophys Geosyst 10:Q10T07. https://doi.org/10.1029/2008GC002354

Stamps DS, Saria E, Kreemer C (2018) A geodetic strain rate model for the east african rift system. Sci Rep 8:732. https://doi.org/10.1038/s41598-017-19097-w

Standish JJ, Sims KWW (2010) Young off-axis volcanism along the ultraslow-spreading Southwest Indian Ridge. Nat Geosci 3:286–292

Sturkell E, Sigmundsson F (2000) Continuous deflation of the Askja caldera, Iceland, during the 1983–1998 noneruptive period. J Geophys Res 105:25671–25684

Takada A (1994) The influence of regional stress and magmatic input on styles of monogenetic and polygenetic volcanism. J Geophys Res 99:13563–13573

Tan YJ, Tolstoy M, Waldhauser F, Wilcox WSD (2016) Dynamics of a seafloor-spreading episode at the East Pacific Rise. Nature 540:261–265

Tapponnier P, Armijo R, Manighetti I, Courtillot V (1990) Bookshelf faulting and horizontal block rotations between overlapping rifts in Southern Afar. Geophys Res Lett 17:1–4

Tarantola A, Ruegg JC, Lepine JC (1979) Geodetic evidence for rifting in Afar. A brittle-elastic model of the behavior of the lithosphere. Earth Planet Sci Lett 45:435–444

Thordarson T, Self S (1993) The Laki (Skaftar Fires) and Grimsvotn eruptions in 1783–1785. Bull Volcanol 55:233–263

Thordarson T, Larsen G (2007) Volcanism in historical time: volcano types, eruptions styles and eruptive history. J Geodyn 43:118–152

Tolstoy M, Cowen JP, Baker ET, Fornari DJ, Rubin KH, Shank TM et al (2006) A sea-floor spreading event captured by seismometers. Science 314:1920–1922

Tolstoy M, Waldhauser F, Bohnenstiehl DR, Weekly RT, Kim WY (2008) Seismic identification of along-axis hydrothermal flow on the East Pacific Rise. Nature 451:181–195

Toomey DR, Hooft EEE (2008) Mantle upwelling, magmatic differentiation, and the meaning of axial depth at fast-spreading ridges. Geology 36:679–682

Trippanera D, Ruch J, Passone L, Jónsson S (2019) Structural mapping of dike-induced faulting in Harrat Lunayyir (Saudi Arabia) by using high resolution drone imagery. Front Earth Sci 7:168. https://doi.org/10.3389/feart.2019.00168

Tryggvason E (1984) Widening of the Krafla fissure swarm during the 1975–1981 volcano-tectonic episode. Bull Volcanol 47:47–69

Tucholke BE, Behn MD, Buck RW, Lin J (2008) Role of melt supply in oceanic detachment faulting and formation of megamullions. Geology 36:455–458

Upcott NM, Mukasa RK, Ebinger CJ, Karner GD (1996) Along-axis segmentation and isostasy in the Western rift, East Africa. J Geophys Res 101:3247–3268

Vanderbeek BP, Toomey DR, Hooft EEE, Wilcock WSD (2016) Segmentation of mid-ocean ridges attributed to oblique mantle divergence. Nat Geosci 9:636–642

Varet J (2018) Geology of Afar (East Africa). Springer International Publishing AG, 345 pp

Vigny C, de Chabalier JB, Ruegg JC, Huchon P, Feigl KL, Cattin et al (2007) Twenty-five years of geodetic measurements along the Tadjoura-Asal rift system, Djibuti, East Africa. J Geophys Res 112: B06410. https://doi.org/10.1029/2004JB003230

Wadge G, Biggs J, Lloyd R, Kendall J-M (2016) Historical volcanism and the state of stress in the East African Rift system. Front Earth Sci 4:86. https://doi.org/10.3389/feart.2016.00086

Waters CL, Sims KWW, Klein EM, White SM, Reagan MK, Girard G (2013) Sill to surface: linking young off-axis volcanism with subsurface melt at the overlapping spreading center at 9°03′N East Pacific Rise. Earth Planet Sci Lett 369–370:59–70

Wernicke B (1985) Uniform-sense normal simple shear of the continental lithosphere. Can J Earth Sci 22:108–125

White SM, Haymon RM, Fornari DJ, Perfit MR, MacDonald KC (2002) Correlation between volcanic and tectonic segmentation of fast-spreading ridges: evidence from volcanic structures and lava flow morphology on the East Pacific Rise at 9°-10°N. J Geophys Res 107:2173. https://doi.org/10.1029/2001JB000571

Williams FM, Williams MAJ, Aumento F (2004) Tensional fissures and crustal extension rates in the northern part of the Main Ethiopian Rift. J Afr Earth Sci 38:183–197

Wilson DJ, Robinson AH, Hobbs RW, Peirce C, Funnell MJ (2019) Does intermediate spreading-rate oceanic crust result from episodic transition between magmatic and magma-dominated, faulting-enhanced spreading? The Costa Rica Rift example. Geophys J Int 218:1617–1641

Wright TJ, Ebinger C, Biggs J, Ayele A, Yirgu G, Keir D et al (2006) Magma maintained rift segmentation at continental rupture in the 2005 Afar dyking episode. Nature 442:291–294

Wright TJ, Sigmundsson F, Pagli C, Belachew M, Hamling IJ, Brandsdottir B et al (2012) Geophysical constraints on the dynamics of spreading centres from rifting episodes on land. Nat Geosci 5:242–250

Xu M, Canales JP, Carbotte SM, Carton H, Nedimovic MR, Mutter JC (2014) Variations in axial magma lens properties along the East Pacific Rise (9°30′N–10°00′N) from swath 3-D seismic imaging and 1-D waveform inversion. J Geophys Res 119:2721–2744

Xu W, Ruch J, Jonsson S (2015) Birth of two volcanic islands in the southern Red Sea. Nat Commun 6:7104. https://doi.org/10.1038/ncomms8104

Xu W, Rivalta E, Li X (2017) Magmatic architecture within a rift segment: Articulate axial magma storage at Erta Ale volcano, Ethiopia. Earth Planet Sci Lett 476:79–86

12 Volcanoes Along Convergent Plate Boundaries

12.1 Introduction

This chapter focuses on the tectono-magmatic relationships along convergent plate boundaries characterized by subduction: these boundaries are the sites of the largest earthquakes and eruptions and constitute the most active, unstable and hazardous areas on Earth. Volcanic arcs are the surface manifestation of the magmatic activity resulting from plate convergence. As anticipated in Chap. 10, their structural setting may vary significantly, mainly depending on the velocity and angle of convergence between the subducting and overriding plates. The arc may in fact be controlled by predominant extensional, strike-slip, contractional, transtensive or transpressive structures, or any combination of these, with transient kinematic variations induced by mega-earthquakes along the plate boundary. As a result of these variable tectonic and structural features, the distribution and composition of the volcanoes, and their erupted styles, volumes and frequencies may also vary. While a volcanic arc usually consists of a focused belt of andesite stratovolcanoes with subordinate calderas, under specific conditions rhyolitic calderas may predominate, fuelling a much more voluminous explosive activity associated with widespread transcrustal magmatism.

The main aims of this chapter are to:

- describe the tectonic and magmatic features of representative volcanic arcs experiencing predominant extensional, contractional and strike-slip kinematics;
- highlight similarities and differences in the magmatic activity of the arcs as a function of their tectonic context, including any transient variation induced by mega-earthquakes;
- discuss the general role of magmatic activity in the evolution of convergent plate boundaries.

12.2 Extensional Arcs

Extensional arcs are characterized by dominant arc-perpendicular extension. As anticipated in Sect. 10.4, extension may result from a weak coupling between the two converging plates, where the downgoing plate does not push tightly against the overriding plate, with the latter experiencing local extension in its frontal portion. In addition, arc extension may be the local expression of a wider area of back-arc opening, resulting from the different convergence rate between the frontal and inner portions of the overriding plate, under the roll-back of the negatively buoyant slab.

V. Acocella, *Volcano-Tectonic Processes*, Advances in Volcanology,
https://doi.org/10.1007/978-3-030-65968-4_12

Table 12.1 Tectono-magmatic features of active volcanic arcs

ARC	R	EC	SS	Q	V_n	V_p	Q_c
Kamchatka	$3 \pm 0.9 \times 10^{-3}$	10 ± 5	5 ± 3	0.67 ± 0.2	74	13	0.85 ± 0.03
Alaska[pa]	$2.6 \pm 0.5 \times 10^{-4}$	-1 ± 1	0 ± 1	-1 ± 0.5	58	6	0.91 ± 0.04
Cascades	$9.5 \pm 0.4 \times 10^{-4}$	2 ± 1	0 ± 1	1 ± 0.5	24	18	0.57 ± 0.05
North Kuril	$3.6 \pm 1 \times 10^{-4}$	-1 ± 1	0.5 ± 0.5	-0.67 ± 0.2	77	12	0.86 ± 0.03
South Kuril	$3.6 \pm 1 \times 10^{-4}$	-3 ± 2	2 ± 2	-0.6 ± 0.2	71	39	0.64 ± 0.02
Less. Antilles[pb]	$4 \pm 0.6 \times 10^{-4}$	4 ± 1 (ap)	3 ± 2	0.57 ± 0.2	12	3	0.8 ± 0.16
Marianas	$1.25 \pm 0.5 \times 10^{-3}$	5 ± 5	0 ± 1	1 ± 0.5	31	21	0.6 ± 0.04
Central Mexico	$1.37 \pm 0.4 \times 10^{-4}$	0.4 ± 0.04	0.14 ± 0.01	0.74 ± 0.3	51	10	0.84 ± 0.04
Central America[c]	$3.1 \pm 0.8 \times 10^{-3}$	12 ± 2 (ap)	11 ± 2	0.52 ± 0.1	73	6	0.92 ± 0.03
NE Japan[d]	$1.5 \pm 0.4 \times 10^{-4}$	-5 ± 3	0 ± 1	-1 ± 0.5	96	32	0.75 ± 0.02
New Zealand[e]	$2 \pm 0.4 \times 10^{-3}$	6.5 ± 0.2	2.6 ± 0.2	0.71 ± 0.2	41	16	0.72 ± 0.04
Izu-Bonin	$4 \pm 0.9 \times 10^{-4}$	1.7 ± 0.3	0.5 ± 0.5	0.77 ± 0.3	55	20	0.73 ± 0.03
Sumatra[pf]	$6.6 \pm 2 \times 10^{-3}$	0 ± 2	23 ± 2	0	41	18	0.69 ± 0.04
East Aleutians	$2 \pm 0.7 \times 10^{-4}$	-5 ± 5	0 ± 1	-1 ± 0.5	65	14	0.82 ± 0.03
South Andes[g]	$1.3 \pm 0.4 \times 10^{-3}$	-1 ± 1	10 ± 8	-0.09 ± 0	75	27	0.73 ± 0.02
SW Japan[ph]	$3 \pm 0.7 \times 10^{-3}$	4 ± 2(ap)	4 ± 2	0.5 ± 0.1	48	23	0.68 ± 0.03

R = volume of erupted magma (km^3) per year per 100 km of length of the arc; *EC* = amount of arc-normal extension (*E*, positive) or compression (*C*, negative), in mm/year; *SS* = amount of arc-parallel slip (mm/year); arc motion $Q = EC/(|EC| + SS)$; V_n = trench-normal component of the convergence vector V_{con} (mm/year); V_p = trench-parallel component of the convergence vector (mm/year); convergence motion $Q_c = V_n/(V_n + V_p)$. Mean errors associated with V_n and V_p are of ±2 mm/year; p = portion of arc; ap = arc-parallel extension; a = Katmai Province; b = central portion; c = South Guatemala and El Salvador; d = northeast Honshu; e = Taupo Volcanic Zone, North Island; f = Toba region; g = Puyehue-Cordón Caulle region; h = Kyushu region (modified after Acocella and Funiciello 2010)

Several magmatic arcs experience predominant extension, including those in Kamchatka (Russia), Cascades (western North America), Marianas, Izu-Bonin (western Pacific), New Zealand and Central Italy (Table 12.1; Acocella and Funiciello 2010, and references therein). The three representative cases described below consider different conditions of extension. These are exemplified by a volcanic arc with low extension rate, such as the Cascade Arc (Sect. 12.2.1), an arc with high extension rate, as the Taupo Volcanic Zone of New Zealand (Sect. 12.2.2) and an arc developed on the margin of an extended back-arc zone, as the Tyrrhenian margin of Central Italy (Sect. 12.2.3).

12.2.1 Cascade Arc

The Cascade Arc strikes north–south for approximately 1200 km from British Columbia (western Canada) to Northern California (western USA). The arc lies above a warm slab convergence zone, where the young (less than 10 Ma) oceanic crust of the Juan de Fuca plate subducts beneath the North American plate north of the Mendocino Triple Junction. At this junction, the tectonic environment shifts from northwest directed shear along the San Andreas Fault to E-W trending convergence along the Juan de Fuca plate (Fig. 12.1; Hildreth 2007 and references therein). Beneath the Cascade Range,

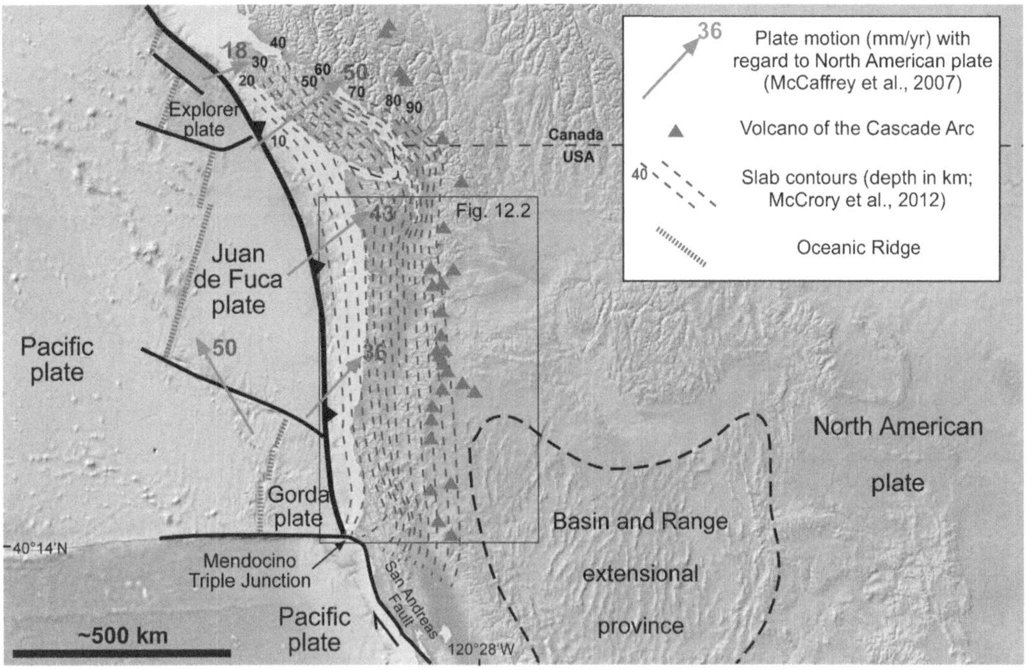

Fig. 12.1 Tectonic setting of the Cascade Volcanic Arc, western North America, showing the major plate boundaries, the slab contours and the vectors of plate motion (modified after McCaffrey et al. 2007). Base DEM provided by GeoMapApp

the Juan de Fuca slab is seismically imaged as a high velocity zone dipping 65°, extending at least to 200 km of depth. This convergence zone is one of the best-documented subduction systems for episodic tremor and slow-slip earthquakes. The crust beneath the Cascades of Oregon is 40 km thick, thinning to 31 km eastward; the 42–47 km thicker crust beneath the Washington Cascades, to the north, may result from underplating. The magmatically most active Oregon portion of the arc shows low seismic velocities at all depths, suggesting widespread melt. The less active Washington portion shows high seismic velocities in the upper mantle and upper crust, possibly due to solidified intrusions. More localized and shallower low velocity zones supportive of $\sim$6% of melt have been imaged below the volcanoes of Mount St. Helens, Mount Adams and Mount Rainier, in the southern Washington Cascades: these are indicative of large-scale basaltic sill emplacement and silicic differentiation, representing the primary reservoir of the region's arc magmatism (Miller et al. 1997; Parsons et al. 1998; Pollitz et al. 2010; Gao et al. 2011; Flinders and Shen 2017).

Northeast-directed subduction of the Juan de Fuca plate results in nearly arc-normal convergence in the northernmost portion of the Cascade Arc, but oblique convergence elsewhere. This imparts a dextral component in the forearc region, from central Washington to California (Hildreth 2007, and references therein). In particular, in the northern Cascades the direction of the maximum principal stress σ_1 changes from margin-normal along the coast to margin-parallel inland. While the former is related to the coupled interface between the North America and Juan de Fuca plates, the latter derives from the motion of the Juan de Fuca plate, which ultimately results in a northward movement at a few mm/yr of the forearc. The Cascades also rotate clockwise with regard to North America at a rate of 0.4–1°/Ma (Fig. 12.2). A southward increase in rotation parallels a change in the arc tectonic regime, from largely contractional in northern Washington to extensional in Oregon. Concomitant is a

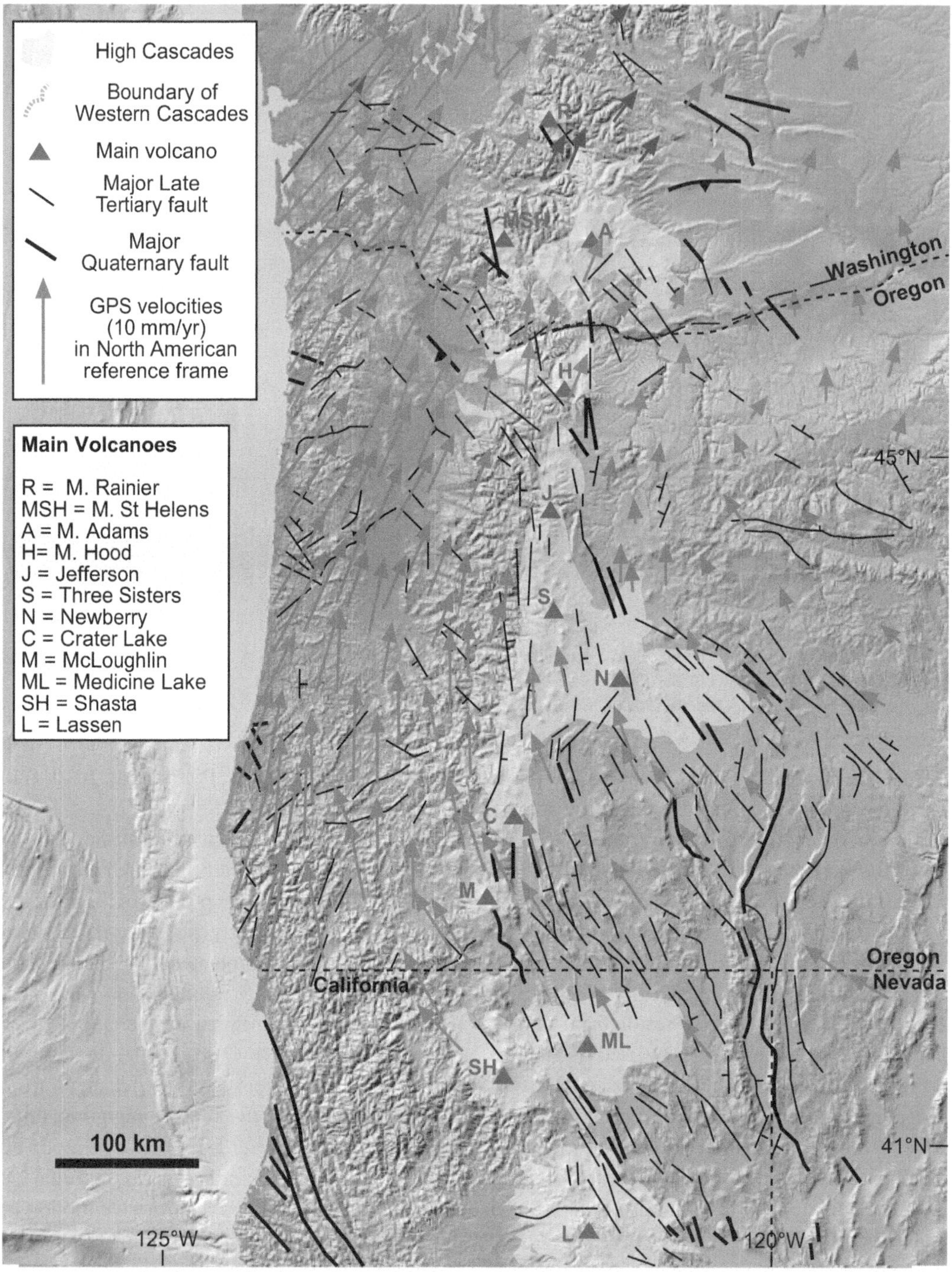

Fig. 12.2 Geology (extent of the Western and High Cascades), structure (late Tertiary and Quaternary faults), present kinematics (GPS motions between 1991 and 2004) and volcanism (main volcanoes) along the central and southern portions of the Cascade Volcanic Arc (Pezzopane and Weldon 1993; Blakely et al. 1997; McCaffrey et al. 2007). Base DEM provided by GeoMapApp

southward increase in the volume of eruptive rocks along the volcanic arc. Extension across the southern part of the arc is also promoted by the buoyant mantle beneath the western Basin and Range Province (Wells 1990; Hildreth 2007; McCaffrey et al. 2007; McCrory et al. 2012). As a result, crustal deformation across Oregon indicates motion of 0.6 cm/year in a N60°W direction, resulting from combined horizontal extension and dextral shear. This motion can account from 10 to 20% of the total Pacific–North American transform motion and much of the lateral component of relative motion between the Juan de Fuca and North American plates. However, only a minor portion of this extension, geodetically undetected (~0.1 cm/year), affects the volcanic arc (Fig. 12.1; Blakely et al. 1997; McCaffrey et al. 2007). Active extension is usually associated with ~N-S trending half-grabens with major offset to the west. In the southernmost portion of the Cascade Arc, the active structures shift from ~N-S trending normal faults to ~NW–SE trending dextral transtensive systems reaching the northern Sierra Nevada (Pezzopane and Weldon 1993; Waldien et al. 2019).

Two distinct volcanic assemblages characterize the central Cascades of Oregon: the Western Cascades and the High Cascades (Fig. 12.2). The Western Cascades were active since ~40 Ma, with volcanism younging eastward during Miocene and Pliocene, towards the High Cascades. The High Cascades have been active since 8–10 Ma in a focused area of extension (Wells and McCaffrey 2013). Eruption rate along the Cascade Arc decreased by a factor of 3 from 35 to 7 Ma, probably due to the slowing of the plate convergence rate and the increase in the obliquity of subduction. Eruption rate then increased in the last 7 Ma, likely because of the new extensional regime promoting mafic volcanism: volcanic production seems therefore a function of convergence rate and upper plate stress regime (Priest 1990; Hildreth 2007; Pitcher et al. 2017).

Of the 2339 Quaternary volcanoes identified in the current Cascade Arc, 37 are andesite-dacite stratovolcanoes and major silicic dome complexes, 110 are shield volcanoes, 340 are lava domes and more than 1850 are monogenic vents. The main Cascade stratovolcanoes exhibit spacings that range from 35 to 170 km. The composition of the erupted products ranges widely and continuously, and includes olivine tholeiites, intraplate basalts, calcalkaline basalts, basaltic andesites, potassic shoshonites, andesites, rhyolites, rhyodacites and dacites. Primitive basalts produce scoria cones, small shields and discrete lava fields, but virtually never erupt centrally at stratovolcanoes, where more evolved compositions dominate as a result of large fractionating reservoirs (Hildreth 2007). The different age of the subducted plate also influences the depth of melting in the mantle wedge beneath the arc and the general composition of the magmas. Younger and hotter oceanic crust undergoes dehydration with minor amount of heating during subduction. The slab may lose much of its volatiles trenchward of the arc, with a lower volatile budget beneath the arc. This occurs in the northern Cascades, where the contribution of the slab fluids in driving arc magmatism is limited and alkali basalts are frequent. Conversely, dehydration reactions are delayed in older oceanic lithosphere and a greater amount of volatiles are released beneath the central-southern Cascade Arc; this results in higher degrees of melting, producing calkalkaline basalt magmas (Green and Harry 1999).

The Quaternary Cascade Arc is 25–100 km wide and its continuity is interrupted by several gaps containing few or no volcanoes. Assuming that the slab dehydrates and the wedge convects and melts continuously along the whole subduction margin suggests that sectors experiencing lower crustal extension locally stall and suppress the intrusion of mantle melts, likely explaining the gaps. On the other hand, higher extension promotes the eruption of the mafic melts. The gaps also define several segments (four to six, depending on the studies) with distinct direction, structure, geochemical composition and melting regimes at depth, reflecting differences not only in tectonic setting, but also in subduction geometry and mantle heterogeneity. While the degree of melting and fluxing is greatest in the southernmost segment, enhanced by a slab window, the central segments show higher and focused extension and heat flow, as well as most of the eruptive vents (Fig. 12.2; Guffanti and Weaver 1988; Pezzopane and Weldon 1993; Hildreth 2007; Schmidt et al. 2008; Pitcher and Kent 2019).

In addition to Mount St. Helens, several volcanoes in the southern part of the arc, including Lassen, Shasta, McLoughlin, seem to lie in localized zones of E-W trending extension controlled by the activity of offset NW–SE trending dextral faults (see also Fig. 6.20a; Weaver et al. 1987; Hildreth 2007). Many volcanoes and volcanic complexes have been recently active in the Cascades. Among these, Lassen Peak (northern California) erupted in 1915, whereas Mount St. Helens (described in Sect. 6.8.2) has been erupting since 1980. In the last decades, deformation studies have identified surface displacements at five of the 13 major Cascade Arc volcanoes (Mount Baker, Mount St. Helens, South Sister, Medicine Lake, and Lassen). No deformation has been detected at five volcanoes (Mount Rainier, Mount Hood, Newberry Volcano, Crater Lake, and Mount Shasta), and there are not sufficient data for a rigorous assessment at the remaining three (Glacier Peak, Mount Adams, and Mount Jefferson; Poland et al. 2017).

12.2.2 Taupo Volcanic Zone of New Zealand

Taupo Volcanic Zone (TVZ) lies in the North Island of New Zealand and is probably the best-known volcanic arc undergoing sustained extension. The NNE-SSW trending TVZ results from convergence between the Australian plate and the westward subducting Pacific plate. The plate interface is offshore and to the east of the North Island of New Zealand, along the Hikurangi Trough (Fig. 12.3; e.g., Stern et al. 2006). The northern offshore continuation of the TVZ arc is the Havre Trough, a young back-arc basin where the original arc occupied a narrow area,

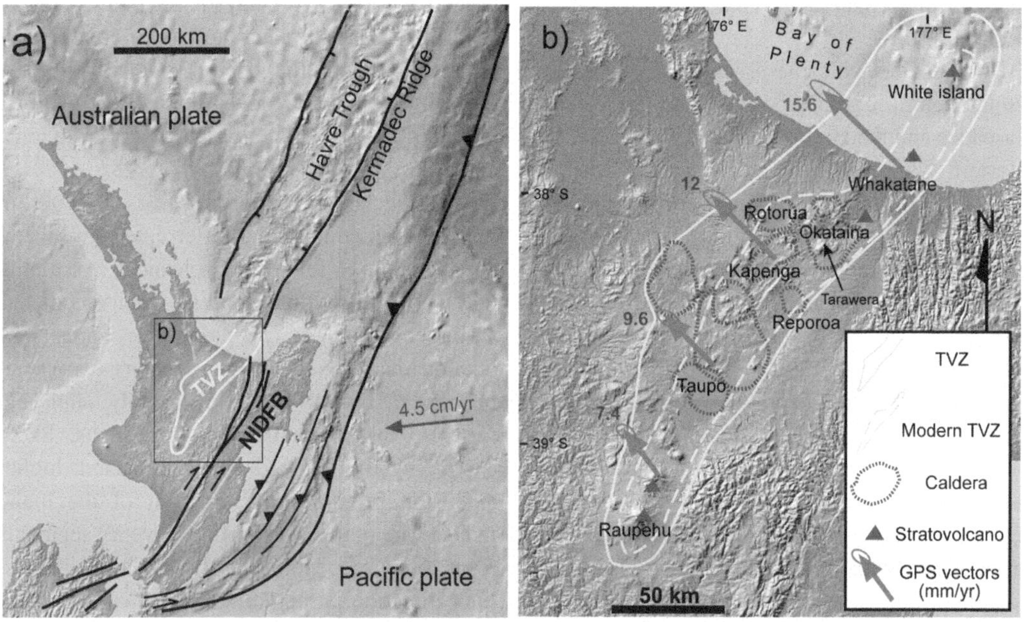

Fig. 12.3 Tectonic setting and main features of Taupo Volcanic Zone (TVZ), New Zealand. **a** Schematic relationship of TVZ to the boundary between the Pacific and Australian plates and fore-arc structures, including the NIDFB (North Island Dextral Fault Belt). **b** Overview of the TVZ, showing the major caldera complexes and stratovolcanoes. Blue arrows show GPS movement of the Australian plate with regard to the fixed Axial Range to the east of TVZ (Wallace et al. 2004). Base DEM provided by GeoMapApp

10–15 km wide, followed by seafloor spreading at ~5.0 Ma, after which arc magmatism dominated again. This rapid sequence of tectono-magmatic regimes is related to the roll-back of the Pacific slab, which may have diverted the arc flux outside the region of seafloor spreading of the back-arc and induced the vertical realignment of surface volcanism with the source of arc melts above the slab (Caratori Tontini et al. 2019). Considered together, the TVZ-Havre Trough system shows an overall northward transition from arc to back-arc magmatism.

At the latitude of TVZ, the mean convergence vector of ~4.5 cm/year between the Australian and Pacific plates is partly oblique, oriented at ~263°. As a result, the area between the trench and the volcanic arc is characterized by an overall constant component of dextral shear, estimated at ~0.25 cm/year. This shear dominates along the North Island Dextral Fault Belt (NIDFB), as the convergence-parallel component passes from dominant contraction in the trench area to negligible contraction along the NIDFB and to dominant extension in the TVZ (Acocella et al. 2003, and references therein).

The current TVZ consists of a rift zone associated with crustal thinning of up to 15 km. Lower crust underplating is significant, with partial melt between 1 and 4%. These features may be related to abundant dehydration fluids coming off the slab at specific locations and migrating upward to the highly fractured volcanic zone. The high mantle flux and rifting are responsible for a remarkably thin (~16 km) and extended crust highly permeable to magma, accounting for the extremely high erupted volumes (Wilson 1996; Bertrand et al. 2012; Gravley et al. 2016; Eberhart-Phillips et al. 2020). Volcanic activity in the last ~2 Ma focused in short periods (25–50 ka) of intense volcanism bracketed by longer periods (100–130 ka) of quiescence, rapidly and asymmetrically narrowing via inward and eastward migration of faulting and also propagating southward. The eastward migration of volcanic activity and faulting followed the roll-back of the Pacific slab at the Hikurangi subduction zone: the correlation in time and space of the loci of voluminous volcanic eruptions and active faulting suggests that a controlling factor in rapid rift narrowing are large shallow crustal heterogeneities, as rhyolitic magma bodies generated by the subduction weakening the crust and localizing deformation (Villamor et al. 2017).

The present site of active volcanism focuses in a NNE-SSW trending narrow zone (20–30 km wide), referred to here as the modern TVZ (Fig. 12.3) and mainly developed in the last ~300 ka, although limited evidence of off-rift magmatism is suggested by seismicity and uplift (Wilson et al. 1995; Hamling et al. 2016). The modern TVZ consists of five main segments displaying an overall graben structure. Each segment shows a variable extension direction, with minor dextral shear resulting from the slightly oblique convergence. GPS measurements and fault slip data suggest a current spreading rate of 0.6–0.8 cm/year across central TVZ, at times with significant transient variations. This extension is associated, in the central portion of TVZ, with overall subsidence of up to 2 cm/year, probably resulting from the cooling and subsequent contraction of shallow magma. This subsidence is confined by a wider zone which coincides with the area of the wider and older TVZ and that is being uplifted of 0.1 cm/year by basaltic melts being injected at depth (Cole 1990; Darby and Meertens 1995; Villamor and Berryman 2001; Hamling et al. 2015; Holden et al. 2015; Houlie and Stern 2017). Seismic activity has been accompanying extension along TVZ, with the most recent significant earthquake (magnitude *M6.5*, in 1987) occurring in the northern part. Shallow microearthquakes may represent up to ~30% of the geodetic deformation, suggesting that the associated small-scale faulting records strains not geologically measured, possibly explaining the disparity between geological and GPS rates of extension across TVZ (Begg and Mouslopoulou 2010; Mouslopoulou et al. 2013).

From a volcanological point of view, the modern TVZ is divided into three parts along its length. (a) The two lateral parts with andesite-dacite stratovolcanoes (Ruapehu-Ngauruhoe-Tongariro volcanoes to the south and

Whakatane-White Island volcanoes to the north). (b) A central part with rhyolitic caldera systems (from Okataina to Taupo calderas) and plutons at depth. This is an extraordinarily productive region of rhyolitic volcanism ($\sim$8800 km^3/Ma) and geothermal fluxes ($\sim$4200 MW), enhanced by vertical low resistivity zones connecting to magmatic sources below the brittle-ductile transition, at 6–7 km. Here the two largest rhyolitic magmatic systems, Taupo and Okataina, erupted thousands of km^3 of ignimbrites in the last 300 ka. Okataina caldera produced the largest historic eruption, a basaltic Plinian, in TVZ along the NE-SW trending 1886 Tarawera fissure, which is still geodetically extending (Fig. 12.4; Nairn and Cole 1981; Spinks et al. 2005; Holden et al. 2015). Taupo caldera has the capability to generate and erupt very rapidly, within 1–100 years, large crustal melts of $\sim$100 km^3 in volume (Barker et al. 2016). The andesitic-rhyolitic-andesitic composition of TVZ is probably related to different water contents in the lower crust (Heise et al. 2007; Deering et al. 2008). Rhyolites derive from diorites accumulated in the mid crust resulting from the differentiation of hot and water rich basaltic magmas produced in the mantle wedge and intruded as sills within the lower metasedimentary crust.

The relationships between regional tectonic extension and magmatic activity have been repeatedly addressed at TVZ. In the lower crust, fast seismic wave speeds suggest that magma has partially filled the stretched crust, with the extended zone coinciding with an area of intruded magma located between 4 and 15 km depth. This implies a feedback relationship influencing the locations of faults and magma: while extension helps magma to rise, the intruded magma heats and weakens the crust to assist more rifting. At the surface, there is an overall along- and across-strike partitioning of extension and volcanism, with the most productive sectors located along the rift axis, feeding polygenetic calderas controlled by the highest amount of regional extension orthogonal to the rift segment (Fig. 12.5; Acocella et al. 2003; Spinks et al. 2005; Gase et al. 2019). Extension at the surface may result from regional tectonics and/or shallow magmatic processes (i.e., diking). The prevalence of one process or the other within a portion of TVZ mainly depends on the availability of magma. While diking assists extension along the

Fig. 12.4 View of part of the $\sim$NE-SW trending 1886 Tarawera eruptive fissure, on the southern portion of Okataina caldera, Taupo Volcanic Zone

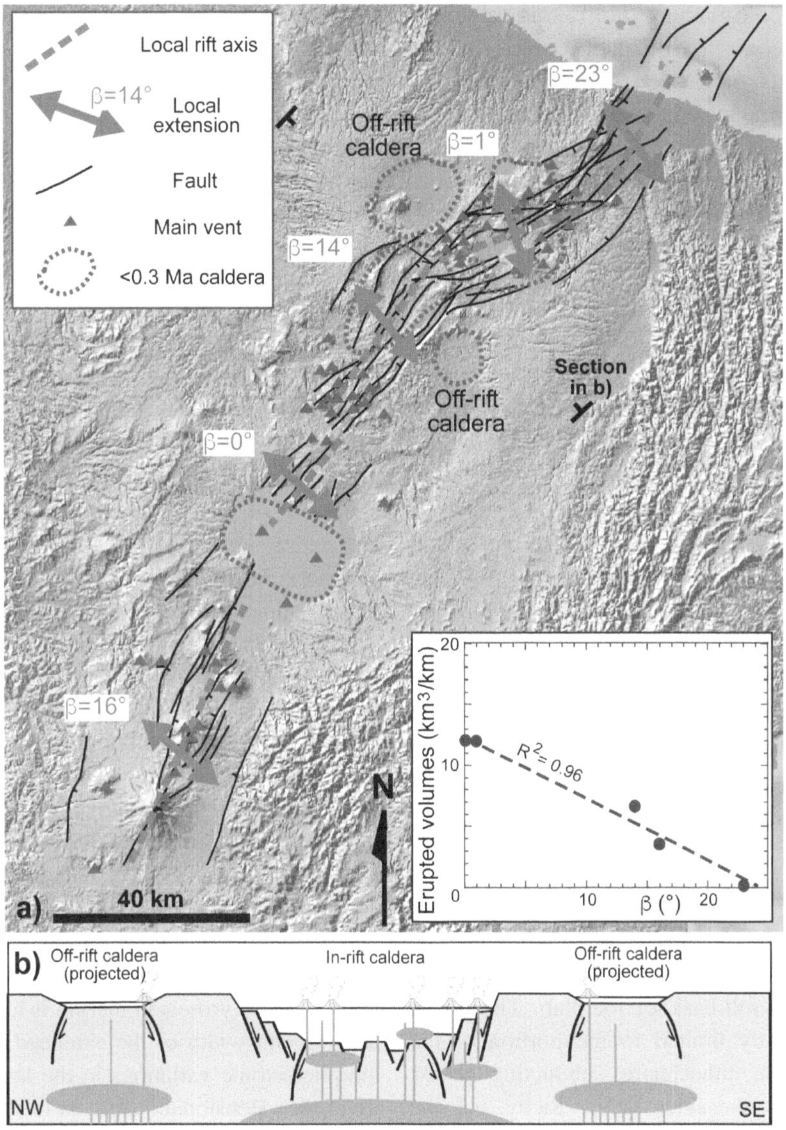

Fig. 12.5 **a** Map of Taupo Volcanic Zone, showing the main volcanoes and the along-rift variations in the dextral component of extension along the rift segments (proportional to the angle β between the opening direction of the rift segment and the perpendicular to the segment strike). The diagram in the inset shows that the magmatically most productive portions undergo orthogonal extension ($\beta \sim 0°$). **b** Across-rift difference in caldera structure: the more productive in-axis polygenetic calderas are deeper and shaped by regional tectonics (see also map view above), whereas off-axis monogenic calderas are less developed and subcircular (Spinks et al. 2005). Base DEM provided by GeoMapApp

most active magmatic portions, regional extension dominates along the magmatically less active portions. For example, field and modelling data across the southern andesitic part of TVZ suggest that the 0.7 cm/year extension is mostly tectonic: here regional faulting accommodates 78–95% of the total extension, while the reminder is accommodated by diking. On the other hand, the 1886 Tarawera event, at Okataina caldera, shows how diking assists extension along the most active magmatic portions. These behaviours imply some complementarity between regional tectonic faulting and diking, although the two processes may be often coupled. In fact, in some parts of TVZ a correlation between increased slip along regional faults and activity from nearby andesitic volcanoes (Ruapehu) or calderas (Rotorua and Ohakuri) has been recognized (Gravley et al. 2007; Villamor et al. 2007, 2011; Seebeck and Nicol 2009; Rowland et al. 2010; Allan et al. 2012; Gomez-Vasconcelos et al. 2017).

12.2.3 Tyrrhenian Margin of Central Italy

The Tyrrhenian margin of Central Italy hosts a volcanic arc located on the northeastern extended portion, or margin, of the back-arc basin of the Tyrrhenian Sea, characterized by significant crustal thinning and incipient oceanization in the southern part. Extensional processes have been affecting the northeastern Tyrrhenian margin since late Miocene as a consequence of back-arc opening. Extension occurred at the back of the eastward migrating Apennine fold and thrust belt, due to the progressive eastward shift of the Apennine subduction (Fig. 12.6a; Malinverno and Ryan 1986; Royden et al. 1987; Patacca et al. 1990). The eastward migration of subduction, and of the associated extension to its rear, is related to the roll-back of the slab. The active slab is currently limited to the portion of the oceanic Ionian lithosphere subducting below Calabria, with the sides below Sicily and the northern and central Apennines experiencing collision between the African and Eurasian continental plates. The product of this collision-subduction-collision configuration is a convergence front that is strongly arcuate and migrating towards the oceanic subducting lithosphere. As a result of this southeastward migration, the southern Tyrrhenian area has been experiencing larger amounts of back-arc extension (5–6 cm/year), associated with incipient oceanization, with regard to the northern Tyrrhenian area (1–2 cm/year; Patacca et al. 1990; Benoit et al. 2011; Savelli and Ligi 2017). Seismic tomography data show gaps within the previously subducted lithosphere below the Tyrrhenian margin of Central Italy, which are interpreted as deep (100–500 km) subvertical tear faults driving post-collision slab breakoff. These tear faults are inferred to control the presence of magmatic activity along the northeastern Tyrrhenian margin (Wortel and Spakman 2000; Rosenbaum et al. 2008; Giacomuzzi et al. 2012).

In Central Italy, the overall NE-SW directed extension associated with NW–SE trending normal faults has migrated eastward from the Tyrrhenian area during Miocene-Pliocene to the Apennines divide during Quaternary, where extension currently manifests through seismogenic faults. Although geodetic or seismological evidence of active extension has not been detected along the northeastern Tyrrhenian margin, here the minimum horizontal stress is currently mainly NE-SW oriented, and locally NW–SE oriented. Extension has produced a thinner crust ($\sim$25 km) in the eastern side of the Apennines with regard to the western part ($\sim$35 km). Associated with crustal thinning is a high heat flux, which reaches 400 mW m^2 in some volcanic areas along the northeastern Tyrrhenian margin and in the Southern Tyrrhenian area (Jolivet et al. 1998; Serpelloni et al. 2005; Di Stefano et al. 2011; Montone et al. 2012; D'Agostino 2014).

The variable amount of stretching along the northeastern Tyrrhenian margin is highlighted by the different width of the extended area and the morphotectonic variations in the lateral continuity of the extensional basins. In fact, the NW–SE trending normal faults extending the margin formed several Plio-Quaternary basins, partly reactivating pre-existing thrust planes generated during the build up of the Apennines orogen. Transverse NE-SW trending faults often interrupt the continuity of these basins, also forming transverse extensional basins, 1–3 km deep, filled by Plio-Quaternary sedimentary and volcanic deposits. The frequency of the transverse structures decreases northeastward, following more limited extension (Fig. 12.6; Mariani and Prato 1988; Faccenna et al. 1994). Surface and subsurface data show that these orthogonal fault systems crosscut each other, suggesting a coeval

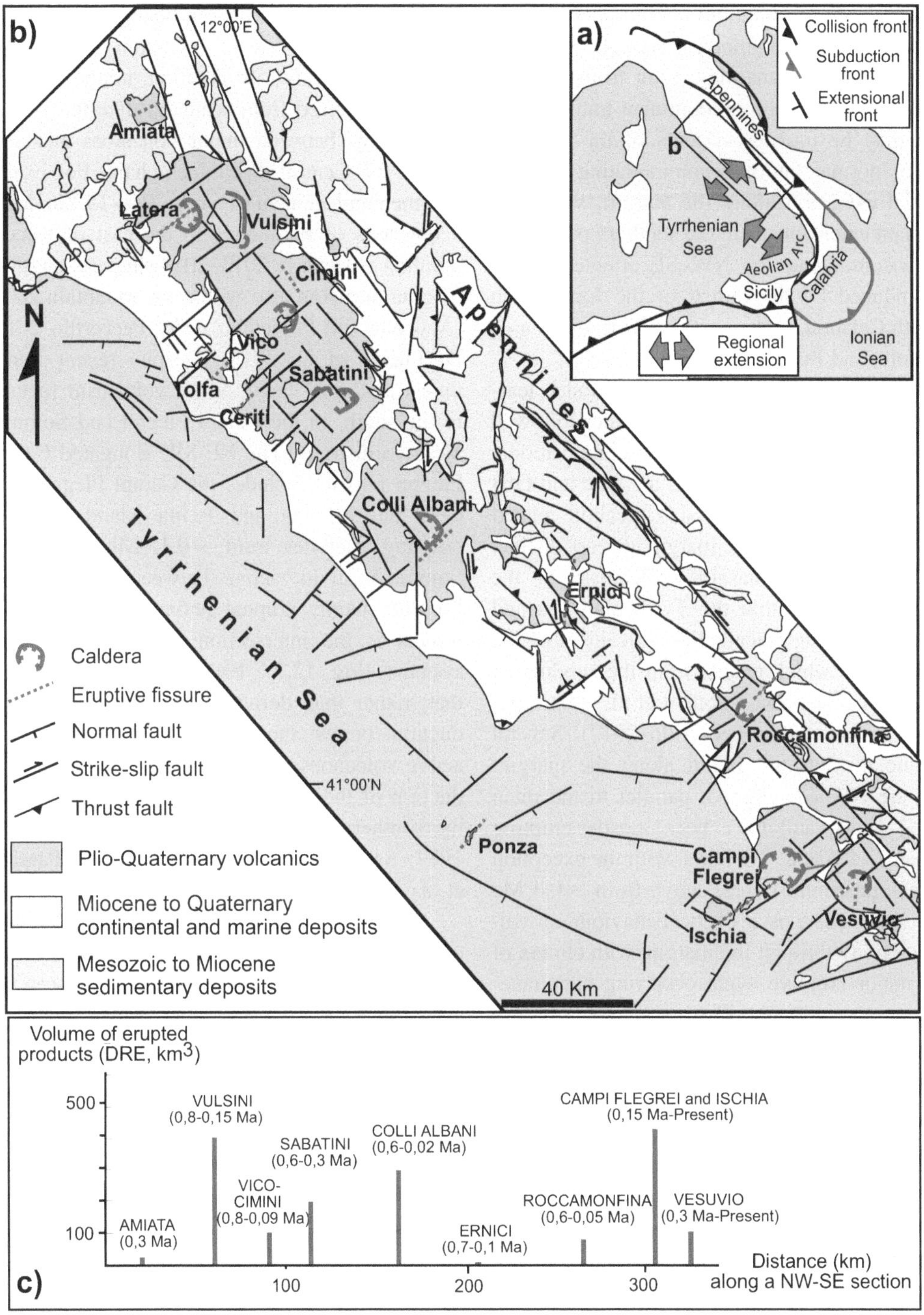

Fig. 12.6 **a** Tectonic setting of the Tyrrhenian margin of Central Italy, showing the orthogonal extension due to the retreat of the Apennines to the northeast (NE–SW extension) and southeast (NW–SE extension) following the arcuate collision-subduction-collision front. **b** Structural setting of the northeastern Tyrrhenian margin, with the main fault systems and the Plio-Quaternary volcanoes. **c** Spacing of volcanoes and related erupted volumes and erupting times (in brackets) projected along a NW–SE trending section (modified after Acocella and Funiciello 2006)

and joint eastward migration. While the margin-parallel faults accommodate crustal extension, the transverse systems may result from two different processes. With predominant transtension, these may be transfer systems of the NW–SE striking normal faults, accommodating the differential extension along the margin. With predominant extension, as in the southern part, these may accommodate the NW–SE oriented extension induced by the retreat of the Ionian slab beneath Calabria, in the southern Tyrrhenian area (Acocella and Funiciello 2006).

Potassic volcanism associated with slab tearing and breakoff beneath the Apennines followed extension along the margin, with compositional signatures transitional between arc type and OIB type magmas. Indeed, the slab breakoff related potassic volcanism of the margin appears distinct from the back-arc basaltic volcanism in the oceanic patches within the Tyrrhenian Sea and from the ongoing calcalkaline volcanism of the Aeolian Arc, which results from the subduction of the Ionian slab (Rosenbaum et al. 2008; Giacomuzzi et al. 2012; Peccerillo 2017). Several volcanic districts are found along the margin: these are NW–SE aligned, parallel to the main normal faults, and have been mostly erupting between ~0.6 and ~0.1 Ma, with the exception of the southernmost area, active from ~0.4 Ma to Present. There is a cyclic behaviour of volcanic activity through the margin, with climax of each major eruptive cycle occurring simultaneously at the different districts every ~48 ka, with most eruptions focusing between 450 and 200 ka (Marra et al. 2004). The northern region, hosting the volcanoes from the Amiata to the Sabatini areas, experienced widespread late Pliocene–Quaternary uplift of a few hundred of metres, probably resulting from repeated magma emplacement (Barberi et al. 1994). The volcanoes mainly consist of caldera complexes with minor relief (Vulsini, Sabatini, Campi Flegrei) or stratovolcanoes with summit calderas (Vico, Colli Albani, Roccamonfina and Vesuvio). Their mean spacing of 42 ± 13 km implies a quite regular distribution. However, this does not correspond to an uniform distribution of the erupted volumes, with most magma being usually erupted at calderas (Vulsini, Sabatini and Campi Flegrei; Fig. 12.6c). In addition, minor volcanic activity occurred from scattered dike-fed monogenic vents between major volcanoes and also within the Central Apennines chain. For example, the emplacement of a dike at ~15 km depth a few tens of kilometres to the east of Roccamonfina volcano in 2013–2014 suggests ongoing magma intrusion also within the mountain chain (Acocella and Funiciello 2006; Peccerillo 2017; Di Luccio et al. 2018). Despite recent minor unrest at Colli Albani, active volcanism focuses to the south, in the Campi Flegrei and Somma-Vesuvio districts. The NE-SW elongated Campi Flegrei District includes the Campi Flegrei Caldera and Procida and Ischia islands, mainly erupting trachytes from ~0.15 Ma to Present. Somma-Vesuvio, active between ~0.3 Ma to Present, mainly erupted phonolites and trachybasalts at the intersection of orthogonal fault systems (Fig. 12.7). Petrological data suggest that, rather than deriving from the extinct subduction below the Central Apennines, these active volcanoes are related to the ongoing subduction of the oceanic Ionian lithosphere below the southern Tyrrhenian area (Rosi and Sbrana 1987; Santacroce 1987; Vezzoli 1988; Passaro et al. 2016; Peccerillo 2017).

Analysis of the fracture systems in the volcanic areas along the margin suggests that the main volcanoes lie at the intersection between the orthogonal fault systems, with the transverse structures predominating over the NW–SE trending structures. This also holds for the major eruptive fissures, which commonly have a transverse direction and moreover erupt the most primitive products (Fig. 12.6). Volcanoes with smaller erupted budget are usually associated with isolated transverse systems (Vico, Vesuvio), whereas volcanoes with higher erupted budget are usually associated with diffuse transverse structures, controlling the shape of calderas and bordering transverse basins usually to the west of the volcano (as at Vulsini, Sabatini, Colli Albani, Campi Flegrei). The onset of sedimentation in these transverse basins is 2–3 Ma older than the

Fig. 12.7 View from the west of the Somma-Vesuvio Volcanic Complex, the last erupting volcano (in 1944) along the Tyrrhenian margin of Central Italy. Mount Somma (to the left) identifies the relict volcano, developed before ~18 ka, separated by the younger arcuate caldera scarp from the active Vesuvio cone (right), developed in the last ~2 ka; the last 1944 lava flow (light colour) lies against part of the caldera scarp. Image courtesy of Giuseppe Vilardo and Laboratorio Geomatica e Cartografia, Osservatorio Vesuviano-INGV

onset of the nearby volcanic activity, suggesting localized pre-volcanic thinning and decompression (e.g., Mariani and Prato 1988; Faccenna et al. 1994).

In synthesis, despite the overall NW–SE orientation of the volcanic arc, there is limited evidence for the activity of NW–SE trending structures controlling magmatism. This is also supported by the fact that active extension due to the NW–SE structures is currently detected within the Apennines ~100 km to the east, and it is likely that such extension already migrated to the east of the volcanic arc during the climax of volcanism between 450 and 200 ka (Devoti et al. 2011; D'Agostino 2014). Therefore, the activity of the NW–SE striking faults along the northeastern Tyrrhenian margin predated volcanic activity, promoting magmatism by thinning the crust and inducing decompression melting. The transverse structures are more closely related to volcanic activity: these may have controlled volcanism in a three-fold way (Fig. 12.8). With a dominant extensional component, as in transverse basins, the transverse structures may localize crustal thinning and decompression, focusing magmatism below the main volcanoes. In addition, any local NW–SE trending extension (especially towards the south) may control dike propagation along a transverse trend. Finally, with a dominant strike-slip component, the transverse structures may provide preferred subvertical crustal paths reactivated by the dikes feeding the major fissure eruptions. The Tyrrhenian case shows how slab breakoff and regional extension promote the overall generation and rise of magma along the margin, but transverse

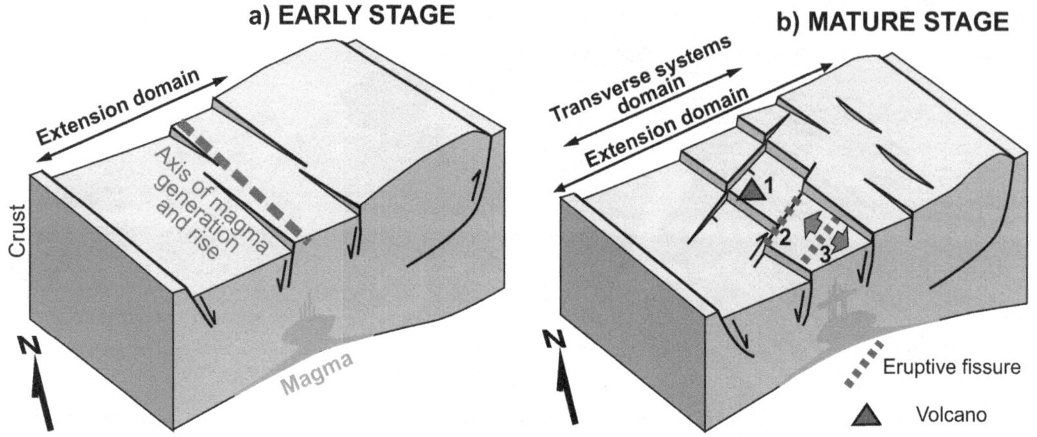

Fig. 12.8 Model of structural control on magmatism along the Tyrrhenian margin of Central Italy. **a** Early stage: extension and decompression due to NW–SE striking normal faults promote magmatism along a NW–SE trending area of partial melting above the torn slab (not shown). **b** Mature stage: extension has migrated to the present Apennines divide. Transverse systems to the back focus the rise of magma in the upper crust, with transverse extensional basins below polygenic volcanoes (1), NE-SW trending dikes intruding subvertical transverse strike-slip faults, (2), or perpendicular to a local NW–SE trending extension, feeding eruptive fissures (3). Not to scale (modified after Acocella and Funiciello 2006)

structures, due to a complex tectonic setting, focus shallower magma rise, emplacement and eruption (Acocella and Funiciello 2006).

12.3 Strike-Slip Arcs

Strike-slip volcanic arcs are characterized by predominant horizontal motion, usually expressed through a principal displacement zone. As described in Sect. 10.4, this condition results from the strain partitioning generated by the oblique convergence between the two plates, with the sense of motion of the trench-parallel component of the convergence vector coinciding with the sense of motion of the strike-slip fault zone along the arc. Strain partitioning, and thus strike-slip arcs, may develop with variable obliquity of the convergence vector: this is expressed by the angle ϕ between the convergence vector and the trench-perpendicular component: the higher the angle, the higher the obliquity (Fig. 10.12). Some strike-slip arcs are related to minor obliquity (of a few tens of degrees), whereas others are related to higher obliquity (several tens of degrees; Table 12.1). While obliquely convergent arcs show a predominant strike-slip motion, a non-negligible contractional or extensional arc-orthogonal component is also commonly found, frequently giving the arc an overall transtensive or transpressive kinematics.

Below two representative cases of strike-slip arcs experiencing minor obliquity (the South Andean Volcanic Zone; Sect. 12.3.1) and major obliquity (Sumatra, Indonesia; Sect. 12.3.2) are described.

12.3.1 The South Andean Volcanic Zone

The N-S trending South Andean Volcanic Zone results from the subduction of the Nazca plate beneath the South American plate in central and southern Chile. To the south of this volcanic zone, after a volcanic gap nearby the Chilean Triple Junction where the spreading Chile Ridge enters the Chile Trench, begins the Austral Volcanic Zone, a young arc related to the Quaternary subduction of the Antarctic plate beneath South America (e.g., Stern 2004).

At the latitude of the South Andean Volcanic Zone, the 0–45 Ma old Nazca plate is being subducted below the continent at 6–9 cm/year in a northeast direction. The dip of the slab increases from $\sim 20°$ at the northern end of the volcanic zone to >25° further to the south. As a consequence, the distance between the trench and the volcanic arc decreases from more than 290 km in the north to less than 270 km in the south. In addition, the depth to the slab below the arc decreases southwards, from 120 to 90 km (Stern 2004). The convergence between the Nazca and South American plates associated with the South Andean Volcanic Zone is moderately oblique, as the angle between the convergence vector and the trench-perpendicular direction is $\sim 22°$. This obliquity is partly partitioned into a contractional forearc and a ~N-S striking domain along the volcanic arc that hosts a large fault system, the Liquine-Ofqui Fault Zone, or LOFZ (Fig. 12.9). This, active since late Oligocene at least, is nearly 1200 km long and,

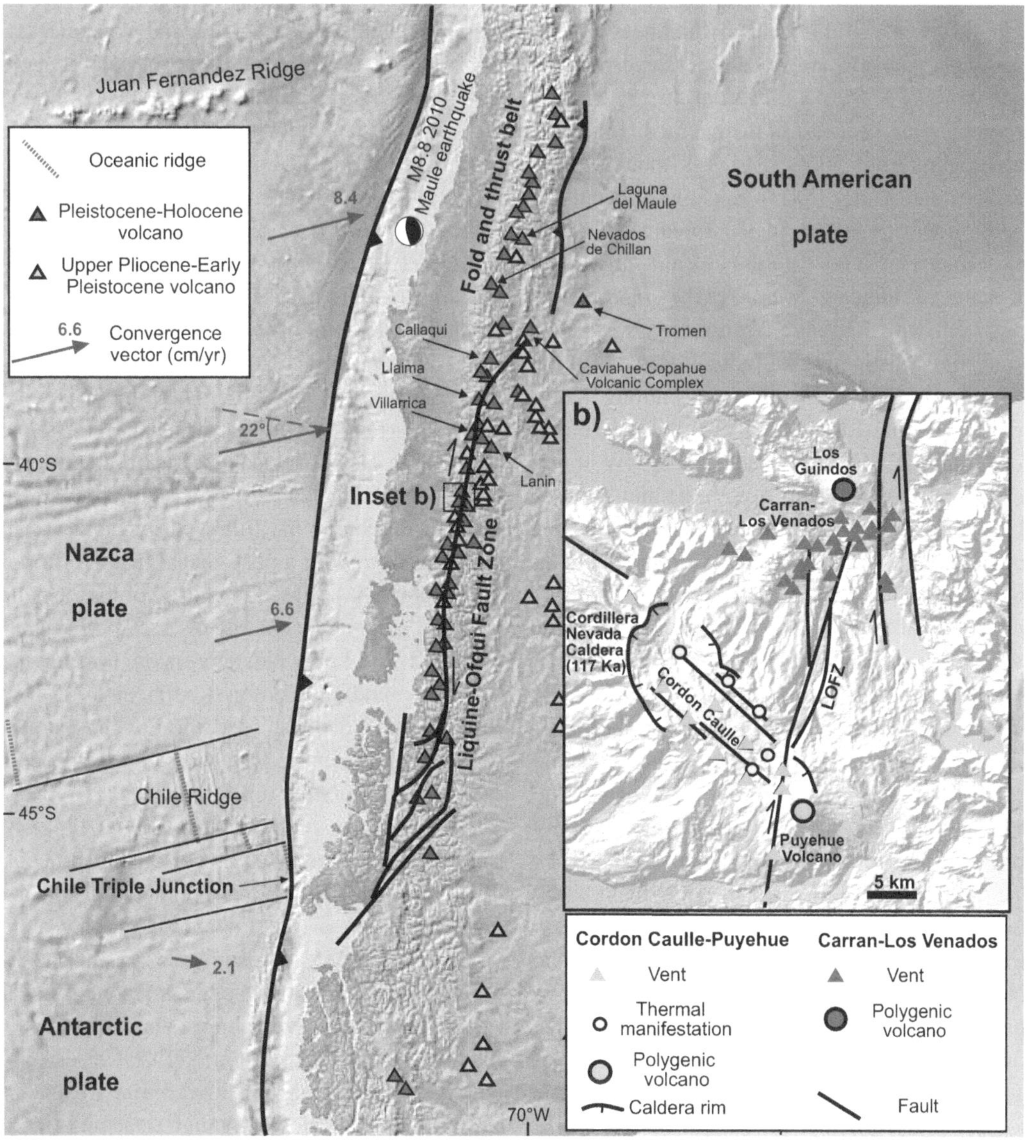

Fig. 12.9 Tectonic setting of the South Andean Volcanic Zone; location and focal mechanism of the 2010 Maule mega-earthquake are also reported. Inset b shows the structure and volcanic features of the Cordón Caulle-Puyehue and Carran-Los Venados volcanic zones (Sepulveda et al. 2005; Melnick et al. 2006; Bucchi et al. 2015). Base DEM provided by GeoMapApp

in addition to its poorly detected Quaternary dextral movement, displays a significant contractional displacement, resulting in overall transpression. The current activity of the LOFZ derives from a Quaternary NE-SW oriented maximum principal stress σ_1, with σ_2 and σ_3 mutually alternating along-strike because of local or regional factors (inherited discontinuities, variations in the convergence angle, topography) and locally producing transpressional or even contractional tectonics and seismicity (Lavenu and Cembrano 1999; Rosenau et al. 2006; Cembrano and Lara 2009; Lara et al. 2010; Legrand et al. 2011). Crustal thickness underneath the volcanic arc decreases steadily from ~50 km at 33 °S to ~35 km at 46 °S, with an accompanying decrease in the average altitude of the orogen, from 5000 m to less than 2000 m (Tassara and Yánez 2003). Seismicity occurs at depths down to 40 km in the forearc and shallower than 12 km beneath the volcanic arc. Focal mechanisms indicate overall strike-slip faulting consistent with ENE-WSW shortening. More in detail, three distinctive latitudinal domains show seismicity consistent with splay faulting along branches of the LOFZ (at its northern termination), along ENE-WSW and ESE-WNW transverse faults (in the central portion) and focused along the master branch of the LOFZ (at its southern part). This indicates a complex strain compartmentalization pattern within the arc, where variable strike-slip faulting dominates over dip-slip motions (Sielfeld et al. 2019).

Magmatic activity focuses along the LOFZ and its intersection with inherited oblique structures, developing the ~N-S trending volcanic arc. The average magma extrusion is here close to 10–13 km^3/km/Ma, although the productivity appears discontinuous (e.g., Volker et al. 2011). The South Andean Volcanic Zone includes at least 60 historically and potentially active main stratovolcanoes, as well as three large silicic caldera systems and numerous minor eruptive centres. The erupted magmas include basalts, basaltic andesites, andesites, dacites and rhyolites. The more evolved rocks have the same isotopic composition of the more primitive rocks, indicating that they formed either by crystal-liquid fractionation without assimilation, or assimilated young, isotopically similar crust, such as Miocene plutonic rocks (Stern 2004, and references therein). At the arc-scale, regional tectonics would control whether basaltic magmas reach the surface or evolve to more differentiated products. In particular, NE-SW striking volcanic alignments contain mainly basaltic to basaltic andesite compositions. Conversely, NW–SE striking alignments contain a wide range of compositions, including rhyolites (Cembrano and Lara 2009).

The central part of the LOFZ, at ~41 °S, displays interesting structural and volcanic features that may be representative of the more general tectono-magmatic relationships along the arc. This area includes the Carrán–Los Venados volcanic field and the neighbour polygenetic Puyehue–Cordón Caulle volcanic complex (Fig. 12.9; Lara et al. 2006; Singer et al. 2008; Bucchi et al. 2015). The Carrán–Los Venados volcanic field is a basaltic to basaltic andesitic volcanic field composed of 65 post-glacial scoria cones and maars and a stratovolcano. The NE-SW elongation of its feeding system, subparallel to the current maximum principal stress σ_1, suggests a zone of local extension. Petrologic data indicate that here magmas differentiate in low crustal reservoirs, followed by rapid ascent to the surface, with a post-glacial magma flux of ~3.1 km^3/ka. The Puyehue–Cordón Caulle volcanic complex consists of a NW–SE elongated 13 × 6 km wide depression hosting the NW–SE trending Cordón–Caulle area of rhyolitic fissure volcanism and the Puyehue stratovolcano to the east. The Cordón Caulle area experienced notable uplift between 2012 and 2015, partly overlapping with the 2011–2012 rhyolitic eruption (Jay et al. 2014; Delgado et al. 2016, 2019). The NW–SE elongation of the feeding system of the Puyehue–Cordón Caulle volcanic complex, perpendicular to the maximum principal stress σ_1, suggests a zone of local contraction. Here the magmas stagnate and differentiate in lower and upper crustal reservoirs, with an average magma flux similar to that of Carrán–Los Venados, although reaching more than twice its value (~9 km^3/ka) during peak eruptive periods. Therefore, at Carrán–Los Venados monogenetic volcanism results from an extensional/transtensional regime that favours rapid magma rise without storage and differentiation in upper crustal reservoirs. Conversely, at Puyehue–Cordón Caulle, contractional deformation and transient high magma flux developed stable upper crustal silicic magma reservoirs feeding polygenic evolved volcanism (Fig. 12.10; Bucchi et al. 2015).

The tectono-magmatic features described for Carrán–Los Venados and Puyehue–Cordón Caulle may be applicable also to other volcanoes along the LOFZ (Folguera et al. 2004). For example, at the northern termination of the Liquine-Ofqui Fault Zone lies the Caviahue-Copahue Volcanic Complex, active within a 20 × 15 km wide depression in the last ~4 Ma. Here magmatic centres and hydrothermal paths have a preferred ENE-WSE alignment, parallel to the direction of the maximum principal stress σ_1. A similar structural setting is found also in the nearby ENE-WSE elongated Callaqui stratovolcano, to the west, and the NE-SW elongated Llaima stratovolcano, to the south (Melnick et al. 2006; De Maisonneuve et al. 2012; Sielfeld et al. 2017). Conversely, other nearby stratovolcanoes are aligned along a NW–SE direction, as for example Villarrica, Quetrupillan and Lanin (Fig. 12.11). All these features suggest recurrent convergence-parallel and convergence-orthogonal structural controls on the volcanoes along the South Andean Volcanic Zone, although more data are needed to generalize any behaviour.

The different tectono-magmatic conditions shown by Carrán–Los Venados and by Puyehue–Cordón Caulle may be explained by the transient variations imposed by seismic cycles associated with mega-earthquakes along the convergent plate boundary. In particular, in the inter-seismic period, the ENE-WSW trending maximum principal stress σ_1 deriving from the convergence is subparallel to the plumbing systems of the NE-SW trending volcanic alignments; this promotes local extension and rapid magma ascent through NE-SW trending feeder dikes. The maximum principal stress σ_1 is also perpendicular to the plumbing system of the NW–SE trending volcanic complexes, promoting local contraction and thus magma stagnation in sills. In the co- and post-seismic periods, that is during and immediately after mega-earthquakes, the situation may

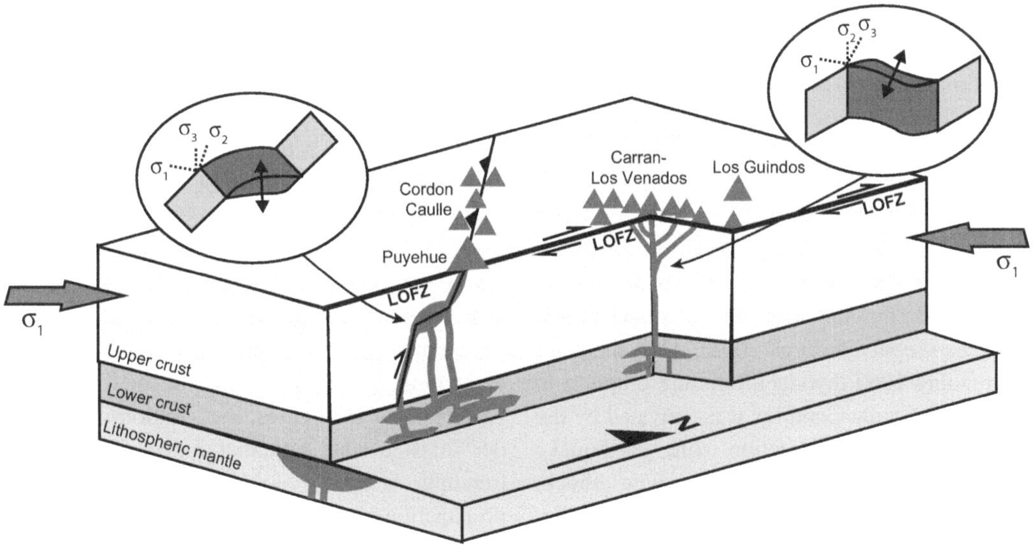

Fig. 12.10 Schematic diagram summarizing the tectono-magmatic features of the Cordón Caulle-Puyehue and Carran-Los Venados volcanic complexes in the South Andean Volcanic Zone (location in Fig. 12.9). The former lies along a NW–SE trending thrust zone, in whose flat parts magma accumulates at depth. The latter lies along a NE-SW extensional zone along the LOFZ, which promotes the rise of magma to the surface (Bucchi et al. 2015)

Fig. 12.11 View of a portion of the South Andean Volcanic Zone showing the Quetrupillan and Lanin (background) volcanoes seen from the top of Villarrica volcano. These three stratovolcanoes form a second-order NW–SE trending alignment within the N–S trending volcanic arc

be reversed, with the upper plate moving trenchward and inducing an overall arc-orthogonal (E-W to NE-SW trending) extension (Walter and Amelung 2007). As a result of this transient extension, a general increase in the eruption rate along the South Andean Volcanic Zone has been recognized in the last centuries soon after major thrust earthquakes along the plate boundary. Eruption locations imply that these effects may manifest up to several hundreds of kilometres beyond the limits of the earthquake rupture zone, with both dynamic and static stress changes affecting eruption-triggering processes over timescales of several months (Watt et al. 2009). This increase in the eruptive rate is explained by the co- and post-seismic transient extension enhancing magma ascent through NW–SE trending dikes feeding NW–SE elongated or aligned volcanoes. The latter condition may justify the Quaternary extension locally experienced by the Puyehue–Cordón Caulle Complex (Sepulveda et al. 2005). Also, the NW-SE trending 1960 rhyodacitic fissure eruption in the Cordón Caulle Complex was activated by the greatest recorded subduction zone earthquake (*M9.5*), starting 38 h after the main shock, 240 km inland (Lara et al. 2004). More recently, the 2010 *M8.8* Maule mega-earthquake induced subsidence of up to 15 cm in five volcanic areas in the South Andean Volcanic Zone within weeks of the earthquake, without detectable thermal changes. This deformation may be related to the co-seismic release of fluids from hydrothermal systems documented at three of the five subsiding volcanic areas. Among these is the Nevados de Chillan Volcanic Complex, which erupted in 2015 along a NW–SE trending fissure, perhaps reactivating a pre-existing basement structure unclamped by the Maule earthquake (Pritchard et al. 2013; Bonali et al. 2015; Lupi et al. 2020).

The northern termination of the transpressive Liquine-Ofqui Fault Zone is connected through a transfer zone to a major N-S trending back-arc thrust system. This contractional back-arc zone is characterized, at a latitude of $\sim 37°$, by several polygenic volcanoes. Among these is Tromen volcano, with the nearby intrusive complex of Cerro Negro de Tricao Malal. While the former is E-W elongated, parallel to the local direction of compression, the latter has an overall N-S elongation, parallel to the axis of the folds in the basement. In both cases, the location of the feeder dikes at the anticlinal hinges of the N-S trending basement folds suggests that dike propagation was controlled by local and shallow stresses related to the stretching of the outer portion of the folds; this mechanism may be responsible for the vertical partitioning of igneous plumbing systems in contractional settings (Galland et al. 2007; Gurer et al. 2016).

Finally, recent studies also consider the possible effect of the climate change-driven deglaciation after the last glacial maximum along the South Andean Volcanic Zone (Mora and Tassara 2019). The resulting changes in pressure at upper crustal levels (<10 km depth) at the scale of several hundred years are of the order of 10–100 MPa. This large decompression may easily surpass the tensile strength of rocks (5–20 MPa), promoting the failure of the reservoir walls, dike propagation and the collapse of the reservoirs accompanying explosive eruptions.

12.3.2 Sumatra, Indonesia

The island of Sumatra (Western Indonesia) exposes part of the highly oblique ($\phi > 45°$) convergence resulting from the Indian plate subducting northward below the Eurasian plate along the NW–SE trending trench. The convergence velocity along the boundary increases towards southeast, from ~5 to ~6 cm/year (Fig. 12.12). This configuration results from the Cenozoic collision of India with Asia, which induced the progressive clockwise rotation of the western portion of the former Sunda subduction zone, whose non-rotational portion remains currently preserved in the island of Java (central

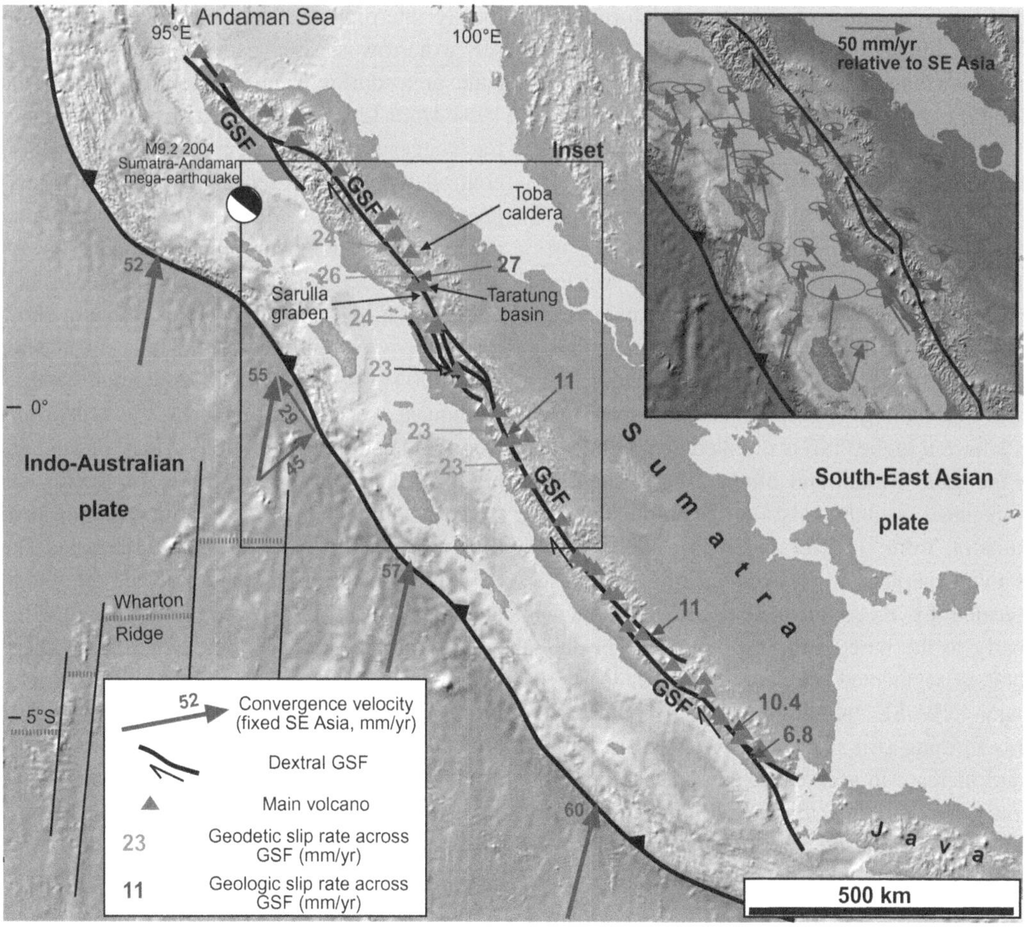

Fig. 12.12 Tectonic setting of Sumatra, Indonesia, with the geologic and geodetic slip rates along the Great Sumatra Fault (GSF) from Natawidja and Triyoso (2007); location and focal mechanism of the 2004 Sumatra–Andaman mega-earthquake are also reported. Inset shows the inter-seismic motion from GPS data between 1989 and 1996 in the central-northern part of the island (McCaffrey et al. 2000). Base DEM provided by GeoMapApp

Indonesia). The rotation increased both the obliquity of convergence and the rate of strike-slip motion along the plate boundary in western Indonesia. The Indo-Australian plate currently subducts beneath Sumatra with a complex geometry. This is highlighted by the steepening of the slab from the north (with a slab dip <60°), where it lies on the 660 km deep upper mantle discontinuity, to the south (dip >70°), where the slab pierces through the upper mantle discontinuity. Also, the slab shows a marked bend in map view in northern Sumatra, approximately below Toba caldera (Pesicek et al. 2008).

While contraction dominates in the offshore portion of the convergent plate boundary, the NW–SE trending dextral Great Sumatra Fault (GSF) focuses most of the inland deformation. The dip-slip motion towards the trench side of the overriding plate and the strike-slip motion along the arc highlight an almost full partitioning of the strain (Fig. 12.12; Bellier et al. 1999; McCaffrey et al. 2000; Bradley et al. 2017). The GSF separates the eastern basin, a region that is part of the Sunda Shelf and undergoes little internal deformation, from the western coastal areas of the forearc sliver. Analysis of well borehole breakouts, focal mechanisms of earthquakes and geologic stress indicators indicates that the regional maximum principal stress σ_1 adjacent to the GSF is oriented at a high angle to the fault (Mount and Suppe 1992). The GSF propagated southwards, in central-southern Sumatra, from ~2 Ma and it is usually active as a vertical plane down to depths of ~15 km, as revealed by the associated seismicity arriving nearly to the base of the brittle crust. The fault zone is segmented and consists of at least 19 major NW–SE trending subparallel branches, also forming strike-slip duplexes, with a northward increase in segment length and in slip rate, from 0.5 to 2.6 cm/year. Overall, most (>80%) of the GSF consists of a narrow (a very few km wide) zone of deformation along a main dextral segment. Limited parts (<20%) of the GSF may show offset segments, at times overlapping. Geomorphic offsets along the fault range as high as 20 km and may represent only the most recent displacement, with other unidentified structures having accommodated the dextral component of oblique convergence in the past. Interaction between dextral segments offset in an en-echelon dextral configuration creates repeated areas of localized extension along the GSF in a context of overall NNE-SSW trending maximum compression (Duquesnoy et al. 1996; Genrich et al. 2000; Prawirodirdjo et al. 2000; Sieh and Natawidjaja 2000; Natawidjaja and Triyoso 2007; Weller et al. 2012; Ito et al. 2016). As mentioned, the activity of the GSF is accompanied by seismicity, locally interrupted in volcanic areas. For example, a narrow seismic gap highlighted the volcanic areas during the aftershock sequence of the 1994 *M6.8* Liwa earthquake, in southern Sumatra, suggesting that the rupture process is also controlled by the presence of magma (Widiwijayanti et al. 1996). More in general, seismicity predominates along the plate boundary, with important effects on the overriding plate. In fact, GPS data in the northern part of Sumatra show that the co- and post-seismic displacement after the 2004 *M9.2* Sumatra–Andaman mega-earthquake, occurred offshore, was characterized by the southwest motion of the volcanic arc region, consisting of several tens of centimetres of co-seismic displacement, plus tens of centimetres in the first year after the earthquake. The 2004 mega-earthquake was followed, one month later, by a seismic swarm and a submarine eruption in the Andaman Sea, along the northern continuation of the Sumatra arc. This seismicity is consistent with the regional tectonic setting and the involvement of fluids, also magmatic, producing arc-parallel normal faults (Subarya et al. 2006; Gahalaut et al. 2008; Shearer and Burgmann 2010; Kamesh Raju et al. 2012; Kundu et al. 2012).

Arc volcanism has been accompanying convergence at least from the Mesozoic. While the late Miocene volcanism in northern Sumatra is scattered, in the southern part it generally overlaps with the Quaternary volcanism, although the latter may be slightly more shifted to the west. Part of the Quaternary volcanism focuses along the GSF (Fig. 12.12). Quaternary volcanics are mainly calcalkaline andesites, dacites and rhyolites, with genetically homogeneous overall composition; relatively primitive rocks, as basalts, are rare, with even the most primitive lavas suffering crustal contamination (Gasparon 2005). The volcanic arc consists of approximately 50 major volcanoes. Most of these are stratovolcanoes or composite volcanic edifices between 600 and 3800 m high, with monogenic vents on their flanks. At times, calderas are present: Toba caldera is the largest and most representative volcanic complex, located above a tear in the bent subducting plate. Toba is the site of Earth's largest Quaternary eruption, ejecting ~3000 km^3 of magma at ~75 ka (Fig. 5.5; Chesner 2012; de Silva et al. 2015; Koulakov et al. 2016). The caldera has a pronounced NW–SE elongation, parallel to the nearby GSF segment, and this configuration has been interpreted as due to a presently inactive stepover between dextral segments. However, the structural relationships between the GSF segment and the caldera structure are not obvious. In fact, the GSF segment lies ~10 km to the southwest of the western caldera rim, and geodetic modelling shows that slip of the GSF segment currently focuses several kilometres southwest of the geologic fault plane. The magma reservoir below the caldera consists of stacked sills down to a depth of 7 km, lying on top of a low velocity anomaly almost extending down to the slab (Detourbet et al. 1993; Genrich et al. 2000; Stankiewicz et al. 2010; Chesner 2012; Jaxybulatov et al 2014). Other volcanic areas, in addition to stratovolcanoes and calderas, host domes and geothermal activity. In the geothermal areas of the Sarulla graben and the Tarutung Basin the fluid pathways are related to pull-apart structures, negative flowers and subvertical splays of the GSF. In the latter case, the geothermal reservoirs are centred along the fault zone, where the highly fractured and hydrothermally altered rocks serve as main conduits for vertical fluid flow from deeper magmatic sources (e.g., Moore et al. 2001).

Despite the apparent spatial coincidence between most volcanoes and the Great Sumatra Fault (Fig. 12.12), closer inspection suggests that the location and the eruption rate of many volcanoes are not really controlled by the activity of the dextral GSF (Fig. 12.13). Rather, the volcanoes mainly lie above a consistent depth of the slab below, at 130 ± 20 km. Indeed, the slab-derived fluid-rich portion of the asthenospheric wedge seems to control the location of both the volcanic arc, through partial melting, and of the GSF, through magma-induced thermal weakening able to localize faulting in the upper crust. The latter feature is supported by the different timing in the development of the GSF along Sumatra. In fact, the more recent central-southern portion of the GSF formed in the last 2 Ma, postdating the older Miocene volcanism. The fact that the latter is found in the same location as the Quaternary volcanism indicates that the Miocene arc was not controlled by the activity of the GSF and suggests that a similar condition applies also to the Quaternary arc. Overall, this implies that the magmatic arc controls the location of plate deformation, although once a major fault zone is established its activity may further promote the rise and shallow emplacement of magma (McCaffrey et al. 2000; Sieh and Natawidjaja 2000; Acocella et al. 2018).

Even though not particularly related to the activity of the GSF, the current volcanic arc also experiences a convergence-orthogonal direction of extension, which is kinematically compatible with the dextral systems of the GSF and explains the common ~NNE-SSW elongation and/or alignment of many volcanoes (Fig. 12.14). However, there is also evidence of subordinate arc-normal extension, as suggested by structural data and volcano elongations, which may be related to a transient co- and post-seismic stress field inversion induced by mega-earthquakes, as observed for the 2004 event. Also, volcanic activity is largely confined within the volcanic

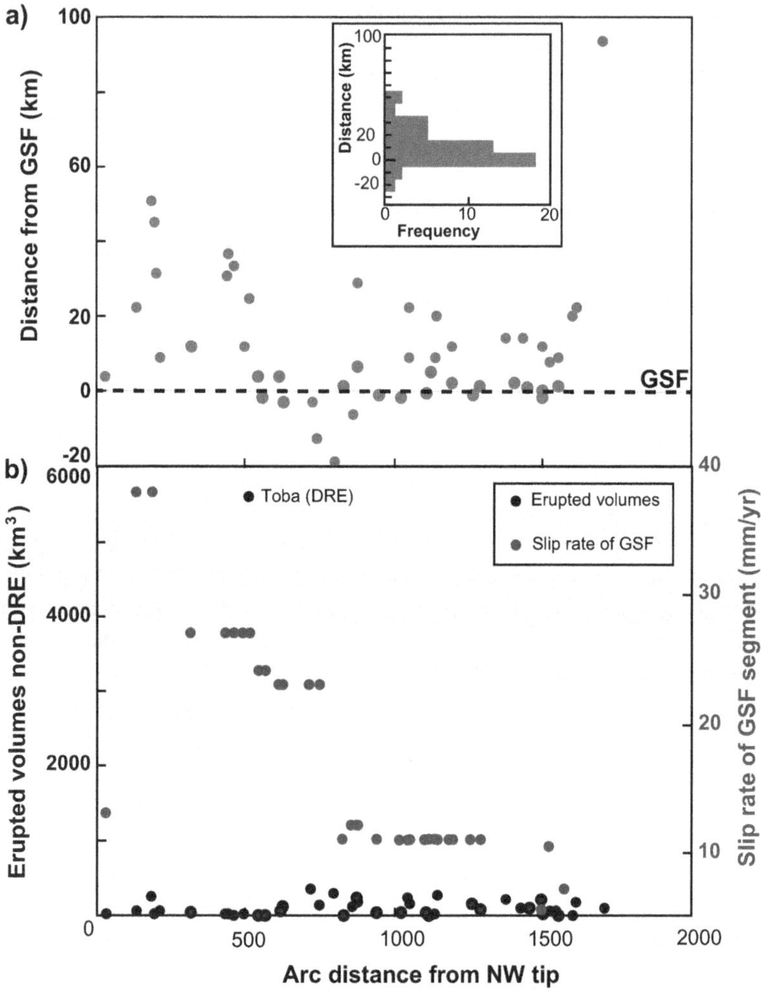

Fig. 12.13 **a** Distribution of volcanoes along Sumatra (from northeast, to the left, to southwest, to the right) as a function of the distance from the Great Sumatra Fault (GSF); **b** Non-DRE (Dense Rock Equivalent, except for Toba) erupted volumes and GSF slip rates along the Sumatra arc, from northeast, to the left, to southwest, to the right (modified after Acocella et al. 2018)

edifices, with limited evidence of monogenic volcanism between the main polygenic volcanoes. This suggests the lack of elongated magmatic systems, implying a central, and not linear (as for example in the Taupo Volcanic Zone), mode of rise of the magma in the crust.

Overall, there is a limited control of the strike-slip structures on volcanism at Sumatra. This supports only in part previous evidence from the deeper structure of extinct and eroded magmatic arcs, where pluton emplacement has been widely associated with local extension created by strike-slip systems (see Sect. 4.5.2). This partial agreement may result from several possibilities: among these, a variable role of the arc structure at depth and at the surface, with a stronger control on magma emplacement at depth and a weaker control on the rise of the magma towards the surface. A similar behaviour has been postulated for the central Aeolian Volcanic Arc of southern Italy. However, conversely to Sumatra, along the central Aeolian Arc there is evidence of magmatism continuing, through eruptive fissures, outside the main volcanoes, suggesting the activity of magmatic systems; these may be induced by a variation in the shallower tectonic regime, which becomes extensional (Ruch et al. 2016; Acocella et al. 2018).

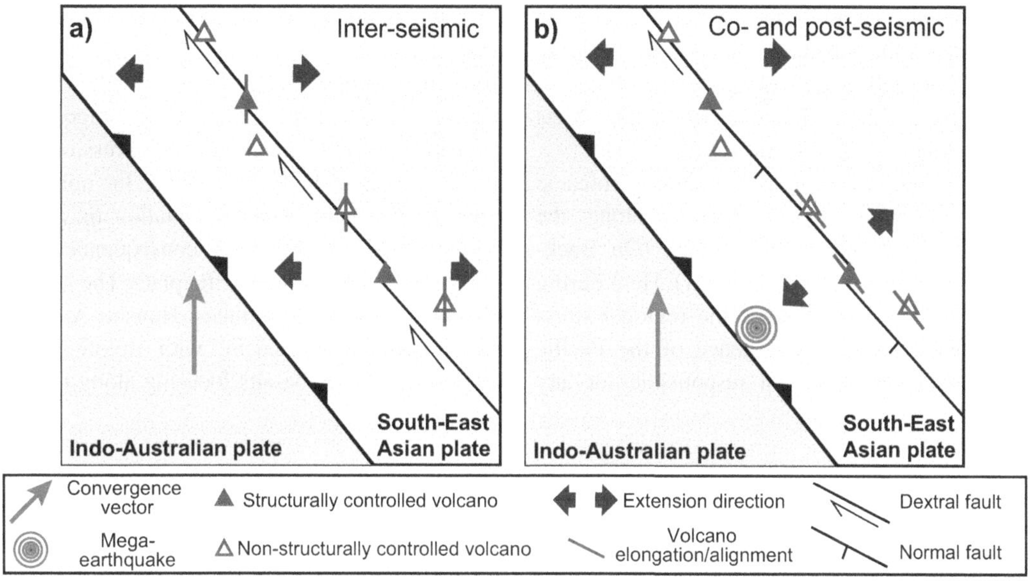

Fig. 12.14 Tectono-magmatic relationships along the Sumatra obliquely convergent volcanic arc. **a** During the inter-seismic period the arc is characterized by the activity of dextral structures associated with strain partitioning. Part of these structures may control the location of the volcanoes ("structurally controlled volcano"). The location of other volcanoes ("non-structurally controlled volcano") away from the strike-slip zone may be explained by other factors, as slab depth or convergence-orthogonal extension, controlling also volcano alignment and elongation. **b** During transient conditions (as in co- and post-seismic periods due to the activity of mega-earthquakes) the arc may undergo arc-normal extension, with the local reactivation of the strike-slip faults as normal faults; the alignment and elongation of volcanoes may partly reflect these transient conditions, becoming subparallel to the arc (modified after Acocella et al. 2018)

12.4 Contractional Arcs

Contractional volcanic arcs are characterized by dominant shortening, resulting from overall arc-orthogonal compression, a condition imparted by orthogonal convergence. As most plate boundaries do not experience pure orthogonal convergence, a frequent and non-negligible strike-slip component is often associated with contractional arcs. Indeed, only very few volcanic arcs experience evident and dominant contraction, and negligible strike-slip motion: among these are the eastern Aleutian Arc, the northeast Japan Arc and, for much of its late Cenozoic-Quaternary evolution, the more complex arc of the Central Andes (Table 12.1; Acocella and Funiciello 2010). Below the focus is on the most-studied and representative volcanic arcs of northeast Honshu (Japan; Sect. 12.4.1) and the Central Andes (Sect. 12.4.2).

12.4.1 Northeast Honshu, Japan

Honshu is the largest island in Japan. The northeast Honshu volcanic arc lies to the west of the N-S trending Japan trench, where the westward Pacific plate subducts with a N115° motion of ~8 cm/year. To the north, the Japan trench is connected to the NE-SW trending Kuril trench. Here the oblique convergence of the Pacific plate induces the arc-parallel southwest motion of the Kuril forearc sliver, colliding in east Hokkaido. To the south, the Japan trench is connected to the N-S trending Izu-Bonin trench, where the Pacific plate subducts below the Philippine Sea plate. To the west lies the Japan Sea, a Miocene back-arc basin associated with widespread basaltic volcanism (Fig. 12.15; Kimura 1986; De Mets 1992; Kimura 1996; Yoshida et al. 2013).

The ~400 km long, N-S trending volcanic arc of northeast Honshu focuses along the uplifted central mountainous range (Ou Backbone Range; Hasegawa et al. 1991). Here during Middle Miocene to Pliocene the regional stress field related to the convergence of the Pacific plate was neutral and not responsible for any distinct structural pattern. However, the southwest indentation of the Kuril sliver, resulting from the oblique convergence with dextral component along the Kuril Arc, imposed a southwest oriented maximum compression and major arc-parallel dextral faults in northeast Honshu. The Kuril sliver indentation may have begun at 10–12 Ma, as a consequence of a change in motion of the Pacific plate. The Middle Miocene to Pliocene northeast Honshu Arc produced bimodal volcanism, with rhyolites and subordinate low K basalts focusing along the Ou

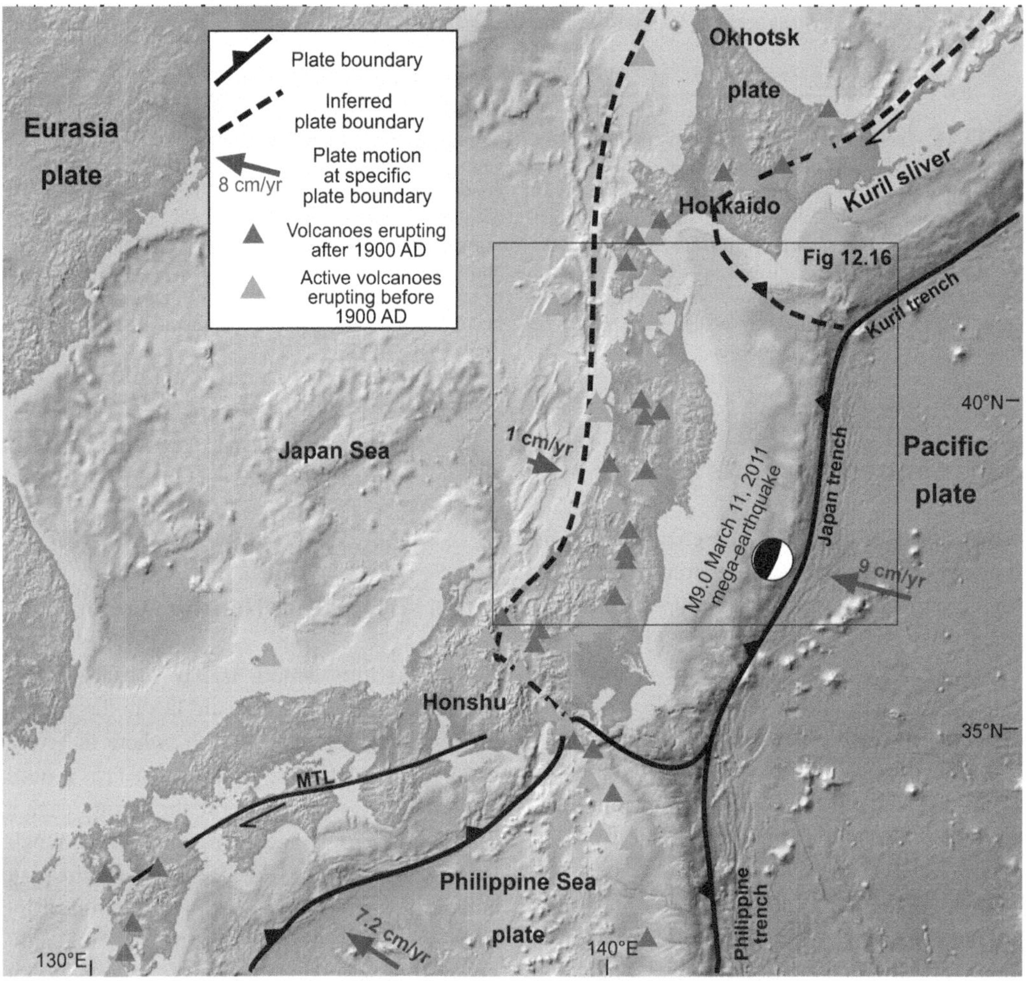

Fig. 12.15 Tectonic setting of northeast Honshu (Japan), with the major active volcanoes. Solid lines: subduction front; dashed lines: inferred boundaries of minor plates. MTL = Median Tectonic Line in southwest Japan. Plate motions from Gripp and Gordon (2002). Location and focal mechanism of the March 11, 2011 Tohoku mega-earthquake are also reported. Base DEM provided by GeoMapApp

Backbone Range, where nearly eighty calderas were fed by felsic plutons following crustal heating and re-melting. The volcanic productivity was approximately 210 km^3 per Ma per 200 km of length of the arc (Fig. 12.16; Sato 1994; Kimura 1996; Yoshida 2001; Acocella et al. 2008; Yoshida et al. 2013). In this period, magma rose and erupted mainly along NE-SW trending areas parallel to the direction of compression imposed by the Kuril sliver. These areas where probably associated in location and orientation with localized extension created by the active arc-parallel dextral faults. These extensional areas also had a different orientation, and were thus uncoupled, from the ~E-W trending paths which have been focusing the rise of melts from the slab in the mantle below the arc in the last 13 Ma (Tamura et al. 2002).

A change in the stress field around the Pliocene–Quaternary boundary resulted in an ~E-W oriented direction of maximum compression, possibly related to the increase in the motion of the Pacific plate (Pollitz 1986). This developed two N-S trending uplifted zones, bounded by reverse faults, in northeast Honshu: the Ou Backbone Range (to the east) and the Dewa Hills (to the west). N-S trending reverse faults become active, also bordering the thermally anomalous part of the crust coinciding with the magmatic arc. The present ~E-W oriented arc-perpendicular compression at crustal levels is supported by various geophysical and geodetic data. This compression is evident also below the arc, were the direction of mantle shear-wave polarization anisotropy is ~E-W oriented, consistent with finger-like mantle diapirs. During Quaternary, the composition of volcanism largely consisted of calcalkaline andesite, erupted from approximately 60 stratovolcanoes, with more evolved compositions being restricted to approximately ten calderas. These stratovolcanoes are also ~E-W aligned and elongated, forming a distinct subpattern with regard to the ~N-S trending one observed at the arc-scale (Fig. 12.16; Ito et al. 2000; Igarashi et al. 2001; Tamura et al. 2002; Nakajima and Hasegawa 2004; Kato et al. 2006; Loveless and Meade 2010). A few of these volcanoes, as Zao, show in the last 2 ka much lower intrusive to extrusive ratios than those estimated for other volcanoes of northeast Honshu, probably following a different thermal and rheological state of the crust (Zellmer et al. 2019).

The Quaternary deformation pattern associated with the volcanoes mostly consists of arc-parallel thrust faults and subordinate arc-perpendicular normal faults, extension fractures and eruptive fissures. These arc-perpendicular extensional structures may result from: a) the shallow propagation and emplacement of E-W trending dikes, parallel to the regional maximum principal stress σ_1; (b) the accumulation of stacked sills, locally increasing the vertical component of the principal stress, which passes from σ_3 to σ_2 and σ_1, and with the arc-parallel oriented σ_2 becoming σ_3 (Fig. 10.24b). Condition b promotes the accumulation of large volumes of magma and is consistent with crustal thickening and the widespread magma-driven uplift along the magmatic arc (Acocella et al. 2008; Yoshida et al. 2013; George et al. 2016). The Quaternary E-W trending structures and the E-W elongation and alignment of volcanoes may be thus considered as the shallowest expression of the ~E–W trending hot mantle fingers imaged at depth, suggesting a Quaternary coupling in the rise of magma from the upper mantle to the upper crust. This possibility is supported by the spatial coincidence between the distribution of the volcanic areas along the Ou Backbone Range and the Dewa Hills and the local negative Bouguer gravity anomalies along the Japan Sea side of the volcanic arc and the low velocity regions in the mantle wedge. The connection between these areas at different depth is confirmed by distinct low velocity regions indicative of partial melting that are continuously distributed from the mantle wedge to the middle crust just below the Ou Backbone Range (Fig. 12.16; Tamura et al. 2002; Acocella et al. 2008; Okada et al. 2010; Yoshida et al. 2013). This coupling leads to a two-dimensional mantle wedge flow, ensuring higher temperatures in the shallow part of the wedge and higher productivity of mafic magma.

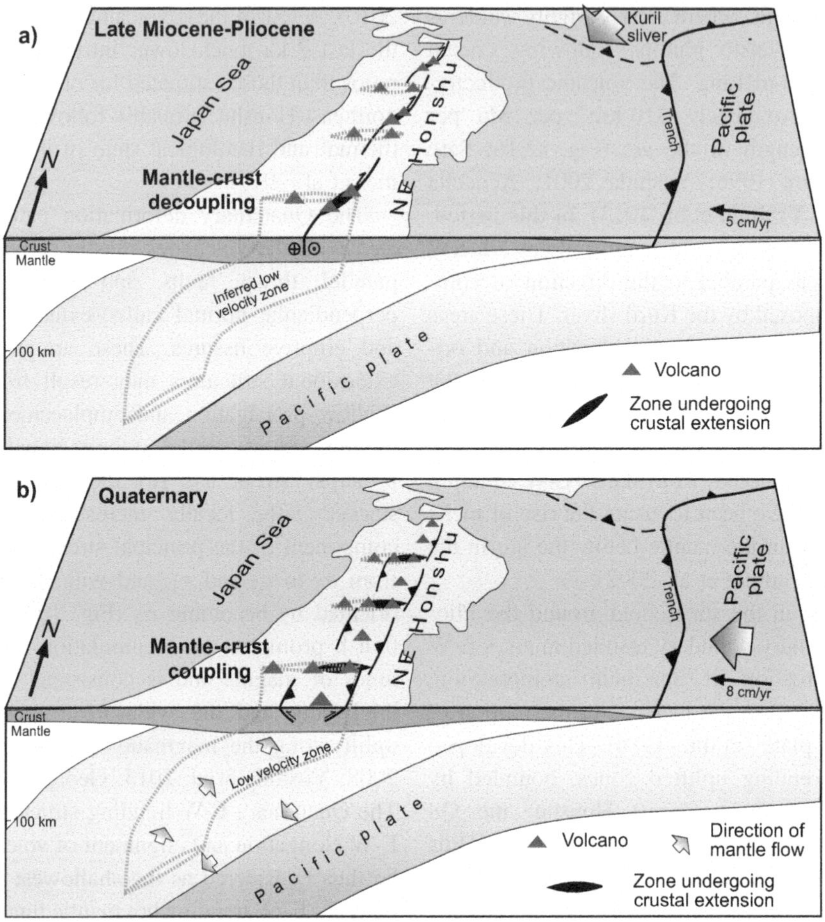

Fig. 12.16 Summary of the tectono-magmatic features of late Miocene-Pliocene (**a**) and Quaternary (**b**) in northeast Japan (modified after Acocella et al. 2008). **a** During late Miocene-Pliocene, the N-S- and NE-SW trending paths for the rise of magma in the crust induced by the indentation of the Kuril sliver (black wedges) are uncoupled with the ~E–W trend of the melts rising in the mantle (dashed orange ellipses); this generates a bimodal volcanism, with lower erupted volumes per Ma. **b** During Quaternary the ~E–W trending paths for the rise of magma in the crust under orthogonal convergence are coupled with those of ~E–W trending hot mantle fingers. The connection between deeper mafic and shallower felsic magmas generates higher volumes of calcalkaline andesites

At crustal levels, the mixing between the deeper mafic and the shallower felsic magma generates the andesites. The Quaternary andesites are associated with larger erupted volumes (approximately 310 km^3 per Ma per 200 km of length of the arc) than the volumes of rhyolites erupted during Miocene-Pliocene. This feature, ultimately attributed to major changes in plate interactions, indicates that pure convergence does not hinder volcanic activity. Rather, under conditions of mantle-crust coupling, magmatic and volcanic activity may be even enhanced (Acocella et al. 2008; Yoshida et al. 2013; Wada et al. 2015; Mahony et al. 2016).

The 2011 *M9.0* Tohoku mega-earthquake caused impressive crustal deformation in northeast Honshu, with significant co-seismic, as well as post-seismic, arc-perpendicular extension across the volcanic arc, reversing the inter-seismic arc-perpendicular contraction (Fig. 12.17). This motion is the transient expression of the trenchward movement of the

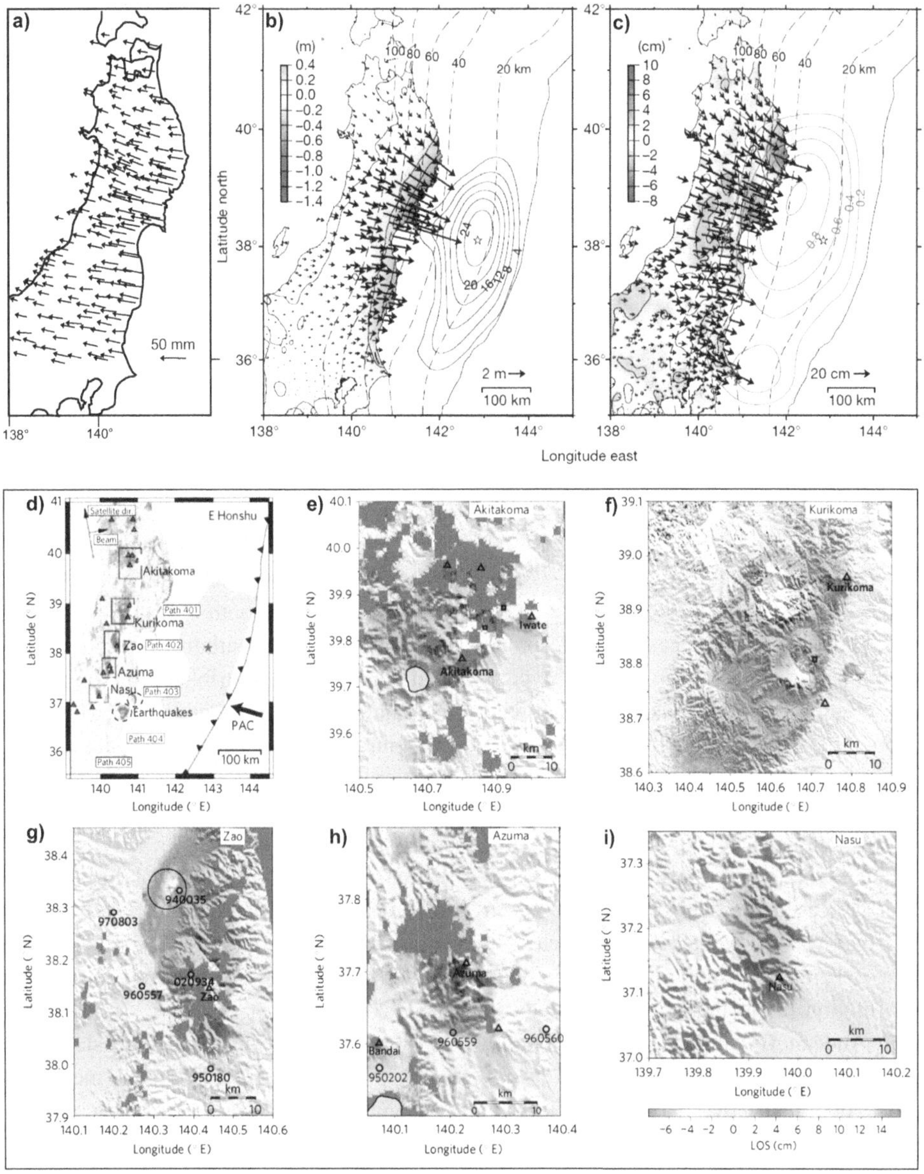

Fig. 12.17 Inter-seismic (**a**, 1997–2001 period), co-seismic (**b**, 10–11 March 2011) and post-seismic (**c**, 12–25 March 2011) GPS motions along the northeast Honshu Arc relative to the March 11, 2011 *M9.0* Tohoku mega-earthquake (Miura et al. 2004; Ozawa et al. 2011). **d** Overview of the subsidence occurred in several volcanic areas (**e–l**) in northeast Honshu in the February-April 2011 period, detected from InSAR data (Takada and Fukushima 2013)

overriding plate during and after the mega-earthquake and has been also confirmed by structural field data, which highlight the coexistence of arc-parallel normal faults and thrust faults along the arc during Quaternary, with the normal faults likely resulting from older seismic events. Despite the sharp co-seismic kinematic reversal in northeast Honshu, a minor inter-seismic component of accelerated trenchward motion was also detected in the decades before the 2011 Tohoku earthquake and related to the increased slip rate on the Japan trench plate interface (Miura et al. 2004; Acocella et al. 2008; Ozawa et al. 2011; Mavrommatis et al. 2014). In the portion of the arc perturbed by the mega-earthquake several volcanoes subsided by 5–15 cm, forming elliptical depressions up to 15–20 km wide and N-S elongated, perpendicularly to the maximum co-seismic extension. The magmatic bodies beneath may have deformed and subsided in response to crustal stress changes or, alternatively, the enhanced crustal permeability may have promoted the escape of fluids from the overlying hydrothermal systems, similarly to what inferred after the 2010 Maule mega-earthquake in southern Chile (Takada and Fukushima 2013; Pritchard et al. 2013).

12.4.2 The Central Andes

The Central Andes provide a case of widespread and exceptional magmatism in an arc experiencing a complex evolution under dominant contraction. The Andes result from the subduction, since late Proterozoic, of the Nazca plate below the South American plate. Magmatism has been accompanying the evolution of the Andes with variable productivity, testified by the evidence that increased regional contraction (as in the Mesozoic) was associated with increase in volcanism, whereas oblique convergence with strike-slip motion (as in the Miocene) enhanced plutonism. Andean volcanism currently focuses above moderately to steeply dipping slabs (dip >25°), delimited by flat slabs portions underlying the non-volcanic parts of the orogen (Jordan et al. 1983; Grocott et al. 1994; McNulty et al. 1998).

The Central Andes magmatism includes the diffuse and voluminous late Cenozoic to Quaternary volcanism that reaches a width of ~300 km in the most thickened portion of the orogen. This volcanism is often defined as **flare-up**, referring to a phase in continental arc evolution characterized by episodic higher flux magmatism producing a large province of calderas and ignimbrites dominated by dacite to rhyolite compositions. Ignimbrite flare-up, observed also in various parts of the western United States and Mexico, requires an elevated heat supply from the mantle to generate a transcrustal magma system fuelling some of the largest silicic eruptions on Earth (Best et al. 2016; de Silva and Kay 2018, and references therein). In the Central Andes, the switch from steady state volcanism (dominantly andesite–dacite composite cones) to flare-up volcanism (dominantly large-scale ignimbrites and caldera complexes) is related to the southward migration and subduction of the oceanic Juan Fernandez Ridge, on the Nazca plate, in the last 25 Ma. The subducting warm and buoyant ridge determined the flattening of the slab below the Central Andes and the temporary lack of volcanism at the surface (Fig. 12.18). This condition coincided with an early Miocene acceleration in the rate of westward drift of South America over the Nazca plate. Both the flat subduction and the accelerated plate motion contributed to the contractional tectonic environment that produced over 300 km of Central Andean shortening, with a 50–70 km thick crust. While the southward migration of the ridge determining the flat slab correlates with a space–time gap in volcanic activity, soon after the passage of the ridge the slab increased its dip, from flat to shallow dipping, finally sweeping volcanism eastward and defining a broad volcanic footprint in the Central Andes (Oncken et al. 2006; Best et al. 2016; de Silva and Kay 2018). The steepening slab then started to roll-back, promoting decompression melting in the overlying enlarging mantle wedge, lowering its viscosity. This condition focused an elevated thermal input, encouraging crustal melting and generating a significant amount of magma. This resulted in the appearance and westward migration of volcanism from ~24 Ma, concurrent with an ignimbrite flare-up, in the Central Andes. In the southern portion of the Central Andes, the flare-up created the Altiplano-Puna Volcanic Complex, active in the last ~10 Ma and characterized by a mantle magma production rate of nearly 20 km^3/km/Ma. This complex, hosted within the Andean plateau at altitudes above 3000 m, is bordered to the west by the Andean

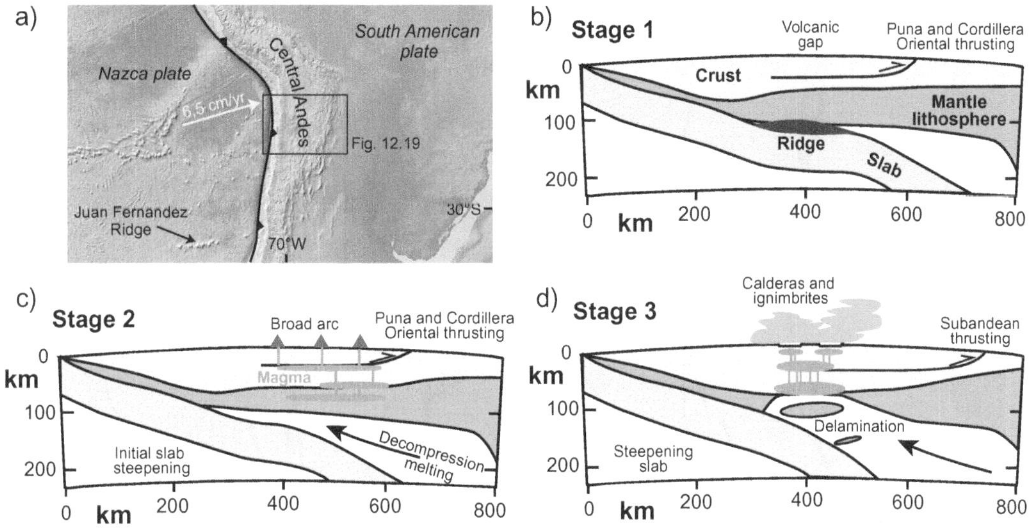

Fig. 12.18 **a** Simplified tectonic setting of the Central Andes, showing also the current location of the southward migrating Juan Fernandez Ridge; base DEM provided by GeoMapApp. **b–d** Main stages of evolution of the Central Andes in the last 25 Ma along an ideal E-W oriented lithospheric section. **b** The subduction of the Juan Fernandez Ridge induces a flat slab underlying a volcanic gap. **c** After the subduction of the ridge the steepening of the slab induces decompression melting in the overlying mantle wedge, promoting volcanism over a broad area. **d** Piecemeal delamination of the lithosphere of the southern part of the Central Andes produces a warmer asthenosphere above the slab, promoting sustained magmatic activity and ignimbrite flare-up of the Altiplano-Puna Volcanic Complex (modified after de Silva and Kay 2018)

forearc and to the east by the Eastern Cordillera and the Subandean foreland (Fig. 12.19). Episodic piecemeal delamination from late Miocene to Quaternary produced peaks in volcanic output of the Altiplano–Puna Volcanic Complex. Delamination may have affected the lower continental crust and the mantle lithosphere, both sinking into and replaced by the underlying hotter and less viscous mantle (Kay and Kay 1993; Kay and Coira 2009; Gioncada et al. 2010). Tomography has repeatedly highlighted high velocity structures beneath the Puna, interpreted as detached continental lithosphere, previously thickened and weakened by orogenic processes. The delamination may have also contributed to the uplift of the Central Andes (Schurr et al. 2006; Bianchi et al. 2013; Liang et al. 2014). Therefore, the exceptional and widespread late Cenozoic volcanic activity in the Central Andes, including the development of the Altiplano–Puna Volcanic Complex, results from decompression melting induced by slab steepening following ridge subduction and a hotter mantle wedge fuelled by warmer asthenosphere following piecemeal delamination. This complex sequence of events is responsible for the ~70 km thick crust supporting a plateau above 3000 m and the exceptional magmatism feeding the Altiplano–Puna Volcanic Complex (Fig. 12.20). Both features are described in more detail below.

The shortening, thickening and uplift of the crust in the Altiplano region began at approximately 25 Ma, following the shallowing of the slab and the increased convergence rate. The latter promoted shortening in the Eastern Cordillera and, subsequently, thrusting in the Subandean foreland to the east. Shortening and uplift in the Puna commenced 5–10 million years later and continued until Quaternary. Known shortening at the surface accounts for only 70–80% of the observed crustal thickening, suggesting that magmatic addition and other processes, such as lower crustal flow or tectonic underplating may contribute significantly to the thickening of the Central Andes (Allmendinger

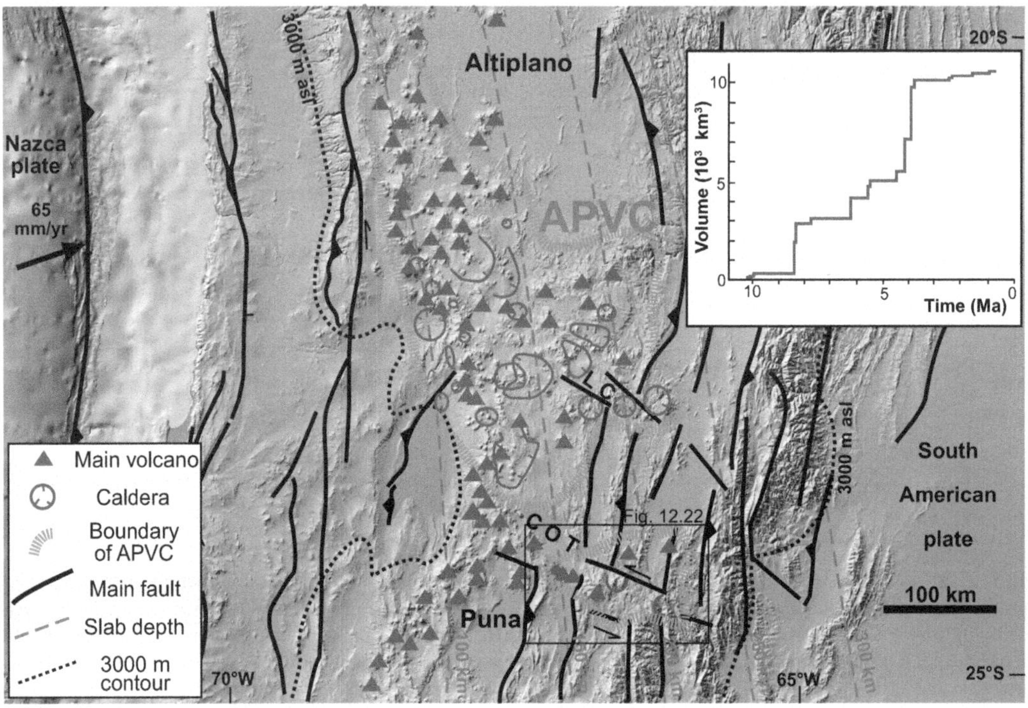

Fig. 12.19 Tectonic setting of the Altiplano–Puna Volcanic Complex (APVC), showing the current slab contours, the major fault systems, including the transverse Lipez-Coranzuli (LC) and Calama-Olacapato-El Toro (COT) fault zones, the major calderas and stratovolcanoes and the 3000 m contour of the Andean plateau (after de Silva and Gosnold 2007). Base DEM provided by GeoMapApp. Inset shows the cumulative volume versus time of major ignimbrite eruptions of the Altiplano-Puna Volcanic Complex (modified after de Silva and Gosnold 2007)

et al. 1997; Hindle et al. 2005). Since 10 Ma, the southern Puna portion of the plateau underwent pure shear deformation, whereas the northern Altiplano portion underwent simple shear. These modes of crustal thickening correlate with changes in lithospheric thickness, modes of isostatic compensation, broad wavelength topography, magmatism and slab steepening (Kay and Kay 1993; Allmendinger and Gubbels 1996; Gioncada et al. 2010).

The Puna and Eastern Cordillera regions experienced NW–SE and NE-SW trending shortening directions during Miocene-Pliocene and during the Pliocene to Quaternary, respectively. The change in the shortening direction has been related to the change in the direction and rate of motion of the South American plate, or to the variation in the orientation and kinematics of prominent fault zones. Neogene deformation also shows orogen-parallel extension propagating southward and eastward, likely induced by mid to lower crustal orogen-parallel channel flow (Marrett and Strecker 2000; Riller et al. 2001). On the Puna, the ~N-S trending high angle reverse faults responsible for a Cenozoic shortening of 10–15% have been reactivated, from late Miocene to Quaternary, with predominant dextral and minor normal motion. Also, much of the Puna has been dominated for the last 1–2 Ma by strike-slip and extensional faulting, in contrast to the protracted earlier history of thrust faulting (Cladouhos et al 1994; Tibaldi et al. 2009). However, there is evidence of ongoing Quaternary contraction along the western and eastern borders of the Puna: while to the west magma intrudes along active reverse faults, the foreland

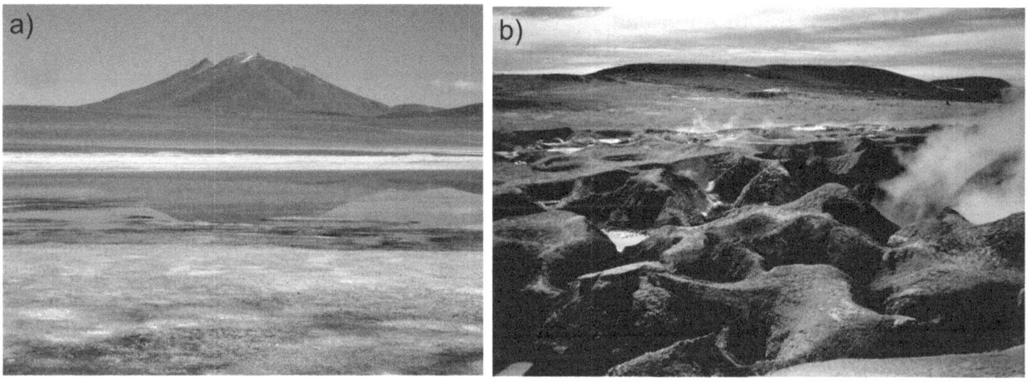

Fig. 12.20 Examples of current volcanic activity in the Altiplano-Puna Volcanic Complex of the Central Andes. **a** View of Aracar stratovolcano, in the southern Puna of Argentina, from the east. The active volcano, 6095 m high, may have last erupted in 1993. **b** View of the Sol de la Manana geothermal area in southern Bolivia, characterized by bubbling mud pots and geysers

to the east is experiencing ~E-W trending compression (Coutand et al 2001; Gonzalez et al. 2009).

Within this tectonic context, a distinct flare-up within the Neogene Central Andean ignimbrite province produced the most voluminous ignimbrite plateau in the southern half of the province from 22 to 24 °S, that is the Altiplano-Puna Volcanic Complex. Approximately 15,000 km^3 of magma were erupted in the last ~10 Ma, focusing in pulses every ~2 Ma, at 8, 6 and 4 Ma, with eruption rates one order of magnitude higher than the average rate for the flare-up of continental silicic provinces The composition of the ignimbrites includes dacites and rhyolites with minor trachydacites, andesites and latites (Fig. 12.19; Best et al 2016, and references therein). Elevated input from the mantle induced crustal melting and assimilation, developing crustal-scale intrusive complexes thermally softening the lower crust (de Silva and Gosnold 2007). A low velocity zone at a depth between 15 and 30 km has been interpreted as a sill-like magma body due to partial melting (~20%) and responsible for magmatic intraplating. This area lies above a 100 to 200 km deep zone of P-wave attenuation, connecting the volcanic areas of the plateau to the seismic zone within the slab: the connecting zone is interpreted as ascent pathway for metamorphic fluids and partial melts (ANCORP Working Group 2003; Schurr et al. 2003; Schilling et al. 2006). More recent seismic images identify a ~200 km wide and ~11 km thick low velocity zone at 4–25 km depth below sea level, interpreted as the plutonic complex that sourced the voluminous volcanics. These images show that the ~500,000 km^3 volume of this plutonic complex, or crustal-scale batholith, is one order of magnitude larger than previous estimates, retaining a significant percentage (up to 25%) of partial melt, most likely in a mush state. These images also allow making one of the best-constrained calculations of a plutonic to volcanic ratio, which is between 20 and 35 (Ward et al. 2014). Geodetic data from the last decades reveal a regional uplift of the entire Altiplano-Puna Volcanic Complex at an average of ~1.0 cm/year, whose fluctuations suggest a non-steady supply of melt and/or volatiles from the partially molten magma body below. Superimposed are more local uplifts over specific volcanic complexes, as Uturuncu and Lazufre, revealing prolonged unrest (Lau et al. 2018; Pritchard et al. 2018; Reath et al. 2019).

Magma emission in the Altiplano-Puna Volcanic Complex focused in ~20 large and complex calderas, along five NW–SE trending transverse magmatic belts, largely recognized from the alignment of magmatic centres,

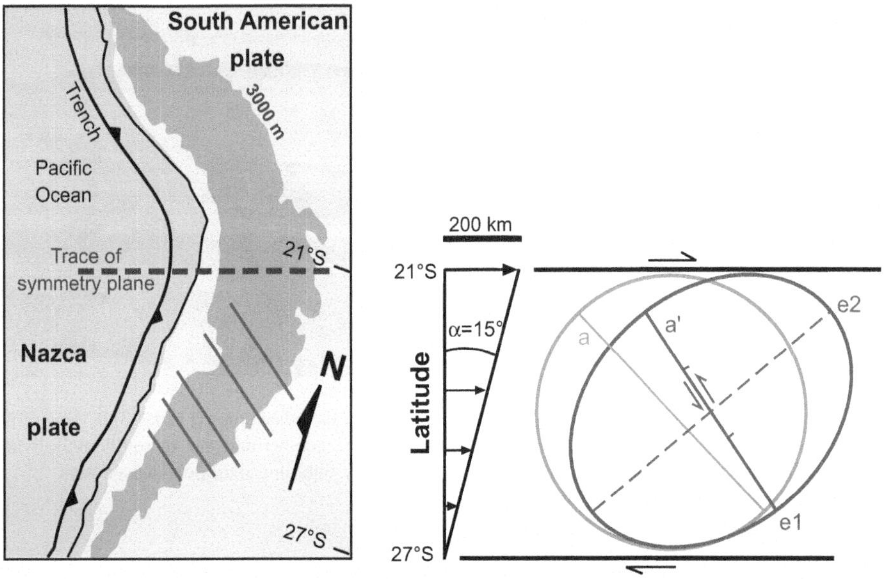

Fig. 12.21 Origin of transverse structures in the Central Andes. Left: bilateral symmetry of the Central Andes. The shaded area is higher than 3 km above sea level. NW–SE striking red lines delineate the traces of major fault zones associated with the transverse volcanic zones. Right: the horizontal sectional strain ellipse, explaining the NW–SE transtensive structures, is deduced from the gradient in transverse shortening in the southern portion of the Central Andes (modified after Riller et al. 2001)

lineaments identified from remote sensing and field analysis (Matteini et al. 2002; Trumbull et al. 2006). These belts are first-order fault zones of the Altiplano-Puna plateau, whose formation and left-lateral transtensive motion results from the gradient in transverse shortening from the central part of the plateau to its southern extent (Fig. 12.21). In addition, there appears to be a local control of pre-existing upper crustal weaknesses, with different orientation and kinematics, on the distribution and activity of polygenic and monogenic volcanoes (Riller et al. 2001; Acocella et al. 2011; Tibaldi et al. 2017).

The largest and best-studied transverse lineament is the ~300 km long Calama-Olacapato-El Toro fault zone. It consists of ~21 major magmatic centres, including explosive vents, calderas, composite stratovolcanoes, monogenetic cones, lava domes and plutons, active since Miocene. The Neogene left-lateral displacement of the Calama-Olacapato-El Toro fault zone is estimated as of ~20 km; in addition, the fault zone is associated with N-S trending extension (Fig. 12.22; Schurr et al. 1999; Riller et al. 2001). Overall left-lateral transtension has facilitated the ascent of magma and caldera formation. The SiO_2 content and $^{87}Sr/^{86}Sr$ ratio of the magmatic rocks suggest that the most evolved products, with upper crustal imprint, focus on the more intensively deformed central fault zone. Conversely, the more primitive, mantle-derived mafic to moderately evolved products focus toward the termini. This points to a genetic relationship between upper-crustal deformation and magmatic activity, leading to encouraged magma storage in the central fault zone (Riller et al. 2001; Petrinovic et al. 2006; Acocella et al. 2011).

12.5 A Synthetic Model for Convergent Plate Boundaries

As presented above, the structural control on arc volcanoes may vary significantly, involving extensional, strike-slip, contractional and oblique systems. The overall structure of the arc is highly

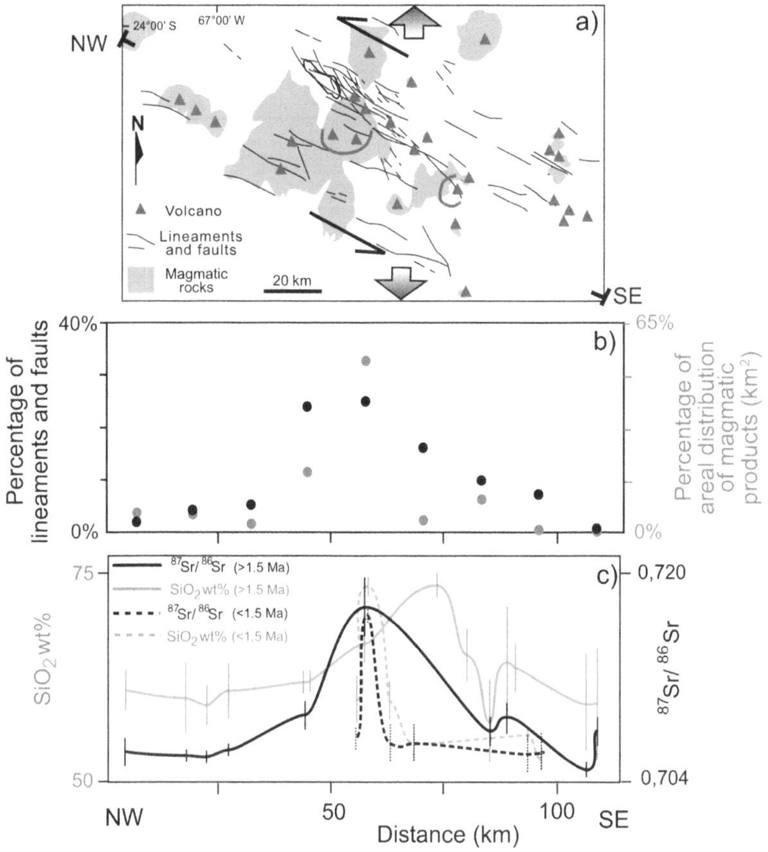

Fig. 12.22 Main tectono-magmatic features of the central and eastern part of the Calama-Olacapato-El Toro (COT) fault zone (location in Fig. 12.19; modified after Acocella et al. 2011). **a** Location of the main volcanoes (triangles), extent of magmatic rocks (in purple), main lineaments and faults (undistinguished). The overall kinematics of the fault zone consists of left-lateral transtension with orogen-parallel extension. **b** Percentages of the lineaments and faults (black) and of the areal distribution of the magmatic products (purple, in km^2) projected as a function of distance along the COT fault zone (see trace of profile in **a**). **c** SiO_2 wt% and $^{87}Sr/^{86}Sr$ ratio of the magmatic products as a function of distance from arc, highlighting more evolved compositions in the most fractured central part of the COT; vertical lines = range of values for each magmatic centre

dependent upon its tectonic setting, as imposed by relative plate motion (Table 12.2). A strike-slip motion along the arc results from a non-negligible trench-parallel component between the converging plates, whereas a dip-slip motion, responsible for extension or contraction, depends on the dominant trench-normal convergence (or subduction) velocity. Extensional conditions along the arc are usually associated with the highest magmatic output (as in the Taupo Volcanic Zone). However, the magmatic output can be important also in contractional arcs, as magma generation above slabs is a steady process. Indeed, contractional arcs may be associated with shallow magma accumulation via stacked sills capable of increasing the vertical stress component, ultimately producing significant volcanism, as in northeast Honshu. Following specific processes (slab steepening and lithospheric delamination, as in the Central Andes) contractional arcs may even produce exceptional amounts of magma.

The type and composition of arc volcanoes are broadly related to their structural setting. While calderas are usually related to felsic (rhyolite) compositions, stratovolcanoes are

Table 12.2 Summary of the relations between the tectonic setting, the structure of the volcanic arc and the amount of erupted magma along the arc (modified after Acocella and Funiciello 2010)

Plate motion (tectonic setting)		Arc motion (structural setting)		Arc production (erupted magma)
Mainly orthogonal	Coupling	Mainly orthogonal	Extension	Higher
			Compression	Lower
Mainly parallel	Coupling	Mainly parallel		Higher
Oblique	Coupling	Oblique	Transtension	Higher
			Transpression	Lower

associated with intermediate (andesite) compositions. Also, while calderas predominate in highly productive and extensional arcs (as TVZ and the Tyrrhenian margin), stratovolcanoes are more frequent in poorly extending (Cascades), strike-slip (Sumatra) and contractional arcs (northeast Honshu). This suggests that arc extension is a requisite to develop the large and long-lived rhyolitic reservoirs below felsic calderas. The Central Andes show a distinct behaviour, as hosting large felsic calderas in an overall contractional arc. Their extremely high eruption rate is roughly equivalent to that of the extensional TVZ, albeit lasting much longer (Houghton et al. 1995). This indicates that in a contractional setting a sequence of external regional conditions (promoting slab steepening and lithospheric delamination), can be equally effective as sustained extension in producing significant volumes of arc volcanism.

The structure of a volcanic arc is not necessarily continuous, with a first-order tectono-magmatic segmentation evident in the distribution of the surface deformation and of monogenic and polygenic volcanoes, as well as their output rates. This segmentation may result from regional stress gradients, associated with microplates, forearc slivers and torn slabs. At a more local scale, the segmentation may also result from the presence of any magmatic system, defined by the continuation of magmatic activity outside a main polygenic volcano, suggesting enhanced conditions for the shallow intrusion of magma. Magmatic systems show a variability depending on the structural setting. Volcanic arcs undergoing evident extension, as TVZ, show a similar architecture to that of mature continental divergent plate boundaries. In fact, volcanic and tectonic activity in both settings is segmented and focused in magmatic systems with similar extension rate and geometry. Major rifting episodes may result from the emplacement of kilometres long mafic dikes, as during the 1886 Tarawera eruption in TVZ or in the early nineteenth century in the northern Main Ethiopian Rift. A small-scale example of volcanic arc experiencing dominant extension and similarly associated with magmatic systems is the Central Aeolian Arc (Italy). Here extension occurs at shallow crustal levels and appears magma-induced, while strike-slip structures are probably active at deeper levels (Ruch et al. 2016). Magmatic systems in volcanic arcs experiencing minor extension (≤ 0.1 cm/year; as the Cascades, Izu-Bonin or the Trans-Mexican Volcanic Belt, Mexico) are less developed or not evident. For example, in 2000 a tens of kilometres long dike propagated laterally out of Miyakejima Island, in the weakly extending Izu-Bonin Arc, suggesting that magmatic activity continues outside the central volcano, although the direction of propagation was subparallel to the direction of convergence (Toda et al. 2002). These features suggest that an extension rate of a few mm/year marks the threshold between arcs with magmatic systems, structurally similar to continental divergent plate boundaries, and arcs with poor or no evidence of magmatic systems. In any case, the structure of extensional arcs is complicated by the fact that the volcanic segmentation, as related to the arc-normal tension, may also result from the stress regime related to the changing dip of the slab and the breadth of the zone coupling with the overriding plate, as highlighted for the Marianas (Andikagumi et al. 2020). Arcs undergoing predominant strike-slip motion also lack evident or recurrent magmatic systems. In the South Andean Volcanic Zone, a few volcanic complexes (as Puyehue–Cordón Caulle, Nevados de Chillan and Laguna del Maule) may resemble magmatic systems, although most polygenic volcanoes lack evidence of volcanic activity continuing outside the edifice. In Sumatra, magmatic systems are lacking and volcanic activity remains, with the exception of Toba caldera, largely punctiform or central. Arc volcanoes undergoing contraction elongate and align parallel to the direction of maximum principal stress σ_1, as in northeast Honshu. Here the Quaternary arc-orthogonal alignment and elongation of volcanoes mimics the direction of "hot mantle fingers" at the sublithospheric scale. However, being volcanic activity restricted to the polygenic volcano, there is no evidence of magmatic systems. Therefore, the linear mode of magmatic accretion of divergent plate boundaries and, partly, extensional arcs is gradually replaced by a dominant central mode of accretion in strike-slip arcs, where magmatic systems are very limited, and an exclusively central mode in contractional arcs, where magmatic systems are absent (Acocella 2014). The Central Andes provide a distinct behaviour, as, for their specific evolution, the predominant central mode of magmatic accretion feeds areal volcanism.

The general transition from linear to central magmatic accretion may be partly reconciled with the average depth of the magma plumbing systems below polygenic arc volcanoes (Fig. 12.23; e.g., Chaussard and Amelung 2012). In fact, the development of a magmatic system, being promoted by shallower magma chambers and dike swarms, depends also upon the average depth of its magma plumbing system. In arc volcanoes, this depth depends upon the arc structure. Arcs undergoing extension and strike-slip motion usually have shallower magma

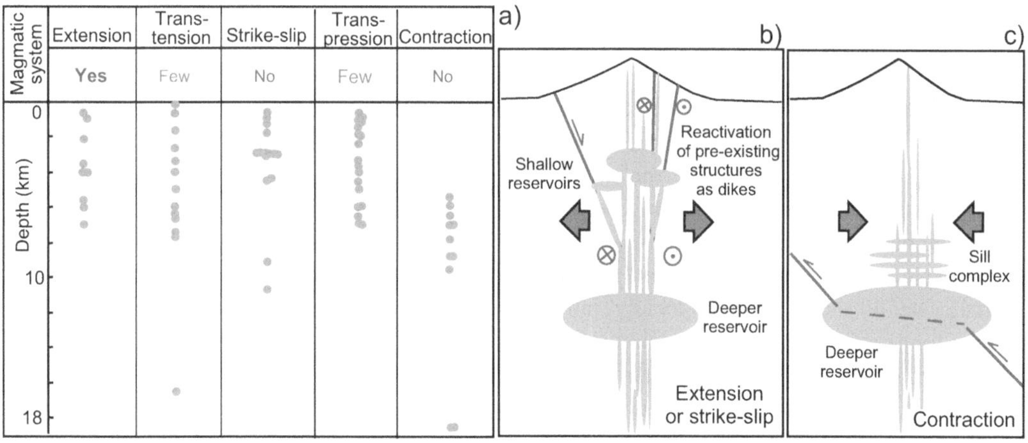

Fig. 12.23 **a** Degree of development of magmatic systems (above) and depths of magma reservoirs as a function of the stress regime in extensional, transtensive, strike-slip, transpressive and contractional volcanic arcs. Reservoir depths are from geodesy, seismology, gravity or petrology. Stress regime is from seismic Centroid Moment Tensor data (CMT) and World Stress Map Project data. **b** and **c** Interpretative cartoons showing in section view the effect of the tectonic setting on magma ascent through the upper crust in volcanic arcs. Extensional and strike-slip settings promote shallow level reservoirs and dikes, while contractional settings promote intermediate-level sills (modified after Chaussard and Amelung 2012)

reservoirs and dike systems, whereas arcs undergoing contraction have deeper reservoirs and dike systems. These structural conditions determine different possibilities for magma to accumulate at shallow crustal levels, propagate vertically and laterally and rise towards the surface, ultimately developing a magmatic system. Therefore, the lack of magmatic systems in contractional arcs may be explained by their deeper magma plumbing systems, which impose a more focused (central) volcanic activity.

Similarly to divergent plate boundaries, magma also affects the evolution of convergent plate boundaries: this may occur through very different mechanisms, which depend on the arc structure. The tectono-magmatic behaviour of some extensional arcs (TVZ) resembles that of continental divergent plate boundaries, where dikes promote shallow opening and magma assists rifting at depth. At the extreme, magma-induced extension may even mask surface evidence of strike-slip structures active in the arc roots, as in the Central Aeolian Arc. In addition, in several arcs the emplacement of magma at depth determines the thermal weakening of the crust, localizing the shallow deformation. This occurs in extensional arcs (Taupo Volcanic Zone) and in strike-slip arcs (Sumatra and South Andean Volcanic Zone), where the fault zones coincide with the core of the volcanic arc (Kendall et al. 2005; Ruch et al. 2016; Villamor et al. 2017; Acocella et al. 2018; Gase et al. 2019). In some arcs with strike-slip component (Kuril Arc and Trans Mexican Volcanic Belt), the weakening may induce an effective decoupling between forearc and back-arc, developing a forearc sliver. Nevertheless, under certain conditions the high thermal state of the crust may locally inhibit faulting: for example, propagating regional faults and related seismicity repeatedly arrested in correspondence of magma chambers, as observed in 1994 after the Liwa earthquake (southern Sumatra) and in 2016 at Aso caldera (western Japan; Widiwijayanti et al. 1996; Lin et al. 2016).

While in obliquely-convergent arcs magma may focus crustal deformation, the produced strike-slip zone may in turn influence magmatic and volcanic activity. In fact, studies on the plutons constituting the roots of eroded volcanic arcs show that pluton growth is promoted by oblique convergence conditions, when strain partitioning activates strike-slip faults creating localised extensional zones, as pull-apart structures, releasing bends and dilational jogs able to promote the rise, emplacement and accumulation of magma (see Fig. 4.20). Similar processes are expected to promote the rise and emplacement of magma and assist the development of magma chambers also below the active volcanoes of Sumatra or the South Andean Volcanic Zone. This may lead to a feedback between magmatic activity and strike-slip faulting along volcanic arcs, where the former localizes the site of active deformation and the latter assists the development of the magma reservoirs constituting the roots of the arcs. Nevertheless, at the surface the control of strike-slip structures on the location and activity of the volcanoes is less evident, as observed at Sumatra or the South Andean Volcanic Zone. In fact, in both cases the location of only a part of the volcanoes may be related to the activity of the strike-slip structures, with another part simply compliant with the responsible regional stress field. Despite the working hypothesis presented in Sect. 12.3.2 of a variable role of the arc structure at depth and at the surface, the limited agreement between these deeper and shallower tectono-magmatic relationships along strike-slip arcs remains poorly understood.

Contractional arcs also show an active role of magma, at times assisted by regional tectonics. In northeast Honshu, pressurized stacked sills may locally increase the vertical component of the principal stress from σ_3 to σ_2 and even to σ_1, promoting arc-parallel extension and arc-perpendicular alignment and elongation of volcanoes. In addition, the co- and post-seismic deformation induced by mega-earthquakes may induce transient stress variations, promoting extension in contractional and strike-slip arcs. This may enhance diking, gas nucleation, magma mixing and ultimately the rise of magma, increasing the eruptive frequency in the post-seismic period. In this case, assisted by regional tectonics, dikes may also intrude and feed

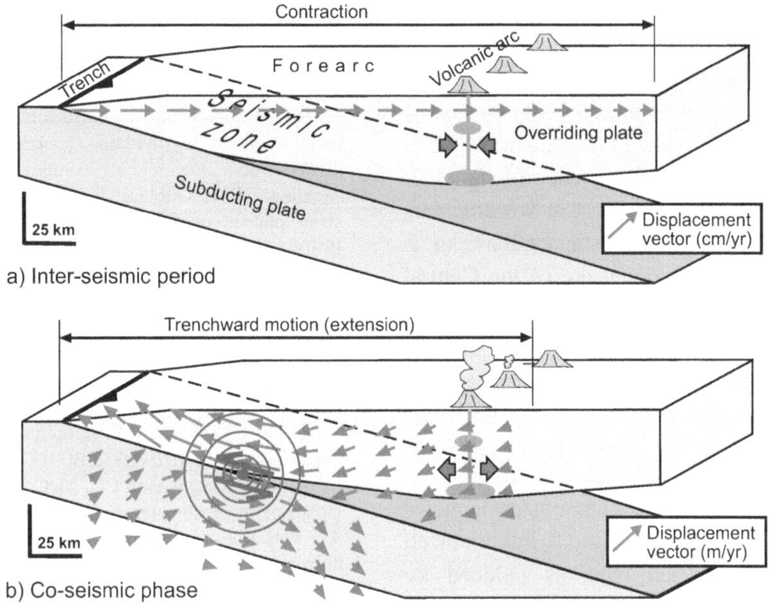

Fig. 12.24 **a** Schematic cross section of a subduction zone in the inter-seismic period, where the overall compressive stress field generates progressively decreasing displacement vectors (blue arrows) pointing towards the inner part of the overriding plate. The volcanoes and their plumbing systems (in orange) experience contraction (purple arrows). **b** Cross section of a subduction zone in the co-seismic phase. The co-seismic displacement field (blue arrows) associated with a megathrust earthquake (red arrows) induces a trenchward movement of the overriding plate. This kinematic inversion extends this portion of the upper plate, promoting volumetric expansion (purple arrows) in the plumbing systems of the volcanoes (orange; modified after Walter and Amelung 2007)

volcanic activity along otherwise non-favourable orientations, as parallel to the arc (northeast Honshu) or perpendicular to the oblique convergence vector (southern Andes; Fig. 12.24; Walter and Amelung 2007; Jonsson 2013).

These different cases from extensional, strike-slip and contractional arcs indicate that magma-driven variations in the stress state or in the structure of a volcanic arc are feasible and common. Therefore, magma may play a dominant role in controlling the evolution of convergent plate boundaries through diking, sill stacking, heat-induced strain localization and stress transients. This implies that a significant part of the structure of a volcanic arc may be magma-induced or assisted, with magmatic processes being largely self-sustained and requiring limited external tectonic contribution.

12.6 Summary

Arc volcanoes may be controlled by extensional, strike-slip, contractional or oblique structures, depending upon the tectonic setting imposed by the converging plates. In particular, the structure of the arc determines the distribution of its volcanism and ultimately the presence and development of the magmatic systems. Arcs experiencing evident extension (several mm/yr, as TVZ) have magmatic systems similar to those of mature continental divergent plate boundaries. Arcs experiencing strike-slip motion (as the South Andean Volcanic Zone or Sumatra) lack dominant magmatic systems, although these may be locally identified. Contractional arcs (as northeast Honshu) lack any evidence of magmatic systems. Therefore, the linear mode of magmatic accretion of divergent plate boundaries and extending arcs is replaced by a dominant central mode of accretion in strike-slip arcs and a completely central mode in contractional arcs. This feature may be a consequence of the shallower magma plumbing systems in volcanic arcs experiencing extension. A distinct behaviour is observed in the contractional arc of the Central Andes, characterized by an areal mode of magmatic accretion resulting from its exceptional productivity.

The structure of a volcanic arc in turn also determines the dominant mechanism(s) through which magma affects the evolution of the convergent plate boundary. These mechanisms include diking, sill stacking, heat-induced weakening and transient stresses induced by mega-earthquakes, suggesting that a significant part of the development of a volcanic arc is magma-induced, with magmatic processes being largely self-sustained, though not insensitive to regional tectonics.

References

Acocella V, Funiciello R (2006) Transverse systems along the extensional Tyrrhenian margin of central Italy and their influence on volcanism. Tectonics 25: TC2003. https://doi.org/10.1029/2005TC001845

Acocella V, Spinks K, Cole J, Nicol A (2003) Oblique back-arc rifting of Taupo Volcanic Zone, New Zealand. Tectonics 22:1045. https://doi.org/10.1029/2002TC001447

Acocella V, Yoshida T, Yamada R, Funiciello F (2008) Structural control on late miocene to quaternary volcanism in the NE Honshu arc, Japan. Tectonics 27:TC5008. https://doi.org/10.1029/2008TC002296

Acocella V, Funiciello F (2010) Kinematic setting and structural control of arc volcanism. Earth Planet Sci Lett 289:43–53

Acocella V, Gioncada A, Omarini R, Riller U, Mazzuoli R, Vezzoli L (2011) Tectono-magmatic characteristics of the eastern Calama-Olacapato-El Toro lineament, Central Andes. Tectonics 30:TC3005. https://doi.org/10.1029/2010TC002854

Acocella V (2014) Structural control on magmatism along divergent and convergent plate boundaries: overview, model, problems. Earth-Sci Rev 136:226–288

Acocella V, Bellier O, Sandri L, Sébrier M, Pramumijoyo S (2018) Weak tectono-magmatic relationships along an obliquely convergent plate boundary: Sumatra, Indonesia. Front Earth Sci 6:3. https://doi.org/10.3389/feart.2018.00003

Allan ASR, Wilson CJN, Millet MA, Wysoczansk RJ (2012) The invisible hand: tectonic triggering and modulation of a rhyolitic supereruption. Geology 40:563–566

Allmendinger RW, Gubbels T (1996) Pure and simple shear plateau uplift, Altiplano-Puna, Argentina and Bolivia. Tectonophysics 259:1–13

Allmendinger RW, Jordan TE, Kay SM, Isacks BL (1997) The evolution of the Altiplano-Puna Plateau of the Central Andes. Annu Rev Earth Planet Sci 25:139–174

ANCORP Working Group (2003) Seismic imaging of a convergent continental margin and plateau in the central Andes (Andean Continental Research Project 1996 (ANCORP'96)). J Geophys Res 108:2328. https://doi.org/10.1029/2002JB001771

Andikagumi H, Macpherson CG, McCaffrey KJW (2020) Upper plate stress controls the distribution o Mariana Arc volcanoes. J Geophys Res 125:e2019JB017391. https://doi.org/10.1029/2019JB017391

Barberi F, Buonasorte G, Cioni R, Fiordelisi A, Foresi L, Iaccarino S et al (1994) Plio-pleistocene geological evolution of the geothermal area of tuscany and latium. Mem Descr Carta Geol Ital XLIX:77–134

Barker SJ, Wilson CJN, Morgan DJ, Rowland JV (2016) Rapid priming, accumulation, and recharge of magma driving recent eruptions at a hyperactive caldera volcano. Geology 44:323–326

Begg JG, Mouslopoulou V (2010) Analysis of late Holocene faulting within an active rift using lidar, Taupo Rift, New Zealand. J Volcanol Geoth Res 190:152–167

Bellier O, Bellon H, Sebrier M, Sutanto P, Maury RC (1999) K-Ar age of the Ranau tuffs: implications for the Ranau caldera emplacement and slip-partitioning in Sumatra (Indonesia). Tectonophysics 312:347–359

Benoit MH, Torpey M, Liszewski K, Levin V, Park J (2011) P and S wave upper mantle seismic velocity structure beneath the northern Apennines: New evidence for the end of subduction. Geochem Geophys Geosyst 12:Q06004. https://doi.org/10.1029/2010GC003428

Bertrand EA, Caldwell TG, Hill GJ, Wallin EL, Bennier SL, Cozens N et al (2012) Magnetotelluric imaging of upper-crustal convection plumes beneath the Taupo Volcanic Zone, New Zealand. Geophys Res Lett 39:L02304. https://doi.org/10.1029/2011GL050177

Best MG, Christiansen EH, de Silva S, Lipman PW (2016) Slab-rollback ignimbrite flareups in the southern Great Basin and other Cenozoic American arcs: a distinct style of arc volcanism. Geosphere 12:1097–1135

Bianchi M, Heit B, Jakovlev A, Yuan X, Kay SM, Sandvol E et al (2013) Teleseismic tomography of the southern Puna plateau in Argentina and adjacent regions. Tectonophysics 586:65–83

Blakely RJ, Christiansen RL, Guffanti M, Wells RE, Donnelly-Nolan JM, Muffler LJP et al (1997) Gravity anomalies, Quaternary vents, and Quaternary faults in the southern Cascade Range, Oregon and California: implications for arc and backarc evolution. J Geophys Res 102:22513–22527

Bonali FL, Tibaldi A, Corazzato C (2015) Sensitivity analysis of earthquake-induced static stress changes on volcanoes: the 2010 Mw 8.8 Chile earthquake. Geophys J Int 201:1868–1890

Bradley KE, Feng L, Hill EM, Natawidjaja DH, Sieh K (2017) Implications of the diffuse deformation of the Indian Ocean lithosphere for slip partitioning of oblique plate convergence in Sumatra. J Geophys Res 122:572–591

Bucchi F, Lara LE, Gutierrez F (2015) The Carrán-Los Venados volcanic field and its relationship with coeval and nearby polygenetic volcanism in an intra-arc setting. J Volcanol Geoth Res 308:70–81

Caratori Tontini F, Bassett D, de Ronde CEJ, Timm J, Wysoczanski R (2019) Early evolution of a young back-arc basin in the Havre Trough. Nat Geosci 12:856–862

Cembrano J, Lara L (2009) The link between volcanism and tectonics in the southern volcanic zone of the Chilean Andes: a review. Tectonophysics 471:96–113

Chaussard E, Amelung F (2012) Precursory inflation of shallow magma reservoirs at west Sunda volcanoes detected by InSAR. Geophys Res Lett 39:L21311. https://doi.org/10.1029/2012GL053817

Chesner CA (2012) The Toba Caldera Complex. Quatern Int 258:5–18

Cladouhos IT, Allmendinger RW, Coira B, Farrar E (1994) Late Cenozoic deformations in the Central Andes: fault kinematics from the northern Puna, northwest Argentina and southwest Bolivia. J South Am Earth Sci 7:209–228

Cole JW (1990) Structural control and origin of volcanism in the Taupo Volcanic Zone, New Zealand. Bull Volcanol 52:445–459

Coutand I, Cobbold PR, de Urreiztieta M, Gautier P, Chauvin A, Gapais D et al (2001) Style and history of Andean deformation, Puna plateau, northwestern Argentina. Tectonics 20:210–234

D'Agostino N (2014) Complete seismic release of tectonic strain and earthquake recurrence in the Apennines (Italy). Geophys Res Lett 41. https://doi.org/10.1002/2014GL059230

Darby DJ, Meertens CM (1995) Terrestrial and GPS measurements of deformation across the Taupo back arc and Hikurangi forearc regions in New Zealand. J Geophys Res 100:8221–8232

De Maisonneuve CB, Dungan MA, Bachmann O, Burgisser A (2012) Insights into shallow magma storage and crystallization at Volcán Llaima (Andean Southern Volcanic Zone, Chile). J Volcanol Geoth Res 211:76–91

DeMets C (1992) Oblique convergence and deformation along the Kuril and Japanese trenches. J Geophys Res 97:17615–17625

DeMets C, Traylen S (2000) Motion of the Rivera plate since 10 Ma relative to the Pacific and North American plates and the mantle. Tectonophysics 318: 119–159

de Silva SL, Gosnold WD (2007) Episodic construction of batholiths: insights from the spatiotemporal development of an ignimbrite flare-up. J Volcanol Geoth Res 167:320–335

de Silva SK, Kay SM (2018) Turning up the heat: high-flux magmatism in the Central Andes. Elements 14:245–250

de Silva SL, Mucek AE, Gregg PM, Pratomo I (2015) Resurgent Toba-field, chronologic, and model constraints on duration, time scales and mechanisms of resurgence at large calderas. Front Earth Sci 3. https://doi.org/10.3389/feart.2015.00025

Deering CD, Cole JW, Vogel TA (2008) Rhyolite compositional continuum governed by lower crustal source conditions in the Taupo Volcanic Zone, New Zealand. J Petrol 49:2245–2276

Delgado F, Pritchard ME, Basualto D, Lazo J, Córdova L, Lara LE (2016) Rapid reinflation following the 2011–2012 rhyodacite eruption at Cordón Caulle volcano (Southern Andes) imaged by InSAR: Evidence for magma reservoir refill. Geophys Res Lett 43:9552–9562

Delgado F, Kubanek J, Anderson K, Lundgren P, Pritchard M (2019) Physicochemical models of effusive rhyolitic eruptions constrained with InSAR and DEM data: A case study of the 2011–2012 Cordón Caulle eruption). Earth Planet Sci Lett 524:115736

Detourbet C, Bellier O, Sébrier M (1993) La caldeira volcanique de Toba et la Grande Faille de Sumatra (Indonesie) vues par l'imagerie SPOT. CR Acad Sci Paris 316:1439–1445

Devoti R, Esposito A, Pietrantonio G, Pisani AR, Riguzzi F (2011) Evidence of large scale deformation patterns from GPS data in the Italian subduction boundary. Earth Planet Sci Lett 311:230–241

Di Luccio F, Chiodini G, Caliro S, Cardellini C, Convertito V, Pino NA (2018) Seismic signature of active intrusions in mountain chains. Sci Adv 4:e1701825

Di Stefano R, Bianchi I, Ciaccio MG, Carrara G, Kissling E (2011) Three-dimensional Moho topography in Italy: New constraints from receiver functions and controlled source seismology. Geochem Geophys Geosyst 12:Q09006. https://doi.org/10.1029/2011GC003649

Duquesnoy TH, Bellier O, Kasser M, Sébrier M, Vigny CH, Bahar I (1996) Deformation related to the 1994 Liwa earthquake derived from geodetic measurements. Geophys Res Lett 23:3055–3058

Eberhart-Phillips D, Bannister S, Reyners M (2020) Attenuation in the mantle wedge beneath supervolcanoes of the Taupo Volcanic Zone, New Zealand. Geophys J Int 220:703–723

Faccenna C, Funiciello R, Bruni A, Mattei M, Sagnotti L (1994) Evolution of a transfer related basin: the Ardea basin (Latium, Central Italy). Basin Res 6:35–46

Flinders AF, Shen Y (2017) Seismic evidence for a possible deep crustal hot zone beneath Southwest Washington. Sci Rep 7:7400. https://doi.org/10.1038/s41598-017-07123-w

Folguera A, Ramos VA, Hermanns RL, Naranjo J (2004) Neotectonics in the foothills of the southernmost central Andes (37–38 S): evidence of strike-slip displacement along the Antiñir-Copahue fault zone. Tectonics 23: TC5008. https://doi.org/10.1029/2003TC001533

Gahalaut VK, Jade S, Catherine JK, Gireedh R, Ananda MB, Dileep Kumar P et al (2008) GPS measurements of postseismic deformation in the Andaman-Nicobar region following the giant 2004 Sumatra-Andaman earthquake. J Geophys Res 113: B08401. https://doi.org/10.1029/2007JB005511

Galland O, Hallot E, Cobbold PR, Ruffet G, de Bremond d'Ars J (2007) Volcanism in a compressional Andean setting: a structural and geochronology study of Tromen volcano (Neuquen province, Argentina). Tectonics 26: TC4010. https://doi.org/10.1029/2006TC002011

Gao H, Humphreys E, Yao H, van der Hilst RD (2011) Crust and lithosphere structure of the northwestern U. S. with ambient noise tomography: Terrane accretion and Cascade arc development. Earth Planet Sci Lett 304:202–211

Gase AC, Van Avendonk HJA, Bangs NL, Luckie TW, Barker DHN, Henrys SA et al (2019) Seismic evidence of magmatic rifting in the offshore Taupo Volcanic Zone, New Zealand. Geophys Res Lett 46:12949–12957

Gasparon M (2005) Quaternary volcanicity. In: Barber AJ, Crow ML, Milsom JS (eds) Sumatra: geology, resources and tectonic evolution. Geol Soc London Memoir London UK 31:120–130

Genrich JF, Bock Y, McCaffrey R, Prawirodirdjo L, Stevens CW, Puntodewo SSO et al (2000) Distribution of slip at the northern Sumatran fault system. J Geophys Res 105:28327–28341

George OA, Malservisi R, Govers R, Connor CB, Connor LJ (2016) Is uplift of volcano clusters in the Tohoku Volcanic Arc, Japan, driven by magma accumulation in hot zones? A geodynamic modeling study. J Geophys Res 121:4780–4796

Giacomuzzi G, Civalleri M, De Gori P, Chiarabba C (2012) A 3D Vs model of the upper mantle beneath Italy: Insight on the geodynamics of central Mediterranean. Earth Planet Sci Lett 335–336:105–120

Gioncada A, Vezzoli L, Mazzuoli R, Omarini R, Nonnotte P, Guillou H (2010) Pliocene intraplate-type volcanism in the Andean foreland at 26°10′S, 64°40′W (NW Argentina): Implications for magmatic and structural evolution of the Central Andes. Lithosphere 2:153–171

Gomez-Vasconcelos MG, Villamor P, Cronin S, Procter J, Palmer A, Townsend D et al (2017) Crustal extension in the Tongariro graben, New Zealand: insights into volcano-tectonic interactions and active deformation in a young continental rift. Geol Soc Am Bull 129:1085–1099

Gonzalez G, Cembrano J, Aron F, Veloso EE, Shyu JBH (2009) Coeval compressional deformation and volcanism in the central Andes, case studies from northern Chile (23°S-24°S). Tectonics 28:TC6003. https://doi.org/10.1029/2009TC002538

Gravley DM, Wilson CJN, Leonard GS, Cole JW (2007) Double trouble: paired ignimbrite eruptions and collateral subsidence in the Taupo Volcanic Zone, New Zealand. Geol Soc Am Bull 119:18–30

Gravley DM, Deering CD, Leonard GS, Rowland JV (2016) Ignimbrite flare-ups and their drivers: A New Zealand perspective. Earth Sci Rev 162:65–82

Green NL, Harry DL (1999) On the relationship between subducted slab age and arc basalt petrogenesis, Cascadia subduction system, North America. Earth Planet Sci Lett 171:367–381

Gripp AE, Gordon RG (2002) Young tracks of hotspots and current plate velocities. Geophys J Int 150:321–361

Grocott J, Brown M, Dallmeyer RD, Taylor GK, Treloar PJ (1994) Mechanisms of continental growth in extensional arcs: an example from the Andean plate-boundary zone. Geology 22:391–394

Guffanti M, Weaver CS (1988) Distribution of Late Cenozoic Volcanic Vents in the Cascade Range's Volcanic Arc Segmentation and Regional Tectonic Considerations. J Geophys Res 93:6513–6529

Gurer D, Galland O, Corfu F, Leanza HA, Sassier C (2016) Structure and evolution of volcanic plumbing systems in fold-and-thrust belts: A case study of the Cerro Negro de Tricao Malal, Neuquén Province, Argentina. Geol Soc Am Bull 128:315–331

Hamling IJ, Hreinsdottir S, Fournier N (2015) The ups and downs of the TVZ: geodetic observations of deformation around the Taupo Volcanic Zone, New Zealand. J Geophys Res 120:4667–4679

Hamling IJ, Hreinsdóttir S, Bannister S, Palmer N (2016) Off-axis magmatism along a subaerial back-arc rift: observations from the Taupo Volcanic Zone, New Zealand. Sci Adv 2:e1600288

Hasegawa A, Zhao D, Hori S, Yamamoto A, Horiuchi S (1991) Deep structure of the northeastern Japan arc and its relationship to seismic and volcanic activity. Nature 352:683–689

Heise W, Bibby HM, Caldwell TG, Nanninster SC, Ogawa Y, Takakura S et al (2007) Melt distribution beneath a young continental rift: the Taupo Volcanic Zone, New Zealand. Geophys Res Lett 34:L14313. https://doi.org/10.1029/2007GL029629

Hildreth W (2007) Quaternary magmatism in the Cascades; geologic perspectives. US Geological Survey Professional Paper 1744,125 p

Hindle D, Kley J, Oncken O, Sobolev S (2005) Crustal balance and crustal flux from shortening estimates in the Central Andes. Earth Planet Sci Lett 230:113–124

Holden L, Wallace L, Beavan J, Fournier N, Cas R, Ailleres L et al (2015) Contemporary ground deformation in the Taupo Rift and Okataina Volcanic Centre from 1998 to 2011, measured using GPS. Geophys J Int 202:2082–2105

Houghton BF, Wilson CJN, McWilliams MO, Lanphere MA, Weaver SD, Briggs RM et al (1995) Chronology and dynamics of a large silicic magmatic system: central Taupo Volcanic Zone, New Zealand. Geology 23:13–16

Houlie N, Stern TA (2017) Vertical tectonics at an active continental margin. Earth Planet Sci Lett 457:292–301

Igarashi T, Matsuzawa T, Umino N, Hasegawa A (2001) Spatial distribution of focal mechanisms for interplate and intraplate earthquakes associated with the subducting plate beneath the northeastern Japan arc: a triple-planned deep seismic zone. J Geophys Res 106:2177–2191

Ito T, Yoshioka S, Miyazaki S (2000) Interplate coupling in northeast Japan deduced from inversion analysis of GPS data. Earth Planet Sci Lett 176:117–130

Ito T, Gunawan E, Kimata F, Tabei T, Meilano I, Agustan et al (2016) Co-seismic offsets due to two earthquakes (Mw 6.1) along the Sumatran fault system derived from GNSS measurements. Earth Planets Space 68:57. https://doi.org/10.1186/s40623-016-0427-z

Jaxybulatov K, Shapiro NM, Koulakov I, Mordret A, Landes M, Sens-Schönfelder C (2014) A large magmatic sill complex beneath the Toba caldera. Science 346:617–619

Jay J, Costa F, Pritchard M, Lara L, Singer B, Herrin J (2014) Locating magma reservoirs using InSAR and petrology before and during the 2011–2012 Cordón Caulle silicic eruption. Earth Planet Sci Lett 395:254–266

Jolivet L, Faccenna C, Goffé B, Mattei M, Rossetti F, Brunet C et al (1998) Midcrustal shear zones in postorogenic extension: example from the northern Tyrrhenian Sea. J Geophys Res 103:12123–12160

JJonsson S (2013) Sunken volcanoes. Nat Geosci 6:891–892

Jordan TE, Isacks BL, Allmendinger RW, Brewer JA, Ramos VA, Ando CJ (1983) Andean tectonics related to geometry of subducted Nazca plate. Geol Soc Am Bull 94:341–361

Kato N, Sato H, Umino N (2006) Fault reactivation and active tectonics on the fore-arc side of the back-arc rift system, NE Japan. J Struct Geol 28:2011–2022

Kay SM, Coira BL (2009) Shallowing and steepening subduction zones, continental lithospheric loss, magmatism, and crustal flow under the central Andean Altiplano-Puna Plateau. In: Kay SM, Ramos VA, Dickinson WR (eds) Backbone of the Americas: shallow subduction, plateau uplift, and ridge and terrane collision. Geol Soc Am Memoir 204:229–260

Kay RW, Kay SM (1993) Delamination and delamination magmatism. Tectonophysics 219:177–189

Kendall JM, Stuart GW, Ebinger CJ, Bastow ID, Keir D (2005) Magma-assisted rifting in Ethiopia. Nature 433:146–148

Kimura G (1986) Oblique subduction and collision: forearc tectonics of the Kuril arc. Geology 14:404–407

Kimura G (1996) Collision orogeny at arc-arc junctions in the Japanese Islands. The Island Arc 5:262–275

Koulakov I, Kasatkina E, Shapiro NM, Jaupart C, Vasilevsky A, El Khrepy S et al (2016) The feeder system of the Toba supervolcano from the slab to the shallow reservoir. Nature Communications 7:12228. https://doi.org/10.1038/ncomms12228

Kundu B, Legrand D, Gahalaut K, Gahalaut VK, Mahesh P, Kamesh Raju KA et al (2012) The 2005 volcano-tectonic earthquake swarm in the Andaman Sea: triggered by the 2004 great Sumatra-Andaman earthquake. Tectonics 31:TC5009. https://doi.org/10.1029/2012TC003138

Lara LE, Naranjo JA, Moreno H (2004) Rhyodacitic fissure eruption in Southern Andes (Cordon Caulle; 40.5°S) after the 1960 (Mw:9.5) Chilean earthquake: a structural interpretation. J Volcanol Geoth Res 138:127–138

Lara LE, Moreno H, Naranjo JA, Matthews S, Perez de Arce C (2006) Magmatic evolution of the Puyehue-Cordón Caulle Volcanic Complex (40°S), Southern Andean Volcanic Zone: From shield to unusual rhyolitic fissure volcanism. J Volcanol Geoth Res 157:343–366

Lara LE, Cembrano J, Lavenu BA (2010) Quaternary vertical displacement along the Liquiñe-Ofqui Fault Zone: differential uplift and coeval volcanism in the Southern Andes? Int Geol Rev 50:975–993

Lau N, Tymofyeyeva E, Fialko Y (2018) Earth Planet Sci Lett 491:43–47

Lavenu A, Cembrano J (1999) Compressional- and transpressional-stress pattern for Pliocene and Quaternary brittle deformation in fore arc and intra-arc zones (Andes of central and Southern Chile). J Struct Geol 21:1669–1691

Legrand D, Barrientos S, Bataille K, Cembrano J, Pavez A (2011) The fluid-driven tectonic swarm of Aysen Fjord, Chile (2007) associated with two earthquakes (Mw=6.1 and Mw=6.2) within the Liquine-Ofqui Fault Zone. Cont Shelf Res 31:154–161

Liang X, Sandvol E, Kay S, Heit B, Yuan X, Mulcahy P et al (2014) Delamination of southern Puna lithosphere revealed by body wave attenuation tomography. J Geophys Res 119:549–566

Lin A, Satsukawa T, Wang M, Mohammadi Asl Z, Fueta R, Nakajima F (2016) Coseismic rupturing stopped by Aso volcano during the 2016 Mw 7.1 Kumamoto earthquake, Japan. Science 354:869–874

Loveless JP, Meade BJ (2010) Geodetic imaging of plate motions, slip rates and partitioning of deformation in Japan. J Geophys Res 115:B02410. https://doi.org/10.1029/2008JB006248

Lupi M, Trippanera D, Gonzalez D, D'Amico S, Acocella V, Cabello C (2020) Megathrust earthquakes and the growth of NW-trending volcanic systems in the Southern Central Andes. Tectonophysics 774:228204

Mahony SH, Sparks RSJ, Wallace LM, Engwell SL, Scourse EM, Barnard NH et al (2016) Increased rates of large-magnitude explosive eruptions in Japan in the late Neogene and Quaternary. Geochem Geophys Geosyst 17:2467–2479

Malinverno A, Ryan WBF (1986) Extension in the Tyrrhenian Sea and shortening in the Apennines as result of arc migration driven by sinking of the lithosphere. Tectonics 5:227–245

Mariani M, Prato R (1988) I bacini neogenici costieri del margine tirrenico: Approccio sismicostratigrafico. Mem Soc Geol Ital 41:519–531

Marra F, Taddeucci J, Freda C, Marzocchi W, Scarlato P (2004) Recurrence of volcanic activity along the Roman Comagmatic Province (Tyrrhenian margin of Italy) and its tectonic significance. Tectonics TC40n https://doi.org/10.1029/2003TC001600

Marrett R, Strecker MR (2000) Response of intracontinental deformation in the central Andes to the late Cenozoic reorganization of South American Plate motions. Tectonics 19:452–467

Matteini M, Mazzuoli R, Omarini R, Cas RAF, Maas R (2002) Geodynamical evolution of the Central Andes at 24°S as inferred by magma composition along the Calama–Olacapato–El Toro transversal volcanic belt. J Volcanol Geother Res 118:205–228

Mavrommatis AP, Segall P, Johnson KM (2014) A decadal-scale deformation transient prior to the 2011 Mw 9.0 Tohoku-Oki earthquake. Geophys Res Lett 41:4486–4494

McCaffrey R, Zwick PC, Bock Y, Prawirodirdjo L, Genrich JF, Stevens CW et al (2000) Strain partitioning during oblique plate convergence in northern Sumatra: geodetic and seismologic constraints and numerical modelling. J Geophys Res 105:28363–28376

McCaffrey R, Qamar AI, King RW, Wells R, Khazaradze G, Williams CA et al (2007) Fault locking, block rotation and crustal deformation in the Pacific Northwest. Geophys J Int 169:1315–1340

McCrory PA, Blair JL, Waldhauser F, Oppenheimer DH (2012) Juan de Fuca slab geometry and its relation to Wadati-Benioff zone seismicity. J Geophys Res 117: B09306. https://doi.org/10.1029/2012JB009407

McNulty B, Farber D, Wallace GS, Lopez R, Palacios O (1998) Role of plate kinematics and plate-slip-vector partitioning in continental magmatic arcs: evidence from the Cordillera Blanca, Peru. Geology 26:827–830

Melnick D, Folguera A, Ramos VA (2006) Structural control on arc volcanism: the caviahue-copahue complex, Central to Patagonian Andes transition (38°S). J S Am Earth Sci 22:66–88

Miller KC, Keller GR, Gridley JM, Leutgert JH, Mooney WD, Thybo H (1997) Crustal structure along the west flank of the Cascades, western Washington. J Geophys Res 102:17857–17873

Miura S, Sato T, Hasegawa A, Suwa Y, Tachibana K, Yui S (2004) Strain concentration zone along the volcanic front derived by GPS observations in NE Japan arc. Earth Planets Space 56:1347–1355

Montone P, Mariucci MT, Pierdominici S (2012) The Italian present-day stress map. Geophys J Int 189:705–716

Moore DE, Hickman S, Lockner DA, Dobson PF (2001) Hydrothermal minerals and microstructures in the Silangkitang geothermal field along the Great Sumatran fault zone, Sumatra, Indonesia. Geol Soc Am Bull 113:1179–1192

Mora D, Tassara A (2019) Upper crustal decompression due to deglaciation-induced flexural unbending and its role on post-glacial volcanism at the Southern Andes. Geophys J Int 216:1549–1559

Mount VS, Suppe J (1992) Present-Day stress orientations adjacent to Active Strike-Slip Faults: California and Sumatra. J Geophys Res 97:11995–12013

Mouslopoulou V, Hristopulos DT, Nicol A, Walsh JJ, Bannister S (2013) The importance of microearthquakes in crustal extension of an active rift: a case study from New Zealand. J Geophys Res 118:1556–1568

Nairn IA, Cole JW (1981) Basalt dikes in the 1886 Tarawera rift. NZ J Geol Geophys 24:585–592

Nakajima J, Hasegawa A (2004) Shear-wave polarization anisotropy and subduction-induced flow in the mantle wedge of northeastern Japan. Earth Planet Sci Lett 225:365–377

Natawidjaja DH, Triyoso W (2007) The Sumatran fault zone—from source to hazard. J Earthquake Tsunami 1:21–47

Okada T, Umino N, Hasegawa A (2010) Deep structure of the Ou mountain range strain concentration zone and the focal area of the 2008 Iwate-Miyagi Nairiku earthquake, NE Japan—seismogenesis related with magma and crustal fluid. Earth Planets Space 62:347–352

Oncken O, Hindle D, Kley J, Elger K, Victor P, Schemmann K (2006) Deformation of the Central Andean Upper Plate System—Facts, fiction, and constraints for plateau models. In: Oncken O, Chong G, Franz G, Giese P, Götze H-J, Ramos VA et al (eds) The andes—active subduction orogeny. Springer, Berlin, pp 3–27

Ozawa S, Nishimura T, Suito H, Kobayashi T, Tobita M, Imakiire T (2011) Coseismic and postseismic slip of the 2011 magnitude-9 Tohoku-Oki earthquake. Nature 475:373–376

Parsons T, Thompson GA, Smith RP (1998) More than one way to stretch: a tectonic model for extension along the plume track of the Yellowstone hotspot and adjacent Basin and Range Province. Tectonics 17:221–234

Passaro S, Tamburrini S, Vallefuoco M, Tassi F, Vaselli O, Giannini L (2016) Seafloor doming driven by degassing processes unveils sprouting volcanism in coastal areas. Sci Rep 6:22448. https://doi.org/10.1038/srep22448

Patacca E, Sartori R, Scandone P (1990) Tyrrhenian basin and apenninic arcs: kinematic relations since late Tortonian times. Mem Soc Geol It 45:425–451

Peccerillo A (2017) Cenozoic volcanism in the tyrrhenian sea region. In: Advances in Volcanology Springer International Publishing, AG, 415 pp

Pesicek JD, Thurber CH, Widiyantoro S, Engdahl ER, DeShon HR (2008) Complex slab subduction beneath northern Sumatra. Geophys Res Lett 35:L20303. https://doi.org/10.1029/2008GL035262

Petrinovic IA, Riller U, Alvarado G, Brod JA, Arnosio M (2006) Bimodal volcanism in a tectonic transfer zone: evidence for tectonically controlled magmatism in the southern Central Andes, NW Argentina. J Volcanol Geoth Res 152:240–252

Pezzopane SK, Weldon RJ (1993) Tectonic role of active faulting on central Oregon. Tectonics 12:1140–1169

Pitcher BW, Kent AJR (2019) Statistics and segmentation: using big data to assess cascades arc compositional variability. Geochim Cosmochim Acta 265:443–467

Pitcher BW, Kent AJR, Grunder AL, Duncan RA (2017) Frequency and volumes of ignimbrite eruptions following the Late Neogene initiation of the Central Oregon High Cascades. J Volcanol Geoth Res 339:1–22

Poland MP, Lisowski M, Dzurisin D, Kramer R, McLay M, Pauk B (2017) Volcano geodesy in the Cascade Arc, USA. Bull Volcanol 79:59

Pollitz FF, McCrory PM, Wilson D, Svarc J, Puskas C, Smith RB (2010) Viscoelastic-cycle model of interseismic deformation in the northwestern United States. Geophys J Int 181:665–696

Pollitz FF (1986) Pliocene change in Pacific-plate motion. Nature 320:738–741

Prawirodirdjo L, Bock Y, Genrich JF (2000) One century of tectonic deformation along the Sumatran fault from triangulation and Global Positioning System surveys. J Geophys Res 105:28343–28361

Priest GR (1990) Volcanic and tectonic evolution of the cascade volcanic arc, central oregon. J Geophys Res 95:19583–19599

Pritchard ME, Jay JA, Aron F, Henderson ST, Lara LE (2013) Subsidence at southern Andes volcanoes induced by the 2010 Maule, Chile earthquake. Nat Geosci 6:632–636

Pritchard ME, de Silva SL, Michelfelder G, Zandt G, McNutt SR, Gottsmann J et al (2018) Synthesis: PLUTONS: investigating the relationship between pluton growth and volcanism in the Central Andes. Geosphere 14:954–982

Kamesh Raju KA, Ray D, Mudholkar A, Murty GPS, Gahalut VK, Samudrala K et al (2012) Tectonic and volcanic implications of a cratered seamount off Nicobar Island, Andaman Sea. J Asian Earth Sci 56:42–53

Reath K, Pritchard ME, Moruzzi S, Alcott A, Coppola D, Pieri D (2019) The AVTOD (ASTER Volcanic Thermal Output Database) Latin America archive. J Volcanol Geoth Res 376:62–74

Riller U, Petrinovic IA, Ramelow J, Grekowiak J, Strecker M, Onken O (2001) Late Cenozoic tectonism, caldera and plateau formation in the central Andes. Earth Planet Sci Lett 188:299–311

Rosenau M, Melnick D, Echtler H (2006) Kinematic constraints on intra-arc shear and strain partitioning in the southern Andes between 38°S and 42°S latitude. Tectonics 25:4013. https://doi.org/10.1029/2005TC001943

Rosenbaum G, Gasparon M, Lucente FP, Peccerillo A, Miller MS (2008) Kinematics of slab tear faults during subduction segmentation and implications for Italian magmatism. Tectonics 27:TC2008. https://doi.org/10.1029/2007TC002143

Rosi M, Sbrana A (1987) Phlegraean fields. CNR Quaderni Ricerca Scientifica Roma 114:167 pp

Rowland JV, Wilson CJN, Gravley DM (2010) Spatial and temporal variations in magma-assisted rifting, Taupo Volcanic Zone, New Zealand. J Volcanol Geoth Res 190:89–108

Royden L, Patacca E, Scandone P (1987) Segmentation and configuration of subducted lithosphere in Italy: an important control on thrust-belt and foredeep-basin evolution. Geology 15:714–717

Ruch J, Vezzoli L, De Rosa R, Di Lorenzo R, Acocella V (2016) Magmatic control along a strike-slip volcanic arc: the central Aeolian arc (Italy). Tectonics 35:407–424

Santacroce R (1987) Somma Vesuvius. CNR Quaderni Ricerca Scientifica Roma, 251 pp

Sato H (1994) The relationship between late Cenozoic tectonic events and stress field and basin development in northeast Japan. J Geophys Res 99:22261–22274

Savelli C, Ligi M (2017) An updated reconstruction of basaltic crust emplacement in Tyrrhenian sea, Italy. Sci Rep 7:18024. https://doi.org/10.1038/s41598-017-17625-2

Schilling FR, Trumbull RB, Brasse H, Haberland C, Asch G, Bruhn G et al (2006) Partial melting in the Central Andean crust: a review of geophysical, petrophysical and petrologic evidence. In: Oncken O, Chong G, Franz G, Giese P, Götze H-J, Ramos VA et al (eds) The Andes—active subduction orogeny. Springer, Berlin, pp 459–474

Schmidt ME, Grunder AL, Rowe MC (2008) Segmentation of the Cascade Arc as indicated by Sr and Nd isotopic variation among diverse primitive basalts. Earth Planet Sci Lett 266:166–181

Schurr B, Asch G, Rietbrock A, Kind R, Pardo M, Heit B et al (1999) Seismicity and average velocities beneath the Argentine Puna plateau. Geophys Res Lett 26:3025–3028

Schurr B, Asch G, Rietbrock A, Trumbull RB, Haberland C (2003) Complex patterns of fluid and melt transport in the central Andean subduction zone revealed by attenuation tomography. Earth Planet Sci Lett 215:105–119

Schurr B, Rietbrock A, Asch G, Kind R, Oncken O (2006) Evidence for lithospheric detachment in the Central Andes from local earthquake tomography. Tectonophysics 415:203–223

Seebeck H, Nicol A (2009) Dike intrusion and displacement accumulation at the intersection of the Okataina Volcanic centre and Paeroa Fault zone, Taupo Rift, New Zealand. Tectonophysics 475:575–585

Sepúlveda F, Lahsen A, Bonvalot S, Cembrano J, Alvarado A, Letelier P (2005) Morpho-structural evolution of the Cordón Caulle geothermal region, Southern Volcanic Zone, Chile: insights from gravity and $^{40}Ar/^{39}Ar$ dating. J Volcanol Geoth Res 148:165–189

Serpelloni E, Anzidei M, Baldi P, Casula G, Galvani A (2005) Crustal velocity and strain-rate fields in Italy and surrounding regions: new results from the analysis of permanent and non-permanent GPS networks. Geophys J Int 161:861–880

Shearer P, Burgmann R (2010) Lessons learned from the 2004 Sumatra-Andaman megathrust rupture. Ann Rev Earth Planet Sci 38:103–131

Sieh K, Natawidjaja D (2000) Neotectonics of the Sumatran fault, Indonesia. J Geophys Res 105:28295–28326

Sielfeld G, Cembrano J, Lara L (2017) Transtension driving volcano-edifice anatomy: insights from andean transverse-to-the-orogen tectonic domains. Quatern Int 438:33–49

Sielfeld G, Lange D, Cembrano J (2019) Intra-arc crustal seismicity: seismotectonic implications for the southern Andes volcanic zone, Chile. Tectonics 38:552–578

Singer BS, Jicha BR, Harper MA, Naranjo JA, Lara LE, Moreno-Roa H (2008) Eruptive history, geochronology and magmatic evolution of the Puyehue-Cordon Caulle volcanic complex, Chile. Geol Soc Am Bull 120:599–618

Spinks K, Acocella V, Cole J, Bassett K (2005) Structural control of volcanism and caldera development in the transtensional Taupo Volcanic Zone, New Zealand. J Volcanol Geoth Res 144:7–22

Stankiewicz J, Ryberg T, Haberland C, Fauzi ND (2010) Lake Toba volcano magma chamber imaged by ambient seismic noise tomography. Geophys Res Lett 37:L17306. https://doi.org/10.1029/2010GL044211

Stern TA, Stratford WR, Salmon ML (2006) Subduction evolution and mantle dynamics at a continental margin: Central North Island, New Zealand. Rev Geophys 44:RG4002

Stern CR (2004) Active Andean volcanism: its geologic and tectonic setting. Revista Geológica De Chile 31:161–206

Subarya C, Chlieh M, Prawirodirdjo L, Avouac JP, Bock Y, Sieh K et al (2006) Plate boundary deformation associated with the great Sumatra-Andaman earthquake. Nature 440:46–51

Takada Y, Fukushima Y (2013) Volcanic subsidence triggered by the 2011 Tohoku earthquake in Japan. Nat Geosci 6:637–641

Tamura Y, Tatsumi Y, Zhao D, Kido Y, Shukuno H (2002) Hot fingers in the mantle wedge: new insights into magma genesis in subduction zones. Earth Planet Sci Lett 197:105–116

Tassara A, Yanez G (2003) Relationship between elastic thickness and the tectonic segmentation of the Andean margin. Rev Geol Chile 30:159–186

Tibaldi A, Corazzato C, Rovida A (2009) Miocene-Quaternary structural evolution of the Uyuni-Atacama region, Andes of Chile and Bolivia. Tectonophysics 471:114–135

Tibaldi A, Bonali FL, Corazzato C (2017) Structural control on volcanoes and magma paths from local- to orogen-scale: the central Andes case. Tectonophysics 699:16–41

Toda S, Stein RS, Sagiya T (2002) Evidence from the AD 2000 Izu islands earthquake swarm that stressing rate governs seismicity. Nature 419:58–61

Trumbull RB, Riller U, Oncken O, Scheuber E, Munier K, Hongn F (2006) The time-space distribution of Cenozoic volcanism in the south-central Andes: a new data compilation and some tectonic implications. In: Oncken O, Chong G, Franz G, Giese P, Götze H-J, Ramos VA et al (eds) The Andes—active subduction orogeny. Springer, Berlin, pp 29–43

Vezzoli L (1988) Island of Ischia. CNR Quaderni Ricerca Scientifica Roma 114, 133 pp

Villamor P, Berryman K (2001) A late Quaternary extension rate in the Taupo Volcanic Zone, New Zealand, derived from fault slip data. NZ J Geol Geophys 44:243–269

Villamor P, Van Dissen R, Alloway BV, Palmer AS, Litchfield N (2007) The Rangipo fault, Taupo rift, New Zealand: an example of temporal slip-rate and single event displacement variability in a volcanic environment. Geol Soc Am Bull 119:529–547

Villamor P, Berryman KR, Nairn IA, Wilson K, Litchfield N, Ries W (2011) Associations between volcanic eruptions from Okataina volcanic center and surface rupture of nearby active faults, Taupo rift, New Zealand: insights into the nature of volcano-tectonic interactions. Geol Soc Am Bull 123:1383–1405

Villamor P, Berryman KR, Ellis SM, Schreurs G, Wallace LM, Leonard GS et al (2017) Rapid evolution of subduction-related continental intraarc rifts: the Taupo Rift, New Zealand. Tectonics 36:2250–2272

Volker D, Kutterolf S, Wehrmann H (2011) Comparative mass balance of volcanic edifice at the southern volcanic zone of the Andes between 33° and 46°S. J Volcanol Geoth Res 205:114–129

Wada I, He J, Hasegawa A, Nakajima J (2015) Mantle wedge flow pattern and thermal structure in Northeast Japan: Effects of oblique subduction and 3-D slab geometry. Earth Planet Sc Lett 426:76–88

Waldien TS, Meigs AJ, Madin IP (2019) Active dextral strike-slip faulting records termination of the Walker Lane belt at the southern Cascade arc in the Klamath graben, Oregon, USA. Geosphere 15:882–900

Wallace LM, Beavan J, McCaffrey R, Darby D (2004) Subduction zone coupling and tectonic block rotations in the North Island, New Zealand. J Geophys Res 109: B12406. https://doi.org/10.1029/2004JB003241

Walter TR, Amelung F (2007) Volcanic eruptions following $M \geq 9$ megathrust earthquakes: implications for the Sumatra-Andaman volcanoes. Geology 35:539–542

Ward KM, Zandt G, Beck SL, Christensen DH, McFarlin H (2014) Seismic imaging of the magmatic underpinnings beneath the Altiplano-Puna volcanic complex from the joint inversion of surface wave dispersion and receiver functions. Earth Planet Sci Lett 404:43–53

Watt SFL, Pyle DM, Mather TA (2009) The influence of great earthquakes on volcanic eruption rate along the Chilean subduction zone. Earth Planet Sci Lett 277:399–407

Weaver CS, Grant WC, Shemeta JE (1987) Local crustal extension at Mount St. Helens, Washington. J Geophys Res 92:10170–10178

Weller O, Lange D, Tilmann F, Natawidjaja D, Rietbrock A, Collings R et al (2012) The structure of the Sumatran Fault revealed by local seismicity. Geophys Res Lett 39:L01306. https://doi.org/10.1029/2011GL050440

Wells RE (1990) Paleomagnetic rotations and the Cenozoic Tectonics of the Cascade Arc, Washington, Oregon, and California. J Geophys Res 95:19409–19417

Wells RE, McCaffrey R (2013) Steady rotation of the Cascade Arc. Geology 41:1027–1030

Widiwijayanti C, Deverchere J, Louat R, Sebrier M, Harjono H, Diament M et al (1996) Aftershock sequence of the 1994, Mw 6.8, Liwa earthquake (Indonesia): seismic rupture process in a volcanic arc. Geophys Res Lett 23:3051–3054

Wilson CJN, Houghton BF, McWilliams MO, Lanphere MA, Weaver SD, Briggs RM (1995) Volcanic and structural evolution of Taupo Volcanic Zone, New Zealand: a review. J Volcanol Geoth Res 68:1–28

Wilson CJN (1996) Taupo's atypical arc. Nature 379:27–28

Wortel MJR, Spakman W (2000) Subduction and Slab Detachment in the Mediterranean-Carpathian Region. Science 290:1910–1917

Yoshida T, Kimura J, Yamada R, Acocella V, Sato H, Zhao D et al (2013) Evolution of the Late Cenozoic Magmatism and the crust-mantle structure in the NE Japan Arc. Geol Soc London Spec Publ 385. https://doi.org/10.1144/SP385.15

Yoshida T (2001) The evolution of arc magmatism in the NE Honshu arc, Japan. Tohoku Geophys J 36:131–249

Zellmer GF, Chen K-X, Gung Y, Kuo B-Y, Yoshida T (2019) Magma transfer processes in the NE Japan Arc: insights from crustal ambient noise tomography combined with volcanic eruption records. Front Earth Sci 7:40. https://doi.org/10.3389/feart.2019.00040

13 Hot Spots

13.1 Introduction

Hot spots are the surface expression of plumes of hotter and lighter material upwelling from the Earth's mantle. The current number of hot spots is estimated to range between 45 and 70: these are mostly in intraplate settings, especially on oceanic lithosphere, and along divergent plate boundaries. Neglecting hot spots along divergent boundaries and shaped by the related far-field tectonics (as with Afar and Iceland, described in Chap. 11), oceanic hot spot volcanoes show a considerable variability in distribution, evolution and activity, as at Hawaii, Galapagos, Easter Island, Reunion, Canary Islands and Azores. This results from the different mantle plume properties (plume configuration and productivity) and local tectonic context (age of the intruded oceanic lithosphere, rate of plate motion, pre-existing structures), which make each hot spot distinctive. Despite this variability, most oceanic hot spot volcanoes also display recurrent structural features, which include overlapping mafic edifices with summit calderas, radial volcanic rift zones and flank instability. Then, there is the less common and more evolved volcanism derived from continental hot spots, of which Yellowstone is the most dramatic example and, at the same time, quite distinct from other less productive continental cases, as for instance Tibesti. All these characteristics make hot spots widely different, stimulating from structural and magmatic perspectives and complicating established models.

This chapter considers examples from oceanic and, subordinately, continental hot spots. In the Pacific Ocean region, three very different types of hot spots are considered: Hawaii, Galapagos and Easter Island. Then, in the Indian Ocean region the productive Reunion hot spot is presented. Subsequently, in the Atlantic Ocean region, the different hot spots of the Canary Islands and the Azores are considered. Finally, the different features of the continental Yellowstone and Tibesti hot spots are presented.

The main aims of this chapter are:

- first to define the primary structural and magmatic characters of hot spot activity, describing representative oceanic and continental cases;
- second to provide a general framework summarizing the structural and magmatic features of hot spot volcanoes, with a working hypothesis for oceanic types.

13.2 Hawaii Hot Spot

The Hawaiian plume is imaged as a continuous low velocity anomaly from the base of the mantle to the surface. Its shallowest part may consist of discrete portions, diapirs, plumelets or filaments, with radii of ~ 10 km and heights ~ 100 km.

V. Acocella, *Volcano-Tectonic Processes*, Advances in Volcanology,
https://doi.org/10.1007/978-3-030-65968-4_13

Ascending mantle diapirs may be sheared by the differential motion between the upper mantle and the lithosphere. The shallow plume contains a melt-rich zone 110–155 km beneath Hawaii Island, whose axial region is a mixture of recycled subducted components and primitive lower mantle materials, implying spatial heterogeneity. The plume outside the head region, inferred to lie beneath the southeast coast of Hawaii Island, is less heterogeneous but distinct from the upper mantle (Wolfe et al. 2009; Farnetani and Hoffmann 2010; Rychert et al. 2013).

The ~6000 km long Hawaiian-Emperor volcanic chain marks the surface trace of the plume, consisting of at least 107 volcanoes with volume of ~1 million km^3 (Fig. 13.1a). Individual seamounts along the chain may be the topographic expression of discrete diapirs from a single deeper mantle plume. The chain is age progressive, with active volcanoes at the southeast end and 75–80 Ma old volcanoes at the northwest end. The bend between the Hawaiian and Emperor chains reflects a major change in Pacific plate motion at ~47 Ma. Age-distance data along the chain also reflect minor deviations from an overall linear trend. Volume calculations highlight an increase in the hot spot activity since 30 Ma and especially during the last 1 Ma (Fig. 13.1b; Vidal and Bonneville 2004; Robinson and Eakins 2006; Sharp and Clague 2006).

The Hawaiian volcanoes undergo progressive subsidence and erosion as they move farther from the hot spot, becoming drowned islands or seamounts. The subsidence results from the load on the lithosphere due to the growing volcanoes, which causes the lithosphere to sag and the Hawaiian Ridge to subside. Superimposed is the subsidence caused by the thermal contraction of the aging oceanic lithosphere, which becomes older westwards. Most volcanoes have subsided 2–4 km since reaching the sea surface, while their bases have subsided 5–8 km, so that about half of the construction of the volcanoes is reduced by subsidence. The bulk of the subsidence is complete ~1 Ma after initiation of volcanism on the seafloor. Overall, the observed rates of vertical motion of the islands adequately record the flexural response of the lithosphere to volcanic loads (Moore 1987; Huppert et al. 2015).

Hawaiian volcanoes contain intrusions estimated at less than 30% by volume, indicating that the edifices are predominately built extrusively. Edifice building has involved four major growth stages. The earliest is the submarine alkalic preshield stage, before the development of a shield volcano. The alkalic composition derives from a nascent magma-transport system and less extensive melting at the plume periphery. The subsequent tholeiitic shield stage accounts for more than 95% of the volume and may be ~1 Ma long. This marks the time when a volcano is near or above the hot spot and its robust magma supply system builds the shield, which mainly consists of tholeiitic basalts and subordinate basalt andesites. Volcanism wanes as the volcano moves away from the hot spot, passing into the post-shield stage. Shallow magma reservoirs (1–7 km depth) cannot be sustained, as magma supply lessens, but reservoirs at 20–30 km depth persist. The rate of extrusion diminishes by a factor of 10 and lavas become more alkalic, eventually evolving to trachytic compositions, as the degree of melting diminishes. Therefore, a cap of alkalic basalt and associated differentiates covers the tholeiitic shield (Lipman and Calvert 2013; Clague and Sherrod 2014). After several million years of erosion, alkalic rejuvenated-stage magma may erupt from isolated vents as the volcanoes override the moat flexure created by loading and subsidence of the new shield volcano lying southeast on the chain. The rejuvenated onshore and offshore volcanism may be partly explained by the enhanced rupture of the oceanic lithosphere in its flexural uplift zone, producing focused magmatic fluxes over an area two to ten times the eruption area. Five Hawaiian volcanoes contain rejuvenated-stage volcanism following quiescent periods that range from 2 to less than 0.5 Ma (Bianco et al. 2005; Thordarson and Garcia 2018, and references therein). Such a prolonged construction of the Hawaiian shields has been also often punctuated by giant and rapid landslides, with seventeen discrete large slides forming around the main islands in the past 5 Ma.

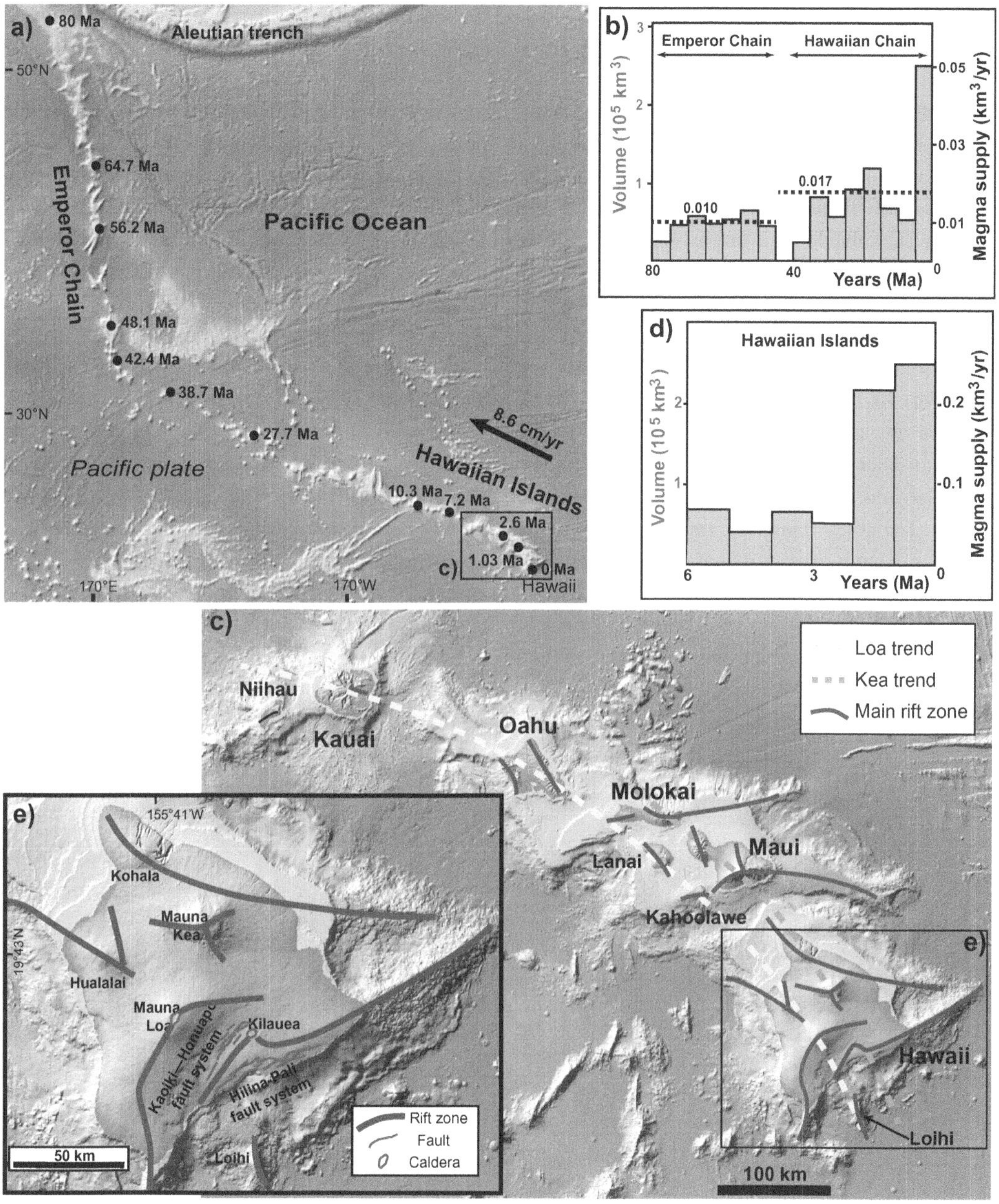

Fig. 13.1 Hawaii hot spot. **a** Overview of the Hawaii-Emperor Chain, resulting from the activity of the Hawaiian plume; plate velocity is relative to the hot spot; **b** related variation of the volcano volumes with time. **c** Overview of the Hawaiian Islands, with the "Kea" and "Loa" trends and the main volcanic rift zones; inset **d** shows the related variation of the erupted volumes with time (Robinson and Eakins 2006). **e** Overview of the subaerial and submarine portions of the island of Hawaii, its volcanoes, rift zones and main fault systems. Digital Elevation Models provided by GeoMapApp

Volcanic centres on the Hawaiian Ridge follow two curved loci roughly parallel to the ridge and ~40 km apart, forming the "Loa" and "Kea" subchains, with geochemical variations reflecting preferential sampling of two distinct sources of deep mantle material. These loci may mark the edges of downbending of the two sides of the lithospheric subsidence trough, where fracturing conducts magma to the surface. The progressive involvement of the "Loa" component, as recorded in the magma erupted since ~6.5 Ma, records the strengthening of the Hawaiian plume through time (Fig. 13.1c; Moore 1987; Weis et al. 2011; Harrison et al. 2017). The volcanoes of the Hawaiian Islands become progressively younger southeastwards, from Kauai (3.9–5.8 Ma) to Hawaii (<1 Ma). Hawaii, the largest of the Hawaiian Islands, has grown at an average rate of ~10^5 km^3/Ma and is presently close to its maximum size. Subaerial Hawaii Island consists of five coalesced and overlapping volcanoes which have formed in the past 1 Ma: Kohala, Hualalai, Mauna Kea, Mauna Loa and Kilauea. Each of these volcanoes requires approximately 600 ka to grow from the ocean floor to the end of shield building stage, reaching the ocean surface about midway (Fig. 13.1e). Kohala completed shield building at ~245 ka, Hualalai and Mauna Kea at 130 ka. Mauna Kea, the tallest volcano, last erupted at ~4.5 ka. It is capped by alkalic lava flows modifying the original shield, and it is uncertain if a summit caldera and well-defined volcanic rift zones existed. Although tall, Mauna Kea may have a modest volume, as its edifice appears perched on the flank of Kohala (Moore and Clague 1992; Lipman and Calvert 2013; Clague and Sherrod 2014; Poland et al. 2014, and references therein). Mauna Loa is the largest volcano, showing shield stage tholeiitic lava mostly issued from the summit and southwest and northeast rift zones, although radial eruptive fissures also exist on its western and northern flanks (Fig. 13.2). Eruptive activity ranges from at least 700 ka to 1984 AD. The current summit magma system includes two interconnected reservoirs, with one centred beneath the south caldera

Fig. 13.2 Hawaii hot spot. Summit of Mauna Loa volcano viewed from the south, showing the caldera with the 1940 eruptive vent. The eruptive fissure and linear vents in the foreground are the proximal part of the Southwest Rift, and the summit of nearby Mauna Kea volcano is partly visible in the background. *Photo credit* U.S. Geological Survey, USGS

margin and the other elongated along the caldera axis (Poland et al. 2014, and references therein). Kilauea is the youngest and most active volcano of Hawaii. It consists of shield stage tholeiitic lava issued from the summit caldera, as well as the east and southwest rift zones since ~100 ka. The present caldera formed at ~1500 AD, with the subsequent 300 years characterized by highly explosive activity (Langenheim and Clague 1987). Whether or not magma is present along and within the deep rift zone, which extends from beneath the shallower rift at ~3 km depth to the base of the volcano at ~9 km depth, remains an open question. It is probable that most magma entering Kilauea passes through the summit reservoir system before entering the rift zones. The connectivity between the distal rift zones and the summit magma chamber of Kilauea has been observed during several eruptions and further confirmed in 2018, when lateral magma propagation and eruption along the East Rift Zone was accompanied by summit caldera collapse (see Sects. 5.10.3 and 6.7; Okubo et al. 1997; Poland et al. 2014; Clague and Sherrod 2014; Neal et al. 2019; Wauthier et al. 2019).

The active submarine volcano of Loihi, 30 km southeast of Hawaii, is the latest activity of the Hawaiian hot spot. The NNW-SSE striking edifice is ~30 km long and 3 km high. The summit has a caldera-like depression, hosting pit craters and a resurgent block above a magma chamber lying between 1 and 2 km of depth. The volcano taps a relatively primitive part of the plume, producing a wide range of magma compositions reflecting increasing degrees of partial melting as it drifts toward the hot spot centre (Clague et al. 2019, and references therein).

Volcanic rift zones are integral structural elements of Hawaiian volcanoes, consisting of narrow (related vents stretch across most less than 2 km) and long (some more than 100 km) portions of focused volcanic and tectonic activity, including spatter cones, spatter ramparts, pit craters, extension fractures, faults, grabens and eruptive fissures (Fig. 13.3).

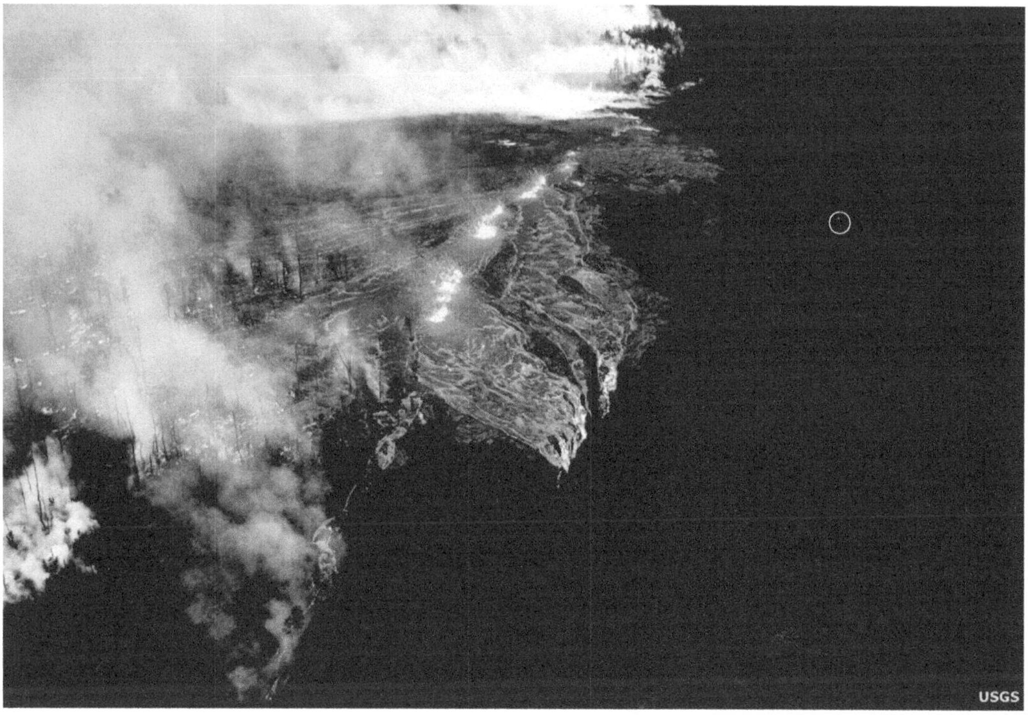

Fig. 13.3 Hawaii hot spot. Detail of eruptive fissure developed along the Eastern Rift Zone of Kilauea on March 2011: lava pours from the fissure just after daybreak and cascades out of sight into a deep crack; geologist in yellow circle for perspective. *Photo credit* U. S. Geological Survey, USGS

At depth, the rift zones consist of a thick, subvertical dike-like magma system, where dikes are usually intruded laterally from a central magma reservoir, propagating up to several tens of kilometres without erupting (Fig. 13.4; Delaney et al. 1990). The orientation of the rift zones depends on the presence of regional structures and the local stress field. The long east trending rift zones of Hawaiian volcanoes may follow regional fractures in oceanic crust activated by arching of the Hawaiian swell in front of the propagating hot spot. At volcanoes that grew on the flanks of an already existing edifice, the orientation of rift zones is influenced by the gravitational stress field resulting from the buttressing conditions of the older volcano. In the case of Kilauea, growing on the flank of Mauna Loa, the rift zones are oriented perpendicular to the direction of least buttressing, imposed by the oceanic abyssal plain to the southeast (Fiske and Jackson 1972; Walter et al. 2006; Lipman and Calvert 2011). Loihi does not follow this behaviour, as its NNW-SSE trending rift zone suggests a formation beyond the influence of the gravitational load of the adjacent older volcanoes, probably influenced by the flexure of the lithosphere (Garcia et al. 2006).

The stresses on the volcano flanks also develop large-scale landslides related to the activity of the rift zones, especially the East Rift Zone of Kilauea. The 1975 magnitude *M7.7* and 2018 *M6.9* earthquakes along the detachment at the base of the southern flank of Kilauea likely resulted from the stress induced by overpressured dikes intruded in the rift zone (Swanson et al. 1976; Chen et al. 2019). Both earthquakes produced an abrupt southward movement of the south flank across the underlying oceanic crust, activating the Hilina-Pali normal fault system, which represents the surface expression of the

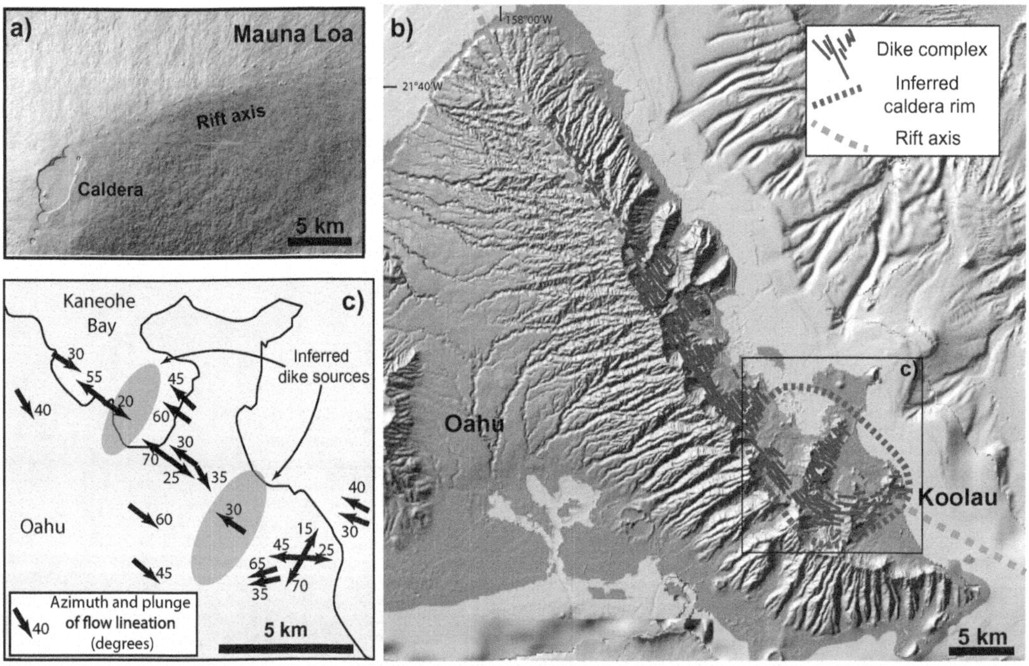

Fig. 13.4 Hawaii hot spot. Comparison between the morphology and structure of active **a** and eroded **b** Hawaiian rift zones, shown at the same scale for comparison. **a** Morphology at the surface of the Northeast Rift of Mauna Loa radiating from the summit caldera and consisting of multiple eruptive fissures at the surface. **b** Deeper structure of the eroded Koolau rift zone, northeast Oahu, at ~1 km depth. **c** Measured directions and plunge of magmatic flow within the dikes along the Koolau rift zone beneath the eroded caldera (Walker 1987). Digital Elevation Models provided by GeoMapApp

instability of this flank (Fig. 13.1e). Such faulting not only provides a means for the flanks to adjust continuously to intrusions, but also contributes to the stress field that encourages dikes to propagate along the rift zone, with dike intrusion and lateral spreading being major contributors to the growth of the volcano (Peterson and Moore 1987; Dieterich 1988; Wang et al. 2019). In addition to Kilauea, Mauna Loa is inferred to have undergone volcano spreading during its early growth, activating the Kaoiki–Honuapo fault system (analogue of the Hilina-Pali fault system on Kilauea; Fig. 13.1e). Cumulative deformation of Mauna Loa's south flank during growth of Kilauea since 200–300 ka involved more than 10 km of seaward spreading, displacing the rift zones of Kilauea, while its deep plumbing system and summit magma reservoir remained nearly fixed; such an evolution also accounts for the arcuate strike of the rift zones of Kilauea (Lipman et al. 2006).

13.3 Galapagos Hot Spot

The Galapagos Archipelago lies in the eastern Pacific Ocean, 200–300 km south of the Galapagos Spreading Ridge that separates the Nazca (to the south) and Cocos plates (Fig. 13.5). The Galapagos Ridge first migrated to the northern edge of the hot spot at 19.5 Ma, overlying the hot spot between 19.5 and 14.5 Ma and then to the south of the hot spot at 14.5 Ma. This split the paleo-Carnegie Ridge into the present Carnegie and Malpelo ridges, with new seafloor separating the two ridges on the Nazca and Cocos plates, respectively. Between 12 and 11 Ma, a later ridge jump shifted the spreading axis north of the hot spot, while spreading between the Carnegie and Malpelo ridges continued until 9.5 Ma (Werner et al. 2003).

Beneath the western Galapagos Archipelago, the thinning of the upper mantle transition zone, between 410 and 660 km depth, within an area

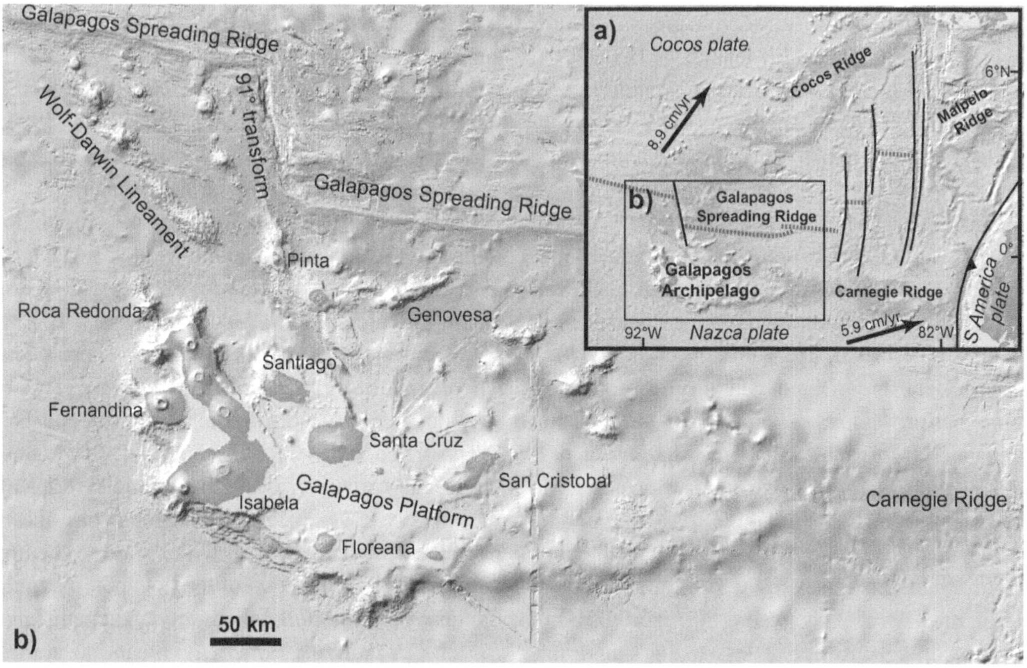

Fig. 13.5 **a** Tectonic setting of the Galapagos hot spot, with the velocities of the plates relative to the hot spot. **b** Overview of the Galapagos Archipelago and the nearby spreading ridge. Digital Elevation Models provided by GeoMapApp

~100 km wide southwest of Fernandina Island, suggests plume upwelling below 410 km. The upper 300 km of the mantle beneath the archipelago contain a low velocity anomaly tilted towards the Galapagos Ridge. At depths of less than 100–120 km, the plume spreads both toward and against eastward plate motion. The plume top is centred near Fernandina (Villagomez et al. 2014). The thickness of the high velocity lithospheric mantle above the plume varies from ~60 km beneath the southwest part of the archipelago to ~45 km below the northeast part. The thicker-than-normal lithosphere to the southwest may result from the residuum from hot spot melting. The thinner lithosphere to the northeast, combined with the lateral deflection of the plume, results in volcanism scattered over a relatively large area. Indeed, the general eastward increase in age of lavas from seamounts, consistent with a hot spot model for their generation, is accompanied by dispersed volcanism, observed also in the distribution of the active volcanoes. In particular, the construction of seven simultaneously active volcanoes indicates that the western Galapagos Islands of Fernandina and Isabela do not result from a simple progression of the lithosphere over a narrowly focused mantle plume. The result is that, while the eruption rate of the Galapagos plume (10^3 km^3/Ma) is smaller than that of the Hawaiian one, active volcanism is more widespread (~40,000 km^2) than in Hawaii (~20,000 km^2; Sinton et al. 1996; Canales et al. 2002; Rychert et al. 2014; Sinton et al. 2018).

The Galapagos Islands stand on a broad volcanic platform lying on young (<10 Ma) oceanic lithosphere. This platform is the shallowest portion of the Carnegie Ridge, whose post-9.5 Ma construction results from the passage of the Nazca plate over the hot spot. The Galapagos platform consists of several tens of large, stepped submarine terraces between 800 and 3500 m depth, sloping outward at less than 2° and surrounded by ~300 m high escarpments with slopes of ~24°. Each terrace results from a sequence of major submarine eruptions, suggesting episodic magma supply. In the western Galapagos, characterized by the currently most active Isabela and Fernandina islands, the platform may have formed in the last ~3 Ma, younging westward and just predating the development of the overlying shield edifices (Geist et al. 2008).

The active volcanoes of the western Galapagos overlying the mantle plume in Isabela and Fernandina islands display a distinctive morphological feature. This consists of the abrupt steepening of their subaerial slope, which passes on average from less than 10° in the distal area to more than 10° and, locally, more than 30°, in the proximal area, giving these volcanoes an "inverted soup plate" profile. From a structural pint of view, each of the volcanoes of the western Galapagos has a summit caldera with circumferential and radial eruptive fissures (Fig. 13.6a).

The circumferential fissures, rarely observed on active volcanoes elsewhere, are high and focused on the upper volcanic edifice, near the caldera rim. The radial fissures are scattered on the mid to lower slopes. This very distinctive fissure pattern results from the stresses due to the load of the volcanic edifice (for the radial fissures) and the unloading induced by the caldera depression (for the circumferential fissures; see Chap. 7; Chadwick and Howard 1991; Mouginis-Mark et al. 1996; Corbi et al. 2015).

The erupted magmas in the western Galapagos shield volcanoes are mainly tholeiitic, although alkali basalts may be locally present, as at Cerro Azul. Also, approximately 1 km^3 of rhyolite was erupted at Alcedo in the last 120 ka from at least three vents and in 2–5 episodes. These rhyolites result from fractional crystallization, which occurs because the volcano has been carried away from the hot spot. More generally, the variability of the magmatic systems of the western Galapagos volcanoes has been explained by their location with regard to the hot spot. At the leading edge of the hot spot, magmas are only partly buffered thermally and chemically (as at Cerro Azul). Once a volcano becomes mature (as Fernandina), magmas transit through a mush pile before residing in a shallow subcaldera sill, which buffers all magmas thermally and chemically and creates a monotonous suite: these volcanoes are the most active. In the dying

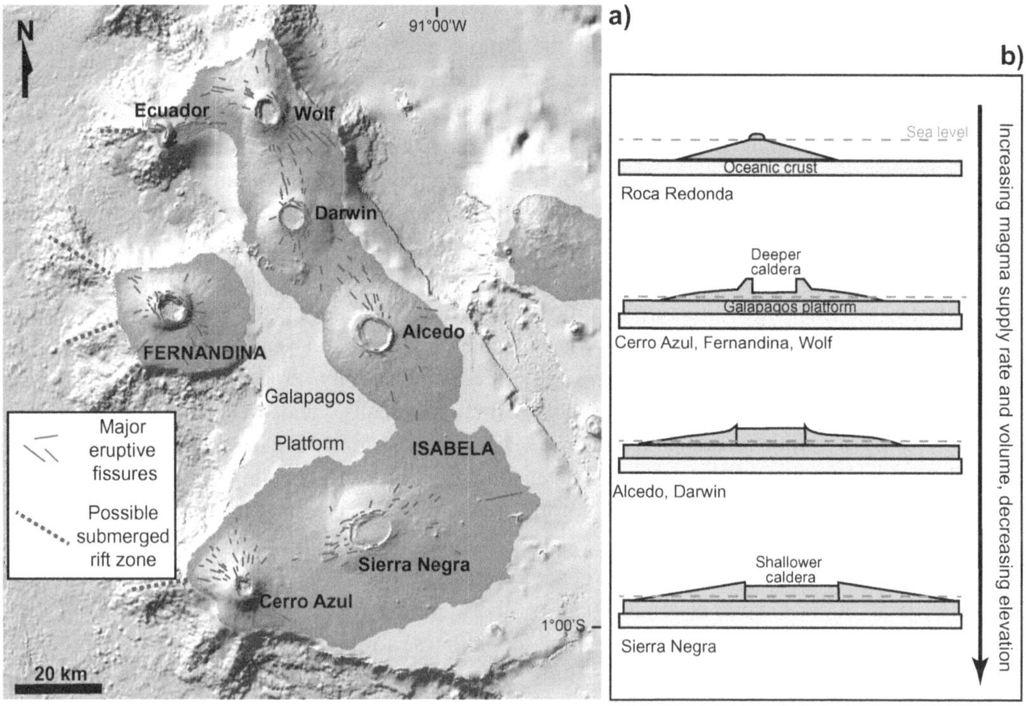

Fig. 13.6 Galapagos hot spot. **a** Overview of Fernandina and Isabela islands, western Galapagos, highlighting the calderas and the main eruptive fissures. Digital Elevation Model provided by GeoMapApp. **b** Model for the effect of magma supply rate on the growing forms of the western Galapagos shield volcanoes (orange, section view), from the lowest supply of the barely emergent conical volcano of Roca Redonda, to the highest supply of large shields (as Sierra Negra), with relatively shallow calderas and gently dipping flanks (modified after Naumann and Geist 2000)

phase, when the volcano is carried away from the hot spot, the mush begins to solidify and silicic magma is also generated (as at Alcedo). The lavas erupted from the eastern Galapagos are on the whole much more primitive (Geist et al. 2014; Harpp and Geist 2018, and references therein). Two discrete regions of magma storage have been often geodetically and petrologically inferred below some western (Wolf, Fernandina, Cerro Azul) and eastern Galapagos volcanoes. These regions consist of a shallower small magma lens and a major zone of magma storage in the lower crust, from which most of the erupted material is sourced (Stock et al. 2018).

Fernandina, ~30 km wide and ~1500 m high above sea level (a.s.l.), is the westernmost volcano, with a summit caldera, well-developed and focused circumferential fissures, and diffuse radial fissures. The caldera results from the last important collapse, reaching a maximum of 350 m in June 1968. A sill-like magma chamber and an oblate-spheroid cavity are postulated at 1.1 and ~4.9 km below sea level, respectively (Simkin and Howard 1970; Chadwick et al. 1991, 2011; Bagnardi and Amelung 2012). The eruptions from radial and circumferential fissures are usually fed by dikes initiated as sills from the shallow magma chamber. The sills propagate laterally, curve upward and steepen to become dikes when feeding circumferential fissures, or twist about a radially oriented axis to feed radial fissures on the volcano flanks (Fig. 7.8; Bagnardi et al. 2013).

Isabela is the largest island in the western Galapagos, resulting from the coalescence of six active volcanoes, each spaced by a few tens of kilometres and together encircling the eastern portion of Fernandina (Fig. 13.6a). From the north, these are: Ecuador, Wolf, Darwin, Alcedo, Sierra Negra and Cerro Azul. Ecuador, at the northwest tip, provides two distinctive features. First, it experienced at least one sector collapse; this removed its western half of the edifice within the past 100 ka. Second, its eastern rift zone, which connects the volcano with Wolf, is constructed of lavas from focused and subparallel linear vents and shows evidence of a feeder conduit; this is the Galapagos structure that most closely resembles a Hawaiian rift zone (Rowland et al. 1994; Geist et al. 2002). The calderas of Alcedo and Sierra Negra volcanoes are both characterized by a minor trapdoor resurgence, with uplift of ~30 and ~100 m, respectively. Interestingly, their short-term asymmetric uplift, as geodetically observed in the last decades, has mimed the longer-term asymmetric resurgence (Chadwick et al. 2006; Galetto et al. 2019).

The observed variations in elevation, slope, volume and area among the western Galapagos volcanoes have been linked to different long-term eruption rates. Volcanoes with shallower calderas, as Sierra Negra, are constructed when radial fissures erupt voluminous lavas to produce a wide, flat shield with minimal, steep, upper flank segments. This behaviour may result from an intermediate to higher supply of magma in more mature volcanoes. Volcanoes with deeper calderas, as Fernandina, develop when numerous small-volume eruptions from circumferential fissures build a steep central summit carapace surmounting gently dipping lower flanks. This behaviour may result from a lower supply of magma in less mature volcanoes (Fig. 13.6b; Naumann and Geist 2000).

The central, northern and eastern Galapagos consist of several smaller islands with older record of activity, eroded inactive volcanoes or their remnants, and evidence of a regional structural control (Fig. 13.7). The northern volcanic province, including the islands of Genovesa and Pinta, results from the combination of excess temperatures, weak lithosphere and regional stresses developed from the interaction between the plume and the Galapagos Ridge. Genovesa formed in the past 350 ka, has ~NE-SW trending fissures parallel to the Genovesa Ridge, a 55 km long volcanic rift zone that is the most prominent submarine rift in the Galapagos, fed by the plume (Harpp et al. 2002, 2003). Pinta lies between Isabela and the Galapagos Ridge, 20 km to the west of the ~N-S trending Galapagos transform fault. It rises from a prominent

Fig. 13.7 Galapagos hot spot. Eroded neck (vertical spine to the right) and vents (in the foreground) accompanied by more recent monogenic volcanism (in the background) on the eastern coast of Santiago Island, central Galapagos

NNW-SSE trending submarine fault scarp, and its elongation reflects the control of this major structure. Regional NNW-SSE trending fracture systems also influenced the fissural volcanism postdating the shield volcano, which became emergent at ~0.7 Ma (Cullen and McBirney 1987). San Cristobal Island, at the eastern end of the archipelago, shows a central shield, formed until 0.6 Ma and, to the northeast, NE-SW trending fissure eruptions active from 0.9 Ma to historical times (Geist et al. 1986).

The differences between the western and the eastern Galapagos volcanoes, in terms of type of volcanic edifice, presence of calderas or rift zones, productivity and composition, have been related to the proximity to the wandering Galapagos Spreading Ridge (Harpp and Geist 2018). In particular, the magmatic flux of the volcanoes negatively correlates to the distance to the ridge: when the ridge was close to the plume, that is when the eastern Galapagos volcanoes formed, most of the plume-derived magma was directed to the ridge. Currently, the active western Galapagos volcanoes are farther from the ridge and most of the plume-derived magma focuses beneath the large young shields. The structure of the volcanoes seems also affected by the proximity to the ridge, with ridge-induced far-field stresses promoting linear volcanism with parallel rift zones in the eastern Galapagos, and plume-induced near-field stresses promoting central volcanoes with calderas and circumferential and radial rift zones in the western Galapagos. Therefore, rather than being controlled by plate motion across the hot spot, as with the Hawaiian volcanoes, the Galapagos volcanoes seem more affected by the changing far-field tectonic environment.

13.4 Easter Island Hot Spot

Easter Island (Chile) lies on the Nazca plate, ~350 km to the east of the East Pacific Rise bordering the Easter microplate, in the western part of the ~3000 km long Easter Seamount Chain. The Easter microplate formed approximately 5 Ma ago during the northward propagation of the eastern spreading segment of the East Pacific Rise and its overlap with the southward propagating western segment (Fig. 13.8a; Naar and Hey 1991; Searle et al. 1993; Rusby and Searle 1995).

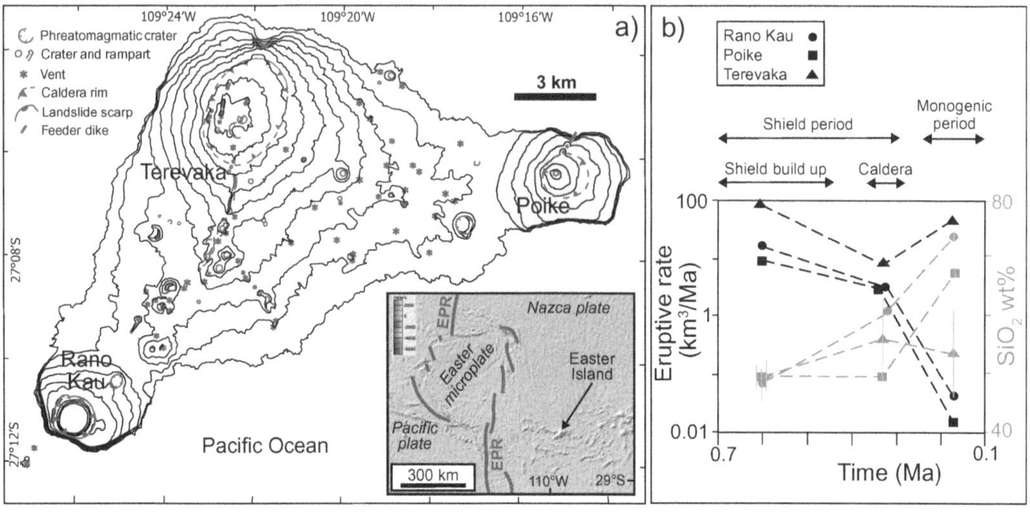

Fig. 13.8 Easter Island hot spot. **a** Structural map of Easter Island, Chile, showing the three volcanic edifices, the buried (dashed lines) and exposed (continuous lines) calderas and the eruptive fissures. Inset shows the regional setting, with the red lines representing segments of the East Pacific Rise. **b** Variation of the eruptive rates (in km^3/Ma, on a log scale; in black) and SiO_2 content of the lavas (in blue) with time considering the periods of shield build up, caldera formation and monogenic volcanism for each of the Easter Island volcanoes (modified from Vezzoli and Acocella 2009)

N-S trending ridges, parallel to the axis of the East Pacific Rise, characterize the oceanic floor of the ~350 km wide region between Easter Island and the East Pacific Rise. This region, characterized by 2–4 Ma old ocean seafloor, hosts several young (0.6–0.2 Ma) submarine volcanic seamounts. The age difference between the oceanic seafloor and its seamounts, as well as the volcanism younger than 0.78 Ma on Easter Island, indicates that the magmatism making up this western portion of the Easter Seamount Chain is intraplate (Naar and Hey 1991; Stoffers et al. 1994; Haase et al. 1997). More generally, radiometric dating along the Easter Seamount Chain shows progressively younger ages westward, confirming that the Easter Seamount Chain results from the motion of the Nazca plate over a mantle plume located immediately to the west of Easter Island and active for at least 30 Ma. Focal mechanisms from earthquakes in the area between Easter Island and the East Pacific Rise show predominant normal motion, with an overall NW–SE direction of extension (O'Connor et al. 1995; Haase 2002; Bird 2003; Ray et al. 2012).

Easter Island is the emerged portion, reaching ~500 m above sea level, of the submarine Rano Kau Ridge, rising for nearly 3000 m from the surrounding ocean floor. This ridge is ~NE-SW elongated, perpendicular to the extension direction revealed by the seismicity west of Easter Island. The emerged portion of Easter Island consists of three overlapping basaltic central volcanoes with a shield-like shape: Rano Kau, Terevaka, and Poike (Fig. 13.8a). Of these, Terevaka is the largest, with Rano Kau and Poike located on its southern and eastern slopes, respectively. These volcanoes range in composition from tholeiitic basalts to rhyolites, and the erupted lavas show considerable crystal fractionation, suggesting shallow magma chambers and a decreasing magma supply (Haase et al. 1997; Gonzalez-Ferran et al. 2004). The three polygenic volcanoes experienced a similar and nearly coeval evolution, characterized by two main stages (Vezzoli and Acocella 2009). The first stage, between ~0.78 and 0.3 Ma, consisted of the build up of the three basaltic shields (Fig. 13.9). This build up was associated with eruption rates of similar order of magnitude for the three volcanoes, although Terevaka grew faster. The SiO_2 content of the erupted magmas remained remarkably constant and low. The final part of the shield stage was characterized by caldera formation, with a significant drop (one order of magnitude) in the eruptive rates, as well as a slight increase in the SiO_2 content of the erupted magmas and the emission of more evolved highly porphyritic lavas (Fig. 13.8b). The second stage, between 0.24 and 0.11 Ma, was characterized by the activity of monogenic vents along the flanks of the three shields, with eruptive fissures showing a dominant NNE-SSW trend, subparallel to the ~NE-SW elongated Rano Kau Ridge. In addition to the preferred NNE-SSW orientation, part of the monogenic

Fig. 13.9 Panoramic view of Easter Island from the summit of Poike volcano, with the Terevaka and Rano Kau shield volcanoes and the post-shield monogenic volcanoes

vents on the Terevaka flanks also show an overall quite dispersed or scattered distribution, resembling poorly-developed or diffuse radial rift zones. During the monogenic stage, the supply to the magmatic systems of Poike and Rano Kau decreased, resulting in minor eruptions of silica-rich, highly viscous, residual and cooler magma. Conversely, the vents on Terevaka, associated with higher eruptive rates and lower SiO_2 content, suggest a more enriched mantle source, with input of new magma (Fig. 13.8b). The input of this new and more primitive magma prevented Terevaka reaching the evolved compositions of the other two volcanoes. The subaerial evolution of Easter Island thus appears controlled by the rise of moderate batches of magma in a context of decreasing supply. This decrease has been responsible for cooling the peripheral magmatic systems (Rano Kau and Poike volcanoes), allowing only the rise of magma through the larger Terevaka magmatic system and its eruption, mainly through radial dikes, along the volcano flanks. Another distinctive feature of Easter Island is the very limited evidence of flank instability and collapse, possibly restricted to a minor northern portion of the Terevaka shield (Vezzoli and Acocella 2009).

Because of its tectono-magmatic features and limited activity, testified by a low eruptive rate, generally diffuse rift zones and very limited lateral collapse, Easter Island may represent an end-member type of oceanic hot spot volcano, opposite to Hawaii.

13.5 Reunion Hot Spot

The Reunion hot spot lies on the African plate, in the southwest Indian Ocean. The products of the 5000 km long ~N-S trending hot spot track highlight a homogeneous, compositionally primitive plume source and include the Deccan basalts (68–66 Ma) and the submarine ridge below the Maldives (55–50 Ma). The most recent hot spot activity has produced the southern Mascarene Ridge, which includes the islands of Mauritius (8.9–0.03 Ma), to the northeast, and Reunion (2.1–0 Ma), to the southwest (Fig. 13.10; Bonneville et al. 1997; O'Neill et al. 2003).

While the present plume conduit, with radius of 100–130 km and upwelling velocity of 5–9 cm/year, lies at an asthenospheric depth at 100–200 km north of Reunion, the hot spot is centred beneath Piton de la Fournaise volcano, in the eastern portion of Reunion Island. The plume may not consist of a single body, as the composition of the Reunion lavas may result from the impingement of two small-scale blobs of plume at the base of the lithosphere (Barruol and Fontaine 2013). A partially molten layer lies at the crust-mantle boundary beneath Reunion. This underplated body is 140 km wide and up to 3 km thick. The oceanic plate around Reunion is not flexurally downwarped toward the island, but instead domed toward southeast. The historical effusive rate of Reunion of $\sim 10^4$ km^3/Ma suggests that its submarine growth began 5–7 Ma ago. Most of the volume edifice is submerged: the volcanic edifice appears as a flattened cone 200–240 km wide, whose subaerial part is only 3% in volume, and whose history spans only the last 2.1 Ma (Charvis et al. 1999; Gallart et al. 1999). Reunion results from at least three volcanic complexes. An inferred older volcano, Les Alizés, below the eastern coast has a large intrusive and cumulate complex, at 1 km depth (Gailler and Lenat 2012). Most of Reunion is constituted by the larger and older Piton des Neiges volcano, to the northwest, and the more recent Piton de la Fournaise volcano, to the southeast, active simultaneously for more than 500 ka (Fig. 13.10). The location and direction of the alignment of both volcanoes and of their connecting rift zone are consistent with an inferred NW–SE trending pre-existing regional structure, whose reactivation may result from the uplift related to the strong underplating (Michon et al. 2007). The magmas erupted at Piton des Neiges and Piton de la Fournaise are mostly olivine basalts, with a composition transitional between tholeiitic and alkali basalts, although progressively evolving towards increasing tholeiitic compositions (Albarede et al. 1997).

Piton des Neiges is a dormant volcano at least 2.08 Ma old. Following a basaltic shield stage

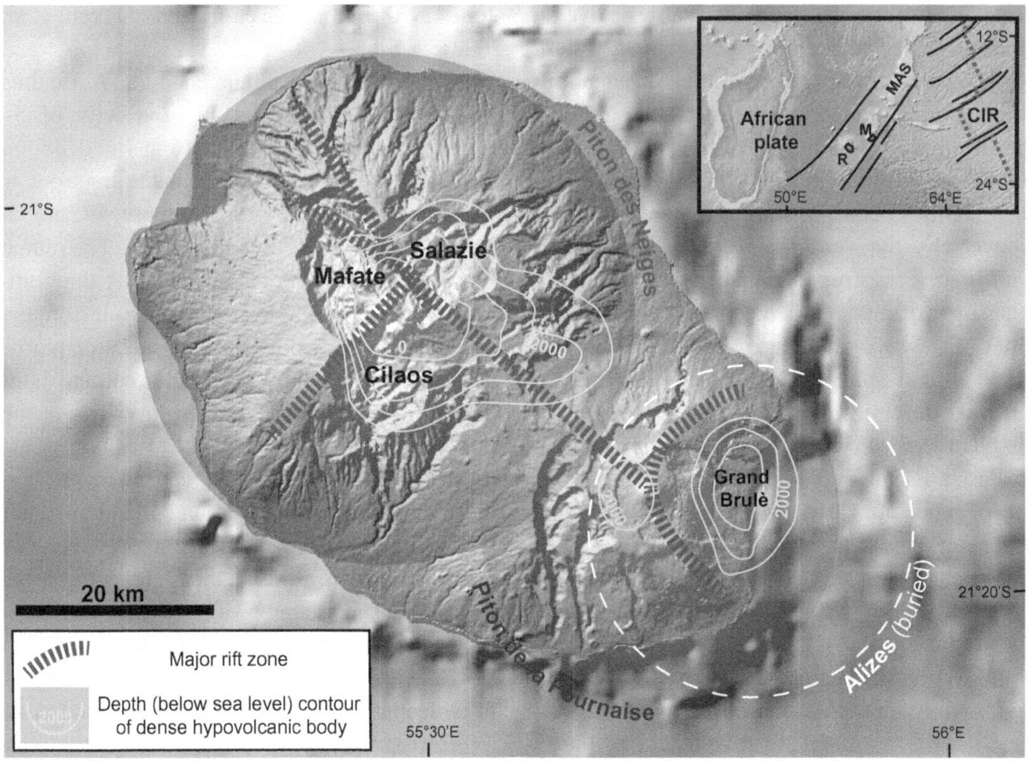

Fig. 13.10 Overview of the island of La Reunion, Indian Ocean, highlighting its three main volcanic edifices, the major volcanic rift zones and the depth contours (in metres below sea level) of buried dense intrusions (from Michon et al 2007; Gailler and Lenat 2012). Inset shows the tectonic setting of the hot spot, along the Mascarene Ridge (MAS); R = Reunion Island; M = Mauritius Island; CIR = Central Indian Ridge. Digital Elevation Models provided by GeoMapApp

(2.08 to 0.43 Ma), it erupted alkaline differentiated lavas (350–12 ka). The volcano is deeply eroded within three main depressions: the Mafate, Salazie and Cilaos Cirques (Fig. 13.10). Minimum eroded volumes of 101 and 105 km^3 have been calculated for the last 180 ka for the Mafate and Cilaos depressions, respectively, giving a minimum long-term erosion rate of 1.2 km^3/ka. The products of the erosion are found on the ocean floor, in large fan-shaped promontories 20–25 km wide at the coastline and 70–150 km across the seafloor. These result from the superimposition of multiple debris avalanche deposits. A hydrothermally altered intrusive complex beneath Piton des Neiges is deeply rooted, to several kilometres below sea level, and is 10–25 km wide (Lenat et al. 2001; Oehler et al. 2004; Gailler and Lenat 2012). Above, three perpendicular intrusion trends (two subvertical and one subhorizontal) coexisted for 2 Ma. Stress accumulation under repeated magma injections resulted in permutations of the principal axes of the stress tensor, reorienting subsequent intrusions. Intrusions cycles started with dike injections in an extensional stress field, reducing the deviatoric stress, then switching the axes of principal stresses and finishing with sill intrusions in a compressional stress field. Sills acted as detachment planes, restoring the extensional stress field and initiating a new cycle of permutations (Chaput et al. 2014).

Piton de la Fournaise has two building stages, described as the Ancient (0.53–0.15 Ma) and Recent Shields (<0.15 Ma). These were separated by a collapse event that shifted the volcano centre eastward by about ~ 15 km, from the present-day Plaine des Sables to the current position of the active summit. Two main subvolcanic intrusions are the Grand Brûlé complex, below the Grand Brûlé depression on the easternmost flank of the volcano, and a complex beneath the Plaine des Sables and the Enclos Fouqué, on the western part of the volcano. The Grand Brûlé complex, N-S elongated and deeply rooted, is the relict of the Alize's Volcano. The subcircular Plaine des Sables-Enclos Fouqué complex was the center of the Ancient Shield before the collapse event, and currently hosts a large and active hydrothermal system (Gailler and Lenat 2012; Dumont et al. 2019). Four U-shaped structures opening seaward are nested on Piton de la Fournaise (Fig. 13.11). The innermost and youngest collapse is the 13 km long and 6–8 km wide Enclos Fouqué-Grand Brûlé depression, E-W elongated and delimited by a horseshoe rim to the west, south and north sides with a continuous cliff, 100–200 m high (Staudacher and Allegre 1993; Gillot et al. 1994; Merle et al. 2010a).

Several hypotheses have been proposed to explain the origin of the Enclos Fouqué-Grand Brûlé depression (Fig. 13.12). A first model suggests that the Enclos Fouqué-Grand Brûlé results from a 4.5 ka old giant landslide partly responsible for the debris avalanche deposits on the submarine flanks (Gillot et al. 1994). A second model suggests that the entire Enclos Fouqué-Grand Brûlé results from vertical collapse, due to the deformation of the hydrothermal system and the downward drag related to the dense intrusive complex of the underlying Alize's volcano (Michon and Saint-Ange 2008). A third set of models suggests that the Enclos depression results from a polyphase caldera collapse, whereas the Grand Brûlé exhibits the scars of lateral landslides with the headwall immediately to the east of the caldera. In this model, the clay-rich altered and pressurized hydrothermal core of the volcano may flow under its own weight, spreading laterally in the Grand Brûlé area and triggering caldera-like collapse in the summit area (Merle and Lenat 2003; Merle et al. 2010b).

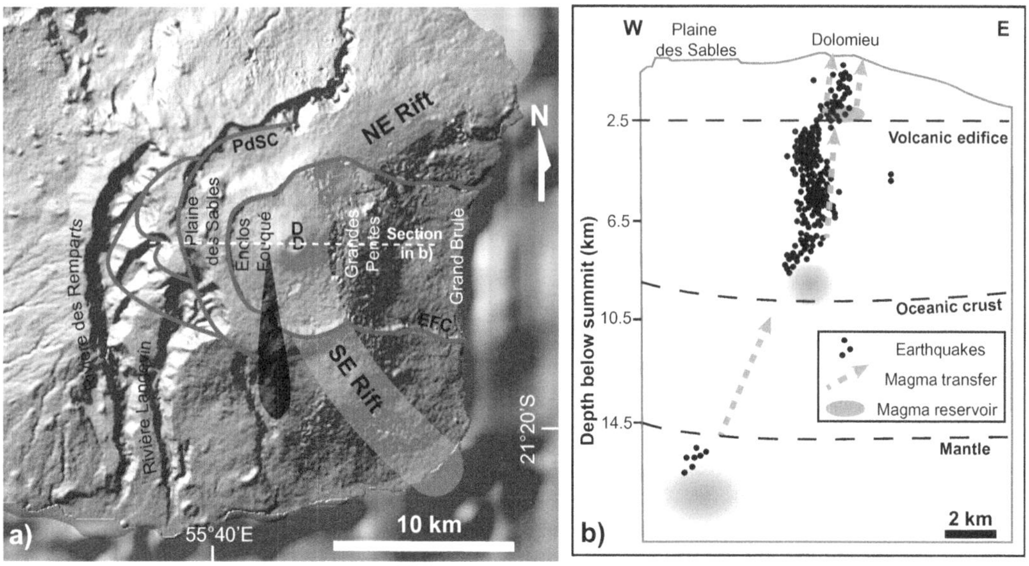

Fig. 13.11 La Reunion hot spot. **a** Simplified structural map of Piton de la Fournaise volcano, including the main rift zones (in orange) and structures (in red); PdSC = Plaine del Sables collapse; EFC = Enclos Fouqué-Grand Brûlé collapse; D = Dolomieu caldera (Merle et al. 2010a; Bonali et al. 2011). Digital Elevation Model provided by GeoMapApp. **b** East–west cross section through Piton de La Fournaise, showing the location of the 1996–2007 earthquakes and magma storage at depth (modified after Peltier et al. 2009)

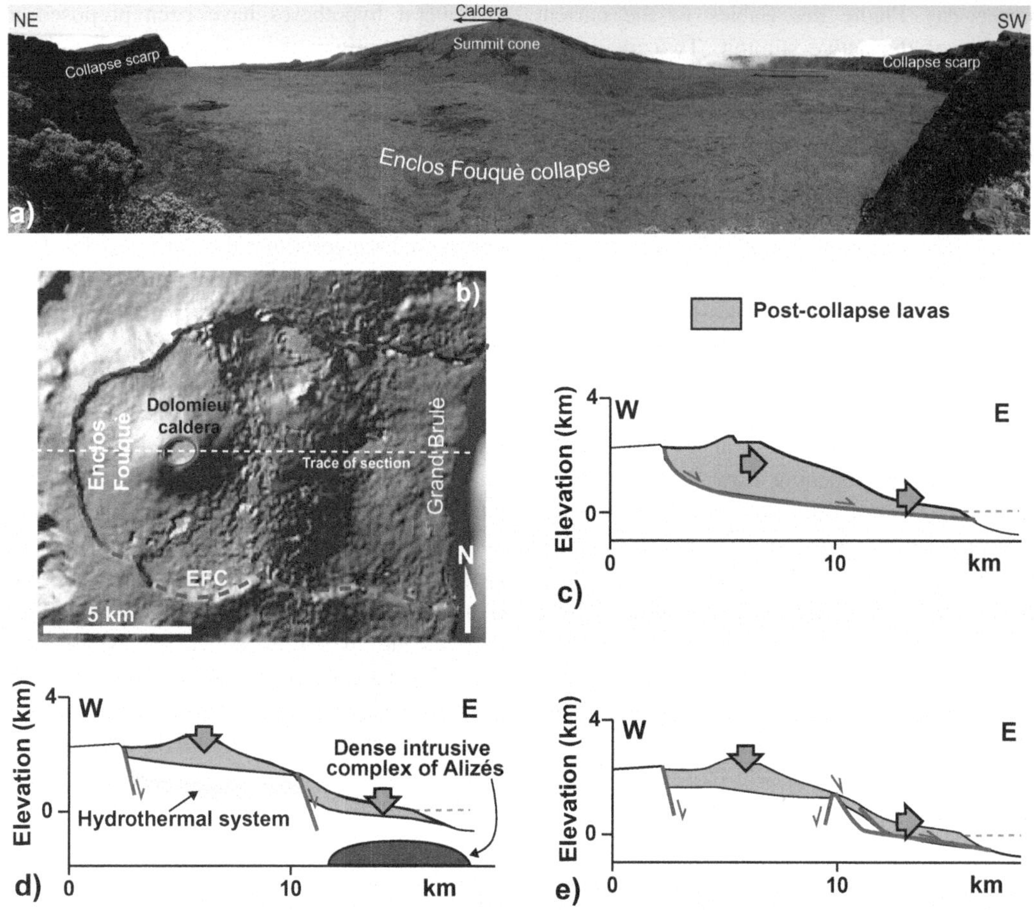

Fig. 13.12 La Reunion hot spot. **a** View from the northwest part of the Enclos Fouquè collapse scarp towards the Dolomieu cone, Piton de la Fournaise volcano. **b** Schematic map view of Piton de la Fournaise and the Enclos Fouqué-Grand Brûlé collapse (yellow dot shows point of view in (**a**). **c–e** Schematic end-member models (in E-W cross section) proposed for the formation of the Enclos Fouqué-Grand Brûlé structure: **c** a single landslide (lateral collapse); **d** a vertical collapse, due to the deformation of the hydrothermal system and the downward drag related to the dense intrusive complex of the underlying Alize's volcano; **e** a polyphase caldera collapse, where the Grand Brûlé is the scar of one or several lateral landslides (Michon and Saint-Ange 2008). DEM in (**a**) provided by GeoMapApp

The Enclos Fouqué-Grand Brûlé depression shares close similarities with the U-shaped collapse structures observed on other hot spot oceanic islands, as at Tenerife (Las Canadas caldera; see Sect. 13.6) and Fogo (Cape Verde). At Fogo, the U-shaped scarp affecting the summit and side of the volcano has been interpreted as a caldera or part of lateral collapse (Fig. 6.1e; Day et al. 1999, and references therein; Masson et al. 2008).

Piton de la Fournaise is a highly active shield, with an average of one eruption every 10 months over the last decades. Present-day summit activity focuses in a 400 m high basaltic cone with caldera (Dolomieu) and pit crater (Bory). Dolomieu caldera has developed by sudden collapses above a 1.5 km wide high velocity plug located above sea level, corresponding to an intrusive, solidified dike and sill complex. Below, two low velocity anomalies, from 1 to 0 km a.s.l. and deeper than 1 km below sea level, correspond to magma storage systems. Dolomieu experienced its last caldera collapse in 2007. To the sides of Dolomieu are the northeast and southeast rift zones, which are among the most developed of the six rift zones of the volcano. These rift zones lack a pronounced topographic expression, as mantling pre-existing topography, and are relatively diffuse (Fig. 13.11; Lenat et al. 1989; Michon et al. 2009; Prono et al. 2009; Bonali et al. 2011).

13.6 Canary Islands Hot Spot

The Canary Islands lie on Jurassic oceanic lithosphere near the west African continental margin. The archipelago consists of seven major islands with east-to-west age progression of ~1.9 cm/year, starting from the oldest subaerial volcanism at ~20 Ma for the eastern islands of Lanzarote and Fuerteventura, to ~15 Ma for Gran Canaria, 8 Ma for Tenerife and less than 2 Ma for the western islands of La Palma and El Hierro (Fig. 13.13).

This progression is compatible with the slow passage of the African plate over a hot spot, currently located beneath the western part of the archipelago. Some features may be difficult to reconcile with a typical hot spot, such as the low melting rate, the tens of Ma long volcanic history at single volcanic centres with long gaps, and the irregular distribution of volcanism. These features suggest that other processes, such as lithospheric intraplate deformation, accompany plume volcanism (Carracedo et al. 1998; Danobeitia and Canales 2000; Geldmacher et al. 2005).

The isotopic signature of the Canary Islands magmas, intermediate between that of the Madeira hot spot to the north and the Cape Verde Islands hot spot to the south, suggests a common, large-scale mantle upwelling off west Africa (Abratis et al. 2002). A low velocity layer, with less than 3% melt, lies at 45–65 km of depth within the thinned lithosphere of the Canary Islands. The Moho deepens eastward, from ~12 to ~25 km. A regional anomaly of ~500 m above the thinned lithosphere, probably related to a sub-lithospheric temperature anomaly of +100 °C, produces a topographic swell partly

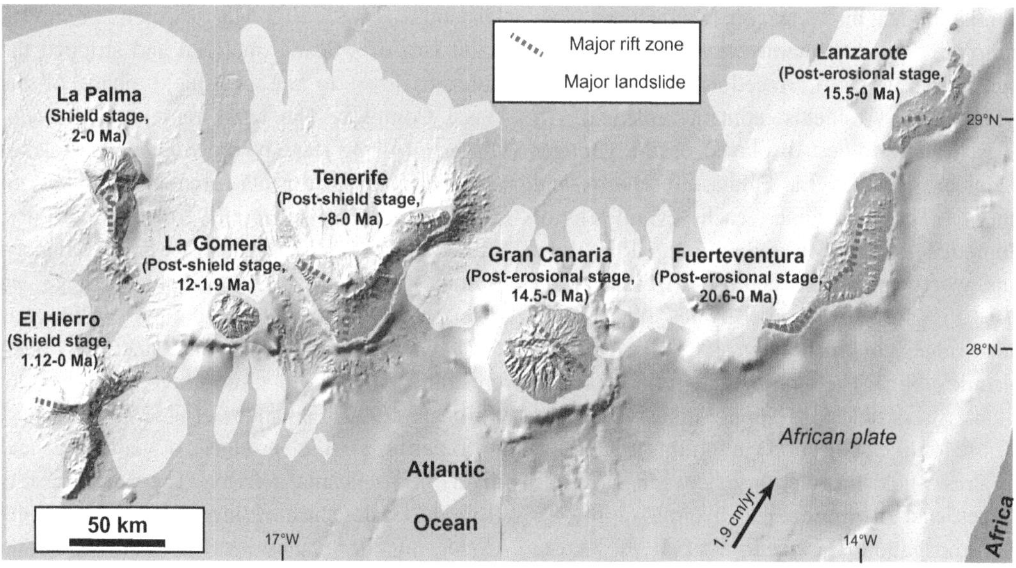

Fig. 13.13 Overview of the Canary Archipelago, highlighting the ages and evolutionary stage of the islands, their main rift zones and submarine landslides; plate velocity is with regard to the hot spot (Carracedo 1999). Digital Elevation Model provided by GeoMapApp

masked by thick sediments and the volcanics of the Canary Islands. Superimposed on this swell is a flexural moat that flanks the Canary Islands, induced by the load of the volcanoes (Collier and Watts 2001; Martinez Arevalo et al. 2013; Fullea et al. 2015). Underplating and crustal thickening by mafic intrusions, usually ranging from 2 to 8 km, occurs along the Moho and the base of the elastic lithosphere beneath Gran Canaria, La Palma, Tenerife, Fuerteventura and Lanzarote. Low velocity zones beneath La Palma and Tenerife are interpreted as due to partial melting (Danobeitia and Canales 2000; Lodge et al. 2012).

Individual islands show initial rapid growth (shield-building stage), followed by quiescence, deep erosion and post-erosional activity. The western islands, La Palma and El Hierro, are the youngest and still in their shield stage. Volcanic activity occurs in these islands and persists in Tenerife and Lanzarote. The composition of the volcanic rocks of the Canary Islands is highly variable, ranging from mafic (transitional tholeiite to melilite nephelinite) to highly evolved (peralkaline rhyolite to trachyte to phonolite). While alkali basalts and tholeiites are the dominant shield-building magma types, highly differentiated magmas, as quartz-trachytes and phonolites, develop mainly during the waning stages (Hoernle and Carracedo 2009).

The Canary shields contain volcanic rift zones, which are best displayed on the younger volcanoes, such as La Palma, El Hierro and Tenerife. These rift zones, each several tens of kilometres long, are usually arranged in three arms, where angles between each rift ideally are 120° (Fig. 13.13). The rift zones range from linear ridges to broad fans at the base of the volcanic edifice. The fanning may reflect: (a) non suitable stress field for trapping dikes within the rift, (b) dike injection migrating laterally, (c) stress field interactions between volcanic edifices, and (d) low rate of magma supply for dike propagation (Carracedo et al. 1998; Acosta et al. 2003). While the concept of three-arm rifts has been accepted for decades as a distinctive feature of the Canary Islands volcanoes, recent studies suggest that these rifts rather result from initial radial eruptive fissures that have been obscured by successive flank collapses (Becerril et al. 2015). The growth of rift zones includes the oversteepening, the instability and the lateral collapse of the flanks to the sides, with dike-induced rifting appearing as the main trigger for flank destabilization and ultimately dismantling the islands. At the shield-stage islands, several Quaternary debris avalanches extend tens of kilometres offshore (Fig. 13.13). There is a higher landslide frequency for the Canary Islands compared to the Hawaiian Islands. The lower stability of the flanks of the Canary Islands may be due to their steeper slopes, a result of the abundance of highly evolved viscous lavas, and the abundance of weak pyroclastic deposits deriving from frequent explosive eruptions due to the elevated volatile contents in the alkalic magmas (Carracedo 1999; Krastel et al. 2001).

The main tectono-magmatic features of some representative islands, from east to west, are summarized below.

Fuerteventura, the oldest island, is distinctive for its long and discontinuous volcanic history, interspersed with major erosional phases. Fuerteventura began and ended its shield stage in the Miocene. The shield was truncated by giant landslides by 17.5 Ma, which removed 3500 km^3 of volcanic material and stripped the volcano down to the pre-shield phase of the Basal Complex. The latter represents the submarine growing stage of the volcanic complexes and the hypabyssal roots (plutons and dikes) of their successive subaerial growing episodes, subsequently affected by long-lived Miocene tectonic activity. After the shield stage an extended period of quiescence and erosion was followed by minor Pliocene and Quaternary post-erosional volcanism (Ancochea et al. 1996; Stillmann 1999; Fernandez et al. 2006).

Tenerife is the product of complex and diversified volcanic activity. The island results from the coalescence of three shields: Roque del Conde to the southwest (between 11.9 and 8.9 Ma), Teno to the northwest (6.2–5.6 Ma) and Anaga to the northeast (4.9–3.9 Ma) (Fig. 13.14). These shields formed the bulk of subaerial Tenerife since 8 ± 4 Ma, with each

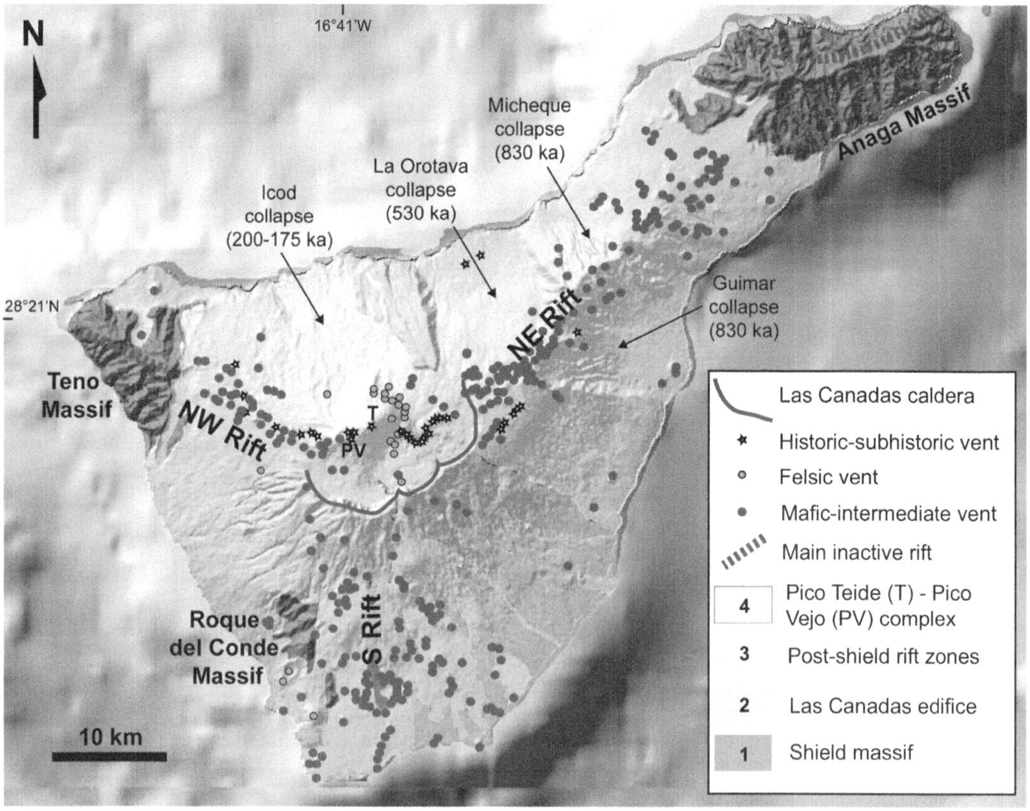

Fig. 13.14 Canary Islands hot spot. Structural map of Tenerife, highlighting the older massifs, the rift zones, the summit caldera and the lateral collapse scarps (modified after Martì et al. 2009)

shield constructed in less than 3 Ma (Guillou et al. 2004). The Anaga basaltic shield volcano was initially dominated by a single rift zone. Destabilization of the northern sector resulted in a collapse and the bending of the rift. Induced by the dilatation of the curved rift, a third rift-arm developed to the south, generating a three-armed rift system (Fig. 6.9; Walter et al. 2005). After the coalescence of the three volcanoes, a significant part of the Tenerife shield was removed by repeated giant landslides. The shield-building stage was thus followed by 2–3 Ma of erosion and volcanic quiescence. Eruptions resumed during the stage of rejuvenation, when the central volcano of Las Cañadas, 40 km wide and ~3000 m high, developed unconformably over the Miocene shield (Fig. 13.14). The central volcano developed three main rift arms (to the south, northwest and northeast), with major collapses in between (Carracedo et al. 2007; Boulesteix et al. 2013). The Las Cañadas volcano summit also collapsed between 200 and 175 ka, creating the Las Cañadas depression (Fig. 13.15). This is explained as the head of a gravitational collapse or, more likely, as the coalescence of three successive caldera collapses due to the emptying of a migrating magma chamber following large eruptions. The twin volcanoes of Pico Teide and Pico Viejo developed within the caldera, marking the most recent eruptive cycle (Martì 2019, and references therein). The oceanic basement dips gently towards the island, whose load forms a flexural moat filled by 2–3 km of sediments. Up to 1.5×10^5 km^3 of magmatic material has been added to the flexured oceanic crust, with magma

Fig. 13.15 Tenerife, Canary Islands: view of the central portion of Las Cañadas caldera (foreground), as seen from the south rim, with Pico Teide on the background

generation rate of 10^3–10^4 km^3/Ma, similar to Reunion and Cape Verde, but at least an order of magnitude less than Hawaii (Watts et al. 1997).

El Hierro is the youngest and westernmost island. The oldest subaerial part is 1.2 Ma old. Eruptive activity has been nearly continuous, growing two successive edifices affected by one or more catastrophic lateral collapses: Tinor (1.2–0.88 Ma) and El Golfo (545–176 ka). After 158 ka, volcanic activity was fed from the present-day rift zones, arranged in the three-armed configuration (Fig. 9.2; Day et al. 1997). Overgrowth instability due to successive rift eruptions has led to several flank failures and giant landslides, the last of which occurred to the north between 130 and 100 ka, resulting in the trilobed shape of the island. The San Andres fault system identifies, along the eastern flank of the northeast rift zone, the failure zone of a giant landslide which has slipped twice between 540 and 430 ka and between 183 and 52 ka (Day et al. 1997; Blahut et al. 2020). Data offshore El Hierro suggest that rifting is not confined to narrow zones. The young Las Hijas volcanic seamounts, 70 km southeast of El Hierro, are consistent with the spacing and timing of propagation of volcanism of the archipelago, and these seamounts may represent future sites of volcanic islands. The last eruption occurred offshore south of the island in 2011 (see Sect. 9.2).

13.7 Azores Hot Spot

Although located next to the Mid-Atlantic Ridge, the Azores Archipelago shows a structure that is mainly controlled by the transtensive motion along the transform plate boundary between the Eurasian and African plates, to the east of the Ridge. This far-field influence motivates the inclusion of the Azores hot spot in this chapter, rather than in Chap. 11, which illustrates the tectono-magmatic features of divergent plate boundaries and includes the Afar and Icelandic hot spots. In particular, the Azores Archipelago straddles the Mid-Atlantic Ridge in the area of a triple junction that includes an oblique and recently formed ultraslow rift, a zone of diffuse seafloor deformation, a major fracture zone and a hot spot (Fig. 13.16).

The archipelago is mostly developed to the east of the Mid-Atlantic Ridge, with only two islands to the west of it. A low velocity anomaly in the shallow mantle (<200 km deep) is imaged under the central group of islands, to the east of the ridge. The low velocity anomaly extends northeastward and downward to connect to a column of low velocity material from $\sim$250 km to at least 400 km of depth. The mantle upwelling velocity is estimated at 3–4 cm/year, smaller than that of Hawaii or Iceland (Yang et al. 2006; Madureira et al. 2014). Plume-ridge interactions and underplating between $\sim$20 and $\sim$7 Ma produced a submarine plateau, on which the Azores Islands lie, shallower than 2000 m, related to a 12–14 km thick crust, thicker than average. The plateau records episodic variations of the hot spot melt production with cyclicity of 3–5 Ma and, following a decrease in the rate of volcanism, it has been rifted by the Mid-Atlantic Ridge since 7 Ma (Gente et al. 2003; Spieker et al. 2018).

The main structures in the Azores area are the Mid-Atlantic Ridge, the Terceira Rift and the East Azores Fracture Zone (Fig. 13.16). The Mid-Atlantic Ridge separates the Eurasian and African plates from the North American plate. The structure and kinematics of the Eurasian-African boundary appears complex. Geodetically, Santa Maria Island, to the southeast, has been following the average African plate movement and Graciosa Island, to the north, has mimicked the average Eurasian plate behaviour, while the other islands display intermediate behaviours. Seismo-tectonic studies locate the main plate boundary between Sao Jorge and Pico, continuing to the east south of Sao Miguel. Other studies suggest that the main plate boundary is the 550 km long ultraslow Terceira Rift, which corresponds to a WNW-ESE trending alignment of segmented and alternating basins and volcanic edifices (Searle 1980; Madeira and

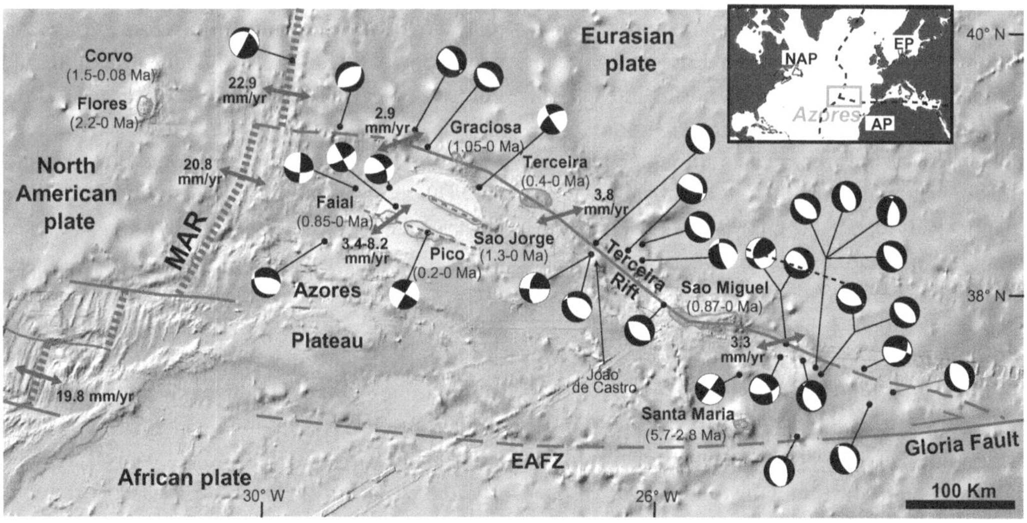

Fig. 13.16 Main features of the Azores Archipelago, including the ages of the islands, the main fault zones (solid and dashed thin red lines) and related opening directions (purple double arrows), and the focal mechanisms of the 1939–2013 M > 5.0 earthquakes. Inset shows the general tectonic setting. Light yellow circle = shallow hot spot location, from −75 to −188 km; MAR = Mid-Atlantic Ridge; EAFZ = East Azores Fracture Zone; NAP = North American plate, EP = European plate; AP = African plate (modified after Trippanera et al. 2014). Digital Elevation Model provided by GeoMapApp

Ribeiro 1990; Vogt and Jung 2004; Fernandes et al. 2006). The Terceira Rift formed ~1 Ma ago and is currently obliquely slowly spreading at 0.2–0.4 cm/year, with WSW-ENE trending extension and dextral component of motion (Fig. 13.16; Pagarete et al. 1998). This motion may result from the different spreading rate of the Mid Atlantic Ridge north (~2.3 cm/year) and south (~1.9 cm/year) of the Azores and the related overall lateral motion between the Eurasian and African plates of ~0.4 cm/year. Four volcanic systems (Sao Miguel, Joao de Castro, Terceira and Graciosa, three of which are islands) along the Terceira Rift are separated by deep non-volcanic basins, similar to ultraslow ridges (see Sect. 11.4.1). The interaction of obliquity, slow spreading rates and thick lithosphere probably prevents larger amounts of melt from reaching the surface. East of the Terceira Rift, the Eurasian-African plate boundary lies along the Gloria Fault, an E-W trending right-lateral transform fault moving at ~0.5 cm/year and reaching Gibraltar (Fig. 13.16; Searle 1980; Vogt and Jung 2004; DeMets et al. 2010). The East Azores Fracture Zone is the westernmost continuation of the Gloria Fault towards the Mid-Atlantic Ridge, to the south of the Terceira Rift. It corresponds to an aseismic and non-volcanic fault system, probably an Eurasian-African boundary abandoned after the northward migration of the triple junction. Therefore, the transtensive Azores region acts simultaneously as ultraslow rift and transfer zone between the Mid-Atlantic Ridge and the Gloria Fault, accommodating the differential shear between the Eurasian and African plates (Lourenço et al. 1998; Vogt and Jung 2004). The Mid-Atlantic Ridge spreading rates do not change abruptly in the area of the Azores triple junction. Rather, they change gradually, suggesting that the Mid-Atlantic Ridge, the Terceira Rift and the East Azores Fracture Zone constitute the boundaries of the kinematically distinct Azores domain, possibly a microplate, with distributed deformation and semi-rigid behaviour (Madeira and Ribeiro 1990; DeMets et al. 2010; Marques et al. 2013).

Seismicity focuses along the Terceira Rift and close to the Faial-Pico Fracture Zone and the Gloria Fault (Fig. 13.16). Focal mechanisms from 1939 to 2013 highlight predominant ~NW–SE oriented normal faults, extending along a NE-SW direction, and minor ~WNW-ESE trending dextral faults and ~NW–SE to NNW-SSE trending sinistral faults. While strike-slip motions predominate in the western portion, the eastern portion undergoes predominant extension, with average slip rate of 0.4 cm/year. Overall, there is a transtensive setting with partition between strike-slip and extensional structures. The estimated magnitudes of the largest paleoearthquakes range from *M6.9* to *M7.1*. Paleoseismological slip rates of 0.1–0.4 cm/year from Faial, Pico and Sao Jorge validate longer-term rates (Lourenço et al. 1998; Madeira and Brum da Silveira 2003; Trippanera et al. 2014). Several sets of faults, with N120 °E, N150 °E and ~N-S directions, define crustal blocks accommodating the interaction between the three main plates. Directions from N110° to N125°, focusing along the Terceira Rift, are interpreted as ancient fracture zones reactivated as transtensional faults. N-S directions are interpreted as former mid-oceanic rift faults reactivated as left-lateral faults (Navarro et al. 2009).

As a result of the activity of the transtensive plate boundary, several of the Azores volcanoes show distinctive morphological, structural and volcanological features. In fact, at least three islands appear strongly elongated (Sao Miguel, Sao Jorge and Pico) parallel to the plate boundary. The elongation is highlighted by transtensive fault systems, fissure eruptions forming focused rift zones and aligned polygenic and monogenic volcanoes. In some cases, as for example at Pico (see Fig. 7.22b), the rift zones extend outside the polygenic volcanoes for tens of kilometres, highlighting strongly elongated zones of magmatic activity, resembling magmatic systems along divergent plate boundaries (see Chap. 11). The development of long and focused rift zones is only partly associated with flank instability, which appears relatively limited on the Azores. In fact, despite frequent recognition of unstable and collapsed volcanoes flanks, as at Pico and Sao Miguel, only a few of the Azores islands show widespread mass-wasting processes affecting their growth, as Flores and Santa Maria (Hildenbrand et al. 2012, 2018; Costa et al. 2014; Sibrant et al. 2015; Marques et al. 2020). Several volcanoes currently host calderas, as Sao Miguel, Terceira, Faial and Corvo, whose formation is usually associated with the emission of more evolved magmas. The magmas erupted in the Azores belong to an alkaline series, exhibiting the full range of compositions, from mafic to felsic. Generally, while the polygenic volcanoes erupt alkali basaltic to trachytic magma, the monogenic volcanoes from the fissure zones erupt basalts and hawaiites; rhyolites have been erupted at the Santa Barbara polygenic volcano, on Terceira Island (Larrea et al. 2018, and references therein).

The tectono-magmatic features of some representative Islands, from east to west, are summarized below.

Sao Miguel, along the eastern Terceira Rift, is the largest island, E-W elongated. Geodetic data suggest that any inferred plate boundary deformation zone at Sao Miguel is narrow, with 75% of the 0.4–0.5 cm/year full plate motion being accommodated over the 10 km width of the island. Sao Miguel consists of six Quaternary loci, which include calderas (Sete Cidades, Agua de Pau and Furnas), stratovolcanoes, shield volcanoes affected by flank collapse and elongated monogenic fields (Moore 1990; Trota et al. 2006; Sibrant et al. 2015).

Sao Jorge, despite its strong WNW–ESE elongation suggesting a far-field tectonic control, is moving neither with Eurasia nor with Africa, presenting two sectors with different behaviour: the northwest sector moves at 0.2 cm/year to N82 °E, while the southeast sector moves at 0.1 cm/year to N109 °E. This velocity field, not explained by slip along the active faults of the island, may be induced by local shallow magmatic processes (Mendes et al. 2013). Horizontal or sub-horizontal flow predominates within the dikes outcropping at Sao Jorge, suggesting that the ascent of magma occurs mainly over localized melting sources, then collected within shallow magma chambers and from here laterally delivered along the island (Silva et al. 2012; Moreira et al. 2015).

Pico and Faial are two nearby WNW-ESE elongated islands with collinear configuration located above the plume head.

Pico, to the east, is the youngest island of the Azores, dominated on its western sector by the stratovolcano of Pico Mountain, active from 57 ka at least, rising $\sim$2.5 km a.s.l. (Fig. 13.17). Several eruptions produced cones and fissures on its western and southeast flanks (Fig. 7.22b). Other major volcanic complexes on Pico include basaltic shield volcanoes (Topo volcano) and monogenic volcanic ridges (the Achada Plateau). Two major collapses affect the northern flank of Pico, where a main debris field larger than

Fig. 13.17 Azores hot spot. The imposing summit of Pico Volcano on the Island of Pico seen from Faial, with the Pedro Miguel Graben in the foreground bordered by two sets of oppositely dipping faults, highlighted by the white (outer fault systems) and yellow (inner fault systems) triangles

10 km^3 and with minimum age of ~70 ka is observed offshore (Nunes et al. 2006; Costa et al. 2014).

Faial, to the west, consists of four main and distinctive units (Fig. 13.18). The oldest volcanic system, to the east, was mainly built around 850 ka, then faulted to form the WNW-ESE-trending Pedro Miguel Graben (Fig. 13.17). An elongated high density body below and parallel to the axis of the Pedro Miguel Graben, interpreted as a solidified dike swarm, suggests that part of the graben is dike-induced. The Caldeira Central Volcano (~130 ka to Present) partly developed within the Pedro Miguel Graben with a summit caldera mainly formed at ~1 ka; a low Vp anomaly at 3–7 km depth images a magma chamber below. The youngest part of the island is the Capelo Peninsula fissure system, west of Caldeira. This consists of a basaltic ridge formed by WNW-ESE aligned vents. Capelo fed the two historical eruptions on Faial: the Cabeço do Fogo eruption in 1672–73 and the Capelinhos Surtseyan eruption in 1957–58 (Romer et al. 2018). The NE-SW trending mean extension rate of the Pedro Miguel Graben has been of 0.3–0.8 cm/year, similar or slightly larger than that of the Terceira Rift. This suggests that Faial, with the nearby Pico Island, where the graben continues buried by recent volcanic deposits, is a major locus of extension within the Azores above the imaged hot spot. The Faial-Pico tectonic segment may thus constitute the offset, westward continuation of magmatic activity to the southwest of the Terceira Rift, or one of the several major structures making up the presently diffuse plate boundary (Madeira and Brum da Silveira 2003; Camacho et al. 2007; Marques et al. 2013; Trippanera et al. 2014).

Flores and Corvo lie west of the Mid-Atlantic Ridge and are both derived from the same mantle

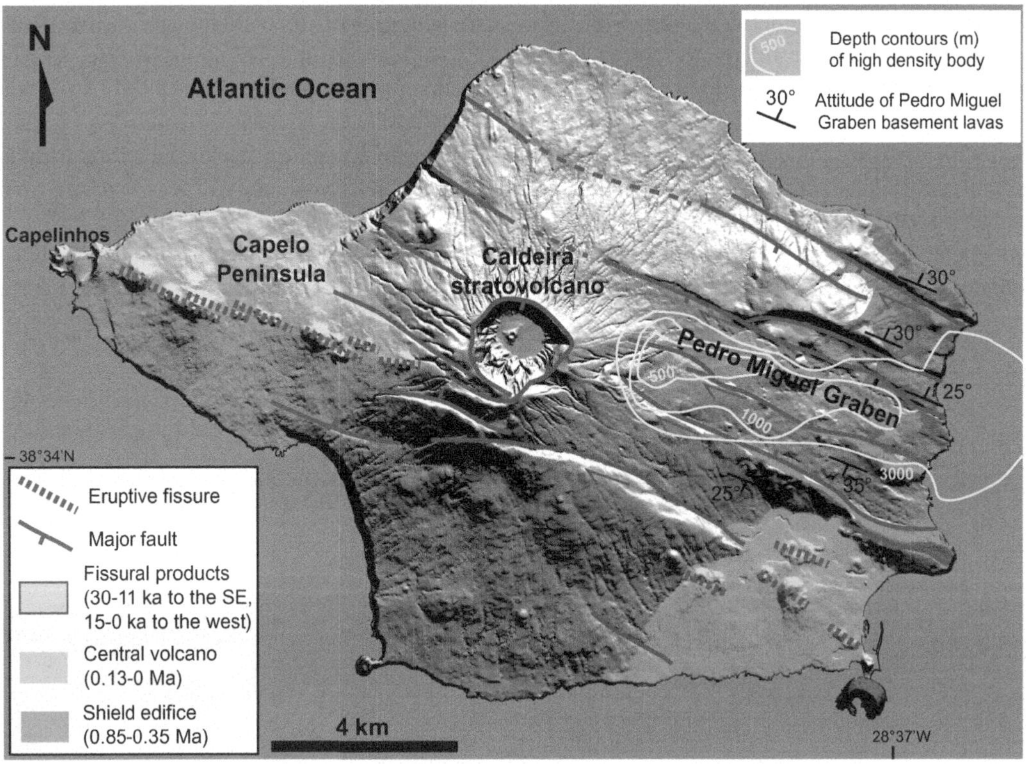

Fig. 13.18 Azores hot spot. Structural map of Faial Island, central Azores, including the Pedro Miguel Graben and its main fault zones, the Caldeira stratovolcano, the Capelo Peninsula and its main fissure eruptions, and the depth contours of the high density body (higher density of >200 kg/m^3) suggesting the presence of a dike complex below the graben (Camacho et al. 2007)

plume responsible for the eastern Azores Islands, although dominated by a source component not evident in the eastern archipelago. Flores consists of overlapping stratovolcanoes, without extended fissure systems. Shallow submarine volcanism occurred between 2.2 and 1.5 Ma. The most voluminous volcanism occurred between 0.7 and 0.5 Ma. Subaerial Corvo has been active between 1–1.5 Ma and 80 ka: its dominant feature is a ~2 km wide and 300 m deep caldera (Genske et al. 2012, and references therein).

13.8 Yellowstone Hot Spot

Yellowstone super-volcano is the most striking and best-known expression of continental hot spot. Large volumes of magma, resulting from the activity of the plume below, were emplaced in western North America to the east of the Juan de Fuca-Farallon subduction system during the last ~20 Ma. Three main magmatic provinces are related to the activity of this plume: the Columbia River Basalts Province, whose magma chambers mark the location of the former plume head, the High Lava Plains of Oregon (or Newberry Hotspot Track), whose western end marks the current plume head, and the Yellowstone Hotspot Track, whose eastern end marks the current plume tail. Such a configuration results from a complex history of plume-slab interaction (Fig. 13.19; Camp and Ross 2004; Wolff et al. 2008).

At ~20 Ma, when the rising Yellowstone plume intersected the overlying eastward-dipping Juan de Fuca-Farallon slab, the plume head flowed westwards, along the upper base of the slab. At ~17 Ma, the plume head punched through the slab and reached the surface, causing volcanism through long and narrow arcuate rifts. Subsequently, the plume head spread out northwards at the base of the lithosphere, probably forced by the thicker cratonic lithosphere boundary to the east, generating the Columbia River Basalts. These consist of ~210,000 km^3 of 17–14 Ma old lava flows and N-S oriented dike swarms. The largest and oldest silicic centres lie along a narrow north-northeast trend at the transition between the craton and accreted oceanic crust. As the North American plate moved southwest, the plume head may have sheared off from the tail and stayed with the plate, trapped by the deeper keel of the older, thicker lithosphere to the east. Contemporaneously, the focused plume tail may have tilted southeast and, by ~15 Ma, propagated northeastward with respect to the North American plate, along the present Eastern Snake River Plain (Xue and Allen 2007; Pierce and Morgan 2009; Obrebski et al. 2010; Coble and Mahood 2012; Camp et al. 2017; Morriss et al. 2020).

The resulting Newberry and Yellowstone tracks show age-progressive volcanism migrating from McDermitt Caldera, which was above the plume head at ~17 Ma. The Newberry Hotspot Track consists of bimodal products with silicic northwest age progression ending in the Newberry volcano (younger than 0.01 Ma), which marks the present location of the plume head, with a propagation rate of 13 km/Ma since 5 Ma. The Yellowstone Hotspot Track is associated with the east-northeast migration of bimodal volcanism along the Eastern Snake River Plain, ending after 800 km at the Yellowstone Caldera, the current tail of the plume (Fig. 13.19; Smith et al. 2009).

Massive rhyolitic volcanism along the Eastern Snake River Plain resulted from basaltic magma differentiation and crustal melting induced by the plume in the last ~16 Ma. The eruptive rate has been fairly uniform since the inception of the Eastern Snake River Plain–Yellowstone volcanic track, between 16 and 17 Ma, with the exception of a gap between 4.5 and 2.1 Ma. In particular, volcanism consisted of frequent (for example peaking every 200–300 ka from 12.7 to 10.5 Ma), high temperature (>1000 °C) and large volume rhyolitic eruptions, with individual ignimbrites reaching 1000 km^3, associated with large calderas. Rhyolites are explained through remelting of their erupted and subvolcanic predecessors on rapid time scales. This volcanism was followed by a similar migration of basaltic lava flows and emplacement of midcrustal mafic intrusions. Finally, the load from the dense intrusions, as well as lithospheric cooling and

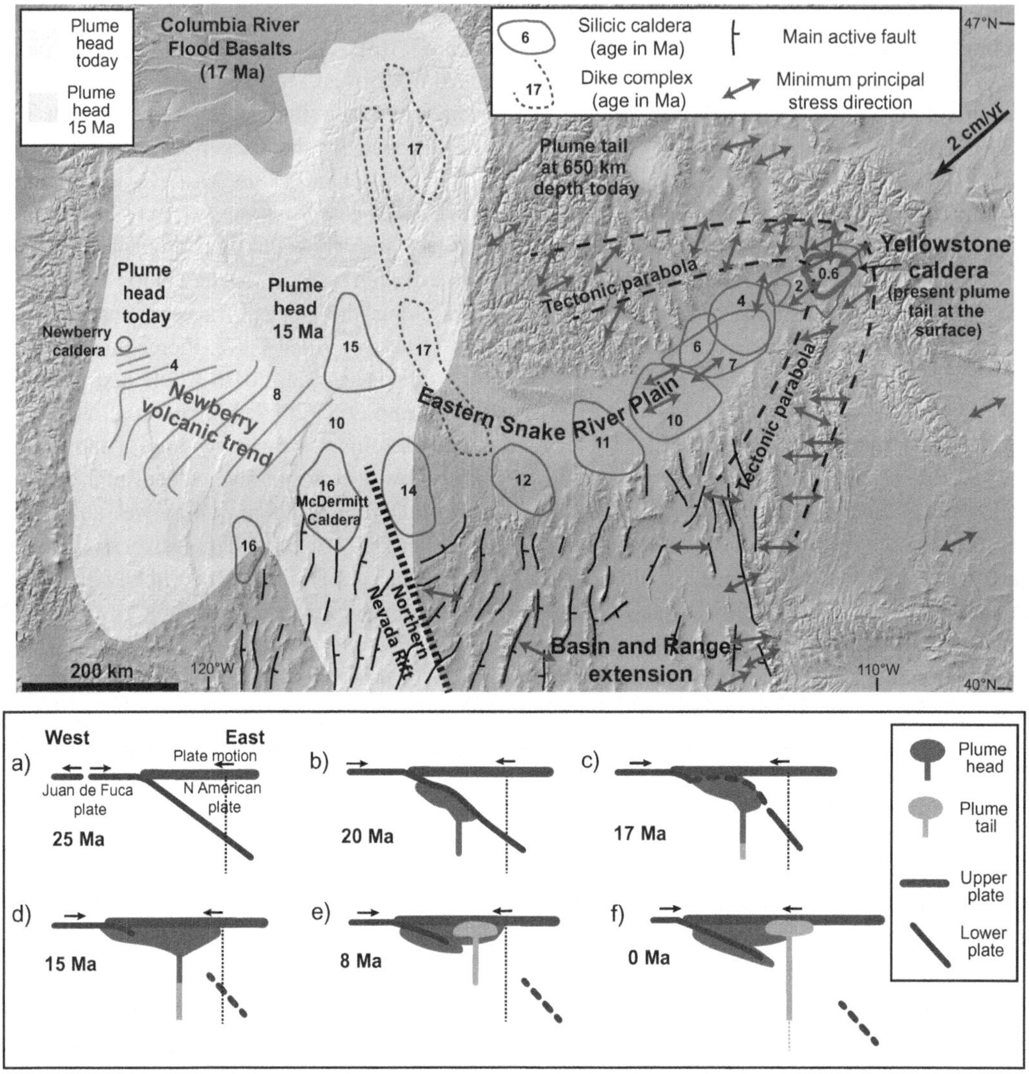

Fig. 13.19 Yellowstone hot spot. Above: overview of the main tectonic and magmatic features of the area affected by the Yellowstone plume, including the distribution of the main volcanic provinces, the current head and tail of the plume and the "tectonic parabola", a region of high seismicity and topography surrounding the Eastern Snake River Plain and the Yellowstone caldera (thick red line). The Newberry volcanic trend highlights local migration of volcanism (blue lines, in Ma). Plate velocity is with regard to the hot spot (modified after Smith et al. 2009). Digital Elevation Model provided by GeoMapApp. Below: proposed model for the interaction between the subducting Juan de Fuca plate and the Yellowstone plume head. **a** 25 Ma, the Yellowstone plume approaches the subducting Juan de Fuca plate; **b** 20 Ma, the plume head has intersected the Juan de Fuca plate and preferentially flowed westwards along the base of the slab; **c** 17 Ma, the plume head has punched through the Juan de Fuca plate, destroyed a larger portion of the slab and caused the volcanism at the surface; **d** 15 Ma, the plume head material spreads out in a larger region at the base of the lithosphere; **e** 8 Ma, the subducting slab drags the remnant plume head material down into the mantle; **f** at present, the hot material from the remnant plume head has been brought to greater depth by the ongoing subducting slab. The vertical dashed line indicates the progression of the current Yellowstone caldera to the west; the orange plume stem represents a hypothetical Yellowstone plume since the arrival of the plume head (shown in red; Xue and Allen 2007)

thermal contraction, has induced, largely before 6.6 Ma, a crustal flexure with 4.5–8.5 km of subsidence along the northeast edge of the plain. Therefore, passage of the North American plate over the melting anomaly has been accompanied by a migrating sequence of events consisting of uplift, regional deformation, massive explosive silicic eruptions and caldera collapse, basaltic volcanism and subsidence (Smith and Braile 1994; McQuarrie and Rodgers 1998; Pierce and Morgan 2009; Knott et al. 2020).

Presently, the Eastern Snake River Plain behaves as a rigid, non-extending block, where much of the minor extension has been accommodated transiently by dikes perpendicular to the plain axis, extending in a NE-SW direction. The Eastern Snake River Plain displays Holocene strain rates one order of magnitude lower than those in the adjacent Basin and Range. The transition zone in between, roughly identified by a parabolic-shaped region ("tectonic parabola"; Fig. 13.19), undergoes dextral and sinistral faulting, associated with seismicity, along the northwest and southeast boundaries of the plain, respectively (Smith and Braile 1994; Parsons et al. 1998; Payne et al. 2012). The ~42 km deep Moho beneath the Eastern Snake River Plain shallows to ~37 km to either side and thickens to ~47 km to the northeast. The lower crust of the plain, generally thickened by a dense underplated layer, shows a low velocity, probably molten layer. This underlies a 4–11 km thick mid crustal high density sill complex of gabbroic lenses interfingering with the granitic upper crust. The amount of mantle-derived mass added to the crust between 11 and 4 Ma is ~340,000 km^3, corresponding to an average mass flux of 0.05 km^3/year, comparable to other hot spots (Peng and Humphreys 1998; McCurry and Rodgers 2009).

Yellowstone caldera constitutes the easternmost active volcanic system fed by the plume tail past the Eastern Snake River Plain. Below Yellowstone caldera, at 60–120 km depth, the plume is sheared southwest by the motion of the North American plate, producing a low velocity layer beneath the thin lithosphere with excess temperature of ~50 to ~200 °C. Below, the plume appears as a ~80 km wide vertical conduit down to 200–300 km depth, where it dips 60° west-northwest, extending to 660 km depth. Upper mantle convection induces eastward flow at ~5 cm/year, that deflects the ascending plume into its west-tilted geometry. The upward deflection of the 660 km deep upper mantle transition zone by 12–18 km over an area ~200 km wide suggests a high-temperature, plume-like upwelling. The upwelling velocity of the plume, characterized by ~2% partial melt, is ~6 cm/year, with a buoyancy flux (0.25 Mg/s) many times smaller than that of oceanic plumes (Xue and Allen 2007; Smith et al. 2009; Schmandt et al. 2012; Tian and Zhao 2012).

The regional signature of the hot spot is currently highlighted by a 600 m high and ~600 km wide topographic bulge centred on Yellowstone caldera, reflecting long-wavelength tumescence (Smith and Braile 1994; Smith et al. 2009). The larger than 15,000 km^3 upper crustal magma reservoir beneath Yellowstone is 6 to 16 km deep, continuing laterally 20 km north of the caldera boundary. Having 2–15% melt, it is assumed as a fluid-saturated porous material consisting of granite and a mixture of rhyolite melt saturated with water and CO_2. In the lower crust, at a depth between 25 and 50 km, a basaltic magma reservoir with a size ~4.5 times that of the upper crustal magma reservoir contains a melt fraction of ~2% and provides the magmatic link between the plume and the upper crustal reservoir (Smith et al. 2009; Huang et al. 2015).

The Yellowstone volcanic field formed during three major caldera-forming rhyolitic eruptions at approximately 2.0, 1.3 and 0.6 Ma, emitting approximately 2500 km^3, 280 km^3 and 1000 km^3 of magma, respectively. The youngest eruption formed the current Yellowstone caldera, followed by ~50 rhyolitic and basaltic events, the youngest at ~70 ka (Fig. 13.20). The composition of the erupted magmas is bimodal, with rhyolitic rocks dominating and basaltic rocks representing a small mostly extra-caldera fraction, and intermediate compositions absent. Silicic magmatism proceeded via shallow-level remelting of earlier erupted and hydrothermally altered source rocks,

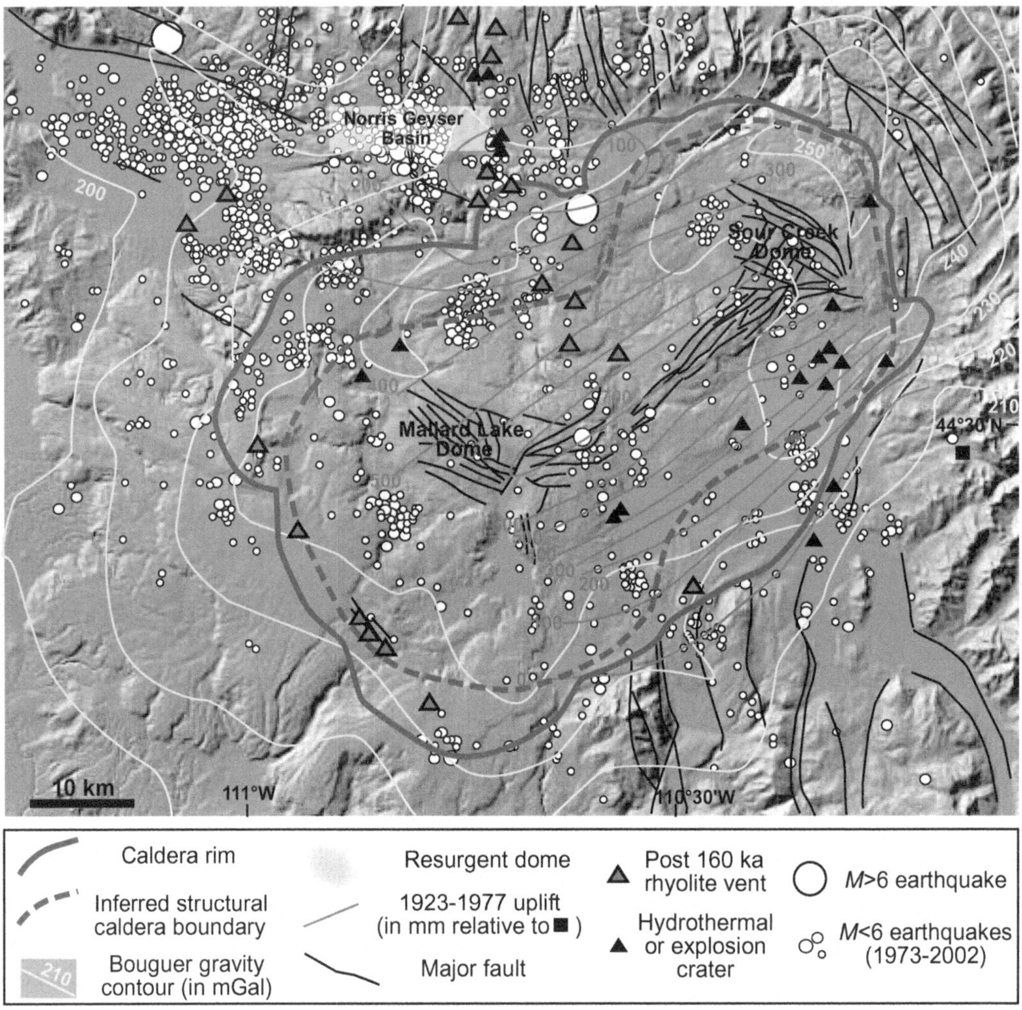

Fig. 13.20 Structure of Yellowstone super-volcano (Wyoming, USA), showing the caldera and its gravity contours, the two resurgent domes, the post-160 ka vents, the uplifted area between 1923–1977 and the 1973–2002 seismicity (after Smith and Braile 1994; Christiansen 2001)

with basaltic magma providing the heat for melting. Analysis of the products of the caldera-forming eruptions reveals a short duration of reservoir assembly, which documents rapid crustal remelting and eruption rates two to three orders of magnitude higher than those of continental arc volcanoes (Smith and Braile 1994; Christiansen 2001; Bindeman et al. 2008; Wotzlaw et al. 2015). The 40 × 70 km wide and ~2 km deep caldera hosts two intensely fractured resurgent domes (Sour Creek and Mallard Lake). The caldera coincides with a −60 mGal gravity anomaly. Focal depths of earthquakes decrease from more than 11 km outside this anomaly to less than 6 km within, reflecting thinning and heating of the seismogenic crust (Miller and Smith 1999). The most striking surface expression of the current activity of the caldera is its huge hydrothermal system, which contains more than 10,000 thermal features, including the world's greatest concentration of geysers, hot springs, mudpots, and steamvents (Fig. 13.21).

Fig. 13.21 Yellowstone hot spot. View of the Grand Prismatic Spring in Midway Geyser Basin at Yellowstone on October 19, 2017. *Photo credit* U.S. Geological Survey, USGS

13.9 Tibesti Hot Spot

The Tibesti hot spot lies in northern Chad, northern Africa. Northern Africa contains several hot spots characterized by anomalous topographic swells and intraplate Cenozoic volcanism, as Hoggar (Algeria), Darfur (Sudan), Cameroon (Cameroon) and Tibesti (Fig. 10.18). Rather than reflecting the involvement of one or several deep mantle plumes, or even a connection with the nearby larger Afar hot spot, these intraplate hot spots are probably related to unconnected plumes and attributed to shallow processes. In particular, the Tibesti hot spot does not show any seismic signature in the upper mantle, suggesting a shallower asthenospheric origin (Pik et al. 2006; Sebai et al. 2006). The low shear velocities at 100–150 km depth in this region can be interpreted in terms of elevated temperatures due to a shallow plume, whose stem appears to migrate approximately 100 km to the west between depths of 100 and 150 km. At shallow levels this plume may also feed the volcanism of the nearby Al Haruj region to the north, in Libya (Keppie et al. 2010). Thermobarometric and tomographic studies suggest that asthenospheric temperatures beneath Al Haruj in the last 5 Ma are more than 50 °C hotter than ambient mantle and 20–40 °C hotter than those predicted to be present beneath the Tibesti region. In addition to the elevated asthenospheric temperature, a contribution from lithospheric melting related to the Cretaceous rifting of the Sirt Basin (Libya), associated with lithospheric thinning, has been inferred for both regions (Cvetkovi et al. 2010; Ball et al. 2019).

The Tibesti swell consists of a ~1000 km wide area with altitudes generally of ~1000 m above the surrounding areas, with several volcanoes with elevation above 2000 m. This positive relief coincides with a geological uplift exposing Precambrian rocks surrounded by Phanerozoic rocks. River profiles suggest that most uplift (0.4–1.2 km) occurred at ~30 Ma, whereas the more recent uplift is estimated at less than 0.4 km. The produced dome is inferred to result from the anomalously hot asthenosphere, which upwells beneath the lithosphere (Keppie et al. 2010; Roberts and White 2010).

The Tibesti hot spot likely lies along a ~6000 km long NW–SE striking lineament (Tibesti Lineament), extending from southwest Algeria to Kenya. This corresponds to a 200–300 km wide fault zone with a possible late Proterozoic origin and reactivated as a transtensive dextral system during hot spot volcanism (Guiraud et al. 2000; Nkono et al. 2018). Despite this probable regional tectonic contribution, the Tibesti region has been aseismic, at least above *M*5.5.

The hot spot-induced Tibesti Volcanic Province (TVP) consists of several shield volcanoes (up to 80 km diameter) with large-scale calderas, extensive lava plateaux and flow fields, and widespread ignimbrite deposits (Fig. 13.22; Permenter and Oppenheimer 2007; Deniel et al. 2015). The composition of magma is mainly bimodal, with basalts representing roughly 60% in volume and felsic rocks, such as rhyolites,

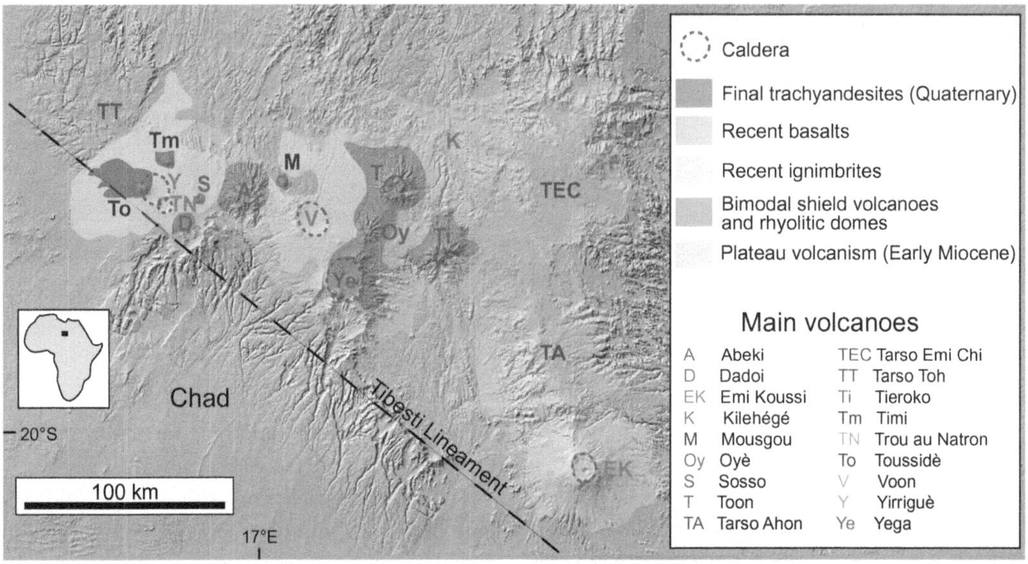

Fig. 13.22 Simplified map of the Tibesti Volcanic Province, Chad (location in inset), with products ranging from the plateau volcanism (approximately 17 Ma) to the final Quaternary volcanism. Labels of major volcanoes have the same colour as the related stage of activity of the province shown above (modified after Gourgaud and Vincent 2004; Deniel et al. 2015)

trachytes, and minor phonolites, approximately 40%. Intermediate compositions, mostly trachyandesites, represent less than 1% volume, and are only encountered in the final volcanic activity. The activity of the TVP began as early as the Oligocene, though the major products that mark its surface date from Lower Miocene to Quaternary. Between at least 17 and 8 Ma widespread alkaline plateau volcanism consisted of flood basalts and silicic lava plugs with intercalated alkaline, peralkaline and rhyolitic ignimbritic sheets. In this period ages of volcanism decrease from northeast to southwest, following the migration of NW–SE trending flexures associated with the Tibesti Lineament and concentrating the feeding dike swarms. Late Miocene large central composite tholeiitic volcanoes, some of which (Toon, Oyé, and Yéga) NNE-SSW aligned, formed in the central part of the TVP. After this initial phase of activity in the central TVP, volcanism migrated to both the eastern and western TVP regions, showing a relatively haphazard spatial development. This is mainly testified by the construction of three large alkaline, peralkaline and rhyolitic ignimbritic volcanoes, associated with significant updoming of the basement, ending with the collapse of large calderas: Voon (5–7 Ma), Emi Koussi (2.4–1.33 Ma) and Yirrigué (0.43 Ma). Basaltic activity, starting at about 5–7 Ma, and essentially consisting of cinder cones and associated lava flows, also occurred to the west and southeast of the TVP (Tarso Tôh, Tarso Ahon, and Tarso Emi Chi). The final volcanic activity is represented by post-Yirrigué caldera activity in the Tarso Toussidé Volcanic Complex and especially Toussidé (the only active volcano in Tibesti), Timi and Mousgou volcanoes (Permenter and Oppenheimer 2007; Deniel et al. 2015, and references therein).

Volcanism of the TVP has been poorly studied in detail, with the above-mentioned Emi Koussi representing one of the best-known volcanoes of the hot spot. This consists of a wide ignimbritic shield-like volcano resulting from the succession of three eruptive sequences, each related to caldera collapse. Its products exhibit two bimodal lava series, silica-saturated and silica-undersaturated, with a wide compositional gap (of ~10% SiO_2). While the silica-saturated

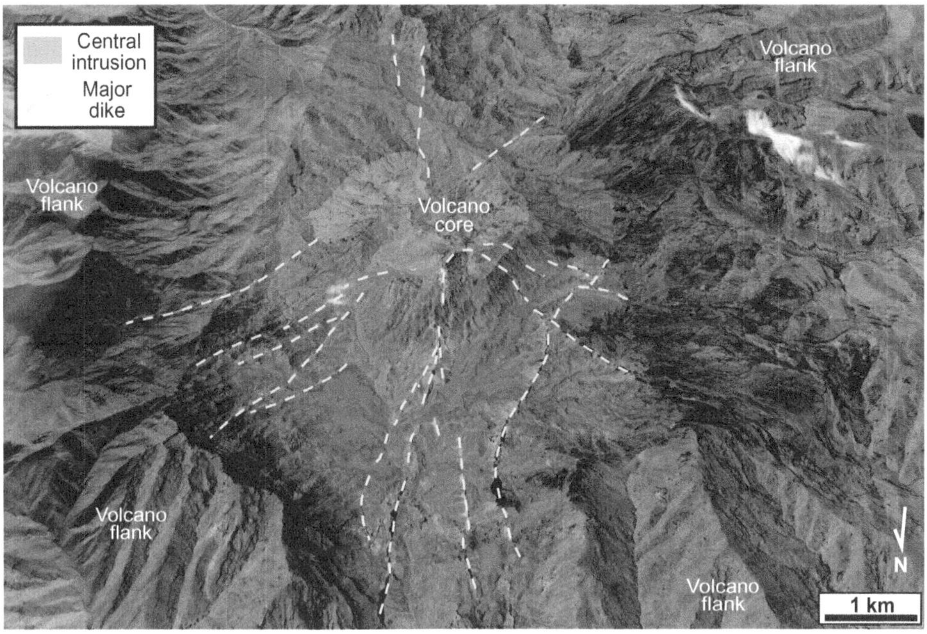

Fig. 13.23 Tibesti hot spot. Google Earth image of the eroded Tieroko volcano, as seen from the north. The eroded central portion of the composite volcano reveals its dike pattern radiating from one or more central intrusions (image source: Google Earth)

suite may be explained by crustal contamination, the undersaturated suite may be explained by crystal fractionation (Gourgaud and Vincent 2004).

The volcanoes of the Tibesti hot spot do not show clear evidence of defined rift zones or significant flank failure, at least from remote sensing data. However, unusual conditions of uplift and erosion currently enable exceptional exposure of the internal structure of some volcanoes. Indeed, the eroded portions of several volcanoes reveal NW–SE trending feeder dikes, associated with flexures, and radial dike patterns. The latter are best developed at Kilehégé, where more than a hundred rhyolitic dikes, 3–15 m thick, converge toward a focal zone with the highest spines and several rhyolitic plugs, and at Tiéroko, where the radial dikes are similarly associated with a central intrusion (Fig. 13.23; Deniel et al. 2015).

As mentioned, a distinctive feature of the Tibesti hot spot is the lack of any clear spatial progression of volcanic centres, though volcanism seems to have initiated in the Central region (e.g. Voon and Toon) and later migrated to the eastern and western regions: this feature is also matched by the lack of a linear hot spot track. In addition, this sparse magmatic activity produced a wide variety of volcanic features, including moderate to large-scale calderas (average diameter of ~12 km), suggesting also significant magma reservoirs below (Permenter and Oppenheimer 2007).

13.10 Comparing Hot Spot Volcanoes

Hot spots display overall common structural and magmatic features, but also variability in terms of plume structure and magma supply, interaction with the lithosphere, distribution, age and type of volcanism, evolution and structure of the volcanoes. These similarities and differences are discussed below, also including knowledge from additional examples of hot spots (Table 13.1). Note that the available information on oceanic hot spot volcanoes mainly concerns their subaerial tips: therefore, the limited knowledge of their submerged portions may affect the considerations made in the following discussion.

A common feature of oceanic and continental hot spots is the geophysical detection of underplated magma, more voluminous than the erupted portion. A broad surface swell is usually, but not systematically, accompanied with magmatic underplating, as at Hawaii, Reunion, the Canary Islands, Yellowstone, Tibesti and in the South Pacific, at both a regional (superswell) and local scale (volcanic chain or cluster; Sichoix et al. 1998; Suetsugu et al. 2009; King and Adam 2014; Park and Rye 2019). The swell may be the surface expression of the flexure induced by the emplacement of the underplated magma and/or the upward pressure exerted by the plume head.

Another common feature of the considered hot spots, with the possible exception of the Canary Islands, is the influence of pre-existing basement structures on volcanic activity. This occurs both on continental and oceanic lithosphere, where any pre-existing discontinuities (faults, fracture zones) may be reactivated and/or intruded by magma. This control of inherited structures, recognized also in the Iceland, Afar and French Massif Central hot spots, indicates that hot spot volcanism is affected by regional structures and/or stresses. The reactivation of regional anisotropies often results in the control of the location, alignment and shape of hot spot volcanoes, including their rift zones and flank instabilities.

Radial volcanic rift zones are common at oceanic hot spots, although their configuration and overall architecture vary widely. On the Hawaiian Islands, the well-developed rift zones consist of long and narrow portions of focused magmatic injection controlled by pre-existing structures, edifice morphology, and flank instability. The best-known Hawaiian rift zones (on Mauna Loa and Kilauea) developed at the head of the unstable flanks of the volcanoes, although the longest rifts extend far beyond the shield edifice, probably also following basement fractures. In contrast, rift zones are poorly developed on the western Galapagos volcanoes or on Easter Island, where diffuse areas of radial dike injection predominate. In between are the defined but relatively wide rift zones of Reunion, which further widen downslope, and the rift zones forming triple arms at 120° mainly on the younger Canary Islands volcanoes, often at the head of collapse structures. In other cases (Azores, the eastern Galapagos Islands), the rift zones are aligned along pre-existing regional trends or anyway highlight the influence of plate boundary stresses. Therefore, the presence and development of rift zones on oceanic hot spots may be reconciled, in addition to the local stress field imposed by the volcanic edifice, with unstable flanks, pre-existing structures and regional stresses. Available evidence (Koolau, Kilauea and Mauna Loa in the Hawaiian Islands, Fernandina in the Galapagos Islands, Tenerife and El Hierro in the Canary Islands, Sao Jorge in the Azores) suggests a predominant lateral propagation of the dikes feeding the rift zones. No evident rift zones have been found in the considered continental hot spots.

Oceanic hot spot volcanoes also show widespread (though not ubiquitous) instability, often resulting in catastrophic collapse. This has been documented in the Hawaiian Archipelago, the Canary Islands, the Azores, Reunion Island and the South Pacific Islands (Tahiti and the Marquesas volcanic chain). In some cases (Canary Islands, Reunion, Marquesas) flank instability results in widespread mass wasting. On the Galapagos Archipelago (with the exception of Ecuador volcano) and Easter Island flank instability is negligible. The general lack of flank instability at the Galapagos volcanoes may be related to their shallow submarine flanks perched on the Galapagos Platform, and/or to the young age of the oceanic lithosphere below the platform, which has hindered the accumulation of clayish oceanic deposits: elsewhere, as below Hawaii, these weak water-saturated deposits may act as decollement level for the volcanic edifices.

Table 13.1 Summary of the regional and local tectonic and magmatic features related to hot spot volcanoes; with the exception of Yellowstone and Tibesti, all listed hot spots are located on oceanic crust

Hot spot	Age of crust (Ma)	Plate velocity (cm/year)	Eruptive rate (km^3/Ma)	Buoyancy flux (Mg/s)	Composition (SiO_2 wt%)	Volcano distribution	Elevation above ocean floor (m)	Rift zones	Summit caldera	Flank instability
Hawaii	80	8.6	10^5	7.10	39–61	Aligned	5000 + 4300	Focused	Yes	Yes
Society	60	10	10^2–10^3	2.72	42–49	Aligned	4000 + 2200	Focused	(Yes)	Yes
Galapagos	8	5.9	10^3	0.71	45–51	Scattered	3500 + 1700	Diffuse	Yes	(No)
Easter	2–4	13.3	0.01–100	0.43	41–74	Aligned	3500 + 500	Diffuse	Yes	No
Réunion	60	1.9	3.4×10^3	0.08	45–70	Aligned	4300 + 3050	(Focused)	Yes	Yes
Canary	160–180	1.9	10^3–10^4	0.13	36–73	Scattered	3600 + 3700	Focused	Yes	Yes
Azores	9–10	2	10^3	0.48	44–65	Scattered	3400 + 2350	Focused	Yes	(Yes)
Madeira	130–140	1.2	10^2–10^3	0.22	41–54	Aligned	4500 + 1850	Focused	No	Yes
Ascension	7	1.7	30	0.09	46–75	Aligned	3500 + 860	Diffuse	No	Minor
Yellowstone	>2500	2	2.5×10^3	0.25	43–79	Aligned	0 + 3400	N/A	Yes	No
Tibesti	>550	1.9	3×10^2	0.51	43–66	Scattered	0 + 3450	N/A	Yes	No

Elevation above seafloor results from depth of ocean (left column) and height above sea level (right column) of the highest volcano in the hot spot. Brackets indicate that the specified feature may not be completely represented (modified from Vezzoli and Acocella 2009; King and Adam 2014)

At Easter Island the lack of flank instability may be attributed to the lower output rate of the hot spot, which builds less imposing volcanic edifices.

There is an overall association between focused volcanic rift zones and unstable volcano flanks (at Hawaii, Society Islands and Canary Islands), as well as between diffuse rift zones and stable flanks (Galapagos, Easter Island). This suggests an interaction where flank instability focuses magmatic activity in rift zones at the head of the unstable flank, preferably of a large volcano, with magmatic activity along rift zones in turn enhancing further instability. At the Azores, the development of the rift zones seems more controlled by the regional structures, rather than flank instability.

The summits of oceanic and continental hot spot volcanoes are also characterized by calderas, as observed on all the above-mentioned examples, as well as in the South Pacific volcanic chains. A few of these summit calderas (Galapagos and, to a much lesser extent, Reunion and Easter Island) are associated with dike-fed circumferential eruptive fissures, rarely documented elsewhere. In addition, the summits of a few oceanic hot spot volcanoes (Tenerife, Reunion and Fogo, Cape Verde) show distinctive horseshoe-shaped scarps continuing on the volcano flanks, probably associated with a complex interplay between vertical and lateral collapse.

Despite these general similarities, hot spot volcanoes also show striking differences in both evolution and structure, with a variability pointing to a growth and architecture that are more complex than those proposed for example for the reference Hawaiian hot spot.

Among the differences in the considered hot spots, a first-order feature is the distribution of localized or delocalized volcanism. In fact, in some cases (Galapagos, Canary Islands, Azores, Tibesti) volcanic activity and volcano growth are synchronous, or scattered, developing several coalesced coeval volcanoes. In others (Hawaii, Reunion, Yellowstone), the activity and growth of the volcanoes are sequential, or aligned, with only a few neighbouring volcanoes active at the same time. Intermediate hybrid types of activity and growth characterize some volcanic chains of the South Pacific. Such an aligned or scattered distribution of volcanism is illustrated by the focused vs. diffuse age-distance relationships shown in Fig. 13.24. These different distributions of the hot spot volcanoes may have various explanations. The distributions may reflect different types of plumes: for example, a superplume fed by different plumelets, as below the Cook-Austral Islands, may generate a scattered distribution of volcanoes more easily than a single plume, as below Reunion (e.g., Koppers et al. 2003). Absolute plate velocity may be an additional feature controlling the duration and persistence of a feeding system below a volcano, with faster plates favouring aligned volcanism (Hawaii, Society Islands in the South Pacific) and slower plates encouraging scattered volcanism (Canary Islands, Azores, Tibesti; Mitchell et al. 2002). A third control may be the age, and the related thickness, of the oceanic lithosphere above the hot spot (Poland 2014). For example, the thicker lithosphere may induce focused volcanism in Hawaii and the thinner lithosphere dispersed volcanism in Galapagos, with both hot spots lying on plates with medium–high velocity. In fact, despite the lower buoyancy flux of the Galapagos plume, a thinner (younger) lithosphere may be more effectively heated and weakened by a mantle plume, developing multiple magmatic paths and thus showing delocalized volcanism. Within an adequate tectonic context, as in proximity to a triple junction, scattered magmatism may develop a microplate with non-rigid behaviour, as observed in the Azores. A fourth factor possibly influencing the aligned or scattered distribution of volcanoes is the magma flux of the hot spot. This is suggested by the more aligned distribution of volcanoes on the less productive Madeira hot spot with regard to that of the nearby and more productive Canary Islands hot spot, both in a similar tectonic setting (Table 13.1 and Fig. 13.24; Geldmacher et al. 2000).

Also, while a swell has been recognized below most hot spot volcanoes, a moat due to the

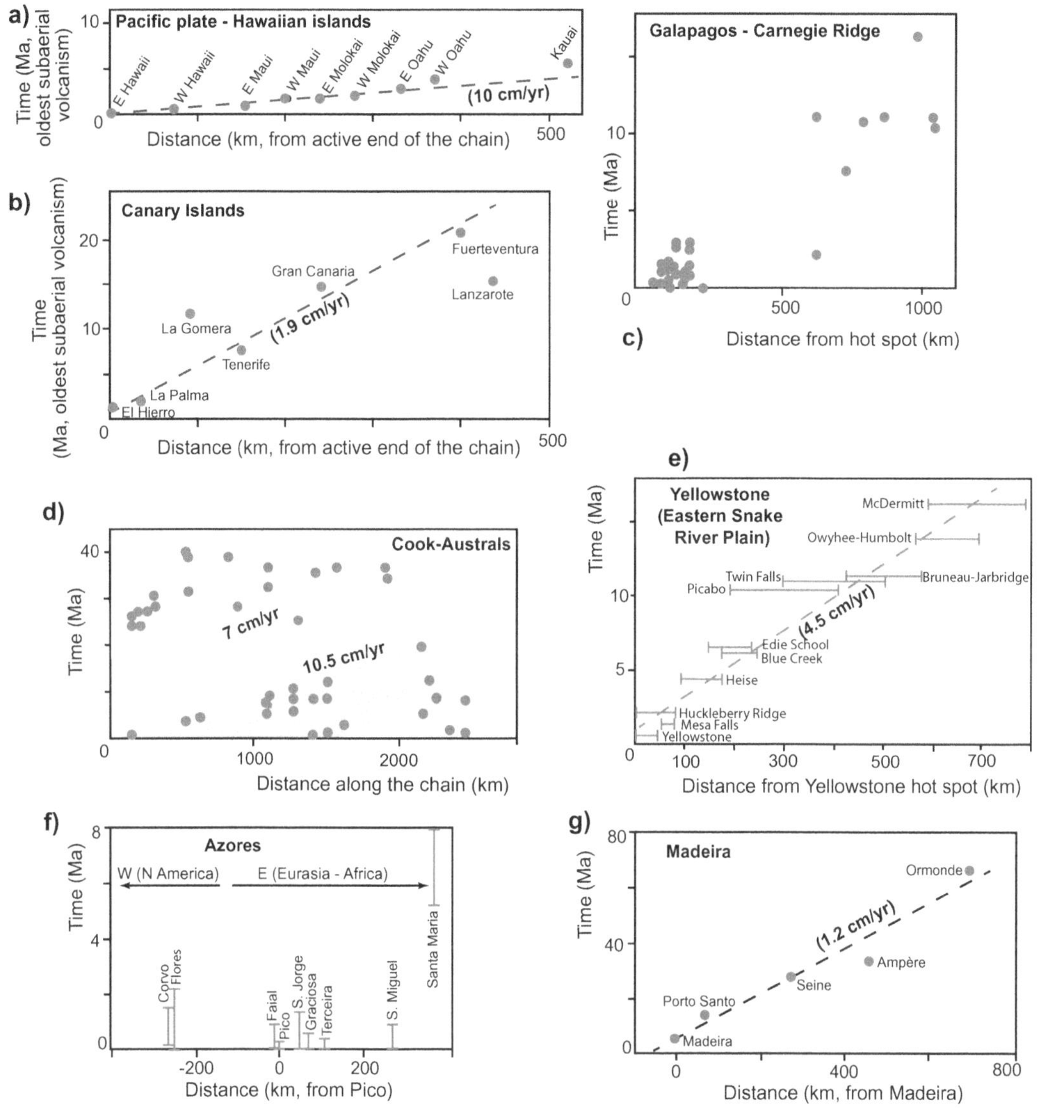

Fig. 13.24 Age of volcanism with regard to the distance from the hot spot location for several hot spots: **a** Hawaiian Islands; **b** Canary Islands; **c** Galapagos-Carnegie Ridge; **d** Cook-Australs; **e** Yellowstone; **f** Azores; **g** Madeira. While several host spots (as Hawaii, Madeira, Yellowstone) show a defined age-distance relationship, implying aligned volcanism, others (as Cook-Australs, Galapagos, Azores) show a more dispersed distribution, implying scattered volcanism

load of the volcanic edifices appears less frequent. A moat has been detected around the volcanoes of Hawaii, Tenerife, and, in part, below the Galapagos, the South Pacific and the Canary Islands: conversely, no moat has been recognized around the Azores and Reunion volcanoes.

A further notable difference in the eruptive style and composition of hot spot volcanoes is related to the nature of their lithosphere. Bimodal compositions, as well as explosive activity related to the evolved magmas, are more common on continental hot spots. At Yellowstone the plume triggers bimodal volcanism, first melting the

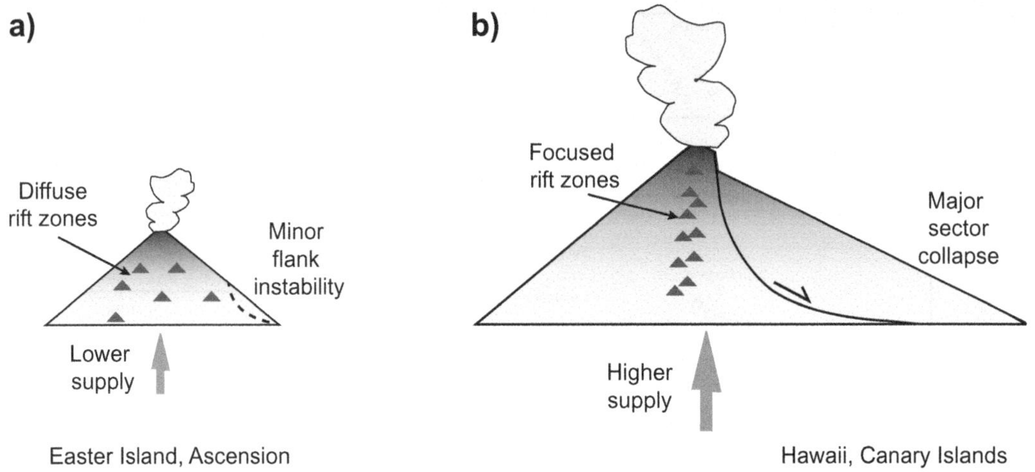

Fig. 13.25 Working hypothesis showing how the structural features (in terms of sector collapses and rift zones to the rear of the collapses) of end-member oceanic hot spots volcanoes depend upon their magma supply and size. **a** Lower magma supplies generate smaller volcanoes with minor flank instability and diffuse rift zones. **b** Higher magma supplies generate larger volcanoes with major sector collapse and focused rift zones. Here only the subaerial portion of the edifice is reported, not to scale

silicic continental crust, producing rhyolites with explosive activity, and then erupting the more primitive products. Tibesti also shows bimodal volcanism, although the rhyolitic compositions are proportionally much less represented than Yellowstone, probably following a more limited crustal melting. Rhyolites and explosive eruptions are usually rare on oceanic hot spots, except in the Canary Islands, Alcedo (Galapagos) and Terceira (Azores). Oceanic hot spot volcanoes are indeed characterized by the predominant effusive activity of mafic magma.

Despite all the differences in the oceanic hot spot volcanoes discussed above, probably the most striking one concerns their extremely variable eruptive rates. Considering the general limitation that estimates for oceanic volcanoes mainly refer to their subaerial tip, the eruptive rates vary over several orders of magnitude, from Hawaii ($\sim 10^5$ km^3/Ma) to Ascension or Easter Island (the latter reaching 10^{-2} km^3/Ma) (Table 13.1). While the highest eruptive rates at Hawaii are associated with focused rift zones and significant lateral collapse, at Easter Island and Ascension the low eruptive rates are mainly accompanied with poorly defined, diffuse rift zones and minor flank instability. These end-member behaviours suggest a complex and not yet fully understood interplay between magma flux, maturity of rift zones and flank instability. A working hypothesis relates these end-member oceanic behaviours to the magma supply and the size of the volcanic edifice. Accordingly, the more productive and larger volcanic edifices generate significant flank instability and focused rift zones, whereas less productive and smaller edifices have minor or no flank instability and diffuse rift zones (Fig. 13.25; Vezzoli and Acocella 2009). The western Galapagos Islands provide an exception, as characterized by high productivity, but lacking flank instability and focused rift zones. This feature may be explained by the relatively small size of the volcanoes and/or their stronger basement, with a thick platform overlying a young oceanic lithosphere lacking the clayish sediments that may promote a basal decollement.

13.11 Summary

Hot spots are the surface expression of plumes upwelling from the Earth's mantle, most often appearing on oceanic lithosphere. Volcanic activity related to representative hot spots shows major recurrent features and differences.

Recurrent features of oceanic and continental hot spot volcanoes include widespread underplating, influence of pre-existing basement structures and the development of volcanic edifices with summit caldera. Oceanic hot spots also show frequent radial rift zones, often accompanied by flank instability. Major differences among hot spot volcanoes, both in oceanic and continental examples, are related to their distribution (aligned or scattered), eruption rate (varying over several orders of magnitude) and composition (with continental examples showing also widespread rhyolites).

The variability of hot spot volcanoes results from differences in the plume structure and supply, and in the age, rate of motion and nature of lithosphere pierced by the plume: the interplay between these features determines different distributions and types of volcanism, as well as different structure and evolution of the volcanoes. In general, the larger the magmatic output of oceanic hot spots, the larger are the islands and the more developed are the instability of the volcanoes flanks and the rift zones, with Hawaii and Easter Island representing end-member behaviours and the Galapagos Islands providing an exception.

References

Abratis M, Schmincke HU, Hansteen TH (2002) Composition and evolution of submarine volcanic rocks from the central and western Canary Islands. Int J Earth Sci 91:562–582

Acosta J, Uchupi E, Smith D, Munoz A, Herranz P, Palomo C et al (2003) Comparison of volcanic rifts on La Palma and El Hierro, Canary Islands and the Island of Hawaii. Mar Geophys Res 24:59–90

Albarede F, Luais B, Fitton G, Semet M, Kaminski E, Upton BGJ et al (1997) The geochemical regimes of Piton de la Fournaise volcano (Reunion) during the last 530,000 years. J Petrol 38:171–201

Ancochea E, Brandle JL, Cubas CR, Hernan F, Huertas MJ (1996) Volcanic complexes in the eastern ridge of the Canary Islands: the Miocene activity of the island of Fuerteventura. J Volcanol Geoth Res 70:183–204

Bagnardi M, Amelung F (2012) Space-geodetic evidence for multiple magma reservoirs and subvolcanic lateral intrusions at Fernandina Volcano, Galápagos Islands. J Geophys Res 117:B10406. https://doi.org/10.1029/2012JB009465

Bagnardi M, Amelung F, Poland MP (2013) A new model for the growth of basaltic shields based on deformation of Fernandina volcano, Galápagos Islands. Earth Planet Sci Lett 377–378:358–366

Ball PW, White NJ, Masoud A, Nixon S, Hoggard MJ, Maclennan J et al (2019) Quantifying asthenospheric and lithospheric controls on mafic magmatism across North Africa. Geochem Geophys Geosyst 20. https://doi.org/10.1029/2019GC008303

Barruol G, Fontaine FR (2013) Mantle flow beneath La Reunion hot spot track from SKS splitting. Earth Planet Sci Lett 362:108–121

Becerril L, Galindo I, Marti J, Gudmundsson A (2015) Three-armed rifts or masked radial pattern of eruptive fissures? The intriguing case of El Hierro volcano (Canary Islands). Tectonophysics 647–648:33–47

Bianco TA, Ito G, Beker JM, Garcia MO (2005) Secondary Hawaiian volcanism formed by flexural arch decompression. Geochem Geophys Geosyst 6: Q08009. https://doi.org/10.1029/2005GC000945

Bindeman IN, Fu B, Kita NT, Walley JW (2008) Origin and evolution of silicic magmatism at yellowstone based on ion microprobe analysis of isotopically zoned zircons. J Petrol 49:163–193

Bird P (2003) An updated digital model of plate boundaries. Geochem Geophys Geosyst 4:1027. https://doi.org/10.1029/2001GC000252

Blahut J, Mitrovic-Woodell I, Baron I, Rene M, Rowberry M, Blard P-H et al (2020) Volcanic edifice slip events recorded on the fault plane of the San Andrés Landslide, El Hierro, Canary Islands. Tectonophysics 776:228317

Bonali FL, Corazzato C, Tibaldi A (2011) Identifying rift zones on volcanoes: an example from La Réunion island, Indian Ocean. Bull Volcanol 73:347–366

Bonneville A, Von Herzen RP, Lucazeu F (1997) Heat flow over Reunion hot spot track: additional evidence for thermal rejuvenation of oceanic lithosphere. J Geophys Res 102:22731–22747

Boulesteix T, Hildenbrand A, Soler V, Quidelleur X, Gillot PY (2013) Coeval giant landslides in the Canary Islands: Implications for global, regional and local triggers of giant flank collapses on oceanic volcanoes. J Volcanol Geoth Res 257:90–98

Camacho AG, Nunes JC, Ortiz E, Franca Z, Vieira R (2007) Gravimetric determination of an intrusive complex under the Island of Faial (Azores): some methodological improvements. Geophys J Int 171:478–494

Camp VE, Ross ME (2004) Mantle dynamics and genesis of mafic magmatism in the intermontane Pacific Northwest. J Geophys Res 109:B08204. https://doi.org/10.1029/2003JB002838

Camp VE, Ross ME, Duncan RA, Kimbrough DL (2017) Uplift, rupture, and rollback of the Farallon slab reflected in volcanic perturbations along the Yellowstone adakite hot spot track. J Geophys Res 122:7009–7041

Canales JP, Ito G, Detrick RS, Sinton J (2002) Crustal thickness along the western Galapagos spreading Center and the compensation of the Galapagos hotspot swell. Earth Planet Sci Lett 203:311–327
Carracedo JC (1999) Growth, structure, instability and collapse of Canarian volcanoes and comparisons with Hawaiian volcanoes. J Volcanol Geoth Res 94:1–19
Carracedo JC, Day S, Guillou H, Rodriguez Badiola E, Canas JA, Perez Torrado FJ (1998) Hotspot volcanism close to a passive continental margin: the Canary Islands. Geol Mag 135:591–604
Carracedo JC, Rodriguez Badiola E, Guillou H, Paterne M, Scaillet S, Perez Torrado FJ et al (2007) Eruptive and structural history of Teide Volcano and rift zones of Tenerife, Canary Islands. Geol Soc Am Bull 119:1027–1051
Chadwick WW, Howard KA (1991) The pattern of circumferential and radial eruptive fissures on the volcanoes of Fernandina and Isabela islands, Galapagos. Bull Volcanol 53:259–275
Chadwick WW, De Roy T, Carrasco A (1991) The September 1988 intracaldera avalanche and eruption at Fernandina volcano, Galapagos Islands. Bull Volcanol 53:276–286
Chadwick WW, Geist DJ, Jonsson S, Poland M, Johnson DJ, Meertens CM (2006) A volcano bursting at the seams: inflation, faulting, and eruption at Sierra Negra volcano, Galápagos. Geology 34:1025–1028
Chadwick WW, Jonsson S, Geist DJ, Poland M, Johnson DJ, Batt S et al (2011) The May 2005 eruption of Fernandina volcano, Galápagos: the first circumferential dike intrusion observed by GPS and InSAR. Bull Volcanol 73:679–697
Chaput M, Famin V, Michon L (2014) Deformation of basaltic shield volcanoes under cointrusive stress permutations. J Geophys Res 119:274–301
Charvis P, Laesanpura A, Gallart J, Hirn A, Lepine JC, de Voogd B et al (1999) Spatial distribution of hotspot material added to the lithosphere under La Reunion, from wide-angle seismic data. J Geophys Res 104:2875–2893
Chen K, Smith JD, Avouac J-P, Liu Z, Song YT, Gualandi A (2019) Triggering of the Mw 7.2 Hawaii earthquake of 4 May 2018 by a dike intrusion. Geophys Res Lett 46:2503–2510
Christiansen RL (2001) The quaternary and pliocene yellowstone plateau volcanic field of Wyoming, Idaho, and Montana. US Geological Survey Professional Paper 729-G 145 pp
Clague DA, Sherrod DR (2014) Growth and degradation of Hawaiian volcanoes. In: Poland MP, Takahashi TJ, Landowski CM (eds) Characteristics of Hawaiian volcanoes. US geological survey professional paper, vol 1801, pp 98–146
Clague DA, Paduan JB, Caress DW, Moyer CL, Glazer BT, Yoerger DR (2019) Structure of Loihi Seamount, Hawai'i and Lava flow morphology from high-resolution mapping. Front Earth Sci 7:58. https://doi.org/10.3389/feart.2019.00058
Coble MA, Mahood GA (2012) Initial impingement of the Yellowstone plume located by widespread silicic volcanism contemporaneous with Columbia River flood basalts. Geology 40:655–658
Collier JS, Watts AB (2001) Lithospheric response to volcanic loading by the Canary Islands: constraints from seismic reflection data in their flexural moat. Geophys J Int 147:660–676
Corbi F, Rivalta E, Pinel V, Maccaferri F, Bagnardi M, Acocella V (2015) How caldera collapse shapes the shallow emplacement and transfer of magma in active volcanoes. Earth Planet Sci Lett 431:287–293
Costa ACG, Marques FO, Hildebrand A, Sibrant AIR, Catita CMS (2014) Large-scale catastrophic flank collapses in a steep volcanic ridge: the Pico-Faial Ridge, Azores Triple Junction. J Volcanol Geoth Res 272:111–125
Cullen A, McBirney AR (1987) The volcanic geology and petrology of Isla Pinta, Galapagos archipelago. Geol Soc Am Bull 98:294–301
Cvetkovi V, Toljic M, Ammar MA, Rundic L, Trish KB (2010) Petrogenesis of the eastern part of the Al Haruj basalts (Libya). J Afr Earth Sc 58:37–50
Danobeitia JJ, Canales JP (2000) Magmatic underplating in the Canary Archipelago. J Volcanol Geoth Res 103:27–41
Day SJ, Carracedo JC, Guillou H (1997) Age and geometry of an aborted rift flank collapse: the San Andres fault system, El Hierro, Canary Islands. Geol Mag 134:523–537
Day SJ, Heleno da Silva SIN, Fonseca JFBD (1999) A past giant lateral collapse and present-day flank instability of Fogo, Cape Verde Islands. J Volcanol Geoth Res 94:191–218
Delaney PT, Fiske RS, Miklius A, Okamura AT, Sako MK (1990) Deep Magma body beneath the summit and Rift Zones of Kilauea Volcano, Hawaii. Science 247:1311–1316
DeMets C, Gordon RG, Argus DF (2010) Geologically current plate motions. Geophys J Int 181:1–80
Deniel C, Vincent PM, Beauvillain A, Gourgaud A (2015) The Cenozoic volcanic province of Tibesti (Sahara of Chad): major units, chronology, and structural features. Bull Volcanol 77:74
Dieterich JH (1988) Growth and persistence of Hawaiian volcanic rift zones. J Geophys Res 93:4258–4270
Dumont M, Peltier A, Roblin E, Reninger P-A, Barde-Cabusson S, Fininzola A et al (2019) Imagery of internal structure and destabilization features of active volcano by 3D high resolution airborne electromagnetism. Sci Rep 9:18280. https://doi.org/10.1038/s41598-019-54415-4
Farnetani CG, Hofmann AW (2010) Dynamics and internal structure of the Hawaiian plume. Earth and Planetary Science Letters 295:231–240
Fernandes RMS, Bastos L, Miranda JM, Lourenco N, Ambrosius BAC, Noomen R et al (2006) Defining the plate boundaries in the Azores region. J Volcanol Geoth Res 156:1–9

Fernandez C, Casillas R, Garcia Navarro E, Gutierrez M, Camacho MA, Ahijado A (2006) Miocene rifting of Fuerteventura (Canary Islands). Tectonics 25:TC6005. https://doi.org/10.1029/2005TC001941

Fiske RS, Jackson ED (1972) Orientation and growth of Hawaiian volcanic rifts. Proc R Soc Lond 329:299–326

Fullea J, Camacho AG, Negredo AM, Fernandez J (2015) The Canary Islands hot spot: new insights from 3D coupled geophysical–petrological modelling of the lithosphere and uppermost mantle. Earth Planet Sci Lett 409:71–88

Gailler L-S, Lenat J-F (2012) Internal architecture of La Réunion (Indian Ocean) inferred from geophysical data. J Volcanol Geoth Res 221–222:83–98

Galetto F, Bagnardi M, Acocella V, Hooper A (2019) Noneruptive unrest at the caldera of Alcedo Volcano (Galápagos Islands) revealed by InSAR data and geodetic modelling. J Geophys Res 124. https://doi.org/10.1029/2018JB017103

Gallart J, Driad L, Charvis P, Sapin M, Hirn A, Diaz J et al (1999) Perturbation to the lithosphere along the hotspot track of La Reunion from an offshore-onshore seismic transect. J Geophys Res 104:2895–2908

Garcia MO, Caplan-Auerbach J, De Carlo EH, Kurz MD, Becker N (2006) Geology, geochemistry and earthquake history of Loihi Seamount, Hawaii's youngest volcano. Chem Erde 66:81–108

Geist DJ, McBirney AR, Duncan RA (1986) Geology and petrogenesis of lavas from San Cristobal Island, Galapagos Archipelago. Geol Soc Am Bull 97:555–566

Geist D, White WM, Albarede F, Harpp K, Reynolds R, Blichert-Toft J et al (2002) Volcanic evolution in the Galapagos: the dissected shield of Volcan Ecuador. Geochem Geophys Geosyst 3:1061. https://doi.org/10.1029/2002GC000355

Geist D, Diefenbach AB, Fornari DJ, Kurz MD, Harpp K, Blusztajn J (2008) Construction of the Galapagos platform by large submarine volcanic terraces. Geochem Geophys Geosyst 9:Q03015. https://doi.org/10.1029/2007GC001795

Geist DJ, Bergantz G, Chadwick WW (2014) Galápagos Magma chambers. In: Harpp KS, Mittelstaedt E, d'Ozouville N, Graham DW (eds) The Galápagos: a natural laboratory for the earth sciences. AGU Geophys Monogr 204:55–69

Geldmacher J, van den Bogaard P, Hoernle K, Schmincke HU (2000) The $^{40}Ar/^{39}Ar$ age dating of the Madeira Archipelago and hotspot track (eastern North Atlantic). Geochem Geophys Geosyst 1:1008. https://doi.org/10.1029/1999GC000018

Geldmacher J, Hoernle K, Bogaard PVD, Duggen S, Werner R (2005) New $^{40}Ar/^{39}Ar$ age and geochemical data from seamounts in the Canary and Madeira volcanic provinces: support for the mantle plume hypothesis. Earth Planet Sci Lett 237:85–101

Genske FS, Turner SP, Beier C, Scaefer BF (2012) The petrology and geochemistry of Lavas from the Western Azores Islands of Flores and Corvo. J Petrol 53:1673–1708

Gente P, Dyment J, Maia M, Goslin J (2003) Interaction between the Mid-Atlantic Ridge and the Azores hot spot during the last 85 Myr: emplacement and rifting of the hot spot-derived plateaus. Geochem Geophys Geosyst 4:8514. https://doi.org/10.1029/2003GC000527

Gillot PY, Lefevre JC, Nativel PE (1994) Model for the structural evolution of the volcanoes of Reunion Island. Earth Planet Sci Lett 122:291–302

Gonzalez-Ferran O, Mazzuoli R, Lahsen A (2004) Geologia del complejo volcanico Isla de Pascua Rapa Nui. Scale 1:30,000. Santiago Chile Centro de Estudios Volcanologicos

Gourgaud A, Vincent P (2004) Petrology of two continental alkaline intraplate series at Emi Koussi volcano, Tibesti, Chad. J Volcanol Geoth Res 129:261–290

Guillou H, Carracedo JC, Paris R, Perez Torrado FJ (2004) Implications for the early shield-stage evolution of Tenerife from K/Ar ages and magnetic stratigraphy. Earth Planet Sci Lett 222:599–614

Guiraud R, Doumnang JC, Baigane M, Carretier S, Dominguez S (2000) Evidence for a 6000 km length NW-SE striking lineament in northern Africa: the Tibesti lineament. J Geol Soc London 157:897–900

Haase KM, Stoffers P, Garbe-Schoenberg CD (1997) The petrogenetic evolution of lavas from Easter Island and neighboring seamounts, near-ridge hotspot volcanoes in the SE Pacific. J Petrol 38:785–813

Haase KM (2002) Geochemical constraints on magma sources and mixing processes in Easter Microplate (SE Pacific): a case study of plume-ridge interaction. Chem Geol 182:335–355

Harpp KS, Wirth KR, Korich DJ (2002) Northern Galapagos Province: Hotspot-induced, near-ridge volcanism at Genovesa Island. Geology 30:399–402

Harpp KS, Fornari DJ, Geist DJ, Kurz MD (2003) Genovesa Submarine Ridge: a manifestation of plume-ridge interaction in the northern Galapagos Islands. Geochem Geophys Geosyst 4:8511. https://doi.org/10.1029/2003GC000531

Harpp KS, Geist DJ (2018) The evolution of Galápagos volcanoes: an alternative perspective. Front Earth Sci 6:50. https://doi.org/10.3389/feart.2018.00050

Harrison LN, Weis D, Garcia MO (2017) The link between Hawaiian mantle plume composition, magmatic flux, and deep mantle geodynamics. Earth Planet Sci Lett 463:298–309

Hildenbrand A, Marques FO, Catalão J, Catita CMS, Costa ACG (2012) Large-scale active slump of the southeastern flank of Pico Island, Azores. Geology 40:939–942

Hildenbrand A, Marques FO, Catalao J (2018) Large-scale mass wasting on small volcanic islands revealed by the study of Flores Island (Azores). Sci Rep 8:13898. https://doi.org/10.1038/s41598-018-32253-0

Hoernle K, Carracedo JC (2009) Canary Islands, Geology. In: Gillespie R, Clague D (eds) Encyclopedia of Islands 140

Huang HH, Lin F-C, Schmandt B, Farrell J, Smith RB, Tsai VC (2015) The Yellowstone magmatic system from the mantle plume to the upper crust. Science 348:773–776

Huppert KL, Royden LH, Perron JT (2015) Dominant influence of volcanic loading on vertical motions of the Hawaiian Islands. Earth Planet Sci Lett 418:149–171

Keppie JD, Dostal J, Murphy JB (2010) Complex geometry of the Cenozoic magma plumbing system in the central Sahara, NW Africa. Int Geol Rev 1:1–17. https://doi.org/10.1080/00206814.2010.496211

King SD, Adam C (2014) Hotspot swells revisited. Phys Earth Planet Inter 235:66–83

Knott TR, Branney MJ, Reichow MK, Finn DR, Tapster S, Coe RS (2020) Discovery of two new super-eruptions from the Yellowstone hotspot track (USA): is the Yellowstone hotspot waning? Geology 48:934–938

Koppers AAP, Staudigel H, Pringle MS, Wijbrans JR (2003) Short-lived and discontinuous intraplate volcanism in the South Pacific: hot spots or extensional volcanism? Geochem Geophys Geosyst 4:1089. https://doi.org/10.1029/2003GC000533

Krastel S, Schmincke HU, Jacobs CJ, Rihm R, Le Bas TP, Alibes B (2001) Submarine landslides around the Canary Islands. J Geophys Res 106:3977–3997

Langenheim VAM, Clague DA (1987) Stratigraphic framework of volcanic rocks of the Hawaiian Islands. US Geol Surv Prof Pap 1350:55–83

Larrea P, Franca Z, Widom E, Lago M (2018) Petrology of the Azores Islands. In: Kueppers U, Beier C (eds) Volcanoes of the Azores. Active volcanoes of the world. Springer, Germany, pp 197–249

Lenat JF, Vincent P, Bachelery P (1989) The off-shore continuation of an active basaltic volcano: Piton de la Fournaise (Reunion Island, Indian Ocean); structural and geomorphological interpretation from sea beam mapping. J Volcanol Geoth Res 36:1–36

Lenat JF, Gibert-Malengrau B, Galdeano A (2001) A new model for the evolution of the volcanic island of Reunion (Indian Ocean). J Geophys Res 106:8645–8663

Lipman PW, Sisson TW, Coombs ML, Calvert A, Kimura JI (2006) Piggyback tectonics: long-term growth of Kilauea on the south flank of Mauna Loa. J Volcanol Geoth Res 151:73–108

Lipman PW, Calvert AT (2011) Early growth of Kohala volcano and formation of long Hawaiian rift zones. Geology 39:659–662

Lipman PW, Calvert AT (2013) Modeling volcano growth on the Island of Hawaii: deep-water perspectives. Geosphere 9:1348–1383

Lodge A, Nippress SEJ, Rietbrock A, Garcia-Yeguas A, Ibanez JM (2012) Evidence for magmatic underplating and partial melt beneath the Canary Islands derived using teleseismic receiver functions. Phys Earth Planet Inter 212–213:44–54

Lourenço N, Miranda JM, Luis JF, Ribeiro A, Mendes Victor LA, Madeira J et al (1998) Morpho-tectonic analysis of the Azores Volcanic Plateau from a new bathymetric compilation of the area. Mar Geophys Res 20:141–156

Madeira J, Ribeiro A (1990) Geodynamic models for the Azores triple junction: a contribution from tectonics. Tectonophysics 184:405–415

Madeira J, Brum da Silveira A (2003) Active tectonics and first paleoseismological results in Faial, Pico and S. Jorge islands (Azores, Portugal). Ann Geophys 46:733–761

Madureira P, Moreira M, Mata J, Nunes JC, Gautheron C, Lorenco N et al (2014) Helium isotope systematics in the vicinity of the Azores triple junction: constraints on the Azores geodynamics. Chem Geol 372:62–71

Marques FO, Catalao JC, DeMets C, Costa ACG, Hildebrand A (2013) GPS and tectonic evidence for a diffuse plate boundary at the Azores Triple Junction. Earth Planet Sci Lett 381:177–187

Marques FO, Hildenbrand A, Costa ACG, Sibrant ALR (2020) The evolution of Santa Maria Island in the context of the Azores Triple Junction. Bull Volcanol 82:39

Martì J (2019) Las Cañadas caldera, Tenerife, Canary Islands: a review, or the end of a long volcanological controversy. Earth Sci Rev 196:102889

Martì J, Ortiz R, Gottsmann J, Garcia A, De la Cruz-Reyna S (2009) Characterising unrest during the reawakening of the central volcanic complex on Tenerife, Canary Islands, 2004–2005, and implications for assessing hazards and risk mitigation. J Volcanol Geoth Res 182:23–33

Martinez Arevalo C, de Lis MF, Hellfrich G, Garcia A (2013) Seismic evidence of a regional sublithospheric low velocity layer beneath the Canary Islands. Tectonophysics 608:586–599

Masson DG, Le Bas TP, Grevemeyer I, Weinrebe W (2008) Flank collapse and large-scale landsliding in the Cape Verde Islands, off West Africa. Geochem Geophys Geosyst 9:Q07015. https://doi.org/10.1029/2008GC001983

McCurry M, Rodgers DW (2009) Mass transfer along the Yellowstone hotspot track I: petrologic constraints on the volume of mantle-derived magma. J Volcanol Geoth Res 188:86–98

McQuarrie N, Rodgers DW (1998) Subsidence of a volcanic basin by flexure and crustal flow: the eastern Snake River Plain, Idaho. Tectonics 17:203–220

Mendes VB, Madeira J, Brum da Silveira A, Trota A, Elosegui P, Pagarete J (2013) Present-day deformation in Sao Jorge Island, Azores, from episodic GPS measurements (2001–2011). Adv Space Res 51:1581–1592

Merle O, Lenat JF (2003) Hybrid collapse mechanism at Piton de la Fournaise volcano, Reunion Island, Indian Ocean. J Geophys Res 108:2166. https://doi.org/10.1029/2002JB002014

Merle O, Mairine P, Michon L, Bachelery P, Smietana M (2010a) Calderas, landslides and paleo-canyons on Piton de la Fournaise volcano (La Réunion Island, Indian Ocean). J Volcanol Geoth Res 189:131–142

Merle O, Barde-Cabusson S, van Wyk deVries B (2010b) Hydrothermal calderas. Bull Volcanol 72:131–147
Michon L, Saint-Ange F, Bachelery P, Villeneuve N, Stadaucher T (2007) Role of the structural inheritance of the oceanic lithosphere in the magmato-tectonic evolution of Piton de la Fournaise volcano (La Reunion Island). J Geophys Res 112:B04205. https://doi.org/10.1029/2006JB004598
Michon L, Saint-Ange F (2008) Morphology of Piton de la Fournaise basaltic shield volcano (La Reunion Island): Characterization and implication in the volcano evolution. J Geophys Res 113:B03203. https://doi.org/10.1029/2005JB004118
Michon L, Villeneuve N, Catry T, Merle O (2009) How summit calderas collapse on basaltic volcanoes: New insights from the April 2007 caldera collapse of Piton de la Fournaise volcano. J Volcanol Geoth Res 184:138–151
Miller DS, Smith RB (1999) P and S velocity structure of the Yellowstone volcanic field from local earthquake and controlled-source tomography. J Geophys Res 104:15105–15121
Mitchell NC, Masson DG, Watts AB, Gee MJR, Urgeles R (2002) The morphology of the submarine flanks of volcanic ocean islands: a comparative study of the Canary and Hawaiian hotspot islands. J Volcanol Geoth Res 115:83–107
Moore JG (1987) Subsidence of the Hawaiian Ridge. US Geol Surv Prof Pap 1350:85–99
Moore JG, Clague DA (1992) Volcano growth and evolution of the island of Hawaii. Geol Soc Am Bull 104:1471–1484
Moore RB (1990) Volcanic geology and eruption frequency, Sao Miguel, Azores. Bull Volcanol 52:602–614
Moreira MA, Geoffroy L, Pozzi JP (2015) Magma flow pattern in dykes of the Azores revealed by anisotropy of magnetic susceptibility. J Geophys Res 120:662–690
Morriss MC, Karlstrom L, Nasholds MWM, Wolff JA (2020) The Chief Joseph dike swarm of the Columbia River flood basalts, and the legacy data set of William H. Taubeneck. Geosphere 16:1–25
Mouginis-Mark P, Rowland SK, Garbeil H (1996) Slopes of western Galapagos volcanoes from airborne interferometric radar. Geophys Res Lett 23:3767–3770
Naar DF, Hey RN (1991) Tectonic evolution of the Easter Microplate. J Geophys Res 96:7961–7993
Naumann T, Geist D (2000) Physical volcanology and structural development of Cerro Azul Volcano, Isabela Island, Galapagos: implications for the development of Galapagos-type shield volcanoes. Bull Volcanol 61:497–514
Navarro A, Lourenco N, Chorowicz J, Miranda JM, Catalao J (2009) Analysis of geometry of volcanoes and faults in Terceira Island (Azores): evidence for reactivation tectonics at the EUR/AFR plate boundary in the Azores triple junction. Tectonophysics 465:98–113
Neal CA, Brantley SR, Antolik L, Babb JL, Burgess M, Calles K et al (2019) The 2018 rift eruption and summit collapse of Kīlauea Volcano. Science 363:367–374
Nkono C, Liegeois J-P, Demaiffe D (2018) Relationships between structural lineaments and Cenozoic volcanism, Tibesti swell, Saharan metacraton. J Afr Earth Sci 145:274–283
Nunes JC, Camacho A, Franca Z, Montesinos FG, Alves M, Vieira R et al (2006) Gravity anomalies and crustal signature of volcano-tectonic structures of Pico Island (Azores). J Volcanol Geoth Res 156:55–70
O'Connor JM, Stoffers P, McWilliams MO (1995) Time-space mapping of Easter Chain volcanism. Earth Planet Sci Lett 136:197–212
O'Neill C, Muller D, Steinberger B (2003) Geodynamic implications of moving Indian Ocean hotspots. Earth Planet Sci Lett 215:151–168
Obrebski M, Allen RM, Xue M, Hung SH (2010) Slab-plume interaction beneath the Pacific Northwest. Geophys Res Lett 37:L14305. https://doi.org/10.1029/2010GL043489
Oehler JF, Labazuy P, Lenat JF (2004) Recurrence of major flank landslides during the last 2 Ma history of Reunion Island. Bull Volcanol 66:585–598
Okubo C, Benz HM, Chouet BA (1997) Imaging the crustal magma sources beneath Mauna Loa and Kilauea volcanoes, Hawaii. Geology 25:867–870
Pagarete J, Pinto JT, Mendes VB, Autunes C, Ribeiro H (1998) The importance of classical geodetic observations for analyzing the geodynamic behaviour of the Azores archipelago. Tectonophysics 294:281–290
Park J, Rye DM (2019) Why is crustal underplating beneath many hot spot islands anisotropic? Geochem Geophys Geosyst 20:4779–4809
Parsons T, Thompson GA, Smith RP (1998) More than one way to stretch: a tectonic model for extension along the plume track of the Yellowstone hotspot and adjacent Basin and Range Province. Tectonics 17:221–234
Payne S, McCaffrey R, King RW, Kattenhorn SA (2012) A new interpretation of deformation rates in the Snake River Plain and adjacent basin and range regions based on GPS measurements. Geophys J Int 189:101–122
Peltier A, Bachelery P, Staudacher T (2009) Magma transport and storage at Piton de La Fournaise (La Réunion) between 1972 and 2007: a review of geophysical and geochemical data. J Volcanol Geoth Res 184:93–108
Peng X, Humphreys ED (1998) Crustal velocity structure across the eastern Snake River Plain and the Yellowstone swell. J Geophys Res 103:7171–7186
Permenter JL, Oppenheimer C (2007) Volcanoes of the Tibesti massif (Chad, northern Africa). Bull Volcanol 69:609–626
Peterson DW, Moore RB (1987) Geologic history and evolution of geologic concepts. Island of Hawaii. US Geol Surv Prof Pap 1350:149–189

Pierce KL, Morgan LA (2009) Is the track of the Yellowstone hotspot driven by a deep mantle plume? review of volcanism, faulting, and uplift in light of new data. J Volcanol Geoth Res 188:1–25
Pik R, Marty B, Hilton DR (2006) How many mantle plumes in Africa? The geochemical point of view. Chem Geol 226:100–114
Poland M (2014) Contrasting volcanism in Hawaii and the Galápagos. In: Harpp KS, Mittelstaedt E, d'Ozouville N, Graham DW (eds) The Galápagos: a natural laboratory for the earth sciences. AGU geophysical monograph, vol 204, pp 5–26
Poland M, Miklius A, Montgomery-Brown EK (2014) Magma supply, storage, and transport at shield-stage Hawaiian volcanoes. In: Poland MP, Takahashi TJ, Landowski CM (eds) Characteristics of Hawaiian volcanoes. US geological survey professional paper, vol 1801, pp 179–236
Prono E, Battaglia J, Monteiller V, Got JL, Ferrazzini V (2009) P-wave velocity structure of Piton de la Fournaise volcano deduced from seismic data recorded between 1996 and 1999. J Volcanol Geoth Res 184:49–62
Ray JS, Mahoney JJ, Duncan RA, Ray J, Wessel P, Naar DF (2012) Chronology and geochemistry of Lavas from the Nazca Ridge and Easter Seamount chain: a 30 Myr Hotspot Record. J Petrol 53:1417–1448
Roberts GG, White N (2010) Estimating uplift rate histories from river profiles using African examples. J Geophys Res 115:B02406. https://doi.org/10.1029/2009JB006692
Robinson JE, Eakins BW (2006) Calculated volumes of individual shield volcanoes at the young end of the Hawaiian Ridge. J Volcanol Geoth Res 151:309–317
Romer RHW, Beier C, Haase KM, Hübscher C (2018) Correlated changes between volcanic structures and magma composition in the faial volcanic system, Azores. Front Ear Sci 6:78. https://doi.org/10.3389/feart.2018.00078
Rowland SK, Munro DC, Perez-Oviedo V (1994) Volcan Ecuador, Galapagos Islands: erosion as a possible mechanism for the generation of steep-sided basaltic volcanoes. Bull Volcanol 56:271–283
Rusby RI, Searle RC (1995) A history of the easter microplate, 5.25 Ma to present. J Geophys Res 100:12617–12640
Rychert CA, Laske G, Harmon N, Shearer PM (2013) Seismic imaging of melt in a displaced Hawaiian plume. Nat Geosci 6:657–660
Rychert CA, Harmon N, Ebinger C (2014) Receiver function imaging of lithospheric structure and the onset of melting beneath the Galápagos Archipelago. Earth Planet Sci Lett 388:156–165
Schmandt B, Dueker K, Humpreys E, Hansen S (2012) Hot mantle upwelling across the 660 beneath Yellowstone. Earth Planet Sci Lett 331–332:224–236
Searle R (1980) Tectonic pattern of the Azores spreading centre and triple junction. Earth Planet Sci Lett 51:415–434
Searle RC, Bird RT, Rusby RI, Naar DF (1993) The development of two oceanic microplates: Easter and Juan Fernadez microplates, East Pacific Rise. Geol Soc London 150:965–976
Sebai A, Stutzamnn E, Montagner J-P, Sicilia D, Beucler E (2006) Anisotropic structure of the African upper mantle from Rayleigh and Love wave tomography. Phys Earth Planet Inter 155:48–62
Sharp WD, Clague DA (2006) 50-Ma initiation of Hawaiian-Emperor bend records major change in Pacific plate motion. Science 313:1281–1284
Sibrant ALR, Hildebrand A, Marques FO, Weiss B, Boulesteix T, Hubscher C et al (2015) Morpho-structural evolution of a volcanic island developed inside an active oceanic rift: S. Miguel Island (Terceira Rift, Azores). J Volcanol Geoth Res 301:90–106
Sichoix L, Bonneville A, McNutt MK (1998) The seafloor swells and Superswell in French Polynesia. J Geophys Res 103:27123–27133
Silva PF, Henry B, Marques FO, Hildebrand A, Madureira P, Meriaux CA et al (2012) Palaeomagnetic study of a subaerial volcanic ridge (Sao Jorge Island, Azores) for the past 1.3 Myr: evidence for the Cobb Mountain Subchron, volcano flank instability and tectonomagmatic Implications. Geophys J Int 188:959–978
Simkin T, Howard KA (1970) Caldera Collapse in the Galapagos Islands, 1968. Science 169:429–437
Sinton CW, Christie DM, Duncan RA (1996) Geochronology of Galapagos seamounts. J Geophys Res 101:13689–13700
Sinton CW, Hauff F, Hoernle K, Werner R (2018) Age progressive volcanism opposite Nazca plate motion: Insights from seamounts on the northeastern margin of the Galapagos Platform. Lithos 310–311:342–354
Smith RB, Braile LW (1994) The Yellowstone hotspot. J Volcanol Geoth Res 61:121–187
Smith RB, Jordan M, Steinberger M, Puskas CM, Farrell J, Waite G et al (2009) Geodynamics of the Yellowstone hotspot and mantle plume: Seismic and GPS imaging, kinematics, and mantle flow. J Volcanol Geoth Res 188:26–56
Spieker K, Rondenay S, Ramalho R, Thomas C, Helffrich G (2018) Constraints on the structure of the crust and lithosphere beneath the Azores Islands from teleseismic receiver functions. Geophys J Int 213:824–835
Staudacher T, Allegre CJ (1993) Ages of the second caldera of Piton de la Fournaise volcano (Reunion) determined by cosmic ray produced ^{3}He and ^{21}Ne. Earth Planet Sci Lett 119:395–404
Stillmann CJ (1999) Giant Miocene landslides and the evolution of Fuerteventura, Canary Islands. J Volcanol Geoth Res 94:89–104
Stock MJ, Bagnardi M, Neave DA, Maclennan J, Bernard B, Buisman I et al (2018) Integrated petrological and geophysical constraints on magma system architecture in the western Galápagos Archipelago: insights from Wolf volcano. Geochem Geophys Geosyst 19:4722–4743

Stoffers P, Hekinian R, Haase KM, Sonne Cruise SO-80 Scientific Party (1994) Geology of young submarine volcanoes west of Easter Island, Southeast Pacific. Mar Geol 118:177–185

Suetsugu D, Isse T, Tanaka S, Obayashi M, Shiobara H, Sugioka H et al (2009) South Pacific mantle plumes imaged by seismic observation on islands and seafloor. Geochem Geophys Geosyst 10:Q11014. https://doi.org/10.1029/2009GC002533

Swanson D, Duffield WA, Fiske RS (1976) Displacement of the South Flank of Kilauea Volcano: the result of forceful intrusion of magma into the rift zones. US geological survey professional paper, vol 963, 44 pp

Thordarson T, Garcia MO (2018) Variance of the flexure model predictions with rejuvenated volcanism at Kılauea point, Kauai, Hawaii. Front Ear Sci 6:121. https://doi.org/10.3389/feart.2018.00121

Tian Y, Zhao D (2012) P-wave tomography of the western United States: insight into the Yellowstone hotspot and the Juan de Fuca slab. Phys Earth Planet Inter 200–201:72–84

Trippanera D, Porreca M, Ruch J, Pimentel A, Acocella V, Pacheco J et al (2014) Relationships between tectonics and magmatism in a transtensive/transform setting: an example from Faial Island (Azores, Portugal). Geol Soc Am Bull 126:164–181

Trota A, Houliè N, Briole P, Gaspar JL, Sigmundsson F, Keigl KL (2006) Deformation studies at Furnas and Sete Cidades Volcanoes (Sao Miguel Island, Azores). Velocities and further investigations. Geophys J Int 166:952–956

Vezzoli L, Acocella V (2009) Easter Island, SE Pacific: an end-member type of hotspot volcanism. Geol Soc Am Bull 121:869–886

Vidal V, Bonneville A (2004) Variations of the Hawaiian hot spot activity revealed by variations in the magma production rate. J Geophys Res 109:B03104. https://doi.org/10.1029/2003JB002559

Villagomez DR, Toomey DR, Geist DJ, Hooft EEE, Solomon SC (2014) Mantle flow and multistage melting beneath the Galápagos hotspot revealed by seismic imaging. Nat Geosci 7:151–156

Vogt PR, Jung WY (2004) The Terceira Rift as hyper-slow, hotspot-dominated oblique spreading axis: a comparison with other slow-spreading plate boundaries. Earth Planet Sci Lett 218:77–90

Walker GPL (1987) The dike complex of Koolau Volcano, Oahu: internal structure of a Hawaiian rift zone. US Geol Surv Prof Pap 1350:961–997

Walter TR, Troll VR, Cailleau B, Belousov A, Schmincke HU, Amelung F et al (2005) Rift zone reorganization through flank instability in ocean island volcanoes: an example from Tenerife, Canary Islands. Bull Volcanol 67:281–291

Walter TR, Klugel A, Munn S (2006) Gravitational spreading and formation of new rift zones on overlapping volcanoes. Terra Nova 18:26–33

Wang K, MacArthur HS, Johanson I, Montgomery-Brown EK, Poland MP, Cannon EC et al (2019) Interseismic quiescence and triggered slip of active normal faults of Kīlauea Volcano's south flank during 2001–2018. J Geophys Res 124:9780–9794

Watts AB, Peirce C, Collier J, Dalwood R, Canales JP, Henstock TJ (1997) A seismic study of lithospheric flexure in the vicinity of Tenerife, Canary Islands. Earth Planet Sci Lett 146:431–447

Wauthier C, Roman DC, Poland MP (2019) Modulation of seismic activity in Kīlauea's upper East Rift Zone (Hawaii) by summit pressurization. Geology 47:820–824

Weis D, Garcia MO, Rhodes JM, Jellinek M, Scoates JS (2011) Role of the deep mantle in generating the compositional asymmetry of the Hawaiian mantle plume. Nat Geosci 4:831–838

Werner R, Hoernie K, Barckhausen U, Hauff F (2003) Geodynamic evolution of the Galapagos hot spot system (Central East Pacific) over the past 20 m.y.: Constraints from morphology, geochemistry, and magnetic anomalies. Geochem Geophys Geosyst 4:1108. https://doi.org/10.1029/2003GC000576

Wolfe CJ, Solomon SC, Laske G, Collins JA, Detrick RS, Orcutt JA et al (2009) Beneath the Hawaiian hot spot. Science 326:1388–1390

Wolff JA, Ramos FC, Hart GL, Patterson JD, Brandon AD (2008) Columbia River flood basalts from a centralized crustal magmatic system. Nat Geosci 1:177–180

Wotzlaw J-F, Bindeman IN, Stern RA, D'Abzac F-X, Schaltegger U (2015) Rapid heterogeneous assembly of multiple magma reservoirs prior to Yellowstone supereruptions. Sci Rep 5:14026. https://doi.org/10.1038/srep14026

Xue M, Allen RM (2007) The fate of the Juan de Fuca plate: Implications for a Yellowstone plume head. Earth Planet Sci Lett 264:266–276

Yang T, Shen Y, van der Lee S, Solomon SC, Hung SH (2006) Upper mantle structure beneath the Azores hotspot from finite-frequency seismic tomography. Earth Planet Sci Lett 250:11–26

Correction to: Calderas

Correction to:
Chapter 5 in: V. Acocella, *Volcano-Tectonic Processes*, Advances in Volcanology, https://doi.org/10.1007/978-3-030-65968-4_5

In the original version of the book, the following belated correction have been incorporated in the chapter "Calderas": Fig. 5.13 has been replaced with the correct figure.

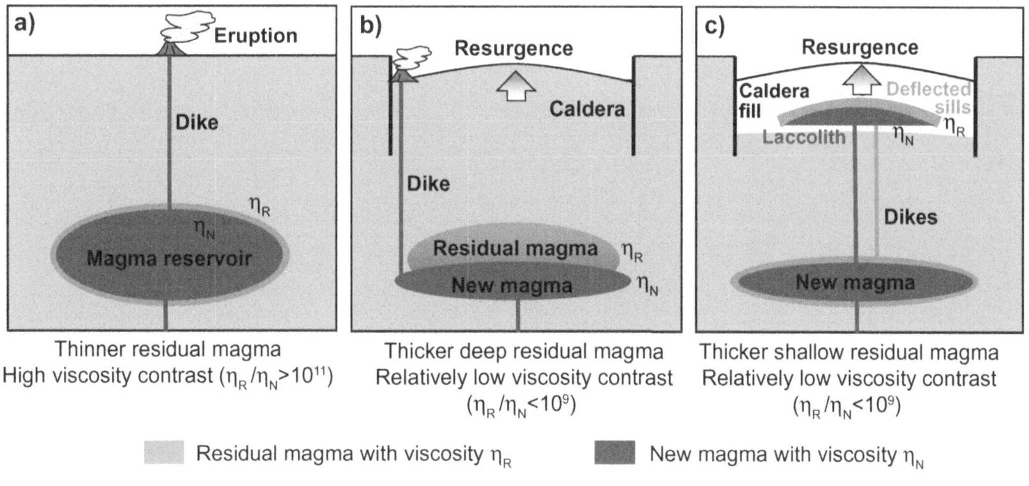

Fig. 5.13 Conditions for resurgence. **a** A thinner layer of viscous residual magma (high viscosity contrasts with the new magma) promotes dike propagation and eruption, without resurgence. **b** A thicker layer of medium viscosity residual magma (with relatively low viscosity contrast with the new magma) hinders dike propagation, promoting stagnation and resurgence; peripheral eruptions may occur. **c** Same as (**b**), but shifted at shallower levels: dikes (orange) may arrest within the altered intracaldera tuff, developing one or more sills (orange) constituting a rheological barrier for successive dikes (red), stagnating in laccoliths and promoting resurgence. Mechanism (**c**) may be alternative to, or combined with, (**b**) (modified after Galetto et al. 2017)

The updated version of this chapter can be found at https://doi.org/10.1007/978-3-030-65968-4_5

V. Acocella, *Volcano-Tectonic Processes*, Advances in Volcanology,
https://doi.org/10.1007/978-3-030-65968-4_14

Index

V. Acocella, *Volcano-Tectonic Processes*, Advances in Volcanology,
https://doi.org/10.1007/978-3-030-65968-4

Q

R

The manufacturer's authorised representative in the EU is Springer Nature Customer Service Centre GmbH, Europaplatz 3, 69115 Heidelberg, Germany. If you have any concerns regarding our products, please contact ProductSafety@springernature.com

Printed and bound by CPI Group (UK) Ltd, Croydon, CR0 4YY
30/09/2026
02225661-0020